3rd International Conference on Mathematics: Pure, Applied and Computation (ICoMPAC 2017)

AA007487

Journal of Physics: Conference Series Volume 974

Surabaya, Indonesia
1 November 2017

ISBN: 978-1-5108-6042-1
ISSN: 1742-6588

Printed from e-media with permission by:

Curran Associates, Inc.
57 Morehouse Lane
Red Hook, NY 12571

Some format issues inherent in the e-media version may also appear in this print version.

Printed by Curran Associates, Inc. (2018)

For permission requests, please contact the Institute of Physics
at the address below.

Institute of Physics
Dirac House, Temple Back
Bristol BS1 6BE UK

Phone: 44 1 17 929 7481
Fax: 44 1 17 920 0979

techtracking@iop.org

Additional copies of this publication are available from:

Curran Associates, Inc.
57 Morehouse Lane
Red Hook, NY 12571 USA
Phone: 845-758-0400
Fax: 845-758-2633
Email: curran@proceedings.com
Web: www.proceedings.com

Table of contents

Volume 974

International Conference on Mathematics: Pure, Applied and Computation

1 November 2017, Surabaya, Indonesia

Accepted papers received: 16 February 2018
Published online: 22 March 2018

Preface

Papers

Welcome Message from the Conference Chair

Bismillahirrohmanirrohiim
The honorable
Rector of ITS,
Keynote Speaker, Director of Research and Community Service,
Invited Speakers,
Dean Of Faculty Of Mathematics, Computing, and Data Science
Head of Mathematics Department
Ladies and Gentlemen,
Assalamu'alaikum warahmatullahi wabarokatuh

On behalf of the ICoMPAC 2017 organizing committee, I am honored and delighted to welcome you to the third International Conference on Mathematics; Pure, Applied and Computation (ICoMPAC 2017) at Wyndham Hotel, Surabaya.

At this year, we are so pleased to accept many papers from Indonesia, Malaysia, Japan, Taiwan, Germany, Thailand, Taiwan, USA, Australia and UK. It is a great pleasure to have 1 keynote speaker and 3 invited speakers with us in this conference to share their knowledge.

This year's conference is themed "Mathematics for Supporting Society Welfare" with the hope that mathematics can take an active role in improving society welfare. The aim of this conference is to provide a forum for researchers, educators, students and industries to exchange ideas, to communicate and discuss research findings and new advancement in mathematics, and to explore possible avenues to foster academic and student exchange, as well as scientific activities. The conference will be a venue to communicate and discuss how mathematics can give contributions to improving society welfare, and give solutions to problems faced by industries.

As a conference chair of ICoMPAC 2017, I realized that the success of this conference depends ultimately on the many people who have worked with us in planning and organizing this conference, in particular for the review process and preparing the technical programs. Recognition should go to the Local Organizing Committee members who

IOP Conf. Series: Journal of Physics: Conf. Series **974** (2018) 011001 doi:10.1088/1742-6596/974/1/011001

have all worked extremely hard for the detail of important aspects of the conference programs.

Last but not least, I would like to thank Institute of Physics (IOP), for the cooperation for publishing papers presented in this conference to their proceedings. I hope this conference will be proven to be an inspiring experience for you. Enjoy your participantion in the ICoMPAC 2017 and we hope that you have a memorable time visiting Surabaya. We also hope you return for the next ICoMPAC with even more colleagues.

Thank You,

Wassalamu'alaikum Wr. Wb.
Mardlijah
Conference Chair

IOP Conf. Series: Journal of Physics: Conf. Series **974** (2018) 011001 doi:10.1088/1742-6596/974/1/011001

The Speech from The Dean of Faculty of Mathematics, Computing, and Data Sciences

Bismillahirrahmanirrahim,
Assalamualaikum warahmatullahi wabarakatuh.
Allahumma Sholli ala saydina Muhammad, wa'ala ali saydina Muhammad, robbis rohli sodri, wayassirli amri, wahlul ukhdatan minlisani yapkohu koili, amma ba'du,

The Honorable Rector of ITS (Prof. Ir. Joni Hermana, M.Sc.ES. Ph.D.), the distinguished speakers, the invited speakers, the participants and the committee of ICoMPAC 2017. We sincerely welcome and thank you for your coming to this event. This event has been going on for the third time. And for this time take the theme of mathematics for supporting the welfare of society.

The increasing number of people is a big problem for the state country in the world, especially developing countries including Indonesia. Indonesia is the country with the fourth largest population in the world after China, India and the United States. Rapid population growth can create complex problems for a country, such as economic, social, educational, cultural and criminal issues.

Economist Alfred Marshall (1842-1924) professor of political science Cambridge University says that poverty is a matter of concern and should be eliminated. For that there needs to be a great effort on the welfare of this society, namely by combining the ability to think logically, thoroughly and mastery of mathematics. This can be done one of them through enhancing the application of basic and applied research results in the field of mathematics. In line with this, it is necessary to work hard and smart in improving the quality and quantity of research in the field of mathematical modeling, operations research, stochastic, biomathematics, actuarial, statistics and finance which focuses on issues in local, national and international scope, either as scientific development or to contribute on problem solving in social life.

ICOMPAC 3 is the right moment to convey the idea to improve the welfare of society based on mathematics. Therefore, in this conference we hope there are efforts to implement the research results of mathematics, statistics and actuary in realizing a society that can meet all needs, whether that is basic need, psychological social or development need that can provide tangible and sustainable contribution for community.

We as the Dean of the Faculty of Mathematics, Computing, and Science Data congratulate the conference, hopefully produce a brilliant idea in the prosperity of society, success and smooth.

IOP Conf. Series: Journal of Physics: Conf. Series **974** (2018) 011001 doi:10.1088/1742-6596/974/1/011001

We do apologize if there is anything less pleased.
Thank you for attention.

Wabillahi Taufiq Walhidayah,
Wasalamualaikum Warahmatullahi Wabarakatuh.

Dean,

Prof. Dr. Basuki Widodo, MSc.

IOP Conf. Series: Journal of Physics: Conf. Series **974** (2018) 011001 doi:10.1088/1742-6596/974/1/011001

The Speech from The Rector of Institut Teknologi Sepuluh Nopember

I would like to convey my sincere congratulation to all involved parties for the successful organization of the third International Conference on Mathematics: Pure, Applied and Computation ICoMPAC 2017, organized by the Department of Mathematics Institut Teknologi Sepuluh Nopember (ITS) Surabaya. The ICoMPAC 2017 is held as part of our 57th ITS Anniversary.

It is a great pleasure and honor for me to welcome to all participants and thank to the keynote speakers and all invited speakers for the worthy time to share your experiences and expertise to all conference participants. I do believe that your participation to this conference is a highlight and give a significant insight to all of us. I expect that your patronage and support towards the advancement of knowledge through this event, will contribute to the future development of Mathematics.

As we know that the role of Mathematics is vital in many aspects of life. There are many problems arise in social, business, economic, environment, and many others that could be solved by Mathematics. I am sure that, ICoMPAC will be the flagship conference for researchers, students, and professionals in the area of Mathematics and its applications to disseminate their research advancements and discoveries, to network and exchange ideas in order to solve more problems.

Last but not least, I wish all participants have a very interesting and learning experiences during the conference. Moreover, I do hope that new collaborations among participants could be established. To our foreign guests, I wish you a memorable stay in Surabaya. We welcome you anytime to visit our university, Institut Teknologi Sepuluh Nopember, Surabaya.

Prof. Ir. Joni Hermana M.Sc.ES. Ph.D
Rector of the Institut Teknologi Sepuluh Nopember (ITS) Indonesia

IOP Conf. Series: Journal of Physics: Conf. Series **974** (2018) 011002 doi:10.1088/1742-6596/974/1/011002

ORGANIZING COMMITTEE

International Scientific Committee
Prof. Basuki Widodo (Institut Teknologi Sepuluh Nopember, Indonesia)
Prof. Erna Apriliani (Institut Teknologi Sepuluh Nopember, Indonesia)
Prof. M. Isa Irawan (Institut Teknologi Sepuluh Nopember, Indonesia)
Prof. Dr. Agus Suryanto (Universitas Brawijaya Malang, Indonesia)
Prof. Dr. Toto Nusantara (State University of Malang, Indonesia)
Dr. Subiono (Institut Teknologi Sepuluh Nopember, Indonesia)
Dr. Hariyanto (Institut Teknologi Sepuluh Nopember, Indonesia)
Drs. Suharmadi, Dipl. Sc, M.Phil (Institut Teknologi Sepuluh Nopember, Indonesia)
Taufik Fuadi Abidin, Ph.D (Universitas Syiah Kuala, Indonesia)
Dr. Suhartono (Institut Teknologi Sepuluh Nopember, Indonesia)
Dr. M. Suhartono (UIN Maulana Malik Ibrahim Malang, Indonesia)
Dr. Arif Muntasa (Universitas Trunojoyo, Indonesia)
Dr. Imam Mukhlash, MT (Institut Teknologi Sepuluh Nopember, Indonesia)
Dr. Budi Setiyono, MT (Institut Teknologi Sepuluh Nopember, Indonesia)
Bernardi Pranggono, Ph.D (Sheffield Hallam University, United Kingdom)
Subchan, Ph.D (Institut Teknologi Sepuluh Nopember, Indonesia)
Dr. Elly Matul Imah (Universitas Negeri Surabaya, Indonesia)
Dr. Intan Muchtadi-Alamsyah (Institut Teknologi Bandung, Indonesia)
Nhat-Tan Le, Ph.D (Mientrung University of Civil Engineering, Viet Nam)
Chaiwat Kosapattarapim, Ph.D (Maejo University, Thailand)
Dr. Mohd Sham Mohamad (Universiti Malaysia Pahang, Malaysia)
Dr. Norazaliza Mohd Jamil (Universiti Malaysia Pahang, Malaysia)
Prof. Hsing-Ko Kenneth Pao (NTUST, Taiwan)
Dr. Hadi Susanto (University of Essex, UK)

Steering Committee

Rector ITS	:	Prof. Ir. Joni Hermana, M.Sc.ES., Ph.D
Vice Rector for Innovation, Cooperation, Alumni, and International Relation Affairs ITS	:	Prof.Dr. Ketut Budi Artana, ST, M.Sc
Director of The Institute for Research and Community Services ITS	:	Prof. Dr. Ir Adi Soeprijanto, MT
Dean of FMKSD ITS	:	Prof. Dr. Drs. Bauki Widodo, M.Sc
Head of Department of Mathematics ITS	:	Dr. Imam Mukhlas, S.Si,MT

Local Organizing Committee
Dian Winda S., S.Si, M.Si (Institut Teknologi Sepuluh Nopember, Indonesia)
Sunarsini, S.Si, M.Si (Institut Teknologi Sepuluh Nopember, Indonesia)
Endah Rokhmati M. P., Ph.D (Institut Teknologi Sepuluh Nopember, Indonesia)
Soleha, S.Si, M.Si (Institut Teknologi Sepuluh Nopember, Indonesia)

Chairman : Dr. Dra. Mardlijah, MT
Co-chairman : Dr. Dieky Adzkia, M.Si

IOP Conf. Series: Journal of Physics: Conf. Series **974** (2018) 011002 doi:10.1088/1742-6596/974/1/011002

IOP Conf. Series: Journal of Physics: Conf. Series **974** (2018) 011003 doi:10.1088/1742-6596/974/1/011003

Peer review statement

All papers published in this volume of *Journal of Physics: Conference Series* have been peer reviewed through processes administered by the proceedings Editors. Reviews were conducted by expert referees to the professional and scientific standards expected of a proceedings journal published by IOP Publishing.

A Mathematical Model Of Dengue-Chikungunya Co-Infection In A Closed Population

Dipo Aldila, Maya Ria Agustin

Department of Mathematics, Universitas Indonesia, Depok 16424, Indonesia

email: aldiladipo@sci.ui.ac.id

Abstract. Dengue disease has been a major health problem in many tropical and sub-tropical countries since the early 1900s. On the other hand, according to a 2017 WHO fact sheet, Chikungunya was detected in the first outbreak in 1952 in Tanzania and has continued increasing until now in many tropical and sub-tropical countries. Both these diseases are vector-borne diseases which are spread by the same mosquito, i.e. the female Aedes *aegypti*. According to the WHO report, there is a great possibility that humans and mosquitos might be infected by dengue and chikungunya at the same time. Here in this article, a mathematical model approach will be used to understand the spread of dengue and chikungunya in a closed population. A model is developed as a nine-dimensional deterministic ordinary differential equation. Equilibrium points and their local stability are analyzed analytically and numerically. We find that the basic reproduction number, the endemic indicator, is given by the maximum of three different basic reproduction numbers of a complete system, i.e. basic reproduction numbers for dengue, chikungunya and for co-infection between dengue and chikungunya. We find that the basic reproduction number for the co-infection sub-system dominates other basic reproduction numbers whenever it is larger than one. Some numerical simulations are provided to confirm these analytical results.

1. Introduction

Dengue disease is one of the most infectious diseases in the world, where more than two over three population in the world are at risk every year [1]. This disease is spread by Aedes mosquitos and is caused by one or more of four serotypes of the DEN virus. Although dengue vaccine has been introduced in several tropical and sub-tropical countries, a medicine to cure infected humans has not been found yet.

Like dengue, chikungunya is also a mosquito-borne disease. Chikungunya is also spread by female Aedes mosquitos, as with dengue disease. Chikungunya was first detected in 1952 in Africa, on the Makonde Plateau [2]. Between 1960 and 1980, chikungunya has been reached its outbreak which was reported in many countries in Asia and Africa. Unlike dengue which is life-threatening, not many reported cases of chikungunya have ended with death.

Since these diseases are spread by the same mosquito, there is a possibility that humans and mosquitos might be infected by these diseases at the same time. For an example, reported in California [3], Central

Africa [4], and Colombian [5]. Therefore, it is very important to understand how these diseases spread among the human and mosquito populations and understand the best prevention or control programs for both of these diseases.

Many mathematical models have been developed to understand dengue [6-10] and chikungunya [11,12], and also some models of co-infection between dengue and chikungunya [13,14]. In contrast to the articles in [13,14], in this article, the effect of co-infection is not only analyzed in the human population, but also in the mosquito population. The mosquito population may also get infected by dengue and chikungunya at the same time.

The paper is organized as follows. In chapter 2, the mathematical model construction will be discussed and followed with mathematical model analysis about the equilibrium points and the basic reproduction number in chapter 3. Numerical experiments to give a better interpretation of the analytical results in chapter 3 are reported in chapter 4. Some conclusions are given in chapter 5.

2. The construction of the mathematical model

In this section, the mathematical models will be described. Firstly, some basic assumptions are given which form the basis of the model construction process, namely :

1. The total human and mosquito populations are constant.
2. There is no increase in the death rate of human and mosquito except from natural death rate.
3. The health status of humans can be seen physically.
4. Humans can be infected by dengue and chikungunya at the same time.
5. The first virus that come into the mosquito's body dominates the other virus, except the mosquito bites the co-infected human.
6. Infection only occurs from direct contact between mosquito and human. Infection cannot occur from human to human or mosquito to mosquito contact.
7. All newborn are susceptible for human and mosquito.

With these assumptions, the construction of the model is given as follows. Let the human population be divided into five compartments, i.e. susceptible humans (x_1), humans infected with dengue (x_2), humans infected with chikungunya (x_3), co-infected humans, or humans infected with both dengue and chikungunya (x_4), and recovered humans (x_5). On the other hand, the mosquito population is divided into four compartments, i.e. susceptible mosquitos (y_1), mosquitos infected with dengue (y_2), mosquitos infected with chikungunya (y_3) and mosquitos co-infected with dengue and chikungunya (y_4). To construct the model, the transmission and transition between each compartment can be seen in Figure 1. More detail about the model construction is given below.

IOP Conf. Series: Journal of Physics: Conf. Series **974** (2018) 012001 doi:10.1088/1742-6596/974/1/012001

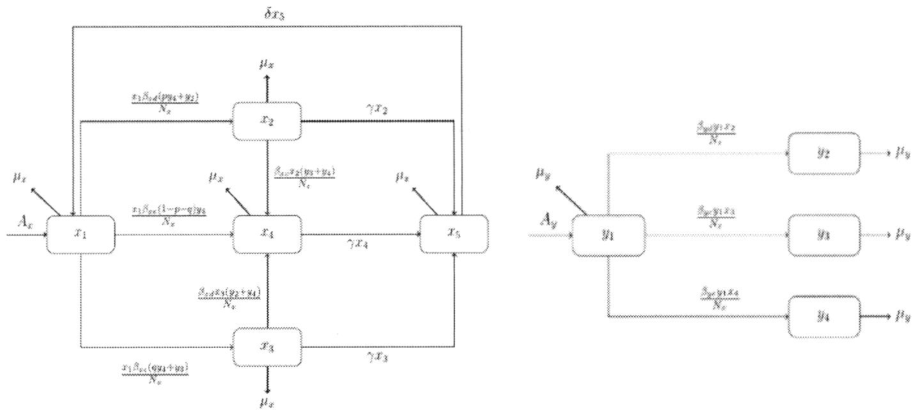

Figure 1. Transmission diagram for the spread of Dengue-Chikungunya Co-Infection

Susceptible humans increase from the newborn stage at a constant rate of A_x and decreases by infection and natural death rate with constant rate of μ_x. Susceptible humans might be infected by mosquitos that carry the DEN virus (y_2 and y_4) with a probability of success infection of β_{xd}, and mosquitos that carry the chikungunya virus (y_3 and y_4) with a probability of success infection of β_{xc}. Since we assume that co-infection between dengue and chikungunya might appear in human and mosquito populations, humans also might be infected by a mosquito carrying both the DEN and chikungunya viruses (y_4) with a probability of success infection of β_{xe}. Therefore, the dynamic of susceptible human respect to time is given by

$$\frac{dx_1}{dt} = A_x - \frac{\beta_{xd}x_1(py_4+y_2)}{N_x} - \frac{\beta_{xc}x_1(qy_4+y_3)}{N_x} - \frac{\beta_{xe}x_1(1-p-q)y_4}{N_x} - \mu_x x_1 + \delta x_5 ,$$

where p and q are the proportion of humans who have contact with y_4 who will be infected by dengue and chikungunya, respectively, and $N_x = x_1 + x_2 + x_3 + x_4 + x_5$ is the total of the human population.

Infected humans with dengue (x_2), chikungunya (x_3) and both dengue and chikungunya (x_4) increase caused by infection process from susceptible human (x_1). Multiple infections might appear in the x_2 and x_3 compartments which allows a possible transition from these compartments into the x_4 compartment. We also assume that infected humans might recover from infection, depending on their natural immune system, with the per capita recovery rate given by constant γ. Therefore, the dynamic of all infected compartments in respect to time is given by

$$\frac{dx_2}{dt} = \frac{\beta_{xd}x_1(py_4 + y_2)}{N_x} - \frac{\beta_{xc}x_2(y_3 + y_4)}{N_x} - \mu_x x_2 - \gamma x_2$$

$$\frac{dx_3}{dt} = \frac{\beta_{xc}x_1(qy_4 + y_3)}{N_x} - \frac{\beta_{xd}x_3(y_2 + y_4)}{N_x} - \mu_x x_3 - \gamma x_3$$

$$\frac{dx_4}{dt} = \frac{\beta_{xe}x_1(1-p-q)y_4}{N_x} + \frac{\beta_{xc}x_2(y_3+y_4)}{N_x} + \frac{\beta_{xd}x_3(y_2+y_4)}{N_x} - \mu_x x_4 - \gamma x_4 .$$

Recovered human increased by transition from infected compartment as a recovery process and decreased caused by natural death rate and as an effect of disappearance of temporary immunity with constant rate δ. Therefore, the dynamic of recovered human compartments with respect to time is given by

$$\frac{dx_5}{dt} = \gamma(x_2 + x_3 + x_4) - \mu_x x_5 - \delta x_5 \,.$$

In the mosquito population, the susceptible mosquitos compartment is increased by constant newborns at the rate of A_y and decreased according to the natural death rate μ_y. This compartment also decreases due to the effect of infection from direct contact with humans infected with dengue y_2, chikungunya y_3 and both y_4 with the probability of successful infection given by β_{yd}, β_{yc}, and β_{ye}, respectively. Therefore, the dynamic of the susceptible mosquitos compartment in respect to time is given by

$$\frac{dy_1}{dt} = A_y - \frac{\beta_{yd} y_1 x_2}{N_x} - \frac{\beta_{yc} y_1 x_3}{N_x} - \frac{\beta_{ye} y_1 x_4}{N_x} - \mu_y y_1 \,.$$

On the other hand, all infected mosquitos (y_2, y_3, y_4) increased by infection term from susceptible mosquitos compartment and decreased by natural death rate μ_y which gave us the dynamic of infected mosquitos is given by

$$\frac{dy_2}{dt} = \frac{\beta_{yd} y_1 x_2}{N_x} - \mu_y y_2$$

$$\frac{dy_3}{dt} = \frac{\beta_{yc} y_1 x_3}{N_x} - \mu_y y_3$$

$$\frac{dy_4}{dt} = \frac{\beta_{ye} y_1 x_4}{N_x} - \mu_y y_4 \,.$$

Therefore, the mathematical model of co-infection between dengue and chikungunya is given by 9-dimensional system of ordinary differential equation which given by system (1) below:

$$\frac{dx_1}{dt} = A_x - \frac{\beta_{xd} x_1 (p y_4 + y_2)}{N_x} - \frac{\beta_{xc} x_1 (q y_4 + y_3)}{N_x} - \frac{\beta_{xe} x_1 (1 - p - q) y_4}{N_x} - \mu_x x_1 + \delta x_5 \quad (1a)$$

$$\frac{dx_2}{dt} = \frac{\beta_{xd} x_1 (p y_4 + y_2)}{N_x} - \frac{\beta_{xc} x_2 (y_3 + y_4)}{N_x} - \mu_x x_2 - \gamma x_2 \quad (1b)$$

$$\frac{dx_3}{dt} = \frac{\beta_{xc} x_1 (q y_4 + y_3)}{N_x} - \frac{\beta_{xd} x_3 (y_2 + y_4)}{N_x} - \mu_x x_3 - \gamma x_3 \quad (1c)$$

$$\frac{dx_4}{dt} = \frac{\beta_{xe} x_1 (1 - p - q) y_4}{N_x} + \frac{\beta_{xc} x_2 (y_3 + y_4)}{N_x} + \frac{\beta_{xd} x_3 (y_2 + y_4)}{N_x} - \mu_x x_4 - \gamma x_4 \quad (1d)$$

$$\frac{dx_5}{dt} = \gamma(x_2 + x_3 + x_4) - \mu_x x_5 - \delta x_5 \quad (1e)$$

$$\frac{dy_1}{dt} = A_y - \frac{\beta_{yd} y_1 x_2}{N_x} - \frac{\beta_{yc} y_1 x_3}{N_x} - \frac{\beta_{ye} y_1 x_4}{N_x} - \mu_y y_1 \quad (1f)$$

$$\frac{dy_2}{dt} = \frac{\beta_{yd} y_1 x_2}{N_x} - \mu_y y_2 \quad (1g)$$

IOP Conf. Series: Journal of Physics: Conf. Series **974** (2018) 012001 doi:10.1088/1742-6596/974/1/012001

$$\frac{dy_3}{dt} = \frac{\beta_{yc} y_1 x_3}{N_x} - \mu_y y_3 \tag{1h}$$

$$\frac{dy_4}{dt} = \frac{\beta_{ye} y_1 x_4}{N_x} - \mu_y y_4 \,. \tag{1i}$$

with initial condition of humans and mosquitoes are given by $x_i(t = 0) = x_{i0} \geq 0$ for $i = 1,2 \dots 5$ and $y_i(t = 0) = y_{i0} \geq 0$ for $i = 1,2 \dots 4$, respectively. Choosing $A_x = \mu_x N_x$ and $A_y = \mu_y N_y$, we have that total of humans and mosquitos population will remain constant all the time, since

$$\frac{dN_x}{dt} = A_x - \mu_x N_x \text{ and } \frac{dN_y}{dt} = A_y - \mu_y N_y \,.$$

3. Mathematical model analysis

In this section, a mathematical model analysis of system (1) will be performed with analyzing the existence and local stability of the equilibrium points, and also their relationship with the basic reproduction number.

3.1. Equilibrium points

System (1) have 4 different equilibrium points which represent the situation that might appear in the field related to the coexistence between dengue and chikungunya. The first equilibrium point is the disease-free equilibrium point (Ω_1) where all infected population equal to 0 in both human and mosquitoes population, which given by

$$\Omega_1 = \left(x_1 = \frac{A_x}{\mu_x}, x_2 = 0, x_3 = 0, x_4 = 0, x_5 = 0, y_1 = \frac{A_y}{\mu_y}, y_2 = 0, y_3 = 0, y_4 = 0 \right).$$

The second and the third equilibrium point is the equilibrium when only dengue infection occur (Ω_2) and only chikungunya infection occur (Ω_3), respectively, which are given by

$$\begin{aligned}
\Omega_2 = \Bigg(x_1 = {} & \frac{N_x \mu_y (\mu_x + \gamma)\left(\delta N_x \mu_x \mu_y + \gamma N_x \mu_x \mu_y + N_x \mu_x^2 \mu_y + \delta A_x \beta_{yd} + A_x \beta_{yd} \mu_x\right)}{\beta_{yd} \mu_x \left(\delta \gamma N_x \mu_y + \delta N_x \mu_x \mu_y + \gamma N_x \mu_x \mu_y + N_x \mu_x^2 \mu_y + \delta A_y \beta_{xd} + \gamma A_y \beta_{xd} + A_y \beta_{xd} \mu_x\right)}, x_2 \\
& = -\frac{\left(\gamma N_x^2 \mu_x \mu_y^2 + N_x^2 \mu_x^2 \mu_y^2 - A_x A_y \beta_{xd} \beta_{yd}\right)(\delta + \mu_x)}{\beta_{yd} \mu_x \left(\delta \gamma N_x \mu_y + \delta N_x \mu_x \mu_y + \gamma N_x \mu_x \mu_y + N_x \mu_x^2 \mu_y + \delta A_y \beta_{xd} + \gamma A_y \beta_{xd} + A_y \beta_{xd} \mu_x\right)}, x_3 \\
& = 0, x_4 = 0, x_5 \\
& = -\frac{\gamma\left(\gamma N_x^2 \mu_x \mu_y^2 + N_x^2 \mu_x^2 \mu_y^2 - A_x A_y \beta_{xd} \beta_{yd}\right)}{\beta_{yd} \mu_x \left(\delta \gamma N_x \mu_y + \delta N_x \mu_x \mu_y + \gamma N_x \mu_x \mu_y + N_x \mu_x^2 \mu_y + \delta A_y \beta_{xd} + \gamma A_y \beta_{xd} + A_y \beta_{xd} \mu_x\right)}, y_1 \\
& = \frac{N_x \mu_x (\delta \gamma N_x \mu_y + \delta N_x \mu_x \mu_y + \gamma N_x \mu_x \mu_y + N_x \mu_x^2 \mu_y + \delta A_y \beta_{xd} + \gamma A_y \beta_{xd} + A_y \beta_{xd} \mu_x)}{\beta_{xd}\left(\delta N_x \mu_x \mu_y + \gamma N_x \mu_x \mu_y + N_x \mu_x^2 \mu_y + \delta A_x \beta_{yd} + A_x \beta_{yd} \mu_x\right)}, y_2 \\
& = -\frac{\left(\gamma N_x^2 \mu_x \mu_y^2 + N_x^2 \mu_x^2 \mu_y^2 - A_x A_y \beta_{xd} \beta_{yd}\right)(\delta + \mu_x)}{\beta_{xd}\left(\delta N_x \mu_x \mu_y + \gamma N_x \mu_x \mu_y + N_x \mu_x^2 \mu_y + \delta A_x \beta_{yd} + A_x \beta_{yd} \mu_x\right)\mu_y}, y_3 = 0, y_4 = 0 \Bigg),
\end{aligned}$$

$$\Omega_3 = \left(x_1 = \frac{N_x\mu_y(\mu_x+\gamma)(\delta N_x\mu_x\mu_y + \gamma N_x\mu_x\mu_y + N_x\mu_x^2\mu_y + \delta A_x\beta_{yc} + A_x\beta_{yc}\mu_x)}{\beta_{yc}\mu_x(\delta\gamma N_x\mu_y + \delta N_x\mu_x\mu_y + \gamma N_x\mu_x\mu_y + N_x\mu_x^2\mu_y + \delta A_y\beta_{xc} + \gamma A_y\beta_{xc} + A_y\beta_{xc}\mu_x)}, x_2 = 0, x_3 \right.$$
$$= -\frac{(\gamma N_x^2\mu_x\mu_y^2 + N_x^2\mu_x^2\mu_y^2 - A_xA_y\beta_{xc}\beta_{yc})(\delta+\mu_x)}{\beta_{yc}\mu_x(\delta\gamma N_x\mu_y + \delta N_x\mu_x\mu_y + \gamma N_x\mu_x\mu_y + N_x\mu_x^2\mu_y + \delta A_y\beta_{xc} + \gamma A_y\beta_{xc} + A_y\beta_{xc}\mu_x)}, x_4$$
$$= 0, x_5$$
$$= -\frac{\gamma(\gamma N_x^2\mu_x\mu_y^2 + N_x^2\mu_x^2\mu_y^2 - A_xA_y\beta_{xc}\beta_{yc})}{\beta_{yc}\mu_x(\delta\gamma N_x\mu_y + \delta N_x\mu_x\mu_y + \gamma N_x\mu_x\mu_y + N_x\mu_x^2\mu_y + \delta A_y\beta_{xc} + \gamma A_y\beta_{xc} + A_y\beta_{xc}\mu_x)}, y_1$$
$$= \frac{N_x\mu_x(\delta\gamma N_x\mu_y + \delta N_x\mu_x\mu_y + \gamma N_x\mu_x\mu_y + N_x\mu_x^2\mu_y + \delta A_y\beta_{xc} + \gamma A_y\beta_{xc} + A_y\beta_{xc}\mu_x)}{\beta_{xc}(\delta N_x\mu_x\mu_y + \gamma N_x\mu_x\mu_y + N_x\mu_x^2\mu_y + \delta A_x\beta_{yc} + A_x\beta_{yc}\mu_x)}, y_2 = 0, y_3$$
$$\left. = -\frac{(\gamma N_x^2\mu_x\mu_y^2 + N_x^2\mu_x^2\mu_y^2 - A_xA_y\beta_{xc}\beta_{yc})(\delta+\mu_x)}{\beta_{xc}(\delta N_x\mu_x\mu_y + \gamma N_x\mu_x\mu_y + N_x\mu_x^2\mu_y + \delta A_x\beta_{yc} + A_x\beta_{yd}\mu_x)\mu_y}, y_4\right).$$

It can be seen that all compartments in Ω_2 and Ω_3 will remain non-negative if and only if $R_{0d} > 1$ and $R_{0c} > 1$, respectively, where $R_{0d} = \frac{\sqrt{\mu_x(\mu_x+\gamma)\beta_{yd}A_y\beta_{xd}A_x}}{\mu_x(\mu_x+\gamma)\mu_yN_x}$ and $R_{0c} = \frac{\sqrt{\mu_x(\mu_x+\gamma)\beta_{yc}A_y\beta_{xc}A_x}}{\mu_x(\mu_x+\gamma)\mu_yN_x}$. The last equilibrium point is the endemic equilibrium point, when all compartment is non-negative (Ω_4) which is not in simple form to be shown in this article. However, the existence of this equilibrium point still could be shown numerically with substituting parameters value in to system (1). Choosing parameters value :

$$data := \left\{\delta = \frac{1}{30}, \gamma = \frac{1}{14}, p = 0.45, q = 0.45, A_x = \frac{40}{949}, A_y = \frac{100}{3}, N_x = 1000, \beta_{xc} = 0.01, \beta_{xd} = 0.02, \beta_{xe} = 0.2, \beta_{yc} = 0.01, \beta_{yd} = 0.02, \beta_{ye} = 0.2, \mu_x = \frac{1}{23725}, \mu_y = \frac{1}{30}\right\},$$

which gave us $R_{0e} = \frac{\sqrt{-\mu_x(\mu_x+\gamma)\beta_{ye}A_y\beta_{xe}A_x(-1+p+q)}}{\mu_x(\mu_x+\gamma)\mu_yN_x} > 1$, we find that the Ω_4 is given by

$$x_1 = 751.8887214, x_2 = 22.05258485, x_3 = 9.903206336, x_4 = 47.05674131, x_5 = 169.0987461,$$

$$y_1 = 770.0938901, y_2 = 10.18953651, y_3 = 2.287919607, y_4 = 217.4286538\,.$$

To analyze the local stability of the equilibrium point, System (1) will be linearized with Jacobian matrix approach. Evaluate the Ω_1 in the Jacobian matrix of system (1), we find that this equilibrium point will locally stable if and only if $R_{0d} < 1, R_{0c} < 1$ and $R_{0e} < 1$. With numerical approach, local stability of each equilibrium points is summarized in Figure 2 below.

IOP Conf. Series: Journal of Physics: Conf. Series **974** (2018) 012001 doi:10.1088/1742-6596/974/1/012001

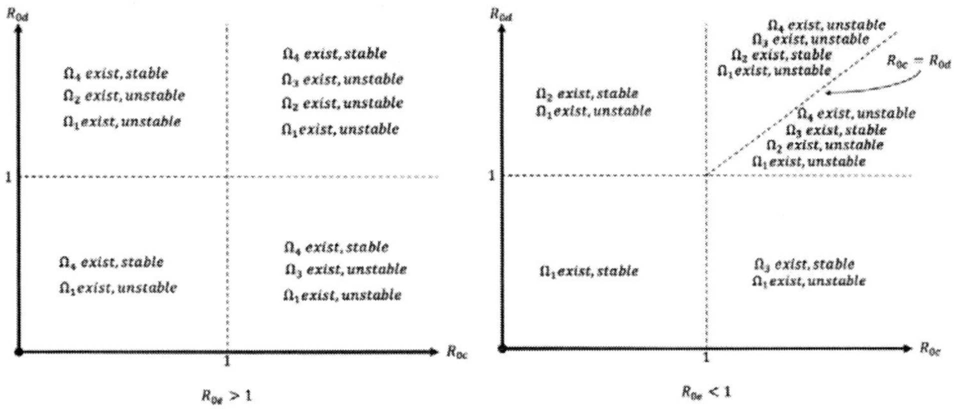

Figure 2. Existence and local stability diagram of equilibrium points $\Omega_i, i = 1,2,3,4$ depend on R_{0c}, R_{0d} and R_{0e}.

Basic Reproduction Number

Basic reproduction number is defined as an expected number of secondary cases caused by one primary case in a virgin population during one infection period [15,16]. Basic reproduction number could be taken from the spectral radius of the next-generation matrix. More detail about the construction of the next-generation matrix could be seen in [15].

The next-generation matrix of system (1) is given by

$$NGM := -\begin{bmatrix} 0 & 0 & 0 & -\frac{\beta_{xd}A_x}{\mu_x N_x \mu_y} & 0 & -\frac{\beta_{xd}A_x p}{\mu_x N_x \mu_y} \\ 0 & 0 & 0 & 0 & -\frac{\beta_{xc}A_x}{\mu_x N_x \mu_y} & -\frac{\beta_{xc}A_x q}{\mu_x N_x \mu_y} \\ 0 & 0 & 0 & 0 & 0 & -\frac{\beta_{xe}A_x(1-p-q)}{\mu_x N_x \mu_y} \\ -\frac{\beta_{yd}A_y}{\mu_y N_x(\gamma+\mu_x)} & 0 & 0 & 0 & 0 & 0 \\ 0 & -\frac{\beta_{yc}A_y}{\mu_y N_x(\gamma+\mu_x)} & 0 & 0 & 0 & 0 \\ 0 & 0 & -\frac{\beta_{ye}A_y}{\mu_y N_x(\gamma+\mu_x)} & 0 & 0 & 0 \end{bmatrix}$$

The element of the next generation read as number of new infection from in column-i caused by one infected person/mosquito in row-j. For example, $NGM_{1,4}$ represent that one infected human by dengue (x_2) produce $\frac{\beta_{xd}A_x}{\mu_x N_x \mu_y}$ new infection in infected mosquito by dengue (y_2). With this NGM in hand, the basic reproduction number of system (1) is given by

$$R_0 = max\{R_{0d}, R_{0c}, R_{0e}\},$$

where R_{0d}, R_{0c}, R_{0e} is given in the previous subsection. Therefore, it can be seen that the existence and local stability of each equilibrium point fully depends on magnitude of the $R_0 = max\{R_{0d}, R_{0c}, R_{0e}\}$, where R_{0d} is the basic reproduction number for dengue infection, R_{0c} is the basic reproduction number for chikungunya infection, and R_{0e} is the basic reproduction number for the co-infection between dengue and chikungunya. In Figure 3 below, we present the dependency of the infection rate respect to the basic reproduction number. It can be seen that smaller the infection rate will reduce basic reproduction number.

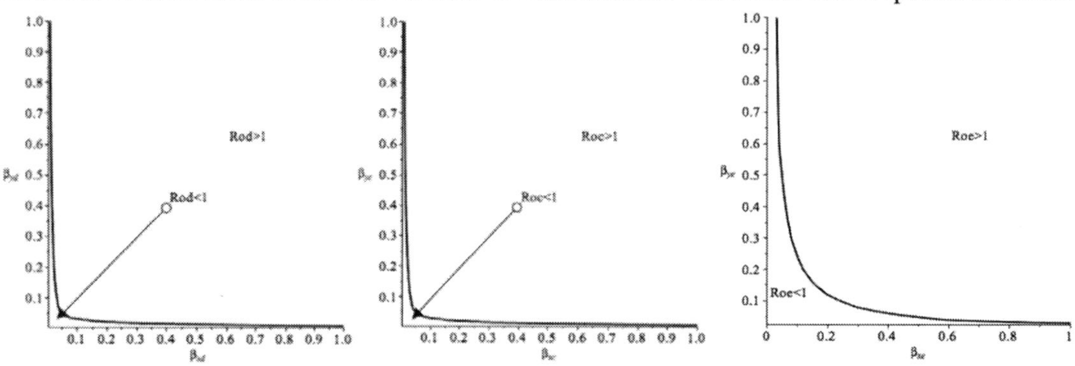

Figure 3. Dependency of infection probability rate for dengue, chikungunya and co-infection (dengue and chikungunya) respect to basic reproduction number from left to right, respectively. Others parameter (except $\beta_{xc}, \beta_{xd}, \beta_{xe}, \beta_{yc}, \beta_{yd}, \beta_{ye}$) are $\delta = \frac{1}{30}, \gamma = \frac{1}{14}, p = 0.45, q = 0.45, A_x = \frac{40}{949}, A_y = \frac{100}{3}, N_x = 1000, \mu_x = \frac{1}{23725}, \mu_y = \frac{1}{30}$.

In the next section, some numerical simulation will be given to give an interpretation of the analytical results.

4. Numerical experiments

Except it is stated, parameters value to run the simulations in this section is given in Table 1 and the initial condition for system (1) is $x_i(0) = (980, 10, 10, 0, 0, 990, 5, 5, 0)$ for $i = 1,2,\ldots,9$.

Table 1. Parameters value for simulations in Figure 4 and 5.

Parameters	A_x	A_y	μ_x	μ_y	γ	p	q	N_x
Value	$\frac{1000}{65 \times 365}$	$\frac{1000}{30}$	$\frac{1}{65 \times 365}$	$\frac{1}{30}$	$\frac{1}{14}$	0.45	0.45	1000

First simulation in this section will be performed in two different scenarios to give a better interpretation of analytical result from the previous section. First case is given when $R_{oe} < 1$ while R_{0d} and R_{0c} are varied. In the other hand, the second case is given when $R_{oe} > 1$ while R_{od} and R_{0c} are varied. Parameters value and the magnitude of the basic reproduction number is given in Table 2 for each case.

Table 2. Basic reproduction number values for simulation in Figure 4 and 5 respected to infection rate β_x and β_y

Case	Color in Figure	R_0 condition	β_{xc}	β_{xd}	β_{xe}	β_{yc}	β_{yd}	β_{ye}	R_0
1 ($R_{0e} < 1$)	Red	$R_{0d} > R_{oc} > 1$	0.1	0.2	0.05	0.1	0.2	0.05	$R_{0c} = 2.04$ $R_{0d} = 4.09$ $R_{0e} = 0.32$
	Green	$R_{0c} > R_{od} > 1$	0.2	0.1	0.05	0.2	0.1	0.05	$R_{0c} = 4.09$ $R_{0d} = 2.04$ $R_{0e} = 0.32$

	Blue	$R_{oc} = R_{od} < 1$	0.03	0.03	0.05	0.03	0.03	0.05	$R_{0c} = 0.61$ $R_{0d} = 0.61$ $R_{0e} = 0.32$
2 ($R_{0e} > 1$)	Red	$R_{0d} > R_{oc} > 1$	0.1	0.2	0.05	0.1	0.2	0.05	$R_{0c} = 2.04$ $R_{0d} = 4.09$ $R_{0e} = 0.32$
	Green	$R_{oc} > R_{od} > 1$	0.2	0.1	0.05	0.2	0.1	0.05	$R_{0c} = 4.09$ $R_{0d} = 2.04$ $R_{0e} = 0.32$
	Blue	$R_{oc} = R_{od} < 1$	0.03	0.03	0.05	0.03	0.03	0.05	$R_{0c} = 0.61$ $R_{0d} = 0.61$ $R_{0e} = 0.32$

In Figure 4, it can be seen that when $R_{0e} < 1$, no matter magnitude of R_{0c} and R_{0d} taken, the co-infection subpopulation both in human and mosquitoes population tend to zero for $t \to \infty$. However, in a beginning time period of simulation, the co-infected subpopulation (x_4, x_9) are increasing and then tends to zero afterward. On the other hand, dengue-infected subpopulation (x_2, x_7) and chikungunya infected subpopulation (x_3, x_8) tend to equilibrium Ω_2 and Ω_3, respectively.

In Figure 5, the other condition of R_{0e} is performed, i.e when $R_{0e} > 1$ while R_{0c} and R_{0d} are varied. It can be seen that when $R_{0e} > 1$, then system (1) will tend to endemic equilibrium of co-infection situation between dengue and chikungunya (Ω_4) no matter the condition of R_{0c} and R_{0d}. It can be seen also that when basic reproduction number for dengue and chikungunya, R_{0c} and R_{0d}, respectively less than 1, number of co-infection subpopulation are much larger in the equilibrium point rather than when one or both of R_{0c} and R_{0d} are worth less than 1.

Second simulation is given in Figure 6 to show how the various value of infection rate for co-infection in human and mosquito population affect the dynamic of all subpopulation in system (1). It can be seen that when β_{xe} and β_{ye} are given for various value from 0.01 to 0.2, susceptible and recovered subpopulation in human and mosquitoes population only slightly different. In the other hand, number of infected human and mosquito (co-infection subpopulation) change significantly. This situation confirms the analytical result that β_{xe} and β_{ye} will dominate the dynamic of system (1) whenever these value increase in to larger than 1 (see Figure (3)).

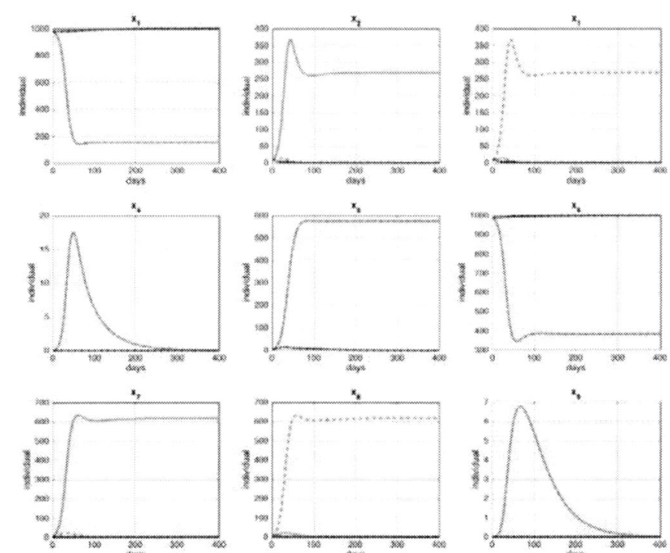

Figure 4. Simulation results for 1st case showing that when $R_{0e} < 1$, then the co infection subpopulation will tend to zero when time is increasing.

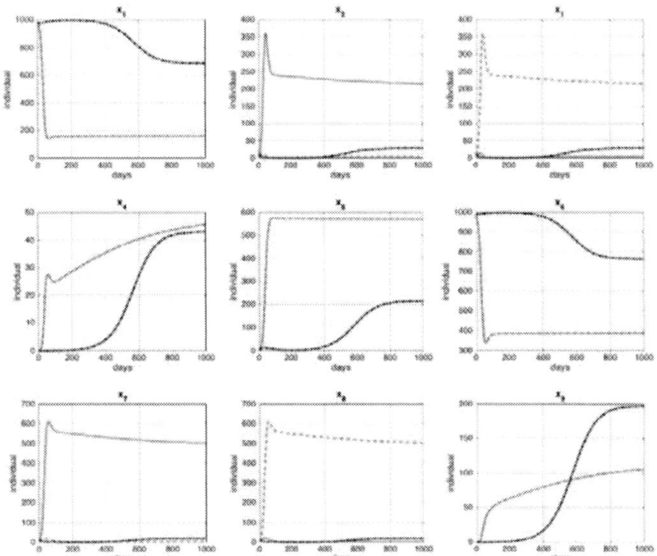

Figure 5. Simulation results for 2nd case showing that when $R_{0e} > 1$, then the co infection subpopulation will tend to the equilibrium point Ω_4 when time is increasing.

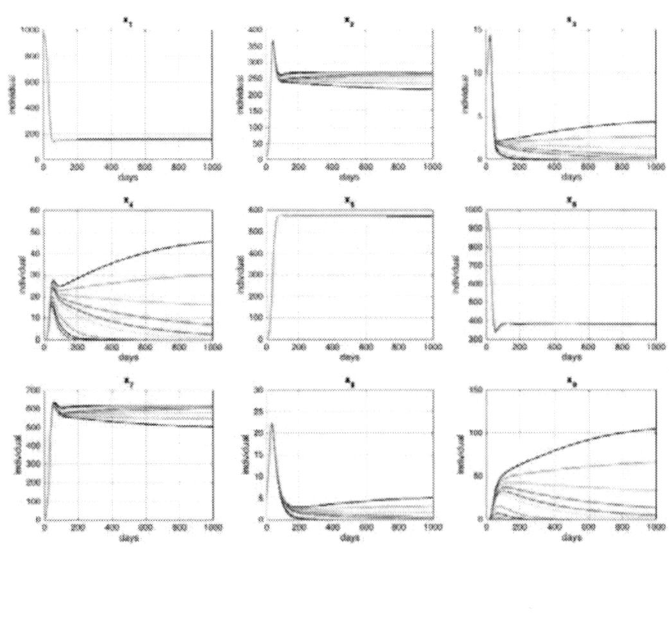

Figure 6. Simulation results for 3rd scenario when R_{0e} is varying while R_{0c} and R_{0d} are remaining the same.

5. Conclusions

In this paper, a comprehensive mathematical model for the co-infection between dengue and chikungunya has been discussed. The mathematical model is developed using a deterministic approach with a nine-dimensional system of ordinary differential equation. Mathematical analysis of the existence and local stability criteria of the equilibrium is given and connected with the basic reproduction number for the related model. We find that the basic reproduction number of the complete system is given by the maximum of three different basic reproduction numbers of three subsystems, i.e. for dengue, for chikungunya, and for co-infection between dengue and chikungunya. From numerical experiments for the local stability criteria, we find that the basic reproduction number for the co-infection sub-system dominates other basic reproduction numbers whenever this value is larger than one. The numerical simulation allowed us to observe this situation, as shown in the last figure in the numerical experiment section.

Acknowledgments

The authors thank to all reviewers for their constructive comments and suggestion. This research is funded by Universitas Indonesia with PITTA research grant 2017 (Project ID: 643/UN2.R3.1/HKP.05.00/2017).

References

[1] *WHO fact sheet about Dengue* [accessed July, 15 2017, available from: http://www.who.int/]
[2] History of Chikungunya Disease [accessed July, 15 2017, available from: http://www.chikungunya.in/]
[3] Dupont-Rouzeyrol M, O'Connor O, Calvez E, Daurès M, John M, Grangeon J P 2015 Co-infection with zika and dengue viruses in 2 patients, New Caledonia, 2014 *Emerg. Infect. Dis.* **21** pp 381-2
[4] Chahar H S, Bharaj P, Dar L, Guleria R, Kabra S K, Broor S 2009 Co-infections with chikungunya virus and dengue virus in Delhi, India *Emerg. Infect. Dis.* [serial on the Internet] **15** pp 1077–80
[5] Villamil-Gómez W E, González-Camargo O, Rodriguez-Ayubi J, Zapata-Serpa D, Rodriguez-Morales AJ 2016 Dengue, chikungunya and zika co-infection in a patient from Colombia *J Infect Public Health* pp 684-6
[6] Y E Putri, S Rozi, H Tasman, and D Aldila 2017 Assessing the effect of extrinsic incubation period (EIP) prolongation in controlling dengue transmission with wolbachia-infected mosquito intervention *AIP Conf. Proc.* **1825** 020019
[7] D Aldila, N Nuraini, E Soewono 2015 Mathematical model in controlling dengue transmission with sterile mosquito strategies *AIP Conf. Proc.* **1677** 030002
[8] D Aldila, E Soewono, and N Nuraini 2012 On the analysis of effectiveness in mass application of mosquito repellent for dengue disease prevention *AIP Conf. Proc.* **1450** 103
[9] D Aldila, N Nuraini, E Soewono and A K Supriatna 2014 Mathematical model of temephos resistance in Aedes aegypti mosquito population *AIP Conf. Proc.* **1589** 460
[10] D Agustine, Y Mahatma, and D Aldila 2015 Effectiveness analysis of fumigation strategy in dengue disease prevention program (Case study: Jakarta province, Indonesia) *AIP Conf. Proc.* **1651** 25
[11] Gerardin P, Guernier V, Perrau J, Fianu A, Le Roux K, *et al.* 2008 Estimating chikungunya prevalence in La Reunion Island outbreak by serosurveys: Two methods for two critical times of the epidemic *BMC Infectious Diseases* **8** p 99

[12] Agusto F B, Shamise Easley, Kenneth Freeman, and Madison Thomas 2016 Mathematical model of three age-structured transmission dynamics of chikungunya virus *Computational and Mathematical Methods in Medicine* **2016** 4320514
[13] Raul Isea, Karl E Lonngren 2016 A preliminary mathematical model for the dynamic transmission of dengue, chikungunya and zika *American Journal of Modern Physics and Application* **3** pp 11-5
[14] Yakob L, Clements A C A 2013 A mathematical model of chikungunya dynamics and control: The major epidemic on Réunion Island *PLoS ONE* **8** 0057448
[15] O Diekmann, J A P Heesterbeek, and Roberts, M G 2010 The construction of next-generation matrices for compartmental epidemic models *J. R. Soc. Interface* **7** pp 873-85
[16] O Diekmann & J A P Heesterbeek 2000 *Mathematical epidemiology of infectious diseases: model building, analysis and interpretation.* (Chichester : Wiley)

IOP Conf. Series: Journal of Physics: Conf. Series **974** (2018) 012002 doi:10.1088/1742-6596/974/1/012002

Reduction of C_d in circular cylinder using two passive control at $R_e = 1000$ and 5000

Chairul Imron[1], **Basuki Widodo**[1], **Tri Yogi Yuwono**[2]

[1] Department of Mathematics, Institut Teknologi Sepuluh Nopember, Surabaya, Indonesia
[2] Department of Mechanical Engineering, Institut Teknologi Sepuluh Nopember, Surabaya, Indonesia

E-mail: imron-its@matematika.its.ac.id

Abstract. There are many ways to reduce the drag force of an object, one of them by adding another smaller object called the passive control. Two passive controls will be placed in front and rear of the main object. The main object is a circular cylinder, the passive control that is in front is the cylinder type-I and the rear is an elliptical or circular cylinder. The distance between the main object with the two passive controls is varied and the Reynolds number used is 1000 and 5000. We want to find the effective distance of the main object with both passive controls so that the drag coefficient of the main object is minimal compared to the non-passive control or with one passive control in front of.

1. Introduction

Research conducted by researchers aims to get new technology, where the new technology is expected to change the behavior of users of these new technologies. Research on the flow of fluids through an object with the aim of reducing the drag force of the object is mostly done by the researchers. Some researchers used one passive control placed in front of various shapes, such as cylindrical cylinders, type-I cylinders, type-D cylinders etc.

Industrial chimneys, the structure of the offshore , the structure of the flyover and others, designing such examples in groups. Circular cylinders, elliptical cylinders or other shapes are commonly used objects. The value of the drag coefficient depending on the geometrical shape of the object, so that a different geometric shape the drag coefficient values are also different. Fluid flow across an object or multiple grouped objects will produce different flow characteristics. This happens because of the interaction between the fluid flow and the object

Fluid flow through the surface of an object then the flow moving around the object will flow slowly even the flow attached to the object will have zero speed. This is caused by the friction between the fluid with the object. While other streams will interact, so the flow rate of the object will move faster. This indicates an increase in shear stress which will affect the flow velocity in each layer, called the boundary layer.

Research has been conducted on the flow of fluids through a single object, such as a single cylindrical circular object[2], or a modified cylinder such as a cylinder type-I or a cylinder type-D[3, 4]. While research has been conducted on fluid flow through more than one object, ie fluid flow through more than one cylinder of various sizes and configurations, fluid flow through circular cylinder with tandem configuration[5, 6, 7, 8] and its elliptical cylindrical side-by-side configuration[9, 10]. The above studies use the boundary layer concept, and the concept of the boundary layer can find the answer to the effect of shear stress having a very important role in the characteristics of the current around the object[1].

If an object is passed by a fluid, then the object will produce a drag force, while the drag force is influenced by several parameters, one of which is the drag coefficient. how to reduce drag on objects passed by fluid, one way is to add a smaller object in front of the main object called passive control. The addition of passive control is carried out to reduce the drag coefficient of 48%[5], also done with different Reynolds numbers and resulting in lower drag coefficient[6]. The cylinder type-*I* is a circular cylinder by cutting the left and right by a certain angle, so the cylinder is shaped like I. The best cutting edge is 53°, this is because the wake occurs wider than the other angle, formed also wider and more annoying flow stronger on the object wall.

In this paper, two passive controls will be used, ie, passive controls in front of the cylindrical type-*I* with the angle is 53° and passive controls behind the circular cylinders and elliptical cylinders. The position of the passive control in front is perpendicular to the flow, while the passive control in the rear is landscape. The distance between the passive control in front and the circular cylinder has varying distances, as well as the distance between the passive control in rear and the circular cylinder. The Reynolds number used is $R_e = 1000$ and 5000. When do these two passive controls effectively decrease the drag coefficient?

2. Numerical Method

The above-mentioned problem can be solved by using the unsteady incompressible fluid equation of Navier-Stokes equations.

$$\frac{\partial \mathbf{v}}{\partial t} + \nabla \cdot \mathbf{v}\mathbf{v} = -\nabla P \quad + \quad \frac{1}{Re}\nabla^2 \mathbf{v} \tag{1}$$

$$\nabla \cdot \mathbf{v} = 0. \tag{2}$$

where Re is the Reynolds number, $\mathbf{v}$ is the velocity, and P is the pressure. SIMPLE algorithm and numerical methods will be used to solve the Navier-Stokes equation. first thing to do is give the initial value for each variable. By ignoring the pressure component we will find the velocity component of the momentum equation, so (1) becomes

$$\frac{\partial \mathbf{v}}{\partial t} = -\nabla \cdot \mathbf{v}\mathbf{v} + \frac{1}{Re}\nabla^2 \mathbf{v} \tag{3}$$

by using the finite difference method, like

$$(f_x)_i = \frac{2f_{i+1} + 3f_i - 6f_{i-1} + f_{i-2}}{6dx} \quad \text{and} \quad (f_x)_j = \frac{2f_{j+1} + 3f_j - 6f_{j-1} + f_{j-2}}{6dy}$$

$$(f_{xx})_i = \frac{f_{i+1} - 2f_i + f_{i-1}}{dx^2} \quad \text{and} \quad (f_{yy})_j = \frac{f_{j+1} - 2f_j + f_{j-1}}{dy^2}$$

afterwards

$$\frac{\partial \mathbf{v}}{\partial t} = \frac{\mathbf{v}^{**} - \mathbf{v}^{*}}{\Delta t} = -\nabla P \tag{4}$$

because equation (2) then equation (4) becomes

$$\frac{\nabla \cdot \mathbf{v}^{*}}{\Delta t} = -\Delta P \tag{5}$$

IOP Conf. Series: Journal of Physics: Conf. Series **974** (2018) 012002 doi:10.1088/1742-6596/974/1/012002

by using *SOR* (*Successive Over Relaxation*)

$$(P_n)_{i,j} = (1-\varepsilon)((P_{n-1})_{i,j}) + \varepsilon(P_n)_{i,j} \tag{6}$$

the result in value P and the last step is

$$\frac{\partial \mathbf{v}}{\partial t} = -\nabla P \tag{7}$$

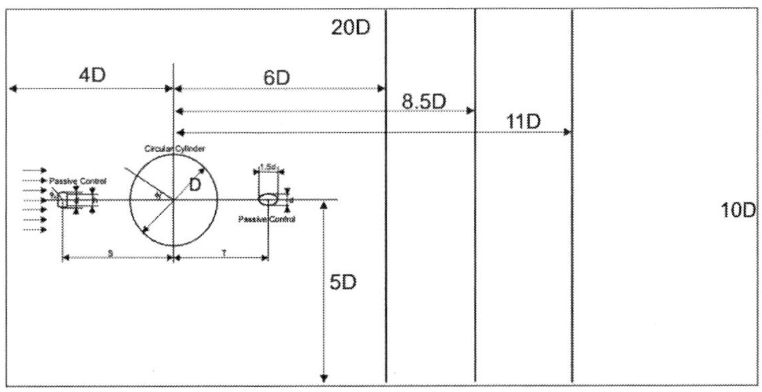

Figure 1. Design research system.

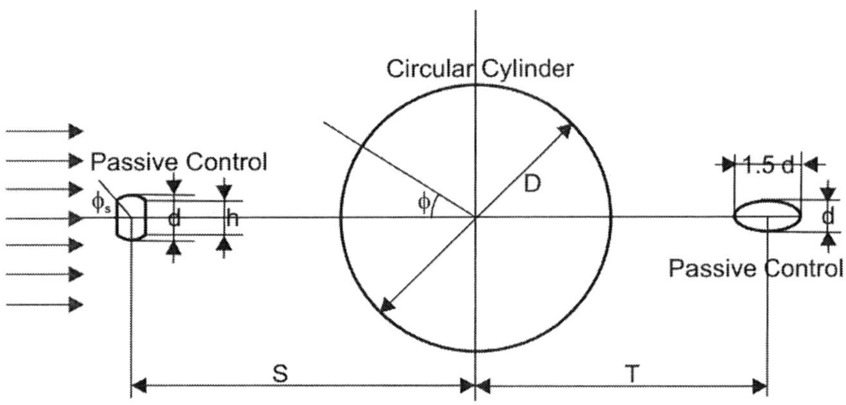

Figure 2. Schematic of two passive controls and circular cylinder.

3. Main Results

Our research system is $10D \times 20D$, where D is the diameter of the circular cylinder and is placed at a distance $4D$ from the front of the system and in the center of the system, like Figure 1.

In this paper, we use two passive controls. The first passive control is an cylinder type-I and is placed in front of a circular cylinder with varying distances, ie $S/D = 0.6, 1.2, 1.8, 2.4$ and 3.0. The second passive control are circular cylindrical and elliptical cylinders. The second passive control is placed rear of the circular cylinder at varying distances, ie $T/D = 0.6, 0.9, 1.2, 1.5, 1.8$ and 2.1 as in Figure 2.

The drag coefficient of a single circular cylinder has been obtained by using the simulation program, the results are compared with experimental results and other simulation programs. We calculate the drag coefficient of a single circular cylinder with $R_e = 100$ is 1,356, while other researchers with the same Reynolds number, such as Zulhidayat is 1.4 and by Five is 1.39[12]. In this paper will be simulated circular cylinder with two passive controls, and the Reynolds number used is 1000 and 5000. The drag coefficient for single circular cylinder with $R_e = 1000$ is 1.21, and for $R_e = 5000$ is 1.51.

Table 1. C_D of a cylinder circular for Re = 1000 and 5000 with difference S/D.

S/D	0.6	1.2	1.8	2.4	3.0
$C_{D_{1000}}$	1.114	0.946	0.896	0.894	0.911
$C_{D_{5000}}$	1.455	1.273	1.221	1.224	1.216

Table 1 is data of the drag coefficient of a circular cylinder with a passive control cylindrical type-I is located at the front with varying distances. From the table it shows that for Reynold number $R_e = 1000$, the best distance to get the minimum drag coefficient is $S/D = 2.4$ with the value of the drag coefficient is 0.894, while for Reynolds number $R_e = 5000$, the best distance for get the minimum drag coefficient is $S/D = 1.8$ or $S/D = 3.0$ with a drag coefficient of $1,221$ or 1.216. The value of the drag coefficient is still smaller than the drag coefficient without passive control.

3.1. Reynold number $R_e = 1000$

Table 2. C_D with passive control cylinder elips for Re = 1000 with diference T/D.

C_{D_E}	S/D				
T/D	0.6	1.2	1.8	2.4	3.0
0.6	1.090	0.923	0.874	0.872	0.889
0.9	1.056	0.892	0.843	0.842	0.860
1.2	1.061	0.896	0.847	0.846	0.863
1.5	1.049	0.885	0.837	0.836	0.853
1.8	1.036	0.874	0.828	0.827	0.844
2.1	1.029	0.870	0.825	0.824	0.841

The drag coefficient of a circular cylinder with two passive controls on the front and the rear. Passive control in front of circular cylinder is cylinder type-I, while passive control in rear of circular cylinder is elliptical cylinder. Data drag coefficient with Reynolds number $R_e = 1000$ and configuration as above, can be seen in Table 2. It appears that the passive control in rear it has a significant effect on the drag coefficient, since the drag coefficient is still smaller than that of a passive control just in front it or without passive control. Minimal drag coefficient of this configuration is 0.824, this occurs is the $S/D = 2.4$ and $T/D = 2.1$.

Passive control in front of the circular cylinder is the cylinder type-I, while the passive control behind the circular cylinder is a small circular cylinder. The data of the drag coefficient with the Reynolds number $R_e = 1000$ and the configuration as above, can be seen in Table 3. It appears that the passive control behind it has a significant effect on the drag coefficient, since the drag coefficient is still smaller than that of a passive control just in front of it or without passive control. the minimum drag coefficient of the configuration is 0.832, this occurs at $S/D = 2.4$ and $T/D = 2.1$.

Table 3. C_D with passive control cylinder circular for Re = 1000 with diference T/D.

C_{D_O}	S/D				
T/D	0.6	1.2	1.8	2.4	3.0
0.6	1.092	0.925	0.876	0.874	0.891
0.9	1.066	0.901	0.852	0.851	0.868
1.2	1.067	0.902	0.853	0.852	0.869
1.5	1.055	0.891	0.843	0.842	0.859
1.8	1.043	0.882	0.835	0.834	0.851
2.1	1.038	0.878	0.833	0.832	0.849

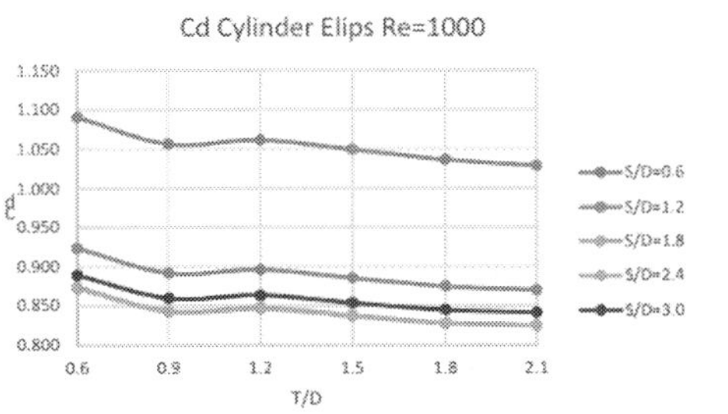

Figure 3. C_D Cylinder Elips $R_e = 1000$.

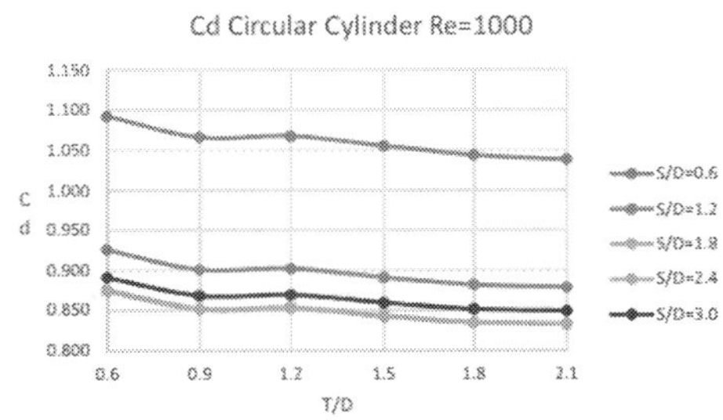

Figure 4. C_D Cylinder Circular $R_e = 1000$.

The graph of Table 2 is Figure 3 and Table 3 is Figure 4. From the two figures shows that the minimum drag coefficient to Reynolds $R_e = 1000$ is the configuration of passive control in the back is circular cylinder with a distance of $S/D = 2.4$ and $T/D = 2.1$. Elliptical cylinder passive controls that are within $S/D = 2.4$ and $T/D = 2.1$ are better than the circular cylindrical passive control that are within $S/D = 2.4$ and $T/D = 2.1$, in other words passive controls in the form of elliptical cylinder is more effective compared with passive control in the form of circular cylinder.

3.2. Reynolds number $R_e = 5000$

Table 4. C_D with passive control cylinder elips for Re = 5000 with diference T/D.

C_{D_E}	S/D				
T/D	0.6	1.2	1.8	2.4	3.0
0.6	1.155	1.005	0.967	0.986	0.982
0.9	1.172	1.016	0.991	0.983	0.991
1.2	1.151	1.004	0.967	0.980	0.976
1.5	1.368	0.941	0.906	0.925	0.957
1.8	1.461	1.225	1.183	1.193	1.193
2.1	1.404	1.229	1.160	1.163	1.157

The drag coefficient of a circular cylinder with two passive controls on the front and the rear. Passive control in front of circular cylinder is cylinder type-I, while passive control in rear of circular cylinder is elliptical cylinder. Data drag coefficient with Reynolds number $R_e = 5000$ and configuration as above, can be seen in Table 4. It appears that the passive control in rear it has a significant effect on the drag coefficient, since the drag coefficient is still smaller than that of a passive control just in front it or without passive control. Minimal drag coefficient of this configuration is 0.906, this occurs is the $S/D = 1.8$ and $T/D = 1.5$.

Table 5. C_D with passive control cylinder circular for Re = 5000 with diference T/D.

C_{D_O}	S/D				
T/D	0.6	1.2	1.8	2.4	3.0
0.6	1.116	1.012	0.973	0.992	0.987
0.9	1.205	1.043	1.015	1.007	1.008
1.2	1.169	1.014	0.977	0.990	0.986
1.5	1.412	1.277	0.916	1.265	1.245
1.8	1.557	1.284	1.225	1.222	1.195
2.1	1.401	1.384	1.220	1.209	1.191

The drag coefficient of a circular cylinder with two passive controls on the front and rear. Passive control in front of the circular cylinder is the cylinder type-I, while the passive control behind the circular cylinder is a small circular cylinder. The data of the drag coefficient with the Reynolds number $R_e = 5000$ and the configuration as above, can be seen in Table 5. It appears that the passive control behind it has a significant effect on the drag coefficient, since the drag coefficient is still smaller than that of a passive control just in front of it or without passive control. the minimum drag coefficient of the configuration is 0.916, this occurs at $S/D = 1.8$ and $T/D = 1.5$

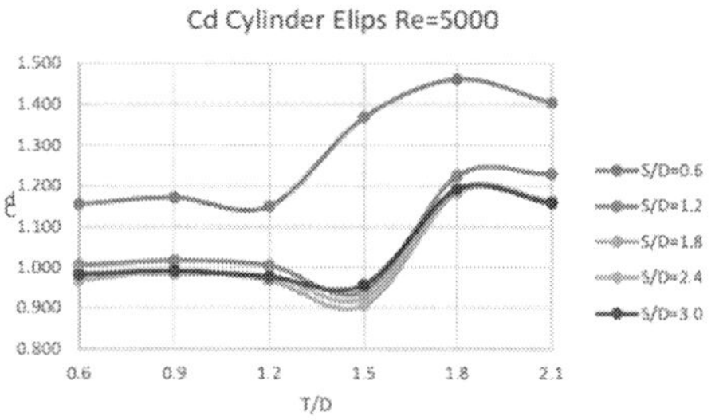

Figure 5. C_D Cylinder Elips $R_e = 5000$.

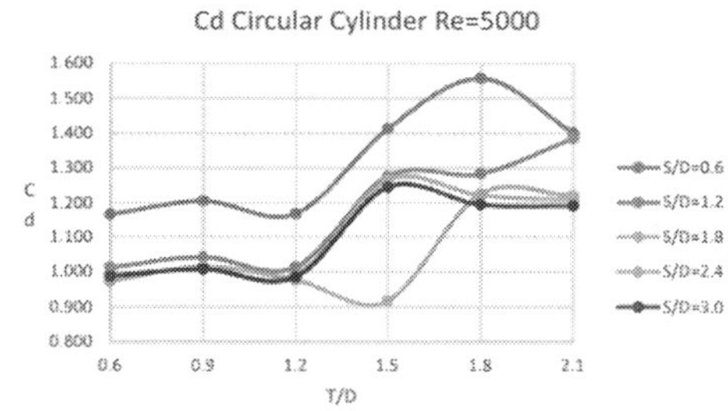

Figure 6. C_D Cylinder Circular $R_e = 5000$.

The graph of Table 4 is Figure 5 and Table 3 is Figure 4. From the two figures shows that the minimum drag coefficient to Reynolds $R_e = 5000$ is the configuration of passive control in the back is circular cylinder with a distance of $S/D = 1.8$ and $T/D = 1.5$. Elliptical cylinder passive controls that are within $S/D = 1.8$ and $T/D = 1.5$ are better than the circular cylindrical passive controls that are within $S/D = 1.8$ and $T/D = 1.5$, in other words passive control in the form of elliptical cylinder is more effective compared with passive control in the form of circular cylinder.

3.3. Wake

We show that the wake occurs for both passive controls with $R_e = 1000$ and 5000. We have shown that the drag coefficient of the main circular cylinder that there is a significant decrease.

We show that the decrease of the drag coefficient also affects the magnitude of the average velocity behind the circular cylinder. The velocity data we took from a distance of $6D$, $8.5D$ and $11D$ from a circular cylinder center or a distance of $10D$, $12.5D$ and $15D$ from the front of the system, like Figure 1.

Table 6. C_D for $R_e = 1000$ and 5000.

R_e	Single	1 PC	%	2 PC			
				Circ	%	Elips	%
1000	1.21	0.896	25.95%	0.832	31.24%	0.824	31.90%
5000	1.51	1.216	19.47%	0.916	39.34%	0.906	40.00%

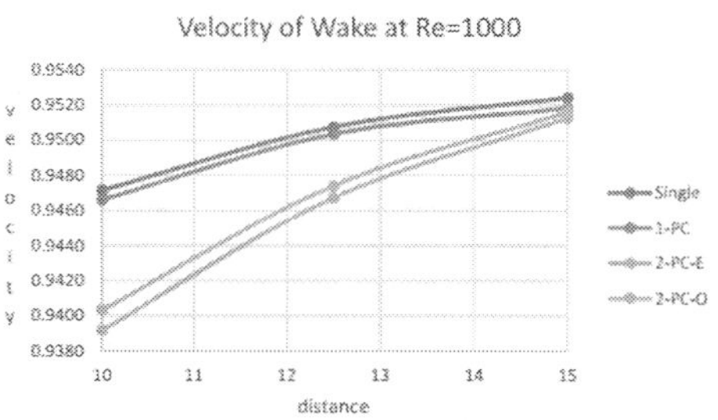

Figure 7. Wake of $R_e = 1000$.

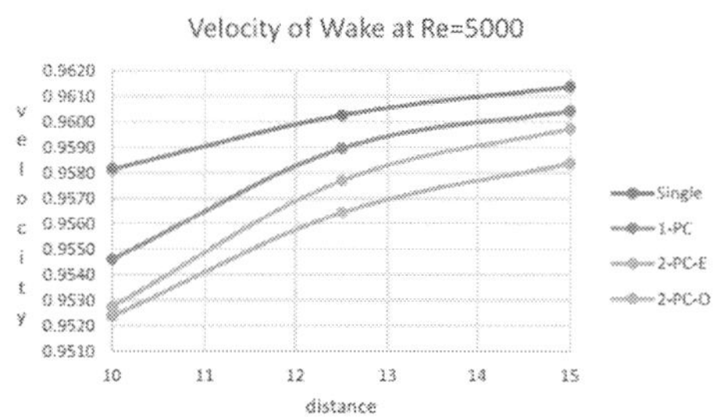

Figure 8. Wake of $R_e = 5000$.

From both figures, the flow velocity near the circular cylinder (ie, $6D$ from the center of the circular cylinder) and will increase as the distance farther away from the center of the circular cylinder and will return equally to the speed without passive control. The decrease in flow velocity behind the circular cylinder corresponds to a decrease in inhibitory coefficient as in Table 6.

4. Conclusion

The reduction of the drag coefficient in circular cylinder can be done by giving passive control. Passive control placement can be placed in front of and or behind. The inhibitory coefficient can be reduced by up to 40% if using passive front pass controls in the form of an cylinder type-I and an ellipse-shaped cylindrical backward control rather than a passive control drag coefficient, this is reinforced by decreasing flow velocity behind the circular cylinder.

5. Acknowledgements

We thank you and appreciation through the head of the Research and Community Service Institut Teknologi Sepuluh Nopember (LPPM ITS) Surabaya, Indonesia, on behalf of the Ministry of Technology and Higher Education (RisTekDikTi) which gave us a research grant under the contract No: 010/SP2H/LT/DRPM/IV/2017, April, 20th 2017, in giving us the opportunity to disseminate our research results to this paper.

References

[1] Widodo, B., (2012), The Influence Of Hydrodynamics On Pollutant Dispersion In The River, International Journal of Contemporary Mathematical Sciences (IJCMS) ISSN 1312-7586, Vol. 7, 2012, no. 45, from 2229 to 2234, HIKARI Ltd Journals and Books Publisher, Bulgaria.

[2] Ladjedel,A.O., Yahiaoui,B.T., Adjlout,C.L., and Imine,D.O., (2011), "Experimental and Numerical Studies of Drag Reduction on Circular Cylinder", World Academy of Sciences, Engineering and Technology, 77, 357-361.

[3] Igarashi T., and Shiba Y., (2006) "Drag Reduction for D-Shape and I-Shape Cylinders (Aerodynamics Mechanism of ReductionDrag)", JSME International Journal, Series B, Vol. 49, No. 4, 1036-1042.

[4] Triyogi Y., and Wawan Aries Widodo, (2010), "Flow Characteristics Around a D-Type Cylinder Near a Plane Wall", Regional Conferences on Mechanical and Aerospace Technology, Bali, Feb 9-10.

[5] Bouak, F. and Lemay, J. (1998), " Passive Control of The Aerodynamic Forces Acting on a Circular Cylinder", Experimental Thermal and Fluid Science, Vol. 16, 112-121.

[6] Tsutsui, T., and Igarashi, T., (2002), "Drag Reduction of a Circular Sylinder in an Air-Stream", Journal of Wind Engineering and Industrial Aerodynamics, Vol. 90, 527-541.

[7] Triyogi Y. and Nuh, M., (2003), "Using of a Bluff Body Cut from a Circular Cylinder as passive Control to reduce Aerodynamics Forces on a Circular Cylinder", The International Conference on Fluid and Thermal energy Conversion 2003, Bali, Indonesia, december 7-11, 2003.

[8] Lee, Sang-Joon., Lee, Sang-Ik., and Park, Cheol-Woo, (2004), "Reducing the Drag on a Circular Cylinder by Upstream Installation of a Small Control Rod", Fluid Dynamic Research, Vol. 34, 233-250.

[9] Imron, C., Yunus, M., (2015), "Mathematical Modeling of Pressure on Cylindrical Ellipse using Side-by-Side Configuration", International Journal of Computing Science and Applied Mathematics, vol 1, No. 1, 10-11, ISSN: 2477-5401, Indonesia.

[10] Imron, C., Apriliani, E., (2016), "Effect of Major Axis Length to the Pressure on Ellips", International Journal of Computing Science and Applied Mathematics, vol 2, No. 3, 41-44, ISSN: 2477-5401, Indonesia.

[11] Aiba, S, and Watanabe, H., (1997), "Flow Characteristics of a Bluff Body Cut from a Circular Cylinder", Journal of Fluids Engineering, Vol. 119, 453-457.

[12] Imron, C., Suharingsih, Widodo, B., & Yuwono, T. (2013), "Numerical Simulation of Fluid Flow Around Circular and I-Shape Cylinder in a Tandem Configuration", Applied Mathematical Sciences, vol 7, No. 114, 5657-5666, HIKARI Ltd Journals and Books Publisher, Bulgaria.

IOP Conf. Series: Journal of Physics: Conf. Series **974** (2018) 012003 doi:10.1088/1742-6596/974/1/012003

Forecasting selected specific age mortality rate of Malaysia by using Lee-Carter model

Halim Shukri Kamaruddin[1] and Noriszura Ismail[2]
[1]Taylor's Business School, Taylor's University, 47500, Subang Jaya, Selangor, Malaysia
[2]Pusat Pengajian Sains Matematik, Fakulti Sains dan Teknologi Universiti Kebangsaan Malaysia, 43600, Selangor, Malaysia

Email: HalimShukri.Kamaruddin@taylors.edu.my

Abstract. Observing mortality pattern and trend is an important subject for any country to maintain a good social-economy in the next projection years. The declining in mortality trend gives a good impression of what a government has done towards macro citizen in one nation. Selecting a particular mortality model can be a tricky based on the approached method adapting. Lee-Carter model is adapted because of its simplicity and reliability of the outcome results with approach of regression. Implementation of Lee-Carter in finding a fitted model and hence its projection has been used worldwide in most of mortality research in developed countries. This paper studies the mortality pattern of Malaysia in the past by using original model of Lee-Carter (1992) and hence its cross-sectional observation for a single age. The data is indexed by age of death and year of death from 1984 to 2012, in which are supplied by Department of Statistics Malaysia. The results are modelled by using RStudio and the keen analysis will focus on the trend and projection of mortality rate and age specific mortality rate in the future. This paper can be extended to different variants extensions of Lee-Carter or any stochastic mortality tool by using Malaysia mortality experience as a centre of the main issue.

1. Introduction

The rise in the life expectancy and gradual decline in the mortality rate across ages in any geographical area has attracted researchers in governments or corporate sector to investigate in further consequences of this situation. Mortality forecasting has become the central focus of study among the academicians and industrial researchers across borderless worldwide. The outcome can be used in different areas mainly in planning of retirement fund, pricing of life insurance contracts, public health or in further medical research interventions. The World Health Organization (WHO) is one of the bodies in the world concerns the movement of worldwide mortality as later is the source of public health and morbidity that most countries can rely on. Hence forecasting future mortality is an important subject for any health organization without exceptional for big organization like WHO in ensuring the stability of health in one nation or specific region in general. Accuracy and reliability forecasting tool play an important role in predicting future mortality and hence estimating the morbidity in specific sub-region. A quite number of mortality models have been proposed since Gompertz model firstly published his law of mortality in 1825 [1]. The methods used in mortality forecast were relatively simple and only involved a fair degree of subjective judgment [2]. Recent methods and tools have been developed by actuaries or demographers involved sophisticated and complicated models. Some of these methods using stochastic processes in the modelling with high outcome expectation in producing more reliable result, which in fact has better significant outcome compared to subjective opinions methods.

This paper presents the fitting of Malaysia mortality data for the past 29 years from the calendar years 1984 to 2012 using original Lee Carter model (1992) and hence from the fitted, Malaysian mortality can be forecasted up to the next ten years. From there, the two demensional fitted and forecasted mortality data will be cross-sectioned and examine for the certain ages. In Lee Carter model, it combines a demographic model with a statistical model of times series in order to forecast mortality rates. Girosi and King [3] views Lee Carter model as a special type of a multivariate process in which the covariance matrix depends on the drift vector and the innovations are inter-temporally correlated. The study on forecasting of mortality has been in the headline of mortality researches when the mortality rate decline dramatically in all developed countries in the twentieth century [4]. According to World Health Ranking 2012 published by WHO, Malaysia's life expectancy in 2011 is 73.9 which a huge jump compared to Malaysia's life expectancy in 1961, which only 54.3.

Apart from concentrating government side impact issues in mortality, insurance companies are also culprit in mortality issues mentioned above. Insurance companies are exposed to uncertainties which lead to whether a surplus or deficit of holding insurance contracts in money term. Life insurance contract in general specifies a stream of payments between the insured and the insurer contingent on some predetermined insurance events. Take whole life insurance as an example, insurer will pay benefits upon the death of the insured. Judging from this situation, the mortality rate and life expectancy are two of main studies should be focused in determining the contract price. Problems have been alarmed to either government or corporate parties in regards to the life insurance issued, mainly due to the inclining of the life expectancy of individuals. Insurance company also concerns on issues regarding insurance products that offer individuals lifetime coverage which highly dependent on the period until when they are alive which in other word individual's life expectancy. Venturing these issues, insurance companies have to make sure a proper lay out on how their funds should be adjusted according to the rapid declining in the mortality rates in Malaysia. In other word, an insurance company should be able to identify and model the sources of risk adequately such that it is be able to price the contract correctly [5]. Hence, a suitable and applicable model or tool of adjustment from the original model is essential for insurance companies so that they can have a clear view in the long run on the mortality rate and life expectancy of Malaysia in guiding them to decide the issued contracts with full accuracy.

Forecasting mortality rate has been developed as early as at the beginning of 1980s which can be sub-divided under three broad approaches: expectation, extrapolation and explanation. In the 1980s and early 1990s, actuaries have relied heavily on the subjective judgments for mortality forecast [2]. Evolution of research in mortality forecasting has drastically changed in the beginning of 1990s, researchers turn to sophisticated extrapolative methods and this forecasting method fully uses the element of statistical method [6]. This approach consists of the traditional and simple extrapolation of aggregate measures for example life expectancy and also more complex methods like the Lee-Carter model. Extrapolative methods assume future trends will essentially be a continuation of the past. In mortality forecasting, this is usually a reasonable assumption because of historical regularities.

Explanation approach makes use of the structural or epidemiological models of certain causes of death involving disease processes and known risk factors, example of this approach is the dependence of lung cancer on tobacco smoking. The main advantage of the explanatory approach is the feedback mechanisms and limiting factors can be taken into account in the modeling. However, this approach is not widely applied in mortality forecasting due to the insufficient study of the relationships between risk factors and mortality, which make them less reliable in forecasting. Many models used in explanatory forecasting are regression-based and therefore they are fit to be used in the GLM methods.

Some situations extrapolation method will not be useful for instance due to certain unpredictable factors, such as fluctuation and stagnation of mortality rate, the researchers began to develop alternative models instead of using simple linear extrapolation. Method that always used for this kind of case is time series due to their stochastic properties and capability to do calculation of probabilistic prediction interval for the estimated value. Time series model can have zero-factor or multiple factors such as time, age and cohort (factor that include correlation between time and age). For time series models with a zero factor, a simple random walk with drift model is developed and applied to age-specific rate. However, the model has a disadvantage which it will create a doubtful age pattern, so the researchers use iterated form of the seemingly unrelated regression estimator to forecast first differences of age-specific log mortality rates based on cross-correlated time series models [1]. For one-factor time series models, some researchers use parameterization functions and Brass relational model, which widely used when encountering poor data situation. Parameterization functions (laws of mortality) are one-factor models signifying the age model of mortality with the advantage of forecasting smoothness across age, familiarity and interpretability. Parameterization method requires repetitious fitting procedure from annual data to obtain a time series for each model parameter using the time series methods [1]. For two-factor time series model, e.g. Lee-Carter model and bivariate regression model which use generalized linear model (GLM) to discover the relationship of mortality rate between age and time (calendar year). Lee-Carter model uses matrix decomposition method to identify independent mortality factors such as age, time and other factors whereas GLM used appropriate time value for forecasting. As a result, time in GLM is considered as explicit covariate while time in Lee-Carter is an independent mortality factor [1]. In conclusion from past research certain extrapolative models and methods will be chosen to allow the discovery of the structural changes which happened among Malaysians as well as the impact of time and age factors towards in constructing fitting mortality rate and hence forecasting the mortality rate in the future which our main concerned focus in Malaysia.

2. Methodology

The Malaysia mortality data for this research are from year 1984 to year 2012 which obtained from Department of Statistics of Malaysia. For each calendar year, it consists of each age group from 0 to 80 year-old, number of Malaysia population as the exposure and number of death for each age. The data obtained are used to fit into the Lee-Carter model for both fitted mortality model and forecasted model. There are three ages selected, 5, 25 and 60 for mortality rate cross-sectional observation. These three ages are chosen to represent mortality pattern for three different groups of ages, i.e. children, young adult and elderly. The fitted mortality rate produced from Lee-Carter model will be cut cross and the trend of mortality will be analysed and observed.

2.1. Lee-Carter model

The basic form of Lee-Carter model is as following:

$$\ln m_{x,t} = a_x + b_x k_t + \varepsilon_{x,t} \tag{1}$$

where

$m_{x,t}$: the age-specific death rate for age x at time t

a_x : the average age-specific mortality

b_x : the rate of improvement of the level of mortality at age x

k_t : the mortality index at time t

$\varepsilon_{x,t}$: the random error for age x at time t

In Lee-Carter model, there is a linear relationship among the parameters and explanatory factors which are the age x, and the time t. The parameters b_x and k_t are estimated by applying the Singular Value Decomposition (SVD) method, assuming the errors are homoscedastic (all the errors have the same finite variance). Using this method, parameters b_x and k_t are subject to the constraints of:

$$\textstyle\sum_x b_x = 1 \tag{2}$$

$$\textstyle\sum_t k_t = 0 \tag{3}$$

Whereas, the a_x is simply the average values over time of the $\ln m_{x,t}$ values for each x. Therefore, a_x is then can be estimated by the following formula:

$$\widehat{a_x} = \frac{1}{n}\sum_{t=1}^{n} \ln(m_{x,t}) \tag{4}$$

At the same time, k_t is estimated by

$$\widehat{k_t} = \sum_x(\ln m_{x,t} - a_x) \tag{5}$$

These two estimated parameters will be used to estimate the b_x using the least squares estimation which the formula as follows:

$$\left(\ln m_{x,t} - \widehat{a_x}\right) = b_x\widehat{k_t} + \varepsilon_{x,t} \tag{6}$$

$$\sum_{x,t}(\ln m_{x,t} - \hat{a}_x - \hat{b}_x\hat{k}_t)^2 \Longrightarrow \widehat{b_x} = \frac{\sum_{t=1}^{n} \hat{k}_t\,(\ln m_{x,t} - a_x)}{\sum_{t=1}^{n} \hat{k}_t^{\,2}} \tag{7}$$

The estimated value of a_x or $\widehat{a_x}$ will give the general pattern of the age-specific death rates, $m_{x,t}$, and estimated value of k_t or $\hat{k}_t$ is the index describes the variation in the level of mortality at t. The $\hat{b}_t$ will describe the tendency of changes of mortality at the specific age x when the general level of mortality $(\hat{k}_t)$ changes. There is high possibility lacking of similarity between the predicted and the actual deaths at the beginning stage of estimation due to the estimation is calculated according to the log of death rates instead of the death rates. In order to overcome the issue and hence to reduce the discrepancies between the fitted death rates and the actual number of deaths, the $\widehat{k_t}$ is re-estimated, by taking the same value of $\widehat{a_x}$ and $\widehat{b_x}$ which estimated from the first step. According to [7], the new $\widehat{k_t}$ denotes as $\ddot{k}_t$, which estimated by an iterative procedure. When the fitted death rates are equal to the actual death rates, then following equation will be fulfilled:

Let

$$q_{x,t} = e^{\left(a_x + b_x\widehat{k_t}\right)} \tag{8}$$

Then

$$\underbrace{\sum d_{x,t}}_{\substack{total\ actual\\ death\ rate}} = \underbrace{\sum N_{x,t}\, q_{x,t}}_{\substack{total\ expected\\ death\ rate}} \tag{9}$$

where

$d_{x,t}$: the actual observed death rate
$N_{x,t}$: the population aged x at time t

The iterative method is done as follows:

(1) First, compare the total expected deaths (*TE*) to the total actual death (*TA*) in each period.
(2) There are three possible states of these comparison:
  (i) If *TE* > *TA*, decrease in *TE* is needed, thus adjusting the $\widehat{k_t}$ so that $\ddot{k}_t$ will be :
    1. $\ddot{k}_t = \widehat{k_t}(1 - d)$, if $\widehat{k_t} > 0$
    2. $\ddot{k}_t = \widehat{k_t}(1 + d)$, if $\widehat{k_t} < 0$
  the d is a small number.
  (ii) If *TE* = *TA*, stop here the iterations.
  (iii) If *TE* < *TA*, increase in *TE* is needed, thus adjusting the $\widehat{k_t}$ so that $\ddot{k}_t$ will be :
    1. $\widehat{\widehat{k_t}} = \widehat{k_t}(1 + d)$, if $\widehat{k_t} > 0$
    2. $\ddot{k}_t = \widehat{k_t}(1 - d)$, if $\widehat{k_t} < 0$
3) Go back to Step 1.

As the k_t index is obtained by the two stages of estimation, it is then able to forecast the index by using ARIMA. Lee and Carter (1992) describes k_t as a random walk with a drift term model such that:

$$\boldsymbol{k_t = k_{t-1} + d + \varepsilon_t} \tag{10}$$

where
d : the drift parameter
ε_t : the uncorrelated term

The term d is estimated by the least square estimator which equal to

$$\boldsymbol{\hat{d} = \frac{1}{n-1}\,(k_n - k_1)} \tag{11}$$

After the k_t index has been forecasted, mortality rate of the projected years is then possible to be obtained. From Andreozzi, Blaconá, & Arnesi [7], the forecast values of k_{n+h} are substituted in the formula:

$$\boldsymbol{\hat{m}_{x,n+h} = \hat{m}_{x,n} e^{\hat{b}_x(\hat{k}_{n+h} - \hat{k}_n)}} \tag{12}$$

where
n: the final year from which data are available
h: the forecast horizon
x: the age

As mention at the beginning of this section the procedure will be run using R-programming and the result and outcome of the data will be discussed in the analysis section.

3. Analysis

The fitted mortality rate from 1984 to 2012 using Lee-Carter portrayed in Figure 1(a) which has been generated by using RStudio. Each coloured lines represent different years and the colours of the lines follow the rainbow colour scheme: red, orange, yellow, green, blue, indigo and violet. The earliest year will be shown in red colour while the latest year will be in violet. Based on the Figure 1(a), the plotted lines generally show the mortality rates increase along with the age in any observed calendar year. However, between the age 0 and 20, the lines formed a rough U pattern. In between these ages, the lines are smoother and less jagged. The lines in Figure 1(a) also show the mortality rate for infants is higher compared to any children aged, young adults in between 15 and 34 and adults before age 40. These findings proved Malaysia has a high birth death rate. But across the calendar year from 1984 to 2012, the infant mortality rate from age 0 to age 5 has sharp decreasing dive with approximately 63.7% in average for these twenty-nine years observation. Government and WHO play an important role in realisation of this improvement. WHO has long term target to reduce two thirds of under five-year-old mortality from 1990 to 2015, which this agenda called Millennium Development Goal (MDA) [8]. From Figure 1(a) mortality rate under five-year-old is declining with average of 8.8% per annum. This statistically implying Malaysia has improved significantly in maternal and child health throughout the years. This drop in the infant mortality rate may due to several different factors. This remarkable improvement in minimizing the child and maternal mortality are the result of hard works from different factor of reasons mainly in Malaysia's policies, strategies and programs that involving socio-economic, cultural, educational, gender, and poverty dimensions [9]. Study in the past also found the mother's educational level has influenced her child's health in Malaysia. Thus, children's chances of surviving have positive correlation with the mother's education [8]. Observation of 29 years in Figure 1(a) also spot a trend of decreasing for mortality rate between age 5 to 10, age 15 to 40 and age 80 and above. However, there is slight increasing in rate of mortality for age in between 45 and 55 throughout 29 years. Moreover, roughly starting from age 30, it can be observed the rate of mortality increase begins to amplify. The Figure 1(a) also can be concluded in general the mortality rate of Malaysia have been lower from one calendar year to another. This may be due to the improvement on the hospital facilities or generally healthcare in Malaysia [10]. Besides that,

Malaysian citizen also pretty much aware over the years the importance of healthy lifestyle, and adopt it in such balance diets and exercises to maintain the healthiness.

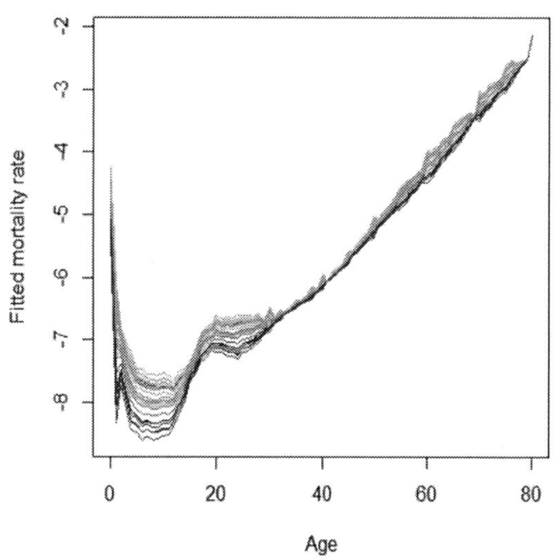

Figure 1(a). Graph of Fitted Line Using Lee-Carter Model

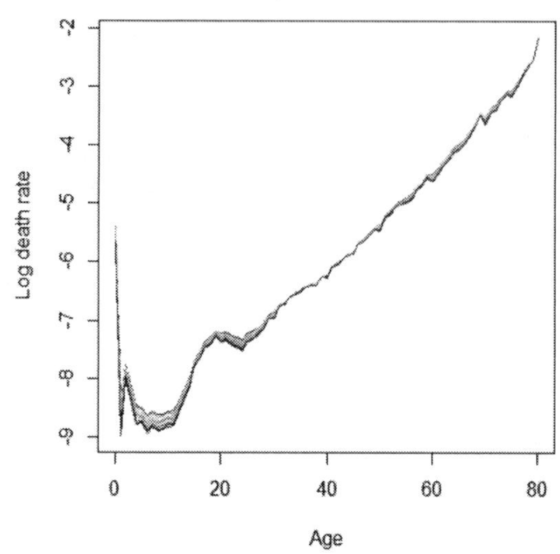

Figure 1(b). Graph of Forecasted Malaysia Mortality from 2013 to 2022.

The Figure 1(b) shows the forecasted mortality rate from year 2013 to 2022. There is no different in colour representing compared to The Figure 1(a). Mortality rate of Malaysia has obvious decreasing trend for infants and teenagers meanwhile the rest of the ages' trends has small increasing tendencies. In general in the future in whatever forecasted year, there is a sharp decrement of mortality rate at the infant age and also slightly decrease in teenager ages. However, a small increment of mortality rate from age 10 to 20. The elder group has the highest log mortality rates throughout the forecasted calendar years as compared to other age groups.

As mentioned in the methodology, further detail analysis of mortality rate can be observed in deferent angle which cross-section of the two dimensional mortality rate at certain selected ages. Figure 2(a), Figure 2(b) and Figure 2(c) are cross-section from different ages, which combination of fitted and forecasted mortality rate from Figure 1(a) and Figure 1(b). Figure 2(a) shows the cross-section of the log mortality rate for age 5. In general the rate has downward trend. The log mortality rate at age 5 declined from -7.51 at year 1984 to -8.35 at year 2012 approximately. However, there are minimum fluctuations within this period of time. Influent by the modelling tool and historical data, the forecasted log mortality rate is continue to decline constantly from -8.35 to -8.75 approximately at year 2022. In overall, there is no large change on log death rate at age 5. Figure 2(b) show the log mortality rate for this age declined from -6.8 at year 1984 to -7.10 at year 1990. The log mortality rate increases inconsistently from -7.10 at year 1990 to peak, -6.58 at year 1996. The log mortality rate then drops to -7.15 at year 2000. Based on this historical data, the forecasted log mortality rate is continue to decline smoothly from -7.15 to -7.38 approximately at year 2022. The higher mortality rate among teenage age is due to smoking related illness result in public health problems. In all three National Health and Morbidity surveys, the prevalence of smoking among adults age 18 and above in Malaysia was 24.8% in 1996 as compared to 21.5% in 1986. It was reported that approximately 10000 deaths due to these illnesses within those few years which making it the primary cause of death in Malaysia [11]. Figure 2(c) shows the log mortality rate for age 60 and declining from -3.75 at year 1984 to -4.37 at year 2012 approximately. However, there are few fluctuations within this period time. Based on this historical data, the forecasted log mortality rate is continue to decline smoothly from -4.37 to -4.63 approximately at year 2022.

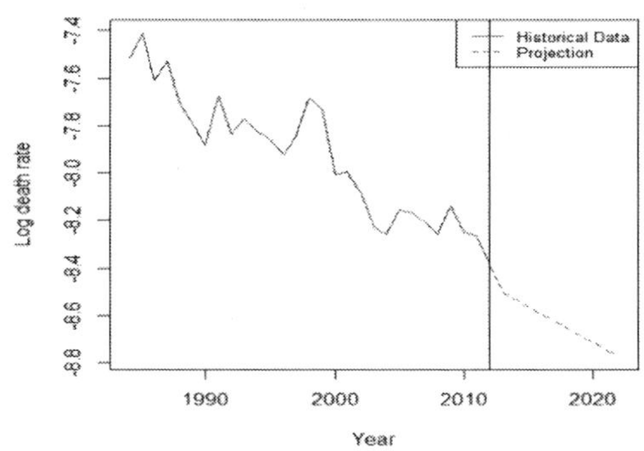

Figure 2(a). Past and forecasted morality rate based on age 5 from year 1984 to 2022.

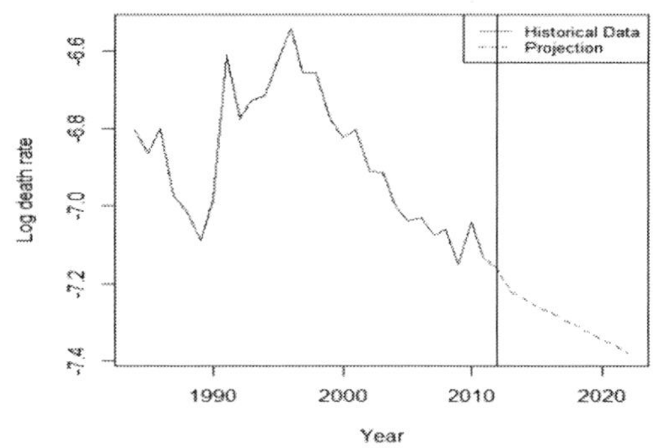

Figure 2(b). Past and forecasted morality rate based on age 20 from year 1984 to 2022.

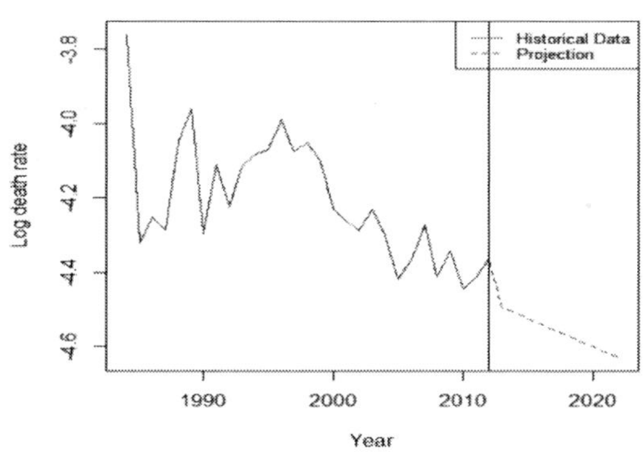

Figure 2(c). Past and forecasted log mortality rate based on age 60 from year 1984 to 2022.

4. Conclusion

Construction of mortality rate for certain ages by cross-section of combination of the fitted and forecasted mortality rate are still not widely done in Malaysia studies. The development of interest in mortality analysis among researcher in Malaysia especially in two-dimensional analysis are still in the beginning phase. The focus of conclusion and discussion are not supposed about the concentration of the models or improvement of the models which most researcher and reviewers lost interest about. The focal point in mortality research is supposed more toward improvement of socio-economics by examining at statistical analysis. Government and semi-government will definitely reap the extra benefits of expending the research matter. Past study by Haberman & Russolillo [7] stated Lee-Carter model is one of the most significant recent developments in the field of mortality forecasting which one of the reasons using this method as the main tool in this research. Besides that, the important feature of Lee-Carter model is for a precision of value of the time index k which enables in defining a complete set of death probabilities in order to eliminate the inconstancies in constructing the life table. Other important feature of the Lee-Carter model is it allows uncertainty in forecasting which is called longevity risk. In addition, the strengths of this method are its simplicity and robustness in the context of linear trends in age-specific death rates. Thus, these features eventually gives Lee-Carter model a good fit to the Malaysia mortality data and hence the keen analysis. This research definitely can be extended to variant different of Lee-Carter extensions or any stochastic mortality tool like Cairns-Blake-Dowd model and Poisson Log-Bilinear model by using Malaysia mortality experience as a centre of the main issue.

5. References

[1] Booth H and Tickle L 2008 *Annals of Actuarial Science* **3** issue 1-2 Mortality modelling and forecasting: A Review of Methods pp 3-43

[2] Pollard J H 1987 *Population Bulletin of the United Nations* **21-22** Projection of age-specific mortality rates pp 55-69

[3] Girosi F and King G 2008 Demographic Forecasting (United Kingdom: Princeton University Press)

[4] Booth H, Maindonald J and Smith L 2002 *Population Studies: A journal of Demography* **56(3)** Applying Lee-Carter under conditions of variable mortality decline pp 325-336

[5] Dahl M 2005 On Mortality and Investment Risk in Life Insurance (Denmark: University of Copenhagen)

[6] *Continuous Mortality Investigation Report* Number 21 2004 (United Kingdom: Institute of Actuaries and Faculty of Actuaries)

[7] Haberman S and Russolillo M 2005 *City University – Actuarial Research Paper* No. 167 Lee-Carter Mortality Forecasting: Application to the Italian Population.

[8] Bujang M A, Hamid A M A, Zolkepali N A, Hamedon N M, Lazim S S M and Haniff J 2012 *Journal of Health Informatics in Developing Countries* **6(2)** Mortality rates by specific age group and gender in Malaysia: Trend of 16 years, 1995–2010 pp 521-522

[9] UNICEF Malaysia - State of the World's Children - Situation in Malaysia. (n.d.) 2014

[10] Efforts aim to improve quality of health service – BorneoPost Online | Borneo, Malaysia, Sarawak Daily News | Largest English Daily In Borneo. (n.d.). 2014

[11] Lim K H, Sumarni M G, Amal N M, Hanjeet K, Rozita W W and Norhamimah A 2009 *Trop Biomed, 26*(1) Tobacco use, knowledge and attitude among Malaysians age 18 and above pp 92-99

Acknowledgment

The authors would like to thank the Department of Statistics Malaysia for their support in supplying the death mortality data.

IOP Conf. Series: Journal of Physics: Conf. Series **974** (2018) 012004 doi:10.1088/1742-6596/974/1/012004

Opinion mining on book review using CNN-L2-SVM algorithm

M F Rozi[1], I Mukhlash[1], Soetrisno[1], and M Kimura[2]

[1] Mathematics Department, Institut Teknologi Sepuluh Nopember, Surabaya, Indonesia
[2] Information Science and Engineering, Shibaura Institute of Technology, 3-5-7 Koto-ku Toyosu, Tokyo, Japan

Email: muhammadfakhrurrozi95@gmail.com, imammukhlash@gmail.com, soetrisno@matematika.its.ac.id, masaomi@shibaura-it.ac.jp

Abstract. Review of a product can represent quality of a product itself. An extraction to that review can be used to know sentiment of that opinion. Process to extract useful information of user review is called Opinion Mining. Review extraction model that is enhancing nowadays is Deep Learning model. This Model has been used by many researchers to obtain excellent performance on Natural Language Processing. In this research, one of deep learning model, Convolutional Neural Network (CNN) is used for feature extraction and L2 Support Vector Machine (SVM) as classifier. These methods are implemented to know the sentiment of book review data. The result of this method shows state-of-the art performance in 83.23% for training phase and 64.6% for testing phase.

1. Introduction

Recently, internet users in this world are increasing every year. It makes internet become large marketplace for people to buy or sell their product on internet. They use electronics and technology infrastructure to make transaction (selling, buying, transfer, or exchanging product, service, and/or information), included in business, business to business interaction, and costumer to business, this action is called E-commerce [1]. However, there are still many problems and enhancements for the implementation that makes many comments, suggestions, and critics emergence in purchasing of product. Review of a product defines the quality of product itself that has been used. That's review can be the opinions that represent good or bad product. But it is so hard to interpret real meaning of that opinion in the implementation. It shows us that it is necessary to extract the meaning of review in order to know the sentiment of that review. The process to extract useful information from user review is called opinion mining. This process is applied by using natural language processing and text analysis methods [2]. A review or opinion is processed in order to know the emotion that is written in that opinion accurately, so that we can get the information of quality of a product that can be classified become good, neutral, or bad. Review of a product has an important role in business. Customer will trust to a product that have good review from other customer. This makes many sellers need positive review for their product in order to be sold well.

There are many methods that have been used in opinion mining. The research from Pang and Lee tried to mine the opinion of movie review by using Naïve Bayes and Support Vector Machine (SVM), besides that they used Based on Minimum Cut for feature selection. That research yielded 86.4% accuracy [3]. Besides that, there is a model which is more modern and enhancing nowadays called Deep Learning Model. One of method that implement Deep Learning model is Convolutional Neural Network (CNN). Generally, CNN is used for image processing as classifier or cluster. Meanwhile, Tang proposes in his research, combination of CNN and L2-Support Vector Machine (L2-SVM) that is tested on image data [4]. His research tried to change activation function of CNN that usually use softmax function become L2-SVM. After that, in 2014 Kim delivered his innovation about implementation of CNN on sentence classification that shown state-of-the art performance [5].

Therefore, this research use combination of CNN and L2-SVM methods for opinion classification. This method will yield two kind of class of opinions, positive and negative. CNN method is used likes Kim's research but activation function is L2-SVM likes Tang's research. Testing of this method is used in book review data that is got from website *amazon.com*. Besides that, performance of method is measured as consideration for next research.

2. Preliminaries

2.1. Related works

There are many methods that had been used by researcher for mining an opinion. It started from a research from Pang and Lee whose were tried to mine opinion from movie review by using Naïve Bayes and Support Vector Machine (SVM) also by using feature selection, Based on Minimum Cut. Their research concluded that their method could reach 86.4% performance [3]. Next, research about opinion mining continued with many kinds of method and combination such as Maximum Entropy, Backpropagation, K-means, that yielded performance 85.4%, 86%, and 78% respectively [3]. Besides that, there is a model which is more modern and enhancing nowadays called Deep Learning Model.

In 2013, Tang introduced his research about implementation of deep learning which is combined with SVM. One of deep learning method that's used by him is Convolutional Neural Network (CNN). In that research, He tried to compare performance between deep learning generally and combination with SVM. He changed activation function of CNN that usually use softmax function become L2-SVM. It was done because Tang thought that there is similarity between softmax classifier and L2-SVM classifier. The result is error of L2-SVM is less than softmax. Because of that, He concluded that changing softmax become L2-SVM can easily help classification process [4].

After that, in 2014, Kim introduced implementation of CNN method on a sentence classification. That research proposed deep learning model that's applied in Natural Language Processing (NLP). Kim made conclusion in his research that simple CNN with one convolution layer can yield state-of-the art performance.

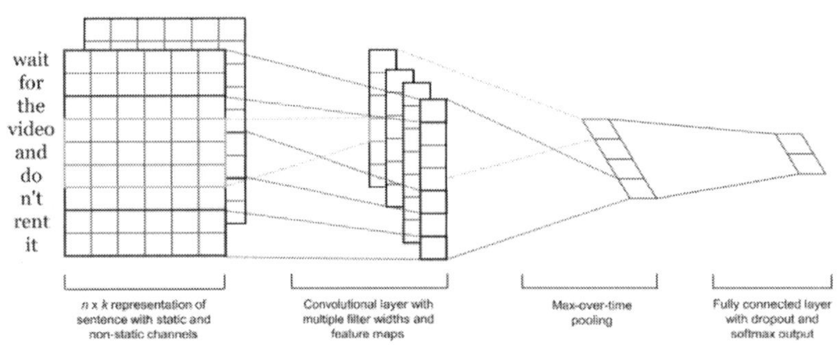

Figure 1. Model architecture with two channel for a sample sentence

2.2. Convolutional neural network

Convolutional Neural Network (CNN) is advanced of Multilayer Perceptron (MLP) that is designed to process two-dimension data. CNN belongs to deep neural network because the depth of network is high and can be applied on many image data. Mechanism of CNN has similarity with MLP, but every neuron in CNN can be represented in two-dimension form that is not likes MLP which every neuron is one-dimension form.

The architecture of CNN consists of two main layers, convolution layer and pooling layer. Implementation of CNN on sentence classification is enough to use two main layers. It can be illustrated on figure 1.

2.2.1. Convolution layer
Convolution layer is first layer in CNN-L2-SVM network architecture. In this layer, convolution process is done to all word vectors of review. figure 2 shows the example of convolution process on a sentence.

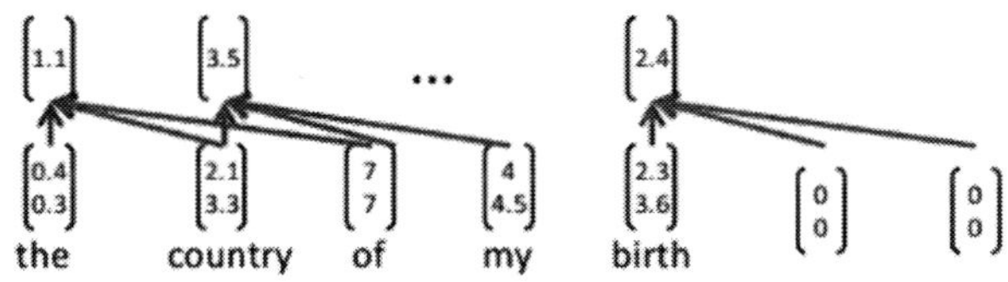

Figure 2. Example of convolution process on a sample sentence

Let the length of a review sentence is n, therefore there are n word vectors with size of dimension is 50 as the input of model. Assume $x_i \in \mathbb{R}^{50}$ is the i-th word vectors of a sentences. If word vectors are concatenated, the review will be able to be represented like below:

$$x_{1:n} = x_1 \oplus x_2 \oplus x_3 \oplus \dots \oplus x_n \quad (1)$$

Where $\oplus$ is concatenation operator. Generally, $x_{i:i+j}$ means the result of word vectors concatenation form i-th indices until $i + j$-th.

$$x_i, x_{i+1}, x_{i+2}, \dots, x_{i+j} \quad (2)$$

Convolution operation us a filter or parameter $w \in \mathbb{R}^{50h}$ where h is the size of window. In order to get feature value, we can get equation likes below:

$$c_i = f(net) \quad (3)$$

$$net = w \cdot x_{i:i+h-1} + b \quad (4)$$

At that equation, there is bias parameter $b \in \mathbb{R}$, $c_i \in \mathbb{R}$, are feature values that is yielded in i-th indices, and f is non-linear function. In this research, non-linear function that is used is Rectified Linear Unit (ReLu) function. Output of that function gives output constraint with positive value. The function can be written as below:

$$ReLu(x) = \max(0, x) \quad (5)$$

So that equations can be represented by:

$$c_i = ReLu(net) \quad (6)$$

$$net = w \cdot x_{i:i+h-1} + b \quad (7)$$

Filter w can be applied to each word window possible in sentence of review $\{x_{1:h}, x_{2:h}, x_{3:h}, \dots, x_{n-h+1:n}\}$. As the result, we can get feature map likes below:

$$c = [c_1, c_2, c_3, \dots, x_{n-h+1}] \tag{8}$$

2.2.2. *Pooling layer*
After we have got feature map that comes from convolution layer, so each of feature map can be processed in pooling layer. In this layer, the most importance value of each feature map is taken. Mathematically pooling operation can be written like below:

$$\hat{c} = \max\{c\} \tag{9}$$

Where $\hat{c}$ is maximum value of feature map c that is correlated with a filter. Because there are m filter, so the result of pooling layer is a vector that consists of maximum values of each feature map and there are m element. Therefore, we can obtain a vector below:

$$z = [\hat{c}_1, \hat{c}_2, \hat{c}_3, \dots, \hat{c}_m] \tag{10}$$

Where z is output vector from pooling layer that will input into the next layer.

2.3. *L2-support vector machine*
Linear support vector machine is firstly formulated for binary classification. Let training data and connected label is $(x_n, y_n), n = 1, \dots, N, x_n \in \{-1, +1\}$, SVM learning consists of constraint optimization likes below:

$$\min_{w,\xi_n} \frac{1}{2} w^T w + C \textstyle\sum_{n=1}^{N} \xi_n \tag{11}$$

$$s.t. w^T x_n t_n \geq 1 - \xi_n \qquad \forall n$$

$$\xi_n \geq 0 \qquad \forall n$$

ξ_n is slack variable which penalizes data point which violate margin requirements. Note that we can include the bias by augment all data vectors x_n with a scalar value of 1. The corresponding unconstrained optimization problem is the following:

$$\min_{w} \frac{1}{2} w^T w + C \textstyle\sum_{n=1}^{N} max(1 - w^T x_n t_n, 0) \tag{12}$$

Objective equation (12) is known as main constraint of L1-SVM, with standard hinge loss. Because L1-SVM is not differentiated, so that the most popular version is known as L2-SVM which minimalize squared loss.

$$\min_{w} \frac{1}{2} w^T w + C \textstyle\sum_{n=1}^{N} max(1 - w^T x_n t_n, 0)^2 \tag{13}$$

L2-SVM can be differentiated and cause more loss for pints which violate margin. To predict class label of testing data x:

$$\underset{t}{argmax}(w^T x)t \tag{14}$$

2.4. Algorithm

This research uses CNN and L2-SVM algorithm in opinion mining on review. CNN algorithm consists of 3 main layer, convolution, pooling, and fully connected layer. The input of this algorithm is a bunch of word vectors that is obtained from pre trained process. That CNN algorithm can be explained in pseudocode below:

```
input  : a data set of opinion sentences that is already represented by word vectors
output : class label of each data
window_size = {h_1, h_2, h_3, ..., h_n},
z = ∅
data = {X_1, X_2, ..., X_m}
for each h in window_size:
    w ← initializeFilter(h, m)
    for X in data:
        s ← size(X)
        c = ∅
        for i in 0: s − n + 1
            x ← concantenate(X^{i:i+h−1})
            temp ← nonLinear(w^T x + b)
            c ← c ∪ temp
        end
        ĉ ← max(c)
        z ← z ∪ ĉ
        probability ← softmax(z)
    end
end
```

Since there is a little bit modification of CNN in this research so that the last layer, fully connected layer is changed by SVM. The result of these algorithms is not a probability but score of class which determine label class of data. Then, L2-SVM loss function is used in training phase that influence the back propagation process to get optimal hyper parameter. The pseudocode of the combination between CNN and L2-SVM is written below:

```
input  : a data set of opinion sentences that is already represented by word vectors
output : class label of each data
window_size = {h_1, h_2, h_3, ..., h_n},
z = ∅
data = {X_1, X_2, ..., X_m}
for each h in window_size:
    w ← initializeFilter(h, m)
    for X in data:
        s ← size(X)
        c = ∅
        for i in 0: s − n + 1
            x ← concantenate(X^{i:i+h−1})
            temp ← nonLinear(w^T x + b)
```

IOP Conf. Series: Journal of Physics: Conf. Series **974** (2018) 012004 doi:10.1088/1742-6596/974/1/012004

$c \leftarrow c \cup temp$
end
$\hat{c} \leftarrow \max(c)$
$z \leftarrow z \cup \hat{c}$
end
end
$w \leftarrow initializeWeight(size(z))$
$skor \leftarrow svm(w^T z + b)$
$if\ skor > 0$
$return\ "positive"$
$else$
$return\ "negative"$

3. Result and discussion

3.1. Result

Dataset that is used for this research is book review dataset from Amazon.com that contain positive and negative review. There are 8 kind of books that have different distribution of positive and negative review. We can see from figure 3 the distribution of each books. Based on figure 3, we can see that the best distribution among other books is "*Gone Girl*". Proportion of positive review is less than others, so that the model that we will build is good enough to predict class label of book review.

After through training phase, we can get model which fit input data. The model is required to be tested the accuracy so that we can conclude that model is suitable model or not. Testing or validation phase is done in order to test data that has been separated before independently. Testing and training data is independent data so that we can see the ability of model to overcome new data. According to implementation result that had been done, model yield the accuracy of each iteration for all kind of data. On figure 4 shows the accuracy of model during training phase that had been done 20 times iteration.

From figure 4 we can see that the accuracy of model with training data is very good. Just 20 times iteration, the accuracy can reach 83.23% and will increase for next iteration. On the other hand, in order to get that result, it takes long times around 76 minutes.

Meanwhile, the accuracy of model with testing data had lower result than training data. It happens because there are not supervising of the result for classification in testing data. figure 5 illustrates the accuracy of training data in 20 times iteration.

We can see figure 5 above that accuracy of testing data is fluctuated if we compare with the result of training data. The accuracy of model with testing data yield final result 61.94%. We can see from figure 5 that the accuracy of testing data decrease from 66% until 60 – 61% at the last iteration. If we see the result of testing and training phase that the graph is in contrary. It means that model has overfitting. Overfitting is an event where model is too good in training data but too bad in testing data. This problem is common in classification with artificial neural network.

3.2. Discussion

Model that has been created is must evaluated. The model need to know the performance for doing specific task. There are several evaluation methods that can be done for classification on a model such as accuracy, precision, recall, and F1 score. In this research, there are four evaluation method are used to know the performance of model. We can see the result in figure 6.

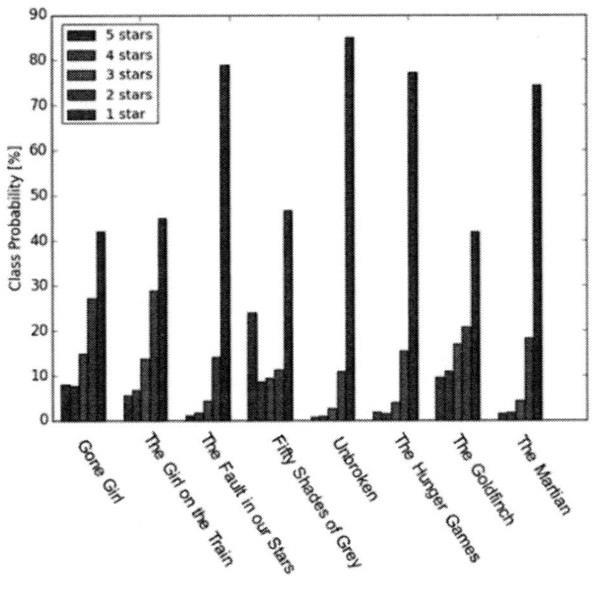

Figure 3. Distribution of book review label

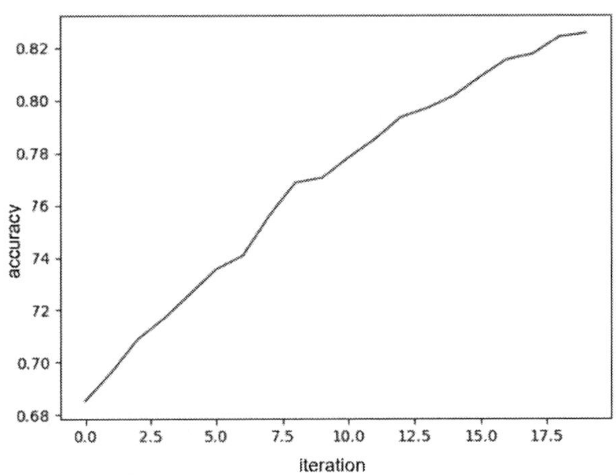

Figure 4. Graph of training data accuracy movement

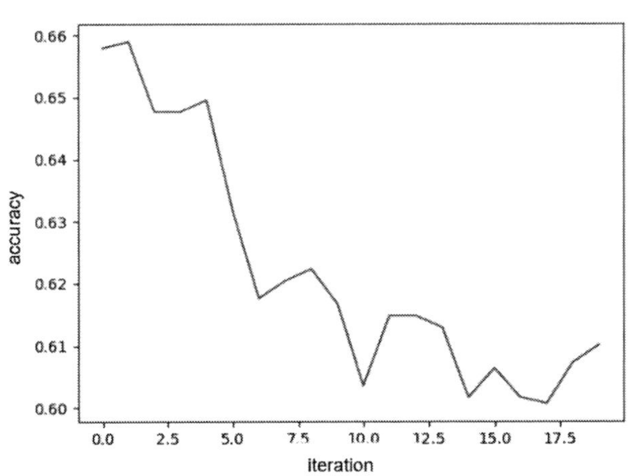

Figure 5. Graph of testing data accuracy movement

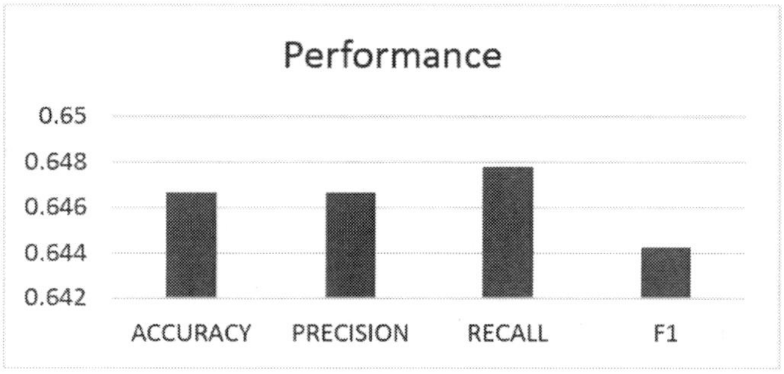

Figure 6. The result of model evaluation

According to that graph, we can see that the accuracy of testing data is around 0.646 or 64.6%. For the precision, we got 0.646 or 64.6% too. Meanwhile, recall value is around 0.647 or 64.7%. That is little bit different with F1 score. We got value around 0.644 or 64.4% which is lower than others. These result shows us that model have good enough performance to classify book review data with accuracy, precision, recall, and F1 score are in same value around 64%.

4. Conclusion

According to the result, it can be concluded that model conduction of Convolutional Neural Network (CNN) and L2 Support Vector Machine (L2-SVM) are able to determine sentiment or opinion of book review. Besides that, we get the performance of model to determine sentiment and opinion of review around 83.23% for training phase and 64.4% for testing phase.

5. References

[1] Manzoor, Amir. 2010. *E-Commerce: An Introduction. Saarbrucken: Lap Lambert*

[2] Ali, Farman.Kwak, Kyung-Sup. dan Kim, Yong-Gi. 2016 *Opinion mining based on fuzzy domain ontology and Support Vector Machine: A proposal to automate online review classification*. Elsevier. 47:235-250

[3] Singh, V. and Dubey, S. K. *Opinion Mining and Analysis: A Literature Review*. IEEE 5th International Conference- Confluence The Next Generation Information Technology Summit. 2014. page 236-237

[4] Tang, Y. 2013 *Deep Learning using Linear Support Vector Machines* International Conference on Machine Learning 2013: Challenges in Representation Learning Workshop. Atlanta USA.

[5] Kim, Y. 2014 *Convolutional Neural Networks for Sentence Classification* Proceedings of the 2014 Conference on Empirical Methods in Natural Language Processing (EMNLP). page 1746-1751

Implementation of multilinear regression in inhibition of thyroid hormone receptor-coactivator

A S Devi[1], **D Sarwinda**[1] **and G F Hertono**[1]
[1] Department of Mathematics, Universitas Indonesia, Depok, 16424, Indonesia
Abstract. Thyroid Hormone Receptors (TRs) play a big role in our body. Upon its nature TRs need coactivator to bind. MSNBAs had claimed to be the inhibitor of thyroid hormone receptor-coactivator interaction. Ninety-five targeted has been obtained by in vivo which classified to active and unspecified compounds. In this study, the active compounds will be tested ex vivo using multilinear regression method to explain the relationship of the physicochemical properties and biological activity – that is Quantitative Structure-Activity Relationship (QSAR) model and to show that the model is an analogue to the synthesis of MSNBAs. It appears that MSNBAs is fit to be an inhibitors of thyroid hormone receptor-coactivator with R-sq > 0,95.

1. Introduction

TRs belong to nuclear hormone receptor superfamily, which regulate development, growth, and metabolism. [1] TR's activity is stimulated *in vivo* by thyroid hormone (T3). [2] TRs contain three functional domains which are an amino terminal transcription activation domain (AF-1), a central DNA binding domain (DBD), and a carboxyl terminal ligand binding domain (LBD) that includes a T3-inducible coactivator binding domain, AF-2.[3] TRs have been classified to two classes, TRα (TRα1) and TRβ (TRβ1 and TRβ2), based on chromosomal localization and amino acid homology. [4] TRα1 is more expressed in bone, gastrointestinal tract, cardiac and skeletal muscle, and the central nervous system. [5] The faulty of TRα makes the thyroid gland can't make enough thyroid hormone to keep the body running normally, causing hypothyroidism. [6] On [7], two scaffolds, β-aminoketones and methylsulfonylnitrobenzoates (MSNBs) act as antagonists of coactivator binding to TRs by competing with NR boxes for binding to the receptor. Though the liabilities make the compounds unstable and give side effect on cardiac ion channel activity. It leads to another experiment on MSNBs with replacement the ester with an amide linkage. The result show that 5 compounds (<50μm) were active to inhibit VDR-SRC2-3, 13 compounds (<50μm) were likely to inhibit the interaction between TRα-SRC2-2, and 5 compounds (<20μm) were active to be an inhibition of TRβ-SRC2-2. [7]

However, in vivo experiments are very expensive and time consuming. At this thought, applied statistic in regression analysis has received attention in various science disciplines. Regression Analysis is performed to determine the correlation between two or more variables having cause-effect relation, to make predictions for the topic by using the relation. It answers some questions such as "is there any relation between dependent and independent variables", "if there are any, what is the power of the relation", "is it possible to make future-oriented predictions regarding the dependent variable", and so on. [8] Regression using one independent variable is called univariate regression analysis while using more than one independent variable is called multivariate regression analysis. The regression model with one dependent variable and more than one independent variable is called multilinear regression or multiple linear regression. [9]

Chemistry, for example, in early and late drug discovery we can use the interaction of drugs with their biological counterparts are determined by intermolecular forces, i.e. by hydrophobic, polar, electrostatic, and steric interactions. QSAR derives models which describe the structural dependence of biological activities either by physicochemical parameters (Hansch analysis), by indicator variables encoding different structural features (Free Wilson analysis), or by three-dimensional molecular property profiles of the compounds (comparative molecular field analysis, CoMFA). QSAR, based on Hansch Analysis is the same as multilinear regression. The physicochemical parameters are group to hydrophobic, polarizability, electrostatic, and steric interactions an analogue to predictors in regression. $Log\ \frac{1}{IC_{50}}$ is the logarithm of the inverse molar dose that produces or prevents a certain biological response. In this case, $Log\ \frac{1}{IC_{50}}$ acts as a dependent parameter or response. The independent parameters are : Log P as the logarithm of the n-octanol/water partition coefficient P, presenting hydrophobic; Molar Refractivity (MR), a polarizability parameter which is used to bind of the substituents to polar surface; Electronic properties describe the influence of a certain group or substituent on electron density distribution, one of them is Hammett Constant (σ); and finally, steric parameter (E_S) mostly use the van der Waals, occurs from geometric consideration or linear free energy relationship [11,12,13].

2. Method

We used multilinear regression to interpret the interaction between biological activity and its physicochemical parameters. It has been successfully found that lipophilicity makes a significant contribution to medicinal chemistry. It offered more comprehensive framework to optimize general absorption, distribution, metabolism and excretion properties, toxicology profiles and ultimately pharmacological response. Refer to QSAR predictors, Ligand Lipophilicity Efficiency (LLE) combines Log P and electrostatic which makes it hard to solve specific problems with [14,15,16]. Ligand Efficiency (LE) assesses the binding affinity in relation to the number of heavy atoms in a molecule [17]. It is like steric parameter on Hansch analysis. Binding Efficiency Index (BEI) and Surface Efficiency Index (SEI) are introduced after the Ligand Efficiency discoveries become hits. These two indices reduce the number of predictors by combining potency with molecular weight and polar surface area – as replacement to polarizability. It is suggested that these indices, either individually or in combination, are useful for effective and efficient drug discovery [18]. We used n (numbers of compound), R-sq (Coefficient of Determination), S (Standard Error of the Regression), F-value (Fischer Statistic) and P-value to evaluate the model.

3. Result

All data is acquired from PubChem AID 738019, AID 738020, AID 738021 (https://pubchem.ncbi.nlm.nih.gov/) and previous paper. Four parameters are used in this model; SEI (Surface Efficiency Index), BEI (Binding Efficiency Index), LE (Ligand Efficiency) and LLE (Ligand Lipophilic Efficiency). The response is $Log\ \frac{1}{IC_{50}}$ represent the biological activity and four parameters as the independent variables or predictors. We use QSAR model by Hansch to find interaction between

$Log\ \frac{1}{IC_{50}}$ and four predictors; SEI, BEI, LE, and LLE. We execute this in statistical software, Minitab 17.

Table 1. Biochemical assay data of Inhibition of VDR

SID	$Log\ \frac{1}{IC_{50}}$	SEI	BEI	LE	LLE
164135898	-1.57978	3.21	8.92	0.17	1.28
164138757	-1.60206	3.19	10.15	0.20	1.98
164138759	-0.84510	3.66	9.50	0.19	2.28
164147066	-1.56820	3.22	9.95	0.20	2.50
164147067	-1.64345	2.97	9.69	0.19	2.45

The equation of Inhibition of VDR:

$$Log\ \frac{1}{IC_{50}} = -9.114 + 1793\ \text{SEI} + 1744\ \text{BEI} - 85.81\ \text{LE} + 0.6324\ \text{LLE} \quad (1)$$

n = 5, R-sq = 100%

Table 2. Biochemical assay data of Inhibition of TRα

SID	$Log\ \frac{1}{IC_{50}}$	SEI	BEI	LE	LLE
164130247	-0.27875	0.00415	0.01364	0.27	3.55
164130253	-1.4624	0.00329	0.0089	0.17	1.08
164133096	-0.23045	0.00419	0.01198	0.23	2.88
164135896	-1.64345	0.00316	0.01179	0.24	3.72
164135897	-1.51851	0.00325	0.01169	0.24	3.59
164135898	-1.51851	0.00325	0.00904	0.17	1.34
164138756	-1.5682	0.00322	0.01093	0.22	2.88
164138757	-1.50515	0.00326	0.01037	0.2	2.07
164141530	-0.62325	0.0039	0.01374	0.27	4.08
164147065	-0.98227	0.00342	0.01152	0.23	3.36
164147066	-1.64345	0.00316	0.00978	0.19	2.43
164147067	-1.39794	0.00314	0.01024	0.2	2.69
164147068	-1.4624	0.00329	0.01037	0.21	2.78

The equation of MSNBA as inhibitor of TRα:

$$Log\ \frac{1}{IC_{50}} = -5.491 + 1364SEI + 144BEI - 11.7LE + 0.195LLE \quad (2)$$

n = 13, S = 0.116716, R-sq = 96.56%, R-sq(adj) = 94.84%, PRESS = 0.239842, R-sq(pred) = 92.43%

Table 3. Biochemical assay data of Inhibition of TRβ

SID	$Log\ \frac{1}{IC_{50}}$	SEI	BEI	LE	LLE
164130247	-0.81954	0.00321	0.00892	0.24	3.01
164133096	-0.5563	0.00319	0.01015	0.22	2.55
164133099	-1.02531	0.00366	0.0095	0.21	2.74
164141530	-0.76343	0.00322	0.00995	0.26	3.94
164147065	-1.20412	0.00297	0.00969	0.22	3.14

The equation of Inhibition of TRβ:

$$Log\ \frac{1}{IC_{50}} = -3{,}301 + 1016\ SEI + 223.1\ BEI - 19.41\ LE + 0.1869\ LLE \quad (3)$$

n = 5, R-sq = 100%

3.1 The goodness of fit

The goodness of fit gives the expression on how well a model fits given set of data, or how well it will predict a future set of observations. There is various way to test the goodness of the model. R-sq measures the proportion of the total variation in $Log\ \frac{1}{IC_{50}}$ explained by the model. R-sq is said to be "good" if R-sq > 60%. Therefore, eq. 1, eq. 2 and eq. 3 show the model with R-sq 100%, 96.56% and 100% are good and can tell the relation between predictors and response. For eq. 2, it gives us more than just R-sq. such as R-sq(adj). R-sq(pred). and S. R-sq(adj) shows the contribution of predictors in the model. It explains that the four predictors in eq. 2 contribute 94.84% of total variance, which is good. R-sq(pred) expresses how well the model will adjust in the future with new observations. It gives 92.43% to overall model, S (Standard Error of the Regression), the smaller error we get, the better the model. The error is 0.116716 which means the average distance of the data from fitted line is 12%.

Table 4. Analysis of Variance of the Model (Eq. 2)

	F-Value	P-Value
Constant	56.11	0.000
SEI	33.84	0.000
BEI	0.33	0.582
LE	1.01	0.344
LLE	1.67	0.232

3.2 The Significance of the Model

The significance of the eq. 2 presents us the F-Value and P-Value. It uses to see which parameters in particular are significant predictors of the response. The F-Value from the ANOVA should be smaller than F-value on F-table with the judgement of $\alpha = 0.01$. From the table. $F_{\alpha.4.8} = 7.01$, which make BEI, LE and LLE are the significant predictors. We should consider removing SEI although the P-Value $< \alpha$. However, the overall performance of model is shown by F-Value (56.11%) is greater than the P-Value. So, we can state that the model is fit better than removing the SEI.
The equation with three predictors which are BEI, LE and LLE :

$$Log\ \frac{1}{IC_{50}} = -5.251 + 1173BEI - 38.1LE - 0.234LLE \quad (4)$$

n = 13, S = 0.251641, R-sq = 82.00%, R-sq(adj) = 76.00%, PRESS = 1.10066, R-sq(pred) = 65.24%

The idea of removing SEI is seen to be bad decision. R-sq to these three predictors is smaller than the previous model. Neither the R-sq(adj), which decrease to 18.84% below, it means the loss of the model is facing caused by removing SEI. It applies also in R-sq(pred) and S. The model barely fits in the future with new observations which is affected by the decrease of R-sq(pred) and the error we get is greater than before. Therefore, we can be sure from previous statement that the model is fit better with four predictors which are SEI, BEI, LE and LLE.

From eq. 2 we can interpret the model to:

- For every 1 μm of SEI increase, it produces/prevents 1364 μm of biological responses.
- For every 1 μm of BEI increase, it decreases 144 μm of molar dose that produces/prevents biological responses.

- For every 10 points of LE increase, it decreases 117 points of biological responses.
- For every 1000 points of LLE increase, it increases 195 points of biological responses.

4. Conclusion

We can conclude that multilinear regression can be used to explain the relationship of physicochemical and biological activity between MSNBAs as inhibitor of thyroid hormone receptors (TRs). In this study, we present the multilinear regression models that we have are an analogue to the result of synthesis of MSNBAs. It gives us an overall model R-sq > 95%.

References

[1] Cheng S Y, Leonard JL and Davis PJ 2010 *Molecular Aspects of Thyroid Hormone Actions* PubMEd
[2] Fraichard A *et al* 1997 *The T3R Alpha Gene Encoding a Thyroid Hormone Receptor is Essential for Post-Natal Development and Thyroid Hormone Production* EMBO Journal p 4412-20
[3] Harvey C B and Williams G R 2002 *Mechanism of Thyroid Hormone Action* PubMed
[4] Mangelsdorf D J *et al* 1995 *The Nuclear Receptor Superfamily: The Second Decade* PubMed
[5] Visser W E, Friesema F C and Visser T J 2011Minireview : *Thyroid Hormone Transporters: The Knowns and The Unknowns* Molecular Endocrinology p 1 -14
[6] www.thyroid.org/hypothyroidism
[7] Hwang J Y 2013 *Synthesis and Evaluation of Methylsulfonylnitrobenzamides (MSNBAS) as Inhibitors of the Thyroid Hormone Receptor-Coactivator Interaction* Bioorg Med Chem Lett
[8] Uyanik G K and Guler N 2013 *A Study on Multiple Linear Regression Analysis* Procedia p 234 – 40
[9] Tabachnick B G and Fidell S L 1996 *Using Multivariate Statistics* 3rd edition (Harper Collins College Publisher, New York)
[10] Kubinyi H 2008 *QSAR: Hansch Analysis and Related Approaches* **1** ed R Mannhold *et al* (Wienhem, Germany) p 1 – 47
[11] Hansen O R 1962 *Hammett Series with Biological Activity* Acta Chemica Scandinavica p 1593- 1600
[12] Hansch C, Maloney P, Fujita T and Muir R M 1962 *Correlation of Biological Activity of Phenoxyacetic Acids with Hammett Substituent Constants and Partition Coefficients* Nature p 178- 80
[13] Hansch C and Fujita T 1964 ρ-σ-π *Analysis, A Method for the Correlation of Biological Activity and Chemical Structure* Journal of the American Chemical Society p 1616-26
[14] Waring M J 2010 *Lipophilicity in Drug Discovery* Expert Opinion on Drug Discovery p 235–48
[15] van de Waterbeemd H 2009 *Drug Bioavailability: Estimation of Solubility, Permeability, Absorption and Bioavailability* Vol 40 ed van de Waterbeemd H and Testa B (Wiley-VCH Verlag GmbH and Co, Weinheim, Germany) p 71–99
[16] Arnott J A and Planey S L 2012 *The Influence of Lipophilicity in Drug Discovery and Design* Expert Opinion on Drug Discovery p 863–75
[17] Hopkins et al, Groom C R and Alex A 2004 *Ligand Efficiency: A Useful Metric for*

IOP Conf. Series: Journal of Physics: Conf. Series **974** (2018) 012005 doi:10.1088/1742-6596/974/1/012005

Lead Selection Science Direct **9** p 430 – 1
[18] Abad-Zapatero C and Metz J T 2005 *Ligand Efficiency Indices as Guideposts for Drug Discovery* Science Direct **10** p 464 - 9

Quaternion wavelet transform for image denoising

Ahmad Khairul Umam[1] and Mahmud Yunus[2]

Mathematics Department, Institut Teknologi Sepuluh Nopember, Sukolilo, Surabaya 60111, Indonesia

E-mail: [1]ahmad_khairul_umam@yahoo.com, [2]yunusm@matematika.its.ac.id

Abstract. Quaternion wavelet transform (QWT) combines discrete wavelet transform (DWT) and quaternion Fourier transform (QFT). QWT has many applications included image processing. In this research, we discuss about construction, characteristics and implementation of QWT on process of image denoising. We construct denoising algorithm with QWT then we do simulation to know performance of algorithm. We use grayscale test images that have size 512×512 pixel with low, medium and high complexity. Experiment removes noise of image successfully. Results of image denoising are used to measure algorithm performance using PSNR (peak signal to noise ratio) value. We compare PSNR values with DWT and QWT for Haar, Biorthogonal, Daubechies and Coiflets wavelet. The method that has the highest PSNR value can be concluded the best performance.

1. Introduction

Quaternion wavelet is a new type of wavelet. Quaternion is proposed by William Rowan Hamilton in 1843. Gai [1] said that quaternion wavelet transform (QWT) is a new multiresolution analysis tool for describing geometric characteristics of the image. Quaternion value is composed of one magnitude and three angle phases. Li [2] said that wavelet transform is based discrete wavelet transform (DWT) and complex wavelet transform (CWT). CWT is a kind of efficient and popular technique for texture representation. Wavelet transform is improvement of Fourier transform. Fourier transform changes a signal domain from spatial domain only to frequency domain. While wavelet transform changes a signal domain from spatial domain to scaling and translation domain.

Yin [3] analyzes quaternion wavelet and applies to image denoising. Firstly, Yin studies standard orthogonal basis of scale space and wavelet space of quaternion wavelet transform (QWT) in spatial $L^2(\mathbb{R}^2)$. Yin proves and presents quaternion wavelet's scale basis function and wavelet basis function concept in spatial scale $L^2(\mathbb{R}^2;\mathbb{H})$. Next, QWT generalized Gauss distribution is used to model QWT coefficient's magnitude distribution. Yin uses Bayesian threshold with soft thresholding method to get threshold value. And then, Yin uses "Farrs" filters at level 1 and Q-shift dual-tree filters at higher levels in quaternion wavelet.

Gai [1] proposes a new hidden markov tree (HMT) model utilizing QWT. And then, Gai applies it to image denoising. Next, Kadiri [4] applies QWT to satellite image with noise. Firstly, Kadiri decomposes the noisy image by QWT. Secondly, Kadiri computes one magnitude and three angle phases $(\boldsymbol{\phi},\boldsymbol{\theta},\boldsymbol{\psi})$. Thirdly, Kadiri computes Neigh-Shrink thresholding of magnitude and phases smoothing of angle phases. The last, Kadiri reconstructs denoised image by QWT.

In this paper, we study QWT for image denoising with Haar, Biorthogonal, Daubechies and Coiflets wavelet. We use hard thresholding with VisuShrink method to get threshold value. We also use mean filter for image denoising. This experiment uses grayscale test images that are corrupted by white Gaussian noise with zero mean and variance σ^2. Test images have size 512×512 pixel with low, medium and high complexity.

2. Preliminaries

2.1. Discrete wavelet transform

Quaternion wavelet transform (QWT) combines some discrete wavelet transforms (DWT). We discuss construction of DWT with multiresolution analysis.

Definition 1. [5] A sequence of closed subspaces $\{\mathcal{V}_j : j \in \mathbb{Z}\}$ of $L^2(\mathbb{R})$ together with a function $\phi \in \mathcal{V}_0$ is called a multiresolution analysis (MRA) if it satisfies the following conditions:

- (Increasing) $\ldots \subset \mathcal{V}_{-1} \subset \mathcal{V}_0 \subset \mathcal{V}_1 \subset \ldots.$
- (Density) $\bigcup_{j\in\mathbb{Z}} \mathcal{V}_j = L^2(\mathbb{R})$.
- (Separation) $\bigcap_{j\in\mathbb{Z}} \mathcal{V}_j = \{0\}$.
- (Scaling) $f(t) \in \mathcal{V}_j$ if and only if $f(2t) \in \mathcal{V}_{j+1}$.
- (Orthonormal basis) There exist a scaling function (of the MRA) $\phi \in \mathcal{V}_0$ whose integer translates $\{\phi(t-n) : n \in \mathbb{Z}\}$ are an orthonormal base for $\mathcal{V}_0$.

■

Wavelet function that constructed by scaling function (father wavelet) is defined:

$$\phi_{j,k}(t) = 2^{\frac{j}{2}}\phi(2^j t - k) \in L^2(\mathbb{R}) \tag{1}$$

where ϕ is scaling function. Scaling function is defined by

$$\phi(t) = \sum_n h_\phi(n)\sqrt{2}\,\phi(2t-n)\,, n \in \mathbb{Z} \tag{2}$$

where $h_\phi(n)$ is scaling function coefficient.

Span of $\{\phi_{j,k}\}$ is defined by

$$\mathcal{V}_j = \overline{\underset{k}{\operatorname{span}}\{\phi_k(2^j t)\}} = \overline{\underset{k}{\operatorname{span}}\{\phi_{j,k}(t)\}} \tag{3}$$

where $\mathcal{V}_j$ is subspace of $L^2(\mathbb{R})$. We can write subspaces of $\mathcal{V}_j$ as

$$\mathcal{V}_0 \subset \mathcal{V}_1 \subset \mathcal{V}_2 \subset \cdots \tag{4}$$

where

$$\lim_{j\to\infty} \mathcal{V}_j = \{L_2(\mathbb{R})\}. \tag{5}$$

Wavelet function that constructed by mother wavelet is defined:

$$\psi_{j,k}(t) = 2^{\frac{j}{2}}\psi(2^j t - k) \in L^2(\mathbb{R}) \tag{6}$$

where ψ is mother wavelet. Mother wavelet is defined by

$$\psi(t) = \sum_n h_\psi(n)\sqrt{2}\,\phi(2t-n)\,, n \in \mathbb{Z} \tag{7}$$

where $h_\psi(n)$ is wavelet function coefficient.

Span of $\{\psi_{j,k}\}$ is defined by

$$\mathcal{W}_j = \overline{\underset{k}{\operatorname{span}}\{\psi_{j,k}(t)\}}. \tag{8}$$

Relationship of spaces $\mathcal{V}_j$ and $\mathcal{W}_j$ can be written by $\mathcal{V}_{j+1} = \mathcal{V}_j \oplus \mathcal{W}_j$, where $\oplus$ is direct sums of two spaces. Spaces $\mathcal{V}_j$ and $\mathcal{W}_j$ also can be written by

$$L^2(\mathbb{R}) = \mathcal{V}_0 \oplus \mathcal{W}_0 \oplus \mathcal{W}_1 \oplus \ldots. \tag{9}$$

Wavelet series expansion is defined by

$$f(t) = \sum_k a_{j_0,k}\phi_{j_0,k}(t) + \sum_{j=j_0}^{\infty}\sum_k b_{j,k}\psi_{j,k}(t) \tag{10}$$

where $j \geq j_0$, $f(t) \in L^2(\mathbb{R})$. If $\{\phi_{j_0,k}(t)\}$ and $\{\psi_{j,k}(t)\}$ are orthonormal in $L^2(\mathbb{R})$, then we can write $a_{j_0,k}$ and $b_{j,k}$ as

$$a_{j_0,k} = \langle f(t), \phi_{j_0,k}(t)\rangle = \int_{\mathbb{R}} f(t)\overline{\phi_{j_0,k}(t)}dt \tag{11}$$

$$b_{j,k} = \langle f(t), \psi_{j,k}(t)\rangle = \int_{\mathbb{R}} f(t)\overline{\psi_{j,k}(t)}dt. \tag{12}$$

Forward discrete wavelet transform is defined by

$$W_{\phi(j_0,k)} = \frac{1}{\sqrt{N}}\sum_t f(t)\phi_{j_0,k}(t) \tag{13}$$

$$W_{\psi(j,k)} = \frac{1}{\sqrt{N}}\sum_t f(t)\psi_{j,k}(t) \tag{14}$$

where $f(t)$ is signal with $t = 0,1,\dots,N-1$ and $j \geq j_0$.

Inverse discrete wavelet transform (IDWT) is defined by

$$f(t) = \frac{1}{\sqrt{N}}\sum_k W_{\phi(j_0,k)}\,\phi_{j_0,k}(t) + \frac{1}{\sqrt{N}}\sum_{j=j_0}^{\infty}\sum_k W_{\psi(j,k)}\psi_{j,k}(t) \tag{15}$$

where $j_0 = 0$, $N = 2^J, j = 0,1,2,\dots,J-1$.

2.2. *Dual tree complex wavelet transform*

Dual tree complex wavelet transform (DTCWT) is based discrete wavelet transforms (DWT). DTCWT decomposes complex signal to real and imaginary. Complex coefficient

$$d_c(j,n) = d_h(j,n) + i d_g(j,n) \tag{16}$$

where $d_g(j,n)$ is Hilbert transform of $d_h(j,n)$.

Definition 2. [6] The Hilbert transform $\mathcal{H}(x(t))$ of a signal $x(t)$ is defined as

$$\mathcal{H}(x(t)) = \frac{1}{\pi}\int_{-\infty}^{\infty}\frac{x(\tau)}{t-\tau}d\tau \tag{17}$$

where τ is dummy.

■

In polar form, complex coefficient is

$$d_c(j,n) = |d_c(j,n)|e^{i\phi} \tag{18}$$

with magnitude

$$|d_c(j,n)| = \sqrt{|d_h(j,n)|^2 + |d_g(j,n)|^2} \geq 0 \tag{19}$$

and phase

$$\phi = \angle d_c(j,n) = \arctan\left(\frac{d_g(j,n)}{d_h(j,n)}\right). \tag{20}$$

A magnitude and angle phase of complex coefficient can be reduced.

3. Results and discussion

3.1. Quaternion wavelet transform

3.1.1. Construction of quaternion wavelet transform

Quaternion wavelet transform (QWT) is based on quaternion algebra, quaternion Fourier transform and Hilbert transform. QWT has characteristics like quaternion Fourier transform (QFT) but QWT about scaling and translation.

QWT has three angle phases so it is suitable for 2-D signal. QWT can be represented in a polar form as

$$\boldsymbol{q} = |\boldsymbol{q}|e^{i\phi}e^{j\theta}e^{k\psi} \tag{21}$$

where $|\boldsymbol{q}|$ is magnitude of $\boldsymbol{q}$ and

$$\phi = \arctan(2(cd + ab)/a^2 - b^2 + c^2 - d^2)/2 \in [-\pi, \pi] \tag{22}$$

$$\theta = \arctan(2(bd + ac)/a^2 + b^2 - c^2 - d^2)/2 \in [-\pi/2, \pi/2] \tag{23}$$

$$\psi = \arcsin\big(2(ad - bc)\big)/2 \in [-\pi/4, \pi/4]. \tag{24}$$

Respectively, $\phi_h(x)$ and $\psi_h(x)$ are 1-D scaling function and wavelet function along the x-axis. Also, $\phi_h(y)$ and $\psi_h(y)$ are 1-D scaling function and wavelet function along the y-axis. Hilbert transforms of $\psi_h(x)$ and $\psi_h(y)$ are

$$\mathcal{H}_x\big(\psi_h(x)\psi_h(y)\big) = \psi_g(x)\psi_h(y) \tag{25}$$

$$\mathcal{H}_y\big(\psi_h(x)\psi_h(y)\big) = \psi_h(x)\psi_g(y) \tag{26}$$

$$\mathcal{H}_{xy}\big(\psi_h(x)\psi_h(y)\big) = \psi_g(x)\psi_g(y) \tag{27}$$

where $\psi_g(x)$ and $\psi_g(y)$ are Hilbert transforms of $\psi_h(x)$ and $\psi_h(y)$, respectively. Also, $\phi_g(x)$ and $\phi_g(y)$ are Hilbert transforms of $\phi_h(x)$ and $\phi_h(y)$.

After processing of Hilbert transform, 2-D a quaternion scaling function and three quaternion wavelet functions become

$$\Phi^q(x, y) = \phi_h(x)\phi_h(y) + i\phi_g(x)\phi_h(y) + j\phi_h(x)\phi_g(y) + k\phi_g(x)\phi_g(y) \tag{28}$$

$$\Psi_h^q(x, y) = \psi_h(x)\phi_h(y) + i\psi_g(x)\phi_h(y) + j\psi_h(x)\phi_g(y) + k\psi_g(x)\phi_g(y) \tag{29}$$

$$\Psi_v^q(x, y) = \phi_h(x)\psi_h(y) + i\phi_g(x)\psi_h(y) + j\phi_h(x)\psi_g(y) + k\phi_g(x)\psi_g(y) \tag{30}$$

$$\Psi_d^q(x, y) = \psi_h(x)\psi_h(y) + i\psi_g(x)\psi_h(y) + j\psi_h(x)\psi_g(y) + k\psi_g(x)\psi_g(y). \tag{31}$$

Definition 3. Function $f(x, y) \in L^2(\mathbb{R}, \mathbb{H})$, we construct

$$c^q = \langle f(x, y), \Phi^q(x, y)\rangle \tag{32}$$

$$d_i^q = \langle f(x, y), \Psi_i^q(x, y)\rangle \tag{33}$$

where $i = h, v, d$. Transformation c^q and d_i^q $(i = h, v, d)$ are called discrete quaternion wavelet transform of $f(x, y)$.

■

3.1.2. Characteristics of quaternion wavelet transform

Quaternion wavelet transform (QWT) is basis like quaternion Fourier transform (QFT) but QWT is specifically scaling and translation basis. Theorem of quaternion Fourier on translation process is

Theorem. [7] Let

$$F^q(\boldsymbol{u}) = \int_{\mathbb{R}^2} e^{-i2\pi ux} f(\boldsymbol{x}) e^{-j2\pi vy}\, d^2\boldsymbol{x} \tag{34}$$

and

$$F_T^q(\boldsymbol{u}) = \int_{\mathbb{R}^2} e^{-i2\pi ux} f(\boldsymbol{x} - \boldsymbol{d}) e^{-j2\pi vy}\, d^2\boldsymbol{x} \tag{35}$$

be the QFT's of a 2D signal f and a shifted version of f, respectively. Then, $F^q(\boldsymbol{u})$ and $F_T^q(\boldsymbol{u})$ are related by shift through $\boldsymbol{d} = (d_1, d_2)^T$

$$F_T^q(\boldsymbol{u}) = e^{-i2\pi ud_1} F^q(\boldsymbol{u}) e^{-j2\pi vd_2}. \tag{36}$$

If we denote the phase of $F^q(\boldsymbol{u})$ by $(\phi(\boldsymbol{u}), \theta(\boldsymbol{u}), \psi(\boldsymbol{u}))^T$, Then, as a result of the shift, the first and the second component of the phase undergo a phase-shift

$$\begin{pmatrix}\phi(\boldsymbol{u})\\ \theta(\boldsymbol{u})\\ \psi(\boldsymbol{u})\end{pmatrix} \mapsto \begin{pmatrix}\phi(\boldsymbol{u}) - 2\pi u d_1\\ \theta(\boldsymbol{u}) - 2\pi v d_2\\ \psi(\boldsymbol{u})\end{pmatrix}. \tag{37}$$

■

QWT is one of QFT, therefore translation process of QWT becomes

$$(\phi(\boldsymbol{u}), \theta(\boldsymbol{u}), \psi(\boldsymbol{u})) \mapsto (\phi(\boldsymbol{u}) - 2\pi u d_1, \theta(\boldsymbol{u}) - 2\pi v d_2, \psi(\boldsymbol{u})) \tag{38}$$

where $\boldsymbol{u} = (u, v)$ is QWT domain and (d_1, d_2) is translation.

3.2. Quaternion wavelet transform for image denoising

Quaternion wavelet transform (QWT) decomposes noisy image into sub-bands. In this paper, we use hard thresholding method with VisuShrink method to reduce detail coefficient magnitudes. And also, we use mean filter to approximation coefficient magnitude.

Hard thresholding method is

$$T_h(y, \lambda) = \begin{cases} y, & |y| \geq \lambda \\ 0, & |y| < \lambda \end{cases} \tag{39}$$

where y is detail coefficient of image and λ is threshold value.

VisuShrink method is

$$\lambda = \sigma\sqrt{2 \log M} \tag{40}$$

where σ is deviation standard of noise and M is image size.

Mean filter 3×3 has mask formula

$$R = \frac{1}{9}\sum_{i=0}^{9} z_i \tag{41}$$

where all number of mask z is 1.

Some steps of QWT for image denoising are:

- Decompose noisy image with QWT.
- Calculate imaginary part of quaternion detail coefficient using hard thresholding with VisuShrink.
- Calculate imaginary part of quaternion approximation coefficient using mean filter.
- Calculate quaternion detail coefficient using hard thresholding with VisuShrink.
- Calculate quaternion approximation coefficient using mean filter.
- Reconstruct the denoised by invers quaternion wavelet transform (IQWT).

3.3. Experimental result of quaternion wavelet transform for image denoising

In the experiment, we use discrete wavelet transform (DWT) and quaternion wavelet transform (QWT) with Haar, Biorthogonal 6.8, Daubechies 9 and Coiflets 5 wavelet. Images are corrupted by white Gaussian noise with zero mean and variance σ^2. White Gaussian noise is one type of noise that is often encountered in digital images. Model of degradation images are

$$g(x, y) = f(x, y) + \varepsilon(x, y) \tag{42}$$

where $f(x, y)$ is clean image, $g(x, y)$ is noisy image and $\varepsilon(x, y)$ is noise.

We use test images with low, medium and hard complexity. Each test image has size 512×512 pixel with grayscale characteristic. We use hard thresholding method with VisuShrink method to reduce detail coefficient magnitudes. And also, we use mean filter to approximation coefficient magnitude. The method that has the highest PSNR (peak signal to noise ratio) value can be concluded the best performance.

IOP Conf. Series: Journal of Physics: Conf. Series **974** (2018) 012006 doi:10.1088/1742-6596/974/1/012006

Denoised image of 100 experiments with $\sigma = 20$ and zero mean are shown in figure 1.

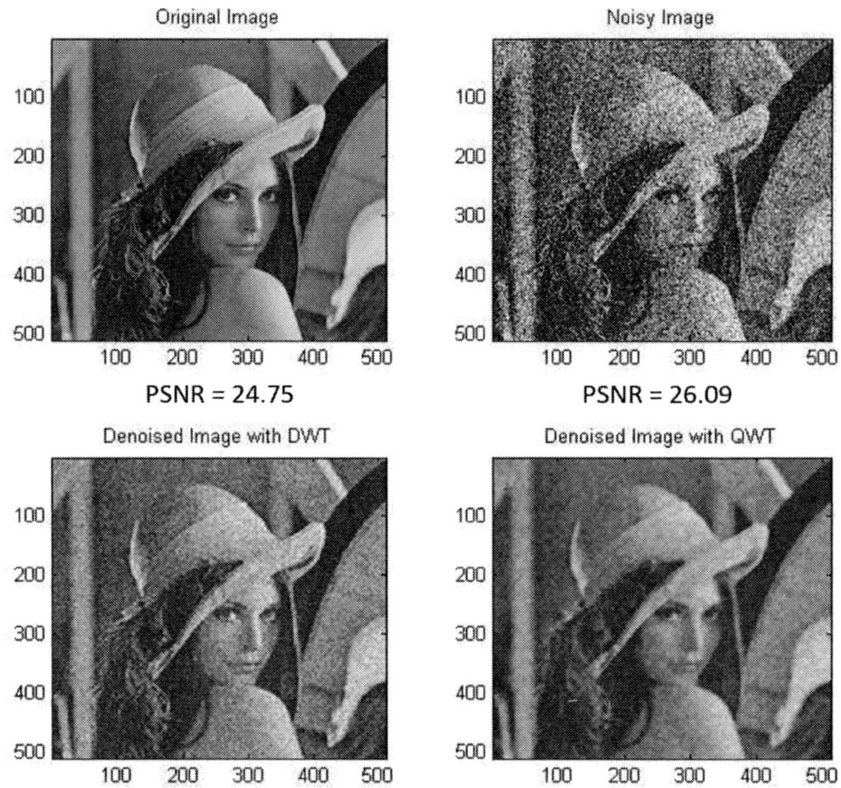

Figure 1. Denoised of Lena image with coiflets 5 DWT and QWT.

PSNR values of 100 experiments with zero mean and some deviation standards are shown in figure 2 and table 1.

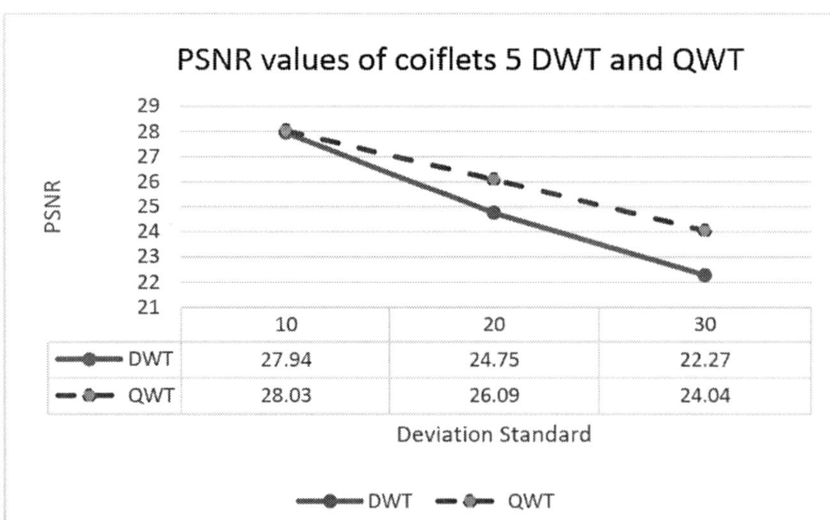

Figure 2. PSNR values of Lena image with coiflets DWT and QWT 5.

Table 1. PSNR values with DWT and QWT for Haar, Biorthogonal 6.8, Daubechies 9 and Coiflets 5.

PSNR (peak signal to noise ratio)										
Images	σ	**Haar**		**Biorthogonal 6.8**		**Daubechies 9**		**Coiflets 5**	
		DWT	**QWT**	**DWT**	**QWT**	**DWT**	**QWT**	**DWT**	**QWT**
Pantai	**10**	31.32	30.02	31.60	34.35	31.61	34.29	31.66	34.38
	20	25.33	25.94	25.35	27.48	25.36	27.46	25.40	27.51
	30	21.44	22.26	21.43	23.00	21.43	22.97	21.45	23.01
Lena	**10**	26.15	26.07	27.93	27.99	27.95	27.96	27.94	28.03
	20	23.71	24.60	24.73	26.10	24.74	26.09	24.75	26.09
	30	21.64	23.06	22.24	24.03	22.24	24.03	22.27	24.04
Mandrill	**10**	22.09	21.95	22.68	22.57	22.70	22.58	22.71	22.60
	20	20.90	21.33	21.42	21.90	21.43	21.93	21.39	21.88
	30	19.80	20.72	20.17	21.16	20.19	21.20	20.14	21.14

We can see from figure 1, DWT and QWT methods remove noise of image successfully. Initially, noisy image can not be seen properly. After it is done image denoising, image can be seen better. We also can see that result of QWT method better than DWT method. We can see from figure 2 and table 1, we compare PSNR values with DWT and QWT for Haar, Biorthogonal 6.8, Daubechies 9 and Coiflets 5 wavelet. Almost all test images with some deviation standards, QWT method has higher PSNR value than DWT method. Performance results of method are stated quantitatively rather than qualitatively. We can say that QWT method has better performance than DWT method.

4. Concluding remarks

Based on the research, we can obtain some conclusions. Quaternion wavelet transform (QWT) can be constructed by multiresolution analysis. QWT can be represented in a polar form with a magnitude and three angle phases. In a polar form, a magnitude and angle phases of quaternion can be reduced. QWT removes noise of image successfully. We compare PSNR (peak signal to noise ratio) values of DWT and QWT with Haar, Biorthogonal, Daubechies and Coiflets wavelet. QWT method has better performance than discrete wavelet transform (DWT) method for image denoising.

Acknowledgments

This work is part of thesis in master's degree mathematics department of Institut Teknologi Sepuluh Nopember.

References

[1] Gai S, Yang G Wan M and Wang L 2014 Hidden Markov Tree Model of Images using Quaternion Wavelet Transform *Computers and Electrical Engineering* **40** p 819.

[2] Li C, Li J and Fu B 2013 Magnitude-Phase of Quaternion Wavelet Transform for Texture Representation Using Multilevel Copula *IEEE Signal Processing Letters* **20** No.8 p 799.

[3] Yin M, Liu W, Shui J and Wu J 2012 Quaternion Wavelet Analysis and Application in Image Denoising *Mathematical Problems in Engineering* **2012** p 1.

[4] Kadiri M, Djebbouri M and Carré P 2014 Magnitude-Phase of The Dual-Tree Quaternionic Wavelet Transform for Multispectral Satellite Image Denoising *EURASIP Journal on Image and Video Processing* **41** p 1.

[5] Bachman G, Narici L and Beckenstein E 2002 *Fourier and Wavelet Analysis* (New York: Springer-Verlag) pp 414-415.

[6] Fletcher P and Sangwine S J 2017 The Development of The Quaternion Wavelet Transform *Signal Processing* **136** p 2.

[7] Bülow T 1999 *Hipercomplex Spectral Signal Representations for the Processing and Analysis of*

IOP Conf. Series: Journal of Physics: Conf. Series **974** (2018) 012006 doi:10.1088/1742-6596/974/1/012006

Images (Ph.D. dissertation Christian Albrechts Univ. Kiel Germany) p 32.

MoREK: The learning media to improve students understanding about electrical circuit in informatics

Indrianto, Meilia Nur Indah Susanti, Rakhmat Arianto

Computer Science Department
Sekolah Tinggi Teknik PLN, Jakarta, Indonesia

Email: indrianto@sttpln.ac.id, meilia@sttpln.ac.id, arianto@sttpln.ac.id

Abstract. The needs for labor in the world is already increasing especially in Indonesia. According to the World Bank, Indonesia is a country that ranks 9th in the world's largest economic growth. To meet that needs, Indonesia needs 55 million workers who are experts in the field of electricity. Therefore, it takes a lot of human resources and has been equipped with knowledge and expertise in the field of electricity. To be able to meet these needs, it takes a better method of learning to increase knowledge and expertise in the field of electricity since college, especially in the field of informatics. Prototype of Electrical Module (The MoREK) requires a Prototype method for the Practicum Module to be created as desired. This method is often used in the real world or it could be said Prototype method is part of the product that expresses the logic and physical external interface that is displayed. For data retrieval is used Pre-experimental method where students will be given pre-test and post-test. The Design of Electrical Module has a purpose to improve the students understanding of Electric Circuit Engineering Courses with the creation of The MoREK so that students are more competent to the course and can meet the needs of manpower or Human Resources (SDM) in the field of electricity. By using The Morek, the score of student learning outcomes increased by 7.8% and informatics students who conduct research in the field of electricity increased to 21%.

1. Introduction

The government has set a plan to build a power plant that can generate 35,000 MW in all parts of Indonesia so that it takes a skilled manpower in the field of electricity with a large number. If the development of electrical energy in a year to reach 5,000 MW then it takes about 20,000 workers consisting of 12,000 skilled workers, 5,000 middle-class workers. "The rest, 1,000-2,000 people are experienced experts [1].

STT-PLN should be a major pioneer in the provision of this workforce. Lesson materials or modules in the field of electricity should be updated immediately tailored to the needs of existing electricity. To produce competent graduated it is necessary lessons in accordance with the needs of the industrial world. When talking about electricity then what is pictured is majoring in Electrical Engineering whereas other majors such as Informatics Engineering can provide graduation to work in the field of electricity. This is because there are many tools connected with computer equipment even computer embedded system.

Based on this research, the researcher average the value of the subjects connected with the literature has decreased almost 80% of the value owned by the informatics students below the value of

50. And this affects the students who take the final project in the field of electricity in STT-PLN informatics majors. Based on the data of informatics students in 2015 who took research in the field of electricity only 6.08% of 205 students and on data informatics students in 2016 showed that students who take research in the field of electricity is only 6.06% of 132 students. This is really worried because STT-PLN should be a pioneer in the field of electricity. For that researchers tried to conduct research in this electrical field. How to make a module in the field of electricity field in the field of informatics (The MoREK) so that graduates will be competent in the field of electricity and will be used in the world of electricity field work so that Human Resources (HR) will be met. Currently, the study of electrical circuit is just a theory and practicum in the electro area but the students were less aware of the implementation of electrical circuits that are combined with the informatics area.

Preparation of practical modules are intended to facilitate students in capturing lessons that researchers teach. "Learning module can be used to improve student learning outcomes" [2]. So that students will be much more understand the subjects taught, will also be easy in carrying out electrical materials because they have learned both in practice and theory. Based on research conducted by Arif Budijanto et al., [3] pointed out that the learning module is an effective method for improving students' psychomotor in learning the course with practical approach". Besides, the researcher tries to develop electricity lesson in order to avoid saturation in electrical subjects.

2. Literature Review

In previous research [4], has successfully developed E-learning Assisted Virtual Laboratory For Practical Basic Physics II in Program Of Study Physical Education FKIP UNSRI which valid, practical and have a potential effect. The method used in this research is a Development Research that use Rowntree model comprising the steps of planning, development and evaluation. The evaluation used is Tesmer formative evaluation consisted of self-evaluation, expert review, a one-to-one, small group and a field test. Data collection techniques with walkthrough, questionnaires and tests. This research has resulted a valid *e-learning* assisted virtual laboratory with validation percentage score 91.67 %, of material aspect, 96.75 % of the media aspects and 77.78 % of instructional design aspects. E-learning Assisted Virtual Laboratory for Practical Basic Physics II developed practicality scores at stage one-to-one stage of 81.30 % and 81.50 % for small group are included in the category of very practical. The test results on the field test phase of 44 students who got an A by 31.82 %, the value of B was 34.09 % and the value of C by 34.09 %. Based on the research results, it can be concluded that Development E-learning Assisted Virtual Laboratory for Practical Basic Physics II has a valid, practical and have a potential effect.

Furthermore, in research [5], tells about teacher's difficulty in explaining the mathematical aspects and the way the phasor diagram depiction of this physical material is one of the causes is the lack of an attractive, flexible and easy-to-use medium to teach the material. Through this research, developed computer-based physics learning media subject of current and electric voltage back and forth for high school students / MA class XII by using Microsoft Excel 2007. Making of this instructional media refers to ADDIE development model that is Analysis, Design, Development, Implementation, and Evaluation. To determine the feasibility of the media, validation is made to several validators consisting of material experts, user experts, and media experts. This validation is in the form of a questionnaire with the criteria / indicators that have been determined. This program has met the eligibility criteria; media (design and program) equal to 78,92% or included in Good (B) category, material (suitability of curriculum subject program of physics subject and electric voltage back and forth SMA / MA class XII) equal to 72,22% or included in good category (B), and users (media and material) of 81.78% included in very good category (SB). From the results of validation analysis can be concluded that the program developed can be utilized and deserve to be used as a medium of physics learning the subject of current and electric voltage back and forth for SMA class XII.

Next, in research [6], the use of network in embedded systems can be done with many kinds of mobile phones, bluetooth, modems, ethernet cards, wireless technology and so on. Using network in embedded system can help people to do remote controlling. On previous research, found that many students have the ability to comprehend the basic concept of embedded systems. They could also make embedded system tools but without network integration. And for that, a development is needed for the embedded system module. The embedded system practicum module design needs a prototype method in order to achieve the desired goal. The prototype method is often used in the real world. Or even, a prototype method is a part of products that consist of logic expression or external physical interface. The embedded system practicum module is meant to increase student comprehension of embedded systems course, and also to encourage students to innovate on technology based tools. It is also meant to help teachers to teach the embedded system concept on the course. The student comprehension is hoped to increase with the use of the practicum course.

Also, in research [7], Observations conducted by researchers showed that in SMK Negeri 2 Bojonegoro student learning is having a module but the module is not used in accordance with the potential students and less according to the standard module, so generally, this study aims to determine the potential of students in SMK Negeri 2 Bojonegoro, quality learning modules, and student learning outcomes. This research development, because in this study developed learning modules and devices. This learning module development refers to the development of 4D (Define, Design, Develop, and Disseminate). The study subjects were 32 students of class X TEI 1. The design of experiments used for data retrieval experimental research design One Group Pretest Posttest with the following results: 1) the results of formal reasoning ability tests concluded that the average student there is at the level of reasoning ability transition, 2) the validation of results result obtained 80,2% for the learning modules and 80% for items we can conclude for use, 3) the result an average value and standard deviation of posttest 64,62 then 12,05 to pretest with an average value of 58,56 and a standard deviation of 12,20. The difference between the results of the study revealed that a significant level of 5%, the value of t_{count} 3,188 while the value of t_{tabel} 2,04, so the value of $t_{count} > t_{table}$ and the probabilities obtained the P Value 0,003, P-Value <0.05 which can be categorized H_1 is accepted and H_0 is rejected, from the statement that the learning outcomes of students after using the learning modules increased compared with students before using the learning modules.

3. Research Method

The method used to develop the module is Prototyping Method so that the module generated can be according to user requirement. This method is often used in the real world or can be said Prototype method is part of a product that expresses the logic and physical external interface is displayed.

IOP Conf. Series: Journal of Physics: Conf. Series **974** (2018) 012007 doi:10.1088/1742-6596/974/1/012007

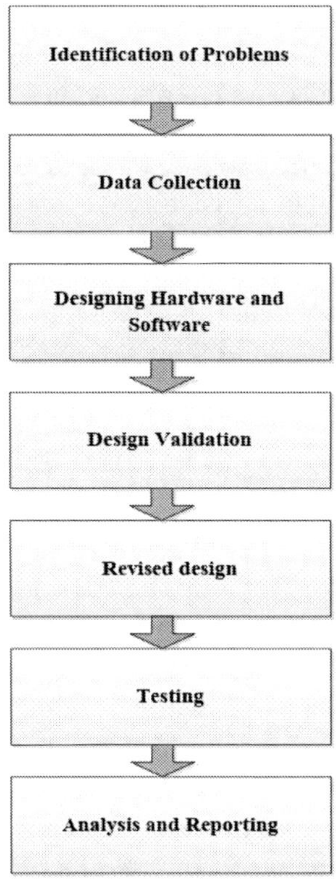

Figure 1. Prototype method

The population used is the students of Informatics Engineering who have taken the course of electrical circuit in the final semester. With a sample consisting of 2 classes of class A and class B consisting of 45 students class A as experimental class given treatment and 45 students of Class B is non-experiment and not given treatment. The design used in this research is pre-experimental design of static group compared.

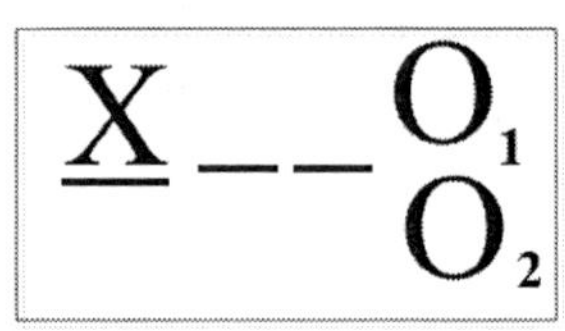

Figure 2. Static group comparation

Class group that is given treatment by using The MoREK is a class of electric class A class Informatics engineering and class of electric class B class Informatics techniques are not given treatments. The technique that is being used for collecting data is by giving test sheets and questionnaires that are analyzed with Likert scale. With criterias such as "very helpful", "helpful", "enough", "unhelpful", "very unhelpful".

Table 1. Criteria of experiment

Percentages	Criterias
0%-20%	Very Unhelpful
21%-40%	Unhelpful
41%-60%	Enough

Percentages	Criterias
61%-80%	Helpful
81%-100%	Very Helpful

Based on the criteria percentages, the MoREK module on helping students in increasing their comprehension of embedded system course is said to be good if it has a percentage above 60 percent.

$$Percentage\ (\%) = \frac{Total\ Actual\ Score}{Total\ Ideal\ Score} x\ 100\% \quad (1)$$

4. Result

Based on the results of the questionnaire calculations conducted on the arbitrariness of what do the student by using The MoREk module. This can be seen from the graph below. With the graph and the questionnaire concluded that students are helping by The MoREK. It is showed by the 75., 22% respondents that said they could understand, learn, and operate the given module.

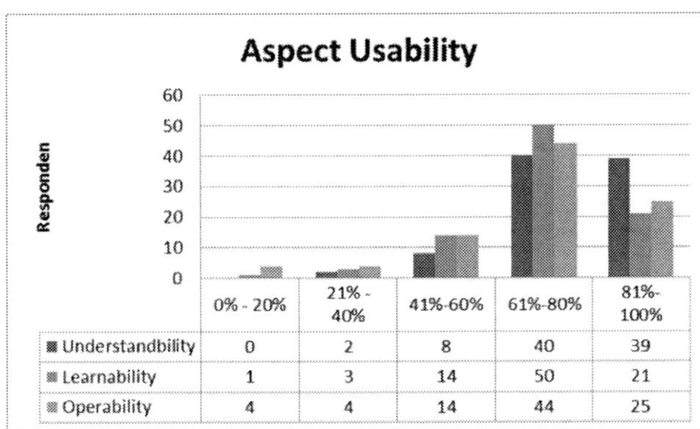

Figure 3. Aspect usability

Based on the results of tests conducted using pre-test and post-test it is seen that the class that provided the MoREK module, has a much better average value compared to that not given the MoREK module. In Class A when done post-test get value equal to 64,7 and after doing post-test get value 72,5.

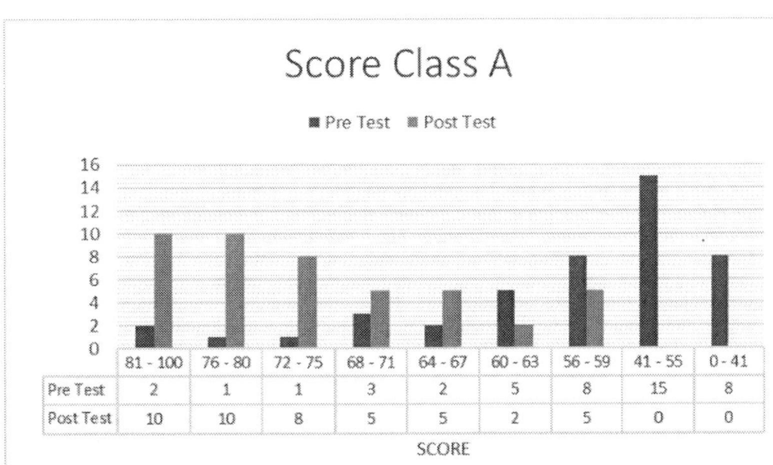

Figure 4. Score of class A

In Class B when done post-test get value equal to 57,4 and after done post-test get value 57,8.

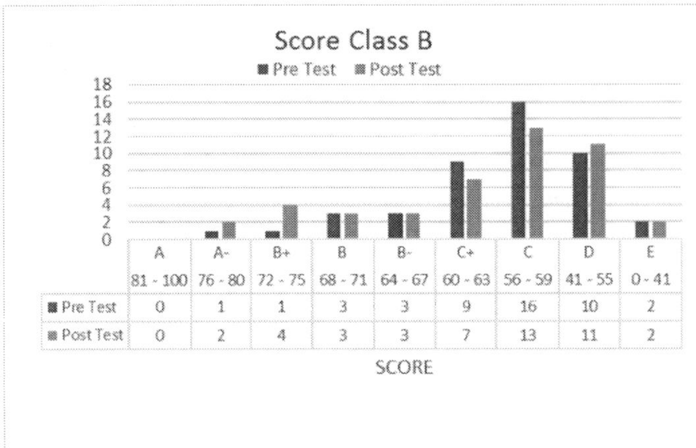

	A	A-	B+	B	B-	C+	C	D	E
	81 - 100	76 - 80	72 - 75	68 - 71	64 - 67	60 - 63	56 - 59	41 - 55	0 - 41
Pre Test	0	1	1	3	3	9	16	10	2
Post Test	0	2	4	3	3	7	13	11	2

Figure 5. Score of class B

This also affects the research generated by students in the field of electricity is visible in the following table.

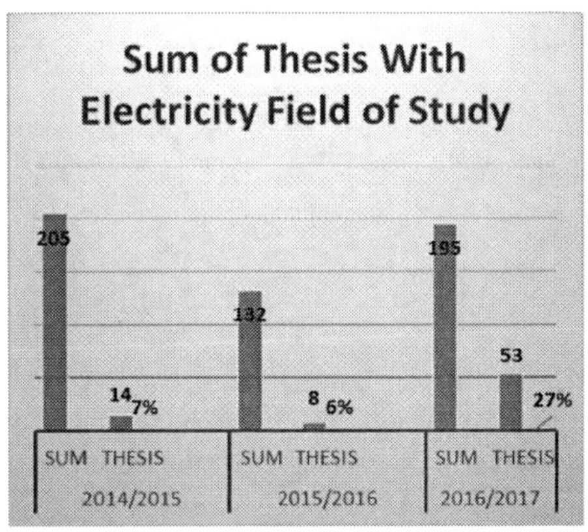

Figure 6. The number of research students in the field of electricity

5. Conclusion

By using the MoREK Module the value obtained by the students increased by 7.8% from the previous value. Student research covering the field of electricity also increased by 21% from the previous year.

References

[1] S. Legino, *Energi Terbarukan*. Surabaya, 2010.

[2] T. D. Sari and S. Isnur, “Pengembangan Modul Pembelajaran Berbasis Proyek Pada Mata Pelajaran Instalasi Penerangan Listrik Untuk Mengetahui Peningkatan Hasilbelajar Siswa Kelas Xi Tiptl Smk Negeri 7 Surabaya,” *J. Pendidik. Tek. Elektro*, vol. 2, no. 2, 2015.

[3] A. Budijanto and A. Shoim, “Prototipe Modul Pembelajaran Embedded System Berbasis Arduino,” in *Seminar Nasional Sains dan Teknologi Terapan Institut Teknologi Adhi Tama Surabaya*, 2015, pp. 1–6.

[4] D. Agustine, K. Wiyono, and M. Muslim, “Pengembangan e-learning berbantuan virtual laboratory untuk mata kuliah praktikum fisika dasar ii di program studi pendidikan fisika fkip unsri,” vol. 1, no. 1, pp. 33–43, 2014.

[5] T. Firdaus, "Pengembangan Media Pembelajaran Arus Dan Tegangan Listrik Bolak-Balik Untuk Sma / Ma," *J. Inov. dan Pembelajaran Fis.*, vol. 2, no. 2, pp. 197–203, 2015.
[6] Indrianto, M. N. I. Susanti, and D. Karina, "RESLanjut: The learning media for improve students understanding in embedded systems," in *AIP Conference Proceedings*, 2017, vol. 1867, no. 1, p. 20047.
[7] M. Puspita and M. Sondang, "Pengembangan Modul Pembelajaran Pada Standar Kompetensi Menerapkan Dasar-Dasar Kelistrikan Di Smk Negeri 2 Bojonegoro," *J. Pendidik. Tek. Elektro*, vol. 2, no. 2, pp. 779–785, 2013.

IOP Conf. Series: Journal of Physics: Conf. Series **974** (2018) 012008 doi:10.1088/1742-6596/974/1/012008

The comparison of proportional hazards and accelerated failure time models in analyzing the first birth interval survival data

Alfensi Faruk

Department of Mathematics, Faculty of Mathematics and Natural Science, Sriwijaya University

E-mail: alfensifaruk@unsri.ac.id

Abstract. Survival analysis is a branch of statistics, which is focussed on the analysis of time-to-event data. In multivariate survival analysis, the proportional hazards (PH) is the most popular model in order to analyze the effects of several covariates on the survival time. However, the assumption of constant hazards in PH model is not always satisfied by the data. The violation of the PH assumption leads to the misinterpretation of the estimation results and decreasing the power of the related statistical tests. On the other hand, the accelerated failure time (AFT) models do not assume the constant hazards in the survival data as in PH model. The AFT models, moreover, can be used as the alternative to PH model if the constant hazards assumption is violated. The objective of this research was to compare the performance of PH model and the AFT models in analyzing the significant factors affecting the first birth interval (FBI) data in Indonesia. In this work, the discussion was limited to three AFT models which were based on Weibull, exponential, and log-normal distribution. The analysis by using graphical approach and a statistical test showed that the non-proportional hazards exist in the FBI data set. Based on the Akaike information criterion (AIC), the log-normal AFT model was the most appropriate model among the other considered models. Results of the best fitted model (log-normal AFT model) showed that the covariates such as women's educational level, husband's educational level, contraceptive knowledge, access to mass media, wealth index, and employment status were among factors affecting the FBI in Indonesia.

1. Introduction

In many practical problems related to time-to-event data, the PH model is widely known as the most used multivariate survival analysis to investigate the effects of several covariates on survival time. The PH model is semi-parametric, which means that the baseline hazard is left unspecified and there is no assumption about the statistical distribution of the survival times. The main assumption of the PH model is that the hazard ratio is constant over time. The violation of this assumption may lead to several statistical problems such as misinterpretation of the estimation results and reducing the power of the related statistical tests [1].

Another model, which also uses PH assumption, is the parametric PH model [2]. This model modifies the PH model by assuming that the baseline hazard function follows one of the statistical distributions. Although it is known that parametric PH model is appropriate to many clinical problems, the parametric PH model cannot be used as an alternative model to cope with non-proportional hazards problem. On the other hand, the accelerated failure time (AFT) model is one of parametric survival models that can be used as an alternative to PH model, especially to overcome the statistical problems due to the violation of PH assumption [3]. The AFT model accounts the effects of the covariates directly on survival times instead of the hazards rate as in the PH model. Moreover, the interpretation of the results in AFT model is easier than the results of PH model because the parameters indicate the effect of the explanatory variables on the mean survival time.

The AFT models have been widely studied by many authors. According to [4], these models fit better to the influenza data than the PH model. The AFT models are also recommended in the analysis of lifetime data, especially in medical research [5]. The comparison of the AFT models and PH model for several numbers of censored observations and the sample size was conducted through simulation study [6]. The similarity among the mentioned works is that the use of AFT models and its comparison to the PH model were conducted in relatively small sample size which is commonly found in clinical research. However, in many real life applications, it is also common that the researchers have to analyze the large data set, such as in demographic and public health studies. Therefore, the main objective of this research is to compare the performance of the Cox PH model and the AFT models in the presence of non-proportional hazards in one of demographical and public health problems, that is the FBI data in Indonesia. The FBI can be defined as the length of time spent by a married couple to get their first child after the day of marriage. The short FBI indicates the high fertility rate of a spouse. In other words, the fertility rate can be reduced by lengthening the FBI. The considered models in this work are limited to the PH model and three types of AFT models, i.e. Weibull AFT model, exponential AFT model, and log-normal AFT model.

2. Materials and methods

The data used in this research were obtained from 2012 IDHS, which was conducted by Statistics Indonesia and United States Agency for International Development (USAID) from May 2012 until June 2012. In order to collect the data in all provinces of Indonesia, they used stratified two-stage cluster sampling technique, where the cluster blocks and the household were the first and the second stage unit sampling, respectively. The cluster blocks were based on the Indonesia population census that was held in 2010. The respondents were asked to answer the questions in the questionnaire. Not all respondents filled the questionnaire completely, therefore only the completed questionnaires were included.

To achieve the research objective, this work employed several steps. In the first step, the log cumulative hazard plots approach and Schoenfeld residual statistical test were applied to the data set in order to examine the existence of non-PH. Next, the PH model, the Weibull AFT model, exponential AFT model, and log-normal AFT model were fitted to the data set. After that, the performance of each model was evaluated by using the AIC values. Finally, as the log-normal AFT model was chosen as the best fitted model, the backward stepwise procedure was applied to select the significant covariates.

3. Basic concepts of the PH model and the AFT models

3.1. PH model

The PH model is a semiparametric regression model which can be used to measure the effects of covariates on the survival time. This model is represented by the relationship of the hazard function, the baseline hazard function, and one or more covariates in the form

$$h(t) = h_0(t)\exp(\boldsymbol{\beta}^t \boldsymbol{X}),$$

where t is survival time, $h(t)$ is the hazard function, $h_0(t)$ is the baseline hazard function which is left unspecified, $\boldsymbol{\beta}$ is a column vector of the regression coefficients, and $\boldsymbol{X}$ is a column vector of the covariates. The Cox regression model assumes that there is the proportionality of the hazard rate between any two individuals in the population. The hazard ratio is defined as the ratio of the hazard functions for two subjects with different values of covariate X_1 and X_2. The formula of the hazard ratio is given by

$$H(t) = \frac{h_0(t)\exp(\beta_2 X_2)}{h_0(t)\exp(\beta_1 X_1)} = \frac{\exp(\beta_2 X_2)}{\exp(\beta_1 X_1)} = \exp(\beta^t(X_2 - X_1)).$$

It can be seen that the hazard ratio $H(t) = \exp(\beta^t(X_2 - X_1))$ is independent of time. In other words, the hazard ratio for any two individuals is constant over time. This property is also known as the PH assumption.

3.2. *Partial likelihood (PL) method*

Cox [7] introduced PL method as the procedure to estimate the regression coefficients of PH model. This method uses the partial likelihood function instead of the likelihood function as in the maximum likelihood estimation (MLE) method [8]. The first step in PL method is by arranging the observed time t_i, $i = 1,2,..,k$ in ascending order so that the times are ordered, that is $0 \leq t_1 \leq t_2 \leq \cdots \leq t_k$. Next, let $\boldsymbol{X}_i$ is a vector of the summation of all p covariates for the individuals who experience the event of interest at time t_i. If there are d_i occurrences of the event at time t_i, the hth element of $\boldsymbol{X}_i$ is $X_{hi} = \sum_{r=1}^{d_i} x_{hir}$, where x_{hir} is the value of the hth covariate for the rth individual of d_i, $h = 1,2,\dots,p$, and $r = 1,2,\dots,d_i$.

The approximation of the partial likelihood function in the case that it allows more than one individual who experiences the event of interest at the observed time t_i is expressed by

$$L_p(\boldsymbol{\beta}) = \prod_{i=1}^{k}\left(\frac{\exp(\boldsymbol{\beta}^t \boldsymbol{X}_i)}{\left\{\sum_{j\in Y_i}\exp(\boldsymbol{\beta}^t \boldsymbol{X}_j)\right\}^{d_j}}\right)^{\delta_i}, \tag{1}$$

where δ_i is an indicator variable whose value equal to one when the event occurs or equal to zero when the event is censored, and Y_i is the number of individuals who experiences the event at or after time t_i, $i = 1,2,..,k$. There are two reasons why equation (1) is called a partial likelihood function, i.e. it is not a full likelihood for $\boldsymbol{\beta}$ and it does not use full data but only use their ranking. The related log-likelihood function of the partial likelihood function (1) is given by

$$\log L_p(\boldsymbol{\beta}) = \sum_{i=1}^{k}\delta_i\left[e^{\boldsymbol{\beta}^t \boldsymbol{X}_i} - d_j \log\left(\sum_{j\in Y_i} e^{\boldsymbol{\beta}^t \boldsymbol{X}_j}\right)\right]. \tag{2}$$

Then, by maximizing the log-likelihood function (2), the estimated value of $\boldsymbol{\beta}$ may be obtained by using a numerical approximation such as Newton-Rhapson method [10].

3.3. *Accelerated failure time model*

Let T is a random variable of survival times and $\boldsymbol{X}$ is a column vector of the covariates $X_1, X_2, \dots, X_p$, the AFT model defines the relationship of survival function for every time $t \in T$, $S(t|\boldsymbol{X})$, and the covariates as follows

$$S(t|\boldsymbol{X}) = S_0[t\exp(\boldsymbol{\beta}^t\boldsymbol{X})], \tag{3}$$

where S_0 is the baseline survival function and $\boldsymbol{\beta}^t = (\beta_1, \beta_2, \dots, \beta_p)$ is a vector of regression coefficients, and $n \in N$. The factor $\exp(\boldsymbol{\beta}^t\boldsymbol{X})$ in equation (3) is known as the accelerated factor which accelerates the survival function with covariate $\boldsymbol{X} = \boldsymbol{0}$. The AFT model assumes that the effects of the covariate are fixed and multiplicative by the accelerated factor on the time scale of t. However, it does not assume that the model holds the constant hazards assumption as in PH model.

The relationship between covariates and the survival time can be also illustrated as a linear relation between the natural logarithm of survival time and the covariate $\boldsymbol{X}$, that is

$$Y = \ln T = \mu + \boldsymbol{\theta}^t\boldsymbol{X} + \sigma W, \tag{4}$$

where μ is the slope, $\sigma > 0$ is an unknown scale parameter, $\boldsymbol{\theta}^t = (\theta_1, \theta_2, \dots, \theta_p)$ is a vector of regression coefficients, $\boldsymbol{\theta} = -\boldsymbol{\beta}$, σ is a scale parameter, and W is a distribution error which is a random variable and assumed to follow a certain parametric distribution. For every distribution of W, there is a related parametric for T. The name for the AFT model come from the distribution of T rather than the parametric distribution of $\ln T$. The commonly parametric distributions, which correspond to the AFT model, are Weibull, exponential, log-normal, gamma, and log-logistic. However, the AFT models that are considered in this research are Weibull AFT model, Exponential AFT model, and log-normal AFT model. The survival function of T_i , $i = 1,2, \ldots, n$ is given by

$$S_i(t) = \Pr(T_i \geq t) = \Pr(\ln T_i \geq \ln t) = \Pr(Y_i \geq \ln t)$$
$$= \Pr(\mu + \boldsymbol{\theta}^t \boldsymbol{X} + \sigma W \geq \ln t) = \Pr\left(W_i \geq \frac{\ln t - (\mu + \boldsymbol{\theta}^t X)}{\sigma}\right).$$

3.4. *Weibull and exponential AFT models*

The Weibull distribution is one of the most popular distributions in the analysis of time-to-event data. It is a flexible model that the hazard rate can be one of the three functions namely monotone increasing, constant, or decreasing function.

The survival function of survival time T for the Weibull distribution with parameters λ and α can be written by $S(t) = \exp(-\lambda t^{\alpha})$, and the hazard function is expressed by $h(t) = \lambda\alpha t^{\alpha-1}$. If the $\boldsymbol{X}$ in equation (4) equal to zero, then it is obtained $Y = \mu + \sigma W$, where $\mu = (-\ln(\lambda)/\alpha)$, $\sigma = \frac{1}{\alpha} = \alpha^{-1}$, and W follows the standard extreme value distribution with probability function

$$f(w) = e^{(w - e^w)}. \tag{5}$$

By integrating the probability density function (5) with respect to variable u on the interval from w and ∞, the survival function of W is $S(w) = \int_w^{\infty} f(u)du = \int_w^{\infty} e^{(u-e^u)}du = e^{(-e^w)}$. Then, the hazard function of W is $h(w) = \frac{f(w)}{S(w)} = \frac{e^{(w-e^w)}}{e^{(-e^w)}} = e^w$.

The univariate probability function $Y = \ln(T)$ is

$$f(y) = \frac{1}{\sigma}\exp[(y-\mu)/\sigma - \exp[(y-\mu)/\sigma]],$$

and survival function for Y is

$$S(y) = \exp\left[-\exp\left(\frac{y-\mu}{\sigma}\right)\right]. \tag{6}$$

Equation (6) can be extended to multivariate case, that is by considering a set of p covariates, $\boldsymbol{X} = (\mathrm{X}_1, \mathrm{X}_2, \ldots, \mathrm{X}_p)^t$ into the models. As the result, the survival function can be expressed by

$$S(y|\boldsymbol{X}) = \exp\left[-\exp\left(\frac{y-(\mu+\boldsymbol{\theta}^t X)}{\sigma}\right)\right]. \tag{7}$$

The exponential distribution can be derived from Weibull distribution, that is by taking $\sigma = 1$ or $\alpha = 1$, so that the equation (6) and equation (7) become $S(y) = \exp[-\exp(y-\mu)]$ and $S(y|\boldsymbol{X}) = \exp[-\exp(y-(\mu+\boldsymbol{\theta}^t\boldsymbol{X}))]$, where $S(y)$ and $S(y|\boldsymbol{X})$ are the survival functions for exponential AFT model in univariate and multivariate case, respectively.

3.5. *Log-normal AFT model*

If the random variable of the survival times T is assumed to follow a log-normal distribution, the baseline survival function $S_0(t)$ and baseline hazard function $h_0(t)$ are expressed by

$$S_0(t) = 1 - \varphi\left(\frac{\ln t - \mu}{\sigma}\right),$$

and

$$h_0(t) = \frac{\varphi\left(\frac{\ln t}{\sigma}\right)}{1 - \Phi\left(\frac{\ln t}{\sigma}\right)},$$

respectively, where μ and σ are parameters, $\varphi(x)$ is density probability function at time x and $\Phi(x)$ is the cumulative distribution function at time x. Therefore, the survival function given a set of covariates $\boldsymbol{X} = (X_1, X_2, \dots, X_p)^t$ for the log-normal AFT model is $S(t) = S_0[t\exp(\boldsymbol{\beta}^t\boldsymbol{X})] = 1 - \varphi((\ln t - (\mu + \boldsymbol{\theta}'\boldsymbol{X}))/\sigma)$, where $t \in T$ is survival time, and $\boldsymbol{\beta}^t = (\beta_1, \beta_2, \dots, \beta_p)$ is a vector of coefficients.

3.6. Parameters estimation procedure of AFT models

Similar to other parametric models, the AFT models are also can be estimated by using MLE. The likelihood function of n observed times $t_1, t_2, \dots, t_n$, with unknown parameters $\boldsymbol{\beta}^t = (\beta_1, \beta_2, \dots, \beta_p), \mu$, and σ, which contain $(n - r)$ right censored data, is expressed by

$$l(t; \boldsymbol{\beta}, \mu, \sigma) = \prod_{i=1}^{n} f_i\,(t_i)^{\delta_i} * S_i(t_i)^{1-\delta_i}, \quad (8)$$

where $0 \le r \le n$, δ_i is indicator variable which is equal to one if t_i is observed and equal to zero if t_i is a censored observation, $f_i(t_i)$ and $S_i(t_i)$ are the density function and survival function at time t_i, respectively. By taking the logarithm of equation (8), it yields the log-likelihood function as follows

$$\log l(t; \boldsymbol{\beta}, \mu, \sigma) = \sum_{j=1}^{n}\{-\delta_i \log(\delta_i t_i + \delta_i \log f_i(t_i) + (1 - \delta_i)\log S_i(W_i))\}, \quad (9)$$

where $W_i = \left\{\log t_i - \frac{\mu + \beta_{1i}X_{1i} + \beta_{2i}X_{2i} + \dots + \beta_{pi}X_{pi}}{\sigma}\right\}$, $\boldsymbol{X} = \{X_{ji}\}$ is a column vector of covariates for the jth subject.

The estimated values of unknown parameters $\boldsymbol{\beta}, \mu$, and σ which maximize the log-likelihood function (9) may be determined by using Newton-Rhapson method [10]. For each parametric distribution, the $f_i(t_i)$ and $S_i(t_i)$ in equation (9) are substituted with the corresponding functions. For example in log-normal AFT model, the $S_i(t_i)$ in equation (9) should be substituted with the $S(t)$ from equation (7).

4. Results

4.1. Description of the data set

In this study, a total of 27488 ever-married women aged 15-49 at the time of interview from 33 provinces in Indonesia were selected for further analysis. Of the total of respondents, there were 25564 (93%) women who gave birth and 1924 (7%) women who did not give birth until the end of the interview. The FBI of ever-married women aged 15-49 years old was defined as the response variable and measured in months.

The research subjects were limited to the women who were not pregnant and did not have a child at the time of their first marriage. It means that the less than nine months and negative FBI were not taken into account. Furthermore, the respondents who did not yet give birth the first child at the time of the interview were categorized as right censored data. Eight covariates which consisted of one continuous covariate and seven categorical covariates were used in this work. These covariates were socioeconomic and demographic characteristics of the respondents (table 1).

4.2. Checking PH assumption by log cumulative hazard plots

In order to identify the presence of non-proportional hazards, it is common to use log cumulative hazard plots. This procedure can be implemented by plotting $log(t)$ versus $log(-log(S(t)))$ for all groups in every categorical covariate, where t and $S(t)$ are survival time and survival function at time t, respectively. If the plot does not yield the parallel curves separated by β over the log(time), then the PH assumption is not correct. In other words, the PH assumption is violated in the model.

The difficulty of this approach will arise if a covariate has too many categories or infinite categories as in continuous covariate. This problem can be solved by dividing the subjects into a few categories. However, this procedure is not applied in this work because the presence of non-proportional hazards can be shown by just checking the categorical covariates. The plots of all categorical covariates based on the data set are depicted in figure 1. These plots were built by using package 'survival' with plot() and survfit() functions in RStudio version 1.1.383.

Table 1. Respondents' characteristics based on 2012 IDHS

Covariate	Categories	Measurement Scale
Age at the first birth	-	Continuous
Place of residence	Rural	Nominal
	Urban	
Women's educational level	No education	Ordinal
	Primary	
	Secondary & above	
Husband's educational level	No Education	Ordinal
	Primary	
	Secondary & above	
Contraceptive knowledge	No	Nominal
	Yes	
Access to mass media	No	Nominal
	Yes	
Wealth index	Poor	Ordinal
	Middle & above	
Employment status	No	Nominal
	Yes	

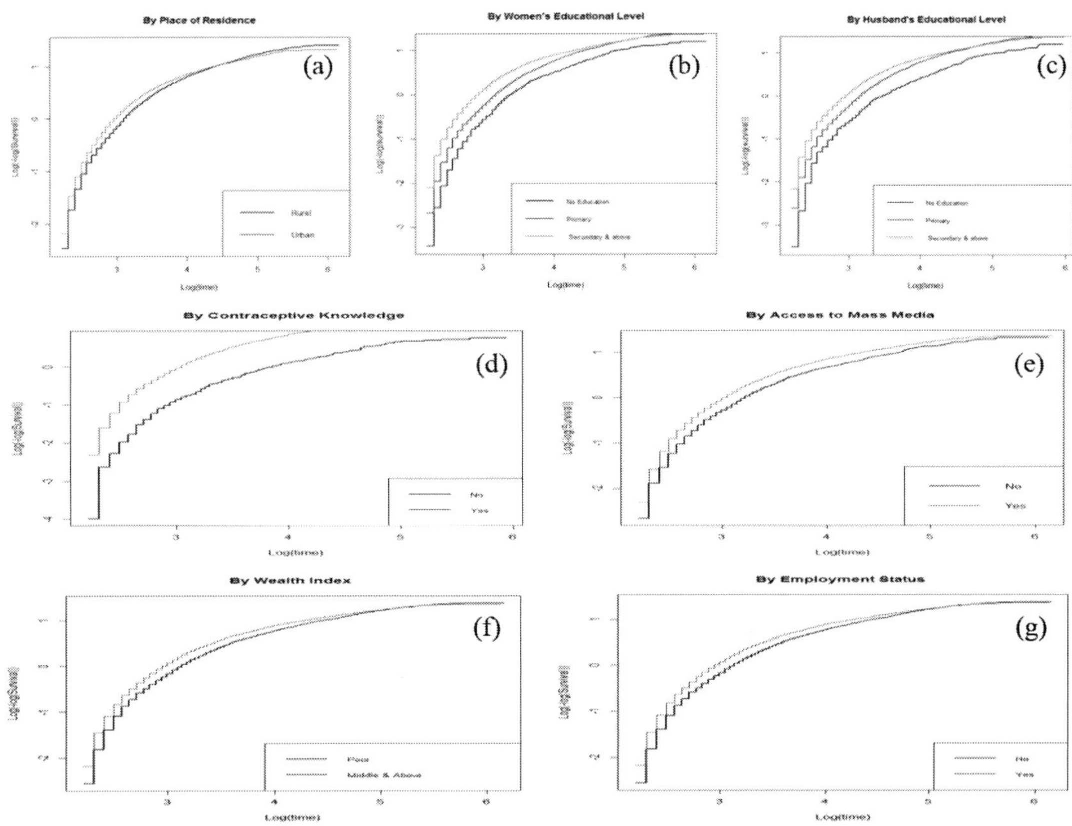

Figure 1. The log cumulative hazard plots of categorical covariates. The curves in (a), (b), (c), (e), (f), and (g) are not parallel. The parallel curves are shown only in (d).

Based on figure 1, the plot that yields two or more parallel curves is just for contraceptive knowledge. Meanwhile, the plots for the other categorical covariates are clearly not parallel because there are at least two curves which intercept each other. It means that by using log cumulative hazard plot, the PH assumption is violated for modelling the FBI data in Indonesia.

4.3. Checking PH assumption by the goodness of fit (GOF) testing approach

The GOF testing approach is very attractive because it provides the statistical tests in order to check the presence of the non-PH. Consequently, the decision about PH assumption is more objective than the log cumulative hazard plots approach which is already discussed in the previous subsection.

One of the most popular GOF testing methods is a GOF test based on the Schoenfeld residual [9]. The basic idea of the test is that if the PH assumption for a covariate is not violated, there is no correlation between the Schoenfeld residuals and the survival time. In other words, it will test the hypothesis $H_0: \rho = 0$ (there is no correlation between the covariate and time). If the hypothesis H_0 is rejected, it can be concluded that the covariate does not hold the PH assumption. In this research, the results of GOF testing with four different forms of survival time for all covariates in the FBI data are obtained by using package 'survival' in RStudio version 1.1.383 (table 2).

Table 2. The GOF test based Schoenfeld residual for the FBI data in Indonesia

	P-value			
	Kaplan-Meier	Log(time)	Time ranking	Time
Age at the first birth	0.00 [a]	0.00 [a]	0.00 [a]	0.00 [a]
Place of residence	0.068	0.039 [a]	0.068	0.086
Women's educational level	1.46e-09 [a]	8.80e-08 [a]	1.35e-09 [a]	0.001 [a]
Husband's educational level	0.003 [a]	0.002 [a]	0.003 [a]	0.01 [a]
Contraceptive knowledge	0.373	0.221	0.380	0.263
Access to mass media	0.481	0.190	0.497	0.041 [a]
Wealth index	0.456	0.448	0.455	0.425
Employment status	0.683	0.553	0.690	0.337
(Global test)	0.00 [a]	0.00 [a]	0.00 [a]	0.00 [a]

[a]: significant at 5% level

In this study, the four different forms of survival time are Kaplan-Meier (KM) transform, log(time) transform, time ranking (or ordering the time), and time (or no transformation). In general, the tests are twofold namely per variable test and global test. From table 2, it can be seen that the per variable tests for all transformation forms consist of at least one significant covariate at 5% level. It means that there is no correlation between the covariate and a specific transformation form. In other words, such covariate does not hold the PH assumption. Moreover, the global tests for all considered survival time transformation forms show the strong evidence of non-PH (p-value = 0.00).

4.4. Fitting the PH model and the AFT models

It has already shown in previous subsections that the PH assumption is violated in the FBI data. Therefore, the AFT models, which has no assumption about PH, are used as the alternative to PH model.

In this research, partial likelihood and MLE approach are employed to estimate the regression coefficients of the PH model and AFT models, respectively. The computation of the estimation of both models for FBI data in Indonesia is performed by RStudio version 1.1.383 with package 'survival' and package 'flexsurv' [10]. The parameters estimation and its corresponding p-values of PH model, Weibull AFT model, exponential AFT model, and log-normal AFT model are given in table 3.

Table 3 shows the estimation results and the p-values of PH model and AFT models in order to assess the effects of each parameter on the survival time. In PH model, eight parameters are significant and two parameters are insignificant at 5% level, that is, the parameters of the place of residence (urban) and access to mass media (yes). Meanwhile, age at the first birth and access to mass media (yes) are insignificant at 5% level for Weibull AFT model and exponential AFT model. The place of residence (urban) and access to mass media (yes) are insignificant at 5% level for log-normal AFT model.

Table 3. Results of fitting the PH model and the AFT models

	Categories	PH model		Weibull AFT model		Exponential AFT model		Log-normal AFT model	
		$e^{(\hat{\beta})}$	p-value	$\hat{\beta}$	p-value	$\hat{\beta}$	p-value	$\hat{\beta}$	p-value
(Intercept)	-	-		4.493	0.00 [a]	4.449	0.00 [a]	3.908	0.00 [a]
Age at the first birth	-	1.004	0.005 [a]	-0.002	6.28e-02	-0.002	2.75e-01	-0.009	5.36e-19 [a]
Place of residence	Rural [b]	-		-		-		-	
	Urban	0.993	0.604	0.039	2.23e-03 [a]	0.035	1.37e-02 [a]	0.006	0.516
Women's educational level	No Education [b]	-		-		-		-	
	Primary	1.079	0.032 [a]	-0.122	1.30e-04 [a]	-0.117	9.16e-04 [a]	-0.079	1.45e-03 [a]
	Secondary & Above	1.267	1.97e-10 [a]	-0.295	1.06e-18 [a]	-0.286	1.19e-14 [a]	-0.197	2.50e-14 [a]
Husband's educational level	No Education [b]	-		-		-		-	
	Primary	1.151	0.001 [a]	-0.209	3.87e-08 [a]	-0.205	1.14e-06 [a]	-0.131	8.76e-06 [a]
	Secondary & Above	1.224	3.14e-06 [a]	-0.277	1.11e-12 [a]	-0.270	3.51e-10 [a]	-0.179	3.14e-09 [a]
Contraceptive knowledge	No [b]	-		-		-		-	
	Yes	1.706	< 2e-16 [a]	-0.652	1.03e-34 [a]	-0.667	6.13e-30 [a]	-0.438	6.56e-30 [a]
Access to mass media	No [b]	-		-		-		-	
	Yes	1.017	0.463	-0.020	3.29e-01	-0.019	3.96e-01	-0.021	0.188
Wealth index	Poor [b]	-		-		-		-	
	Middle & Above	1.045	0.002 [a]	-0.027	3.91e-02 [a]	-0.032	3.01e-02 [a]	-0.027	9.11e-03 [a]
Employment status	No [b]	-		-		-		-	
	Yes	0.938	5.53e-07 [a]	0.095	3.96e-16 [a]	0.089	3.51e-12 [a]	0.047	1.77e-07 [a]

[a] Significant at 5% level
[b] Reference category

In many cases, it is also important to compare the efficiency of different models. AIC is one of the common criteria to select the best model. AIC can be computed by

$$\text{AIC} = -2\log(likelihood) + kp, \qquad (10)$$

where $\log(likelihood)$ is the loglikelihood function, p is the number of parameters, and k is the predetermined constant (we shall take as 2). The AIC values based on equation (10) of the PH model and the three AFT models for the data set are computed by using RStudio version 1.1.383 with package 'survival' (table 4).

Table 4. Comparison of the PH model and the AFT models using AIC criteria

Model	AIC Value
PH model	477950.2
Weibull AFT model	221728.4
Exponential AFT model	222249.5
Log-normal AFT model	207266.9

The AIC values in table 4 are obtained in the case of the full model which means that all covariates are included in the model. According to table 4, the AIC value of PH model is 477950.2 and it doubles the average AIC values of all proposed AFT models. It proves that the AFT models are much better than the PH model to fit the data set. In addition, it can be seen also in table 4 that the log-normal AFT model has the lowest AIC value. In other words, the log-normal AFT model is the best fitted model among the other considered models.

4.5. Variable selection using backward stepwise

In this subsection, the backward stepwise procedure is applied to select the best combination variables for log-normal AFT model. This procedure is implemented by using package 'survival' with step() function in RStudio version 1.1.383 (table 5). In addition, the computation also yields that the AIC value of the log-normal AFT model in table 5 is 207265, which is less than the full log-normal AFT model (AIC= 207266.9) as shown in table 4. It means that the log-normal AFT model in table 5 is better than the full model in table 4.

Table 5. The best log-normal AFT model based on the backward stepwise procedure

	The Estimate of Coefficients							
(Intercept)	Women's educational level (primary)	Women's educational level (secondary & above)	Husband's educational level (primary)	Husband's educational level (secondary & above)	Contraceptive knowledge (yes)	Access to mass media (yes)	Wealth index (middle & above)	Employment status (yes)
3.903	-0.083	-0.203	-0.133	-0.181	-0.445	-0.027	0.046	-0.009

5. Discussion

Published studies in various practical problems tend to use the PH model rather than the AFT models. However, there are only a few of the studies which investigated the required PH assumption [11]. The violation of the PH assumption can cause the model to be unreliable and biased [12]. Therefore, the AFT models can become one of the solutions to deal with the presence of non-PH in the data set. Moreover, several works have proved that the AFT models better than the PH model to fit the data set with non-PH [3], [4], [6].

In order to check the violation of PH assumption in the FBI data set, this work applied two methods namely log cumulative hazard plots and the GOF test based Schoenfeld residual. The results of both methods showed that the non-PH existed in the data set (figure 1 and table 2). This finding is similar to [13], which also showed the presence of non-PH in the data set.

In this study, the results of the PH model and the AFT models were compared to analyze the FBI data in Indonesia. To compare these models, this work used the AIC value. The estimation of the full models (table 3) showed that the PH model and the AFT models had no much differences in testing the significance of the covariates. However, the AIC value of PH model (AIC= 477950.2) approximately doubles the AIC values of each considered AFT models, i.e. Weibull AFT model (AIC= 221728.4), exponential AFT model (AIC=222249.5), and log-normal AFT model (AIC=207266.9). It means that, in this circumstance, the performance of the AFT models is much better than the PH model. Based on the AIC value, it can be seen also that the log-normal is the best fitted model among the other considered models. These results are consistent with the results of the other works [5], [14], which showed that the log-normal AFT model was the best fitted model to their data set.

The further analysis of the best fitted model (the log-normal AFT model) showed that the variables of women's educational level (primary), women's educational level (secondary & above), husband's educational level (primary), husband's educational level (secondary & above), contraceptive knowledge (yes), access to mass media (yes), wealth index (middle & above), and employment status (yes) (table 5). This result is different with the result by [15], which concluded that age at the first birth is significant at 10% level. However, the significance of employment status, contraceptive knowledge, and women education are consistent with [16].

6. Conclusions

In this paper, the PH model and three AFT models were applied to the FBI data in Indonesia. Based on the log cumulative hazard plots and GOF test, it was clear that the constant hazards assumption for the FBI data in Indonesia was violated. Although the procedure of the log cumulative hazard plots was relatively simple, it was difficult to make the decision about the PH assumption for any covariate which consists of many categories. This drawback could be overcome by applying GOF test which provides more powerful results using a statistical test. The results showed that the log-normal AFT model had the lowest AIC value comparing to the PH model, Weibull AFT model, and exponential AFT model. It means that the log-normal AFT is the best fitted model for the FBI data in Indonesia among the other considered models.

References

[1] Schemper M 1992 Cox analysis of survival data with non-proportional hazard functions *J. R. Stat. Soc. Series D The Stat.*, **41** 455-465

[2] Lee E T and Wang J W 2003 *Statistical methods for survival data analysis* (New Jersey- John Wiley & Sons, Inc.)

[3] Wei L J 1992 The accelerated failure time model: a useful alternative to the Cox regression model in survival analysis *Stat. Med.* **11** 1871-79

[4] Kay R and Kinnersley 2002 On the use of the accelerated failure time model as an alternative to the proportional hazards model in the treatment of time to event data : a case study in influenza *Drug Inf. J.* **36** 571-79

[5] Khanal S P, Sreenivas V and Acharya S K 2014 Accelerated failure time models : an application in the survival of acute liver failure patients in India *Int. J. Sci. Res.* **3** 161–66

[6] Orbe J, Ferreira E and Núñez-Antón V 2002 Comparing proportional hazards and accelerated failure time models for survival analysis *Stat. Med.* **21** 3493–3510

[7] Cox D R 1972 Regression models and life-tables *J. R. Stat. Soc. Ser. B* **34** 187–220

[8] Faruk A 2014 Estimasi parameter data tersensor tipe I berdistribusi log-logistik menggunakan maximum likelihood estimate dan iterasi Newton-Rhapson *Proc. Seminar Nasional MIPA 2014* Palembang pp 21-25

[9] Schoenfeld D 1982 Partial residuals for the proportional hazards regression model *Biometrika* **69** 239–41

[10] Jackson C H 2016 Flexsurv: a platform for parametric survival modeling in R *J. Stat. Softw.* **70** 1–33

[11] Altman D G, De Stavola B L, Love S B and Stepniewska K A 1995 Review of survival analyses published in cancer journals *Br. J. Cancer* **72** 511–18

[12] Zare A, Hosseini M, Mahmoodi M, Mohammad K, Zeraati H and Naieni K H 2015 A comparison between accelerated failure-time and cox proportional hazard models in analyzing the survival of gastric cancer patients *Iran. J. Public Health* **44** 1095–1102

[13] Rahman M A 2015 Fitting time to first birth using extended Cox regression model in presence of non-proportional hazard *Dhaka Univ. J. Sci.* **63** 25-30

[14] Adelian R, Jamali J, Zare N, Ayatollahi S M T, Pooladfar G R and Roustaei N 2017 Comparison of Cox's regression model and parametric models in evaluating the prognostic factors for survival after liver transplantation in Shiraz during 2000–2012 *Int. J. Org. Transplant Med.* **6** 119-125

[15] Faruk A 2017 Aplikasi regresi Cox pada selang kelahiran anak pertama di provinsi Sumatera Selatan *J. Mat.* **7** 19-29

[16] Gebeyehu A 2015 *Survival analysis of time-to-first birth after marriage among women in Ethiopia : application of parametric shared frailty model* (Jimma- Jimma University)

Dynamic Segmentation Of Behavior Patterns Based On Quantity Value Movement Using Fuzzy Subtractive Clustering Method

Iriansyah Sangadji, Yozika Arvio and Indrianto
Sekolah Tinggi Teknik PLN, Jakarta
iriansyah@sttpln.ac.id, yozika@sttpln.ac.id, indrianto@sttpln.ac.id

Abstract. to understand by analyzing the pattern of changes in value movements that can dynamically vary over a given period with relative accuracy, an equipment is required based on the utilization of technical working principles or specific analytical method. This will affect the level of validity of the output that will occur from this system. Subtractive clustering is based on the density (potential) size of data points in a space (variable). The basic concept of subtractive clustering is to determine the regions in a variable that has high potential for the surrounding points. In this paper result is segmentation of behavior pattern based on quantity value movement. It shows the number of clusters is formed and that has many members.

1. Background.

In order to understand by analyzing the pattern of changes in value movements that can dynamically vary over a given period with relative accuracy, an equipment is required based on the utilization of technical working principles or specific analytical method. This will affect the level of validity of the output that will occur from this system.

There are many techniques and methods in order to analyze the pattern of dynamically changing value movements as mentioned above. One of them is the method used based on the use of cluster to do the analysis. Clustering can also be done by utilizing various known models such as Self Organizing Mapping, Linear vector Quantitation, K Means and so on. Of the many such methods, not many are using fuzzy subtractive clustering in Clustering and classification in published research.

In this paper, it will use fuzzy subtractive clustering method in order to understand by analyzing the pattern of change in value movements of sales data in particularly. In this experiment will utilize the actual time series sales data as raw data to be simulated then analysis of simulation output results based on fuzzy subtractive clustering method.

The data used as raw data is the rough data of sales of various products within a company whose business scope is of a special characteristic.

The table 1 below is the total value of annual sales volume of the data to be simulated.

Table 1. Example times series annual sales data

No	Year	Volume
1	2011	Rp. 14,566,101,153
2	2012	Rp. 18,222,610,140
3	2013	Rp. 26,112,686,484
4	2014	Rp. 28,620,582,707
5	2015	Rp. 49,603,557,816

Segmentation can be intended as a division into different groups (heterogeneous) into homogeneous groups, where each group can be exposed to patterns of changes in varied value movements. This can be done by looking at changes in the formation of clusters according to the steps contained in the fuzzy subtractive clustering method. Number of clusters will be formed without specified number of clusters first. The algorithm with the input of a specified radius distance will determine the number of clusters to be formed.

The above will explain that the smaller the defined radius, the more clusters will be formed. On the other hand, that the larger the radius is defined, the less cluster will be formed as well. The pattern of dynamically changing value movements can be defined as cluster segmentation. Each set of resulting clusters, can contain the number of set members that can not be the same.

2. Objective

The objective of this paper is; Applying substractive clustering fuzzy method in forming cluster in time series data and Conducting segmentation process in order to get picture of behavior pattern based on movement of value of quantity.

Segmentation is a division of groups that have different characteristics, or different behaviors within a certain value. Segmentation can also be interpreted as identifying difference analysis. Segmentation can be intended as a distinct (heterogeneous) division into homogeneous groups. The purpose of segmentation is to make it easier to distinguish, meaning every value / cluster produced is easier to distinguish.

Clustering is a process of classifying data in a class or clusters so that data in a cluster has a high degree of similarity with each other but very different from the data in other clusters. Clustering is one of unsupervised learning techniques where we do not need to practice the method.

The purpose of this method is to group with number of data or objects into the cluster so that each cluster will be filled as closely as possible data. Clustering model analysis is a technique of multivariable analysis used to group objects (variables or data) so as to produce an information to assist the implementation of testing of the object and can present a hypothesis based on the relation. The principle is to maximize homogeneity in one group and to maximize the heterogeneity (inequality) between groups.

Fuzzy clustering will determine the cluster center by marking the average location for each cluster. In the initial conditions, the cluster center is still not accurate, but each data point has a membership degree for each cluster. By fixing the cluster center and the degree of membership of each data point repeatedly it will be seen that the center of the cluster will move towards the right location.

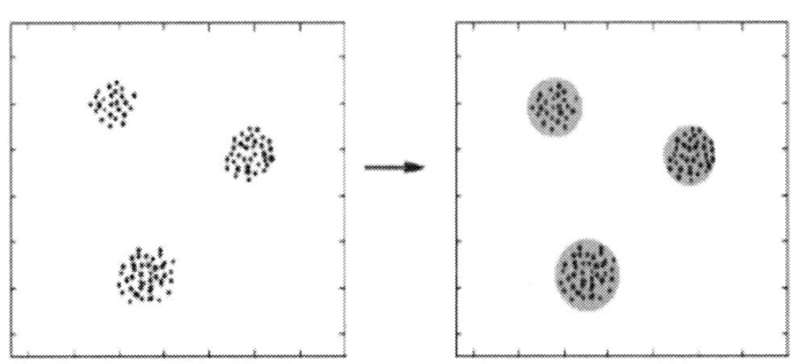

Figure 1. Example clustering process

3. Theoretical Study

in this paper, focuses on the implementation of the subtractive clustering fuzzy method. In theory that fuzzy subtractive clustering is based on the density (potential) size of data points in a space (variable). The basic concept of subtractive clustering is to determine the regions in a variable that has high potential for the surrounding points as shown in figure 1. The points with the most number of neighbors will be selected as cluster centers. The point that has been selected as the center of this cluster will then be reduced its potential. Then the algorithm will choose another point that has the most neighbors to be the center of another cluster.

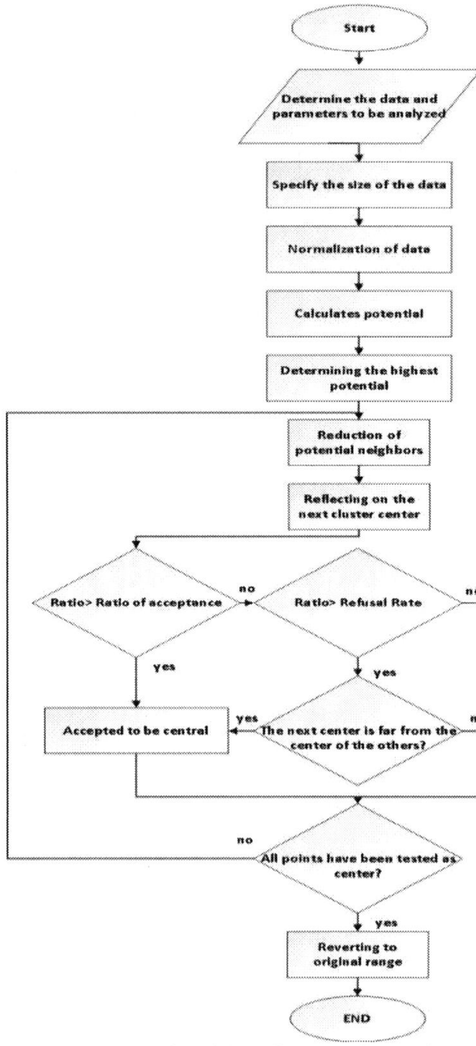

Figure 2 Algorithm fuzzy subtractive clustering flowchart

This will be done repeatedly until all the points are tested. The Fuzzy Subtractive clustering algorithm is :

Matrix X which is the data to be in the cluster, the size of i x j, with i = the amount of data to be clustered and j = number of variables / attributes (criteria)

$$\begin{bmatrix} X_{11} & X_{12} & \dots & X_{1j} \\ X_{21} & X_{22} & & X_{2j} \\ \vdots & \vdots & \ddots & \vdots \\ X_{i1} & X_{i2} & \cdots & X_{ij} \end{bmatrix} \tag{1}$$

Determine:

a. rj (the radius of each data attribute); j = 1,2, .., m.

b. q (squash factor).

c. accept ratio.

IOP Conf. Series: Journal of Physics: Conf. Series **974** (2018) 012009 doi:10.1088/1742-6596/974/1/012009

d. reject ratio.
e. Xminj (min allowed data in each data attribute); j = 1,2, .., m.
f. Xmaxj (max allowed data in each attribute of data); j = 1,2, .., m.

Normalize

$$X_{ij} = \frac{X_{ij} - X\min_j}{X\max_j - X\min_j} \qquad (2)$$

i = 1,2, ..., n; j = 1,2, .., m

Where :
Xij = data to the j-attribute.
Xmin j = minimum data on j-attribute.
Xmax j = maximum data on j-attribute.

Determine the initial potential of each data point
a. i = 1; i is data to.
b. Do it up to i = n,
   i. Tj = Xij (Tj is the i th jth attribute data); j = 1,2, ..., m

   ii. Count:

$$Dist_{kj} = \left(\frac{T_j - X_{kj}}{r_j}\right) \qquad (3)$$

   iii. Calculate initial potential:

$$\text{if } m = 1\text{, Then}$$
$$D_i = \sum_{k=1}^{n} e^{-4\left(Dist_{k1}^2\right)}$$
$$\text{if } m > 1\text{, Then}$$
$$D_i = \sum_{k=1}^{n} e^{-4\left(\sum_{j=1}^{m} Dist_{kj}^2\right)} \qquad (4)$$

To calculate the degree of membership

$$\mu_{ki} = e^{-\sum_{j=1}^{m}\frac{(X_{ij}-C_{kj})^2}{2\sigma_j^2}} \qquad (5)$$

4. Data

The data variables used in this research are:
a) Variable 1: data of quantity
b) Variable 2: net value data
c) Variable 3: data cogs

Process requirements here are variables and parameters used to process data in general. The specifications of software and hardware that the authors use in this study are Variables:
X1 = Quantity
X2 = Net Value
X3 = COGS
The parameter value used is the result of the experiment that has been done so that obtained the maximum parameter value.

a. r (radius) = 0.001
b. q (squash factor) = 0.2
c. accept ratio = 0.8
d. reject ratio = 0.2
e. X1Min = 1, X1Max = 90
   X2Min = 46162800, X2Max = 1140000000
   X3Min = 63649002, X3Max = 1014124250

5. Result and Discussion

In the raw data groupings of a year data, the radius used is 0.8 resulting in 1 cluster, 0001 yielding 2 clusters and 0.0001 resulting in 3 clusters.

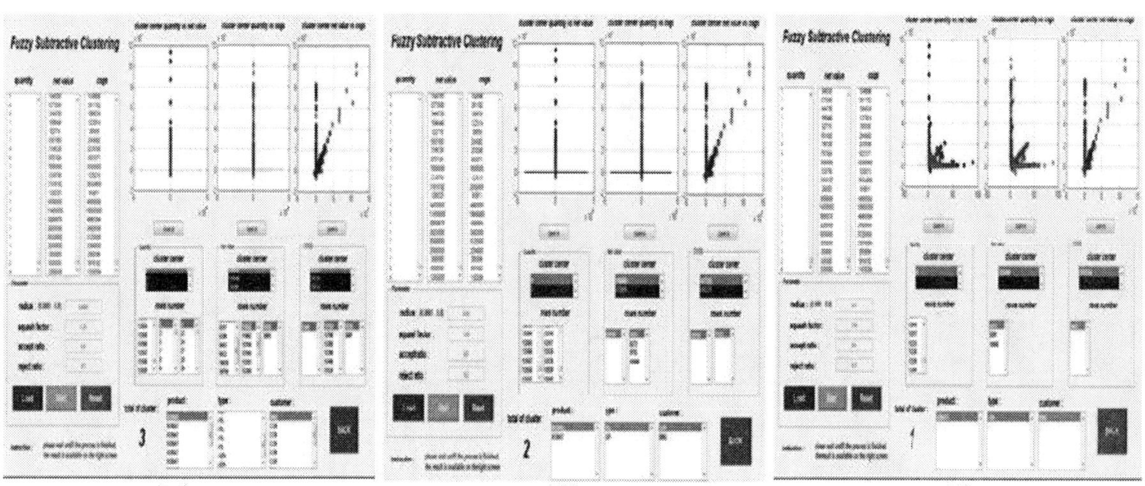

(a) (b) (c)

Figure 3. Example of gui's clusters

Gui's clusters has been designed using matlab as shown Figure 3 above. A year data testing is done by dividing into four quarters result data. 3-dimensional correlation of data (X1,X2,X3) can be seen also entirely in figure 3a,b,c.

Table 2. Quantity data

Cluster Center	Number of Cluster Members	Total data	Quarters	Clusters
1	7589	12561	First Quarter	2
2	2663	12561		
1	7589	12561	Second Quarter	2
2	2663	12561		
1	7589	12561	Thirth Quarter	3
1	7589	12561		
2	2663	12561		

1	7589	12561		
1	7589	12561	Fourth Quarter	3
2	2663	12561		
1	7589	12561		
1	7589	12561	Total A Year	3
2	2663	12561		

Table 2 shows that there is a significant change in the behavior of the data increase pattern in the third and fourth quarters

Table 3. Net value data

Cluster Center	Cluster Members	Number of Cluster members	Quarters	Clusters
78	217,646	2	First Quarter	2
148	92,373	2		
32	1721	1	Second Quarter	2
77200	4279	1		
196480	192	1	Thirth Quarter	3
1480410	46,49,68,110,138,271,274,299, 302,305,320,383,417,432,448,465, 472,481,498,530,566,567,759,761, 796,811,855,888,889,895,911,982, 983,984,1031,1093,1108,1126	38		
122268	573	1		
67191	424,547	2	Fourth Quarter	3
263464	45	1		
52800	1385	1		
25400	5400,6317,6399,8451, 8533,10630,10704	7	Total A Year	3
116008	11856,11863,12184, 12190,12202,12208, 12466,12524	8		
30200	5636,5881	2		

Table 3 showed that there was a significant increase in behavior change in the third quarter. Although there is also the same thing in the fourth quarter, but not as high as third quarter

Table 4. Cogs data

Cluster Center	Cluster Members	Number of Cluster members	Quarters	Clusters
546840	217	1	First Quarter	2
864672	92	1		
219247	1721	1	Second Quarter	2
162779	4279	1		
106965	192,731	2	Thirth Quarter	3
759814	35, 46,49,68,110,138,182,271,274,299, 302,305,320,383,417,432,448,465,472, 481,498,530,566,567,759,761,796,811, 855,888,889,895,911,982,983,984,986, 1031,1093,1108,1126,1135,1136,1137	95		

63629	573	1	Fourth Quarter	3
48242	14,547,2466	2		
146998	45,323,355,381,594,749, 1323,1491,1600,1633,1760	11		
45435	1385	1		
25400	5400,6317,6399,8451, 8533,10630,10704	7	Total A Year	3
116008	11856,11863,12184, 12190,12202,12208, 12466,12524	8		
30200	5636,5881	2		

Table 4 the same thing as table 3, shows that there is a significant jump in behavior change in the data pattern in the third quarter. Although there is also the same thing in the fourth quarter, but not as high as third quarter.

6. Conclusion

The data grouped by the fuzzy subtractive clustering method begins with the input data to be clustered, which begins by setting the value of cluster parameters consisting of: the radius of each attribute data (r), squash factor (q), Accept ratio, reject ratio, Xmin (minimum allowed data), Xmax (minimum allowed data), data normalization, determine the initial potential of each data point, locate the highest potential point, determine the cluster center locate the new cluster center candidate and calculate the sigma cluster value. It can produce the number of clusters according to the data being implemented.

There was a significant dynamic segmentation in the third quarter where the value of the changes was quite high

References

[1] A.Taupik Rahman.,Nasrun Hariyanto.,& Sabat Anwari. (2014). Peramalan Beban Puncak Jangka Pendek Khusus Hari Libur Nasional Berbasis Algoritma Fuzzy Subtractive Clustering, Studi Kasus di Jawa – Bali. *Jurnal Reka Elkomika*. 2337-439X April 2014.
[2] Away, A. (2010). *Matlab Programming*. Bandung : Informatika
[3] Gita Febrina Wulandari. (2014). Segmentasi Pelanggan Menggunakan Algoritma K-Means Untuk Customer Relationship Management (CRM) Pada Hijab Miulan.
[4] Helmy Yulianto Hadi.,R. Rizal Isnanto.,& Budi Setiyono. (2006). Pembagian Kelas Kuliah Mahasiswa Menggunakan Algoritma Pengklasteran Fuzzy. Makalah Seminar Universitas Diponegoro.
[5] Heru Dibyo Laksono.,& Muhammad Hafis. (2013). Aplikasi Fuzzy Clustering Dengan Menggunakan Algoritma
[6] Subtractive Clustering Untuk Perkiraan Kebutuhan Energi Listrik Jangka Panjang Di Provinsi Sumatra Barat Dari Tahun 2012-2021. *Jurnal Teknologi Informasi & Pendidikan*. 2086 – 4981.
[7] I Gede Oka Artawan. (2013). Penentuan Lokasi SMP Baru di Kabupaten Klungkung dengan Algoritma Fuzzy Subtractive Clustering. *Jurnal Matematika*. 1693-1394
[8] Kotler, Philip dan Keller, Kevin L. (2009), Manajemen Pemasaran edisi ke-13 jilid 1, Penerbit Erlangga, Jakarta
[9] Kusumadewi, Sri. (2002). Analisis dan Desain Sistem Fuzzy Menggunakan Toolbox Matlab. Yogyakarta: Graha Ilmu.
[10] Kusumadewi, Sri, Hari Purnomo. (2004). Aplikasi Logika Fuzzy Untuk Pendukung keputusan. Graha Ilmu, Yogyakarta.

Application of FKNN on Positioning of Potential Salt in Coastal South Beach of Madura

Tony Yulianto[1*], Rica Amalia[2], Kuzairi[3]

[1*, 2, 3)] Department of Mathematics, Mathematics and Science Faculty, Islamic University of Madura, Indonesia

Email: [1*)]toniyulianto65@gmail.com

Abstract. The sea holds a lot of potential to be utilized, including salt which is one of the most important needs in everyday life. In terms of quality domestic salt production is still not meet the health requirements, especially salt produced from salt farmers. So, there is an effort to optimize salt production by improving the quality of sea water, especially in terms of salinity. Madura is one of the areas in Indonesia that can be developed as a center of national salt, especially the southern region of Madura. That opportunities can be developed considering that until now Indonesia is still imported salt, whereas the potential of salt in Madura is very large. Therefore, in this study used FKNN method to simplify the process of classification for decision making about the position of salt content in the beach of Madura that has the potential to be a quality salt. From all the south beach of Madura, 12 data were taken (each district had 3 coastal data) with each data is tested as many as 8 test. Then from 12 data is taken 11 data for the training and the rest in testing using FKNN. So that, the results are obtained that almost all beaches on the south beach of Madura are in the category of Kw III, except for 2 beaches that are Pang Pajung beach (located at latitude -7,207,387 and Coordinates longitude 113,006943) located in the category Kw I and Gading beach (located at latitude -7,185039 coordinates and longitude coordinates 112,915988) located in the category Kw II and the two beaches are located in Bangkalan district.

1. Introduction

Indonesia as an archipelagic country with a long coastline of 81,000 kilometers is a coastal and marine area that has a huge variety of biological and non-biological resources. With the oceans constituting 70% of the total area of the archipelago, the sea has a lot of potential to be utilized, among others are salt (Yulianto, Komariyah, & Ulfaniyah, 2017).

Salt is one of the most important needs in everyday life. The making of salt is mostly done traditionally by smallholders besides by industrial salt companies. In terms of quality of domestic salt production is still not meet health requirements, especially salt produced from salt farmers, because the quality of salt is generally below the quality of II according to SNI / SII specifications No.140-76 (Hasan, 2011).

Madura is one of the areas in Indonesia that can be developed as a national salt center. That opportunities can be developed considering that until now Indonesia is still imported salt, whereas the potential of salt in Madura is very large. Of the 2.4 million tons of national salt demand, currently only 1.4 million tons are supplied. Of that amount, Madura's salt contribution is 600,000 ton. This data shows that almost half of the national salt demand is supplied by Madura, optimize salt production by improving seawater quality especially in terms of its salt content. The phenomenon of most of the traditional salt processing on the coast of Madura does not see the proportion of salt content in the region (Hasan, 2011). The research uses the south coast of Madura, because the southern coast becomes the main crossing point for frequent transportation by many vehicles, thus becoming the main access in industry.

The background of the use of the method among others is based on previous research stated that k-nearest neighbor fuzzy method (FKNN) proved to consider the ambiguous nature of neighbors, where ambiguous neighbors do not have an important role in the classification process, this is due to the membership value of each class giving an instance to a class (Putri, Cholissodin, & Setiawan, 2015). Therefore, in this study, the authors will use the FKNN method in the classification process for decision making about the position of salt content in the coast of Madura that has the potential to be a quality salt.

2. Preliminaries

This section presents briefly the basic concept related to salt and FKNN method and implementation of FKNN.

2.1. *Definition of Salt*

Salt is a crystal-shaped white solid object that is a collection of compounds with the largest portion of Sodium Chloride (> 80%) as well as other compounds, such as Magnesium Chloride, Magnesium Sulphate, and Calcium Chloride. Natural salt sources derived from seawater, salt lake water, in-ground deposits, salt mines, groundwater sources. Components - these components have an important role for the human body, so it takes salt consumption with the right size to support human health (Hasan, 2011).

According to Desrosier (1988), there are three main sources of salt, namely (Mohi, 2014):

1. Solar salt is salt obtained by evaporation from brine either from the sea or from the saline lake of the land.
2. A salt or salt mine is a salt normally declared as a salt rock, derived from mining operating as deep as a thousand feet or more below the earth's surface.
3. The salt obtained from the evaporation of the sun contains chemical waste and salt-tolerant halophililis microbes. Mineral salt or salt sources are generally free from contamination of these organisms.

2.2. Quality of Salt Based on SNI Salt

Salt can be divided into several categories such as; very good category, good and medium. It is said to be very good if it contains NaCl> 95%, good NaCl content 90-95%, and moderate NaCl between 80-90% but the preferred is the salt content above 95% (Mayasari & Lukman, 2004).

Salt quality based on NaCl content can be mapped (see table 1).

Table 1 Salt Quality based on NaCl (Usman, 2017)

No.	Substance	Ks
Quality I	NaCl > 98%	Maximum water content of 4%
Quality II	94,4% < NaCl <98%	Maximum water content of 5%
Quality III	NaCl < 94%	Water content of > 5%

Every salt producer is required to obtain Indonesian National Certificate (SNI) before trading (see table 2). According to the Indonesian National Certificate (SNI) no. 01-3556 of the year 1994 and the Decree No. 77/1995 the salt used must contain iodine of 30-80 ppm (Hartati, Supriyo, & Zainuri, 2014).

Table 2 Standard of Salt SNI (Hartati, Supriyo, & Zainuri, 2014)

No.	Parameter	Unity	SNI Standard
1.	Flavors	-	Salty
2.	Smell	-	Normal
3.	Color	-	White
4.	NaCl (kitchen salt)	w/w	Min 94,73
5.	Water (H_2O)	w/w	Max 7
6.	Fe_2O_3 (iron oxide/sludge/sand)	mg/kg	Max 100
7.	Ca (calcium) atau Mg (magnesium)	w/w	Max 1,00
8.	SO_4 (sulphate)	w/w	Max 1,00
9.	Impurity	w/w	Max 2,00
10.	Iodium as KIO_3	w/w	30-80
11.	Pb (lead)	mg/kg	Max 10
12.	Cu (copper)	mg/kg	Max 10
13.	Hg (mercury)	mg/kg	Max 0.1
14.	As (arsen)	mg/kg	Max 0.1

2.3. Fuzzy Set

The firm set (crisp) A is defined by the items in the set. In the set of crisp, the membership value of an item x in a set A ($\mu_A(x)$) has two possibilities, namely (Yulianto, Komariyah, & Ulfaniyah, 2017):

1. One (1), which means that an item becomes a member in a set.
2. Zero (0), which means that an item is not a member in a set.

The Fuzzy set is based on the idea of extending the range of characteristic functions such that the function will include real numbers at intervals [0,1]. The membership value indicates that an item in the universe of speech is not only 0 or 1, but also the value that lies between them. In other words, the truth value of an item is not just right or false. Value 0 indicates false, value 1 indicates true, and there are still values that lie between right and wrong (Yulianto, Komariyah, & Ulfaniyah, 2017).

2.4. Data Mining

Data mining is a term used to describe the discovery of knowledge in a database. Data Mining is a process that uses statistical techniques, mathematics, artificial intelligence and machine learning to identify useful information and knowledge assembled from various large databases (Wijaya, Mardji, & Furqon, 2015).

In this study used normalization min-max. Normalization of min-max is calculated (see equation 1). According to Wijaya et al (2015) normalization min-max has the advantage of maintaining relationships in the data. And has a function that unites units of various attributes (Wijaya, Mardji, & Furqon, 2015):

$$V' = \frac{V - \min_i A}{\max_i A - \min_i A} \quad (1)$$

with:

V'	:	The result of normalization whose value ranges between 0 and 1
V	:	The value of attribute A to be normalized
$\min_i A$	:	The minimum value of an attribute, A
$\max_i A$	:	The maximum value of an attribute, A

2.5. Fuzzy K-Nearest Neighbor

Fuzzy K-Nearest Neighbor (FKNN) is a Fuzzy set theory that generalizes the classical K-NN theory by defining the membership value of a data in each class. The formula used to obtain the results of FKNN with the form (Selly, Ridok, & Muflikhah, 2013):

$$u(x, y_i) = \frac{\sum_{j=1}^{K} w(x_j, y_i)^* d(x, x_j)^{\frac{-2}{(m-1)}}}{\sum_{j=1}^{K} d(x, x_j)^{\frac{-2}{(m-1)}}} \tag{2}$$

with $u(x, y_i)$ is the membership value of x class c. K is the number of nearest neighbors used. $w(x_j, y_i)^*$ is the value of neighboring member's membership in the neighboring K in the y_i class, the value 1 if the training data x_k belongs to class c_i or 0 if not belong to class c_i, for $d(x, x_k)$ is the distance from data x to x_k data in the nearest neighbor's k, m is the weight exponent of magnitude $m>1$ (Selly, Ridok, & Muflikhah, 2013).

The membership value of a data in a class is greatly influenced by the distance of that data to its nearest neighbor, the closer to its neighbor the greater the membership value of that data to its neighboring class, and vice versa. The distance is measured by N dimension (feature) data. The measurement of distance (inapplication) of two data used in FKNN (see equation 2) is generalized by (Putri, Cholissodin, & Setiawan, 2015):

$$d(x_i, x_j) = \left(\sum_{l=1}^{N} |x_{il} - x_{jl}|^p\right)^{\frac{1}{p}} \tag{3}$$

with N is the dimension (number of features) of data. While p is the distance determinant used, if $p = 1$ then the distance used is Manhattan, if $p = 2$ fed the distance used is Euclidean, if $p = \infty$ then the distance used is Chebychev. Although FKNN uses membership values to declare membership data on each class, but to provide final output, FKNN must still provide the final class of predicted results. For this purpose, FKNN selects the class with the largest membership value in the data (Putri, Cholissodin, & Setiawan, 2015).

Prediction Algorithm with FKNN (Putri, Cholissodin, & Setiawan, 2015):

1. Normalize the data using the largest and least values of data on each feature (see equation 1).
2. Find the nearest neighbor's K for the test data x using equation (3).
3. Calculate the membership value $u(x, y_i)$ using equation (2) for each i with $1 \le i \le C$.
4. Take the largest value $v = u(x, y_i)$ for all $1 \le i \le C$, with C is the number of classes.
5. Give the class label v to the data u in x that is y_i.

2.6. Data Collection

The data collected from all districts in Madura located on the south coast of Madura as much as 12 data per each point (beach) with sampling per each district there are 3 samples. We just use 12 data not more, because we have no much cost to do more and variable test tools also require a large cost. Because of that reason, we can only retrieve the data is not too much and we limit the data retrieval south coast in each district only as much as 3. The data collected in the form of seawater on the coast by taking a bottle of drinking water and then tested in the lab for the test level of each concentration. The concentrations of the test were 8 concentrations: Mg (magnesium), Cu (copper), SO4 (sulfate), As (arsenic), Hg (mercury), Pb (lead), NaCl (kitchen salt), and Ca (calcium).

2.7. Salt Quality Criteria

Based on coastal water data from laboratory test result, to be able to determine the quality of salt contained in sea water because it has not done the processing process in this study does not directly use the existing SNI salt, but using the approach approach value almost close to the actual SNI Salt. So the salt quality criterion data is determined first to facilitate the calculation process using the FKNN method which then can be obtained coastal location containing adequate salt quality which is almost equal to the best salt quality on the market. The SNI Salt is used for salt that has been done processing and distillation of seawater in several stages, so in this data do not use SNI Salt directly. The salt quality criterion data to the selected concentration (see table 3).

Table 3 Data on Salt Quality Criteria

Quality	NaCl	Mg	Ca	SO_4	Pb	Cu	Hg	As
Kw I	10	300	200	0	0	0	0	0
Kw II	27	400	300	1	5	5	0,05	0,25
Kw III	30	1000	400	2	10	10	0,1	0,5

2.8. FKNN Process

As for the FKNN algorithm, so we take 11 data from 12 data to do the training test. So the first step by entering the value of m. In this research is inserted value $m = 5$, then after that do normalization of data by using equation (1), for example:

a. At Mg test in 12 beaches it is known that the smallest value of data is 270 and the biggest value is 1200 so that the normalization result of Mg Patoroman beach (south coast of Bangkalan) is worth $\frac{330-270}{1200-270} = \frac{60}{930} = 0{,}064516129$,.
b. On the NaCl test at 12 beaches it is known that the smallest value of data is 4 and the largest value is 30 so that the normalization result of NaCl beach Patoroman (south coast of Bangkalan) is worth $\frac{29-4}{30-4} = \frac{25}{26} = 0{,}9615384615$, and so on.
c. For weighted values also performed normalization of data also, for example on the test Mg on Kw I obtained $\frac{10-10}{30-10} = \frac{0}{20} = 0$, on Kw II obtained value $\frac{27-10}{30-10} = \frac{17}{20} = 0{,}85$ and on Kw III we get value $\frac{30-10}{30-10} = \frac{20}{20} = 1$, and also done for other weight values.

Then look for the distance using equation (3) and the value $w(x_k, y_i)^*$ from equation (2). From the result of normalization then do a distance search such as used $p = 2$ (Euclidean distance) and $w(x_k, y_i)^*$. Example:

a. For the Mg test when y_1 (on Kw I) on Patoroman beach the distance value:

$$d_{\text{Patoroman}} = \sqrt{(330-330)^2 + (330-1200)^2 + (330-960)^2 + \ldots} \quad (4)$$

Equation (4) is calculated continuously to 12 points then obtained value $d_{\text{Patoroman}}$.

b. Whereas to obtain $w(x_k, y_i)^*$ from equation (2) for test Mg y_1 (at Kw I) on Patoroman beach is: $w^* = 0$.

The result of the value u of equation (2) on the south coast of Madura is obtained by combining the values of w^* and d from all points. After finish, we can do testing for the remaining data in the same way.

3. Result And Discussions

The simulation result after testing on all the data using matlab for south beach of Madura using FKNN method obtained the largest u result (Rank u) (see table 4).

Table 4 Ranking Results u Using FKNN On the South Beach

Beach Name	Latitude Coordinate	Longitude Coordinate	Rank u
A. Bangkalan District			
1. Patoroman Beach	-7,214059	113,040249	3
2. Pang Pajung Beach	-7,207387	113,006943	1
3. Gading Beach	-7,185039	112,915988	2
B. Sampang District			
1. Taddan Beach	-7,215940	113,283964	3
2. Camplong Beach	-7,218862	113,319411	3
3. Tanjung Beach	-7,217936	113,394611	3
C. Pamekasan District			
1. Branta Beach	-7,223800	113,449749	3
2. Jumiang Beach	-7,236570	113,544467	3
3. Talang Siring Beach	-7,141684	113,593849	3
D. Sumenep District			
1. Prenduan Beach	-7,110188	113,676640	3
2. Aing Panas Beach	-7,110020	113,689512	3
3. Kalianget Beach	-7,050465	113,932520	3

The result (see table 4) shows that almost all beaches on the south beach of Madura are in the category of Kw III, except for 2 beaches that are Pang Pajung beach (located at latitude -7,207,387 coordinates and longitude coordinates 113,006943) located in the Kw I and Gading beach (located at latitude -7,185039 coordinates and longitude coordinates 112,915988) located in the Kw II category and both beaches are located in Bangkalan district. From the data we can get as much 83.33% of beaches are in cluster 3 (Kw III), 8.33% of beaches are in cluster 1 (Kw I) and 8,33% another of beaches are in cluster 2 (Kw II).

4. Conclussions

The conclusion of this research (see table 4) shows that almost all beaches on the south beach of Madura are in the category of Kw III, except for 2 beaches that are Pang Pajung beach (located at latitude -7,207,387 coordinates and longitude coordinates 113,006943) located in the Kw I and Gading beach (located at latitude -7,185039 coordinates and longitude coordinates 112,915988) located in category Kw II and second beach located in Bangkalan district. Because the data is still small, so the future can be done quite a lot to the validity of the data can be processed better.

5. Acknowledgments

Praise the presence to Allah SWT and the author also thanked Ristek Dikti who has allocated research funding for later can be utilized in this research. In addition, the authors also thank the Dean of Mathematics and head of Mathematics department of UIM who has provided support in the form of initial assistance to conduct research, as well as dean of FKIP UIM who has assisted in the provision of laboratory for testing the concentration of sea water data in this study so that this study to run smoothly.

References

[1] Hartati, R., Supriyo, E., & Zainuri, M. (2014, April). Yodisasi Garam Rakyat Dengan Sistem Screw Injection. *GEMA TEKNOLOGI, 17*(4), 160-163.

[2] Hasan, T. I. (2011). Identifikasi Sosial Ekonomi Dan Ketenagakerjaan Petani Garam Di Kabupaten Bireuen. *Sains Riset, 1*(2).

[3] Mayasari, V. A., & Lukman, R. (2004). *Studi Peningkatan Mutu Garam Dengan Pencucian.* Surabaya: Fakultas Teknologi Industri Jurusan Teknik Kimia Institut Teknologi Sepuluh Nopember.

[4] Mohi, R. A. (2014). Analisis Potensi Pengembangan Tambak Garam Di Desa Siduwonge Kecamatan Randangan Kabupaten Pohuwato. Dalam *Tesis* (hal. 12-20). Gorontalo: Universitas Negeri Gorontalo.

[5] Putri, I. R., Cholissodin, I., & Setiawan, B. D. (2015). Optimasi Metode Adaptive Fuzzy K-Nearest Neighbor Dengan Particle Swarm Optimization Untuk Klasifikasi Status Sosial Ekonomi Keluarga. *Repositori Jurnal Mahasiswa PTIIK UB, 5*(3).

[6] Selly, Y., Ridok, A., & Muflikhah, L. (2013). Perbandingan K-Nearest Neighbor Dan Fuzzy K-Nearest Neighbor Pada Diagnosis Penyakit Diabetes Melitus. *Repositori Jurnal Mahasiswa PTIIK UB, 2*(10).

[7] Usman, M. A. (2017). Studi Eksperimen Penggunaan Air Garam Sebagai Sumber Energi Alternatif. Dalam *Skripsi* (hal. 15-18). Kendari: Jurusan Teknik Mesin, Fakultas Teknik, Universitas Halu Oleo.

[8] Wijaya, H. T., Mardji, & Furqon, M. T. (2015). Penerapan Fuzzy K-Nearest Neighbor (FKNN) Untuk Diagnosa Penderita Liver Berdasarkan Indian Liver Patient Dataset (ILPD). *Repository Jurnal Mahasiswa PTIIK UB, 5*(9).

[9] Yulianto, T., Komariyah, S., & Ulfaniyah, N. (2017). Application of Fuzzy Inference System by Sugeno Method on Estimating of Salt Production. *AIP Conference Proceedings* (hal. 1-7). Surabaya: American Institute of Physics. doi:10.1063/1.4994442

IOP Conf. Series: Journal of Physics: Conf. Series **974** (2018) 012011 doi:10.1088/1742-6596/974/1/012011

Numerical investigation of the flow around circular cylinder with two passive controls

Amirul Hakam[1], Basuki Widodo[1], Tri Yogi Yuwono[2] and Chairul Imron[1]

[1] Department of Mathematics, Institut Teknologi Sepuluh Nopember, Surabaya, Indonesia

[2] Department of Mechanical Engineering, Institut Teknologi Sepuluh Nopember, Surabaya, Indonesia

E-mail: imron-its@matematika.its.ac.id

Abstract. Investigation is done on incompressible flow around circular cylinder with two passive controls. The first passive control has a cylinder type I, placed in front of the cylinder with a fixed distance. The second passive control investigated with two types are ellipse and circle, placed behind the cylinder at varying distances. Reynolds numbers used are 100, 500 and 1000. Variations of the passive control's shapes, the distance between the passive control and the cylinder, and the variation of the Reynolds number influence the drag coefficient received passive control. The smallest drag coefficient received passive control occurs in the Reynolds number 1000 and the distance $T/D=1.8$. The drag coefficient received passive control ellipse or circular shape has very small difference, but for comparison investigation, passive control ellipse receives drag coefficient smaller than passive control circle shape.

1. Introduction

The phenomenon of movement of fluid flow across the object plays an important role in the development of the industrial world and technique. One of them is the flow of fluid through the cylinder, as a representation of a pile building or pipes on the offshore oil drilling process. The flow of fluid through the cylinder will stagnate, boundary layer, separation and wake behind the cylinder. The larger the wake of fluid flow, the greater difference in force received between the front of cylinder and beside cylinder. As a result the drag coefficient received by the cylinder is also getting bigger[1]. The drag coefficient received by the cylinder is an attractive phenomenon. In this case is closely related to the durability and strength of the cylinder. If a large drag coefficient can be reduced then the cylinder's durability can last a long time.

In general, to reduce the drag coefficient received the cylinder can be done by giving the passive controller placed around the cylinder with varying shapes, the distance with the cylinder or the number of Reynolds used. Research has been done by using single passive control cylinder I and D with variation of the Reynolds number $>2.3\times10^4$. The result is the drag coefficient received by the cylinder is reduced to 50% of the main cylinder without passive control[2].

In this study, the main cylinder will be given by two passive controls. The first passive control is a cylinder type-I with angle 53° at the upright position that is placed in front of the main cylinder with a fixed distance, while the second passive control, there are two types to be studied are the ellipse and the circle shape placed behind the cylinder with varying distances. Passive control scheme and main cylinder can be seen in the figure 1. Reynolds number used 100, 500 and 1000. Furthermore, will be investigated and compared the influence of distance variation and Reynolds number to the drag coefficient received ellipse and circle passive control.

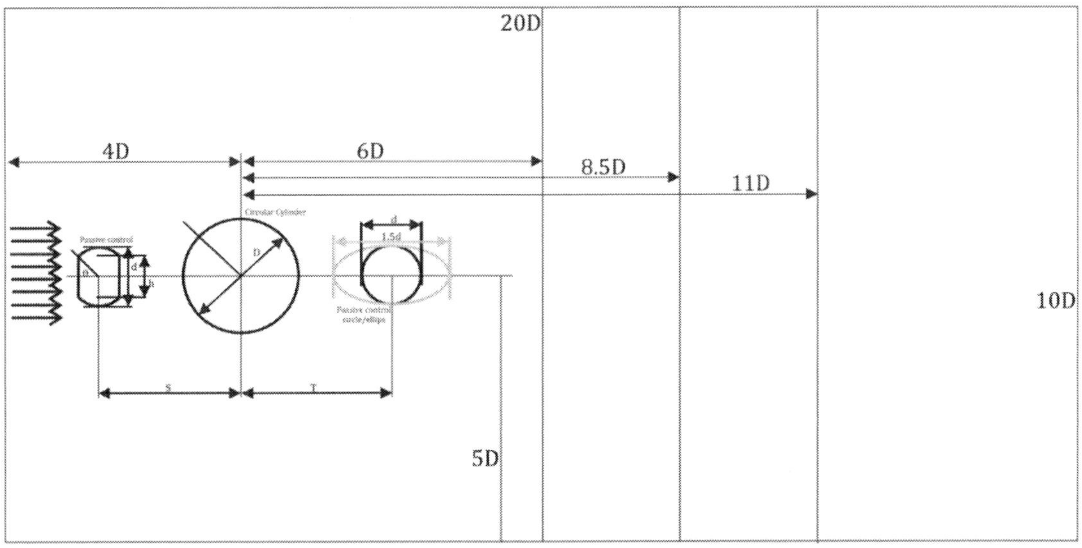

Figure 1. Design research with varies of distances

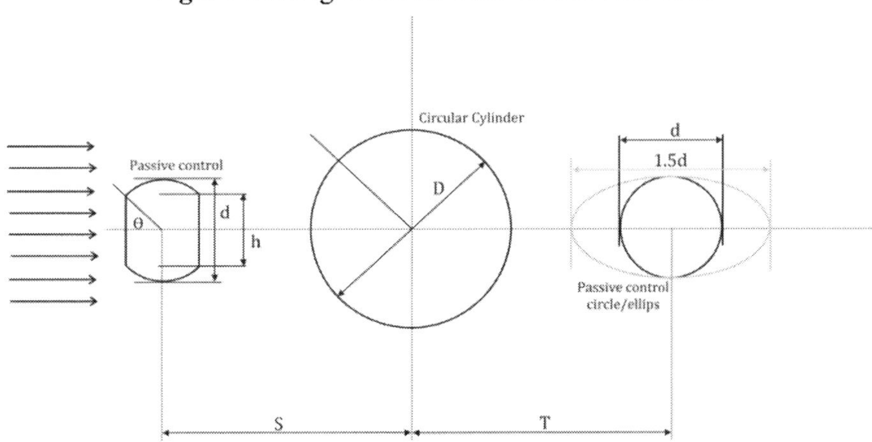

Figure 2. Passive control scheme

2. Numerical Method

The equations used for unsteady incompressible fluid is the Navier-Stokes equations[3]

$$\frac{\partial u}{\partial t} + \nabla uu + \nabla p = \frac{1}{\text{Re}} \nabla^2 u \quad (1)$$

$$\nabla u = 0 \quad (2)$$

where u is the velocity, Re is the Reynolds number and p is the pressure. Numerical method and SIMPLE algorithm is used to solve the Navier-Stokes equations. The first step is set initial value for each variable, then find the velocity component (u^*) of the momentum equation without the pressure component.

$$\frac{\partial u}{\partial t} + \nabla uu = \frac{1}{\text{Re}} \nabla^2 u \quad (3)$$

Equation (3) can be solved using finite difference, like

$$(u_x)_i = \frac{2u_{i+1} + 3u_i - 6u_{i-1} + u_{i-2}}{6dx} \quad \text{and} \quad (u_y)_j = \frac{2u_{j+1} + 3u_j - 6u_{j-1} + u_{j-2}}{6dy}$$

$$(u_{xx})_i = \frac{u_{i+1} - 2u_i + u_{i-1}}{dx^2} \quad \text{and} \quad (u_{yy})_j = \frac{u_{j+1} - 2u_j + u_{j-1}}{dy^2}$$

Then, second step is find the pressure component using equation

$$\nabla^2 p = \frac{1}{\Delta t} \nabla u^* \tag{4}$$

Equation (4) can be solved using finite difference and completed using the SOR method. So, the convergence of the solution can be obtained quickly. SOR method can be expressed by

$$(p_n)_{i,j} = (1-\omega)\left((p_{n-1})_{i,j}\right) + \omega (p_n)_{i,j}$$

for choises of ω with $\omega > 1$ [4]. The last step is correction velocity component using equation as follows :

$$\frac{\partial u}{\partial t} = -\nabla p$$

3. Result and Discussion

The fluid flow across the circular cylinder with two passive controls simulated in the 10Dx20D area, D is the circular cylinder diameter. Circular cylinder is placed in 4D position from front and 5D from above. While the first passive control position is placed in front of the cylinder with fixed distance at the ratio of S/D = 2.1 and the second passive control is placed behind the circular cylinder with varying distances at the ratio of T/D = 0.6, 1.2, 1.8, 2.4, 3. Reynolds numbers used are 100, 500 and 1000. While the fluid flow profile can be known with the wake profile at varying distance of 6D, 8.5D and 11D from the center of circular cylinder.

3.1. Validation code

To determine the accuracy of calculations performed by the computer code, we identify with the calculation of the drag coefficient received by a single circular cylinder without passive control. The identification of the drag coefficients is done by Reynolds number 100, then compared with the result of the calculation ever done.

Table 1. Comparison with other research

Re		Zulh[5]	Lima[6]	Ding[7]
100	1.35837	1.4	1.39	1.356

Based on the table 1, the results of the calculation drag coefficient obtained is not much different from other researchers. Thus, the computer code that has been created can be used for simulation in this study.

3.2. Pressure distribution

Different shapes of the passive control will obtain different pressure distributions. Based on the Reynolds number used, the effect of the distance variation between the passive control and the main cylinder to the pressure distribution on the passive control is shown in figure 3 and figure 4.

Based on figure 3 and figure 4, the pressure distribution that occurs for all types of variations produces a symmetrical graph. The minimum pressure coefficient occurs at an angle of 50-100 and 250-300. The smallest coefficient occurs when the Reynolds number 1000. The greater of Reynolds number is obtained the smallest drag coefficient. For the variation of distance between the passive control and the main cylinder T/D, the smallest pressure distribution is obtained when the distance T/D=1.2.

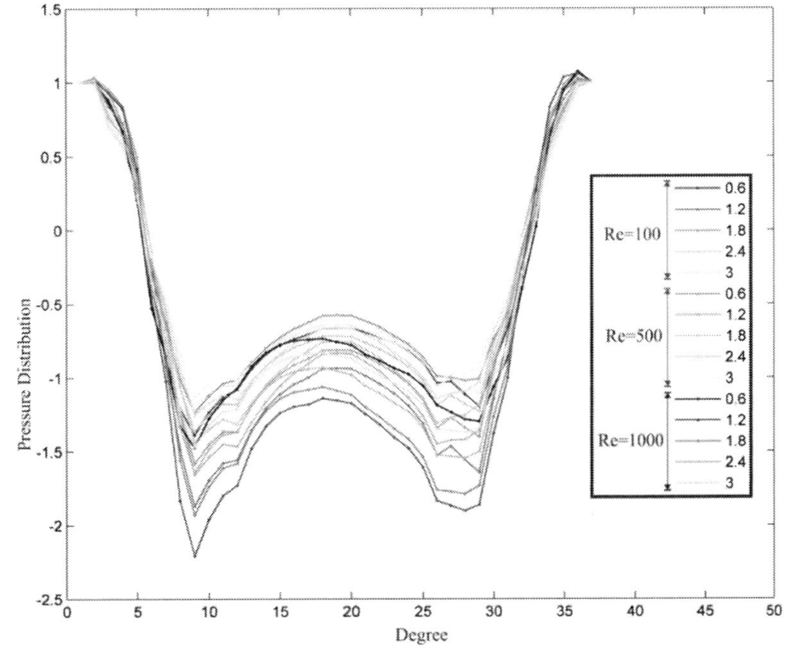

Figure 3. Pressure distribution of ellipse passive control

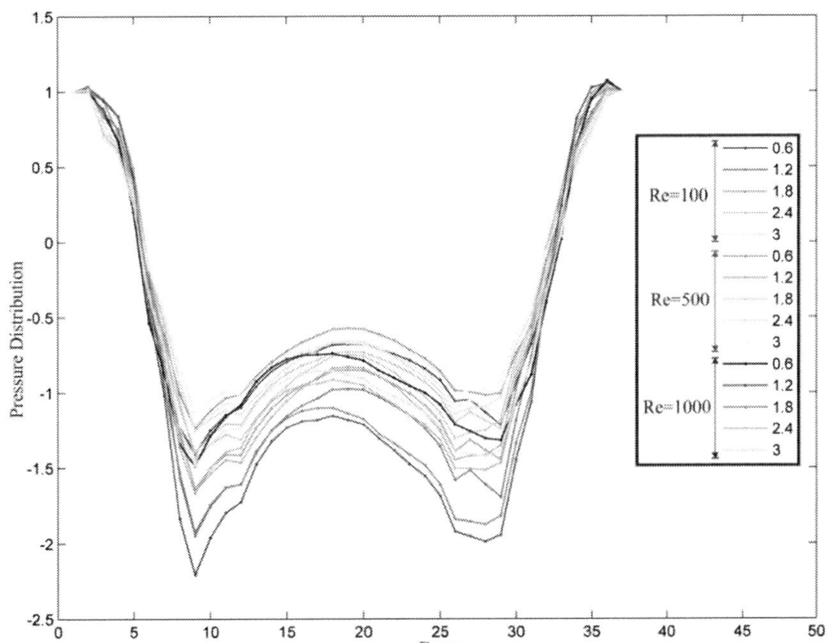

Figure 4. Pressure distribution of circle passive control

The ratio of pressure distribution that occurs in the ellipse or circle passive control is not much different. Passive control elliptical shape accepts a smaller pressure coefficient compared to a circular passive control. To be more accurate, the difference of pressure distribution received by passive control ellipse and circle at T/D=1.2 can be shown in figure 5. Passive control elliptical shape accepts a smaller pressure coefficient compared to a circular passive control.

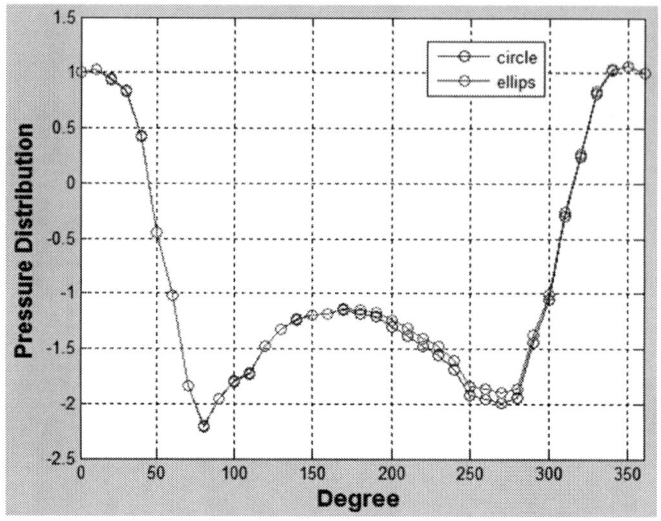

Figure 5. Comparison elliptical and circle control passive

3.3. Drag coefficient

The drag coefficient on a single cylinder with Re 100 has been validated in Table 1. Whereas the Reynolds number 500 and 1000 are shown in Table 2.

Table 2. Drag coefficient in the Reynolds Number 100, 500, 1000

Reynolds number	100	500	1000
Cd	1.356	1.231	1.210

Based on the Reynolds number, the greater the Reynolds number the smaller the drag coefficient on the cylinder. Furthermore, with a single passive control with a type I cylinder placed in front of the main cylinder, the smallest drag coefficient is obtained at a distance S/D=2.1[3]. Furthermore, with variation of passive control and main cylinder distance, the drag coefficient received by the second passive control is the shape of ellipse or circle shown in table 3.

Table 3. Drag coefficient in varies distances and the Reynolds number

	Circle			ellipse		
T/D	Reynolds number			Reynolds number		
	100	500	1000	100	500	1000
0.60	1.169	1.108	1.038	1.161	1.109	1.029
1.20	0.986	0.938	0.878	0.977	0.938	0.870
1.80	0.941	0.911	0.833	0.932	0.912	0.825
2.40	0.944	0.923	0.832	0.935	0.924	0.824
3.00	0.969	0.953	0.849	0.960	0.953	0.841

The drag coefficient received by passive control of elliptical shape is generally smaller than the passive control of the circle shape. The drag coefficient on Reynolds number 1000 is smaller than the drag coefficient on the Reynolds number 500. And the drag coefficient on the Reynolds number 500 is smaller than the drag coefficient on the Reynolds number 100. Thus the smallest drag coefficient occurs in the Reynolds number 1000 at T/D=1.8. This applies to both forms of passive control (ellipse and circle).

3.4. Wake

The fluid flow profile behind the passive control is known through the wake tested on the Reynolds 1000 number with the passive control distance and the main cylinder T/D=1.8. This test is arranged in three variations of the wake position are the distance of 6D, 8.5D and 11D which is measured from the center of the diameter of the cylinder. The wake profile at that distance can be seen in the figure 6 and figure 7.

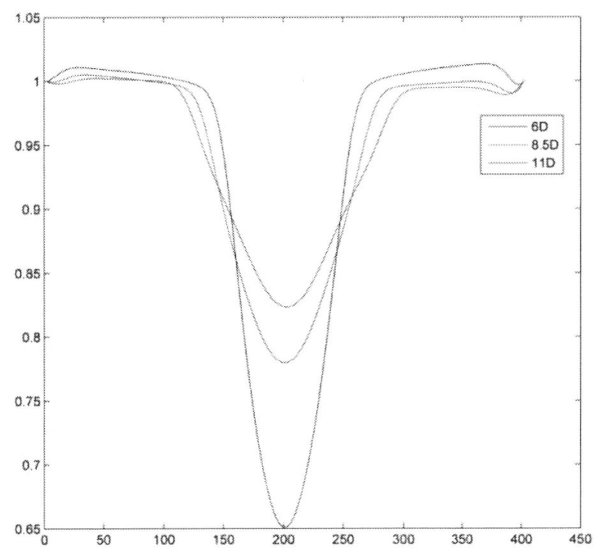

Figure 6. Wake of Ellipse passive control

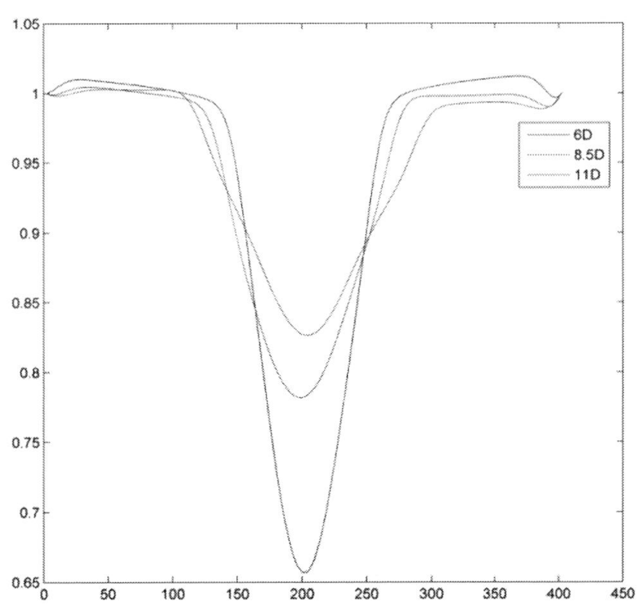

Figure 7. Wake of Ellipse passive control

4. Conclusion

Investigating fluid flow across the circular cylinder and two passive controls, the object of observation is the passive control placed behind the cylinder with two types of ellipses and circles. The results of the investigation obtained :

- For all variations of distance (T/D) and Reynolds numbers, the average pressure distribution received by the passive control of the ellipse is smaller than the passive control of the circular shape.
- The drag coefficient received by the ellipse passive control is smaller than the passive control of the circle, although the drag coefficient received by the ellipse passive control and the cylinder is almost the same.
- The appearance of wake when the smallest drag coefficient is at the time of Re 1000 and T/D=1.8 occurs in the circular cylinder shows symmetrical results.

Acknowledgements

We appreciate through the head of research institute and community service of Institut Teknologi Sepuluh Nopember (LPPM ITS) Surabaya Indonesia, on behalf of the Ministry of Technology and Higher Education who gave us a research grant under contract No: 010/SP2H/LT/DRPM/IV/2017, April, 20th 2017, in giving us to disseminate our results to this paper.

References

[1] Tista, S.P.G.G., (2010), "*Pengaruh Penempatan Penghalang Berbentuk Silinder Pada Posisi Vertikal Dengan Variasi Jarak Horisontal Di Depan SIlinder Utama Terhadap Koefisien Drag*", Department of Mechanical Engineering , Universitas Udayana, Indonesia.

[2] Igarashi T., Shiba Y., (2006), "*Drag Reduction for D-Shape and I-Shape Cylinders (Aerodynamic Mechanism of Reduction of Drag)*", *JSME international Journal,* SeriesB, Vol **49**, No. 4. Japan.

[3] Imron, C., Suhariningsih, Widodo, B. & Yuwono, T., (2013), "*Numerical Simulation of Fluid Flow Around Circular and I-Shape Cylinder in a Tandem Configuration*", *Applied Mathematical Sciences,* vol **7** &, No. 114, 5657-5666, HIKARI Ltd Journal and Books Publisher, Bulgaria.

[4] Carroll, J. (2011), "*Iterative Techniques in Matrix Algebra, Relaxation Techniques for Solving Linear Systems*", *Numerical Analysis (9th Edition)*, Dublin City University.

[5] Zulhidayat, D.N., M-J. C. & Horng, T.-L. (2009), "*An Immersed Boundary Method to Solve Fluid-Solid Interction Problems*", *Computational Mechanics* **44**, 447-453.

[6] A.L.F. Lima E. Silva, A. S.-N. & Damasceno, J. (2003), "*Numerical Simulation of Two Dimensional Flows Over a Circular Cylinder Using the Immersed Boundary Method*",*Journal of Computational Physics* ***189***, 351-370.

[7] H. Ding, C. Shu, K. Y. & Xu, D. (2007), "*Numerical Simulation of Flows Around Two Circular Cylinders by Mesh-Free Least Square-Based Finiter Difference Methods*", *International Jounal for Numerical Methods in Fluids* **53**, 305-332.

Student's scheme in solving mathematics problems

Nining Setyaningsih[1,3], Dwi Juniati[2,4], and Suwarsono[2]

[1]Mathematics Education, Universitas Muhammadiyah Surakarta, Indonesia
[2]Mathematics Education, Universitas Negeri Surabaya, Indonesia

[3]Email: ns259@ums.ac.id , [4]dwi_juniati@yahoo.com

Abstract. The purpose of this study was to investigate students' scheme in solving mathematics problems. Scheme are data structures for representing the concepts stored in memory. In this study, we used it in solving mathematics problems, especially ratio and proportion topics. Scheme is related to problem solving that assumes that a system is developed in the human mind by acquiring a structure in which problem solving procedures are integrated with some concepts. The data were collected by interview and students' written works. The results of this study revealed are students' scheme in solving the problem of ratio and proportion as follows: (1) the content scheme, where students can describe the selected components of the problem according to their prior knowledge, (2) the formal scheme, where students can explain in construct a mental model based on components that have been selected from the problem and can use existing schemes to build planning steps, create something that will be used to solve problems and (3) the language scheme, where students can identify terms, or symbols of the components of the problem.Therefore, by using the different strategies to solve the problems, the students' scheme in solving the ratio and proportion problems will also differ.

1. Introduction

Mathematics is a science that has a relatively strict structure. So, when we will study mathematical concepts, we must go through certain sequences. It means, in the early part of the study of mathematical concepts, it generally begins with a definition that is the basis for understanding the topics of the next part which generally consists of axioms or theorems. The basic objects of mathematics in the form facts, concepts, operations or relations, and principles are mental objects or mind objects [1]. Because of facts, concepts, operations and principles are mental objects, so information processing will occurs and they will be stored in memory [2].

Scheme is a collection of prior knowledge that provides a context for interpreting new information [3][4]. Howard defines the scheme as an abstract knowledge structure that organizes some information into meaningful systems [5]. From the explanation of the scheme, it can be derived that the scheme is a concept structure that is stored in memory and helps to interpret the situation or new information.

The scheme can be done through the learning process [6,7]. Students can form a new scheme of a new experience and students can add new components in the old scheme. In the learning process, students face changes to the scheme either by adding components, refining, developing or altering the old scheme [8], [9]. The scheme that students have can be developed through the giving of problems. To sharpen students' skills in mathematics, problems need to be addressed, where the problem is not a routine problems [10]. However, students cannot solve the mathematics problem due to lack of students in understanding the problem and low ability in computing [11][12]. Meanwhile, Silver is more interested in discussing the contents and arrangements of student knowledge (problem solver) [13]. He stated that by knowing the background information (knowledge) students can help to explain why the instruction is not effective. Based on the explanation, this study aims to investigate student's scheme in solving mathematics problems.

2. Content knowledge scheme

Rumelhart and Norman said that Schemes are data structures for representing the generic concepts stored in memory [14,15]. There are schemes for generalized concepts underlying objects, situation, events, sequences of events, action and sequences of actions [14,15]. Schemes in some sense represent the stereotypes of these concepts. Howard explained that a scheme is a mental representation of some aspect of the world [5,16]. It has slots that are related to each other in prescribed ways and that are filled by stimuli to create an instantiation of the scheme

Carrel, Adams and Collins define the scheme is as an abstract knowledge structure [17,18]. Furthermore, they claim that the acquired knowledge is called the knowledge background of the solver. They explain the types of schemes, namely (1) content scheme, (2) formal scheme, and (3) linguistic scheme[17,18]. Content scheme relates to the prior knowledge background of the text content. Content schemes contain conceptual knowledge or information about what is happening, and how relationships occur to each other to form a logical relationship [17,18]. Formal scheme is related to a formal and structured knowledge background [17,18]. And also, formal scheme is related to knowledge that discusses how stories or problems are built that are expected to improve understanding. Thus, the formal scheme is a process of constructing and testing of a text or problem based on the knowledge possessed. Language scheme deals with knowledge of terms, grammar, vocabulary etc. The language scheme plays an important role in understanding a text or information [17,18].

3. Content knowledge in solving mathematics problems

The part of mathematics topics which have been learned by students are ratio and proportion. The concepts of ratio and proportion is very important because ratio and proportion are commonly used in daily life [11,19,20]. However, in general, students' achievement on this topic is not good enough [21]. The possible cause is that because students find some difficulties in learning this topic [21].

The main parts of the problem to find are: what to look for?, how is the data known?and how is it? While, the problem proves to be a problem to indicate whether a statement is true or false, or not both [21]. This can be done by answering the question of whether the statement is true or false. The main part of this problem is the hypothesis and conclusion of a theorem that must be verified. Thus, a mathematical question can be a problem for a person, depending on how to solve the problem. Therefore it is necessary to think complex from the person to think about how to solve the problem right.

Problem solving is defined as the process by which individuals combine prior knowledge to deal with new situations [22]. This means problem-solving is a process that a person does in combining prior knowledge to accomplish a task that is not yet known of the settlement procedure. Mathematical problem solving is a complex mental process that requires visualization, imagination, manipulation, analysis, abstraction and unification of ideas [6,7]. So, mathematical problem solving cannot be separated from one's knowledge of the substance of the problem.

Scheme is related to problem solving in that it assumes that a system is developed in the memory by acquiring a structure in which problem solving procedures are integrated with some concept. The scheme is responsible for providing a structure for understanding a problem solving situation as well as for guiding problem solving procedures. To solve a problem, student is first assumed to encode information in the cognitive domain. A system of production of information is then applied to transform its encoded form to intermediate form of the problem solving procedures.

4. Methods

This is a qualitative study which aims to investigate students' scheme in solving mathematics problems. The data were collected through interview and test about the ratio and proportion . The data collected by those instruments helped researcher in interpreting and enable us to make data triangulation.

This study were conducted at Junior High School Muhammadiyah Program Khusus in Surakarta, Indonesia. The subject of this study were students from grade seven. The result of student work to the two questions were analysed to identify the knowledge applied in solving the problems. This study uses data analysis with the following stages: 1) reduction, 2) presentation, and 3) the conclusions or verification [23]. Data reduction of test results and interviews conducted with students. Then the data is presented in the form of a narrative text.

5. Results and discussion

To know the scheme in solving mathematics problem, subjects were given two tasks about ratio and proportion. Response of subject for each task and interview result presented as follows.

5.1. Task: Ratio

Mr. Rahmad plans to paint the walls of the house. But the color of paint sold in the store is less in accordance with the wishes of Mr. Rahmad. To get the desired color, Mr. Rahmad will mix the blue paint and white color. The A mixture consists of 3 blue paint cans and 2 white paint cans, while the B mixture consists of 4 blue paint cans and 3 white paint cans. Each paint can is the same size. Of the two mixtures, Mr. Rahmad wants an older blue color. Among mixes of A and mix B, which mixture yields the older blue color according to the wall color that Mr. Rahmad wants? Give the reason!

The results of written answers and interviews of subject S1 are presented in Figure 1.

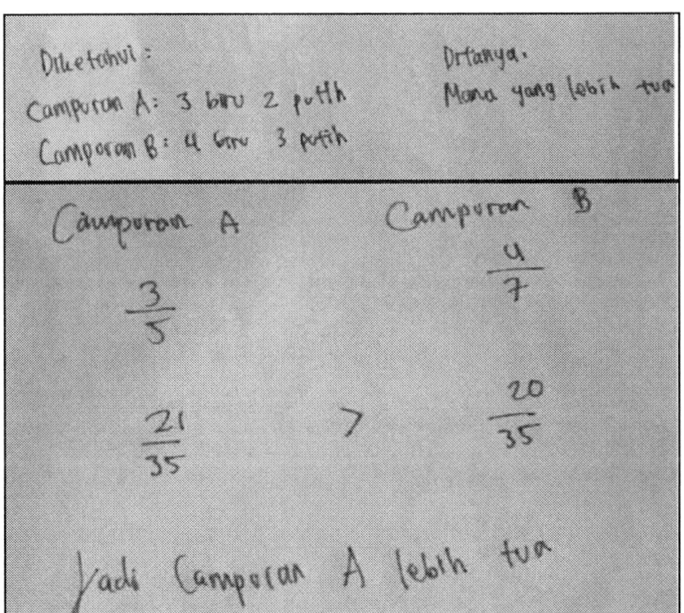

Figure 1. Subject S1's written answer.

Q: Have you had any problems like this before?
S1: not yet
Q: Now, solve this problem
S1: (nod, pause for a moment and then start doing)
Q: Can you explain what is known and what is asked from this problem?
S1: known: A mixture of 3 blue and 2 white and mix B consists of 4 blue and 3 white.
Asked: Which mixture gives the older blue color?
Q: What do you mean with $\frac{3}{5}$?
S1: Blue color compared with mixed colors = 3: 5
Q: Why not 2: 5?
S1: because of the blue color question.
Q: What do you mean with $\frac{4}{7}$?
S1: Blue color compared with mixed color = 4: 7
Q: Why $\frac{3}{5}$ *tobe* $\frac{21}{35}$?
S1: the denominator are equated for $\frac{3}{5}$ and $\frac{4}{7}$, that is $5 \times 7 = 35$, so $\frac{3}{5} = \frac{3x7}{5x7} = \frac{21}{35}$
Q: why $\frac{4}{7}$ *to be* $\frac{20}{35}$?
S1: yes the same wih above, the denominator is equated , so $\frac{4}{7} = \frac{4x5}{7x5} = \frac{20}{35}$
Q: why do you choose the mixture A older than the mixture B
S1: because $\frac{3}{5}$ is larger than $\frac{4}{7}$

Meanwhile, the results of written answers and interviews of subject S2 are presented in Figure 2. .

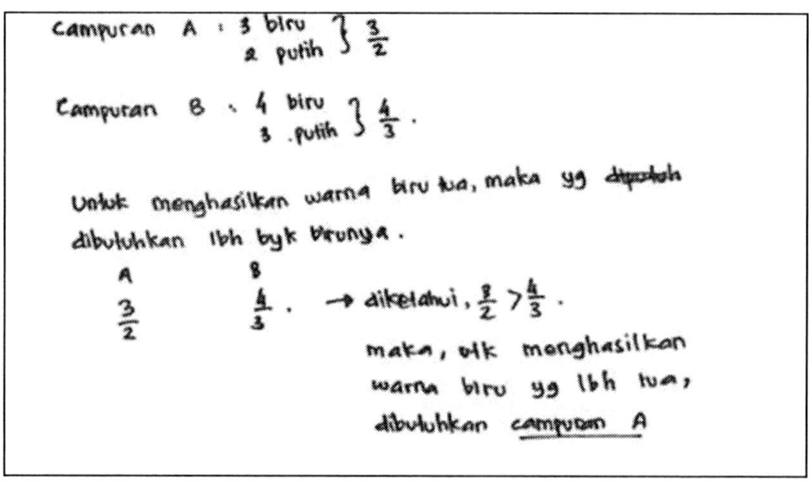

Figure 2. Subject S2's written answer.

Q: Have you had any problems like this before?
S2: not yet
Q: Now, try to solve it
S2: (nod, pause for a moment and then start doing)
Q: Can you explain what is meant in the question?
S2: Mr. Rahmad wants to do the house wall, but the color is not available in the store. He made the mixture A consists of 3 blue
Q: What do you mean with $\frac{3}{2}$?
S2: Blue color compared with white color = 3: 2
Q: Why not 2 : 3?
S2: because the question for blue color.
Q: What do you mean with $\frac{4}{3}$?
S2: Blue color compared with white color = 4 : 3
Q: why do you choose the mixture A has a blue color that is older than mixture B

IOP Conf. Series: Journal of Physics: Conf. Series **974** (2018) 012012 doi:10.1088/1742-6596/974/1/012012

paint cans and 2 white paint cans and the mixture B consists of 4 blue paint cans and 3 white paint cans. Which mixture gives the older blue color.

Q: From the problem, can you explain what is known and what is asked of that problem?

S2: known: the mixture A consists of 3 blue and 2 white and the mixture B consists of 4 blue and 3 white.

S2: because $\frac{3}{2}$ is larger than $\frac{4}{3}$

Q: why $\frac{3}{2}$ is larger than $\frac{4}{3}$

S2: the denominator are equated for $\frac{3}{2}$ and $\frac{4}{3}$, that is $2 \times 3 = 6$, so $\frac{3}{2} = \frac{3x3}{2x3} = \frac{9}{6}$ and $\frac{3}{2} = \frac{4x2}{3x2} = \frac{8}{6}$, so $\frac{3}{2} = \frac{9}{6} > \frac{4}{3} = \frac{8}{6}$

From the results of the analysis of S1 and S2's written answers and interviews, it show that the content scheme owned by subject S1 and S2 are shown by their ability in identifying the problem, they are able to answer what components are known and what is asked. As for the formal scheme demonstrated subject S1 and S2's ability to plan knowledge and execute knowledge. From the interview also obtained data that subject S1 is able to explain the components on the matter of the ratio problem or in other words they are able to identify the knowledge. In addition, they are also able to form a mental model of the components that have been determined or in other words they are able to elaborate his prior knowledge about ratio to solve the problem. Furthermore, they are able to make plans to solve problems and to execute the plans to answer the problem.

The scheme built by S1 and S2 is almost the same. The difference lies in the reasoning of the ratio formulation. S1 look for ratio for blue color with mixed color, while S2 looks for ratio for blue color and white paint. The scheme of subject S1 in solving ratio problem is shown in Figure 3.

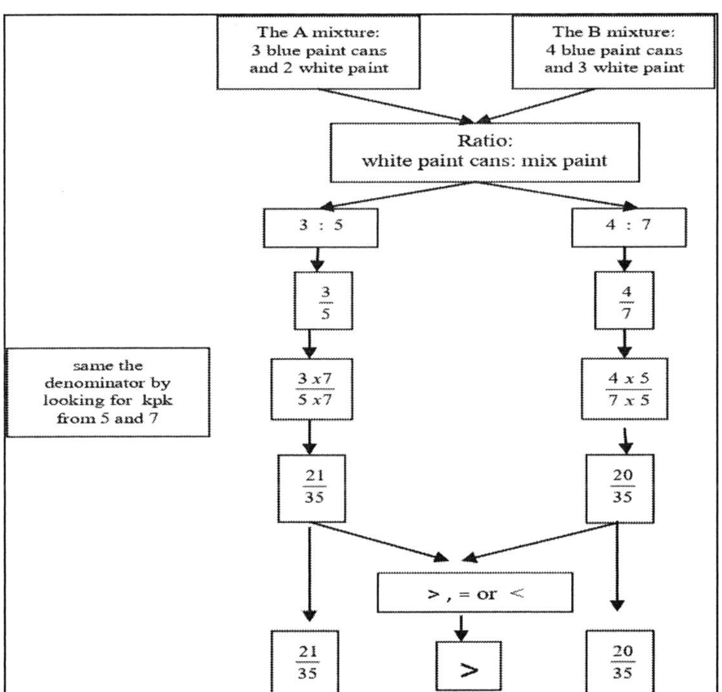

Figure 3. Subject S1' s scheme in solving ratio problem

5.2. Task 2: Proportion

Hidayat likes cats. He has 3 cats. For 4 days, Hidayat gave food to his three cats with 4 kg of cat food. One day, Pak Anwar, his uncle Hidayat, bought 9 cats for Hidayat. Using the same size, how many kg of cat food does Hidayat need to feed all his cats for a week?

The results of written answers and interviews of subject S1 are presented in Figure 4.

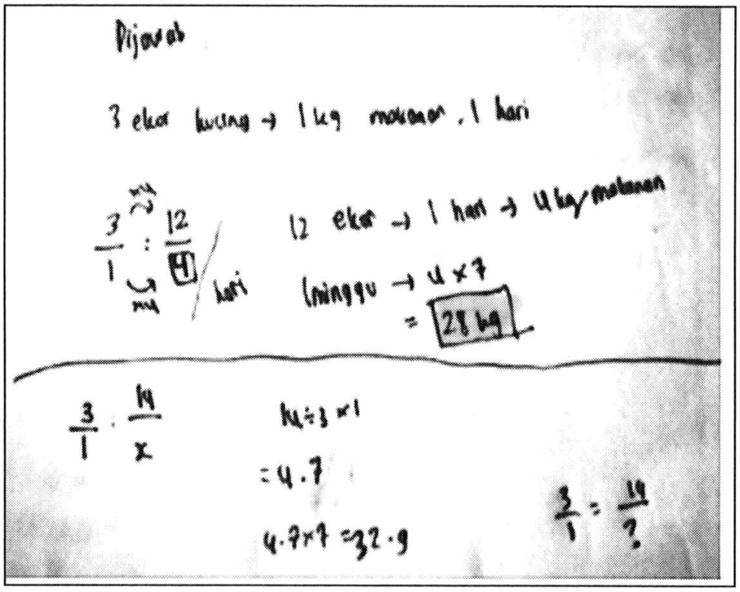

Figure 4. Subject S1's written answer

Q: Have you had any problems like this before?
S1: Not yet
P: Now, try to finish it
S1: (nod, pause for a moment and then start doing)
P: already
S1: (nods)
Q: Can you explain what is meant in the question?
S1: Hidayat has 3 cats, for 4 days he gives cat food 4 kg. Now Hidayat has 12 cats, how many kg of food is needed for 1 week?
Q: From the problem, it can be explained what is known and asked.
S1: Known 3 cats --- 4kg of food, for 4 days. Asked : how many kg of food for 12 cats for 1 week?.
Q: What do you mean 3 cats - 1 kg of food, 1 day
S1: known to 3 cats --- 4kg of food, 4 days. So 1 day for 3 cats need 1 kg of food
Q: What do you mean with $\frac{3}{1}=\frac{12}{4(?)}$?
S1: 3 cats, 1 kg of food. 12 cats, 4 kg of food
Q: How to get it?
S1: 3 multiplied 4 = 12, the bottom should also be multiplied by 4, so ? = 4
Q: what does it mean?
S1: 12 cats need 4 kg of food for one day.
Q: For 1 week how many kg of food for 12 cats?
S1: 4 x 7 = 28 kg of food

From the results of the analysis of written answers and interview results, showed that the content scheme owned subject S1 shown with his ability in identifying the problem, that he is able to answer what components are known and what is asked. As for the formal scheme demonstrated S1's ability to plan knowledge and execute knowledge. On the other hand, from the interview also obtained data that S1 able to explain the components on the problem of proportion or in other words he was able to identify knowledge. In addition, he is also able to form a mental model of the components that have been determined or in other words he is able to elaborate knowledge. Furthermore, he is able to make plans to solve problems and simultaneously execute knowledge.

In explaining the solution of problem-solving proportions, S1 completed in three stages. The first stage, he looks for how many kg of food for 3 cats for one day by using the unit rate strategy [24]. The second stage he looks for how many kg of food for 12 cats for one day and the third stage he looks for how many kg of food for 12 cats for 1 week by using the cross product strategy [24]. Thus, he already has a content scheme, which demonstrates his ability in identifying the components in the matter. In addition, he also has a formal scheme, which demonstrated his ability to elaborate components in problems, plan and execute knowledge. He also has a language scheme, which demonstrates his ability to write the mathematical symbols used in solving problems.

Figure 5 and Figure 6 below show the schemes of S1 and S2 in solving the proportion problem.

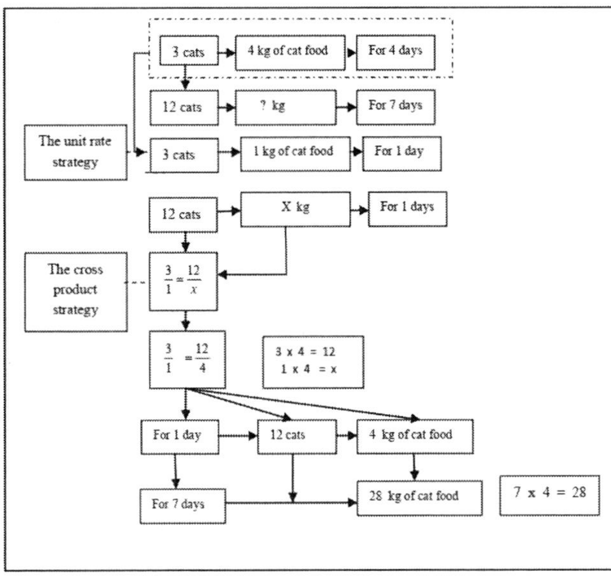

Figure 5. Subject S1's scheme in solving proportion problem

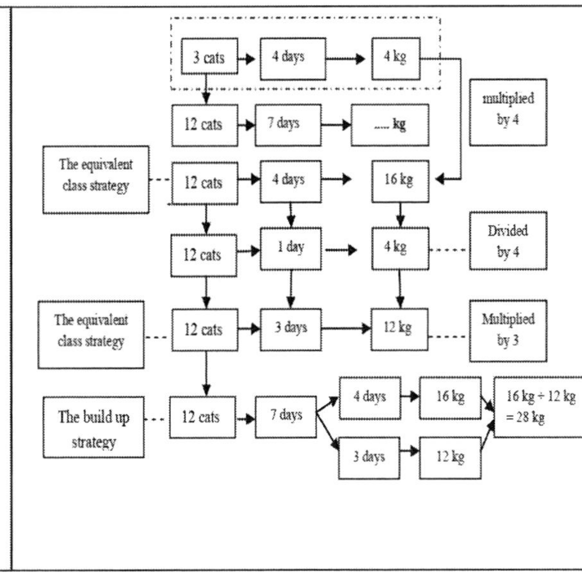

Figure 6. Subject S2's scheme in solving proportion problem

6. Conclusion

The scheme of the research subjects (S1and S2) in solving mathematical problems, especially the material ratios and proportions can be explained as follows: (1) the content scheme, in which the subject describes the information or components that have been selected in accordance with the initial knowledge possessed, (2) the formal scheme, in which the subject is able to explain in constructing a mental model based on the components that have been selected from the problem and using the scheme to build the planning steps, creating something that will be used to solve the problems encountered and (3) the language scheme, in which the subject is able to identify terms, symbols of the components of the ratio problem and proportion. Also informed that the research subjects using different strategies in solving the problem of ratios and proportions, so the scheme was built also differently.

7. Acknowledgment

This paper is a part of the results of dissertation research funded by Universitas Muhammadiyah Surakarta (UMS). Therefore, we acknowledge UMS for the funding. Moreover, we also thanks to SMP Muhammadiyah Program Khusus Kottabarat Surakarta as the partners for this program.

References

[1] Bell F H 1987 *Teaching and Learning Mathematics* (USA: Wm.C. Brown Company)
[2] Hintzman D L 1986 "Scheme Abstraction" in A Multiple Trace Memory Model *Psychol. Rev.* **93** 411–28
[3] Anderson J R 1984 *Cognitive Psychology and Its Implication* (San Fransisco: W.H. Freeman and Company)
[4] Bartlett F C 1932 *Remembering: A Study in Experimental and Social Psychology* (Cambridge: Cambridge University Press)
[5] Howard R W 1987 *Concept and Schemeta* (London: Cussel Educational, Artillery House)
[6] Marshall S P 1995 *Shemas in Problem Solving* (New York: Cambridge University Press)
[7] Hershkotz S and Nesher P 2003 The Role of Shemes in Solving World Problem. The Mathematical Educator *Math. Educ.* **7** 1–24
[8] Piaget J and Cook M 1952 *The Origin of Intelligence in Children* (New York: International University Press)
[9] Anon 1971 *The Psychology of Intelligence* (London: Routledge & Kegen Paul)
[10] Ball D L 1990 Prospective Elementary and Secondary Teacher's Understanding of Division *J. Res. Math. Educ.* **21** 132–44
[11] Lo J-J and Watanabe T 1997 Developing Ratio and Proportion Schemes: A Story of a Fifth Grader *J. Res. Math. Educ.* **28** 216–36
[12] Rejeki S, Setyaningsih N and Toyib M 2017 Using LEGO for learning fractions, supporting or distracting? *AIP Conference Proceedings* vol 1848
[13] Kheong F H 1999 Strategic Models For Solving Ratio and Proportion Problems *Math. Educ.* **4** 34–52
[14] Rumelhart D E 1980 Schemeta: The Buiding Blocks of Cognitive *W.F. Brewe (ed) Theoretical Issues in Reading Comprehesion* (Hisdalle, New Jersey: Lawrence Erlbain Assosiation)
[15] Rumelhart D E and Norman D A 1985 *Representation and Knowledge. In A.M. Aitkenhead & J.M. Slack (Ed) Issues in Cognitive Modelling* (Hisdalle, New Jersey: Lawrence Erlbain Assosiation)
[16] Harel G and Showder L 1998 Students' Proof Shemes : Result from Exploratory Studies *CBMS Issues Math. Educ.* **7** 234–83
[17] Adamsand and Collins 1979 Scheme Theoretic View of Reading *Fredolle, R.O. Discourse Processing Multi Disiplinary Perpevtive* (Norwood, New Jersey)
[18] Carrel P L, Devine J and Eskey D E 1988 *Interactive Approaches to Second Language Reading* (Cambridge: Cambridge University Press)
[19] Sukro A and Filliz D 2015 The Effect of Project Based Learning in Ratio and proportion On Mathematics Success and Attitude *Eur. J. Sci. Math. Educ.* **3** 1–13
[20] Singh P 2000 Understanding The Concepts of Proportion and Ratio Constructed by Two Grade Six Students *Educ. Stud. Math.* **43** 271–92
[21] Ledesma E F R 2011 Primary and Secondary Teachers' knowledge, Interpretation and Approaches to Student Error about Ratio and Proportional Topic *Creat. Educ.* **2** 264
[22] Mayer R E 1982 The Psychology of Mathematical Problem Solving *Mathematical Problem Solving: Issues in Research* (Philadelphia: The Franklin Institute Press)
[23] Merriam S B 1998 *Qualitative Research and Case Study Applications in Education* (San Fransisco: Jossey-Bass Publication)
[24] Cramer K, Post T and Currier S 1993 Learning and Teaching Ratio and Proportion: Research Implications *Research Ideas for the Classroom (Middle Grades Mathematics)* ed D Owens (New York: Macmillan Publishing Co, Inc) pp 159–78

Reducing a congestion with introduce the greedy algorithm on traffic light control

Puji Catur Siswipraptini, Wisnu Hendro Martono, Dian Hartanti

Informatics Department
Sekolah Tinggi Teknik PLN, Jakarta-Indonesia

Email : pujicatur@sttpln.ac.id; wisnuhendro@sttpln.ac.id; dianhartanti@sttpln.ac.id

Abstract. The density of vehicles causes congestion seen at every junction in the city of jakarta due to the static or manual traffic timing lamp system consequently the length of the queue at the junction is uncertain. The research has been aimed at designing a sensor based traffic system based on the queue length detection of the vehicle to optimize the duration of the green light. In detecting the length of the queue of vehicles using infrared sensor assistance placed in each intersection path, then apply Greedy algorithm to help accelerate the movement of green light duration for the path that requires, while to apply the traffic lights regulation program based on greedy algorithm which is then stored on microcontroller with Arduino Mega 2560 type. Where a developed system implements the greedy algorithm with the help of the infrared sensor it will extend the duration of the green light on the long vehicle queue and accelerate the duration of the green light at the intersection that has the queue not too dense. Furthermore, the design is made to form an artificial form of the actual situation of the scale model or simple simulator (next we just called as scale model of simulator) of the intersection then tested. Sensors used are infrared sensors, where the placement of sensors in each intersection on the scale model is placed within 10 cm of each sensor and serves as a queue detector. From the results of the test process on the scale model with a longer queue obtained longer green light time so it will fix the problem of long queue of vehicles. Using greedy algorithms can add long green lights for 2 seconds on tracks that have long queues at least three sensor levels and accelerate time at other intersections that have longer queue sensor levels less than level three.

1. Introduction

Base on daily data of Kompas, data Departement Transportation of DKI in 2015, the number of vehicles in Jakarta is 5.4 million private vehicles, 84,891 vehicles with 9.5% growth this year. As more and more vehicles are increasing every year causing frequent traffic jams, especially on the road junction. Meanwhile, losses due to congestion are estimated to reach 12.8 trillion rupiah as measured by time, fuel and health. According to a traffic light survey at Grogol intersection, West Jakarta, at the intersection of Kedoya West Jakarta, at the junction of Pasar Baru Gunung Sahari, Central Jakarta, the Fatmawati road junction in South Jakarta can be inferred if there is an imbalance of volume, speed, time and length of queue compared to availability roads cause congestion that is difficult to solve.

This case study was conducted at the intersection of four matramans. The intersection of four matramans was chosen from the four intersections surveyed, at the intersection of four things that still apply conventional traffic light systems, not based on length of queue of vehicles, and also four intersections are one of four intersections Having vehicle density causing congestion due to queue of vehicles the long one. A special problem with traffic light crossings is the period of red and green lights that remain/does not change at any time/static and can only be changed green period and red/off by the operator. In solving the problem of vehicle density causing congestion will be used a model of smart traffic lights with Greedy Algorithm equipped with infrared detection sensor with Arduino Mega 2560 microcontroller as its processor, which allows reading the length of the queue of vehicles as a determinant of the Green light Time.

2. Literature Review

Researchers have been attempting to model a new algorithm in order to come up with the optimization of the traffic control.Control design such as a decentralized and self-organizing control strategy for a traffic signal network adapting to dynamical environmental changes based on only local traffic information [1], design of a new urban traffic control system which simulated the Ant Colony Optimization (ACO) based system as an adaptive path length and path traffic [2], a smart traffic control system based on the wireless sensor network [3] and an alerting system for red light crossing scenario to alert the drivers on other sides to save their lives based on the queue length of thevehicles on the traffic lights [4]. The studies in [5][6][7] pose the traffic signal control as an optimization problem such as a combination of reinforcement learning to maximize local traffic at each intersection and global optimization by Evolutionary Computational Method [5][8], minimization of the total delay experienced by all network traffic at each section in a given network using a dynamic programming technique [6] and adjustment of green period by reinforcement learning [7].

However, we have found no studies concerned. with adjusting all the three parameters (cycle time, spilt and offset) based on the completely decentralized approach.

3. Research Method

The objective of this paper is designing traffic control management to obtain models of traffic control devices using Greedy Algorithm. This research is in the form of a model 4 arm -shaped intersection with 6 lanes consist of one lane bus way, 3 fast lane, 2 slow lane. In this paper, first we describe the real data involving such kind of some intersection in Jakarta to intents model, second we describe a kind of a board model of traffic light with some detector, while the detector output will be used as a computer model. Output from a number of bound detector will race with other adjacent output detector in the same intersection. The first process analyzed with TSP algorithm to get light exchange in normal condition and the second analyzed with Greedy algorithm to get up normal time condition in order to win the race (scheduling). And the third, we develop intersection traffic light control system with 6 lanes consist of one lane busway, 3 fast lane, 2 slow lane according the model analizing TSP and Greedy algorithm.

Based on the description above, a design of smart traffic light that has an infrared sensor as a control system for the adjustment based on the density of vehicles the roads is required. It is expected that traffic control system will be well-adjusted with the density that is detected by the infrared sensor at every intersection. This smart traffic light system is not constant (real-time) and will be followed by changes of the density of the vehicles at the intersection. This sensor send traffic density input to controlling unit, according to these input traffic pole timer works. [9]. The principle of the control system at traffic lights will be using infrared sensors as a vehicle density level control system with media control utilizing AVR microcontroller ATmega16, because this microcontroller is able to access program and data simultaneously . The algorithm used in this design of smart traffic light is a greedy algorithm that is expected will be able to determine which path has the longest time delay and which path has the shortest and it will purposely prevent and decrease bottlenecks in order to increase the effectiveness of the traffic lights. The design of smart traffic light system is going to reduce the number of traffic/bottlenecks that usually happens at crossroads/intersection with the design that is very dynamic and real-time (duration of green light is adjusted with density of vehicles on the roads).

IOP Conf. Series: Journal of Physics: Conf. Series **974** (2018) 012013 doi:10.1088/1742-6596/974/1/012013

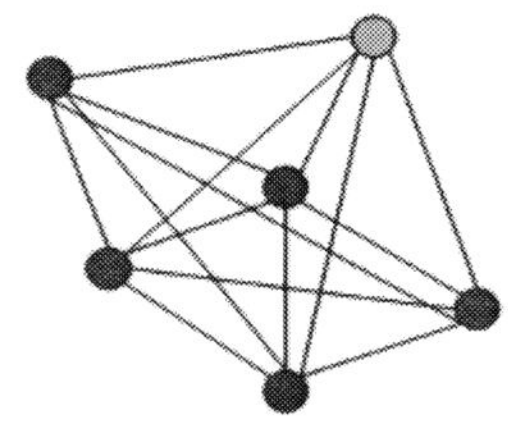

Figure 1. The graph of greedy algorithm and tsp

Generally traffic lights are located at the crossroads such: T Junctions, Multi Junction, etc that is why a mechanism to control the sequences of each lamp is required in order to prevent a congetion and colliding. Travelling Salesman problem above could be solved with the Greedy Algorithm [10]:

$g(j, j+1) = 1, \quad \text{for } j = 1, \ldots, m-1$

$g(m, 1) = M \quad \text{(for some large number M)}$

$g(j,k) = 2$, otherwise Optimal trip as shown in Figure 3 [9]

Figure 2. Travelling salesman problem

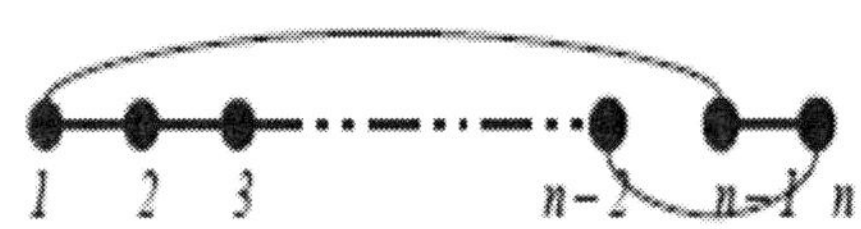

Figure 3. Optimal trip

The solution of Greedy Algorithm is in figure 4

Figure 4. Greedy algorithm optimal solution

B set R = $\{b_1,\ldots,b_n\}$ of n activities that wish to use a resource.

activity b_i : start time r_i, finish time f_i

Activities b_i and b_j are compatible if

$(r_i, f_i) \cap (r_j, f_j) = \emptyset$

Goal : Select a maximum-size subset of mutually compatible activities.

Assumption:

Activities are sorted order of non-decreasing of finish time:

$f_1 \leq f_2 \leq f_3 \leq \ldots \leq f_n$

Algorithm *Greedy-Activity-Selector* [11]

Input: *n* activities represented by intervals

$[r_i, f_i)$, $1 \leq i \leq n$, where $f_i \leq f_i+1$

Output: b maximum-size set of mutually compatible activities

```
B1= {b1}
last = 1 /* last indexes the activity added most recently */
for i = 2 to n do
  if ri<flast
    then Bi= Bi-1
    else /* ri≥ flast*/
      Bi= Bi-1 ∪{bi}
      last = i
return Bn
```

Running time: $O(n)$

Figure 5. The activity-selection problem

4. System Architecture

Our research has the concept of a traffic light regulation system based on dynamically operating greedy algorithm, meaning the green light on this traffic system is based on the length of its queue of vehicles at each intersection. If the intersection is not queue of vehicles, then the traffic light at the intersection operates normally, otherwise if at the intersection there is a long queue, resulting in the duration of the green light increases more. To find out the length of the vehicle queue on the models is read from the infrared sensor.

In the scale model of simulator device that is built equipped with five sensors equipped and installed equipment within a distance of 0 cm - 10 cm, and 10 cm - 25 cm for each intersection and two sensors dedicated to the busway lane. Placement of this sensor will be able to detect the length of the queue of vehicles that have been determined by researchers. The determination of the green light duration of the traffic system on the scale model of simulator not only on the queue difference, but also applied greedy algorithms that will increase the performance of traffic lights at the intersection.

Figure 6. Diagram of research

4.1 Components diagram

Each components works as bellow:

- Car is an object to be detected by an infrared sensor.
- The infrared consist of a receiver and transmitter emits infrared light. If the light is blocked by an object, then the infrared as receiver will detect the object and the signal will be on. Arduino Mega 2560 microcontroller as the controller will process the input received from an infrared sensor, which will then be processed and applied to the greedy algorithm to be forwarded to the LED lighting circuit components.
- On the set of LED components, will accept the process of Arduino Mega 2560 and adding the application of greedy algorithm used as outputs. This output will set the length of time which the green LED was on.

5. Design of greedy traffic light simulator model

Design of greedy traffic light simulator model (GTLS) as shown in Figure 6, New traffic light scale model are built using Arduino processors and Greedy algorithms equipped with sensor devices, which provide a more accurate response.

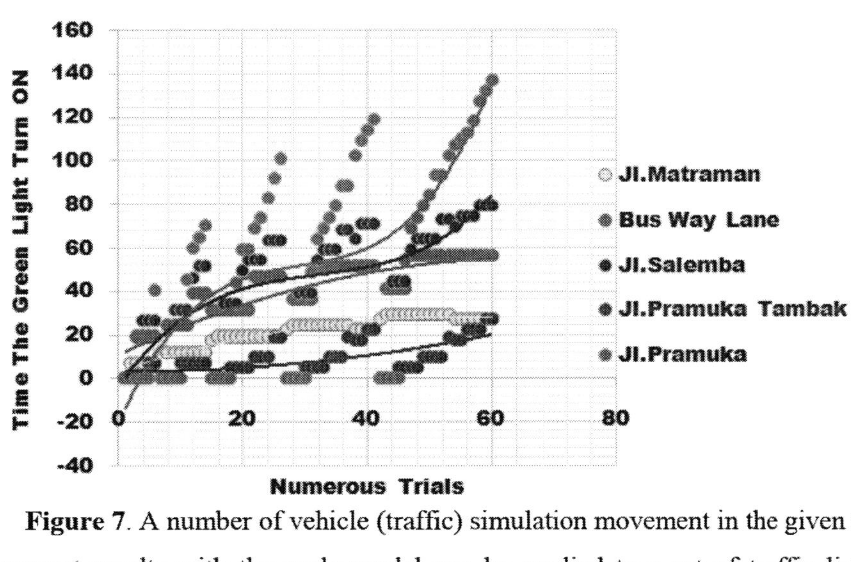

Figure 7. A number of vehicle (traffic) simulation movement in the given time

The measurement results with the scale model can be applied to a set of traffic lights that are applicable at the Matraman intersection. At each junction is given the value of five sensors that represent the distance each has been determined by the researchers

IOP Conf. Series: Journal of Physics: Conf. Series **974** (2018) 012013 doi:10.1088/1742-6596/974/1/012013

Table 1. Level sensor for distance

Level Sensor	Distance (m)
1	0 - 25
2	25 - 50
3	50 - 100
4	100 - 200
5	200 - 300

Table 1. represents the distance of each sensor. In GTLS, each sensor is provided with a distance of 10 cm. Only on the busway lane is equipped with a set of sensors, while on the path to another vehicle is equipped with four sets of sensors.

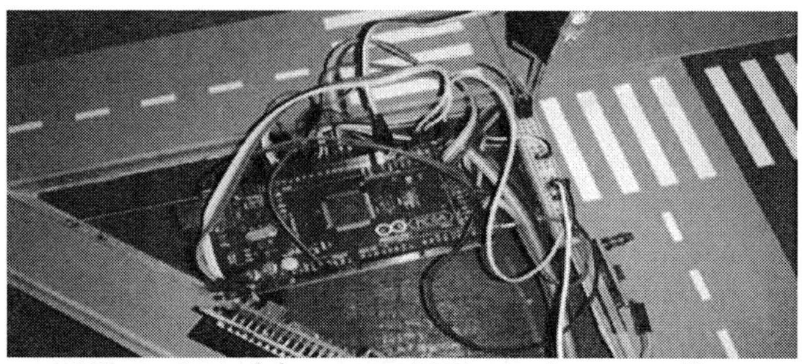

Figure 8. Arduino mega 2560 as microcontroller

Figure 9. Greedy traffic light simulator

Figure 8 and 9 are design of GTLS which equipped with hardware devices include LED that are used as traffic lights, that are infrared sensors on each lane to detect the length of queues of vehicles simulators, Arduino Mega 2560 as mikrokontroller, and 20x04 LCD character to indicate the state of the traffic. The junction has a length of 50 cm, thenthe middle path length is 37 cm, total size scale model simulators is 137 cm x 137 cm.

6. Greedy algorithm for smart traffic light

The above flow chart is a flow of Greedy algorithm as a complement to smart traffic light processor. The sensor will read the traffic density and the data will be sent to Arduino. What is interpreted as a sensor reads the length of the vehicle queue, greater or equal to X.

- X is a sensor that is greater or equal to level three. Then the sensor puts the value into the resolution circuit, is there an N path in the solution set?

- N is a value greater or equal to the three that are in the set of solutions, then the next step is to add time to a green light like P.

P is the amount of time that will be added to the green light at the intersection that has a queue of vehicles on three or more sensors, and will reduce the green light time like P at another intersection. We can see in figure 10

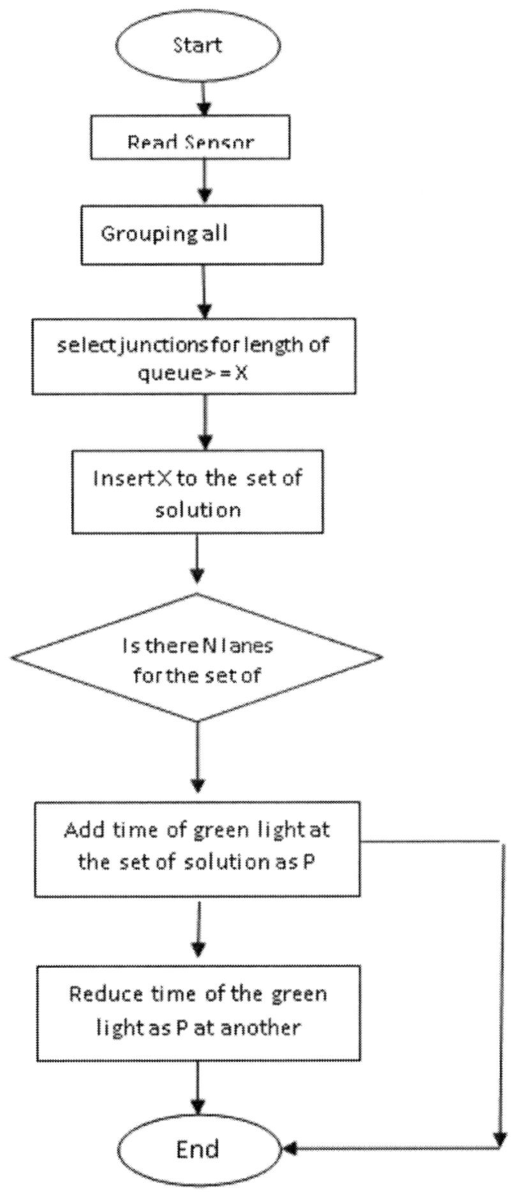

Figure 10. Greedy algorithm for smart traffic light

IOP Conf. Series: Journal of Physics: Conf. Series **974** (2018) 012013 doi:10.1088/1742-6596/974/1/012013

7. Optimising traffic light control using greedy algorithm

The greedy algorithm yields an optimal solution, it proof show that for all $1 \leq j \leq m$ the following holds, there exists an optimal solution S* with [11] :

$S^* \cap \{a_1, \ldots, a_j\} = S_i$

$j = 1$:

Let $S^* \subseteq \{a_1, \ldots, a_m\}$ be some optimal solution, $S^* = \{a_{j1}, \ldots, a_{jk}\}$

$S^* = a_{j1}, a_{j2}, a_{j3}, \ldots, a_{jk}$

$j - 1 \rightarrow j$:

Let $S^* \subseteq \{a_1, \ldots, a_m\}$ be some optimal solution with $S^* \cap \{a_1, \ldots, a_{j-1}\} = S_{j-1}$

Consider $R = S^* \setminus S_{j-1}$

Figure 11. The greedy algorithm yields an optimal solution

The Observation, R is an optimal solution to the problem of finding a maximum-size set of activities in $\{a_j, \ldots, a_m\}$ that are compatible with all activities in S_{i-1}.

Case 1: $s_j < f_{last}$

$S_{j-1} = s_j, f_1, \ldots a_{last}, f_{last}$

S_j is not compatible with S_{j-1}

S_j is not contained in S^*

$S^* \cap \{a_1, \ldots, a_j\} = S_{j-1} = S_i$

Case 2: $s_j \geq f_{last}$

$S_{j-1} = s_j, f_1, \ldots a_{last}, f_{last}$

S_j is compatible with S_{j-1}

There is: $R \subseteq \{a_j, \ldots, a_n\}$

$R = b_1, b_2, b_3, \ldots b_j \; a_j$

$B^* = S_{j-1} \cup (R \setminus \{b_1\}) \cup \{a_j\}$ is optimal

$B^* \cap \{a_1, \ldots, a_j\} = S_{j-1} \cup \{a_j\} = S_j$

Figure 12. Optimal solution of finding maximum size

8. Result

From the simulation results, it can be obtained that the quantity can be used as a decision in the optimization of traffic light control by using greedy algorithm as shown in Figure 5 about GTLS performance. From 118 simulation experiments on scale model of simulator, it was found that greedy algorithms can add long green lights for 2 seconds on tracks that have long queues at least three sensor levels at 100 m of infrared sensor distance, and accelerate time at other intersections that have longer queue sensor levels less than level three, seen in Figures 13 and 14.

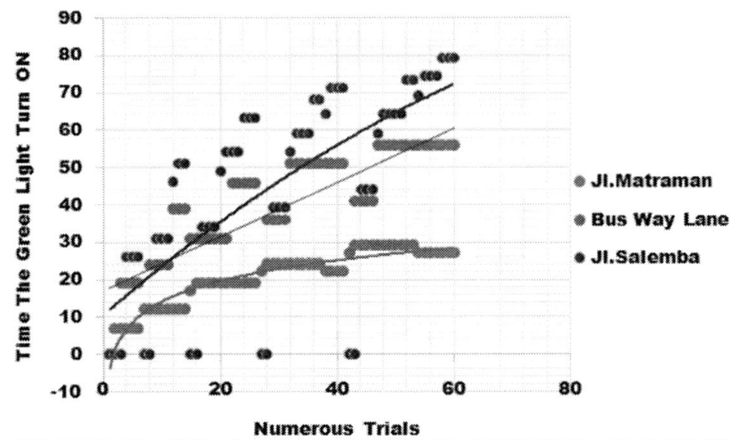

Figure 13. Numbers of vehicles in any given time between Jl. matraman, bus way lane and Jl. salemba

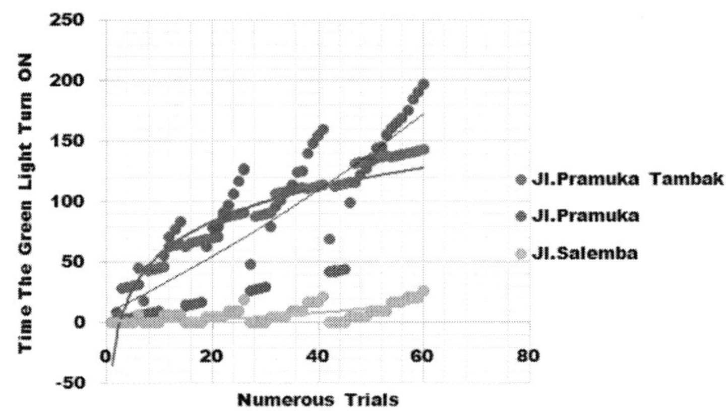

Figure 14. Optimum time for green light on at jl.Pramuka Tambak, jl Pramuka and jl. Salemba

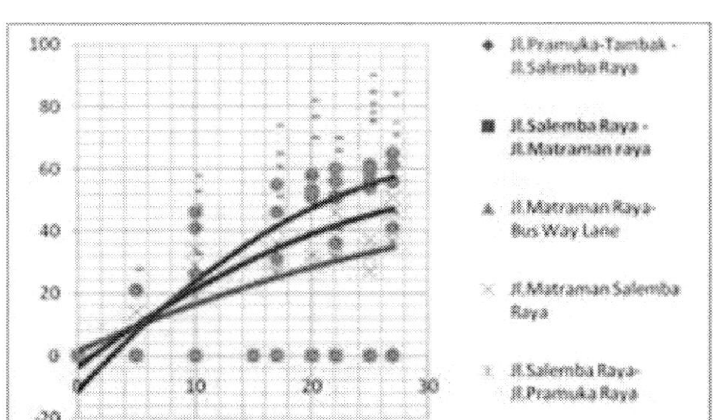

Figure 15. Optimum time for the 5 junction branch

From figure 15, it shows that optimum time for green light is reaches at $12<X<18$. The meeting point of the red line, blue and green will cause traffic jam.

A<= 12:18 <= B

A = means stagnancy

B = means smooth traffic

C = means optimum time for green light

Figure 16. Consideration

9. Conclusion

In this paper, we describe a new architecture of the control system to reduce the duration of fixed traffic lights by using Greedy and Infra Red sensors as well as Arduino. The result of scale model simulator by using traffic signal control modification based on Greedy method can be applied to handle simple and complex traffic light system in Indonesia. The Greedy algorithm works successfully on the traffic system in the case of this research by accelerating or increasing the green light at each intersection of a certain level.

References

[1] W. Jatmiko, A. Azurat, A. Wibowo, H. Marihot, and M. Wicaksana, "Self-Organizing Urban Traffic Control Architecture With Swarm-Self Organizing Map In Jakarta : Signal Control System And Simulator," *Int. J. Smart Sens. Intell. Syst.*, vol. 3, no. 3, pp. 463–465, 2010.

[2] R. Foroughi, A. Montazer, and R. Sabzevari, "Design of a New Urban Traffic Control System Using Modified Ant Colony Optimization Approach," *Iran. J. Sci. Technol. Trans. B Eng.*, vol. 32, pp. 167–173, 2008.

[3] W. Zang and G. Tan, "Traffic Congestion Evaluation and Signal Control Optimization Based on Wireless Sensor Networks: Model and Algorithms," *Hindawi Publ. Corp. Math. Probl. Eng.*, 2012.

[4] A. Faisal, A. Nasser, and H. Rowaihy, "Simulation of Dynamic Traffic control system based on Wireless sensor network," in *IEEE Symp on Computers & Informatics*, 2010, pp. 40–45.

[5] M. Sadayoshi and K. Yukinori, "Genetic Reinforcement Learning for Cooperative Traffic Signal Control," *IEEE Conf Evol. Comput.*, pp. 223–228, 1994.

[6] I. Porche, R. Segupta, M. Sampath, and Y. L. Chen, "A Decentralized Scheme for Real-Time Optimization of Traffic Signals," in *IEEE Int Conf on Control Application*, 1996, pp. 582–589.

[7] M. Tadnobu, K. Haruhiko, H. Sadaki, and O. Nobuyasu, "Multi Agent-Based Traffic Signal Control with Reinforcement Learning," in *Trans, IEICE*, 2000, pp. 478–486.

[8] N. Ikuku, I. Takeshi, and S. Kazutoshi, "No TitleImprovements of the Traffic Signal Control by Complex-Valued Hopfield Networks," in *IEEE Int Conf World Congress of Evolutionary Computational*, 2006, pp. 1186–1191.

[9] D. Mayee, L. Jindal, A. K. Yadav, and Chintoo, "Smart Traffic Light Signals," *Int. J. Sci. Res. Dev.*, vol. 4, no. 2, pp. 1919–1921, 2016.

[10] P. C. Siswipraptini, W. H. Martono, and D. Hartanti, "New Smart Traffic Light Using Travelling Salesman problem and Greedy Algorithm," in *The 3rd Asia Future Conference Proceeding*, 2016, pp. 1412–1419.

[11] S. Albers, "10-Greedy algortihm." Technische Universitait Munchen, 2008.

Securing Digital Audio using Complex Quadratic Map

Suryadi MT, Tjandra Satria Gunawan and Yudi Satria

Department of Mathematics, Faculty of Mathematics & Natural Sciences, Universitas Indonesia, Depok, 16424, Indonesia

E-mail: yadi.mt@sci.ui.ac.id, tjandra.satria@sci.ui.ac.id, ysatria@sci.ui.ac.id

Abstract. In This digital era, exchanging data are common and easy to do, therefore it is vulnerable to be attacked and manipulated from unauthorized parties. One data type that is vulnerable to attack is digital audio. So, we need data securing method that is not vulnerable and fast. One of the methods that match all of those criteria is securing the data using chaos function. Chaos function that is used in this research is complex quadratic map (CQM). There are some parameter value that causing the key stream that is generated by CQM function to pass all 15 NIST test, this means that the key stream that is generated using this CQM is proven to be random. In addition, samples of encrypted digital sound when tested using goodness of fit test are proven to be uniform, so securing digital audio using this method is not vulnerable to frequency analysis attack. The key space is very huge about 8.1×10^{31} possible keys and the key sensitivity is very small about 10^{-10}, therefore this method is also not vulnerable against brute-force attack. And finally, the processing speed for both encryption and decryption process on average about 450 times faster that its digital audio duration.

1. Introduction

The needs for securing information especially digital audio has become important and critical. For fulfilling that we need good encryption process, one of them is using chaos function.

There are many previous research discussing about securing digital multimedia using chaos function, some of them are: securing digital image that already discussed by many researcher [1,2], on securing digital audio [3,4], and securing digital video [5,6].

Chaos function [7] is usually easier to be implemented and based on research by Kocarev & Lian [8] on Table 14, encryption method using chaos function is significantly faster compared to other type of encryption like AES and DES.

Common chaos function used for many encryption process is Logistic Map that usually implemented on digital image encryption.

In this research, we are discussing other chaos function that is complex quadratic map (CQM). This function have one complex parameter c and this function map a complex number z to other complex number z' using this equation: $z' = z^2 + c$.

Because both c and z are complex numbers, the parameter domain in CQM will be more general than the parameter domain in Logistic Map. So we predict that the key space is very big so it will not vulnerable against brute-force attack.

In this paper, we will develop new algorithm for securing digital audio using CQM function and analyzing its performance.

2. Research Method

This CQM function mapping some complex number to other complex number using this equation [9]:

$$g_c(z) = z^2 + c \tag{1}$$

where $z, c \in \mathbb{C}$, but finding some parameter c such that this CQM function to be chaotic and bounded is not easy. This is CQM function in recurrence form:

$$z_{n+1} = z_n^2 + c \tag{2}$$

where $z_0 = 0 + 0i$ and $c \in \mathbb{C}$. Complex parameter c can be decomposed into two real parameter $a, b \in \mathbb{R}$ using this equation: $c = a + bi$. In this paper we only consider $|a| \leq 2$ and $|b| \leq 2$ as candidate paremeter.

In Equation (2), if we have c we will get the sequence of complex number $z_1, z_2, z_3, \ldots$, but because we use digital audio with integer samples, that sequence can't be used for key stream directly. Each element in that sequence z_n should be transformed first to k_n which is integer number.

We define the transformation from z_n to k_n as follows:

- z_n will be represented as 2 real numbers a_n and b_n using this equation $z_n = a_n + b_n i$
- Both a_n and b_n are represented as 64 bit floating point number using double precision floating point format (DPFPF) that is defined as follows [10]:

$$-1^{sign}\left(1 + \sum_{i=1}^{52} prec_{52-i} \cdot 2^{-i}\right) \times 2^{expo-1023} \tag{3}$$

- Because k_n is integer and it's used to encrypt or decrypt digital audio with max 32 bit integer, we use 32 bit unsigned integer to represent k_n.
- The first 16 bit (from bit 1 to bit 16 inclusive) on k_i are taken from bit 33 to bit 48 inclusive on binary representation of a_i
- The last 16 bit (from bit 17 to bit 32 inclusive) on k_i are taken from bit 33 to bit 48 inclusive on binary representation of b_i

Using that rule to transform z_i to k_i will always make k_i become 32 bit integer. To encrypt or decrypt digital audio with bit depth b bit, and to make the resulting digital audio also with bit depth b bit, then we first should convert k_i to become b bit integer x_i with this equation: $x_i = k_i \; mod \; 2^b$ where mod denoting modulo operation. This x_i will be XOR-ed with corresponding audio sample value on digital audio that to be either encrypted or decrypted.

Equation (2) is used for generating key stream that satisfy chaotic condition. That key stream is used indirectly to encrypt plain audio using bit exclusive or (XOR) operation and some preprocessing with some parameter and the resulting data is called cipher audio. That resulting cipher audio containing information that is too different than original plain audio and it's very hard to get that original plain audio without knowing the parameter used to generate the key stream. To get the original plain audio back, we should do decryption process that is inverse of encryption process and we should use the exact same key parameter to generate key stream.

This encryption algorithm performance is tested based on: key sensitivity, key stream randomness based on 15 NIST test [11], ergodicity is tested using Goodness of Fit test [12], algorithm efficiency based on computation time, and key space analysis based on IEEE 754 double precision binary floating point format.

Restriction in this paper are: digital audio that is used for encryption and decryption process has (*.wav) format and pulse code modulation (PCM) data type with bit depth per sample varies from 8 bit to 32 bit (8, 16, 24, and 32 bit). The programming language that is used on this research is C language, and the program run single thread on Kubuntu 17.04 with Core i7-4710HQ @2.50 GHz processor.

3. Result and Analysis

Test data that is used on this research is digital audio that having two channels left and right (stereo), and each channel having 44100Hz sample rate, more detailed info are given on Table 1.

Table 1. Test data used in this research

Number	Sound Type	File Name	Duration	Number of Samples	Bit Depth
1	Music taken from No Copyright Sound (NCS)	Record.wav	9.95 seconds	438663 samples ×2 channel	8
2					16
3					24
4					32
5	Short audio recording	Ricochet.wav	3 minutes 42.03 seconds	9791479 samples ×2 channel	8
6					16
7					24
8					32
9	Mozart melody (Created using superposition of some sine wave)	Mozart.wav	9 minutes 20.39 seconds	24712979 samples ×2 channel	8
10					16
11					24
12					32

3.1. Key Sensitivity

We use $c = -0.47 + 0.54i$ as a key, and we use that key to encrypt test data number 10 (Mozart.wav) and then we try to decrypt the file but with different but almost equal key $c = -0.47 - 10^{-x} + 0.54i$. Analysis result for some value of x are given on Table 2.

Table 2. Key sensitivity analysis result

X	Qualitative	Channel	χ^2	P_{value}
18	Same as original	Left	5903479841	0
		Right	9238326871	0
17	There are a little noise	Left	164174628.8	0
		Right	254073224.8	0
16	There are a little noise	Left	106923069.1	0
		Right	152171262.5	0
15	There are some noise	Left	38619867.08	0
		Right	42425652.77	0
14	There are many noise	Left	8940486.307	0
		Right	8919213.488	0
13	Not recognized after 2 minutes	Left	205585.0038	0
		Right	207153.1011	0
12	Not recognized after 12 seconds	Left	66442.89612	0.006242993
		Right	66646.60339	0.001125034
11	Not recognized after 1 second	Left	65185.36123	0.832887257
		Right	65583.51007	0.445989412
10	Completely not recognized	Left	65535.43138	0.498790016
		Right	65879.51358	0.170604590
Not Decrypted		Left	65805.95026	0.226862604
		Right	65594.53662	0.433983646

On Table 2, if we use $c = -0.47 + 0.54i$ for encrypting data, then we need $x \leq 10$ or error that is larger than or equal 10^{-10} to keep the decrypted data completely not recognized. So we conclude that key sensitivity on this CQM function is about 10^{-10}.

3.2. Key Stream Randomness Analysis (NIST Test)

NIST test result with level of significance $\alpha = 0.01$ on gey stream that is generated and converted to 32 bit integer are given on Table 3.

Table 3. NIST test result on CQM function with key parameter $c = -0.47 + 0.54i$

Statistical Test	Repetition	P_{value}	Proportion
Frequency	1	0.534146	128.00/128
Block Frequency	1	0.275709	127.00/128
Cumulative Sums	2	0.323322	127.50/128
Runs	1	0.378138	127.00/128
Longest Run	1	0.070445	124.00/128
Rank	1	0.170294	127.00/128
FFT	1	0.834308	127.00/128
Non Overlapping Template	148	0.408235	126.68/128
Overlapping Template	1	0.057146	126.00/128
Universal	1	0.090936	126.00/128
Approximate Entropy	1	0.422034	127.00/128
Random Excursions	8	0.242362	107.25/108
Random Excursions Variant	18	0.319938	106.44/108
Serial	2	0.612282	126.00/128
Linear Complexity	1	0.819544	128.00/128

It can be seen from Table 3 that CQM function with key parameter $c = -0.57 + 0.54i$ is generating completely random key stream based on all 15 NIST test. This is because for each test the P_{value} are larger than the level of significance that is 0.01 and passing proportion are all larger than minimum pass rate.

The minimum pass rate for each statistical test with the exception of the random excursion (variant) test is approximately 123 for a sample size 128 binary sequences. The minimum pass rate for the random excursion (variant) test is approximately 103 for a sample size 108 binary sequences.

3.3. Ergodicity Analysis

For measuring ergodicity level, we use Goodness of Fit test on cipher audio for each test data and we use level of significance $\alpha = 0.01$. The result are given on Table 4.

Table 4. Goodness of Fit test result

Number	Bit Depth	Degree of Freedom	Left Statistical Test	Left P_{value}	Right Statistical Test	Right P_{value}
1	8	255	291.098	0.05963155	222.170	0.93207627
2	16	65535	65506.12642	0.53105557	65357.02580	0.68800259
3	24	65535	65215.39514	0.81121859	65202.84559	0.82047453
4	32	65535	65853.33069	0.18951083	66256.11154	0.02349425
5	8	255	278.002	0.15416357	294.036	0.04680851
6	16	65535	65626.19250	0.39989747	65209.97125	0.81525438
7	24	65535	66160.38751	0.04237252	65675.26680	0.34863661
8	32	65535	65617.66541	0.40901475	65597.15755	0.43113867
9	8	255	265.180	0.31761973	258.303	0.43049635
10	16	65535	65805.95026	0.22686260	65594.53662	0.43398365
11	24	65535	65374.86500	0.67033449	66085.37948	0.06452742
12	32	65535	65000.59374	0.93033663	65269.76015	0.76784941

It can be seen from Table 4 that for all 12 test data used on this research, all P_{value} are larger than level of significance that is 0.01. That means that the cipher audio samples are uniformly distributed or satisfy the ergodicity condition.

3.4. Measuring Computation Time

For testing with key parameter $c = -0.47 + 0.54i$, we get the measured time for encryption and decryption process that is given on Table 5.

Table 5. Computation time for encryption and decryption process using CQM function

Number	Bit Depth	Sound Duration (seconds)	Encryption Time (seconds)	Ratio: Sound Duration per Encryption Time	Decryption Time (seconds)	Ratio: Sound Duration per Decryption Time
1	8	9.95	0.0235722	422.1073977	0.0241974	411.2012034
2	16		0.0225165	441.8981636	0.0224950	442.3205157
3	24		0.0222062	448.0730607	0.0221952	448.2951269
4	32		0.0223089	446.0103367	0.0223861	444.4722395
5	8	222.03	0.4679249	474.4992199	0.4656103	476.8580077
6	16		0.4972977	446.4730080	0.4945643	448.9406130
7	24		0.4700998	472.3039661	0.4692973	473.1116075
8	32		0.4704180	471.9844904	0.4709581	471.4432133
9	8	560.39	1.2373781	452.8850155	1.2450933	450.0787210
10	16		1.1953703	468.8003374	1.2046227	465.1996015
11	24		1.2612918	444.2984566	1.2885842	434.8881509
12	32		1.2697750	441.3301569	1.2550427	446.5107044

It can be seen on Table 5 that the computation time for encryption and decryption process are nearly equal. Digital audio bit depth is not affecting the processing speed for both encryption and decryption process that is on average about 450 faster that its digital audio duration.

3.5. Key Space Analysis

Based on Equation (3) the bit value of $expo$ is not independence because based on Equation (2) all key parameter which is used on CQM function that is a and b are all satisfy $|a| \leq 2$ and $|b| \leq 2$, therefore the variable that is independence are only $sign$ and $prec_1, prec_2, \ldots, prec_{52}$ so there are $2^{53} \approx 9 \times 10^{15}$ different numbers for each parameter.

There are 2 real parameters on CQM function that is a and b so based on DPFPF there are total $(2^{53})^2 = 2^{106} \approx 8.1 \times 10^{31}$ possible keys.

4. Conclusion

Based on analysis and simulation on test data that is used on previous section, we get conclusion as follows:

a. Key sensitivity on CQM function is about 10^{-10}
b. CQM Function is passing all 15 NIST test with $P_{value} > 0.01$. So the key stream that is generated using this CQM function are proven to be random.
c. Generated key stream by CQM function can make the cipher audio having uniform distribution (ergodic) on both channel based on Goodness of Fit test with $P_{value} > 0.01$.
d. Computation time for encryption and decryption process are nearly equal. Digital audio bit depth is not affecting the processing speed for both encryption and decryption process that is on average about 450 faster that its digital audio duration.
e. Large key space that is $\approx 8.1 \times 10^{31}$ possible keys.

Acknowledgment
The authors would like to thank Kemenristekdikti RI and Directorate of Research and Community Service (DRPM) Universitas Indonesia for the grant PTUPT 2017 who support us in this research.

References

[1] Suryadi MT, Eva Nurpeti, and Dhian Widya. (2014), Performance of Chaos-Based Encryption Algorithm for Digital Image, *Journal Telecommunication Computing Electronics and Control*, **Vol. 12**, No. 3, pp. 675-682, September 2014, ISSN: 1693-6930.

[2] Yohan Suryanto, Suryadi MT, and Kalamullah Ramli. (2016), A Secure and Robust Image Encryption Based on Chaotic Permutation Multiple Circular Shrinking and Expanding, *Journal of Information Hiding and Multimedia Signal Processing*, **Vol. 7**, No. 4, pp. 697-713, July 2016, ISSN: 2073-4212.

[3] Vishakha B. Pawar, Pritish A. Tijare, Swapnil N. Sawalkar. (2014). A Review Paper on Audio Encryption, *International Journal of Research in Advent Technology*, **Vol. 02**, No. 12, pp. 45-48, December 2014, E-ISSN: 2321-9637.

[4] Sangeeta, Vikas Dhankhar, and Amita Dhankhar. (2013), Secure LSB Based Audio Steganography, International Journal of Engineering, *Applied and Management Sciences Paradigms*, **Vol. 05**, Issue 01, pp. 131-134, July 2013, ISSN (Online): 2320-6608.

[5] Suryadi MT. (2013), New Chaotic Algorithm for Video Encryption. *4th The International Symposium on Chaos Revolution in Science, Technology and Society* 2013, Jakarta, August 28-29. 2013.

[6] Brahim B, Mohamed M.L, and Salah B. (2011), *Multimedia Chaos-based Encryption Algorithm*, Department of Electronics, Faculty of Eng. Sciences, University Badji Mokhtar of Annaba, 23000 Annaba, Algeria.

[7] Devaney, R.L. (1989). *An introduction to chaotic dynamical systems* (2nd ed.). Addison-Wesley Publishing company, Inc.

[8] Kocarev, L and Lian, S. (2011), *Chaos-Based Cryptography: Theory, Algorithm and Applications*, Springer-Verlag, Berlin.

[9] Sarah N, Kabes. (2012), *The Transition Between the Complex Quadratic Map and the Hénon Map*, Department of Mathematics and Statistics, University of Minnesota, Duluth, MN 55812, July 2012.

[10] David Goldberg. (1991), *What Every Computer Scientist Should Know About Floating-Point Arithmetic*, ACM Computing Surveys, Vol 23, No 1, March 1991.

[11] Lawrence E Bassham III. (2010). *A Statistical Test Suite for the Validation of Random Number Generator and Pseudo Random Number Generators for Cryptographic Applications*, NIST Special Publication 800-22 Revision 1a.

[12] Walpole, R.E., Myers, R. H., Myers, S. L., Ye, K. (2012). *Probability and Statistics for Engineers & Scientists* (9th edition). Boston: Prentice Hall.

Turbulent flow in a vessel agitated by side entering inclined blade turbine with different diameter using CFD simulation

N N Fathonah[1], T Nurtono[1], Kusdianto[1], and S Winardi[1*]

[1] Laboratory of Fluid Mechanic and Mixing, Department of Chemical Engineering, Institut Teknologi Sepuluh Nopember
Kampus ITS Keputih, Surabaya Indonesia 60111

*Email: swinardi@chem-eng.its.ac.id

Abstract. Single phase turbulent flow in a vessel agitated by side entering inclined blade turbine has simulated using CFD. The aim of this work is to identify the hydrodynamic characteristics of a model vessel, which geometrical configuration is adopted at industrial scale. The laboratory scale model vessel is a flat bottomed cylindrical tank agitated by side entering 4-blade inclined blade turbine with impeller rotational speed N=100-400 rpm. The effect of the impeller diameter on fluid flow pattern has been investigated. The fluid flow patterns in a vessel is essentially characterized by the phenomena of macro-instabilities, i.e. the flow patterns change with large scale in space and low frequency. The intensity of fluid flow in the tank increase with the increase of impeller rotational speed from 100, 200, 300, and 400 rpm. It was accompanied by shifting the position of the core of circulation flow away from impeller discharge stream and approached the front of the tank wall. The intensity of fluid flow in the vessel increase with the increase of the impeller diameter from d=3 cm to d=4 cm.

1. Introduction

One of the mixer widely used is agitated tank by side entering impeller which an agitator is mounted horizontally on the sides of the tank wall, near the bottom of the tank. The main advantages of side entering agitated vessel includes large volume capacity, low capital and operating cost, and simple installation as no mounting support on the top of the tank is needed [1,2]. One of the main drawbacks of side entering agitated vessel is the poorer mixing performance with respect to baffled vessels, due to the smaller pumping rate generated by the impeller. As a consequence, limited attention has been devoted so far to the fundamental mixing characterization of these mixers. Nevertheless, vessel with side entering mixer are gaining interest as they may provide significant advantages in a number of applications including biochemical, oil and pulp and paper processes.

Mixing performance in side entering agitated vessel can be studied experimental and computational approaches. Over the past years a number of works have been carried out to investigate mixing performance by side entering impeller using experimental and simulation approaches [3,4,5,6]. However, almost of all simulation studies has been conducted for large-scale mixing system without quantitative validation with experimental data. This is because due to experimental work is a laborious and expensive process when carried out in large scale side entering agitated vessel. In this study, therefore, the hydrodynamics and characteristics of side entering agitated vessel will be studied by numerical simulations using CFD software package ANSYS Fluent 17.1. Impeller used are 4-bladed inclined blade turbine (IBT) with diameter 3 and 4 cm. The influence of impeller diameter on turbulent fluid flow will investigated. The simulation results will be validated with experimental data.

2. Geometry of mixing system

A side entering agitated vessel is used in this work includes the flat bottomed cylindrical vessel (D=40cm), 4-blade IBT with diameter d=3 and 4 cm (d/D=0.075 and 0.1). The length of the impeller shaft (s) and the distance of impeller shaft to the tank bottom (h), are equal to the impeller diameter (d) respectively. The horizontal angle of impeller shaft (β) is constant at 0o. The geometry and dimension of the vessel and impeller are shown in figures 1 and 2, schematically.

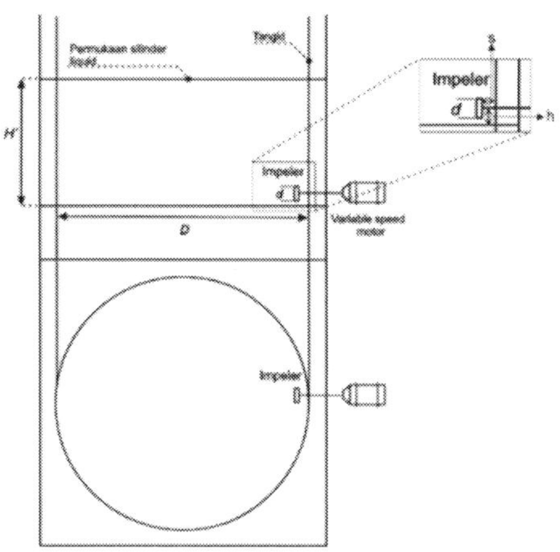

Figure 1. Vessel Geometry

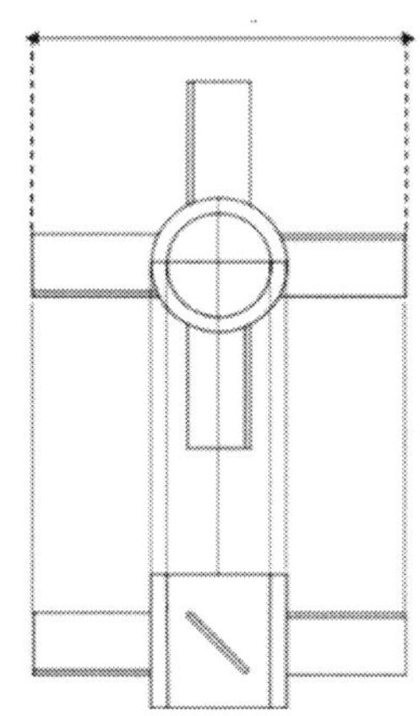

Figure 2. 4-blade IBT geometry

3. Model development

3.1. Assumption

The assumption of operating condition for this system is set as fluid flow is three dimensional and unsteady state. Working fluid in vessel is incompressible, isothermal, and Newtonian fluid. The model is single phase without accounting for multiphase interactions.

3.2. Description of computational model

The computational model is based on solving the conservation equations which describe the flow in the tank [7]. For the mean flow of an incompressible liquid, the equation of mass conservation is expressed as,

$$\frac{\partial U_i}{\partial x_i} = 0 \quad (1)$$

U_i denotes the mean velocity in direction x_i. Under steady-state laminar flow conditions the momentum balance is given by:

$$\rho U_i \frac{\partial U_i}{\partial x_i} = \frac{\partial p}{\partial x_i} + \frac{\partial \tau_{ij}}{\partial x_j} + \rho g_i \quad (2)$$

The stress tensor τ_{ij} is given by:

$$\tau_{ij} = \mu\left(\frac{\partial U_i}{\partial x_i} + \frac{\partial U_j}{\partial x_j}\right) \quad (3)$$

When the flow is turbulent an additional term appears in the momentum balance for the mean flow:

$$\rho U_i \frac{\partial U_i}{\partial x_j} = -\frac{\partial p}{\partial x_i} + \frac{\partial}{\partial x_j}\left(\tau_{ij} + \rho \overline{u_i u_j}\right) + \rho g_i \quad (4)$$

u_i is the fluctuating, turbulent, velocity component in direction x_i. The cross correlation term denotes the Reynolds stress tensor. Although analytical equations for the Reynolds stress tensor can be derived, these equations again contain higher order cross correlation terms, and the set of equations is not closed. For modeling the Reynolds stresses various so-called turbulence models are available. An often used model is the *k*-ε model which models the Reynolds stresses as being proportional to gradients in the mean flow.

$$\rho\overline{u_i u_j} = -\mu_t\left(\frac{\partial u_i}{\partial x_j} + \frac{\partial u_j}{\partial x_i}\right) + \frac{2}{3}\delta_{ij}k \qquad (5)$$

The turbulent viscosity μ_t is given by:

$$\mu_t = c_\mu \rho \frac{k^2}{\varepsilon} \qquad (6)$$

The model equations for the conservation of the turbulent kinetic energy density k and the turbulent energy dissipation rate density ε are:

$$U_k \frac{\partial k}{\partial x_k} = \frac{\partial}{\partial x_k}\left(\left(v + \frac{v_t}{\sigma_k}\right)\frac{\partial k}{\partial x_k}\right) + P_k - \varepsilon \qquad (7)$$

$$U_k \frac{\partial \varepsilon}{\partial x_k} = \frac{\partial}{\partial x_k}\left(\left(v + \frac{v_t}{\sigma_\varepsilon}\right)\frac{\partial \varepsilon}{\partial x_k}\right) + C_{\varepsilon 1}\frac{\varepsilon}{k}P_k - \varepsilon \qquad (8)$$

Equation (5) for the Reynolds stresses is similar to equation (3) for the laminar shear stresses. The turbulent viscosity μ_t plays the same role as the molecular viscosity μ. As a result the form of the turbulent momentum equations is similar to the form of the laminar momentum equation except that μ is replaced by an effective viscosity μ_{eff}:

$$\mu_{eff} = (\mu + \mu_t) \qquad (9)$$

3.3. CFD simulation

The simulation were performed using software ANSYS® ver. 17.1 that included Design Modeler as geometry modeler, ANSYS Meshing as grid generator, and ANSYS Fluent as the CFD solver. The simulations were started from steady-state condition using the combination of k-ε as turbulence model and multiple reference frame (MRF) as impeller motion model. The calculation domain were divided into two zones that are depicted in Figure 3 below.

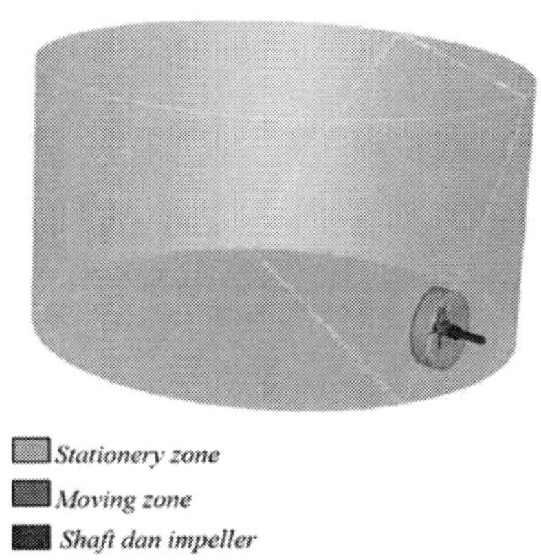

Figure 3. The calculation domain with two zones

The display of the flow field vector in each plane inside the agitated vessel was recorded every time step to make flow pattern animations. The flow pattern recognition and characterization were obtained from vertical and horizontal planes passing through the impeller shaft.

4. Result and discussion

4.1. Influence of impeller rotational speed

Fluid flow patterns in side entering agitated vessel is essentially characterized by the phenomena of macro-instabilities, i.e. the flow patterns change in space and time with large scale. These phenomena always accompanied with strong velocity fluctuations in the entire of the vessel. In this study, for simplification, changes in fluid flow patterns only observed in a vertical and horizontal plane passing through the impeller shaft. This phenomena of macro-instabilities shown in figure 4 below. This flow patterns are recorded in 60, 100, 120, 300, 400 and 600 second. From these figures, it can be seen that macro-instabilities phenomena occur throughout the mixing process runs at rotational speed 300 rpm and agitated with IBT 3 cm of diameter. Flow patterns formed are different every time.

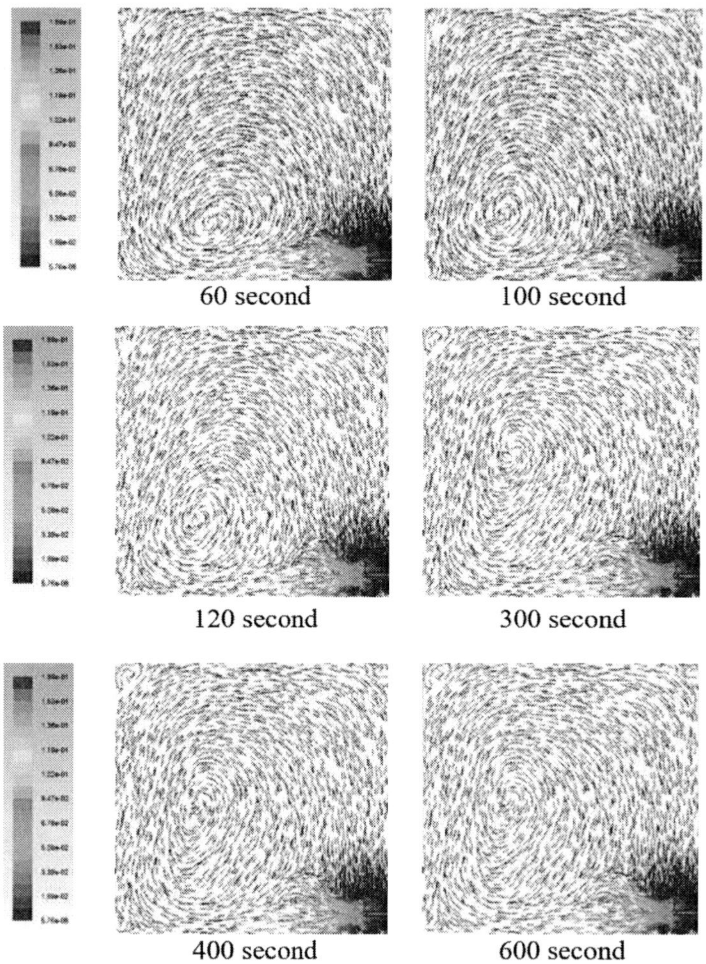

Figure 4. Macro-instabilities phenomena in tank stirred by IBT with diameter of d=3 cm

Effect of impeller rotational speed to fluid flow pattern in tank stirred by IBT with diameter of d=3 cm are shown in Figure 5. From side view observation, at impeller rotational speed N=100 rpm, discharge stream from impeller will flow toward the wall of tank, then most discharge stream impingement wall, and along the wall rises up towards the liquid surface. Another part of the discharge stream turned upward to the liquid surface then flow return back to the impeller in short time. It can be seen that the large circulation flow is formed with the core located close to the impeller discharge stream. On the right side of the area near the liquid surface formed a stagnant region. From the top view observation, it can be found that the circulation flow also occurs horizontally on the area on the left side and the right side of the impeller.

When the impeller rotational speed is increased in the range N=200-400 rpm, from the side view observation shows that the intensity of circulation flow increase with the increase of rotational speed of the impeller. It was accompanied by shifting the position of the core of circulation flow away from impeller discharge stream and approached the front of the tank wall. From the top view observation, it can be seen that the circulation flow position shifted away from the impeller area and approached the tank wall.

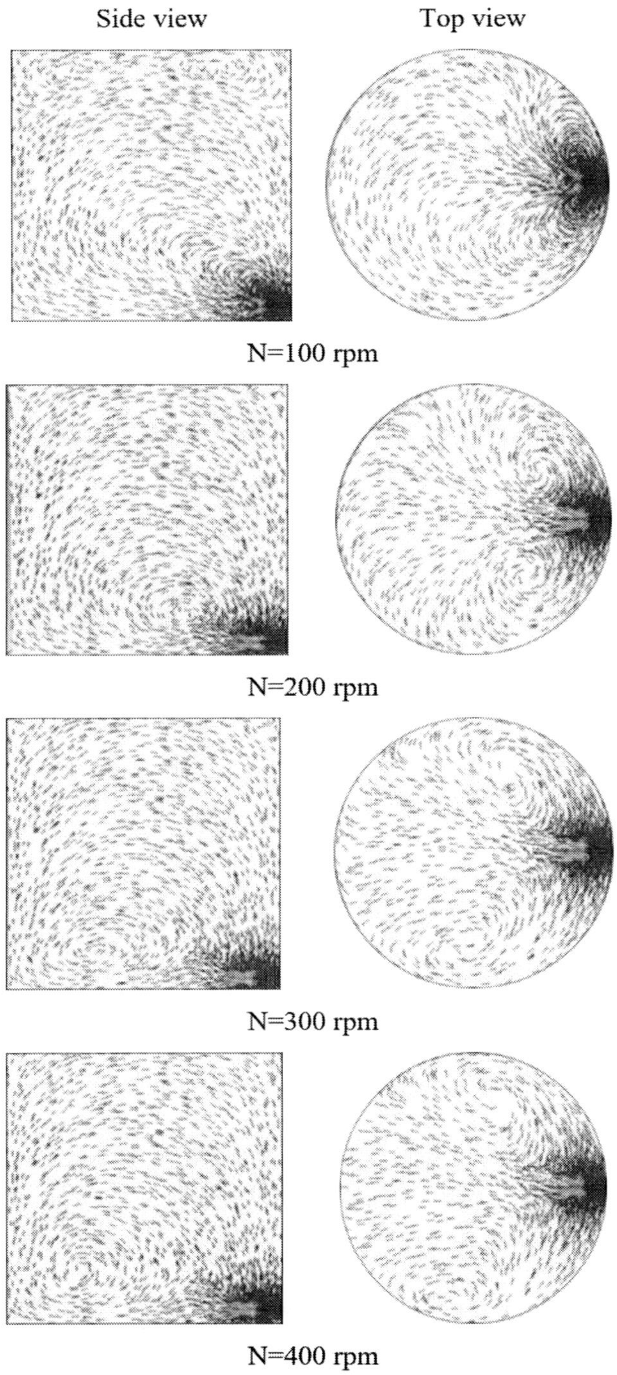

Figure 5. Fluid flow patterns in tank stirred by IBT with *d*=3 cm in different impeller rotational speed

IOP Conf. Series: Journal of Physics: Conf. Series **974** (2018) 012015 doi:10.1088/1742-6596/974/1/012015

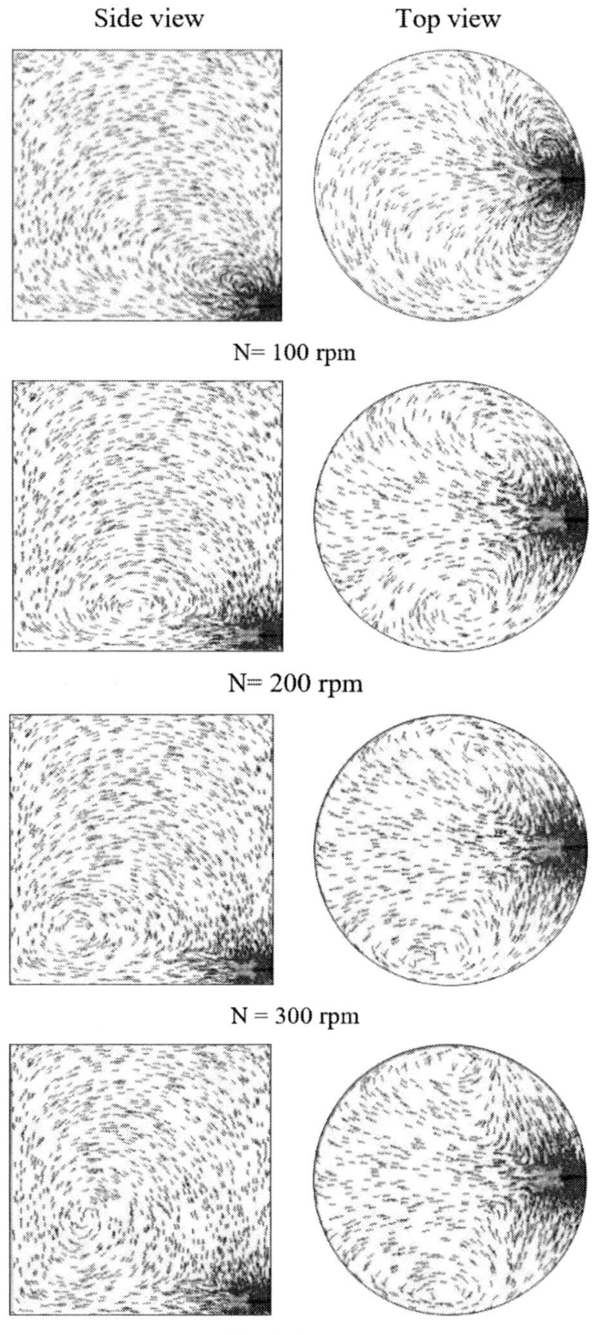

Figure 6. Fluid flow patterns in tank stirred by IBT with *d*=4 cm in different impeller rotational speed

Effect of the impeller speed to the fluid flow patterns in the tank agitated by IBT with a diameter d=4 cm are shown in Figure 6. From the side view and top view observations, it can be seen that the circulation flow intensity increase with the increase in impeller rotational speed from N=100-400 rpm as seen with impeller diameter of 3 cm. This was indicated by a shift in the position of the core of circulation flow, which is further away from the impeller discharge stream and approached the tank wall in front of them.

4.2. Influence of impeller diameter

As shown in Figures 5 and 6 above, the fluid flow patterns in the tank produced by IBT showed the same tendency when the impeller rotational speed increase. However, the intensity of fluid flow increases with the increase of the impeller diameter from d=3 cm to d=4 cm as shown in Figure 6.

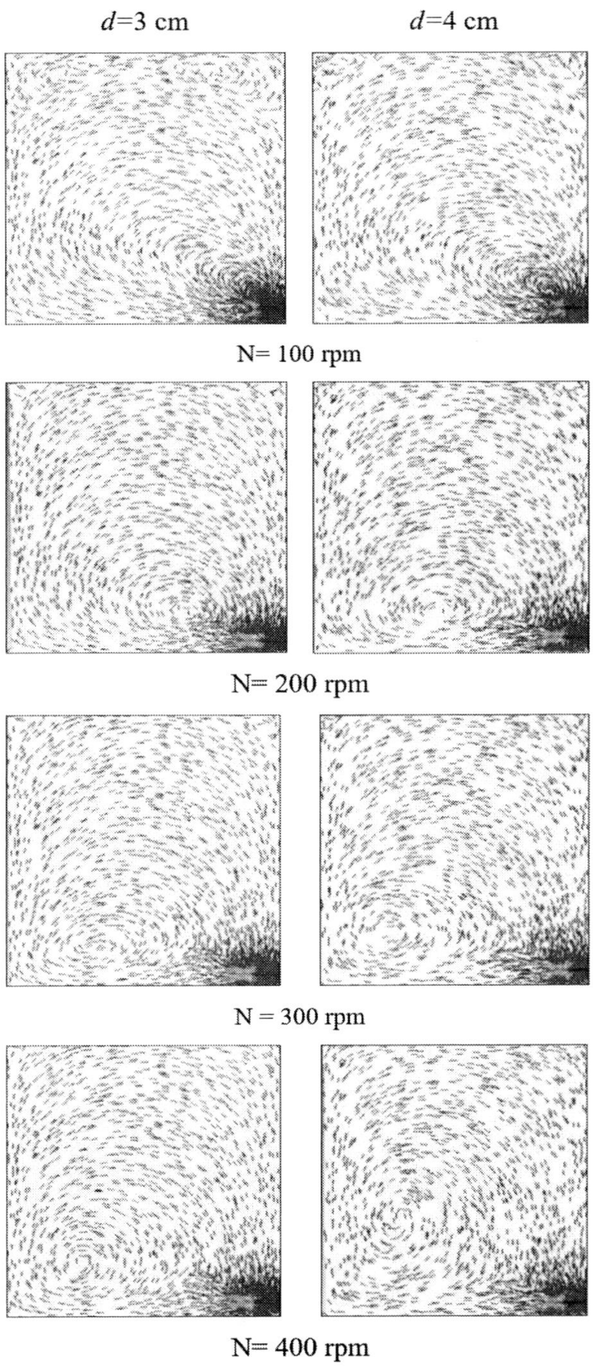

Figure 7. Influence of impeller diameter to fluid flow pattern in side entering agitated vessel

At a rotational speed N=100 rpm, the intensity of the fluid flow generated by the impeller d=4 cm is relatively large compared with those produced by the impeller with d=3 cm. This is shown by the size

IOP Conf. Series: Journal of Physics: Conf. Series **974** (2018) 012015 doi:10.1088/1742-6596/974/1/012015

of the core of a larger circulation flow and stagnation region in the top right corner regions near the surface of the liquid does not appear.

4.3. Comparison with experimental data

Computational data needs to be validated with experimental data. Because the simulation results is depend on several factors such as geometry of the system used, meshing structure, definition and properties of the fluid used, the turbulence model and impeller motion model selected [8].

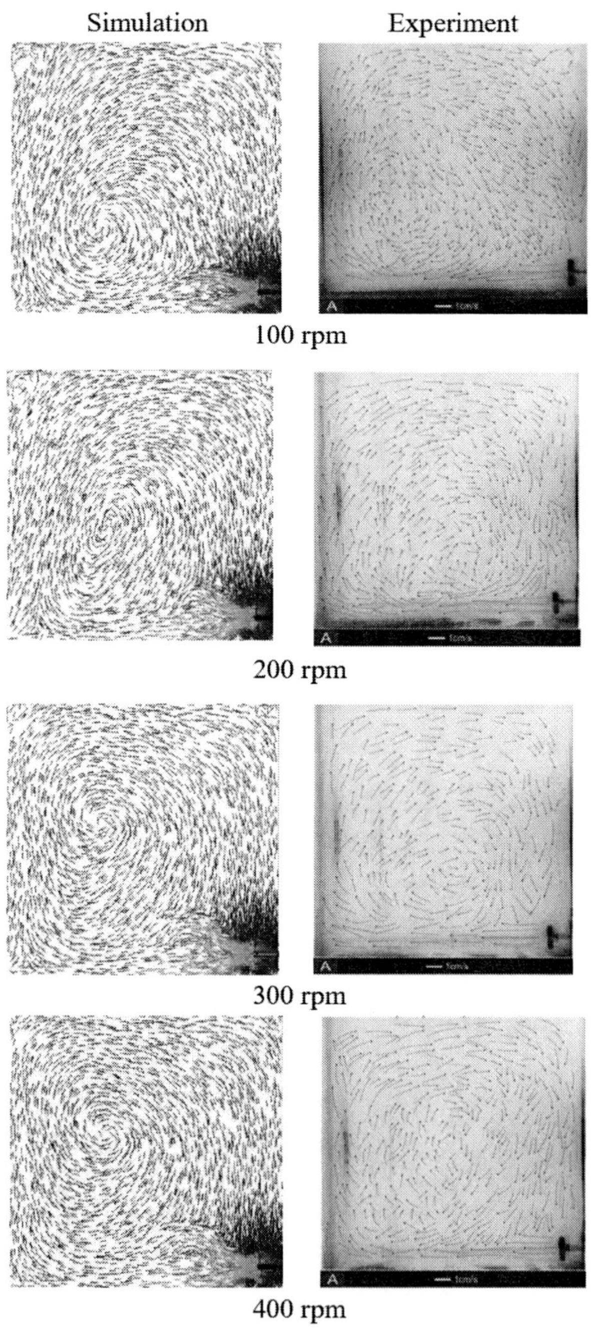

Figure 8. Flow pattern comparison between simulation and experimental results

Figure 8 above shows the comparison of flow pattern produced by simulation and experimental results. Flow patterns produced by axial flow impeller has main characteristic one loop circulation. Additional flow pattern might arise due to the interaction of the structures and the vessel wall and have been explained in macro instabilities phenomena section 4.1 [9]. From this figure we can conclude that simulation results has a good agreement with experimental data, that can be seen from flow pattern produced tend to be similar.

5. Conclusion

The fluid flow patterns in the vessel equipped with side entering 4-blade IBT is essentially characterized by the phenomena of macro-instabilities, i.e. the flow patterns change with large scale in space and low frequency. The intensity of fluid flow in the tank with IBT increase with the increase of impeller rotational speed from 100, 200, 300, 400 rpm. It was accompanied by shifting the core position of circulation flow away from impeller discharge stream and approached the front of the tank wall. Also the intensity of fluid flow increase with the increase of impeller diameter from d=3 cm to d=4 cm. Simulation results has a good agreement with experimental data, that can be seen from flow pattern produced tend to be similar.

Acknowledgment

The authors are grateful for financial support provided by a grant from Directorate of Research and Public Service, Directorate General of Research Strengthening and Development, Ministry of Research, Technology and Higher Education of the Republic of Indonesia.

References

[1] Vincent W. Uhl Joseph B. Gray 1966 *Mixing Theory and Practice I* (New York: Academic Press)
[2] James W. Oldshue 1983 *Fluid Mixing Technology* (New York: McGraw-Hill Publications Co.)
[3] Johannes A. Wesselingh 1975 *Mixing of Liquids in Cylindrical Storage Tanks With Side-Entering Propellers* (Chemical Engineering Science)
[4] Klausdieter K. 1984 *Suspension by Side Entering Agitators* (Chemical Engineering Process) vol. 18 p 233
[5] Asghar A. D. Masoud R. 2004 *CFD Simulation of Homogenization in Large Scale Crude Oil Storage Tanks* (Petroleum Science & Engineering/ Elsevier) vol. 43 p 151-161
[6] Masoud R. 2005 *The Effect of Impellere Layout on Mixing Time in Large-Scale Crude Oil Storage Tank* (Petroleum Science & Engineering/ Elsevier) vol. 48 p 161-170
[7] H K Versteeg W. Malalasekera 1995 *An Introduction to Computational Fluid Dynamics: The Finite Volume Method* (London: Prentice Hall)
[8] C. Gomez C.P.J. Benington F. Taghipour 2010 *Investigation of the Flow Field in a Rectangular Vessel Equipped with a Side-entering Agitator* (Journal of Fluid engineering/ ASME) vol. 132
[9] Jaime S.E Fariborz T. 2015 *Computational Simulation of Mixing Flow of Shear Thinning non-Newtonian Fluids with Various Impellers in a Stirred Tank* (Chemical Engineering and Processing: Process Intensification/ Elsevier)

IOP Conf. Series: Journal of Physics: Conf. Series **974** (2018) 012016 doi:10.1088/1742-6596/974/1/012016

Digital Sound Encryption with Logistic Map and Number Theoretic Transform

Yudi Satria, Gabe Rizky P.H. and Suryadi MT

Department of Mathematica, Faculty of Mathematica and Natural Sciences, Universitas Indonesia, Kampus UI, Depok, 16424

Email: ysatria@sci.ui.ac.id, gabe.rizky@sci.ui.ac.id, and yadi.mt@sci.ui.ac.id

Abstract. Digital sound security has limits on encrypting in Frequency Domain. Number Theoretic Transform based on field (GF $2^{521}-1$) improve and solve that problem. The algorithm for this sound encryption is based on combination of Chaos function and Number Theoretic Transform. The Chaos function that used in this paper is Logistic Map. The trials and the simulations are conducted by using 5 different digital sound files data tester in Wave File Extension Format and simulated at least 100 times each. The key stream resulted is random with verified by 15 NIST's randomness test. The key space formed is very big which more than 10^{469}. The processing speed of algorithm for encryption is slightly affected by Number Theoretic Transform.

1. Introduction

Various efforts to secure digital data has been done using cryptograph application. There are two main process in cryptograph, namely encryption and decryption process. The main reliability measure of the Encryption algorithm is the durability against various attack and the processing speed.

The encryption algorithm commonly used is DES, AES, and RSA. However, because relating to their durability and or processing speed, currently many of developing encryption are based on chaos function. This is based on the fact that the usage of chaos function in encrypting has more durability against attack and or better processing speed [1, 2].

Several studies related to encryption algorithm based on chaos function in digital data security are encryption algorithm using Logistic Map [3-8], other chaos function [9-14] and non-chaotic mode [15-17]. Each of this studies has its own performance associated with the level of resulted key stream, the key space and the processing time (processing speed).

Digital data security done in previous studies are mostly done to secure digital image while security related to digital sound has not many yet explored. In this study, the improvement for algorithm encryption to secure digital sound will be developed. The developed algorithm is using chaos function, in this case Logistic Map, and combined Number Theoretic Transform (NTT). The development is conducted to provide alternative digital sound encryption algorithm which has better durability against attack and faster processing speed.

2. Research Method

Chaos function in this study will act as the random number generator which will generate a random key stream sequence. This study using Logistic Map as chaos function and combined by NTT. The sequence of key stream will generated by the Logistic Map in this paper is using the recursive form in Equation (1) as follow [2]:

$$x_{i+1} = rx_i(1 - x_i) \tag{1}$$

Where $x_0 \in (0,1)$, $r \in (0,4]$ for $i = 0,1,2,3, ...$

Sound wave not only can be interpreted by wave super position based on the primitive wave frequency (Frequency Domain) but also can be interpreted by wave amplitude sequence (Time Domain). Randomization at Frequency Domain has limitation in discrete form due to loss of information as the consequence of transformation from continuous form to discrete form and inversely.

The transformation, which to be the bridged both Time Domain and Frequency Domain, is using Fourier Transformation variation called Number Theoretic Transform. It was based on the Fourier Transformation's formulation [18].

$$w = e^{-\frac{2*i*\pi}{N}} \tag{2}$$

$$f_k = \sum_{n=0}^{N-1} t_n * w^{-k*n} \tag{3}$$

$$t_k = \frac{1}{N}\sum_{n=0}^{N-1} f_n * w^{k*n} \tag{4}$$

But the different about the variation is about using the field where $w, f_k t_k \in f(*', +')$ and $w^k \neq 1$ and $w^N = 1, \forall k = 1 \dots N-1$.

The design of developed improvement encryption algorithm in this study is shown in Figure 1.

Furthermore, the viewed performance of developed algorithm is based on the randomness analysis of randomness through the 15 NIST's test [19], the key space and the measurement of processing speed.

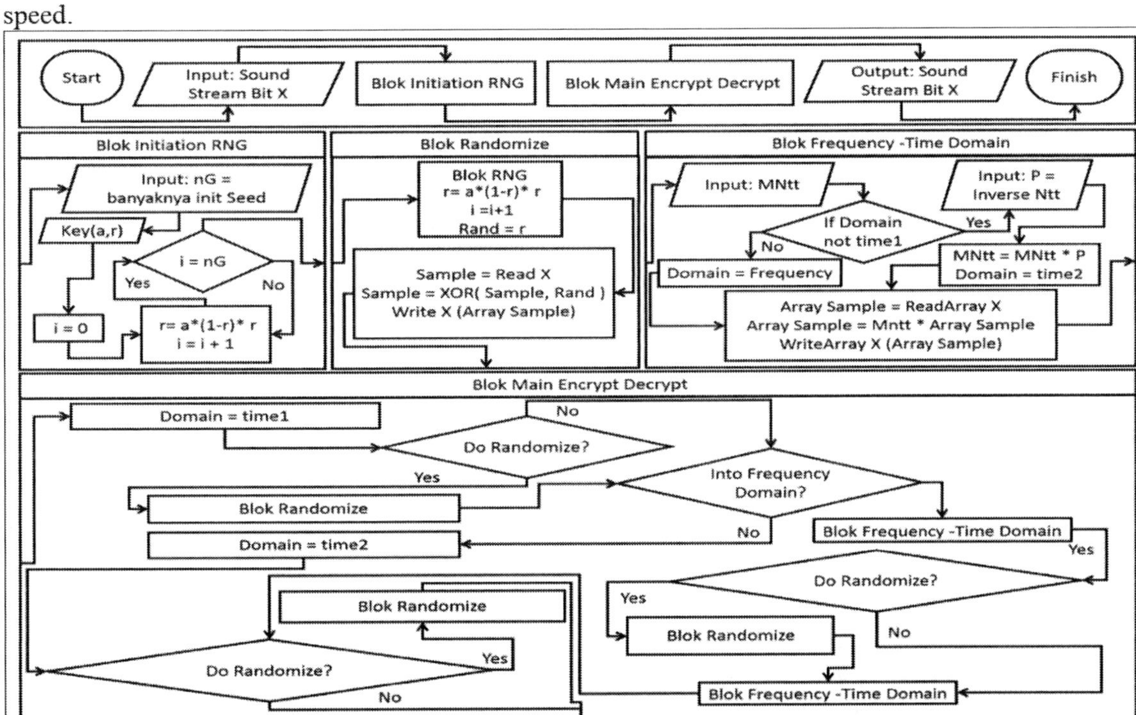

Figure 1. Encryption Algorithm

3. Result and Analysis

Encryption algorithm developed in this study will be implemented using C and java programing language and tested using trials data as showed in Table 1. The performance of the trials and simulations result is explained in next section.

Table 1. Data Tester

Name of File	Size of File (MB)	Duration (Seconds)	Bit Rate (KB/s)
Lagu1.wav	2.66	31	705
Lagu2.wav	20.00	123	1411
Lagu3.wav	37.90	225	1411
Lagu4.wav	44.50	264	1411
Lagu5.wav	798.00	4746	1411

3.1. Key Randomness Analysis (by 15 NIST's test)

The result with significant level are $\alpha = 0{,}01$ on generated key stream was showed on Table 2.

Table 2. 15 NIST's Randomness Test Result

STATISTICAL TEST	PROPORTION	P-VALUE
Frequency	32/32	0.739918
Block Frequency	32/32	0.671779
Cumulative Sums	32/32	0.573505
Runs	32/32	0.862344
Longest Run	32/32	0.178278
Rank	32/32	0.534146
FFT	32/32	0.739918
Non Overlapping Template	31/32	0.436885
Overlapping Template	32/32	0.468595
Universal	31/32	0.862344
Approximate Entropy	32/32	0.862344
Random Excursions	23/24	0.267745
Random Excursions Variant	23/24	0.301477
Serial	32/32	0.762807
Linear Complexity	29/32	0.299251

In the table 2, all Resulting P-Value values for all 15 NIST's tests are greater than significant level (0.01) which means passing the NIST's test. It can be deducted the random stream which was generated by Logistic map, was random and verified by 15 NIST's test.

3.2. Analyst key space

There are two core parameters in Logistic map function as initial value in Real domain but with modified in the field (GF $2^{521} - 1$) then the key space in every parameter will be same as the norm set of the field. In other word, the key space in a Logistic map which based on the field will be $(2^{521} - 1)^2 > 2^{1040} \cong 10^{313}$.

There are one core parameter in Fourier Transformation as initial value in Real Domain and it just constant and the value was depending on the size (or order or N). By using NTT over the field, the parameter still will be one core parameter but there has restriction on the order but the good point is all element can be used as the key which is better than just only using NTT over a ring like normally use in integer 8-bit, 32-bit, 64-bit or may be 1024-bit. In the other word, the key space in a NTT over the field was $(2^{521} - 1) > 2^{520} \cong 10^{156}$.

In the summary, with the combining Logistic Map and NTT, it can make the key space larger and help to reach $(2^{521} - 1)^3 > 2^{1560} \cong 10^{469}$. The result is better than only using Logistic Map as algorithm with IEEE Floating Point Format which only can reach 10^{30} [7]. It can improve the Logistic Map 15.6 time in exponential to fight against Brute Force Attack.

3.3. *Processing time*

The simulation process of encryption and decryption process to get the average time process was tested with 4 different key and at least 100 times for every key for every data tester. Table 3 shows the average time of encryption and decryption process.

Table 3. Average Time process for encryption and decryption

Data tester	Logistic map		Logistic map + NTT	
	Encryption(s)	Decryption(s)	Encryption(s)	Decryption(s)
Lagu1.wav	4.711863	5.640585	4.90875	5.938621
Lagu2.wav	12.84444	13.78578	15.32002	16.47847
Lagu3.wav	20.38666	21.43264	25.28059	26.38297
Lagu4.wav	23.51062	24.40229	29.08023	30.02419
Lagu5.wav	358.6138	358.0492	468.8384	462.2297

The average time for encryption and decryption process is relatively similar. The format of relative key streams also does not have any significant impact on the speed of processing time. The file size is highly impacting the speed both encrypting and decrypting. Other than those, The NTT has slightly affected.

4. Conclusion

Based on the discussion, trials and simulation used in previous segment. We can conclude the following:

A. The key streams generated by Logistic Map are random which was verified by NIST's tests.
B. The key space reach 10^{469} which has high durability against Brute Force Attack.
C. The average encryption time is relatively faster than decryption time.
D. The usage key in Logistic map will not affected the time.
E. The NTT has slightly affected the processing time.

Acknowledgment

This study is supported by Kementerian Ristek Dikti RI through Direktorat Riset dan Pengabdian Masyarakat Universitas Indonesia, in Hibah Penelitian Terapan Unggulan Perguruan Tinggi Tahun 2017 scheme.

References

[1] Devaney, R.L. 2.5(1989). *An introduction to chaotic dynamical systems (2nd ed.)*. Addison-Wesley Publishing company, Inc.
[2] Kocarev, L., & Lian, S. (2011). *Chaos-based cryptography*. Berlin Heidelberg : Springer-Verlag.
[3] Pareek, N.K., Patidar, V., Sud, K.K. (2006). Image encryption using chaotic logistic map. *Journal of Image and Vision Computing*, **24**, 926-934.
[4] Patidar, V., Pareek, N.K., Sud, K.K. (2009). A new subtitution-diffusion based image cipher using chaotic standard and logistic maps. *Journal of Commun Nonlinear Sci Numer Simulat*, **14**, 3056-3075.
[5] Miki Yasushi, (2013). Using the Logistic Map to Generate Scratching Sound. *The Mathematica Journal*, Wolfram Media, Inc., **8**, 1-27

[6] Suryadi M.T., (2013). *New Chaotic Algorithm for Video Encryption*, 4th the International Symposium on Chaos Revolution in Science, Technology and Society 2013, Jakarta, 28-29, August, 2013.

[7] Suryadi M.T., Eva Nurpeti, and Dhian Widya. (2014), Performance of Chaos-Based Encryption Algorithm for Digital Image, *Journal Telecommunication Computing Electronics and Control*, **12**(3), 675-682.

[8] Sonis, M., (1996). Once more on Hénon map: Analysis of bifurcations, *Journal of Chaos, Solitons & Fractals*, **7**(12), 2215–2234.

[9] Zhang, W., Wong, K., Yu, H., Zhu, Z. (2013). An image encryption scheme using reverse 2-dimensional chaotic map and dependent diffusion. *Journal of Commun Nonlinear Sci Numer Simulat*, **18**, 2066-2080.

[10] Ye, Ruisong and Zhao, Haiying., (2012), An Efficient Chaos-based Image Encryption Scheme using Affine Modular Maps, *International Journal Computer Network and Information Security*, **7**, 41-50.

[11] Ansari, S., Gupta, N and Agrawal, S., (2012), An Image Encryption Approach Using Chaotic Map in Frequency Domain, *International Journal of Emerging Technology and Advanced Engineering*, **2**(8), 287 – 291.

[12] Liu, S., Sun, J. and Xu, Z., (2009), An Improved Image Encryption Algorithm based on Chaotic System, *Journal of Computer*, Academy Publisher, **4**(11), 1091-1100.

[13] Abu Zaid, Osama M., El-Fishawy, Nawal A., Nigm, E.M., (2013). Cryptosystem Algorithm Based on Chaotic System for Encrypting Colored Image, *International Journal of Computer Science Issues*, **10**, Issue 4, No 2, 215-224.

[14] Gao, H., Zhang, Y., Liang, S., Li, D. (2006). A new chaotic algorithm for image encryption. *Journal of Chaos, Solutions and Fractals*, **29**, 393-399.

[15] Liu, L., Zhang, Q., & Wei, X. (2012). A RGB image encryption algorithm based on DNA encoding and Chaos Map. *Computers and Electrical Engineering design*, **38**(5), 1240-1248.

[16] Yohan Suryanto, Suryadi M.T., and Kalamullah Ramli. (2016), A Secure and Robust Image Encryption Based on Chaotic Permutation Multiple Circular Shrinking and Expanding, *Journal of Information Hiding and Multimedia Signal Processing*, ISSN 2073-4212, **7**(4), 697-713.

[17] Yohan Suryanto, Suryadi M.T., and Kalamullah Ramli. (2017), A new image encryption using color scrambling based on chaotic permutation multiple circular shrinking and expanding, *Journal Multimedia Tools and Application*, Springer, **76**, 16831–16854

[18] Makwana, Vinita and Parmar, Neha (2013). Analysis of Performance of Fast Fourier Transformation of an Audio File, *International Journal of Application or Innovation in Engineering and Management*, (IJAIEM), **2**, (11), 68-71.

[19] Rukhin A., et.all, (2010). *A Statistical Test Suite for Random and Pseudorandom Number Generator for Cryptographic Application. National Institute of Standards and Technoloogy* (Revision 1a), Technology Administration U.S. Depertament of Commerce.

Preventing Superinfection in Malaria Spreads with Repellent and Medical Treatment Policy

Fanny Fitri, Dipo Aldila*

Department of Mathematics, Universitas Indonesia, Depok 16424, Indonesia

E-mail: *aldiladipo@sci.ui.ac.id

Abstract. Malaria is a kind of a vector-borne disease. That means this disease needs a vector (in this case, the *anopheles* mosquito) to spread. In this article, a mathematical model for malaria disease spread will be discussed. The model is constructed as a seven-dimensional of a non-linear ordinary differential equation. The interventions of treatment for infected humans and use of repellent are included in the model to see how these interventions could be considered as alternative ways to control the spread of malaria. Analysis will be made of the disease-free equilibrium point along with its local stability criteria, construction of the next generation matrix which followed with the sensitivity analysis of basic reproduction number. We found that both medical treatment and repellent intervention succeeded in reducing the basic reproduction number as the endemic indicator of the model. Finally, some numerical simulations are given to give a better interpretation of the analytical results.

1. Introduction

Malaria is a disease caused by a type of *Plasmodium* parasite. This parasite enters the human body through the bite of the female anopheles mosquito and the parasite breeds asexually in the human body [1]. Generally, the incubation period of malaria or the time between a mosquito bite and the appearance of clinical symptoms depends on the number of parasites that enter the human body, but is in the range of about 7-30 days. The clinical symptoms are fever, chills, general malaise, sweats, headaches, body aches, nausea and vomiting. There are specific symptoms such as repetitive fever that appear in three stages: a cold stage (sensation of cold, shivering), a hot stage (fever, headaches, vomiting; seizures in young children), and finally a sweating stage (sweating, return to normal temperature, tiredness) [2]. Severe malaria occurs when infections are complicated by serious organ failures or abnormalities in the patient's blood or metabolism. Based on the epidemiological side, this disease is one of the three major infectious diseases in the world, along with *acquired immunodeficiency syndrome* (AIDS) and *tuberculosis* (TB) [3].The disease is life-threatening to the people who live in tropics and subtropics countries [4] such as Asia Pacific [5], Africa [6], and Latin America [7].

In 2003, malaria was spread over 30 provinces, 226 districts, and 6,053 villages in Indonesia [8]. The condition is exacerbated in developing areas which have resistance to drugs that have been used for malaria. The spread of malaria can be combatted by educating the public about the dangers of malaria, and by using mosquito nets or repellents while sleeping [9, 10, 11].However, the ministry of health imported anti-malaria medicine from China. The medicine is made from vegetable matter containing an artemisinin derivatives combination [12]. Malaria parasites have a complex life cycle. Therefore, the malaria vaccine is given in different ways at each stage.

For the asexual exoerythrocytic stage, the vaccine is the circumsporozoite protein (CSP), for the asexual erythrocytes stage the vaccine is the merozoite surface protein (MSP), and for the sexual stage, the vaccine is Pfs 28 and Pfs 25 [13].

Some researchers have developed a mathematical model of the spread of malaria i.e. an age-structured mathematical model of malaria transmission in which acquired immunity can act in three ways (immunity functions): reducing the probability of clinical disease, speeding the clearance of parasites, and increasing tolerance to sub-patent infections [14]. The other authors that also discuss about the spread of malaria model is the transmission and maintenance of malaria, paying attention both to the overall dynamics and to the population biology of the infection in human hosts and mosquito vectors [15]. Spreading drug resistance makes malaria control even more difficult. Therefore, JC Koella and R Antia developed epidemiological models for the spread of anti-malarial resistance based on the Macdonald-Ross model of malaria transmission. In our simplest model, resistance does not spread if the proportion of infected individuals treated is less than a threshold value; if drug treatment exceeds this threshold, resistance will eventually become fixed in the population. In more complex models, where hosts can be infected by multiple parasite strains or where treatment varies spatially, resistance is generally not fixed, but rather some level of sensitivity is often maintained in the population [16]. Some authors have explained about malaria transmission model for different levels of acquired immunity and temperature-dependent parameters (vector). A mathematical model was developed based on the following parameters: human host immunity, assuming the existence of acquired immunity and immunological memory, which boosts the protective response upon reinfection; mosquito vector, taking into account that the average period of development from egg to adult mosquito and the extrinsic incubation period of parasites (transformation of infected but non-infectious mosquitoes into infectious mosquitoes) are dependent on the ambient temperature [17]. Other authors also discuss how large-scale climate simulations and infectious disease systems may be modeled and analyzed and how these methods may be applied to predicting changes in the basic reproduction number of malaria across Tanzania. This research has concluded that disease emergence, extinction, and transmission all depend strongly on climate. Mathematical models offer powerful tools for understanding geographic shifts in incidence as climate changes [18].

Different from the models above, the superinfection process in malaria spread discuss about two types of symptom, namely symptomatic and asymptomatic. It also considers humans that use repellents. Medical treatment is only given to symptomatic infected humans. This article is constructed as follows. Sections 1 and 2 describe the introduction and the mathematical model construction, respectively. Mathematical model analysis concerning the equilibrium point and the basic reproduction number is given in section 3. Finally, a numerical simulation for various conditions, and some conclusions are given in sections 4 and 5.

2. Mathematical model

Let the human and mosquito population be divided into five and two compartments, respectively. Let us assume that the human population can be differentiated explicitly into susceptible (S_h), exposed (E_h), infected with symptomatic malaria (D_h), infected with asymptomatic malaria (A_h), and asymptomatic with undetectable parasite (U_h). In the other hand, the mosquitoe population can only divided into susceptible and infected mosquitoes, denoted by S_m and I_m, respectively. We assume that total of human population is given by

$$N_h(t) = S_h(t) + E_h(t) + D_h(t) + A_h(t) + U_h(t),$$

while total of mosquitoes population is given by

$$N_m(t) = S_m(t) + I_m(t).$$

The compartment of susceptible human (S_h) is increased by the recruitment rate from newborns (B_h) (please note that we assume that malaria is not transmit vertically). This compartment also increased due to the recovery rate of infected compartments D_h, A_h, U_h with rates of γ_D, γ_A and γ_U, respectively. To construct the recovery process, we assume that not all those infected with symptomatic and asymptomatic malaria could recovered from this disease. We assume only f proportion of infected with symptomatic malaria compartment will get medical treatment that will increase their natural recovery rate from γ_D into γ_T. Therefore, total of individual in $D_h(t)$ who recovered from malaria is $f\gamma_T D_h$ for they who got medical treatment and $(1-f)\gamma_D D_h$ for those who do not get medical treatment. If the recovery rate is succeeded, then they will enter the susceptible population with the rate of $\xi_T f \gamma_T D_h$ and $\xi_D(1-f)\gamma_D D_h$ for D_h who get and do not get medical treatment, respectively. The rest of it, $1-\xi_D$ and $1-\xi_T$, will transmit into infected with asymptomatic compartment. The susceptible human compartment will decrease caused by infection from infected mosquitoes at the rate of $\beta_h S_h I_m$ where β_h is the probability of successful transmission rate. Since we assume that portion p of total of human population are using repellent to prevent them from direct contact with infected mosquitoes, then the total of susceptible human who gets infected by infected mosquito is given by $\beta_h(1-p)S_h I_m$ and $r\beta_h p S_h I_m$ for human who do not and use repellent, respectively. Please note that r is the reduction of β_h as an effect of repellent use. Therefore, we have that

$$\frac{dS_h}{dt} = B_h(N_h) - \beta_h(1-p)S_h I_m - r\beta_h p S_h I_m + \xi_T f \gamma_T D_h + \xi_D(1-f)\gamma_D D_h + \gamma_U U_h - \mu_h S_h.$$

Exposed human compartment (E_h) is generated by infection rate from susceptible human compartment (S_h). The individual will remain in this compartment for mean duration of $\frac{1}{\delta_h}$ and will transfer into symptomatic and asymptomatic compartments with fraction of ϕ and $1-\phi$, respectively. This compartment also decreased caused by natural death rate of μ_h. Thus, we have

$$\frac{dE_h}{dt} = \beta_h(1-p)S_h I_m + r\beta_h p S_h I_m - \delta h E_h - \mu_h E_h.$$

There are two types of infected compartment, symptomatic (D_h) and asymptomatic (A_h). Both compartments are generated by transition from exposed compartments, i.e $\phi\delta_h E_h$ and $(1-\phi)\delta_h E_h$, respectively for D_h and A_h. Some proportion (f) of D_h will receive get medical treatment which increase the recovery rate from γ_D into γ_T, where $\gamma_T > \gamma_D$. On the other hand, D_h also generated from superinfection from asymptomatic compartment with a rate of $\beta_h A_h I_m$. Thus, we have :

$$\frac{dD_h}{dt} = \phi\delta_h E_h - (1-f)\gamma_D D_h - f\gamma_T D_h + \phi r \beta_h p A_h I_m + \phi\beta_h(1-p)A_h I_m - \mu_h D_h.$$

The asymptomatic compartments increased by transition from exposed compartment, failure of natural recovery rate with rate of $(1-\xi_D)(1-f)\gamma_D D_h$ and $(1-\xi_T)f\gamma_T D_h$, for symptomatic human who get and do not get medical treatment, respectively. This compartment also will be decreased by superinfection process, natural recovery rate into asymptomatic with undetectable parasites (γ_A) and also natural death rate. Therefore,

$$\begin{aligned}\frac{dA_h}{dt} &= (1-\phi)\delta_h E_h + (1-\xi_D)(1-f)\gamma_D D_h + (1-\xi_T)f\gamma_T D_h - \phi r \beta_h p A_h I_m \\ &- \phi\beta_h(1-p)A_h I_m - \gamma_A A_h - \mu_h A_h.\end{aligned}$$

The last, asymptomatic undetectable parasite (U_h) is generated by natural recovery rate from A_h at a rate of γ_A and decreased by natural recovery rate γ_U and natural death rate. Thus,

$$\frac{dU_h}{dt} = \gamma_A A_h - \gamma_U U_h \mu_h U_h.$$

Susceptible mosquito (S_m) is generated by a recruitment rate (B_m) and decreased by infection with symptomatic and asymptomatic human compartments (D_h, A_h, U_h) at the rate of β_m and natural death rate (μ_m). On the other hand, an infected compartment (I_m) is generated by infected individuals from susceptible compartment and decreased by natural death rate. Since life expectancy of mosquitoes is relatively short, about one month, recovered population is not included in to the model. Thus,

$$\begin{aligned}\frac{dS_m}{dt} &= B_m(N_m) - r\beta_m S_m p(A_h + D_h + U_h) - \beta_m S_m(1-p)(A_h + D_h + U_h) - \mu_m S_m \\ \frac{dI_m}{dt} &= r\beta_m S_m p(A_h + D_h + U_h) + \beta_m S_m(1-p)(A_h + D_h + U_h) - \mu_m I_m.\end{aligned}$$

Description of transmission process of the model that will be constructed can be seen in Figure 1, where yellow arrows for the transmission process and blue arrows for the transition process.

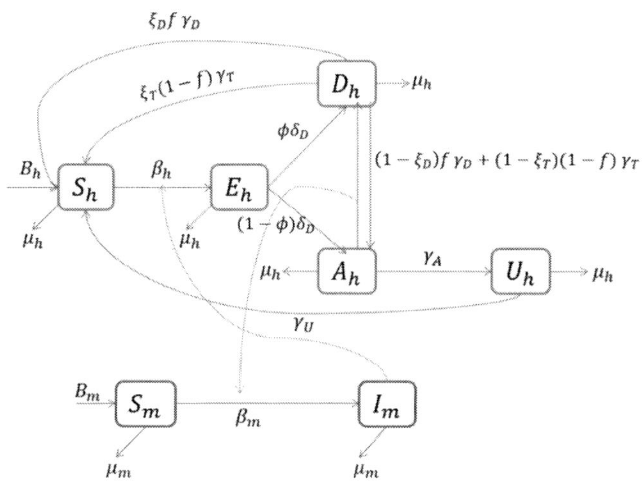

Figure 1. Transmission diagram for the spread of malaria in system 1Mathematical modelequation.2.1.

Thus, the model for the malaria transmission with superinfection process is given by the following deterministic system of non-linear differential equations :

$$\begin{aligned}\frac{dS_h}{dt} &= B_h(N_h) - \beta_h(1-p)S_h I_m - r\beta_h p S_h I_m + \xi_T f\gamma_T D_h + \xi_D(1-f)\gamma_D D_h + \gamma_U U_h - \mu_h S_h, \\ \frac{dE_h}{dt} &= \beta_h(1-p)S_h I_m + r\beta_h p S_h I_m - \delta h E_h - \mu_h E_h, \\ \frac{dD_h}{dt} &= \phi\delta_h E_h - (1-f)\gamma_D D_h - f\gamma_T D_h + \phi r\beta_h p A_h I_m + \phi\beta_h(1-p)A_h I_m - \mu_h D_h, \\ \frac{dA_h}{dt} &= (1-\phi)\delta_h E_h + (1-\xi_D)(1-f)\gamma_D D_h + (1-\xi_T)f\gamma_T D_h - \phi r\beta_h p A_h I_m \\ &- \phi\beta_h(1-p)A_h I_m - (\gamma_A + \mu_h)A_h, \\ \frac{dU_h}{dt} &= \gamma_A A_h - \gamma_U U_h - \mu_h U_h, \\ \frac{dS_m}{dt} &= B_m(N_m) - r\beta_m S_m p(A_h + D_h + U_h) - \beta_m S_m(1-p)(A_h + D_h + U_h) - \mu_m S_m, \\ \frac{dI_m}{dt} &= r\beta_m S_m p(A_h + D_h + U_h) + \beta_m S_m(1-p)(A_h + D_h + U_h) - \mu_m I_m,\end{aligned} \tag{1}$$

with $S_{h0}, E_{h0}, D_{h0}, A_{h0}, A_{h0}, U_{h0}, S_{m0}$, and I_{m0} are the initial condition for above system are given for $t = 0$. Please note that the change rate of total of human population respect to time is given by :

$$\begin{aligned} \frac{dN_h}{dt} &= \frac{d\left(S_h + E_h + D_h + A_h + U_h\right)}{dt} \\ &= B_h(N_h) - \mu_h\left(S_h + E_h + D_h + A_h + U_h\right) \\ &= B_h(N_h) - \mu_h N_h. \end{aligned}$$

If we choose that total of human population is constant for all time $t > 0$, then we have that recruitment rate of human population is given by $B_h(N_h) = \mu_h N_h$, which also constant for all time $t > 0$.

Similar with human population, in mosquitoes population we have that

$$\begin{aligned} \frac{dN_m}{dt} &= \frac{d(S_m + I_m)}{dt} \\ &= B_m(N_m) - \mu_m(S_m + I_m) \\ &= B_m(N_m) - \mu_m N_m. \end{aligned}$$

Again, if total of mosquitoes population is constant for all time $t > 0$, we have that $B_m(N_m) = \mu_m N_m$, which also mean that recruitment rate of mosquitoes population are constant for all time $t > 0$.

3. Model analysis

In this section, mathematical model analysis about the existence of equilibrium points and the local stability criteria will be discussed analytically and numerically. Basic reproduction number as the endemic criteria of the model in system 1Mathematical modelequation.2.1 will be constructed along with the parameter sensitivity.

3.1. Disease free equilibrium point and it local stability criteria

The disease-free equilibrium point is a condition when there is no disease present in a population or when all individuals are susceptible. Disease-free equilibrium point in the system 1Mathematical modelequation.2.1 is obtained when E_h=D_h=A_h=U_h=I_m=0. Setting the right-hand side of system 1Mathematical modelequation.2.1 equal to 0, we find that the disease-free equilibrium point is obtained as follows.

$$DFE = \left\{ A_h = 0, D_h = 0, E_h = 0, In_m = 0, S_h = \frac{B_h}{\mu_h}, S_m = \frac{B_m}{\mu_m}, U_h = 0 \right\}. \tag{2}$$

Using the Jacobian matrix approach, the local stability of DFE will be analyzed.

The Jacobian matrix of system 1Mathematical modelequation.2.1 when it is evaluated in DFE is given by :

$$J = \begin{bmatrix} -\mu_h & 0 & C4 & 0 & \gamma_U & 0 & -C1 \\ 0 & -\delta_h - \mu_h & 0 & 0 & 0 & 0 & C1 \\ 0 & \phi\,\delta_h & C5 & 0 & 0 & 0 & 0 \\ 0 & (1-\phi)\,\delta_h & C3 & -\gamma_A - \mu_h & 0 & 0 & 0 \\ 0 & 0 & 0 & \gamma_A & -\gamma_U - \mu_h & 0 & 0 \\ 0 & 0 & -C2 & -C2 & -C2 & -\mu_m & 0 \\ 0 & 0 & C2 & C2 & C2 & 0 & -\mu_m \end{bmatrix}, \tag{3}$$

IOP Conf. Series: Journal of Physics: Conf. Series **974** (2018) 012017 doi:10.1088/1742-6596/974/1/012017

where

$$
\begin{aligned}
C1 &= -\frac{\beta_h(1-p)B_h}{\mu_h} - \frac{r\beta_h p B_h}{\mu_h} \\
C2 &= \frac{r\beta_m B_m p}{\mu_m} + \frac{\beta_m B_m(1-p)}{\mu_m} \\
C3 &= (1-\xi_D)(1-f)\gamma_D + (1-\xi_T) f\gamma_T \\
C4 &= f\gamma_T\xi_T + \xi_D(1-f)\gamma_D \\
C5 &= -(1-f)\gamma_D - f\gamma_T - \mu_h.
\end{aligned}
$$

The eigenvalues of above matrix is taken from the root of characteristic polynomial

$$
-(\lambda+\mu_m)(\lambda+\mu_h)\left(\sum_{i=0}^{4} a_i\lambda^i\right) = 0, \tag{4}
$$

where

$$
\begin{aligned}
a_0 &= (x_1x_2\beta_h + x_3)\,\delta_h + x_4 \\
a_1 &= \left(-{\mu_h}^4\mu_m + x_5 + x_6 + x_7 - x_8\right)\delta_h + \mu_m(x_9 + x_{10} + x_{11})\,\mu_h \\
a_2 &= x_{12} + 2\,(x_{13} f - \gamma_A\gamma_U + x_{14})\,\mu_m{\mu_h}^2 + \mu_m(x_{15} + x_{16})\,\mu_h + x_{17} \\
a_3 &= 3\,(x_{18} + x_{19} + x_{20})\,\mu_m\mu_h \\
a_4 &= ((f-1)\gamma_D - f\gamma_T - 4\,\mu_h - \mu_m - \delta_h - \gamma_A - \gamma_U)\,\mu_m\mu_h.
\end{aligned}
$$

and

$$
\begin{aligned}
x_1 &= -B_m\beta_m\,(1+(r-1)\,p)^2\,B_h \\
x_2 &= -(\phi\,\xi_D - 1)(\gamma_A + \gamma_U + \mu_h)(f-1)\,\gamma_D + \gamma_T\,(\gamma_A + \gamma_U + \mu_h)(\phi\,\xi_T - 1)\,f - {\mu_h}^2 + (-\gamma_A - \gamma_U)\,\mu_h - \gamma_A\gamma_U\phi \\
x_3 &= (\gamma_A + \mu_h)((f-1)\,\gamma_D - f\gamma_T - \mu_h)\,{\mu_m}^2\mu_h\,(\gamma_U + \mu_h) \\
x_4 &= (\gamma_A + \mu_h)((f-1)\,\gamma_D - f\gamma_T - \mu_h)\,{\mu_m}^2{\mu_h}^2\,(\gamma_U + \mu_h) \\
x_5 &= (-3\,\mu_m + (f-1)\,\gamma_D - f\gamma_T - \gamma_A - \gamma_U)\,\mu_m{\mu_h}^3 \\
x_6 &= (((2\,f-2)\,\gamma_D - 2\,f\gamma_T - 2\,\gamma_A - 2\,\gamma_U)\,\mu_m + (\gamma_A + \gamma_U)(f-1)\,\gamma_D - \gamma_T\,(\gamma_A + \gamma_U)\,f - \gamma_A\gamma_U)\,\mu_m{\mu_h}^2 \\
x_7 &= \left(((\gamma_A + \gamma_U)(f-1)\,\gamma_D - \gamma_T\,(\gamma_A + \gamma_U)\,f - \gamma_A\gamma_U)\,{\mu_m}^2 + \gamma_A\,((f-1)\,\gamma_D - f\gamma_T)\,\gamma_U\mu_m + 2\,B_m\beta_h\beta_m\,(1+(r-1)\,p)^2\,B_h\right)\mu_h \\
x_8 &= B_m\beta_h\beta_m\,(1+(r-1)\,p)^2\,(-(\phi\,\xi_D - 1)(f-1)\,\gamma_D + \gamma_T\,(\phi\,\xi_T - 1)\,f - \gamma_A - \gamma_U)\,B_h \\
x_9 &= -{\mu_h}^4 + (-4\,\mu_m + (f-1)\,\gamma_D - f\gamma_T - \gamma_A - \gamma_U)\,{\mu_h}^3 \\
x_{10} &= (((3\,f-3)\,\gamma_D - 3\,f\gamma_T - 3\,\gamma_A - 3\,\gamma_U)\,\mu_m + (\gamma_A + \gamma_U)(f-1)\,\gamma_D - \gamma_T\,(\gamma_A + \gamma_U)\,f - \gamma_A\gamma_U)\,{\mu_h}^2 \\
x_{11} &= ((2\,(\gamma_A + \gamma_U)(f-1)\,\gamma_D - 2\,\gamma_T\,(\gamma_A + \gamma_U)\,f - 2\,\gamma_A\gamma_U)\,\mu_m + \gamma_A\,((f-1)\,\gamma_D - f\gamma_T)\,\gamma_U)\,\mu_h + \gamma_A\,((f-1)\,\gamma_D - f\gamma_T)\,\gamma_U\mu_m \\
x_{12} &= -4\,{\mu_h}^4\mu_m + 3\,\mu_m\,(-2\,\mu_m - \delta_h + (f-1)\,\gamma_D - f\gamma_T - \gamma_A - \gamma_U)\,{\mu_h}^3 \\
x_{13} &= (-3/2\,\delta_h + (3/2\,f - 3/2)\,\gamma_D - 3/2\,f\gamma_T - 3/2\,\gamma_A - 3/2\,\gamma_U)\,\mu_m \\
x_{14} &= ((f-1)\,\gamma_D - f\gamma_T - \gamma_A - \gamma_U)\,\delta_h + (\gamma_A + \gamma_U)(f-1)\,\gamma_D - \gamma_T\,(\gamma_A + \gamma_U) \\
x_{15} &= (((f-1)\,\gamma_D - f\gamma_T - \gamma_A - \gamma_U)\,\delta_h + (\gamma_A + \gamma_U)(f-1)\,\gamma_D - \gamma_T\,(\gamma_A + \gamma_U)\,f - \gamma_A\gamma_U)\,\mu_m \\
x_{16} &= ((\gamma_A + \gamma_U)(f-1)\,\gamma_D - \gamma_T\,(\gamma_A + \gamma_U)\,f - \gamma_A\gamma_U)\,\delta_h + \gamma_A\gamma_U\,((f-1)\,\gamma_D - f\gamma_T) \\
x_{17} &= B_h\beta_m\beta_h\,(1+(r-1)\,p)^2\,B_m\delta_h \\
x_{18} &= (f-1)\,(\mu_h + 1/3\,\mu_m + 1/3\,\delta_h + 1/3\,\gamma_A + 1/3\,\gamma_U)\,\gamma_D - \gamma_T\,(\mu_h + 1/3\,\mu_m + 1/3\,\delta_h + 1/3\,\gamma_A + 1/3\,\gamma_U)\,f \\
x_{19} &= -2\,{\mu_h}^2 + (-4/3\,\mu_m - \delta_h - \gamma_A - \gamma_U)\,\mu_h + (-1/3\,\delta_h - 1/3\,\gamma_A - 1/3\,\gamma_U)\,\mu_m \\
x_{20} &= (-1/3\,\gamma_A - 1/3\,\gamma_U)\,\delta_h - 1/3\,\gamma_A\gamma_U.
\end{aligned}
$$

Using the Routh-Hurwitz criterion [19] for the 4 degree polynomial, all the roots of above polynomial will be negative if and only if $a_i > 0$ for $i = 1, 2, 3, 4$ and also $a_1a_2a_3 > a_3^2 + a_1^2a_4$.

3.2. Basic reproduction number ($\mathcal{R}_0$)

Many analytical behaviors for the epidemic models, like the local and global stability of disease-free equilibrium point and endemic equilibrium point, are determined by the basic reproduction number, denoted by $\mathcal{R}_0$ [23, 24, 25, 26, 27]. The basic reproduction number is defined as the expected number of secondary cases caused by one primary case in a virgin population during one infection period [20, 21, 22]. The term of "virgin" population here is defined as a population in which, in initial condition, there is no single infection at all in a whole population. If $\mathcal{R}_0 = 2$, then one primary infected individual will produce two new infected individuals. On the other hand, if $\mathcal{R}_0 = 0.5$, it needs two infected individuals in the initial condition to produce one secondarily infected individual in the next generation. If $\mathcal{R}_0 = 1$, then a number of infected individuals will always be the same from generation to generation. Therefore, the endemicity of the disease in the case of $\mathcal{R}_0 = 1$ is determined by the initial condition.

Using the next-generation matrix approach, the basic reproduction number of system 1Mathematical modelequation.2.1 is given by

$$\mathcal{R}_0 = \sqrt{da + db + dc} \tag{5}$$

with :

$$\begin{aligned}
a &= \frac{\phi\, B_h \delta_h \beta_h \,(1 + (r-1)\,p)}{\mu_h\,(\mu_h + (1-f)\,\gamma_D + f\gamma_T)\,(\delta_h + \mu_h)} \\
b &= -\frac{B_h \delta_h \beta_h\,((1-\phi)\,\mu_h + (\phi\,\xi_D - 1)\,(f-1)\,\gamma_D - f\gamma_T\,(\phi\,\xi_T - 1))\,(1 + (r-1)\,p)}{\mu_h\,(\gamma_A + \mu_h)\,(\delta_h + \mu_h)\,(-\mu_h + (f-1)\,\gamma_D - f\gamma_T)} \\
c &= -\frac{\delta_h \gamma_A \beta_h B_h\,((1-\phi)\,\mu_h + (\phi\,\xi_D - 1)\,(f-1)\,\gamma_D - f\gamma_T\,(\phi\,\xi_T - 1))\,(1 + (r-1)\,p)}{\mu_h\,(\gamma_A + \mu_h)\,(\gamma_U + \mu_h)\,(\delta_h + \mu_h)\,(-\mu_h + (f-1)\,\gamma_D - f\gamma_T)} \\
d &= \frac{1}{\mu_m}\left(\frac{r\beta_m B_m p}{\mu_m} + \frac{\beta_m B_m\,(1-p)}{\mu_m}\right).
\end{aligned}$$

The disease-free equilibrium will locally asymptotic stable if and only if $\mathcal{R}_0 < 1$, and unstable otherwise. To see the dependency of each parameter, respect to the magnitude of basic reproduction number, one alternative method is using the level set of basic reproduction number respect to some free parameters, with setting the other parameters to remain constant. Except it is stated, the parameter's value that is using in this section are given in Table 2.

First, analysis is given in Figure 2 to perform the dependency of the proportion of treatment (f) respect to recovery rate caused by the treatment intervention (γ_T). It can be seen that as it was expected, more massive the treatment given to infected human (larger value of f), the basic reproduction number will become smaller. Similar behavior also is shown in the recovery rate caused by treatment intervention. More effective the treatment implemented to infected human (larger γ_T, which means that the duration of infection become shorter) will reduce basic reproduction number. It can be seen also that γ_T is more sensitive rather than f. Therefore, it is more effective to accelerate the effect of treatment to reduce basic reproduction number, rather than implementation of treatment into infected human.

Next sensitivity analysis is given in Figure 3 to show how basic reproduction number will reduced depend on natural recruitment rate of human and mosquitoes. It can be seen that larger the magnitude of recruitment rate of human or mosquitoes, then the magnitude of basic reproduction number will increase.

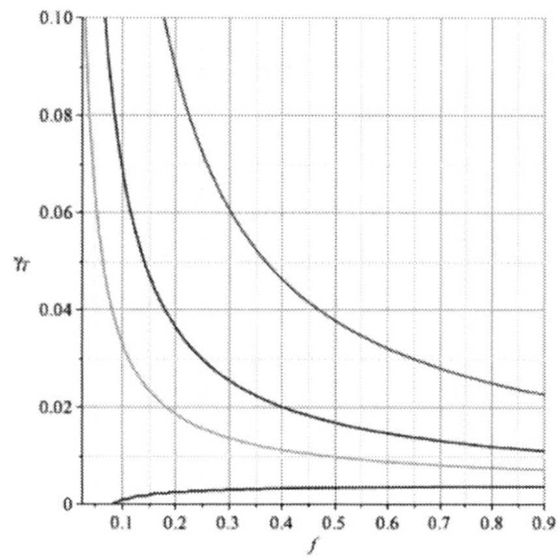

Figure 2. Sensitivity analysis of basic reproduction number ($\mathcal{R}_0$) respect to treatment proportion (f) and the recovery rate caused by treatment intervention (γ_T). Please note that black, green, blue and red curve is the curve for $\mathcal{R}_0$ equals to 0.89, 0.87, 0.85 and 0.8, respectively.

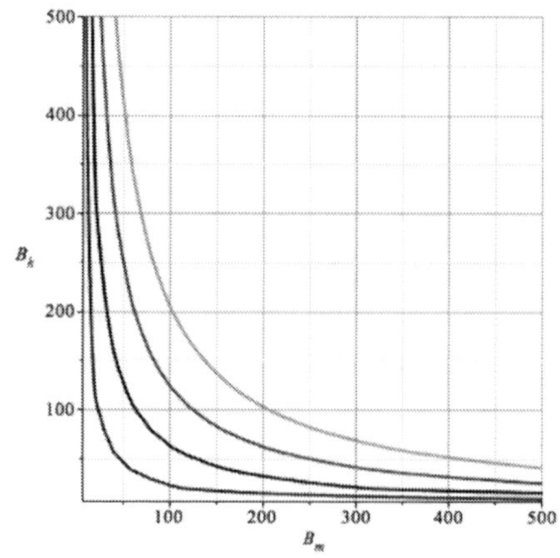

Figure 3. Sensitivity analysis of basic reproduction number ($\mathcal{R}_0$) respect to recruitment rate of human B_h and mosquitoes B_m. Please note that blue, black, red and green curve is the curve for $\mathcal{R}_0$ equals to 0.6, 1, 1.4 and 1.8, respectively.

4. Numerical simulations

In this section, some numerical simulation will be conducted to describe the effect of how the variation of magnitude of parameters might affect the change of the dynamics of system 1Mathematical modelequation.2.1. To perform the following simulations, the initial condition is given in Table 1 and value of each parameter (except it stated differently) is given in Table 1Initial Condition of system 1Mathematical modelequation.2.1table.1.

Table 1. Initial Condition of system 1Mathematical modelequation.2.1.

initial condition	S_h	E_h	D_h	A_h	U_h	S_m	I_m
value	990	2	3	5	0	990	10

IOP Conf. Series: Journal of Physics: Conf. Series **974** (2018) 012017 doi:10.1088/1742-6596/974/1/012017

Table 2. Parameters Value for simulation in section 3 and 4.

Parameters	f	p	ϕ	r	B_h	B_m	β_h	β_m	ξ_D
Value	0	0	0.9	0.4	$\frac{1000}{65\times365}$	$\frac{1000}{30}$	$\frac{0.1}{1000}$	$\frac{0.2}{1000}$	0.45

Parameters	δ_h	γ_A	γ_D	γ_T	γ_U	μ_h	μ_m	ξ_T
Value	$\frac{1}{15}$	$\frac{1}{180}$	$\frac{1}{180}$	$\frac{1}{21}$	$\frac{1}{180}$	$\frac{1}{65\times365}$	$\frac{1}{30}$	0.9

4.1. Numerical simulation for effect of p

In this section, variation of p which present the presentation of human who is using repellent is given to see how these parameters give an impact to the dynamic of system 1Mathematical modelequation.2.1. The value of p is given for $p = 0.1n$, with $n = 0, 1, \ldots, 10$, and the decreasing of $\mathcal{R}_0$ as an impact of these various value of p are given in Table 3. It can be seen that with $p > 0.5$, which present that more than 50% human are using repellent to avoid direct contact to infected mosquito could not reduce $\mathcal{R}_0$ which is smaller than 1. This is mean that the disease-free equilibrium point will locally stable, as shown in Figure 4.

From Figure 4Dynamic of infected individuals respect to the change of $p$figure.4, it can be seen that increasing number of human who is using repellent in fact not make the system tends to the disease-free equilibrium as $t \to \infty$. Also, it can be seen that increasing value of p will delay the rapid increase of infected individuals in both human and mosquito population, and also reduce the level of endemicity (number of infected individuals when $t \to \infty$).

Table 3. Change of $\mathcal{R}_0$ respect to different values of p.

p	0	0.1	0.2	0.3	0.4	0.5	0.6	0.7	0.8	0.9	1
$\mathcal{R}_0$	1.49	1.40	1.31	1.22	1.13	1.04	0.95	0.86	0.77	0.69	0.69

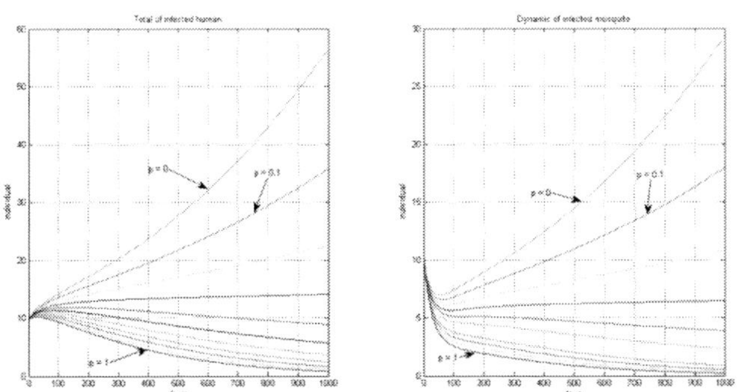

Figure 4. Dynamic of infected individuals respect to the change of p.

4.2. Numerical simulation for effect of f

In this section, variation of f which present the symptomatic infected human who gets medical treatment is given to see how these parameters give an impact to the dynamic of system 1Mathematical modelequation.2.1. Similar with the previous scenario, the value of f is given for $f = 0.1n$, with $n = 0, 1, \ldots, 10$, and the decreasing of $\mathcal{R}_0$ caused by these various values of f are given in Table 4. It can be seen that with more than 20% of infected human following medical treatment, can reduce $\mathcal{R}_0$ which is smaller than 1. This is mean that the disease-free equilibrium point will locally stable, as shown in Figure 5.

From Figure 5, it can be seen that increase in the number of symptomatic infected human who gets medical treatment, in fact, make the system tends to the disease-free equilibrium as $t \to \infty$. Also, it can be seen that the intervention of medical treatment is more sensitive to suppress the dynamic of infected human and mosquito when $f < 0.4$, since when $f > 0.4$, it is true that the dynamic is tend to the disease-free equilibrium point, but the difference is not significant anymore. This result confirms our result in the previous section (Figure 2) that f only significantly change the magnitude of $\mathcal{R}_0$ in the beginning of intervention value.

Table 4. Change of $\mathcal{R}_0$ respect to different values of f.

f	0	0.1	0.2	0.3	0.4	0.5	0.6	0.7	0.8	0.9	1
$\mathcal{R}_0$	1.49	1.18	1.03	0.94	0.88	0.83	0.80	0.77	0.75	0.73	0.72

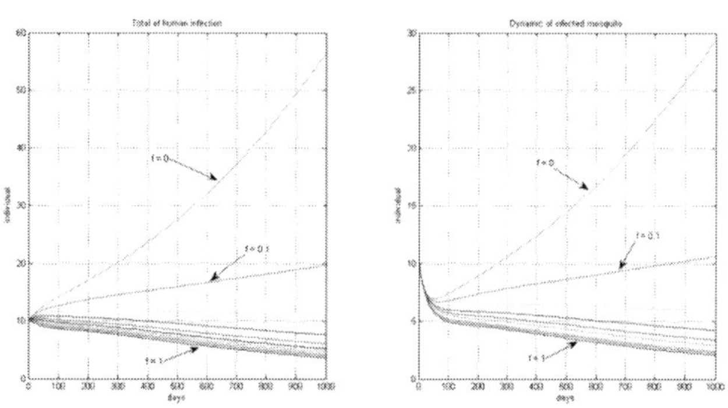

Figure 5. Dynamic of infected individuals respect to the change of f.

5. Conclusions

Malaria is a vector-borne disease which is caused by *Plasmodium* and spread by female *anopheles* mosquitoes. In this article, a mathematical model has been constructed to understand how Malaria might spreads. The intervention of repellent to avoid or to minimize contact between mosquitoes and humans included in the model along with medical treatment to cure people from Malaria.

From the analysis about the equilibrium point and the basic reproduction number, the latter denoted by $\mathcal{R}_0$, we find that repellent and medical treatment is succeeding to reducing the occurrence of disease. This is can be seen from the sensitivity analysis and numerical simulation of the dynamics of infected humans and mosquitoes, which tend to disease-free equilibrium when repellent and medical treatment is given in some specific values. We also find that an intervention using repellent and medical treatment is very sensitive to the reduction of the basic reproduction number in the early values of them, please see section 3 and tables 3 and 4. Therefore, if the government wants to control the spread of malaria, intervention in a not very massive policy could be an option if there is a limited budget.

This study could be continued by making the control parameters of repellent and medical treatment control variables which will depend on time. This approach is used to minimize the size of the infected population while using the minimum budget. One option to handle this problem is to reconstruct it as a control optimum problem. Please see [28, 29] for some references to the application of optimal control in some different forms of disease spread.

Acknowledgments

We thank all the reviewers for their supportive comments and suggestions. This research is funded by Universitas Indonesia with PITTA 2017 Research Grant Scheme (Project ID: 643/UN2.R3.1/HKP.05.00/2017)

References

[1] *WHO fact sheet about malaria* [accessed September,3 2017, available from: http://www.who.int/]
[2] *Malaria disease* [accessed September,3 2017, available from http://www.cdc.gov/malaria/about/disease.html/]
[3] *Malaria summary* [accessed September,3 2017, available from: vault.gakken-idn.id/content/files/malaria-summary-full-text.pdf]
[4] U F Achmadi 2005. Manajemen Penyakit Berbasis Wilayah **3** p 228-248
[5] C E G Mercado, N Ekapirat, A M Dondorp and R J Maude 2017 An assessment of national surveillance systems for malaria elimination in the Asia Pacific. *Malaria Journal* **16** p 127
[6] S Litsios 2015 Re-imagining the control of malaria in tropical Africa during the early years of the World Health Organization. *Malaria Journal* **14** p 178
[7] L R Cruz, T Spangeberg, M V G Lacerda, T N C Well 2013 Malaria in South America: a drug discovery perspective. *Malaria Journal* **12** p 168
[8] *Malaria:Pemeriksaan parasit malaria secara mikroskopik.* (Depkes RI 2004, Direktorat Jendral Pemberamtasan Parasit Menular dan Penyehatan Lingkungan Permukiman)
[9] A O Spjeldnæs, A Y Kitua and B Blomberg 2014 Education and knowledge helps combating malaria, but not degedege: a cross-sectional study in Rufiji, Tanzania. *Malaria Journal* **12** p 168
[10] N H Ntonifo and S Veyufambom 2016 Assessing the effective use of mosquito nets in the prevention of malaria in some parts of Mezam division, Northwest Region Cameroon. *Malaria Journal* **15** p 390
[11] Insect repellents help prevent malaria and other diseases spread by mosquitoes [accessed September,4 2017] available from: http://www.cdc.gov/
[12] *Penggunaan artemisin untuk atasi malaria di daerah yang Resisten klorokuin.*(DEPKES RI. Smart T. HIV/malaria.2004) [accessed September,4 2017 available from :www.depkes.go.id]
[13] V C Saputra 2014 Laporan praktek kerja profesi apoteker di apotek kimia farma, slipi jakarta barat periode 10 maret 2014 - 19 april 2014
[14] J A N Filipe, E M Riley, C J Darkeley, C J Sutherland, and A C Ghani 2007 Determination of the processes driving the acquisition of immunity to malaria using a mathematical transmission model. *PLoS Comp Biol*, **3** pp 2569-79
[15] J L Aron and R M May 1982 The population dynamics of malaria. *In Population Dynamics of Infectious Disease.* Edited by: Anderson RM. London: Chapman and Hall **50** pp 139-79
[16] J C Koella, R Antia 2003 Epidemiological models for the spread of anti-malarial resistance. *Malar J* **2** p 3
[17] H M Yang 2000 Malaria transmission model for different levels of acquired immunity and temperature-dependent parameters (vector). *Revista de Saude Publica,* **34** pp 223-31
[18] P E Parham and E Michael 2010 Modeling the effects of weather and climate change on malaria transmission. *Environ Health Perspect* **118** pp 620-26 0901256
[19] J L Willems 1970 *Stability theory of dynamical system*
[20] O Diekmann and J A P Heesterbeek and M G Robert 2010 The construction of next-generation matrices for compartmental epidemic models. *J. R. Soc. Interface,* **7** pp 873-85
[21] O Diekmann and J A P Heesterbeek 2000 *Mathematical epidemiology of infectious disease : model building, analysis and interpretation.* **30** pp 303
[22] E Soewono and D Aldila 2015 A survey of basic reproductive ratios in vector-borne disease transmission modeling. *AIP Conference Proceedings* **1651** pp 12
[23] H M Yang 2014 The basic reproduction number obtained from jacobian and next generation matrices – A case study of dengue transmission modelling. *BioSystems* **126** pp 52-75

[24] K Dietz 1993 The estimation of the basic reproduction number for infectious diseases. *Sage Journal* **2** pp 23-41 096228029300200103

[25] M B Barongo, K Ståhl, B Bett, R P Bishop, E M Fèvre, T Aliro, E Okoth, C Masembe, D Knobel and A Ssematimba 2015 Estimating the Basic Reproductive Number (R0) for African Swine Fever Virus (ASFV) Transmission between Pig Herds in Uganda *(Plos One)* **10** 0125842

[26] H Inaba and H Nishiura 2008 The Impact of Population Growth Rate on the Eradication Threshold. *The Basic Reproduction Number of an Infectious Disease in a Stable Population* **3** pp 194-228

[27] P V D Driessche and J Watmough 2002 Reproduction numbers and sub-threshold endemic equilibria for compartmental models of disease transmission. *Mathematical Biosciences* **180** pp 29–48

[28] D Aldila, T Gotz, E Soewono 2013 An optimal control problem arising from a dengue disease transmission model, *Mathematical Biosciences* **242** pp 9–16

[29] D Aldila, N Rarasati, N Nuraini, and E Soewono 2014 Optimal Control Problem of Treatment for Obesity in a Closed Population. *International Journal of Mathematics and Mathematical Sciences* **2014** 273037

IOP Conf. Series: Journal of Physics: Conf. Series **974** (2018) 012018 doi:10.1088/1742-6596/974/1/012018

Implementation of Kalman filter algorithm on models reduced using singular pertubation approximation method and its application to measurement of water level

Vimala Rachmawati [1], Didik Khusnul Arif [2] and Dieky Adzkiya [3]

Department of Mathematics, Faculty of Mathematics, Computing, and Data Sciences, Institut Teknologi Sepuluh Nopember, Kampus ITS Sukolilo-Surabaya 60111, Indonesia

Email : [1] vimalarachmawati@gmail.com, [2] didik@matematika.its.ac.id , [3] dieky@matematika.its.ac.id

Abstract. The systems contained in the universe often have a large order. Thus, the mathematical model has many state variables that affect the computation time. In addition, generally not all variables are known, so estimations are needed to measure the magnitude of the system that cannot be measured directly. In this paper, we discuss the model reduction and estimation of state variables in the river system to measure the water level. The model reduction of a system is an approximation method of a system with a lower order without significant errors but has a dynamic behaviour that is similar to the original system. The Singular Perturbation Approximation method is one of the model reduction methods where all state variables of the equilibrium system are partitioned into fast and slow modes. Then, The Kalman filter algorithm is used to estimate state variables of stochastic dynamic systems where estimations are computed by predicting state variables based on system dynamics and measurement data. Kalman filters are used to estimate state variables in the original system and reduced system. Then, we compare the estimation results of the state and computational time between the original and reduced system.

1. Introduction

River is one source of water that holds and drain the flow of water. River can be viewed as a system which is a combination of several components that work together to achieve a certain objective. In this case, the river level system is used to keep the flow of river water flow in the desired condition. The system from this river level can be represented in the form of mathematical models.

The systems contained in the universe often have a large order. Thus, the mathematical model has many state variables that affect the computation time. Therefore, we need a simplification of large order systems into smaller orders without significant errors. Simplification of this system is called model reduction [1]. Currently, there are many developed methods of model reduction such as balanced truncation methods [2,3,7,8,9,11] and singular perturbation approximation [10]. In [10], we have developed the algorithm of singular perturbation approximation. In this paper, we apply the Singular Perturbation Approximation (SPA) method to shallow water equations.

In large systems, generally not all variables are known, so a useful estimate is needed to measure the magnitude of the system that cannot be measured directly. Kalman filter is a recursive algorithm for estimating state variables of a stochastic dynamic system. Estimation using this method is done by predicting state variables based on system dynamics and measurement data [4].

In [1], Kalman filter algorithm has been developed in the reduced model and applied to the heat conduction distribution problem. The simulation results show that the Kalman filter estimation on the reduced system is more accurate and faster compared to Kalman filter on the original system. In this paper, we estimate the model of shallow water equations that has been reduced using Singular Perturbation Approximation (SPA) method by implementing Kalman Filter algorithm to measure water level of the river.

2. Shallow Water Equations

In this section, we discuss the shallow water equation that describe the flow of water in the river [5] :

$$\begin{aligned} \frac{\partial h}{\partial t} + D\frac{\partial u}{\partial x} &= 0 \\ \frac{\partial u}{\partial t} + g\frac{\partial h}{\partial x} + C_f u &= 0 \end{aligned} \qquad (1)$$

with initial conditions :

$$h(x,0) = 2 + \sin(2\pi x),\; u(x,0) = 0 \qquad (2)$$

and boundary conditions :

$$\begin{aligned} h(0,t) &= h(x-1,t),\; h(L,t) = h(2,t) \\ u(0,t) &= u(x-1,t),\; u(L,t) = u(2,t) \end{aligned} \qquad (3)$$

where $h(x,t)$ is the water level above the reference plane at position x and time t, $u(x,t)$ is the average current velocity at position x and time t, t is the time variable, x is the position along the river, D is the water depth above the reference point, g is the gravitational acceleration and C_f is a friction constant.

3. Discretization

In this section, we discretization the Shallow Water Equations with Lax Wendroff Scheme so that we can obtain a discrete-time system that is suitable for model reduction. First, from equation (1) is discretized with FTCS (*Forward Time Centered Space*) Scheme as follows

$$\begin{aligned} h_i^{k+1} &= h_i^k - \tfrac{D\Delta t}{2\Delta x}\left(u_{i+1}^k - u_{i-1}^k\right) \\ u_i^{k+1} &= \left(1 - C_f\Delta t\right)u_i^k - \tfrac{g\Delta t}{2\Delta x}\left(h_{i+1}^k - h_{i-1}^k\right) \end{aligned} \qquad (4)$$

The Lax Wendroff Scheme is a combination of Lax Friedrichs Scheme and Leapfrog Scheme. The Lax Friedrichs Scheme works as follows replace h_i^k and u_i^k in equation (4) respectively with $\frac{1}{2}(h_{i+1}^k + h_{i-1}^k)$ and $\frac{1}{2}(u_{i+1}^k + u_{i-1}^k)$. So we obtain :

$$\begin{aligned} h_i^{k+1} &= \tfrac{1}{2}\left(h_{i+1}^k + h_{i-1}^k\right) - \tfrac{D\Delta t}{2\Delta x}\left(u_{i+1}^k - u_{i-1}^k\right) \\ u_i^{k+1} &= \left(\tfrac{1-C_f\Delta t}{2}\right)\left(u_{i+1}^k + u_{i-1}^k\right) - \tfrac{g\Delta t}{2\Delta x}\left(h_{i+1}^k - h_{i-1}^k\right) \end{aligned} \qquad (5)$$

IOP Conf. Series: Journal of Physics: Conf. Series **974** (2018) 012018 doi:10.1088/1742-6596/974/1/012018

The Leapfrog Scheme works as follows replace Δt with $2\Delta t$ in equation (4) with the aim that $g\Delta t$ or $D\Delta t$ has a smaller value than Δx to achieve the desired accuracy. So we obtain :

$$\begin{aligned} h_i^{k+1} &= h_i^k - \tfrac{D\Delta t}{\Delta x}\left(u_{i+1}^k - u_{i-1}^k\right) \\ u_i^{k+1} &= \left(1 - 2C_f\Delta t\right)u_i^k - \tfrac{g\Delta t}{\Delta x}\left(h_{i+1}^k - h_{i-1}^k\right) \end{aligned} \qquad (6)$$

Substitution u_i^{k+1} in equatiom (5) to h_i^{k+1} in equation (6) and h_i^{k+1} in equation (5) to u_i^{k+1} in equation (6). Then we obtain :

$$\begin{aligned} h_i^{k+1} &= h_i^k - a\left(u_{i+1}^k - u_{i-1}^k\right) + c\left(h_{i+1}^k - 2h_i^k + h_{i-1}^k\right) \\ u_i^{k+1} &= du_i^k - b\left(h_{i+1}^k - h_{i-1}^k\right) + c\left(u_{i+1}^k - 2u_i^k + u_{i-1}^k\right) \end{aligned} \qquad (7)$$

with

$$a = \tfrac{D\Delta t}{\Delta x}\left(1 - C_f\Delta t\right), b = \tfrac{g\Delta t}{\Delta x}, c = \tfrac{Dg\Delta t^2}{2\Delta x^2}, d = \left(1 - 2C_f\Delta t\right)$$

Generate terms (7) for $i = 0,1,2,..,N-1$. So if equations (7) is written in the form of state space system, we obtain

$$x(k+1) = Ax(k) + Bu(k) \qquad (8)$$

where

$$x(k+1) = \begin{bmatrix} h_1 \\ u_1 \\ h_2 \\ u_2 \\ h_3 \\ u_3 \\ \vdots \\ h_{N-1} \\ u_{N-1} \end{bmatrix}^{k+1}, x(k) = \begin{bmatrix} h_1 \\ u_1 \\ h_2 \\ u_2 \\ h_3 \\ u_3 \\ \vdots \\ h_{N-1} \\ u_{N-1} \end{bmatrix}^{k}, u(k) = \begin{bmatrix} h_0 \\ u_0 \\ h_N \\ u_N \end{bmatrix}^{k}$$

Then, we construct a measurement equation at time k that is

$$y(k) = Cx(k) + Du(k) \qquad (9)$$

Furthermore, a simulation is done to determine the water level of the river by using the parameter values as follows

$$D = 10\ m, g = 9{,}8\frac{m}{s^2}, C_f = 0{,}0002, x = 1, t = 0{,}2$$

So we obtained the matrix for the original system (A, B, C, D) with the order $n = 20$ as follows

$$A = \begin{bmatrix} 0{,}02 & 0 & 0{,}49 & -0{,}5 & 0 & 0 & \cdots & \cdots & \cdots & 0 & 0 & 0 \\ 0 & 0{,}02 & -0{,}49 & 0{,}49 & 0 & 0 & \ddots & \ddots & \ddots & 0 & 0 & 0 \\ 0{,}49 & 0{,}5 & 0{,}02 & 0 & 0{,}49 & -0{,}5 & \ddots & \ddots & \ddots & 0 & 0 & 0 \\ 0{,}49 & 0{,}49 & 0 & 0{,}02 & -0{,}49 & 0{,}49 & \ddots & \ddots & \ddots & 0 & 0 & 0 \\ 0 & 0 & 0{,}49 & 0{,}5 & 0{,}02 & 0 & \ddots & \ddots & \ddots & 0 & 0 & 0 \\ 0 & 0 & 0{,}49 & 0{,}49 & 0 & 0{,}02 & \ddots & \ddots & \ddots & 0 & 0 & 0 \\ \vdots & \ddots & \ddots & \ddots & \ddots & \ddots & \ddots & \ddots & \ddots & \ddots & \ddots & \vdots \\ \vdots & \ddots & \ddots & \ddots & \ddots & \ddots & \ddots & \ddots & \ddots & \ddots & \ddots & \vdots \\ \vdots & \ddots & \ddots & \ddots & \ddots & \ddots & \ddots & \ddots & \ddots & \ddots & \ddots & \vdots \\ 0 & 0 & 0 & 0 & 0 & 0 & \ddots & \ddots & \ddots & 0{,}02 & -0{,}49 & 0{,}49 \\ 0 & 0 & 0 & 0 & 0 & 0 & \ddots & \ddots & \ddots & 0{,}5 & 0{,}02 & -0{,}49 \\ 0 & 0 & 0 & 0 & 0 & 0 & \cdots & \cdots & \cdots & 0{,}49 & 0 & 0{,}02 \end{bmatrix}_{20\times 20}$$

$$B = \begin{bmatrix} 0{,}49 & 0{,}49 & 0 & 0 & \cdots & \cdots & \cdots & \cdots & 0 & 0 & 0 & 0 \\ 0{,}5 & 0{,}49 & 0 & 0 & \cdots & \cdots & \cdots & \cdots & 0 & 0 & 0 & 0 \\ 0 & 0 & 0 & 0 & \cdots & \cdots & \cdots & \cdots & 0 & 0 & 0{,}49 & -0{,}49 \\ 0 & 0 & 0 & 0 & \cdots & \cdots & \cdots & \cdots & 0 & 0 & -0{,}5 & 0{,}49 \end{bmatrix}^{T}_{4\times 20}$$

$$C = \begin{bmatrix} 1 & 0 & 0 & 0 & 0 & 0 & 0 & 0 & 0 & 0 & 0 & 0 & 0 & 0 & 0 & 0 & 0 & 0 & 0 & 0 \\ 0 & 0 & 1 & 0 & 0 & 0 & 0 & 0 & 0 & 0 & 0 & 0 & 0 & 0 & 0 & 0 & 0 & 0 & 0 & 0 \\ 0 & 0 & 0 & 0 & 1 & 0 & 0 & 0 & 0 & 0 & 0 & 0 & 0 & 0 & 0 & 0 & 0 & 0 & 0 & 0 \\ 0 & 0 & 0 & 0 & 0 & 0 & 1 & 0 & 0 & 0 & 0 & 0 & 0 & 0 & 0 & 0 & 0 & 0 & 0 & 0 \\ 0 & 0 & 0 & 0 & 0 & 0 & 0 & 0 & 1 & 0 & 0 & 0 & 0 & 0 & 0 & 0 & 0 & 0 & 0 & 0 \\ 0 & 0 & 0 & 0 & 0 & 0 & 0 & 0 & 0 & 0 & 1 & 0 & 0 & 0 & 0 & 0 & 0 & 0 & 0 & 0 \\ 0 & 0 & 0 & 0 & 0 & 0 & 0 & 0 & 0 & 0 & 0 & 0 & 1 & 0 & 0 & 0 & 0 & 0 & 0 & 0 \\ 0 & 0 & 0 & 0 & 0 & 0 & 0 & 0 & 0 & 0 & 0 & 0 & 0 & 0 & 1 & 0 & 0 & 0 & 0 & 0 \\ 0 & 0 & 0 & 0 & 0 & 0 & 0 & 0 & 0 & 0 & 0 & 0 & 0 & 0 & 0 & 0 & 1 & 0 & 0 & 0 \\ 0 & 0 & 0 & 0 & 0 & 0 & 0 & 0 & 0 & 0 & 0 & 0 & 0 & 0 & 0 & 0 & 0 & 0 & 1 & 0 \end{bmatrix}_{10\times 20}$$

$$D = [0]$$

The requirement for model reduction is that the system must be asymptotically stable, controllable and observable. So it is necessary to analyze the original system (A, B, C, D) above whether it is asymptotically stable, controllable and observable.

4. Model Reduction

Model Reduction is an attempt to replace large models or systems with simpler models without significant errors. Model reduction can be done with several methods. One of them is the method of Singular Perturbation Approximation. Model Reduction is done by establishing a balanced system first.

4.1. Balanced Systems

The balanced system $(\tilde{A}, \tilde{B}, \tilde{C}, \tilde{D})$ is a new system derived from the original system (A, B, C, D) with controllability gramian $\widetilde{W}$ and observability gramian $\widetilde{M}$ which is the unique solution of the Lyapunov equations. Thus it satisfies $\widetilde{W} = \widetilde{M} = \Sigma = \mathrm{diag}(\sigma_1, \sigma_2, \ldots, \sigma_n), \sigma_1 \geq \cdots \geq \sigma_r \geq \cdots \geq \sigma_n > 0$. With σ_i is the Hankel singular value of the system (A, B, C, D) which can be defined as $\sigma_i = \left|\sqrt{\lambda_i(WM)}\right|$, $i = 1, \ldots, n$, where λ_i is the eigen values of WM. The balanced system is obtained by transforming the original system of equations (8) and (9) to the matrix of transformation T which satisfies [6] :

$$x_k = T\tilde{x}_k \tag{10}$$

Thus we obtained the matrix for the balanced system $(\tilde{A},\tilde{B},\tilde{C},\tilde{D})$ by order $n=20$ with $\tilde{A}=T^{-1}AT, \tilde{B}=T^{-1}B, \tilde{C}=C, \tilde{D}=D$ that is asymptotically stable, controllable and observable as follows

$$\tilde{A}=\begin{bmatrix} 0.7942 & 2.0188e-15 & \cdots & \cdots & \cdots & 5.7600e-15 & 0.1254 \\ -1.5623e-14 & 0.2265 & \ddots & \ddots & \ddots & -0.0482 & 3.0281e-15 \\ \vdots & \ddots & \ddots & \ddots & \ddots & \ddots & \vdots \\ \vdots & \ddots & \ddots & \ddots & \ddots & \ddots & \vdots \\ \vdots & \ddots & \ddots & \ddots & \ddots & \ddots & \vdots \\ -9.1001e-16 & 0.0226 & \ddots & \ddots & \ddots & -0.1770 & -5.0922e-15 \\ -0.0357 & -1.1540e-15 & \cdots & \cdots & \cdots & -6.7856e-15 & -0.2743 \end{bmatrix}_{20\times 20}$$

$$\tilde{B}=\begin{bmatrix} 0.4565 & -0.6739 & -0.4899 & \cdots & \cdots & \cdots & \cdots & \cdots & \cdots & \cdots & 0.0465 & 0.0508 \\ 0.4616 & -0.6815 & -0.4955 & \ddots & \ddots & \ddots & \ddots & \ddots & \ddots & \ddots & 0.0428 & 0.0473 \\ 0.4565 & 0.6739 & -0.4899 & \ddots & \ddots & \ddots & \ddots & \ddots & \ddots & \ddots & -0.0465 & 0.0508 \\ -0.4616 & -0.6815 & 0.4955 & \cdots & \cdots & \cdots & \cdots & \cdots & \cdots & \cdots & 0.0428 & -0.0473 \end{bmatrix}^{T}_{4\times 20}$$

$$\tilde{C}=\begin{bmatrix} 0.2084 & -0.3456 & \cdots & \cdots & \cdots & -0.4641 & -0.4376 \\ 0.2040 & -0.0857 & \ddots & \ddots & \ddots & 0.0842 & 0.0738 \\ \vdots & \ddots & \ddots & \ddots & \ddots & \ddots & \vdots \\ \vdots & \ddots & \ddots & \ddots & \ddots & \ddots & \vdots \\ \vdots & \ddots & \ddots & \ddots & \ddots & \ddots & \vdots \\ 0.2040 & 0.0857 & \ddots & \ddots & \ddots & -0.0842 & 0.0738 \\ 0.2084 & 0.3456 & \cdots & \cdots & \cdots & 0.4641 & -0.4376 \end{bmatrix}_{10\times 20}$$

$$\tilde{D}=[0]$$

4.2. Singular Pertubation Approximation Methods

Having obtained the balanced system $(\tilde{A},\tilde{B},\tilde{C},\tilde{D})$ with the same controllability gramian $\tilde{W}$ and observability gramian $\tilde{M}$ that equals the diagonal matrix Σ. Then the system $(\tilde{A},\tilde{B},\tilde{C},\tilde{D})$ is partitioned according to $\Sigma = diag(\Sigma_1, \Sigma_2)$, where $\Sigma_1 = diag(\sigma_1,\sigma_2,\cdots.\sigma_n)$ and $\Sigma_2 = diag(\sigma_{r+1},,\sigma_{r+2,}\cdots.\sigma_n)$. Thus, the system realization $(\tilde{A},\tilde{B},\tilde{C},\tilde{D})$ can be written as follows :

$$\begin{bmatrix}\tilde{x}_1(k+1)\\ \tilde{x}_2(k+1)\end{bmatrix}=\begin{bmatrix}\tilde{A}_{11} & \tilde{A}_{12}\\ \tilde{A}_{21} & \tilde{A}_{22}\end{bmatrix}\begin{bmatrix}\tilde{x}_1(k)\\ \tilde{x}_2(k)\end{bmatrix}+\begin{bmatrix}\tilde{B}_1\\ \tilde{B}_2\end{bmatrix}u(k)$$
$$\tilde{y}(k)=\begin{bmatrix}\tilde{C}_1 & \tilde{C}_2\end{bmatrix}\begin{bmatrix}\tilde{x}_1(k)\\ \tilde{x}_2(k)\end{bmatrix}+\tilde{D}u(k) \qquad (11)$$

with $\tilde{x}_1(k)\in\mathbb{R}^r$ and $\tilde{A}_{11}\in\mathbb{R}^{rxr}$ corresponds to gramian Σ_1, and $\tilde{x}_2(k)\in\mathbb{R}^{n-r}$ corresponds to Σ_2. The next step, we take $\tilde{x}_2(k+1)=0$ and we assume $\tilde{A}_{22}$ is a nonsingular matrix. So that the reduced system of r corresponding to gramian Σ_1 is as follows :

$$\tilde{x}_1(k+1)=\tilde{A}_{11}\tilde{x}_1(k)+\tilde{B}_1\,u(k)$$
$$\tilde{y}(k)=\tilde{C}_1\,\tilde{x}_1(k)+\tilde{D}u(k) \qquad (12)$$

for $k=0,1,2,\cdots$, with $\tilde{x}_1(k)\in\mathbb{R}^r$, $u(k)\in\mathbb{R}^s$ and $\tilde{y}(k)\in\mathbb{R}^t$ then we obtained

$$\tilde{A}_r = \tilde{A}_{11} - \tilde{A}_{12}\tilde{A}_{22}{}^{-1}\tilde{A}_{21}, \quad \tilde{C}_r = \tilde{C}_1 - \tilde{C}_2\tilde{A}_{22}{}^{-1}\tilde{A}_{21}$$
$$\tilde{B}_r = \tilde{B}_1 - \tilde{A}_{12}\tilde{A}_{22}{}^{-1}\tilde{B}_2, \quad \tilde{D}_r = \tilde{D} - \tilde{C}_2\tilde{A}_{22}{}^{-1}\tilde{B}_2$$

Thus we obtained a reduced system $(\tilde{A}_r, \tilde{B}_r, \tilde{C}_r, \tilde{D}_r)$ of size r and expressed in the form of :

$$\begin{aligned} \tilde{x}_{r_{k+1}} &= \tilde{A}_r\tilde{x}_{r_k} + \tilde{B}_r\tilde{u}_k \\ \tilde{y}_{r_k} &= \tilde{C}_r\tilde{x}_{r_k} + \tilde{D}_r\tilde{u}_k \end{aligned} \tag{13}$$

From model reduction with the Singular Perturbation Approximation (SPA) method on the asymptotically stable, controllable and observable, $(\tilde{A}, \tilde{B}, \tilde{C}, \tilde{D})$ systems produced by the reduced system $(\tilde{A}_r, \tilde{B}_r, \tilde{C}_r, \tilde{D}_r)$ with $r < n$ order that is asymptotically stable.

For example, if we reduced the original system by order $n = 20$ to order $r = 8$, then we get the matrix for reduced system $(\tilde{A}_r, \tilde{B}_r, \tilde{C}_r, \tilde{D}_r)$ as follows :

$$\tilde{A}_r = \begin{bmatrix} -0.2848 & -1.4130e-11 & \cdots & \cdots & 1.1471e-12 & 0.5900 \\ -1.3615e-11 & 0.5768 & \ddots & \ddots & -0.0568 & 1.0024e-11 \\ 0.9040 & 1.5027e-11 & \ddots & \ddots & 7.4797e-14 & 0.1567 \\ 1.2735e-11 & -0.7229 & \ddots & \ddots & -0.1769 & 1.2704e-13 \\ 2.6574e-12 & -0.2175 & \ddots & \ddots & -0.2041 & 6.1157e-13 \\ -0.1784 & -2.2115e-12 & \ddots & \ddots & -2.7623e-15 & -0.0259 \\ 3.1440e-13 & -0.0653 & \ddots & \ddots & -0.3026 & 3.1992e-15 \\ -0.1299 & -2.1658e-12 & \cdots & \cdots & -5.9488e-15 & -0.2961 \end{bmatrix}_{8\times 8}$$

$$\tilde{B}_r = \begin{bmatrix} 0.4471 & -0.6071 & 0.1357 & -0.3188 & -0.0694 & -0.0156 & -0.1725 & -0.0939 \\ 0.4528 & -0.6135 & 0.1365 & -0.3232 & -0.0700 & -0.0159 & -0.1749 & -0.0952 \\ -0.4471 & -0.6071 & -0.1357 & -0.3188 & -0.0694 & 0.0156 & -0.1725 & 0.0939 \\ 0.4528 & 0.6135 & 0.1365 & 0.3232 & 0.0700 & -0.0159 & 0.1749 & -0.0952 \end{bmatrix}^T_{4\times 8}$$

$$\tilde{C}_r = \begin{bmatrix} 0.3242 & -0.3205 & 0.1012 & -0.1717 & 0.0204 & -0.0271 & 0.1363 & -0.0615 \\ -0.0938 & -0.2004 & 0.3330 & 0.2593 & -0.0084 & 0.0198 & 0.2535 & -0.0407 \\ 0.0355 & -0.0797 & -0.0269 & 0.1329 & 0.2710 & 0.3324 & 0.1249 & -0.0177 \\ -0.0719 & -0.0759 & -0.0179 & -0.0629 & 0.3043 & -0.0233 & 0.2433 & -0.2477 \\ 0.0489 & -0.0775 & -0.0586 & -0.0250 & 0.1606 & -0.0013 & 0.5051 & -0.1272 \\ -0.0489 & -0.0775 & 0.0586 & -0.0250 & 0.1606 & -0.0013 & 0.5051 & 0.1272 \\ 0.0719 & -0.0759 & -0.0179 & -0.0629 & 0.3043 & 0.0233 & 0.2433 & 0.2477 \\ -0.0355 & -0.0797 & -0.0269 & 0.1329 & 0.2710 & -0.3324 & 0.1249 & 0.0177 \\ 0.0938 & -0.2004 & -0.3330 & 0.2593 & -0.0084 & -0.0198 & 0.2535 & 0.0407 \\ -0.3242 & -0.3025 & -0.1012 & -0.1717 & 0.0204 & 0.0271 & 0.1363 & 0.0615 \end{bmatrix}_{10\times 8}$$

$$\tilde{D}_r = \begin{bmatrix} -0.1468 & -0.1281 & -0.1010 & \cdots & \cdots & 0.0544 & 0.0438 & 0.0460 \\ -0.1437 & -0.1293 & -0.1019 & \ddots & \ddots & 0.0550 & 0.0443 & 0.0466 \\ 0.0460 & 0.0438 & 0.0544 & \ddots & \ddots & -0.1010 & -0.1281 & -0.1468 \\ -0.0466 & -0.0443 & -0.0550 & \cdots & \cdots & 0.1019 & 0.1293 & 0.1437 \end{bmatrix}^T_{4\times 10}$$

5. Estimation

In this section, we discuss about estimation the original system and reduced system. Then, we compare the estimation results of the state and computational time between the original system and reduced system.

5.1 Kalman Filter Algorithm

Kalman filter is one method to estimate state variable from stochastic dynamic system first introduced by Rudolf E. Kalman in 1960. Estimation using this method is done by predicting state variable based on system dynamics and measurement data [4]. In system modelling, there is no mathematical model of a perfect system. This can be caused by the noise factor that affects the system. Therefore, it is necessary to add stochastic factors to the deterministic system of equations (8) and (9) in the form of noise system and noise measurement, thus becoming the stochastic dynamic system as follows:

$$\begin{aligned} \boldsymbol{x}_{k+1} &= \boldsymbol{A}_k\boldsymbol{x}_k + \boldsymbol{B}_k\boldsymbol{u}_k + \boldsymbol{G}_k\boldsymbol{w}_k \\ \boldsymbol{z}_k &= \boldsymbol{H}_k\boldsymbol{x}_k + \boldsymbol{v}_k \end{aligned} \tag{14}$$

Kalman Filter algorithm consists of 4 parts. The first and second sections provide the model system and measurement model as well as the initial value (initialization), while the third and fourth sections are the prediction and the correction phase.

a. System models and measurement models

$$x_{k+1} = A_k x_k + B_k U_k + G_k w_k$$
$$z_k = H_k x_k + v_k$$
$$x_0 \sim (\overline{x_0}, P_{x_0}), \qquad w_k = (0, Q_k), \ v_k = (0, R_k)$$

b. Initialization

$$\widehat{x_0} = \overline{x_0}, \qquad P_0 = P_{x_0}$$

c. Time Update
Covariance error : $P_{k+1}^- = A_k P_k A^T + G_k Q_k {G_k}^T$
Estimation : $\hat{x}_{k+1}^- = A_k \hat{x}_k + B_k U_k$

d. Measurement Update
Kalman Gain : $K_{k+1} = P_{k+1}^- {H_{k+1}}^T (H_{k+1} P_{k+1}^- {H_{k+1}}^T R_{k+1})^{-1}$
Covariance error : $P_{k+1} = (I - K_{k+1} H_{k+1}) P_{k+1}^-$
Estimation : $\hat{x}_{k+1} = \hat{x}_{k+1}^- + K_{k+1}(z_{k+1} - H_{k+1}\hat{x}_{k+1}^-)$

5.2 Simulation

In this section, we will do the model reduction on the original system (A, B, C, D) beginning with a balanced system $(\tilde{A}, \tilde{B}, \tilde{C}, \tilde{D})$. Next, we focus on model reduction using the Singular Pertubation Approximation method to obtain a reduced system $(\tilde{A}_r, \tilde{B}_r, \tilde{C}_r, \tilde{D}_r)$. Based on the reduction procedure in the original system, it can be concluded that the system can be reduced from order 1 to 19 which satisfies $\|G_s - G_r\|_\infty \leq 2(\sigma_{r+1} + \cdots + \sigma_n)$, with G_s and G_r respectively is a system transfer function (A, B, C, D) and its reduced system.

Furthermore, estimation is done with the implementation of the kalman filter algorithm on the original system and reduced system from the order 8 to 19 using MATLAB software. From the simulation results, we get the graphic estimations in Figure 1 and 2 and the comparison of error values in Table 1.

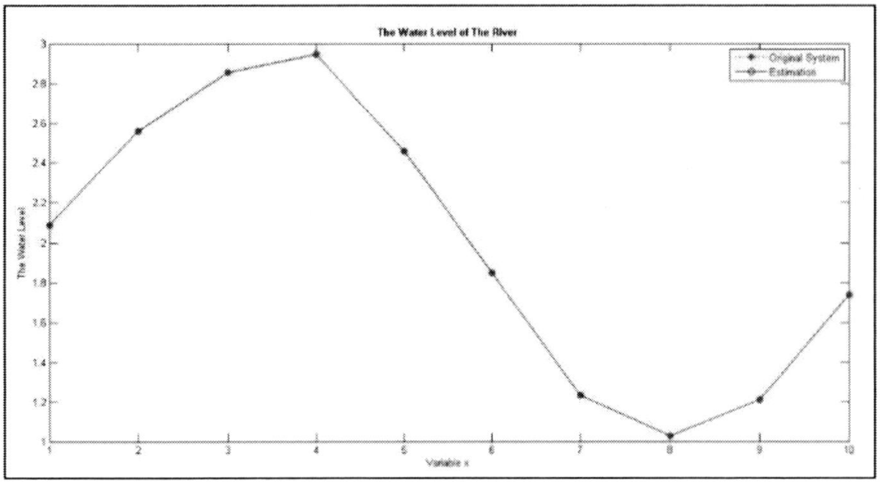

Figure 1. Estimation of original system by using Kalman filter

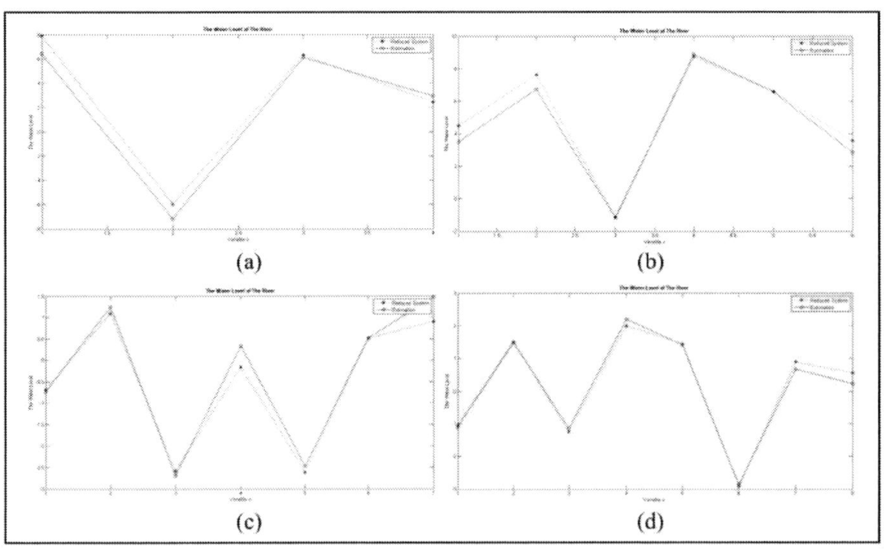

Figure 2. Estimation of the reduced order system 8,11,14 and 16 by using Kalman Filter

Based on Figure 1, the graphic between the original system and the estimated value is almost the same so as to produce a small error. While in Figure 2, the graphic shown between the reduced system at the order of 8,11,14 and 16 with estimation result have slightly different value at some point so that it yield big error value. For more details about error values in each system can be seen from the following table :

Table 1. Eror value by using Kalman Filter

Experiment	Original	8th Order	9th Order	10th Order	11th Order	12th Order	13th Order	14th Order	15th Order	16th Order	17th Order	18th Order	19th Order
1	1,63E-04	2,0945	1,8717	0,0199	1,0363	3,2058	0,0605	0,4596	7,37E-04	1,07E-05	0,5243	0,0154	0,1357
2	6,26E-06	23,7105	0,0183	0,0356	0,9372	0,2381	0,5023	0,3160	0,0095	0,0011	1,2355	0,0092	0,0610
3	1,99E-06	0,6903	0,0999	0,0452	0,0457	0,0048	0,6105	0,0242	0,1725	0,0401	0,9412	0,0014	0,0046
4	4,39E-06	10,3773	0,6125	0,0864	0,1522	0,0023	0,0651	0,0576	0,0737	0,0065	0,6803	0,0135	0,0058
5	2,86E-06	0,0335	0,0496	0,0192	0,3528	0,5858	0,0075	0,0933	0,2205	0,0010	1,0084	0,0519	0,0108
Mean Value	**3,58E-05**	**7,3812**	**0,5304**	**0,0413**	**0,5048**	**0,8074**	**0,2492**	**0,1901**	**0,0954**	**0,0097**	**0,8779**	**0,0183**	**0,0436**

Based on Table 1, the original system has the smallest average error compared to the others. Whereas for comparison in the reduced system, the order 16 has the smallest error value. For computational time comparison is shown in Table 2 as follows:

Table 2. Computational time to estimate the systems by Kalman Filter

Order of reduced systems	Computational time (Seconds)
Original	0,01015
8th order	0,01913
9th order	0,00685
10th order	0,00724
11th order	0,00711
12th order	0,00655
13th order	0,00679
14th order	0,00665
15th order	0,00652
16th order	0,00645
17th order	0,00645
18th order	0,00668
19th order	0,00717

According to Table 2, the reduced system has faster computation time than the original system and those are reduced system order 16 and 17.

6. Conclusions

In this work, we have studied estimation of the river flow models that have been reduced by the singular pertubation approximation method. Our approach is as follows. First, we discretized the differential equation using Lax Wendroff Scheme to obtain a discrete-time linear-time-invariant system for the original system. Then, we construct the balanced system. After that, we apply model reduction using singular pertubation approximation method. Next, we estimate the original system and reduced system using Kalman filter algorithm. Finally, we compare the results of the state and computational time between the original system and reduced system. According to our simulation results, the original system has the smallest estimation error value and the reduced system has the fastest computation time.

References

[1] Arif, D.K, et al. (2014). *Construction of the Kalman Filter Algorithm on the Model Reduction*. International Journal Control and Automation, Vol. 7, No. 9, pp. 257-270

[2] Mustaqim, K., Arif, D.K., Apriliani, E. and Adzkiya, D. (2017). *Model reduction of unstable systems using balanced truncation method and its application to shallow water equations*. Journal of Physics: Conference Series, Vol. 855, No. 1, p. 012029.

[3] Rochmah, M., Fatmawati, dan Purwati, U.D. (2015). *Model Tereduksi Sistem Linier Waktu Diskrit dengan Metode Singular Perturbation Approximation*. Surabaya : Universitas Airlangga.

[4] Lewis, F. L. (1986). *Optimal Estimation With An Introduction Stochastic Control Theory*. USA

[5] Verlaan M. (1998). *Efficient Kalman Filtering Algorithms for Hydrodynamic Models*. Belanda : University of Delft.

[6] Skogestad,S., Postlethwaite,I. (2001). *Multivariable Feedback Control Analysis and Design*. Chichester: John Wiley and Sons.

[7] Lesnussa, T.P., Arif, D.K., Adzkiya, D. and Apriliani, E. (2017). *Identification and Estimation*

of state variables on reduced model using balanced truncation method. Journal of Physics: Conference Series, Vol. 855, No. 1, p. 012023.

[8] Sari, Y.I., Arif, D.K., Apriliani, E. and Adzkiya, D. (2017). *A study on model order reduction of stable discrete-time linear-time-invariant systems using balanced truncation methods*. AIP Conference Proceedings, Vol. 1867, No. 1, p. 020049.

[9] Kartika, D.A., Arif, D.K., Apriliani, E. and Adzkiya, D. (2017). *A study on model order reduction of stable continuous-time linear-time-invariant systems using balanced truncation methods*. AIP Conference Proceedings, Vol. 1867, No. 1, p. 020050.

[10] Arif, D.K., Fadhilah, H.N., Adzkiya, D., Rochmah, L.N., Yoga, N.A.W., Setiawan, H., Kamilah, R. and Cahyani, R.R.N. (2017). *Accelerating simulation time of unstable discrete-time systems using singular perturbation approximation*. Proceedings of the International Conference on Instrumentation, Control and Automation, pp. 7-12.

[11] Arif, D.K., Adzkiya, D., Apriliani, E. and Khasanah, I.N. (2017). *Model reduction of non-minimal discrete-time linear-time-invariant systems*. Malaysian Journal of Mathematical Sciences, Vol. 11, No. 3, pp. 377-391.

IOP Conf. Series: Journal of Physics: Conf. Series **974** (2018) 012019 doi:10.1088/1742-6596/974/1/012019

Improving a two-equation eddy-viscosity turbulence model to predict the aerodynamic performance of thick wind turbine airfoils

Galih Bangga[1,2], Tri Kusumadewi[3], Go Hutomo[4], Ahmad Sabila[5], Taurista Syawitri[6], Herlambang Setiadi[7], Muhamad Faisal[8,9], Raditya Wiranegara[10], Yongki Hendranata[11], Dwi Lastomo[12], Louis Putra[13] and Stefanus Kristiadi[14]

[1] Institute of Aerodynamics and Gas Dynamics, University of Stuttgart, Stuttgart, Germany
[2] Mechanical Engineering Department, Institut Teknologi Sepuluh Nopember, Surabaya, Indonesia
[3] School of Manufacturing System and Mechanical Engineering, Sirindhorn International Institute of Technology, Thammasat University, Thailand
[4] Toyota SO Department, Astra International, Ltd., Indonesia
[5] Civil Engineering Department, Universitas Brawijaya, Malang, Indonesia
[6] Mechanical Engineering Department, Universitas Muhammadiyah Surakarta, Surakarta, Indonesia
[7] School of Information Technology & Electrical Engineering, The University of Queensland, Brisbane, Australia
[8] Non-Installation Top Range Department, Berca Schindler Lifts, Ltd., Indonesia
[9] Graduate Institute of Automation and Control, National Taiwan University of Science and Technology, Taiwan
[10] School of Aerospace, Transport and Manufacturing, Cranfield University, United Kingdom
[11] Mechanical Engineering Department, Texas A&M University, College Station, United States
[12] PUI-PT Mechatronics and Industrial Automation Research Center, Intitut Teknologi Sepuluh Nopember, Surabaya, Indonesia
[13] Mechanical Engineering Department, Politecnico di Milano, Milan, Italy
[14] Aeronautics Department, Imperial College of Science, Technology and Medicine, London, United Kingdom

E-mail: `bangga@iag.uni-stuttgart.de` and `galih.bangga90@gmail.com`

Abstract. Numerical simulations for relatively thick airfoils are carried out in the present studies. An attempt to improve the accuracy of the numerical predictions is done by adjusting the turbulent viscosity of the eddy-viscosity Menter Shear-Stress-Transport (SST) model. The modification involves the addition of a damping factor on the wall-bounded flows incorporating the ratio of the turbulent kinetic energy to its specific dissipation rate for separation detection. The results are compared with available experimental data and CFD simulations using the original Menter SST model. The present model improves the lift polar prediction even though the stall angle is still overestimated. The improvement is caused by the better prediction of separated flow under a strong adverse pressure gradient. The results show that the Reynolds stresses are damped near the wall causing variation of the logarithmic velocity profiles.

IOP Conf. Series: Journal of Physics: Conf. Series **974** (2018) 012019 doi:10.1088/1742-6596/974/1/012019

1. Introduction

Flow separation has become the subject of interest for many years due to its importance in practical engineering problems especially for lifting structures or airfoils. Unfortunately, these lifting devices often attain optimum performance at the condition of "onset of separation" which implies that separation phenomena must be well understood if the analysis is aimed at practical applications [1]. The separation process can be explained by the change of pressure gradient occurring on the airfoil surface. It happens when the angle of incidence of the incoming flow increases, which results in an augmentation of the pressure gradient especially on the suction side of the airfoil. This causes flow detachment resulting in a significant loss of the aerodynamic performance. Rhie and Chow [1] described that this condition involves a strong interaction between the viscous flow and the inviscid flow outside of the boundary layer. On the other hand, in the linear lift region where the flow is attached, the interaction is significantly weaker. This indicates that viscous effects are important and need to be considered for the flow experiencing a large adverse pressure.

Wind turbines often operate at over-rated wind speed or pitch-fault conditions. These issues increase the local angle of attack and pressure gradient occurring on the surface. As a result, stronger separation is expected to occur. These phenomena become more severe if the inboard blade area is considered because the local inflow angle is usually very high due to limitation of the twist angle by structural constraints [2]. Furthermore, as the rotor size increases vastly nowadays, thicker airfoils need to be used from the inboard area extending further outboard longer than more conventional rotors which even enhance the adverse pressure gradient occurring on the blade sections. It is clear that the rotor becomes more vulnerable towards flow separation, and this implies that accurate predictions of flow separation are important for wind turbine aerodynamics.

Due to complexity of the issue, one effective way to bring the physics into numerical approach is to employ a large scale computation using the Navier-Stokes equations. These computational fluid dynamics methods (CFD) have been applied in many engineering problems involving the prediction of flow over wind turbines [3–7]. However, due to the lack of robust CFD methods, most CFD studies were done focusing only on the validation of the CFD codes rather than the physical phenomena of the flow. It is already well known that the Reynolds Averaged Navier - Stokes (RANS) method, commonly used to treat the flow turbulence, is inaccurate for massively separated flows. The maximum lift coefficient and post stall behaviour of airfoils were reported not to be sufficiently captured [8]. Applying eddy-resolving models like Large Eddy Simulations (LES) or Detached Eddy Simulations (DES) has been shown to improve the situation [8–11]. It has been demonstrated that RANS had difficulty in predicting the stall characteristics of some airfoils, while the LES based simulations were able to deliver accurate predictions. However, due to the 3D nature of eddies, computations should be carried out in three dimensional configuration. For this purpose, two dimensional simulations can not be used and this can be quite an issue if a lot of computations need to be performed like in generating airfoil databases for preparation of the Blade Element Momentum (BEM) model [12]. It is always possible to develop the airfoil databases using DES only for the high angle of attack cases where separation presents (using 3D grid) while the standard RANS approach or even XFOIL can be applied for the lower angle of attack (2D grid is applicable). However, the the grids should be sufficiently small to resolve the inertial scales of turbulence. The problem arises when this approach is used for the inboard wind turbine airfoil because the relative thickness of the airfoil is greater than 35%. As a consequence, no linear region presents and separation occurs for the whole range of the angle of attack.

In the URANS approach, the turbulent closure is modeled by assuming the "eddy" viscosity as a function of the "laminar" viscosity. In fact, actually eddy viscosity is not physical and is generated based on a simple relation called Boussinesq hypothesis. The idea is to relate the turbulence stresses to the mean flow in order to close the URANS equations by modeling the Reynolds stress terms. Therefore, turbulence models based on this theorem is usually called the eddy-viscosity turbulence models (EVTM). It has been agreed that it is too dangerous to assume validity of a turbulence model for a wide variety of flows. Many authors have developed improved models for the RANS approach. Myong and Kasagi [13] extended the original $k-\varepsilon$ model considering the two characteristic length scales for the dissipation rate, one very near the wall and the other remote from the wall, which are related to the transfer of the turbulent momentum. The new approach was able to correctly model the wall-limiting behaviour of the major turbulence quantities for the case of fully developed turbulent pipe and channel flows. One of the most striking contribution was given by Menter [14] in combining the Wilcox $k-\omega$ model [15] with the standard $k-\varepsilon$ model according to Launder and Spalding [16]. The model reduces the freestream sensitivity of the Wilcox model while maintaining a good prediction for wall bounded flows. In 2001, Nagano and Hattori [17] developed an improved model based on a low Reynolds number non-linear $k-\varepsilon$ model flow in a rotating channel. Chitsomboon and Thamthae [18] and Bangga et al. [19, 20] have documented that modifications of the eddy viscosity parameter can lead to a more accurate flow prediction, especially if massive flow separation is encountered. In fact, most of turbulence models tune the relation of the eddy viscosity parameter in their development. Based on these insights, the present studies aim to improve CFD predictions of wind turbine airfoils by modifying the turbulent viscosty of a two-equation Unsteady-RANS (URANS) turbulence model. In this case, unsteady simulations are chosen because the resulting flow field of airfoils under massive separation is always unsteady.

A damping factor is introduced in the turbulent viscosity equation for the Menter SST turbulence model [14]. This model is chosen because it is one of the most commonly used turbulence models in industry, and reasonably accurate to predict flows experiencing a strong adverse pressure gradient. The main idea behind the modification is to reduce the contribution of the Reynolds stress terms to the flow near wall region. The modification incorporates the ratio between the the turbulent kinetic energy to its specific dissipation for separation detection. The goal is to activate the modification for separated flow while minimizing the impact for the attached flow case.

The paper is organized as follows. The test case and numerical procedures are given in Section 2. Section 3 presents the results, and the conclusion is made in Section 4.

2. Methods

2.1. Adjustment of the turbulence model

The present section describes the governing equation of the adjusted turbulence model. The Menter SST [14] turbulence model is selected because the model is well known to provide reasonable accuracy for RANS based simulations under strong adverse pressure gradient flows. The original model was proposed by Menter and has become the most used EVTM since then. The approach combines the "standard" $k-\varepsilon$ [16] in the free shear flows and the Wilcox $k-\omega$ [21] in the wall bounded area to reduce the sensitivity of the $k-\omega$ model towards the inflow conditions. This is done by transforming the $k-\varepsilon$ equations to the equivalent $k-\omega$ based through the blending functions F_1 and F_2. They are unity in the near-wall area activating the standard k- ω model, and zero far away from the surface enforcing the transformed k- ε model to be activated. In addition, the SST model introduces a damped cross-diffusion derivative term for the ω equation and revises the definition of the turbulent viscosity to account for the turbulent shear stress transport. These features make the SST model more accurate and reliable than the Wilcox $k-\omega$ in many cases.

The model is governed by two transport equations, one for turbulent kinetic energy (k) and one for its specific dissipation rate (ω), which can also be thought of as the ratio of ε to k. It is a semi-empirical model combining the exact equation for k and non-exact formulation for ε. The

IOP Conf. Series: Journal of Physics: Conf. Series **974** (2018) 012019 doi:10.1088/1742-6596/974/1/012019

equation for the specific dissipation rate is governed using physical reasoning and bears little resemblance to its mathematically exact counterpart. The transport equations for k and ω are given in Equations (1) and (2), respectively.

$$\frac{\partial}{\partial t}(\rho k) + \frac{\partial}{\partial x_i}(\rho k u_i) = \frac{\partial}{\partial x_j}\left(\Gamma_k \frac{\partial k}{\partial x_j}\right) + \tilde{G}_k - Y_k + S_k \tag{1}$$

$$\frac{\partial}{\partial t}(\rho \omega) + \frac{\partial}{\partial x_i}(\rho \omega u_i) = \frac{\partial}{\partial x_j}\left(\Gamma_\omega \frac{\partial \omega}{\partial x_j}\right) + G_\omega - Y_\omega + D_\omega + S_\omega \ . \tag{2}$$

G_ω and $\tilde{G}_k$ represent the generations of ω and the turbulence kinetic energy due to mean velocity gradients, respectively, while Y_k and Y_ω represent the dissipations of k and ω due to turbulence, respectively [22]. The blending of the standard $k-\varepsilon$ model with the Wilcox $k-\omega$ models is facilitated by the blending functions, F_1 and F_2, which introduce a cross-diffusion term D_ω. The effective diffusivities are given by

$$\Gamma_k = \mu + \frac{\mu_t}{\sigma_k} \tag{3}$$

$$\Gamma_\omega = \mu + \frac{\mu_t}{\sigma_\omega} \ , \tag{4}$$

where σ_k and σ_ω are the turbulent Prandtl numbers for k and ω, respectively [22]. μ is the dynamic viscosity and μ_t is the turbulent viscosity which is defined as

$$\mu_t = \frac{\rho k}{\omega} \frac{1}{\max\left[\frac{1}{\alpha^*}, \frac{S F_2}{a_1 \omega}\right]} . \tag{5}$$

S defines the strain rate magnitude. For high Reynolds number flows, a_1 is equal to 0.31 and α^* is unity .

It has been shown by Chitsomboon and Thamthae [18] and Bangga et al. [19, 20] that the modification of the turbulent viscosity near the wall can improve CFD predictions of airfoils. The modifications employed a damping function implemented in the eddy viscosity formulation. Both the approaches in [18] and in [19, 20] assumed that μ_t depends on the wall distance. The modification of μ_t by Chitsomboon and Thamthae [18] is:

$$\mu_t^* = \left\{0.1 + \left[1 - 0.1\tanh\left((0.03y^+)^4\right)\right]\left[0.9 + 0.1\tanh\left((0.005y^+)^8\right)\right]\right\}\mu_t, \tag{6}$$

and according to Bangga et al. [19, 20] (for the $k-\varepsilon$ turbulence model) is

$$\mu_t = f_b \rho C_\mu \frac{k^2}{\varepsilon}, \tag{7}$$

$$f_b = f C_b, \tag{8}$$

where

$$C_b = \frac{1000}{A_0 + A_1 \frac{kU^*}{\varepsilon}} \tag{9}$$

with

$$U^* = \sqrt{S_{ij} S_{ij} + \tilde{\Omega}_{ij} \tilde{\Omega}_{ij}} \tag{10}$$

$$\tilde{\Omega}_{ij} = S_{ij} - 2\varepsilon_{ijk}\omega_k \tag{11}$$

$$S_{ij} = \frac{1}{2}\left(\frac{\partial u_j}{\partial x_i} + \frac{\partial u_i}{\partial x_j}\right). \tag{12}$$

The value of f is damped on the buffer zone around $5 < y^+ < 30$ and is unity on the other zones. Further details of the modification can be seen in [19, 20].

The main objective of the turbulent viscosity modification is to reduce the Reynolds stresses contribution of the turbulence model. By doing so, the ability of the flow to withstand separation reduces. This is helpful to facilitate the simulation in predicting stronger separation (that is usually underestimated). However, it shall be noted that the modified model shall not destroy the basic calibration case for flows without adverse pressure gradient like the flow over a flat plate. Based on the similar idea as in [18–20], the present study also introduces a simple modification by adding a damping factor for the turbulent viscosity equation. The equation for the turbulent viscosity is redefined as

$$\mu_t^* = f_b \mu_t \tag{13}$$

In Equation (13), μ_t^* is the newly defined turbulent viscosity that is damped by parameter f_b with the condition

$$f_b = \begin{cases} < 1, & \text{if } \sqrt{k}/(0.09\omega y) > w_d \\ 1, & \text{otherwise} \end{cases} \tag{14}$$

In Equation (14), y represents the wall distance. The idea is that the damping factor is less than unity, *i.e.*, reducing the turbulent viscosity, if the dimensionless property of the inverse wall distance, $\sqrt{k}/(0.09\omega y)$, is larger than a specific wall distance constant defined as w_d. Physically, this means that the turbulent viscosity is being reduced by Equation (14) approaching the wall. It is shown from Equation (14) that f_b acts as a step fuction, similar to those employed in [18–20]. The discussion about the impact of this parameter will be given in section 3.3. The present modification is indeed simpler than the one employed by Bangga et al. [19,20] for the $k-\varepsilon$ turbulence model. This is done to facilitate faster convergence than the previous implemented approaches. Furthermore, the definition of the wall distance is not directly obtained from y^+ but instead employing the connection between the $\sqrt{k}/(0.09\omega y)$ ratio. This usage is intended for a better detection of separation (based on k and ω) rather than purely y^+. It shall be noted that noticeable magnitude of $\sqrt{k}/\omega$ occurs within the separation area.

2.2. *Computational mesh and numerical methods*

Two dimensional numerical simulations have been carried out using a CFD code, FLOWer, from the German Aerospace Center (DLR) [23–25] employing the Unsteady Reynolds-Averaged Navier-Stokes (URANS) approaches. During the last years, the code was continuously extended at the Institute of Aerodynamics and Gas Dynamics - University of Stuttgart for wind turbine applications [26, 27]. The numerical procedure of the FLOWer code is based on structured meshes. The spatial discretization scheme used in the present study is a central cell-centered finite volume formulation because it provides high robustness and is well-suited for parallel applications [24]. The method employs the Jameson-Schmidt-Turkel (JST) [28] approach for flux computations. The scheme is second order accurate in space on smooth meshes. The method utilizes central space discretization with artificial viscosity and explicit hybrid 5-stage Runge-Kutta time-stepping schemes. Dual time-stepping according to Jameson [29] with second-order accuracy in time, multi-grid level 3 and the implicit residual smoothing with variable coefficients

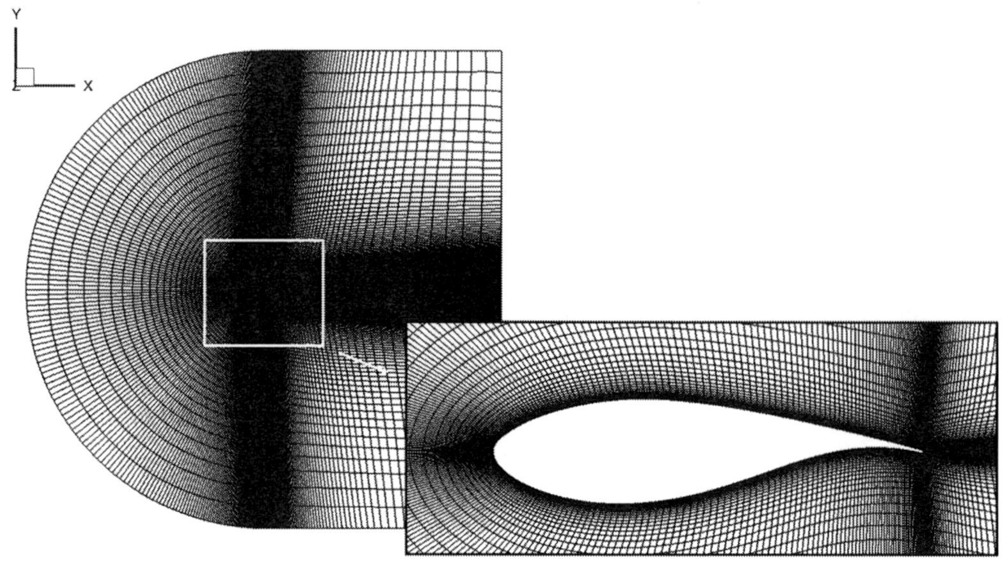

Figure 1: Computational mesh used in the present CFD calculations.

according to Radespiel *et al.* [30] were applied. All simulations were performed by considering the flow unsteadiness with the time step value less than 2% of the time required for a fluid particle convecting over the airfoil, defined as

$$\Delta t < \frac{2}{100}\frac{c}{U_\infty}\,, \tag{15}$$

where U_∞ is the freestream velocity and c is the chord length. Basic sensitivity studies concerning the temporal and spatial resolutions will be given in Section 3.1.

The mesh of the airfoil was generated using an automated script developed at IAG using the grid generator IGG. The domain is C-H type extending 50 times chord length in all directions. The structured mesh is intensively refined near the airfoil wall, see Figure 1. The distance of the first cell layer was set to meet the non-dimensional wall distance of $y^+ < 1$. As no wall function is used, maintaining y^+ lower than unity is important to properly resolve the laminar sub-layer region near the wall. The resolution of the grid consists of 280 and 128 cells in streamwise and wall normal directions, respectively, and among them 32 cells are located within the boundary layer.

2.3. The airfoils

In the present studies, four different wind turbine airfoils designed at the Delft University of Technology, namely DU 91-W2-250 (25% relative thickness), DU 97-W-300 (30%), DU 00-W2-350 (35%) and DU 00-W2-401 (40%) are considered. The description of the airfoil nomenclature was given in [31]. DU stands for Delft University and followed by the year in which the airfoil was designed. The letter W represents that the airfoil is used for wind energy application where the additional number following it indicates that there has been more than one airfoil design with the same thickness that year. The last digits of the nomenclature represent 10 times of the relative thickness percentage. A good review of the airfoils performance and their sensitivity towards surface roughness can be found in [31,32].

The airfoils were measured in the Delft University wind tunnel (LST) and in the laminar wind tunnel at IAG. The measurement was carried out at a Reynolds number of $Re = 3$ million. In this

Table 1: Operating condition of the measurements.

Airfoil name	Re [million]	Trip location [%]
DU 91-W2-250	3.0	5% SS
DU 97-W300	3.0	5% SS, 20% PS
DU 00-W2-350	3.0	2% SS, 10% PS
DU 00-W2-401	3.0	2% SS, 10% PS

campaign, turbulators were employed on the airfoil surfaces to evaluate its performance under soiled conditions. Detailed information of the trip locations is given in Table 1. In the CFD simulations, the transition was prescribed at turbulator position, *i.e.*, turbulence production was switched-on at this location. For the DU 91-W2-250 airfoil, the boundary layer tripping was introduced only on the suction side at $x/c = 5\%$. The transition location on the pressure side was estimated using XFOIL [33] for each angle of attack, resulting in 20 different transition location data. For the airfoil with 30% relative thickness, DU 97-W300 airfoil, the location of trip in the CFD simulation is at the relative chordwise positions of 5% and 20% on the suction and pressure sides, respectively. The relative thicknesses of the examined thicker DU airfoils, DU 00-W2-350 and DU 00-W2-401 airfoils, are 35% and 40%, respectively. On these airfoils, the boundary layer tripping at $x/c = 2\%$ and 10% on the suction and pressure sides, respectively, was applied. In addition to that, CFD computations for fully turbulent boundary layer were also performed for comparison.

3. Results and discussion

3.1. Temporal and spatial discretization studies

Before the examinations are carried out, it is necessary to verify the consistency of the employed CFD model. In this section, basic sensitivity studies for the temporal and spatial discretizations are presented. The test was performed on the DU 91-W2-250 airfoil operating at a very high Reynolds number (Re) of 17.6 million. This high Re condition was purposely chosen because this airfoil is commonly used for the application of large wind turbines in range of 10-20 MW like the AVATAR rotor [34]. The employed turbulence model is the Menter SST $k-\omega$ and no laminar-turbulence prediction/prescription was applied in this grid study, *i.e.*, fully turbulent computations.

The time step size for the present 2D simulations is Δ_t 0.00035 s, corresponding to 1.66% of the convective time of a fluid particle passing the airfoil. The definition of the convective time can be seen in Equation (15). To demonstrate that this time step size is sufficient, an additional CFD simulation at $\alpha = 14°$ was carried out employing a smaller relative time step size of $\Delta_t =$ 0.4%. Figure 2 shows that reducing the time step does not bring a significant influence on the results. This implies that the CFD simulations using the basic time step size of $\Delta_t = 1.66\%$ are plausible.

A grid study was performed in order to show that the numerical results are independent of the number of grid cells. Three different streamwise grid densities were examined; Grid 1: 200 cells, Grid 2: 280 cells, Grid 3: 360 cells. The results of the grid studies are presented in Figure 3. It is shown that the increased grid resolutions hardly produce any differences in both the lift and drag polars for a wide range of angles of attack (α), inferring that the grid is spatially converged. All the grid resolutions generally are in a good agreement within the linear lift region but Grid 1 slightly underestimates the results in the post stall regimes, both in positive and negative stall angles. Considering the accuracy and computational cost, Grid 2 with the streamwise resolution of 280 cells is used in all further simulations.

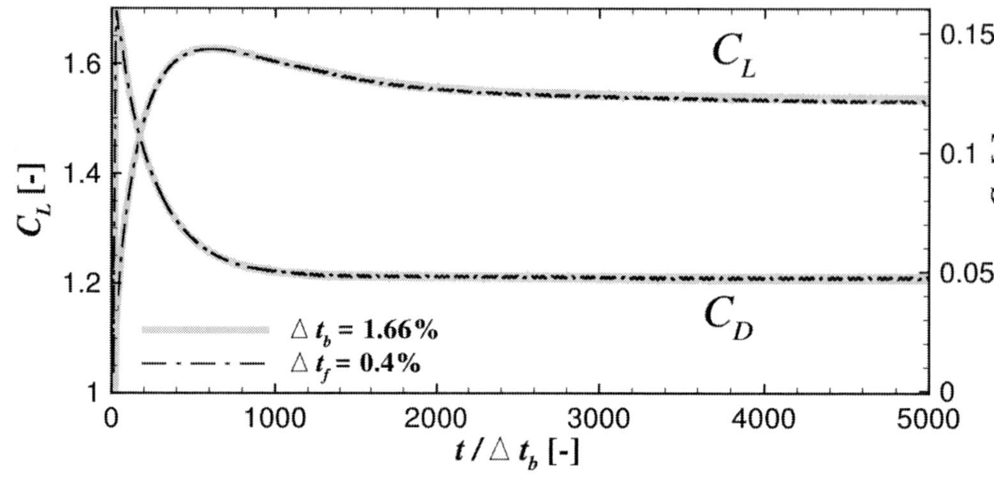

Figure 2: The effect of temporal discretization on the predicted aerodynamic coefficients. The simulations were carried out for the DU 91-W2-250 airfoil, at $\alpha = 14°$ and $Re = 17.6\text{x}10^6$, for the baseline ($\Delta t_b = 1.66\%$) and fine ($\Delta t_f = 0.4\%$) time steps.

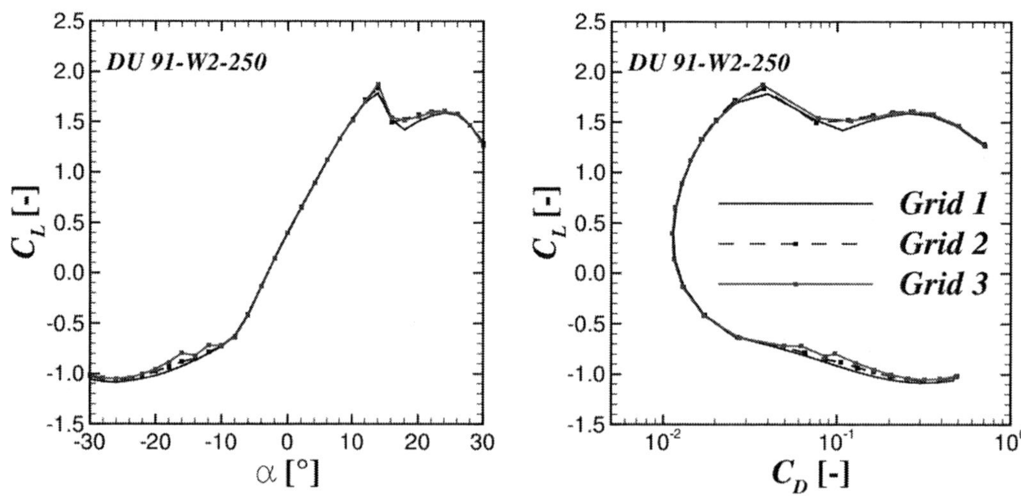

Figure 3: Computed lift and drag polars using three examined grid densities. The increased grid density hardly influences the CFD results.

3.2. Effects of the Reynolds number and transition location

Despite the use for large turbine blades, the measurement data for this airfoil is only available at *Re* of 3 million [32]. The computation at this condition was also carried out using Grid 2 to see the influence of *Re*. The results for lift and drag coefficients are depicted in Figures 4a and 4b, respectively. Black-squared symbol represents the measurement at $Re = 3\text{x}10^6$, dashed-dotted line is for the CFD computation at $Re = 3\text{x}10^6$ by applying the turbulator described in Section 2.3, dashed line is for the CFD computation at $Re = 3\text{x}10^6$ assuming a fully turbulent boundary layer and solid line is for the CFD computation at $Re = 17.6\text{x}10^6$ assuming a fully turbulent boundary layer. It is shown that all the computations over-estimate the maximum

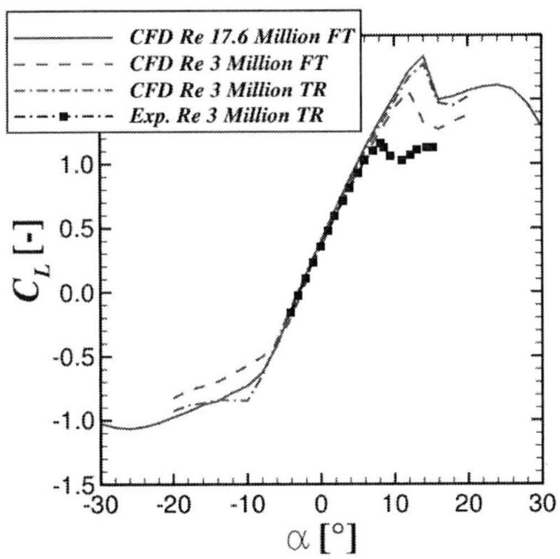

(a)

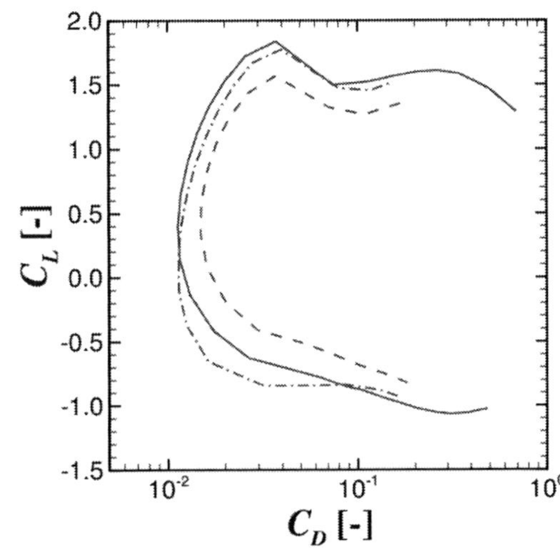

(b)

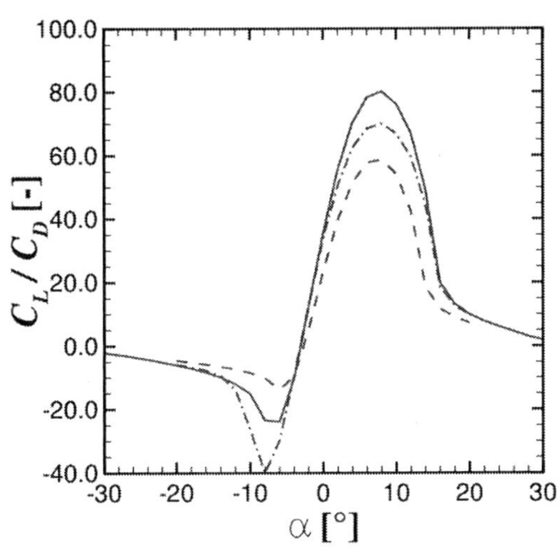

(c)

Figure 4: The effect of Reynolds number for the DU 91-W2-250 airfoil at fully turbulent (FT) and tripped (TR) conditions. The maximum lift coefficient reduces at smaller Re for fully turbulent boundary layer flow.

lift coefficient, showing the inaccuracy of the original SST model to simulate the aerodynamic characteristics in the post-stall region. It is observed that a higher attained maximum lift can be achieved by the airfoil operating at a larger Reynolds number. For example, the maximum lift coefficient for the case employing $Re = 17.6\text{x}10^6$ is about 1.85 at $\alpha = 13.9°$, while it is only about 1.58 at $\alpha = 11.5°$. This indicates that increasing Re can lead to a higher maximum lift and to a noticeable delay of the lift stall. In terms of drag, a remarkable reduction of the drag coefficient is observed for the case operating at a larger Re. This effect seems to occur for the whole range of the studied angles of attack. From Figure 4c, it becomes evident that the airfoil performance enhances considerably with increasing Re. It is shown that the ratio of lift to drag

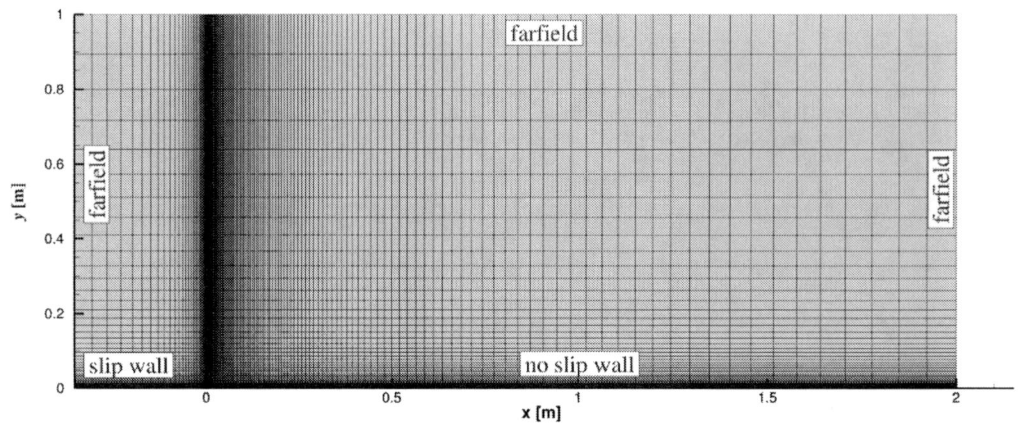

Figure 5: 2D mesh for parametric studies on a flat plate flow. The employed boundary conditions are embedded on the figure.

becomes larger with Re especially approaching its maximum value. It is shown that maximum C_L/C_D for the case at $Re = 3\text{x}10^6$ (FT) is 59 while it reaches 80 for $Re = 17.6\text{x}10^6$.

In Figure 4, it is shown that the location of transition point has a little influence on C_L for the linear lift region. The deviation starts to occur when the angle of attack is higher than 13.9°, corresponding to the stall angle for the case assuming the fully turbulent boundary layer. It is shown that earlier transition location results in the lower attained maximum lift as well as a smaller lift magnitude in the post-stall area. The transition location seems to have a stronger influence for the drag coefficient. Similarly, the ratio of C_L/C_D also increases for the case employing transitional flows. It is noticed that even though the boundary layer tripping is so close to the leading edge (< 5%), the laminar arc length of the airfoil is long enough to influence the flow field which confirms that thick airfoils are more sensitive towards the location of laminar-turbulent transition.

3.3. Damping factor influence

In this section, parametric studies on the damping effects are given. The modification involves two main parameters: the damping magnitude (f_b) and the wall distance constant (w_d) where the damping is applied. Both parameters give an influence on the resulting flow field. In order to ensure that the modification does not destroy the basic calibration of the original SST model, a careful study needs to be carried out for a problem with a well known solution like for flat plate flows. To do so, an additional mesh was prepared for this purpose as shown in Figure 5. The resolutions of the mesh are of 161 and 97 grid points in streamwise and normal flow directions, respectively. The employed boundary conditions are embedded as well in Figure 5. It shall be noted that the entrance of the flow was set as a slip wall to facilitate better convergence and to allow a better observation of the results downstream this area where a no-slip wall boundary condition was applied. In the interface between these two different wall-type boundary conditions, a significant mesh refinement was conducted allowing a better observation of the flow at a smaller Reynolds number. Despite that, a fully turbulent boundary layer was assumed. Thus, the observation is focused on high Reynolds number flow greater than 1e5. This value is reasonable as the intended purpose of the study is for wind turbine applications. Steady simulations were carried out as no flow separation occurs for a flat plate.

Figure 6 shows the results of the simulations in terms of the skin friction coefficient (C_f) as

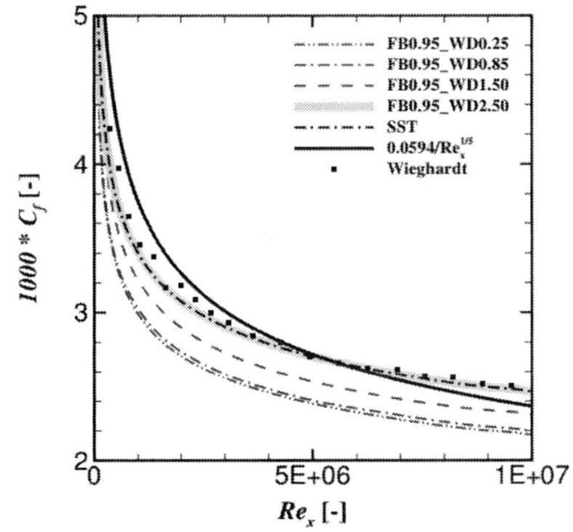

(a)

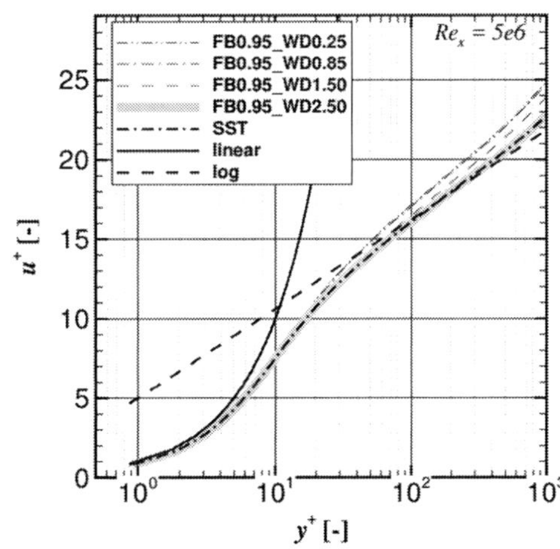

(b)

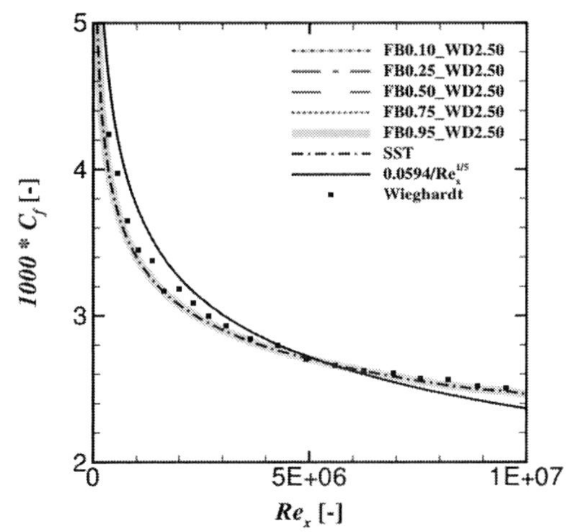

(c)

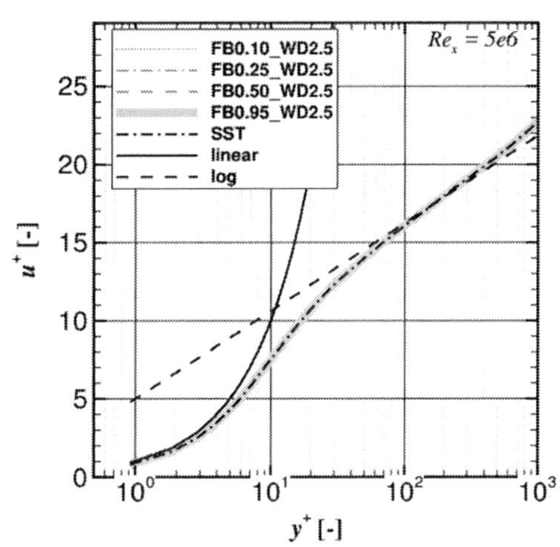

(d)

Figure 6: Sensitivity of modified parameters on flat plate flow. Measurement data for C_f is obtained from [35].

a function of the local streamwise Reynolds number and the logarithmic velocity plot is also presented. In Figures 6a and 6b, the effects of w_d on the flow are assessed. The value of f_b was set to constant ($f_b = 0.95$) in these two figures. It can be seen that C_f is underestimated if w_d is too small. This means that the affected wall distance above the wall is too large. A similar behaviour is observed as well for u^+ within $y^+ > 10$. The model becomes closer to the reference data as w_d increases, and it coincides to each other with the original SST model when w_d reaches 2.5. Thus, this value was set as a reference in assessing the effects of f_b in Figures 6c and 6d. As a consequence, the model still well resembles the original SST results even when f_b is as small as 0.1, please see Figures 6c and 6d.

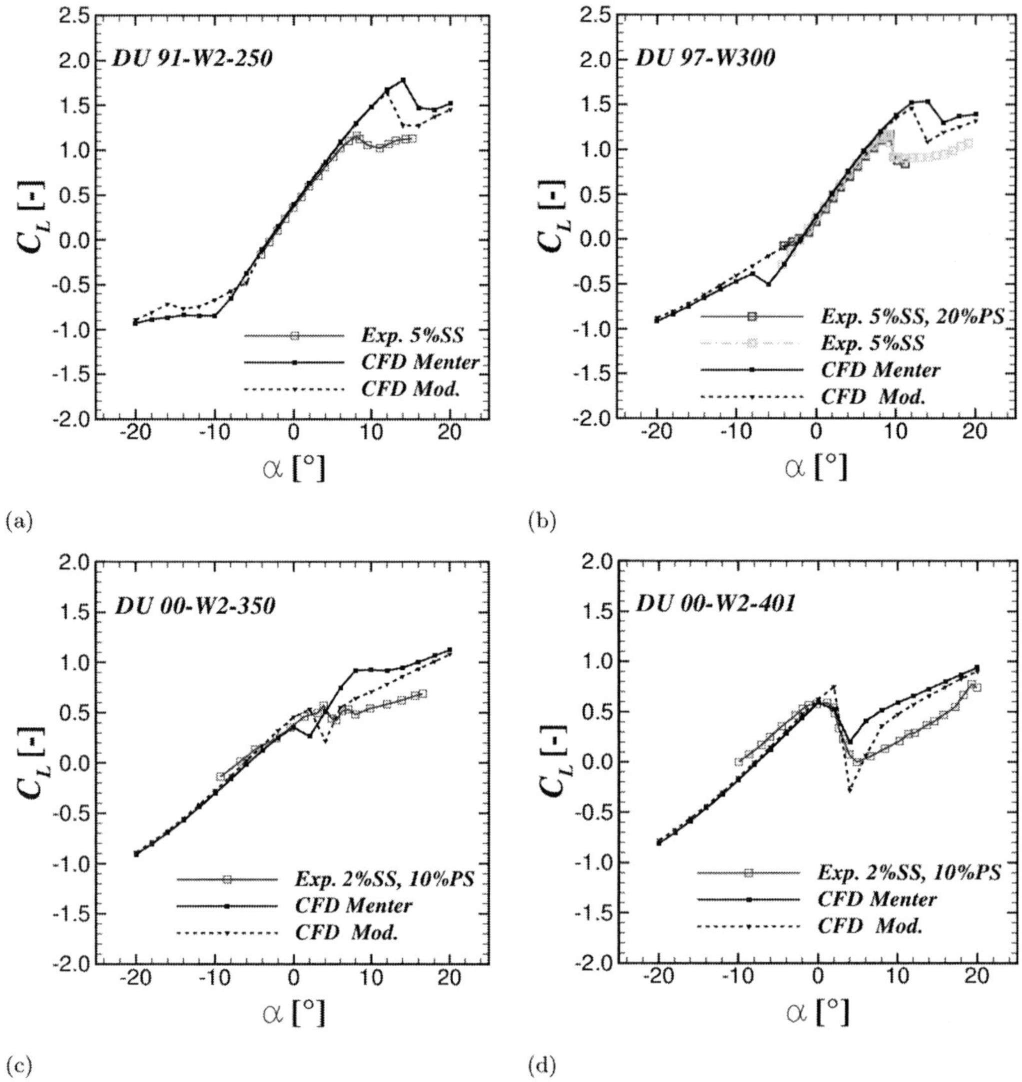

Figure 7: Lift coefficient polar of the four examined DU-airfoils. The adjustment of the turbulent viscosity improves the CFD prediction for these thick wind turbine airfoils.

3.4. Performance of the modified model

The performance of the adjusted turbulence model is examined in this section. The damping factor of $f_b = 0.1$ (with $w_d = 2.5$) was selected according to the consideration described in the previous section. The predicted lift polars from the studied airfoils are compared with measurement data [32] and CFD simulations using the Menter SST turbulence model, depicted in Figure 7. All the numerical computations were carried out according to the measurement data in Table 1. The transition location was prescribed according to the turbulator position in the experiment. Downstream of this location the turbulent generation term was switched-on, ramp-up to 1 within two grid cells. Detailed description on the boundary layer tripping applied in each studied airfoil can be seen in Section 2.3. The studies were performed for a relatively wide angles of attack range, from -20° up to 20°, capturing both the positive and negative stalls.

For each airfoil, the step of α in the simulations is 2°. This implies that 20 individual simulations are required for each airfoil in each studied turbulence model. In total, 160 simulations were carried out for these four different airfoils.

The results of the CFD computations in terms of the lift coefficient are presented in Figure 7. Symbolized, solid and dashed lines represent the measurement, CFD results with the Menter SST model and with the modified model, respectively. All the CFD models show similar results for the linear lift regime. At higher angles of attack, deviations start to occur as depicted in Figure 7. The modified model seems to reduce the maximum lift coefficient and results in earlier stall, especially for the DU 91-W2-250, DU 97-W-300 and DU 00-W2-350 airfoils, but is less dominant for the DU 00-W2-401 airfoil section. In fact, the maximum lift for the thickest studied airfoil increases slightly as shown in Section 3.3. However, in the post-stall regime, the present model shows consistent results for all the studied airfoils. It can be seen that the addition of the damping factor reduces the lift coefficient predicted by the CFD computations. In general, it is shown that the proposed modification improves the CFD prediction compared to the original model, even though all the models are failed in mimicking the experimental results. Three main reasons may explain this behavior. (1) Flow separation on airfoils at high angles of attack is a complex 3D-unsteady flow phenomenon. Two dimensional URANS computations may not be able to model the flow properly, and 3D computations are expected to improve the situation [36]. (2) As all the turbulent eddies are modeled using URANS, the accuracy of CFD computations depends strongly on the employed model. This implies that the ultimate accuracy of the CFD simulations is the accuracy of the turbulence model itself [37]. The use of eddy resolving approaches like Large Eddy Simulations (LES) or Detached-Eddy Simulations (DES) are believed to better mimic the experiment. (3) The overestimation of the maximum lift seems to be contributed as well by the strong unsteadiness of the flow. It can be clearly seen in Figure 8 that predicted C_L from the modified model drops with increasing simulation time up to around $t/\Delta t = 3000$. Then, C_L increases again as the lift fluctuation presents. This indicates that the accurate prediction can be hindered by a strong unsteady effect. Despite that, the modified model clearly improves the prediction as the lift fluctuation at high angle of attack is captured in comparison to the original SST model which shows no indication of unsteadiness

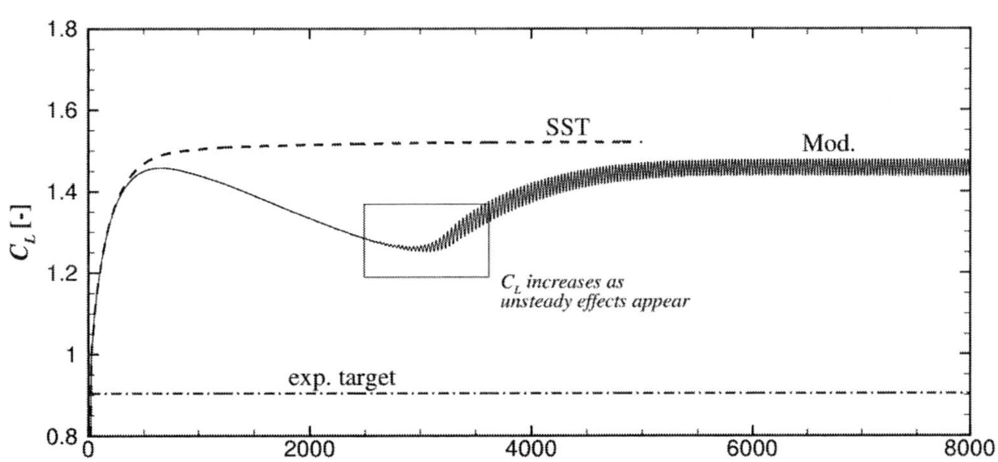

Figure 8: Unsteady effects at maximum lift. The simulations were carried out for the DU 97-W300 airfoil at $\alpha = 12°$.

up to $t/\Delta t = 5000$.

Figures 9a and 9b show the pressure coefficient distributions of the DU 91-W2-250 airfoil at $\alpha = 6°$ and at $\alpha = 14°$. The predicted Cp distribution of the present model coincides with the Menter SST model for the small angle of attack while it deviates strongly at a higher α. It is shown that, for $\alpha = 14°$, the minimum pressure peak becomes more positive in the modified model. This results in the pressure level increase downstream of this position. A slightly lower pressure compared to the Menter SST model is observed only within a small area near the trailing edge. On the pressure side of the airfoil, a small deviation is also observed. It can be seen that the pressure level reduces (becomes more negative) compared to the original model. As a consequence, the lift coefficient reduces in the present approach. It seems that the model is strongly influenced by the pressure gradient. For example, at $\alpha = 14°$ the deviation starts to occur downstream of the minimum Cp not only within the separated flow region.

In Figures 9c and 9d, the corresponding skin friction coefficient distributions are shown. It can be observed that the model again resembles the original SST model at a small angle of attack where the pressure gradient is small and no separation occurs (Figure 9c). From Figure 9d, a strong deviation of C_f starts to occur just downstream of the transition point while this is

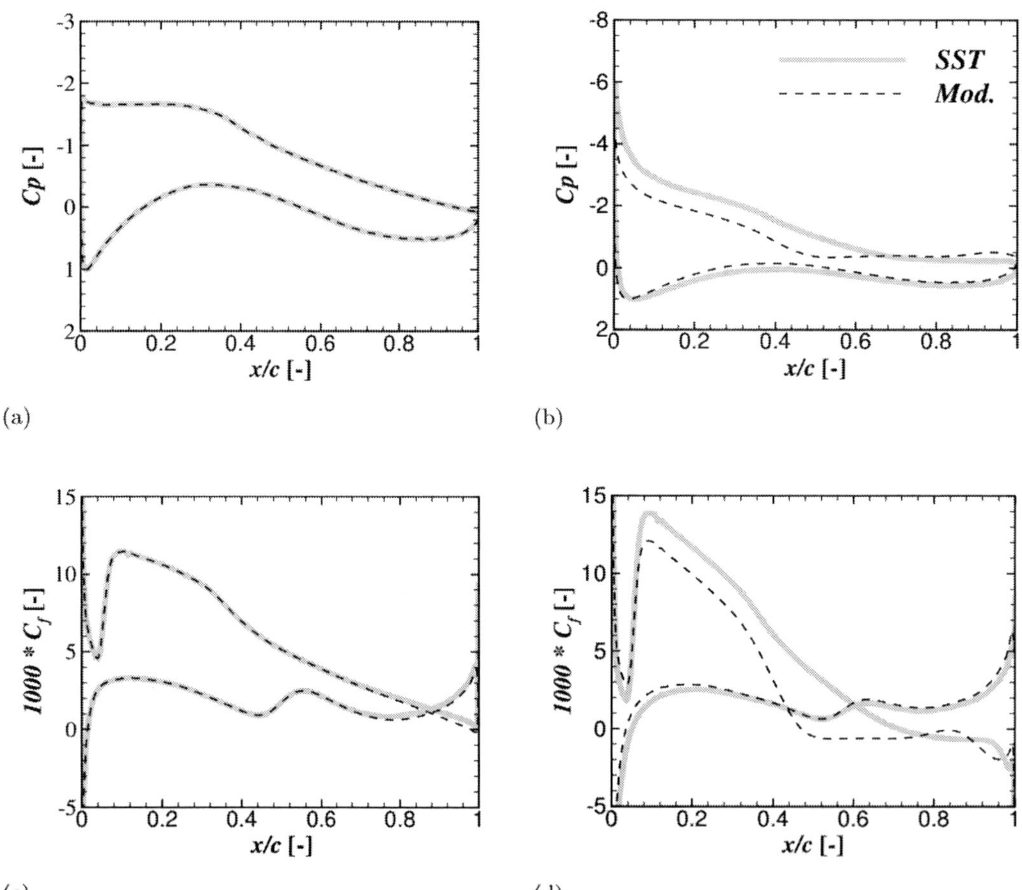

(a) (b) (c) (d)

Figure 9: Predicted pressure and skin friction coefficients of the DU 91-W2-250 airfoil at $\alpha = 6°$ (left) and at $\alpha = 14°$ (right).

less prominent for the laminar portion of the flow ($x/c < 5\%$). The modified model reduces the magnitude of friction contribution compared to the original model. Furthermore, significantly earlier separation is observed as well (where $C_f < 0$). Within the separated flow area, the wall shear stress seems to be constant over a relatively large portion of the chord especially on the suction side.

Figure 10 presents the flow field around the airfoil at $\alpha = 14°$. It is shown clearly that stronger separation is observed for the present model compared to the Menter SST. The separated flow area starts from $x/c \approx 0.5$ for the present model, while it is further downstream for the Menter SST model from $x/c \approx 0.7$. The predicted separated flow area obtained by the present modification is expected to better represent the experiment because the error in the lift prediction is only 13.6%, compared to the original model that reaches 58.7% at this specific angle of attack.

To better understand the origin of the improvement, the velocity profile comparison is shown in Figures 11 and 12 for $x-$ and $y-$velocity components, respectively. The profiles are extracted from the velocity field in Figure 10 at five chordwise positions, namely $x/c = 0.1$, 0.25, 0.5, 0.75 and 0.9. It can be seen that the streamwise velocity profile well resembles the original SST model upstream of the separation point in Figures 11a and 11b. However, as separation starts to occur at $x/c = 0.5$, the velocity profile deviates showing a significant reduction of u^+ up to $y^+ < 253$. It shall be noted that no separation takes place for the original SST model at this position. The strongest deviation between the two models is observed at $x/c = 0.75$ in Figure 11d. Further downstream, the Menter SST model predicts strong flow separation that the difference between the models reduces.

A similar behaviour is observed also for the logarithmic $y-$velocity profiles in Figure 12. No deviation of the velocity profile is observed before separation takes place. A positive $y-$velocity component is observed for the modified model starting from $x/c = 0.5$ due to backflow. This phenomenon is observed also at $x/c = 0.75$ and 0.9 in Figures 12d and 12e, respectively. The positive v^+ magnitude is shown for the Menter SST model from $x/c = 0.75$. Despite that, the backflow is not as strong as for the modified model.

Figure 13 displays the Reynolds stress components from the turbulence models at the

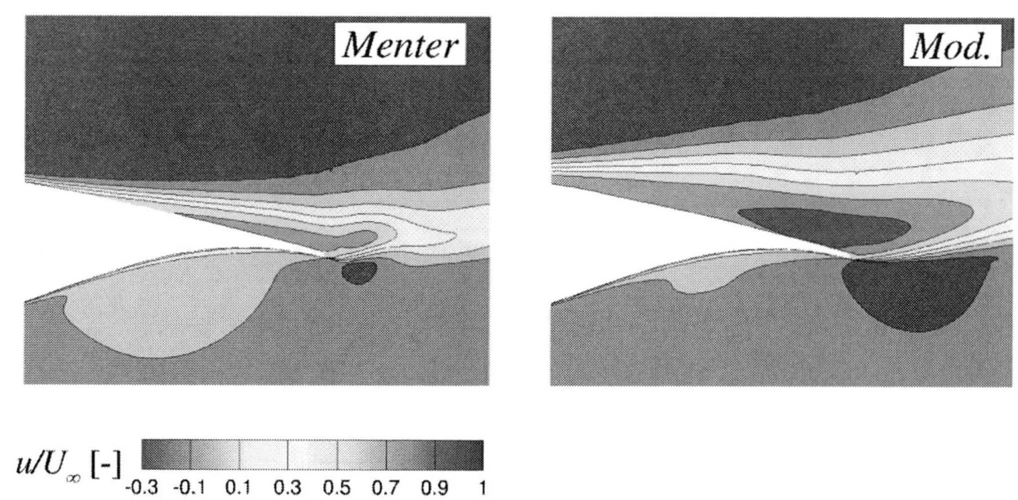

Figure 10: Predicted velocity field of the DU 91-W2-250 airfoil at $\alpha = 14°$. Stronger separation is predicted by the present model.

corresponding positions as the velocity profiles in Figures 11 and 12. Three different components are observed: $- < u'u' >$, $- < u'v' >$ and $- < v'v' >$. The Reynolds stresses are estimated by:

$$- < u_i' u_j' >= \mu_t \left(\frac{\partial u_i}{\partial x_j} + \frac{\partial u_j}{\partial x_i} \right). \tag{16}$$

It shall be noted that the contribution of w remains zero since the simulations are two dimensional.

The $- < u'u' >$ plots are illustrated in Figures 13a-13d, $- < u'v' >$ in Figures 13e-13h and $- < v'v' >$ in Figures 13i-13l. The variable is normalized by the square of friction velocity (u^{*2}).

The plots again support the observation for the velocity profiles. It can be seen that both

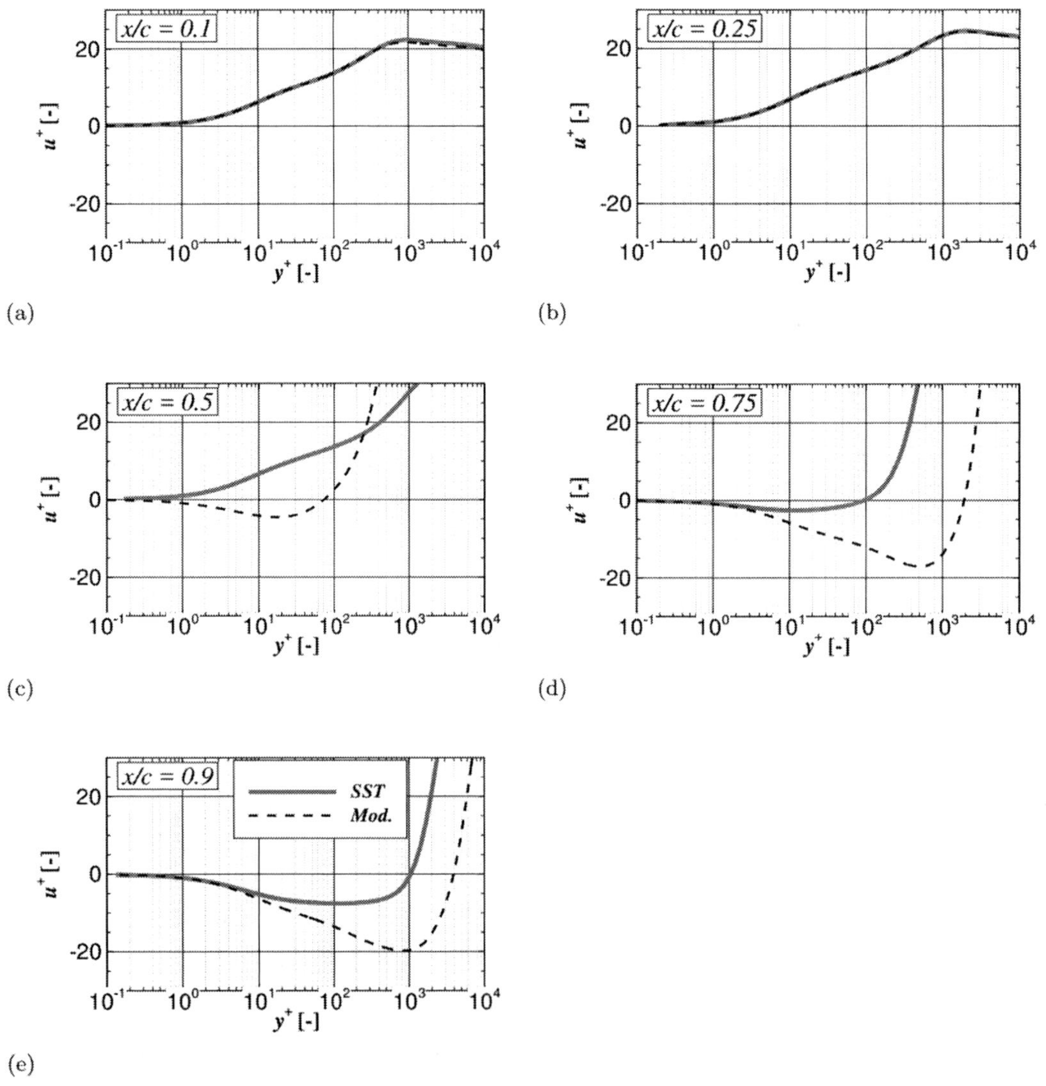

Figure 11: Nondimensional logarithmic $x-$velocity profiles on the suction side of the DU 91-W2-250 airfoil at $\alpha = 14°$. Dashed and solid lines represent the modified and the Menter SST models, respectively.

IOP Conf. Series: Journal of Physics: Conf. Series **974** (2018) 012019 doi:10.1088/1742-6596/974/1/012019

turbulence models have the similar prediction at $x/c = 0.25$ where no separation occurs. A small deviation is shown starting from $x/c = 0.5$. It becomes notable that the cross-contribution between $x-$ and $y-$velocity ($- < u'v' >$) is stronger than $- < u'u' >$ and $- < v'v' >$. Downstream of this position, the Reynold stresses are considerably different. The Reynolds stresses are damped near the wall up to a certain y^+ value for the modified model compared to the fuller profile for the original SST model. As a consequence, the stability of the flow to withstand the adverse pressure gradient becomes smaller, resulting in stronger separation. The damped area seems to increase with increasing x/c. For example, the Reynolds stresses are damped up to $y^+ = 1000$ for $x/c = 0.75$ but it reaches $y^+ = 2500$ for $x/c = 0.9$. It is shown, however, that the Reynolds stresses from the modified model are stronger at a larger y^+ value

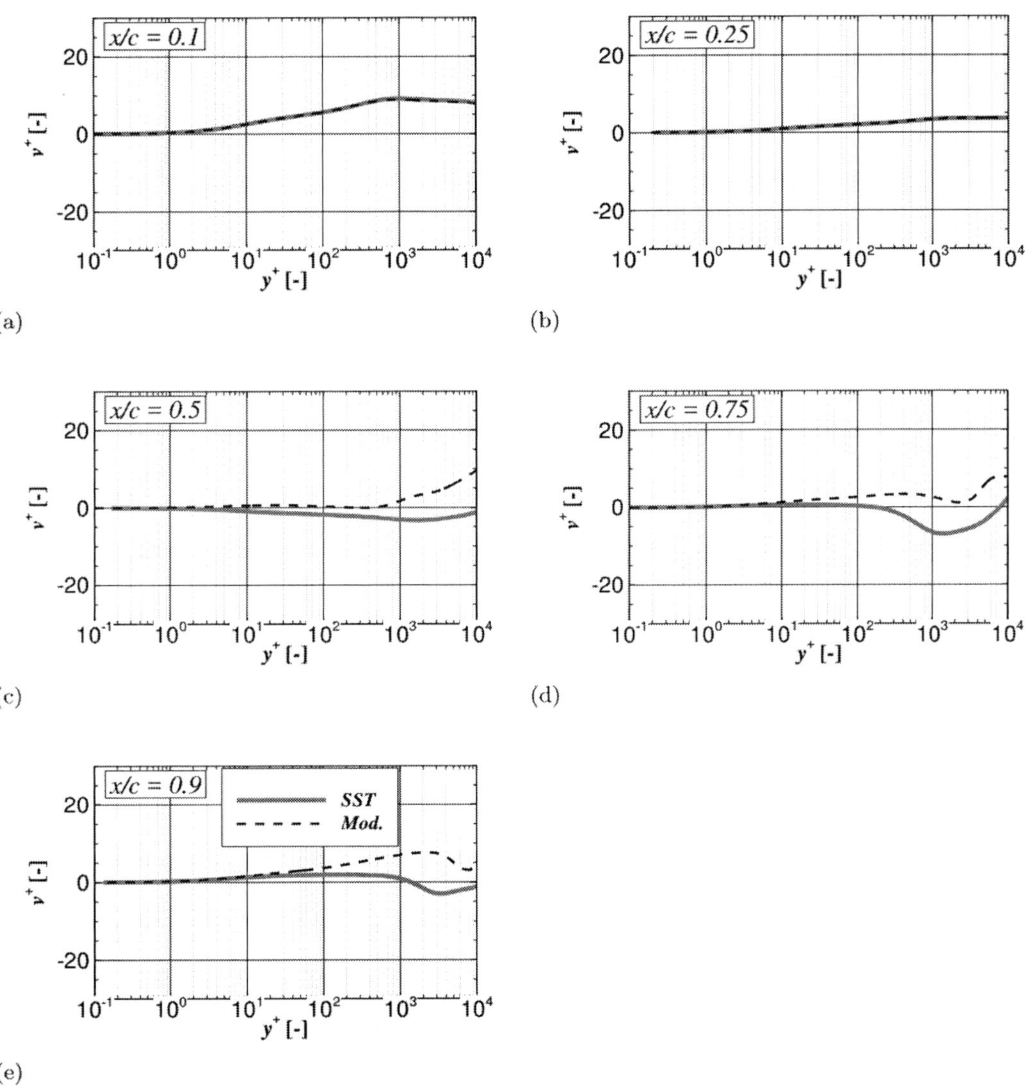

Figure 12: Nondimensional logarithmic $y-$velocity profiles on the suction side of the DU 91-W2-250 airfoil at $\alpha = 14°$. Dashed and solid lines represent the modified and the Menter SST models, respectively.

IOP Conf. Series: Journal of Physics: Conf. Series **974** (2018) 012019 doi:10.1088/1742-6596/974/1/012019

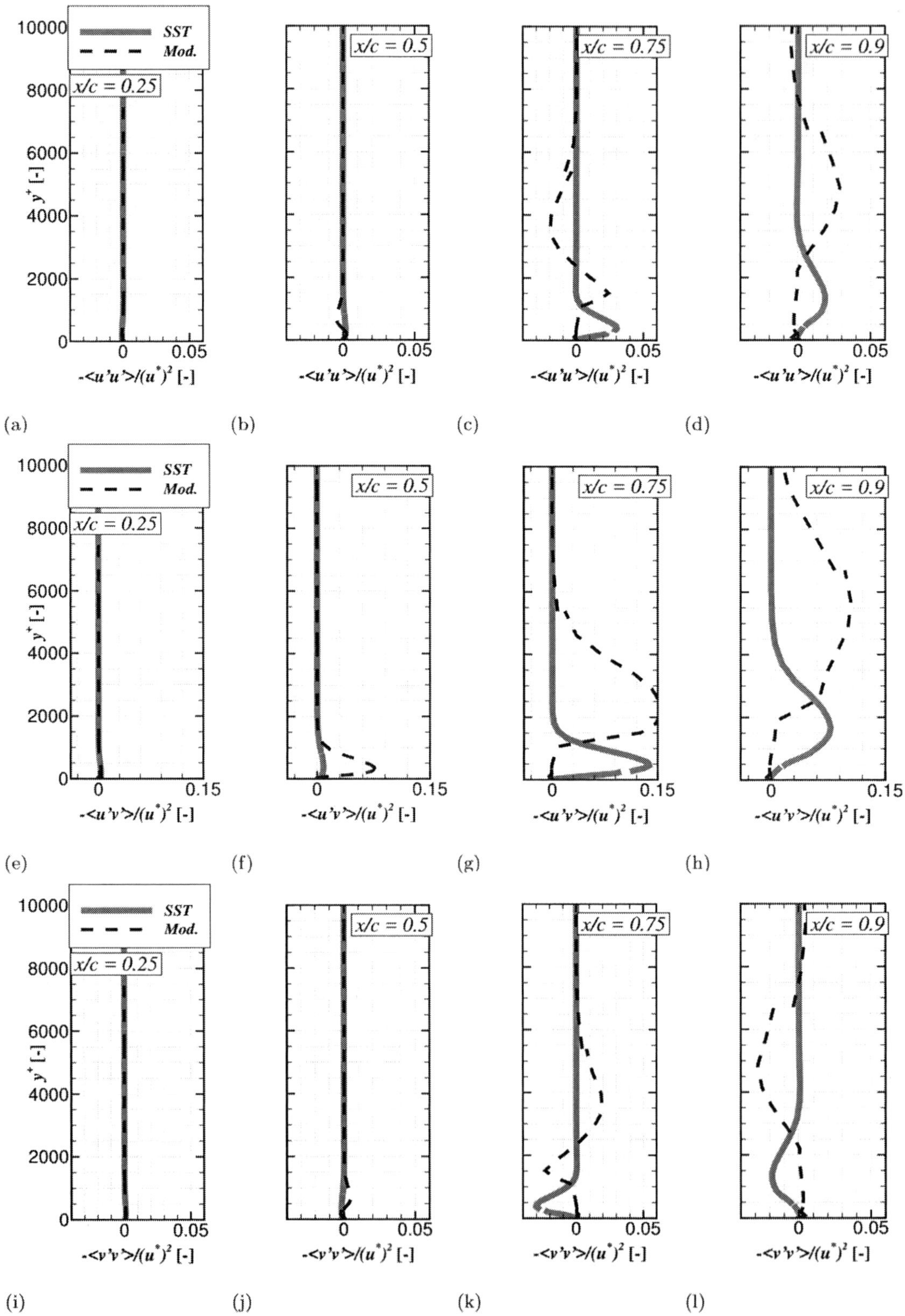

Figure 13: Reynolds stresses from the computed by turbulence models.

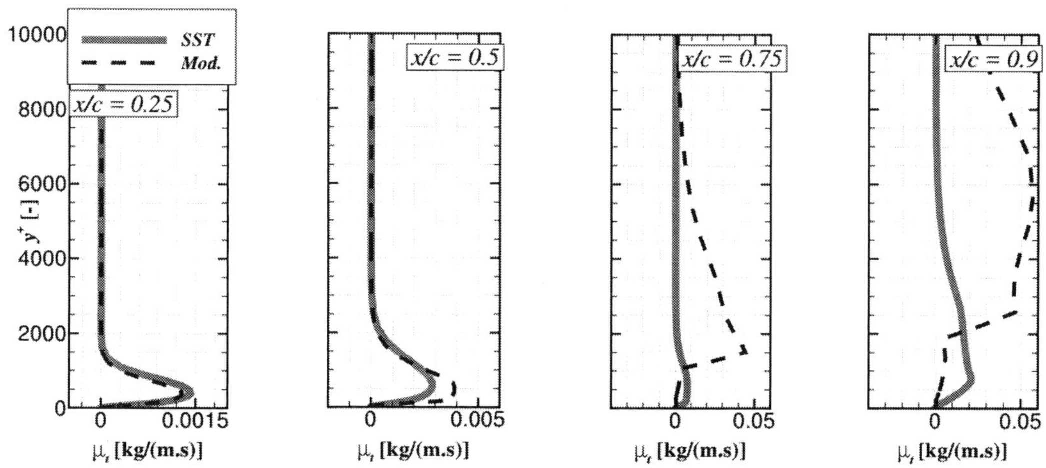

(a) (b) (c) (d)

Figure 14: Reynolds stresses from the computed by turbulence models.

than the damped area. This seems to be related to the displacement effect of the flow due to separated vortex up to the boundary layer edge. For example, the Reynolds stress terms increase up to around $y^+ = 9500$ (wall distance of about $0.18c$) at $x/c = 0.75$ while the boundary layer edge is at around $y^+ = 7000$.

The behaviour of the Reynolds stresses are strongly influenced by the turbulent viscosity magnitude. It is shown that the turbulent viscosity resembles the original SST model as long as no separation occurs. The effects of the damping factor are clearly displayed especially in Figures 14c and 14d. A sharp change of μ_t is shown since a step function model is used in the present works. But the influence of this sharp change seems to be minor because the Reynolds stresses profiles are still relatively smooth as already shown in Figure 13.

4. Conclusion and outlook

Computational Fluid Dynamics (CFD) studies have been performed on four thick wind turbine airfoils, namely DU 91-W2-250 (25% relative thickness), DU 97-W-300 (30%), DU 00-W2-350 (35%) and DU 00-W2-401 (40%). The simulations were carried out by employing the adjusted Menter SST turbulence model. The equation of the turbulent viscosity was redefined by adding a damping factor (f_b) to damp the turbulence. The damping factor was applied at regions near the wall controlled by a wall distance constant (w_d). Parametric studies were carried out to evaluate the sensitivity of the CFD results towards variation of the damping factor and wall distance constant. The results were evaluated and compared with the original Menter SST turbulence model and available measurement data to assess the accuracy of the model.

The evaluations show that the modified model is able to improve the CFD prediction especially at high angles of attack. The maximum lift coefficient and post stall characteristics are better captured even though the stall angle is not accurately predicted, which is expected due to three dimensionality and unsteadiness of the flow. The use of 3D CFD computations or eddy resolving technique like Large Eddy Simulations (LES) and Detached-Eddy Simulations (DES) may improve the situations. Nevertheless, both the models agree fairly well in the linear lift regime. At high angles of attack, the minimum pressure coefficient becomes more positive in the present model, increasing the pressure level on the suction side. On the other hand, the pressure level reduces on the pressure side of the airfoil. As a result, the lift coefficient alleviates, reducing the lift error with the measurement data to 13.6% compared to the SST model that reaches 58.7%. Stronger separation is observed for the present model. The logarithmic velocity profile of the modified model well resembles the original SST model in the region where no separation occurs. The deviation between the the velocity profiles predicted by both models are noticeable within the separation area, indicating that the modification is activated. These phenomena are caused by the a strong reduction of the Reynolds stresses contribution near the wall that makes the flow becomes more vulnerable towards the adverse pressure gradient especially at high angles of attack.

In the present studies, only two dimensional simulations were carried out. It is suggested to examine the performance of the model for three dimensional flow. Applying the model using the full 3D Navier-Stokes computations on wind turbine rotors are strongly recommended. Furthermore, the present model can be used as the near wall RANS model for the DES approach to improve the prediction of separation within the boundary layer flows. The model is, however, not yet tested for rotating coordinate system since the objective of the present studies is to improve the prediction of the model for generating airfoil databases used by BEM simulations. Further investigations regarding this aspect are recommended.

Acknowledgements

The authors gratefully acknowledge the Ministry of Research, Technology and Higher Education of Indonesia for the funding through Directorate General of Higher Education (DGHE) scholarship. This paper is dedicated to F.F.S. Lesy who passed away on June 12, 2017 while accompanying the author G. Bangga in finishing the works. At last but not least, the authors would like to specially mention GFriend, BlackPink, Twice, AOA, IOI, Sistar and SNSD for their assistance in creating a suitable working environment during the completion of the studies.

References

[1] Rhie C and Chow W 1983 *AIAA journal* **21** 1525–1532
[2] Bangga G, Lutz T, Jost E and Krmer E 2017 *Journal of Renewable and Sustainable Energy* **9** 023304 URL `http://dx.doi.org/10.1063/1.4978681`
[3] Duque E P, Burklund M D and Johnson W 2003 *Journal of Solar Energy Engineering* **125** 457–467
[4] Pape A L and Lecanu J 2004 *Wind Energy* **7** 309–324
[5] Bangga G, Lutz T and Krämer E September 23-25, 2015 *EAWE PhD Seminar 11* (Stuttgart, Germany)
[6] Bangga G, Lutz T and Krämer E May 19-20, 2015 *Proceedings of German Wind Energy Conference 12, DEWEK 2015* (Bremen, Germany)
[7] Kim Y, Jost E, Bangga G, Weihing P and Lutz T 2016 *Journal of Physics: Conference Series* vol 753 (IOP Publishing) p 032047 URL `https://doi.org/10.1088/1742-6596/753/3/032047`
[8] Shur M, Spalart P, Strelets M and Travin A 1999 *Engineering turbulence modelling and experiments* **4** 669–678
[9] Bangga G, Weihing P, Lutz T and Krämer E 2017 *Journal of Mechanical Science and Technology* **31** 2359–2364 URL `http://dx.doi.org/10.1007/s12206-017-0432-6`
[10] Bangga G, Lutz T and Krämer E 2017 *Journal of Mechanical Science and Technology* **31** 3839–3844 URL `http://dx.doi.org/10.1007/s12206-017-0728-6`
[11] Bangga G, Weihing P, Lutz T and Krämer E 2018 *Hybrid RANS/LES simulations of the three-dimensional flow at root region of a 10 MW wind turbine rotor. New Results in Numerical and Experimental Fluid Mechanics XI* (Springer) pp 707–716 URL `http://dx.doi.org/10.1007/978-3-319-64519-3_63`
[12] Bangga G, Guma G, Lutz T and Krämer E 2018 *Wind Engineering* (To be published)
[13] Myong H K and Kasagi N 1990 *JSME international journal. Ser. 2, Fluids engineering, heat transfer, power, combustion, thermophysical properties* **33** 63–72
[14] Menter F R 1994 *AIAA journal* **32** 1598–1605
[15] Wilcox D C *et al.* 1998 *Turbulence modeling for CFD* vol 2 (DCW industries La Canada, CA)
[16] Launder B E and Spalding D B 1972
[17] Nagano Y and Hattori H 2001 *TSFP DIGITAL LIBRARY ONLINE* (Begel House Inc.)

[18] Chitsomboon T and Thamthae C 2011 *Proceedings of the World Renewable Energy Congress* vol 15 pp 4114–4120
[19] Bangga G 2013 *Assessment of modified two-equations U-RANS turbulence model to predict the onset of dynamic stall* Master's thesis Institut Teknologi Sepuluh Nopember
[20] Bangga G and Sasongko H 2017 *Journal of Applied Fluid Mechanics* **10** 1–10 URL `https://doi.org/10.18869/acadpub.jafm.73.238.26391`
[21] Wilcox D C 1993 *AIAA journal* **31** 1414–1421
[22] Ansys A F 2011 *ANSYS inc* 218–221
[23] Kroll N, Rossow C C, Becker K and Thiele F 2000 *Aerospace Science and Technology* **4** 223–237
[24] Aumann P, Bartelheimer W, Bleecke H, Kuntz M, Lieser J, Monsen E, Eisfeld B, Fassbender J, Heinrich R and Kroll N, Mauss M, Raddatz J, Reisch U, Roll B, Schwarz T 2008 *FLOWer installation and user manual* Deutsches Zentrum fur Luft- und Raumfahrt
[25] Schwarz T, Spiering F and Kroll N 2010 *Second Symposium of Simulation of Wing and Nacelle Stall*
[26] Bangga G, Kim Y, Lutz T, Weihing P and Krämer E 2016 *Journal of Physics: Conference Series* **753** 022026 URL `https://doi.org/10.1088/1742-6596/753/2/022026`
[27] Bangga G, Lutz T, Dessoky A and Krämer E 2017 *Journal of Renewable and Sustainable Energy* **9** 053303 URL `http://dx.doi.org/10.1063/1.5003772`
[28] Jameson A, Schmidt W, Turkel E *et al.* 1981 *AIAA paper* **1259** 1981
[29] Jameson A 1991 *AIAA paper* **1596** 1991
[30] Radespiel R, Rossow C and Swanson R 1990 *AIAA journal* **28** 1464–1472
[31] Timmer W and Van Rooij R 2003 *Journal of Solar Energy Engineering* **125** 488–496
[32] Van Rooij R and Timmer W 2003 *Journal of solar energy engineering* **125** 468–478
[33] Drela M 1989 *Low Reynolds number aerodynamics* (Springer) pp 1–12
[34] Lekou D, Chortis D, Chaviaropoulos P, Munduate X, Irisarri A, Madsen H, Yde K, Thomsen K, Stettner M and Reijerkerk M, Grasso F, Savenije R, Schepers G, Andersen CF 2015 Avatar deliverable d1.2 reference blade design Tech. rep. ECN Wind Energy
[35] Wieghardt K and Tillmann W 1951
[36] Bangga G, Hutomo G, Wiranegara R and Sasongko H 2017 *Journal of Mechanical Science and Technology* **31** 261–267 URL `http://dx.doi.org/10.1007/s12206-016-1228-9`
[37] Fröhlich J and von Terzi D 2008 *Progress in Aerospace Sciences* **44** 349–377

IOP Conf. Series: Journal of Physics: Conf. Series **974** (2018) 012020 doi:10.1088/1742-6596/974/1/012020

Enhancing BEM simulations of a stalled wind turbine using a 3D correction model

Galih Bangga [1,2], Go Hutomo[3], Taurista Syawitri[4], Tri Kusumadewi[5], Winda Oktavia[6], Ahmad Sabila[7], Herlambang Setiadi[8], Muhamad Faisal[9,10], Yongki Hendranata[11], Dwi Lastomo[12], Louis Putra[13], Stefanus Kristiadi[14] and Ilmi Bumi[15]

[1] Institute of Aerodynamics and Gas Dynamics, University of Stuttgart, Stuttgart, Germany
[2] Mechanical Engineering Department, Institut Teknologi Sepuluh Nopember, Surabaya, Indonesia
[3] Toyota SO Department, Astra International, Ltd., Indonesia
[4] Mechanical Engineering Department, Universitas Muhammadiyah Surakarta, Surakarta, Indonesia
[5] School of Manufacturing System and Mechanical Engineering, Sirindhorn International Institute of Technology, Thammasat University, Thailand
[6] Project Support and Cost Control Division, Indoturbine, Ltd., Jakarta, Indonesia
[7] Civil Engineering Department, Universitas Brawijaya, Malang, Indonesia
[8] School of Information Technology & Electrical Engineering, The University of Queensland, Brisbane, Australia
[9] Non-Installation Top Range Department, Berca Schindler Lifts, Ltd., Indonesia
[10] Graduate Institute of Automation and Control, National Taiwan University of Science and Technology, Taiwan
[11] Mechanical Engineering Department, Texas A&M University, College Station, United States
[12] PUI-PT Mechatronics and Industrial Automation Research Center, Intitut Teknologi Sepuluh Nopember, Surabaya, Indonesia
[13] Mechanical Engineering Department, Politecnico di Milano, Milan, Italy
[14] Aeronautics Department, Imperial College of Science, Technology and Medicine, London, United Kingdom
[15] Materials Engineering Department, Institut Teknologi Sepuluh Nopember, Surabaya, Indonesia

E-mail: `bangga@iag.uni-stuttgart.de` and `galih.bangga90@gmail.com`

Abstract. Nowadays wind turbine rotors are usually employed with pitch control mechanisms to avoid deep stall conditions. Despite that, wind turbines often operate under pitch fault situation causing massive flow separation to occur. Pure Blade Element Momentum (BEM) approaches are not designed for this situation and inaccurate load predictions are already expected. In the present studies, BEM predictions are improved through the inclusion of a stall delay model for a wind turbine rotor operating under pitch fault situation of -2.3° towards stall. The accuracy of the stall delay model is assessed by comparing the results with available Computational Fluid Dynamics (CFD) simulations data.

1. Introduction

The increased fossil fuel depletion in past decades has led to a shortage in electricity production. It causes an increasing need in finding alternatives to fossil energy sources. Wind energy has been identified as one of the most promising sources for renewable energy industry [1, 2]. Currently, there are two categories of modern wind turbines, namely Horizontal Axis Wind Turbines (HAWTs) and Vertical Axis Wind Turbines (VAWTs), which are used mainly for electricity generation. HAWTs are more commercially dominant since they have received significant research and development works over the decades [3].

Several control mechanisms have been investigated and applied in wind turbine rotors, mainly to maintain a stable power production and to alleviate occurring unsteady aerodynamic loads on the rotor. For example, the development of an active pitch control system can reduce the tendency of massive flow separation on wind turbine blade at high wind speed cases [4]. It is agreed that blade pitch control system is critical for turbine operation, as pitching is an important action for enhancing energy capture, mitigating operational load, stalling and aerodynamic braking [5]. In pitch fault condition, a wind turbine rotor may be imposed to strong aerodynamic loads that can be dangerous to the structural stability of this rotating machine. Bangga et al. [6–9] documented that massive separation occurs in this situation and can cause three dimensional (3D) effects or rotational augmentation (RA).

Blade Element Momentum (BEM) approaches are not designed for the aforementioned conditions since the methods are based only on the 2D airfoil polar data and no three-dimensionality can be captured [10]. Even so, the methods are, ironically, widely used in the early design phase and load prediction of wind turbines because of its robustness and low computational cost. Thus, inaccurate prediction of wind rotor loads under pitch-faulted conditions is already expected. BEM often underpredicts the rotor loads in the blade root area due to the absence of RA in these simple approaches [6].

The main purpose of the present studies is to improve BEM predictions for a wind turbine rotor operating under pitch-fault condition of -2.3° towards stall. This is done through the inclusion of a stall delay model to correct the 2D airfoil polar data. The investigations are carried out on the generic MEXICO (Model Experiments in Controlled Conditions) rotor [11] studied in the MEXNEXT project. CFD simulation data obtained from Bangga et al. [8, 12] is used to verify the BEM computations. The following sections are divided into: description of BEM and the employed stall delay model in Sections 2 and 3, respectively; results of the simulations and discussion in Section 4; and all conclusions are summarized in Section 5.

2. Blade element momentum theory

BEM theory (BEMT) has its origins in momentum theory and the development from this to BEMT is well explained in many texts [13–15]. The mathematical descriptions of this approach are given in the following discussions, divided into two sections. The first one is to use momentum balance on a rotating annular stream tube passing on a turbine (momentum theory). The second is to examine the forces generated by the airfoil sections at various sections along the blade (blade element theory). These two methods then give a series of equations that can be solved iteratively (BEMT).

2.1. One dimensional momentum theory and the momentum transfer

In this section, one dimensional momentum theory as the foundation of the BEM theory is described. The approach assumes the flow to be steady, inviscid, incompressible and axisymmetric. The rotor in this case is modelled as a frictionless permeable actuator disc which is assumed to impart no rotational velocity to the flow [13]. The momentum theory basically consists of a control volume for conservation of mass, axial and angular momentum balances, and energy conservation [16]. This surrounds the actuator disc and is bounded by a

IOP Conf. Series: Journal of Physics: Conf. Series **974** (2018) 012020 doi:10.1088/1742-6596/974/1/012020

stream-tube, with two cross sections far upstream and far downstream of the disc [13]. The used general assumption is that the stream-tube has no interaction with the fluid flow outside of the boundary. The actuator disc removes energy from the stream-tube by providing a drag (axial) force that produces a pressure drop in the fluid just downstream of the disc. Both upstream and downstream surfaces of the stream-tube are assumed at ambient static pressure and so the flow speed must drop downstream to satisfy Bernoulli's equation.

Using Equations (1) and (2), the axial force can be defined in terms of the pressure differential, upstream-and-downstream-vicinity of the actuator disc. By assuming the cross sectional areas of the stream tubes just upstream and downstream of the disc are effectively the area of the disc, A_{disc}, the axial force can be rewritten as in Equation (3)

$$p_{amb} + \frac{1}{2}\rho U^2 = p_{ud} + \frac{1}{2}\rho U_{disc}^2, \tag{1}$$

$$p_{dd} + \frac{1}{2}\rho U_{disc}^2 = p_{amb} + \frac{1}{2}\rho U_1^2, \tag{2}$$

$$F_A = \frac{1}{2}\rho A_{disc}\left(U^2 - U_1^2\right). \tag{3}$$

The axial force can also be calculated using the conservation of momentum as follows:

$$F_A = \dot{m}\left(U - U_1\right). \tag{4}$$

Equating Equations (3) and (4) yields

$$U_{disc} = \frac{U + U_1}{2}. \tag{5}$$

Introducing the axial induction or interference factor, a, as the fractional reduction in flow speed betweeen free-stream and the actuator disc.

$$a = \frac{U - U_{disc}}{U} \tag{6}$$

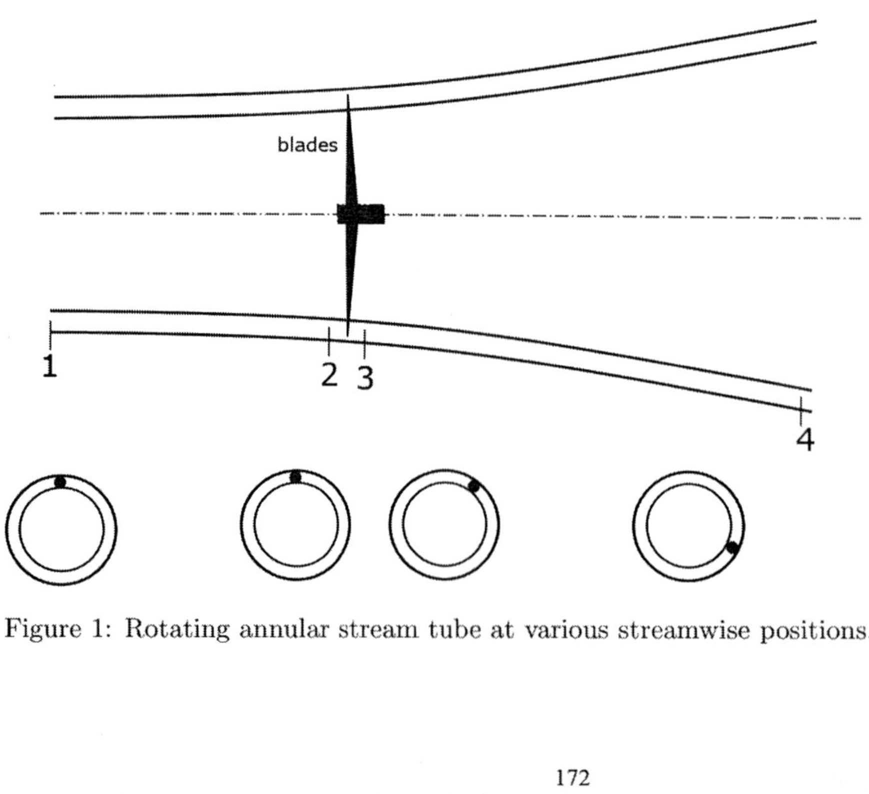

Figure 1: Rotating annular stream tube at various streamwise positions.

$$U_{disc} = U\left(1-a\right) \tag{7}$$

$$U_1 = U\left(1-2a\right) \tag{8}$$

The axial force in Equation (3) then can be rewritten as in Equation (9)

$$F_A = 2\rho A_{disc} U^2 a\left(1-a\right) \tag{9}$$

Using the axial induction factor with Equation (3), one may obtain the following equation for power extraction:

$$P = 2\rho A_{disc} U^3 a\left(1-a\right)^2. \tag{10}$$

According to one dimensional momentum theory, there is a pressure drop (energy lost) when the wind flows through the rotor. Some of the energy loss from this axial flow is converted into rotational momentum of the stream-tube, as a reaction to the rotational torque imparted to the turbine rotor thus rotating annular stream tube is then introduced.

Rotating annular stream tube is shown in Figure 1. Four stations are shown in the diagram, some way upstream of the turbine (1), just before the blades (2), just after the blades (3) and some way downstream of the blades (4). As the wind passes between stations 2 and 3, the motion of the turbine causes the wind to rotate in the wake downstream of the turbine. The blade wake rotates with an angular velocity ω and the blades rotate with an angular velocity of Ω. Recalling the conservation of angular momentum, the torque calculation of a rotating annular element of fluid at a radius r can be written as

$$dT = d\dot{m}\left(\omega r\right)r = \rho U_{disc} 2\pi r dr\left(\omega r\right)r \tag{11}$$

by introducing the angular (tangential) induction factor $b = \frac{\omega}{2\Omega}$, the elemental torque can then be rewritten as

$$dT = 4b(1-a)\rho U\Omega r^2 \pi r dr \tag{12}$$

2.2. Blade element theory

As described earlier in Section 2, the blade element theory divides the blades into discrete span-wise elements (airfoils). This theory relies on two key assumptions: (1) There are no aerodynamic interactions between the 2D airfoil elements in different sections, and (2) the forces acting on the elements are solely determined by the lift and drag characteristics of the airfoils shape.

In Figure 2, the definitions of the velocity vector and the lift (L) and drag (D) acting on the blade section are illustrated. The total angle between the circumferential direction to the relative inflow (V) is denoted by ϕ. This angle is composed by the local angle of attack (α) and the sum of the pitch and twist angle (θ). It shall be noted that U_∞ defines the inflow wind speed and Ω is the rotational speed of the turbine. The variable U_∞ is the same as U used in Section 2.

The axial force and torque acting on the blade element can be obtained using the relations for lift and drag, that are represented by the lift (C_L) and drag (C_D) coefficients assuming the chord length (c) of the blade element is known from the geometrical model. These are formulated as follows

$$dF_A = N\frac{1}{2}\rho V^2 c(C_L\cos\phi + C_D\sin\phi)dr \tag{13}$$

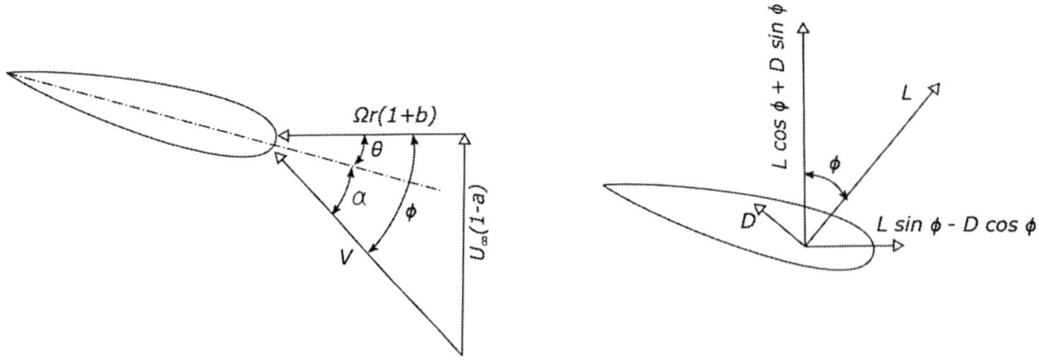

Figure 2: Velocity vector and forces acting on the blade element.

$$dT = N\frac{1}{2}\rho V^2 cr(C_L \sin\phi - C_D \cos\phi)dr. \tag{14}$$

Furthermore, by introducing a new parameter defining the ratio between the tangential relative to the inflow velocity, $\lambda = \frac{r\Omega}{U_\infty}$, the total flow angle can be redefined as [13]

$$\phi = \tan^{-1}\left(\frac{U(1-a)}{r\Omega(1+b)}\right) = \tan^{-1}\left(\frac{(1-a)}{\lambda(1+b)}\right) \tag{15}$$

3. Stall delay model

In this section, the mathematical formulation of the employed stall delay model is described. As mentioned briefly in Introduction, the BEM method is not designed to account for flow three-dimensionality as no interaction between each blade section to the neighbouring sections is assumed in the model. Thus, the approach depends strongly on the supplied aerodynamic polars. Due to this limitation, the 3D effects shall be stimulated in the given polars. In the present works, the 3D correction model based on Chaviaropoulos and Hansen [17] is used.

The model was developed by adopting the idea from Snel [18] to express the 3-D correction of the lift coefficient (C_L) as a fraction of the difference (ΔC_L) between the inviscid ($C_{L,INV}$) and the corresponding 2-D value ($C_{L,2D}$). In this model, the influence of twist angle is introduced in the ΔC_L multiplier. The model can be formulated as follows

$$C_{L,3D} = C_{L,2D} + a(c/r)^h \cos^n(\text{twist})(C_{L,INV} - C_{L,2D}) \tag{16}$$

where $a = 2.2$, $h = 1.3$ and $n = 4$. Chaviaropoulos and Hansen [17] also introduced a correction for drag coefficient (C_D) increase. However, Bangga et al. [7] pointed out that the 3D behaviour for drag is not general, and even decreases for the AVATAR rotor. The drag reduction was also observed for the MEXICO rotor in [8]. Thus, no drag correction is applied in the present studies. The same correction model has been applied in [19].

4. Results and discussion

The present section describes the results of the BEM simulation with and without the inclusion of the stall delay model. The computations were carried out using the QBlade code [20] which has been developed at the Technical University of Berlin.

4.1. Polar correction

Figure 3 presents the lift coefficient polar after being corrected by the stall delay model described in Section 3 compared to the corresponding 2D insviscid and 2D viscous lift coefficient values.

Four different blade sections are presented, namely $0.2R$, $0.3R$, $0.4R$ and $0.6R$. It shall be noted that the value of the chord to radius ratio (c/r) decreases as the radial distance from the center of rotation increases.

It can be clearly seen that the 3D effects the lift coefficient magnitude in the blade inboard area compared to the 2D viscous C_L. The corrected 3D lift coefficient is even close to the corresponding 2D inviscid polar at $0.2R$ in Figure 3a. This is reasonable since a strong radial flow occurs within the blade area where separation occurs. Despite that, the lift augmentation is mainly observed only in the inboard blade area where c/r is large. This becomes evident from Figure 3 as the lift augmentation becomes weaker at larger radial distances. According to Bangga et al [7], the augmentation becomes less pronounced at $c/r < 0.1$, and this corresponds to $r/R > 0.6$ for the MEXICO rotor. This behaviour is clearly observed in the Polar plots in Figure 3. A similar characteristic was also observed in the CFD studies in [8].

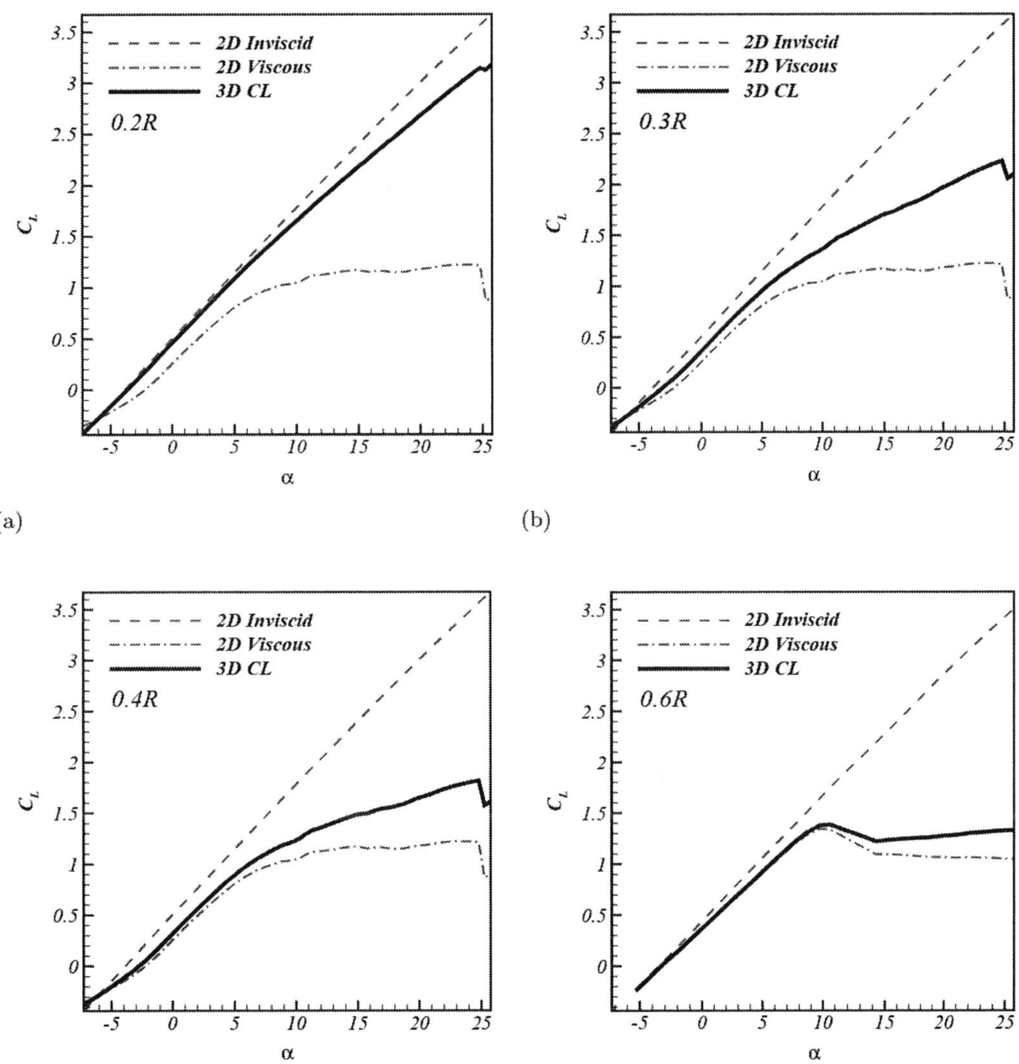

(a) (b) (c) (d)

Figure 3: 3D corrected polar in comparison to the 2D inviscid and viscous polars.

IOP Conf. Series: Journal of Physics: Conf. Series **974** (2018) 012020 doi:10.1088/1742-6596/974/1/012020

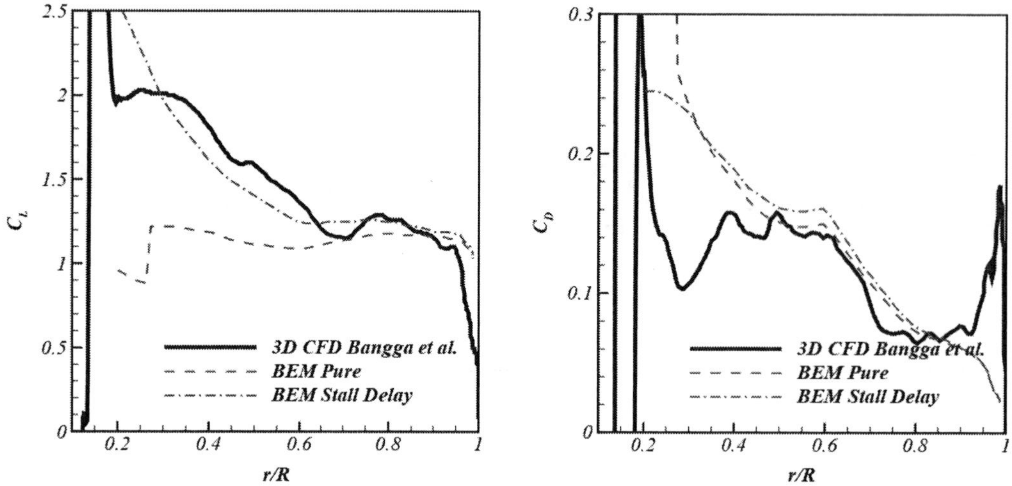

(a) (b)

Figure 4: Lift and drag coefficients along the blade span.

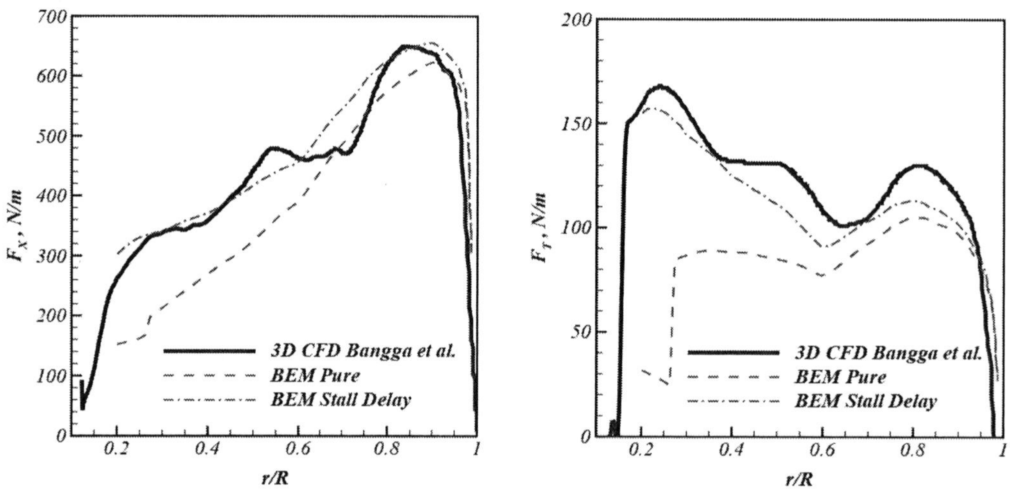

(a) (b)

Figure 5: Sectional loads acting on the blade.

4.2. Sectional blade forces

In this section, the results of the BEM simulation in terms of sectional loads are presented. The sectional lift and drag coefficients along the blade span are given in Figures 4a and 4b, respectively. The effects for lift is clear as demonstrated in Section 4.1. The lift coefficient increases in the inboard area compared to the 2D situation. Thus, the 3D CFD results from Ref. [8] are substantially higher than the pure BEM simulation without any 3D correction model. By the inclusion of the 3D model from haviaropoulos and Hansen [17], the BEM prediction is improved, becoming closer to the CFD data. It shall be noted, however, that there might be uncertainty provided by the angle of attack estimation, where the azimuthal averaging technique [21] was employed for the CFD results.

In Figure 4b, the drag coefficient distribution along the blade span is presented. The drag predicted by BEM with and without the stall delay model agrees reasonably well with the CFD data even though an overestimation is observed within $0.2 < r/R < 0.4$. Two reasons may cause this phenomenon: (1) the uncertainty in the angle of attack estimation and/or (2) the local downwash or upwash effect may alleviate the drag force as explained in [7]. It is worth mentioning that no drag correction was applied in the BEM simulation but the prediction for drag is different. This is mainly caused by the deviation in the prediction of the local induction factor of the blade section which affects the local angle of attack. As a result, drag is influenced since the direction of the vector is different.

Figure 5 presents the sectional blade loads in terms of the axial force (Fig. 5a) and tangential force (Fig. 5b) relative to the rotor plane. It is observed that the BEM simulation employing the stall delay model is able to mimic the CFD results fairly well for both forces. On the other hand, the pure BEM computation fails to predict the aerodynamic loads not only on the magnitude, but also on the trends and slope as well especially for the tangential force. This indicates that the power prediction using pure BEM simulation is inaccurate as the rotor driving moment is represented by the tangential force.

5. Conclusion and outlook

Numerical studies employing the Blade Element Momentum (BEM) approach have been performed and the results were compared to available CFD data from the literature. It was shown that the employed stall delay model augments the 3D lift coefficient polars at the inboard blade sections. The observed 3D effects becomes weaker as the ratio of the chord to radius (c/r) reduces. The observation confirmed that the limit of observable lift augmentation is for $c/r > 0.1$. The inclusion of this stall delay model is able to improve the prediction of the aerodynamic loads acting on the blade. The lift and drag distributions from the CFD data were reasonably mimicked by the BEM approach employing the stall delay model. In terms of the tangential load, a strong deviation was shown between the pure BEM computation compared to the CFD data, not only on the magnitude but also on the trends and slope. In contrast, the BEM simulation employing the stall delay model was able to mimic the behaviour of the tangential load accurately in comparison to the CFD data. This indicates that the inclusion of a 3D correction model is important for accurate load prediction using BEM. For future works, a new correction model will be developed and studied further.

Acknowledgements

The authors gratefully acknowledge the Ministry of Research, Technology and Higher Education of Indonesia for the funding through Directorate General of Higher Education (DGHE) scholarship. This paper is dedicated to F.F.S. Lesy who passed away on June 12, 2017 while accompanying the author G. Bangga in finishing the works. At last but not least, the authors would like to specially mention Via Vallen and Nella Kharisma for their assistance in creating a suitable working environment during the completion of the studies.

References

[1] Bangga G, Hutomo G, Wiranegara R and Sasongko H 2017 *Journal of Mechanical Science and Technology* **31** 261–267 URL `http://dx.doi.org/10.1007/s12206-016-1228-9`

[2] Bangga G, Lutz T, Dessoky A and Krämer E 2017 *Journal of Renewable and Sustainable Energy* **9** 053303 URL `http://dx.doi.org/10.1063/1.5003772`

[3] Hutomo G, Bangga G and Sasongko H 2016 *Applied Mechanics and Materials* **836** 109–114 URL `https://doi.org/10.4028/www.scientific.net/AMM.836.109`

[4] Bossanyi E 2003 *Wind energy* **6** 119–128

[5] Lu B, Li Y, Wu X and Yang Z 2009 *Power Electronics and Machines in Wind Applications, 2009. PEMWA 2009. IEEE* pp 1–7

[6] Bangga G, Kim Y, Lutz T, Weihing P and Krämer E 2016 *Journal of Physics: Conference Series* **753** 022026 URL `https://doi.org/10.1088/1742-6596/753/2/022026`
[7] Bangga G, Lutz T, Jost E and Krämer E 2017 *Journal of Renewable and Sustainable Energy* **9** 023304 URL `http://dx.doi.org/10.1063/1.4978681`
[8] Bangga G, Lutz T and Krämer E 2017 *Journal of Mechanical Science and Technology* **31** 3839–3844 URL `http://dx.doi.org/10.1007/s12206-017-0728-6`
[9] Bangga G, Weihing P, Lutz T and Krämer E 2018 *Hybrid RANS/LES simulations of the three-dimensional flow at root region of a 10 MW wind turbine rotor. New Results in Numerical and Experimental Fluid Mechanics XI* (Springer) pp 707–716 URL `http://dx.doi.org/10.1007/978-3-319-64519-3_63`
[10] Shen W Z and Sørensen J N 1999 *Jornal of Computational Physics* **150** 518–548
[11] Boorsma K and Schepers J 2014 New mexico experiment Tech. rep. ECN Wind Energy, Netherlands
[12] Bangga G, Weihing P, Lutz T and Krämer E 2017 *Journal of Mechanical Science and Technology* **31** 2359–2364 URL `http://dx.doi.org/10.1007/s12206-017-0432-6`
[13] Masters I, Chapman J, Willis M and Orme J 2011 *Journal of Marine Engineering & Technology* **10** 25–35
[14] Moriarty P J and Hansen A C 2005 *AeroDyn theory manual* (Citeseer)
[15] Manwell J F, McGowan J G and Rogers A L 2010 *Wind energy explained: theory, design and application* (John Wiley & Sons)
[16] Sørensen J N 2016 *General Momentum Theory for Horizontal Axis Wind Turbines* (Springer) pp 99–121
[17] Chaviaropoulos P and Hansen M O 2000 *Journal of Fluids Engineering* **122** 330–336
[18] Snel H, Houwink R, Bosschers J, Piers W J, van Bussel G J W and Bruining A 1993 *EWEC 3959*
[19] Bangga G, Guma G, Lutz T and Krämer E 2018 *Wind Engineering* (To be published)
[20] Marten D, Wendler J, Pechlivanoglou G, Nayeri C and Paschereit C 2013 *IJETAE* **3** 264–269
[21] Hansen M, Srensen N, Srensen J and Michelsen J 1997 499501

IOP Conf. Series: Journal of Physics: Conf. Series **974** (2018) 012021 doi:10.1088/1742-6596/974/1/012021

Slow and fast dynamics model of a Malaria with Sickle-Cell genetic disease with multi-stage infections of the mosquitoes population

Shanti Dewi Siawanta
Dept. of Mathematics, Faculty of Mathematics and Natural Sciences, Universitas Gadjah Mada, Sekip Utara Yogyakarta 55281, INDONESIA.
E-mail: shantidewisiawantasds@gmail.com

Fajar Adi-Kusumo
Dept. of Mathematics, Faculty of Mathematics and Natural Sciences, Universitas Gadjah Mada, Sekip Utara Yogyakarta 55281, INDONESIA.
E-mail: f_adikusumo@ugm.ac.id

Aluicius Irwan Endrayanto
Dept. of Mathematics, Faculty of Mathematics and Natural Sciences, Universitas Gadjah Mada, Sekip Utara Yogyakarta 55281, INDONESIA.
E-mail: endrayanto@ugm.ac.id

Abstract. Malaria, which is caused by Plasmodium, is a common disease in tropical areas. There are three types of Plasmodium i.e. Plasmodium Vivax, Plasmodium Malariae, and Plasmodium Falciparum. The most dangerous cases of the Malaria are mainly caused by the Plasmodium Falciparum. One of the important characteristics for the Plasmodium infection is due to the immunity of erythrocyte that contains HbS (*Haemoglobin Sickle-cell*) genes. The individuals who has the HbS gene has better immunity against the disease. In this paper, we consider a model that shows the spread of malaria involving the interaction between the mosquitos population, the human who has HbS genes population and the human with normal gene population. We do some analytical and numerical simulation to study the basic reproduction ratio and the slow-fast dynamics of the phase-portrait. The slow dynamics in our model represents the response of the human population with HbS gene to the Malaria disease while the fast dynamics show the response of the human population with the normal gene to the disease. The slow and fast dynamics phenomena are due to the fact that the population of the individuals who have HbS gene is much smaller than the individuals who has normal genes.

1. Introduction

Malaria is one of the most common disease which is transmitted by female Anopheles Mosquitos and happens mostly on the tropical areas and also some subtropical areas. According to WHO's data, it is reported that there are 214 million clinical cases of malaria in the world with 438.000 fatal cases, and about 3.2 billion people at risk. There are more than 30 of 400 species of

IOP Conf. Series: Journal of Physics: Conf. Series **974** (2018) 012021 doi:10.1088/1742-6596/974/1/012021

Anopheles plays role as the vector of Malaria. The transmission intensity depends on some factors such as the parasite, the vector, the human host and the environment, see [13].

One of the interesting phenomena in Malaria infection is that the individuals who have the Sickle-cells anaemia genetic disease have better endurance against the disease. One of the uniqueness of the Sickle-cell anaemia is that when the Plasmodium falciparum take the actin fibre from erythrocyte to build an intracellular-bridge for transporting protein that is made by the parasite into cell surface, the Sickle-cell will cut that bridge so it can not be used by the Plasmodium, see [1, 9, 12].

In this paper we consider a model of the Malaria transmission which involves the interactions between three populations i.e. the human population with normal genes (AA), the human population with Sickle-cells anaemia genes (AS), and the vector populations. The model is motivated by the study in [5] and [10]. The study in [5] is focused to analyze the interaction between individuals who have normal genes (AA), the individuals who have Sickle-cell genes (AS), and the vector population. However, the analysis in [5] did not consider the transmission process of the disease on the mosquitos population. The authors consider the slow-fast dynamics of the system that show the interactions between the populations.

The transmission process of the disease in the mosquitos population that has been studied in [10] is focused on the interaction dynamics between the human population with normal genes (AA) and the mosquitos population, which are separated into three compartments adopting the classical SIR epidemic model.

In this paper, we combine the human-mosquitos population model in [10] with the HbS cases of the human population. The model contains seven compartments, i.e. the uninfected and infected humans from malaria with genotype AS, the uninfected and infected humans from malaria with genotype AA, the mosquitos which are susceptible with Plasmodium parasites, mosquitos which have Plasmodium parasite in gametocyte form which can not transmit malaria, and mosquitos which can transmit malaria (the Plasmodium parasite is in sporozoite form).

We follow the assumptions in [5], i.e., the malaria infection rate and the malaria-induced death rate of AA individuals are higher than the one in AS individuals, the recovery rate from malaria in AA individuals are lower than in AS individuals, the background mortality rate of AS individuals is higher than AA individuals, the malaria epidemic occurs on a much faster timescale than their progress in sickle-cell gene frequency, and that SS individuals are never born.

2. Construction of the model

The interaction between the individuals with AA and AS genes are each classified into two compartments, i.e. the uninfected ($u_i, i = 1, 2$) and the infected ($v_i, i = 1, 2$) individuals by the plasmodium. The mosquitos population is classified into three compartments, i.e., the susceptible mosquitos (z_1), the mosquitos which has plasmodium but do not have ability to transmit the plasmodium (z_2), and the mosquitos which has ability to transmit the plasmodium (z_3). The total population of the mosquitos is $N_z = z_1 + z_2 + z_3$. The transfer diagram of our system is shown on Figure 1.

We assume that the number of individuals in the human population is N with $N = \sum_{i=1}^{2}(u_i + v_i)$ and the density dependency of the birth function is $B(N) = b(1 - \frac{N}{K})$ where the b is the birth rate of the human populations. The frequencies of the AA and AS genes are $p = \frac{2u_1+2v_1+u_2+v_2}{2N}$ and $q = \frac{u_2+v_2}{2N}$ and the death rate of the human with AA and AS genes are $m_1 = \mu$ and $m_2 = \mu + \vartheta$ where μ is the natural death rate of the human and the ϑ is the extra death rate of the human with AS gene. Lastly, $P_1 = p^2$ and $P_2 = 2pq$ show the proportion of the AA genes and AS genes in the human populations.

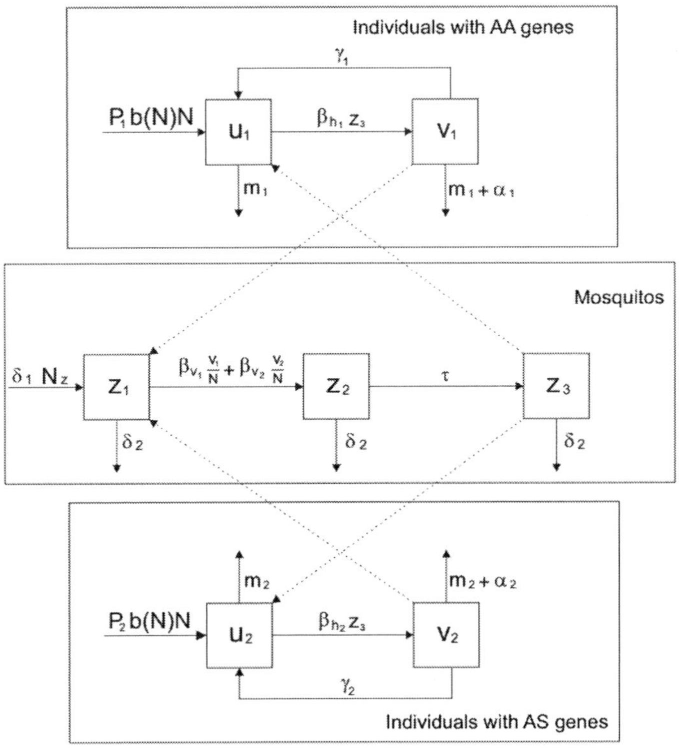

Figure 1. The Transfer Diagram

Based on the transfer diagram in Figure 1, we derive the mathematical model as follows.

$$\begin{aligned}
\frac{du_i}{dt} &= P_i B(N)N - m_i u_i - \beta_{h_i} z_3 u_i + \gamma_i v_i, \\
\frac{dv_i}{dt} &= \beta_{h_i} z_3 u_i - (m_i + \gamma_i + \alpha_i) v_i, \quad i = 1, 2, \\
\frac{dz_1}{dt} &= \delta_1 N_z - (\beta_{v_1} \frac{v_1}{N} + \beta_{v_2} \frac{v_2}{N}) z_1 - \delta_2 z_1, \\
\frac{dz_2}{dt} &= (\beta_{v_1} \frac{v_1}{N} + \beta_{v_2} \frac{v_2}{N}) z_1 - \tau z_2 - \delta_2 z_2, \\
\frac{dz_3}{dt} &= \tau z_2 - \delta_2 z_3,
\end{aligned}$$

see the Table 1 on the Appendix for the biological meaning of the parameters in our system.

3. Fast and slow dynamics

In this section, we introduce new variables and parameters transformation. By these transformations, we consider two time scales of the system which show the fast dynamic when the mosquitos population has interaction with individuals with normal genes and slow dynamics when the mosquitos has the interaction with individuals who have Sickle-cell genes.

The fast and slow dynamics on this paper is due to the fact that the individuals who have HbS gene are much smaller than the individuals with normal genes. This situation implies that the population dynamics of the individuals who have HbS gene is much slower compare with the population dynamics of the normal individuals.

Following [5], we assume that $m_1 < m_2$, $\gamma_1 \leq \gamma_2$, $\beta_{h_1} \geq \beta_{h_2}$, and $\alpha_1 \geq \alpha_2$. Due to the connection with realistic biological parameters the parameters $m_i, \alpha_i, i = 1,2$ and b are assumed much smaller than the other epidemiological parameters, see [5]. So that we can rescale those parameters, $m_i = \varepsilon \tilde{m_i}$, $\alpha_i = \varepsilon \tilde{\alpha_i}$, $b = \varepsilon \tilde{b}$, where ε is a small positive parameter. In [5], the authors define new variables that show the proportion of the human with AA and AS genes, i.e., $x_i = \frac{u_i}{N}$, $y_i = \frac{v_i}{N}$, $w = x_2 + y_2$, $i = 1,2$, where x_i, y_i show human populations of the respective types and w is the total frequency of genotype AS. We notice that $x_1 + y_1 + x_2 + y_2 = 1, x_1 = 1 - y_1 - w$, and $x_2 = w - y_2$. In this case we also consider two time-scales, i.e. the original time t which is referred as the fast time variable, and $\iota = \varepsilon t$ which is referred to the slow time variable. Then, we denote '.' $as\frac{d}{dt}$ and '/' $as\frac{d}{d\iota}$.

By using the the new variables and the rescaled parameters, we have the transformed systems of System (1) that shows the fast and slow dynamics of the model, i.e.,

$$
\begin{aligned}
\frac{dy_1}{dt} &= \beta_{h_1} z_3(1 - y_1 - w) - \gamma_1 y_1 - \varepsilon y_1[(\tilde{m_1} - \tilde{m_2})w + \tilde{\alpha_1}(1 - y_1) - \tilde{\alpha_2} y_2 + (P_1 + P_2)\tilde{b}(1 - \frac{N}{K})], \\
\frac{dy_2}{dt} &= \beta_{h_2} z_3(w - y_2) - \gamma_2 y_2 - \varepsilon y_2[(\tilde{m_1} - \tilde{m_2})(w - 1) - \tilde{\alpha_1} y_1 \\
&\quad + \tilde{\alpha_2}(1 - y_2) + (P_1 + P_2)\tilde{b}(1 - \frac{N}{K})], \\
\frac{dz_1}{dt} &= \delta_1 N_z - (\beta_{v_1} y_1 + \beta_{v_2} y_2) z_1 - \delta_2 z_1, \\
\frac{dz_2}{dt} &= (\beta_{v_1} y_1 + \beta_{v_2} y_2) z_1 - \tau z_2 - \delta_2 z_2, \\
\frac{dz_3}{dt} &= \tau z_2 - \delta_2 z_3, \\
\frac{dw}{dt} &= \varepsilon \ (\ ((1 - w)P_2 - wP_1)\tilde{b}(1 - \frac{N}{K}) + (\tilde{m_1} - \tilde{m_2})w(1 - w) \\
&\quad + \tilde{\alpha_1} w y_1 - \tilde{\alpha_2}(1 - w) y_2), \\
\frac{dN}{dt} &= \varepsilon N(\ (P_1 + P_2)\tilde{b}(1 - \frac{N}{K}) - \tilde{m_1}(1 - w) - \tilde{m_2} w \\
&\quad - \tilde{\alpha_1} y_1 - \tilde{\alpha_2} y_2).
\end{aligned} \tag{1}
$$

System (1) shows the fast dynamics, and the system

$$
\begin{aligned}
\frac{\varepsilon dy_1}{dt} &= \beta_{h_1} z_3(1-y_1-w) - \gamma_1 y_1 - \varepsilon y_1[(\tilde{m}_1-\tilde{m}_2)w + \tilde{\alpha}_1(1-y_1) \\
&\quad - \tilde{\alpha}_2 y_2 + (P_1+P_2)\tilde{b}(1-\frac{N}{K})], \\
\frac{\varepsilon dy_2}{dt} &= \beta_{h_2} z_3(w-y_2) - \gamma_2 y_2 - \varepsilon y_2[(\tilde{m}_1-\tilde{m}_2)(w-1) - \tilde{\alpha}_1 y_1 \\
&\quad + \tilde{\alpha}_2(1-y_2) + (P_1+P_2)\tilde{b}(1-\frac{N}{K})], \\
\frac{\varepsilon dz_1}{dt} &= \delta_1 N_z - (\beta_{v_1} y_1 + \beta_{v_2} y_2) z_1 - \delta_2 z_1, \\
\frac{\varepsilon dz_2}{dt} &= (\beta_{v_1} y_1 + \beta_{v_2} y_2) z_1 - \tau z_2 - \delta_2 z_2, \\
\frac{\varepsilon dz_3}{dt} &= \tau z_2 - \delta_2 z_3, \\
\frac{dw}{dt} &= (\ ((1-w)P_2 - wP_1)\tilde{b}(1-\frac{N}{K}) + (\tilde{m}_1-\tilde{m}_2)w(1-w) \\
&\quad + \tilde{\alpha}_1 w y_1 - \tilde{\alpha}_2(1-w)y_2), \\
\frac{dN}{dt} &= N(\ (P_1+P_2)\tilde{b}(1-\frac{N}{K}) - \tilde{m}_1(1-w) - \tilde{m}_2 w \\
&\quad - \tilde{\alpha}_1 y_1 - \tilde{\alpha}_2 y_2).
\end{aligned}
\tag{2}
$$

shows the slow dynamics of System (1). We note that y_1, y_2, z_1, z_2, z_3 are the fast variables and w, N are the slow variables.

In the next section, the fast dynamics of System (1) that is for $\varepsilon = 0$ will be discussed. We will also analyze the basic reproduction ratio of the system that determine the important parameters which influence the spreads of the disease. In the Numerical Simulation section, we will show the dynamics in full system, that is for $\varepsilon \neq 0$ and also the slow dynamics of the system, that is the dynamics for $\varepsilon = 0$ of System (2).

4. The basic reproduction ratio for the fast dynamics

For $\varepsilon = 0$, we get the new system from the system (1),

$$
\begin{aligned}
\frac{dy_1}{dt} &= \beta_{h_1} z_3(1-y_1-w) - \gamma_1 y_1, \\
\frac{dy_2}{dt} &= \beta_{h_2} z_3(w-y_2) - \gamma_2 y_2, \\
\frac{dz_1}{dt} &= \delta_1 N_z - (\beta_{v_1} y_1 + \beta_{v_2} y_2) z_1 - \delta_2 z_1, \\
\frac{dz_2}{dt} &= (\beta_{v_1} y_1 + \beta_{v_2} y_2) z_1 - \tau z_2 - \delta_2 z_2, \\
\frac{dz_3}{dt} &= \tau z_2 - \delta_2 z_3.
\end{aligned}
\tag{3}
$$

System (3) show the fast dynamics of the Malaria infection processes for the human with AA and AS genes.

By a straightforward calculation, the System (3) has the equilibrium point which represents the stedy state condition $({y_1}^*, {y_2}^*, {z_1}^*, {z_2}^*, {z_3}^*)$ with

$$\begin{aligned}
{y_1}^* &= \frac{T_{h_1}{z_3}^*}{1+T_{h_1}{z_3}^*}(1-w), \\
{y_2}^* &= \frac{T_{h_2}{z_3}^*}{1+T_{h_2}{z_3}^*}w, \\
{z_1}^* &= \frac{C_{11}{z_3}^*\delta_1}{(\tau+\delta_2)(T_{v_1}{y_1}^*+T_{v_2}{y_2}^*+1)-\delta_1(C_{11}+T_{v_1}{y_1}^*+T_{v_2}{y_2}^*)} \\
{z_2}^* &= \frac{T_{v_1}{y_1}^*+T_{v_2}{y_2}^*}{C_{11}}{z_1}^*, \\
{z_3}^* &= \frac{-k_1+\sqrt{\Delta}}{2k_0}
\end{aligned} \tag{4}$$

where $s = \frac{\tau}{\delta_2}$, $C_{11} = 1+s$, $T_{v_i} = \frac{\beta_{v_i}}{\delta_2}$, $T_{h_i} = \frac{\beta_{h_i}}{\gamma_i}$, $R_i = T_{v_i}T_{h_i}, i = 1,2$, $k_0 = R_1(1-w)T_{h_2} + R_2wT_{h_1} + T_{h_1}T_{h_2}$, $k_1 = R_1(1-w)+R_2w+T_{h_1}+T_{h_2}$, and $\Delta = {k_1}^2 - 4k_0$.

We employ the next generation matrix methods to compute the basic reproduction number of the System (3) and we have

$$R_0 = \frac{\delta_1}{\delta_2}. \tag{5}$$

For $R_0 > 1$, the equilibrium point (4) is asymptotically stable. In this case, the disease goes to the steady state conditions.

The basic reproduction number in our case is different with the one in [5]. In our case the ratio depends only on the birth and death rates of the mosquitos population. The differences of the ratio is due to the appearance of the interaction dynamics of the malaria infection on the mosquitos population.

5. Numerical Simulations for the slow-dynamics

In this section we consider the slow dynamics of our model which is represented by System (2). For the numerical simulation, we use the similar parameter that being used in [5],

$$\begin{aligned}
\beta_{h_1} &= 0.12, & \beta_{h_2} &= 0.11, & \beta_{v_1} &= 0.1, & \beta_{v_2} &= 0.15, & \delta_1 &= 0.1, \\
\delta_2 &= 0.005, & \mu &= 0.00004, & \vartheta &= 0.00001, & \gamma_1 &= 0.01, & \gamma_2 &= 0.01, \\
\alpha_1 &= 0.0001, & \alpha_2 &= 0.00008, & b &= 0.00004, \\
K &= 10000, & \tau &= 0.0001, & N_z &= 1000.
\end{aligned} \tag{6}$$

where the time variable is on daily scale.

On Figure 2 we show that the solutions will be absorbed by the slow manifold and approach to the equilibrium in very slow time.

The situation can be interpreted as follows. The population of the individuals who has Sickle-cell genetic disease will grow or decay very fast into a certain condition that we call the slow-manifold. In the slow-manifold the population will increase or decrease very slow to reach the steady state condition. The situation is due to the fact that the individuals who have Sickle-cell genes has ability against the plasmodium falciparum infection. On this figure we show the interaction between infected individuals with AA genes and the frequency of the individuals with AS genes.

In those case we found that the solution will tend to the stable equilibrium point. From the position of the equilibrium point, we can conclude that the infected individuals with the normal genes has higher risk for the Plasmodium infection than the individuals with Sickle-cells anaemia

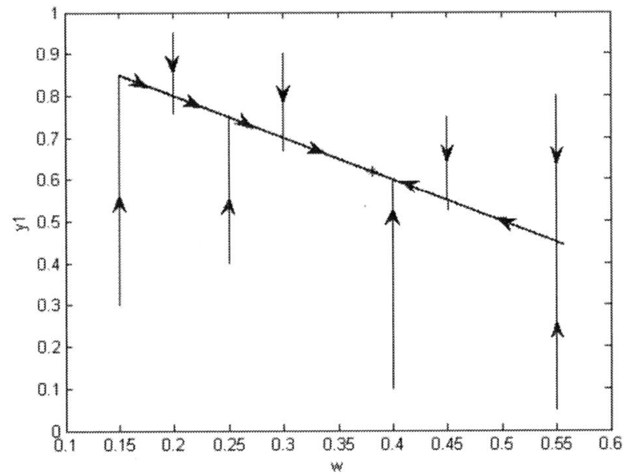

Figure 2. The dynamics on the slow manifold for System (1) that shows the interaction between the individuals with Sickle-cell genes with the total frequency of the AS genotype for $\varepsilon = 0.01$.

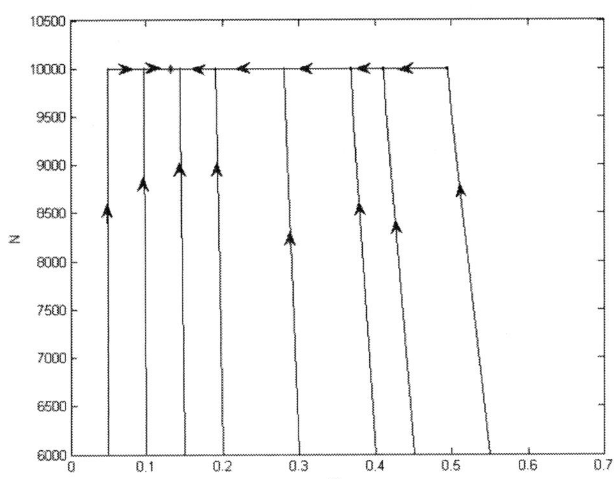

Figure 3. The interaction between the total population of the human and the total frequency of AS

genes. The spreads of the disease in AS population grow slowly to reach the equilibrium points, see Figure 3.

In Figure 3, we show that the human population grow rapidly until the carrying capacity where the slow-manifold related to the total frequency of AS genotype appear. The Figure 2 and Figure 3 show the long term behavior of the disease which is represented by the stable equilibrium point. This situation can be interpreted as follows. If the total number of the individuals reaches the carrying capacity, the number of individuals who has HbS genes play role to reduce the spreads of the plasmodium and the disease will be isolated into a certain number of individuals.

6. Concluding remarks

The sickle cells has the gene which prevents the growth of Malaria plasmodium. The slow-manifold in the system can be interpreted as follows. The individuals who has sickle cells anaemia disease has ability to survive from Malaria. So that, the growth of the Malaria in this population is much slower than the normal individuals population. In our model we show the steady state conditions of the spreads of the malaria on those populations that represents the longterm behavior of this disease in such populations. In this case, the number of individuals who has HbS genes play role to reduce the spreads of the plasmodium and the disease will be isolated into a certain number of individuals.

Acknowledgements

The authors would like to thank to Ministry of Research, Technology, and Higher Education Republic of Indonesia and the Dept. of Mathematics, Universitas Gadjah Mada (UGM) for the support during the research. The special thanks also for some colleagues in UGM for the discussions during the research.

References

[1] Anwar, C., Musral, Y., 1997, *Atlas Parasitologi Kedokteran*, Hipokrates, Jakarta.
[2] Anton, H., Rorres C., 2000, *Elementary Linear Algebra*, Eight Edition, John Wiley and Sons, Inc., New York.
[3] Diekmann, O., Heesterbeek, J.A.P, 2000, *Mathematical Epidemiology of Infectious Diseases : Model Building, Analysis, and Interpretation*, John Wiley and Sons, Inc., New York.
[4] Diekmann, O., Heesterbeek, J.A.P, Roberts, M.G., 2010, The Construction of Next-generation Matrices for Compartmental Epidemic Models, *J. R. Soc. Interface*, 7, 873-885.
[5] Feng, Z., Yi, Y. , Zhu, H. , 2004, Fast and Slow Dynamics and the S-gene Frequency, *J Dyn Diff Equat*, 16:869 .
[6] Fuller, G., Tarwater, D. , 1986, *Analytic Geometry*, Sixth Editions, Addison-Wesley Publishing Company, Inc, USA.
[7] Kocak, H. and Hole, J.K., 1991, *Dynamic and Biffurcation*, Springer-Verlag, New York.
[8] Perko, L., 1991, *Differential Equation and Dynamical System*, Springer-Verlag New York, Inc., USA.
[9] Porth, C. M., 2011, *Essentials of Pathophysiology Concepts, Third Edition*, Wolters Kluwer Health - Lippincott Williams & Wilkins, China.
[10] Chitnis, N., Cushing, J. M., Hyman, J. M., 2006, Bifurcation Analysis of A Mathematical Model for Malaria Transmission, *SIAM J. Appl. Math*, Vol 67, No. 1, pp.24-45.
[11] Verhulst, F. , 1939, *Nonlinear differential equations and dynamical systems*, Springer-Verlag New York, Inc., USA.
[12] Wahlgren, M., Perlmann, P., 2005, *Malaria : Molecular and Clinical Aspects*, Harwood Academic Publishers, Singapore.
[13] WHO 2016, *Malaria : The Fact Sheet* , WHO Media Centre, http://www.who.int/mediacentre/factsheets/fs094/en/

Appendix.
The variables and parameters which is used in the system has been described in Table 1

Symbol	Description	Symbol	Description
u_1	the number of uninfected people with genotypes AA.	p	the frequencies of the AA gene.
u_2	the number of uninfected people with genotypes AS.	q	the frequencies of the AS gene.
v_1	the number of infected people with genotypes AA.	P_1	the proportion of the genotypes AA.
v_2	the number of infected people with genotypes AS.	P_2	the proportion of the genotypes AS.
N	the total population size of humans.	$B(N)$	a density dependent birth function.
z_1	the fraction mosquitos that are susceptible with Plasmodium parasites	β_{h_i}	the malaria infection rate of a human of type i.
z_2	the fraction mosquitos that have Plasmodium parasite (the form is gametocyte) but cant transmit malaria.	β_{v_i}	the malaria infection rate of a mosquito from bitting people of type i.
z_3	the fraction mosquitos that can transmit malaria (the Plasmodium parasite's form is sporozoit)	τ	the growth rate per capita of mosquitos from the form parasite is gametocyte into sporozoit.
α_i	the additional death rate of infected individuals due to malaria.	$\frac{1}{\gamma_i}$	the average time until a victim of malaria recovers.
μ	the natural death of human.	δ_1	the birth rate per capita of mosquitos.
ϑ	the extra death rate of AS individuals due to S gene.	δ_2	the death rate per capita of mosquitos.
K	carrying capacity.	N_z	the total population size of mosquitos.

Table 1. Variables and parameters which are used on the Compartment diagram 1

Profile of Metacognition of Mathematics and Mathematics Education Students in Understanding the Concept of Integral Calculus

La Misu[1,3], I Ketut Budayasa[2,4], and Agung Lukito[2,5]

[1]Department of Mathematics Education Universitas Halu Oleo, Kendari-Indonesia
[2]Department of Mathematics Universitas Negeri Surabaya-Indonesia

Email: [3]lamisuhamid@yahoo.co.id, [4]ketutbudayasa@unesa.ac.id, [5]agunglukito@unesa.ac.id

Abstract. This study describes the metacognition profile of mathematics and mathematics education students in understanding the concept of integral calculus. The metacognition profile is a natural and intact description of a person's cognition that involves his own thinking in terms of using his knowledge, planning and monitoring his thinking process, and evaluating his thinking results when understanding a concept. The purpose of this study was to produce the metacognition profile of mathematics and mathematics education students in understanding the concept of integral calculus. This research method is explorative method with the qualitative approach. The subjects of this study are mathematics and mathematics education students who have studied integral calculus. The results of this study are as follows: (1) the summarizing category, the mathematics and mathematics education students can use metacognition knowledge and metacognition skills in understanding the concept of indefinite integrals. While the definite integrals, only mathematics education students use metacognition skills; and (2) the explaining category, mathematics students can use knowledge and metacognition skills in understanding the concept of indefinite integrals, while the definite integrals only use metacognition skills. In addition, mathematics education students can use knowledge and metacognition skills in understanding the concept of both indefinite and definite integrals.

1. Introduction

Integral calculus is one of the compulsory courses programmed by students of Mathematics Department and Mathematics Education Department. When viewed from the graduates of the Department of Mathematics Education is to produce a bachelor candidate for high school mathematics teacher, so it is expected to be able to comprehend integral calculus material, in addition to apply it to everyday life, and other relevant courses, also can teach it to students in high school. While the graduates of Department of Mathematics are producing pure mathematics scholars, so it is expected to be able to comprehend the material of integral calculus, and can apply it to everyday life, as well as other relevant subjects [4].

Based on the difference of interest of students in the selection of Mathematics Department and Mathematics Education Department, it is possible to differentiate students' metacognition in understand mathematical concepts. This can be seen from the difference of mathematics learning outcomes of students in high school and in the first semester of Mathematics Department or Mathematics Education Department. When viewed from the average value of the national exam of high school mathematics, for students better mathematics education than math students (63.58> 59.0), while the average achievement index first semester, mathematics students better than mathematics education students (3.17> 3.07).

The metacognition profile is a natural and intact description of a person's cognition that involves his own thinking in terms of his or her knowledge, and the ability to plan and monitor his thinking process, and evaluate the process and outcome of one's thinking when understanding a concept. According to Koriat, in Goh [8], Metacognition refers to what people know about cognition in general and about their own cognitive processes, in particular, as well as how they use this knowledge to adjust their informational processes and behavior. While cognitive understanding is a process that occurs internally within the central nervous system at the time of human thinking (Gagne [7]). Then Neisser [14] says that cognition is the acquisition, arrangement, and use of knowledge.

The concept of integral calculus that is examined in this paper is the concept of integral indeterminate and integral of course. To understand the concept of integral calculus using the Bloom theory developed by Anderson et al. [1], and Mayer [12], and can be seen in Table 1.

Table 1: Categories in understand concepts by Anderson et al. and Mayer

Categories	Indicator	Definition
1. Interpreting	Clarifying, Paraphrasing , Representing , and Translating.	Changing from one form of representation to another
2. Exemplifying	Illustrating, and Instantiating	Finding a specific example or illustration of a concept or principle.
3. Classifying	Categorizing and Subsuming	Determining that something belongs to a category
4. Summarizing	Abstracting, and generalizing	Abstracting a general theme or major point(s)
5. Inferring	Concluding, Extrapolating, Interpolating, and Predicting	Drawing a logical conclusion from presented information
6. Comparing	Contrasting, Mapping, and Matching	Detecting similarities and differences between two or more objects, events, ideas, problems, or situations
7. Explaining	Constructing models	Constructing a cause and effect model of a system

This paper discusses only two categories of conceptual understanding, summarizing and explaining. The metacognition profile in understanding the concept of integral calculus especially the summarizing and explaining category can be seen from 2 components of metacognition, ie cognitive knowledge consists of declarative knowledge, procedural knowledge, and conditional knowledge (Flavell [6]) and metacognition skills consisting of planning, monitoring and evaluating (Dawson [5], Livingston [10], and CCD Malaysia [3]).

The students' metacognition profile review was distinguished between mathematics students and mathematics education students. Thus, the purpose of this study was to produce the metacognition profile of mathematics and mathematics education students in understanding the concept of integral calculus.

2. Method

This research method is the explorative method with a qualitative approach. This explorative research is intended to explore the metacognition of mathematics students and mathematics education in understanding the concept of integral calculus so as to obtain a large profile of students' metacognition. The subjects of study (RS) are students of the mathematics department and Mathematics Education department who are studying integral calculus, consisting of 1 mathematics student and 1 student of mathematics education. Selection of the subject of this study based on the highest score of the Mathematics Ability Test (score >70), and consider the value of student achievement. The main data collection of this research was obtained by using interview technique. In addition, there is supporting data which is the result of written work of the research subject in understand integral calculus task. The data analysis process follows analysis model of Miles and Huberman[13], consisting of: (1) data reduction, data presentation, and (3) conclusion. Interpretation of data carried out simultaneously with data presentation activities. In this activity, researchers do an interpretation of research data (answers RS at the time of interview). Making conclusions and recommendations based on the results of data processing to answer research questions, namely finding the profile of students' metacognition in understanding the concept of integral calculus.

3. Results and Discussion

The results of this study reveal the metacognition profile of mathematics and mathematics education students in understanding the concept of integral calculus, especially the categories of summarizing and explaining. To reveal the metacognition profile of the students, first, look at the understanding of mathematics and mathematics education students about the concept of integral calculus based on the students' answers in completing the task of understanding the concept of integral. Second, through interviews can reveal the metacognition of students in understanding the concept of integral calculus.

3.1. Understanding of mathematics and mathematics education students on the concept of integral calculus on the Summarizing Category

The answers of mathematics and mathematics education students in understanding the concept of integral calculus in the summarizing category can be seen in Table 2.

Table 2: The Answers to mathematics and mathematics education students in understanding the concept of integral calculus in the Summarizing category

Form of task	Answers of Students	
	Mathematics	Mathematics Education
a. $D_x[x^4]=4x^3$; $D_x[x^4+3]=4x^3$ and $D_x[x^4-2]=4x^3$ Define $\int 4x^3\ dx$	$\int 4x^3dx = x^4 + C$	$\int 4x^3dx = x^4 + C$
b. Given $\int_0^2 \lvert x\rvert\ dx = \int_0^2 x\,dx = F(2) - F(0) = 2$ where F is the antiderivative of x. Given $\int_{-2}^0 \lvert x\rvert\ dx = \int_{-2}^0 (-x)\ dx = 2$ Define $\int_{-2}^2 \lvert x+2\rvert\ dx$	$\int_{-2}^2 \lvert x+2\rvert dx =$ $\int_{-2}^0 -(x+2)dx + \int_0^2 (x+2)dx =$ - (½(0)²+2(0) – (1/2(-2)² +2(-2)) + ½ (2)²+2(2)) = 2-4+2+4 = 4	$\int_{-2}^2 \lvert x+2\rvert dx =$ $\int_{-2}^0 -(x+2)dx + \int_0^2 (x+2)dx =$ - (½(0)²+2(0) – (1/2(-2)² +2(-2)) + ½ (2)²+2(2)) = 2-4+2+4 = 4

Based on the results of the answers in Table 2, mathematics and mathematics education students can summarize the main points given in the form of indefinite integrals so that the answer is correct. Meanwhile, the main points given in the form of definite integrals cannot be summarized so the answer is wrong.

3.1.1. Results of mathematics student interviews about understanding the concept of integral calculus on the summarizing category

Results of mathematics student interviews about understanding the concept of integral calculus on the category of summarizes consists of the concept of indefinite integrals and definite integrals:

3.1.1.1. The Indefinite Integral Concepts

The results of mathematics student interviews in understanding the concept of indefinite integrals in the Summarizing category as follows: (1) Students of mathematics can summarize the main points given in the form of improper integrals correctly ie $\int 4x^3\ dx = x^4 + C$, (2) The mathematics students can give the reason that $\int 4x^3\ dx = x^4 + C$ is based on the pattern that some x^4 functions are added with a different number but the derivatives are the same. (3) Student of mathematics can use procedure to get $\int 4x^3\ dx = x^4 + C$, that is $\int 4x^3$ anti derived from function x^4 although have different konstata, so result x^4 in added C, (4) Students of mathematics have not been able to determine the hook elements about the results obtained from the indefinite integral whether a function or not a function, (5) Students of mathematics can explain easily from the summary of the main points given in the form of indefinite integrals, and (6) Students of mathematics can emphasize that summarizing the main points given in the form of indefinite integrals to $\int 4x^3\ dx = x^4 + C$ is true.

Thus, results of mathematics student interviews in understanding the indefinite integral concepts indicate that they have used metacognition knowledge and metacognition skills in understanding the concepts of indefinite integrals in the summarizing category. Only metacognition skills in the planning component are not yet visible.

3.1.1.2. The Definite Integral Concepts
The results of mathematics student interviews in understanding the definite integral concepts on summarizing category as follows: (1) Students of mathematics cannot summarize the main points given in form the definite integrals ie $\int_{-2}^{2}|x+2|\,dx = \int_{-2}^{0}-(x+2)\,dx + \int_{0}^{2}(x+2)\,dx = 4$ is wrong (2) The mathematics student can not give the reason that the area of the function $|x + 2|$ on the interval [-2,2] is above the x-axis, (3) Math students are less able to use the procedure to obtain the area of function $|x + 2|$ at the interval [-2,2], (4) Math students have not been able to review their understanding that the area of function $|x + 2|$ on the interval [-2,2] is above the x-axis, (5) The mathematics student has not been able to explain the steps of calculating the area of the function $|x + 2|$ at the interval [-2.2], ie $\int_{-2}^{2}|x+2|dx = \int_{-2}^{2}(x+2)\,dx$, and (6) Students of mathematics have not mastered the procedure to obtain results from $\int_{-2}^{2}|x+2|dx = 8$.

Thus, the Results of mathematics student interviews in understanding the concept of integral certainly in the Summarizing category has not shown metacognition knowledge and metacognition skills.

3.1.2. Results of mathematics education student interviews about understanding the concept of integral calculus on the summarizing category
Results of mathematics education student interviews about understanding the concept of integral calculus on the category of summarizes consists of the concept of indefinite integrals and definite integrals:

3.1.2.1. The Indefinite Integral Concepts
The results of mathematics education students interviews in understanding the concept of indefinite integrals in the summarizing category as follows: (1) Students of mathematics education can summarize the main points given in the form of improper integrals correctly ie $\int 4x^3\,dx = x^4 + C$, (2) The mathematics education students can give the reason that $\int 4x^3\,dx = x^4 + C$ is based on the pattern that some x^4 functions are added with a different number but the derivatives are the same. (3) Student of mathematics education can use procedure to get $\int 4x^3\,dx = x^4 + C$, that is $\int 4x^3$ anti derived from function x^4 although have different konstata, so result x^4 in added C, (4) Mathematics education students can determine the hook elements about the results obtained from the indefinite integral is a relation, because the value of C is not single, (5) Students of mathematics education can explain easily from the summary of the main points given in the form of indefinite integrals, and (6) Students of mathematics education can emphasize that summarizing the main points given in the form of indefinite integrals to $\int 4x^3\,dx = x^4 + C$ is true.

Thus, the result of the mathematics education student's interview in understanding the indefinite integrals concepts indicate that it has used metacognition knowledge and metacognition skills in understanding the concept of indefinite integrals in the summarizing category.

3.1.2.2. The Definite Integral Concepts
The results of interviews of mathematics students in understanding the concept of definite integrals on summarizing category as follows: (1) Students of mathematics education cannot summarize the main points given in form the definite integrals ie $\int_{-2}^{2}|x+2|\,dx = \int_{-2}^{0}-(x+2)\,dx + \int_{0}^{2}(x+2)\,dx = 4$ is wrong (2) The mathematics education students cannot give the reason that the area of the function $|x + 2|$ on

IOP Conf. Series: Journal of Physics: Conf. Series **974** (2018) 012022 doi:10.1088/1742-6596/974/1/012022

the interval [-2,2] is above the x-axis, (3) Students of math education are less able to use procedures to obtain the area of function $|x + 2|$ at the interval [-2,2], (4) Students of mathematics education can review their understanding that the area of function $|x + 2|$ on the interval [-2, 2] is above the x-axis, (5) Students of mathematics education can explain the steps of calculating the area of function $|x + 2|$ at the interval [-2, 2], ie $\int_{-2}^{2}|x+2|dx = \int_{-2}^{2}(x+2)dx$, and (6) Students of mathematics education can master the procedure to obtain results from $\int_{-2}^{2}|x+2|dx = 8$.

So, the result of student interviews mathematics education in understanding the concept of definite integrals on summarizing category does not show knowledge of metacognition. After being reminded again of the concepts of definite integrals, students can use their metacognitive skills to understanding the concepts of definite integrals that.

3.2. Understanding of mathematics students and mathematics education on the concept of integral calculus on the explaining category

The answers of mathematics and mathematics education students in understanding the concept of integral calculus in the explaining category can be seen in Table 3.

Table 3: Answers to mathematics and mathematics education students in understanding the concept of integral calculus in the explaining category

Form of task	Answers of Students	
	Mathematics	Mathematics Education
a. Is known for the linearly properties of the definite integrals, ie if f and g have antiderivatives, then $\int [f(x)+g(x)]dx = \int f(x)dx + \int g(x)dx$ Based on your knowledge of the polynomial function, and if f the polynomial function then try to explain whether $\int [f(x)]\, dx$ apply this trait.	Let $f(x)$ be a polynomial function with the form $f(x) = x^2 + 2x + 3$. To integrate the function of this polynomial can be done one by one ie: $\int (x^2+2x+3)dx =$ $\int x^2 dx + \int 2x\, dx + \int 3dx$	Let $f(x)$ be a polynomial function with the form $f(x) = a_n x^n + a_{n-1}x^{n-1} + \ldots + a_1 x + a_0$. To integrate the function of this polynomial can be done one by one ie: $\int (a_n x^n + \ldots + a_1 x + a_0)dx =$ $\int a_n x^n dx + .. + \int a_1 x dx + \int a_0 dx$
b. Is known of property of additivity for definite integrals, ie if f is integral to an interval containing three dots a, b, and c respectively, then $\int_a^c f(x)dx = \int_a^b f(x)dx + \int_b^c f(x)dx$. Based on your knowledge of the definitions of definite integrals, and the sequence properties of the real numbers, try to explain whether this property applies also to certain intervals containing four consecutive points a, b, c, and d when f is integrated at that interval.	If f on the interval [a, d] contains four dots a, b, c, and d respectively, illustrated  Then L = $\int_a^b f(x)dx + \int_b^c f(x)dx +$ $\int_c^d f(x)dx$	If f on the interval [a, d] contains three dots a, b, and c respectively, illustrated  Then L = $A_1 + A_2$, so $L = \int_a^b f(x)dx + \int_b^c f(x)dx$ If f on the interval [a, d] contains four dots a, b, c, and d respectively, illustrated  Then L = $A_1 + A_2 + A_3$, so $L = \int_a^b f(x)dx + \int_b^c f(x)dx + \int_c^d f(x)dx$

Based on the results of the answers in Table 3, the mathematics student can explain a model of the consequences of a given property of either indefinite integrals or definite integrals, but the explanation is very brief and not detailed. While of mathematics education students can explain a model of the consequences of a given property of either indefinite integrals or definite integrals, with a very detailed explanation.

3.2.1. Results of mathematics student interviews about understanding the concept of integral calculus on the explaining category

Results of mathematics student interviews about understanding the concept of integral calculus on the category of explaining consists of the concept of indefinite integrals and definite integrals:

3.2.1.1. The Indefinite Integral Concepts

The results of the mathematics student interviews in understanding the concept of indefinite integrals in the explaining category as follows: (1) Students of mathematics can explain a model of an indefinite integral with a polynomial function based on the linearly properties, ie $\int(x^2+2x+3)dx$ be $\int x^2dx+\int 2x\,dx+\int 3dx$, (2) Students of mathematics can give the reason that the polynomial function applies the linearly properties so that $\int(x^2+2x+3)dx$ can be changed to $\int x^2dx+\int 2x\,dx+\int 3dx$, (3) Students of mathematics can use the procedure to explain $\int(x^2+2x+3)dx$ changes to $\int x^2dx+\int 2x\,dx+\int 3dx$, (4) Students of mathematics can show that the polynomial function applies the linearly properties, (5) Students of mathematics can show results from $\int(x^2+2x+3)dx$ same with result $\int x^2dx+\int 2x\,dx+\int 3dx$, and (6) Students of mathematics can conclude that the function of polynomial is a linear function then apply the linearly properties, so that $\int(x^2+2x+3)dx$ = $\int x^2dx+\int 2x\,dx+\int 3dx$.

Thus, the results of mathematical student interviews in understanding the concepts of indefinite integrals indicate that it has used metacognition knowledge and metacognition skills in understand the concept of indefinite integrals in the explaining category.

3.2.1.2. The Definite Integral Concepts

The results of interviews of mathematics students in understanding the concept of definite Integrals in the Explaining category as follows: (1) The mathematics student can explain a model of the definite integrals of function f at intervals $[a, d]$ based on the sequence properties of the real numbers with f can be integrated at $[a, d]$, ie $\int_a^b f(x)dx=\int_a^b f(x)dx+\int_b^c f(x)dx+\int_c^d f(x)dx$, (2) The mathematics student can give the reason that the interval $[a, d]$ contains four dots a, b, c and d respectively and f can be integrated at $[a, d]$ so apply $\int_a^b f(x)dx+\int_b^c f(x)dx+\int_c^d f(x)dx$, (3) Students of mathematics can use the procedure with the help of graphs to obtain the area of each interval $[a, b]$, $[b, c]$, and $[c, d]$, (4) The mathematics student has not been able to explain that if f can be integrated into $[a, d]$, then f can also be integrated into each interval $[a, b]$, $[b, c]$, and $[c, d]$, (5) The mathematics student has not been able to explain the integral form at each interval $[a, b]$, $[b, c]$, and $[c, d]$ ie $\int_a^b f(x)dx+\int_b^c f(x)dx+\int_c^d f(x)dx$, and (6) The mathematics student can conclude that a function f on the interval $[a, d]$ contains four dots a, b, c and d sequentially with f can be integrated into $[a, d]$, is $\int_a^d f(x)dx=\int_a^b f(x)dx+\int_b^c f(x)dx+\int_c^d f(x)dx$.

Thus, the results of the mathematics student's interview in understanding the definite integral concepts of the explaining category indicate that they have used metacognition knowledge, but have not used metacognition skills especially in planning and monitoring.

3.2.2. Results of mathematics education students interviews about understanding the concept of integral calculus on the explaining category

Results of mathematics education student interviews about understanding the concept of integral calculus on the category of explaining consists of the concept of indefinite integrals and definite integrals:

IOP Conf. Series: Journal of Physics: Conf. Series **974** (2018) 012022 doi:10.1088/1742-6596/974/1/012022

3.2.2.1. The Indefinite Integral Concepts

The results of the mathematics education student interviews in understanding the concept of indefinite integrals in the explaining category as follows: (1) Students of mathematics education can explain a model of an indefinite integral with a polynomial function based on the linearly properties, ie $\int(a_n x^n + ... + a_1 x + a_0)dx$ be $\int a_n x^n dx + .. + \int a_1 x dx + \int a_0 dx$, (2) Students of mathematics education can give the reason that the polynomial function applies the linearly properties so that $\int(a_n x^n + ... + a_1 x + a_0)dx$ can be changed to $\int a_n x^n dx + .. + \int a_1 x dx + \int a_0 dx$,(3) Students of mathematics education can use the procedure to explain $\int(a_n x^n + ... + a_1 x + a_0)dx$ changes to $\int a_n x^n dx + .. + \int a_1 x dx + \int a_0 dx$, (4) Students of mathematics education can show that the polynomial function applies the linearly properties, (5) Students of mathematics education can show results from $\int(a_n x^n + ... + a_1 x + a_0)dx$ same with result $\int a_n x^n dx + .. + \int a_1 x dx + \int a_0 dx$, and (6) Students of mathematics education can conclude that the function of polynomial is a linear function so that it behaves the linearly properties, so apply $\int(a_n x^n + ... + a_1 x + a_0)dx = \int a_n x^n dx + .. + \int a_1 x dx + \int a_0 dx$.

Thus, the result of the mathematics education student interview in understanding the indefinite integral concepts indicate that it has used metacognition knowledge and metacognition skills in understand the concept of indefinite integrals in the explaining category.

3.2.2.2. The Definite Integral Concepts

The result of interview of mathematics education students in understanding the concept of definite integral in Explaining category as follows: (1) The mathematics education student can explain a model of the definite integrals of function *f* at intervals [*a*, *d*] based on the sequence properties of the real numbers with *f* can be integrated at [*a*, *d*], ie $\int_a^b f(x)dx = \int_a^b f(x)dx + \int_b^c f(x)dx + \int_c^d f(x)dx$, (2) The mathematics education students can give the reason that the interval [a, d] contains four dots *a*, *b*, *c* and *d* respectively and *f* can be integrated at [*a*, *d*] so apply $\int_a^b f(x)dx + \int_b^c f(x)dx + \int_c^d f(x)dx$, (3) Students of mathematics education can use the procedure with the help of graphs to obtain the area of each interval [*a*, *b*], [*b*, *c*], and [*c*, *d*], (4) Students of mathematics education can explain that if *f* can be integrated into [*a*, *d*] then *f* can also be integrated into each of the intervals [*a*, *b*], [*b*, *c*], and [*c*, *d*], (5) Students of mathematics education can write integral forms at each interval [*a*, *b*], [*b*, *c*], and [*c*, *d*] ie $\int_a^b f(x)dx + \int_b^c f(x)dx + \int_c^d f(x)dx$, and (6) Students of mathematics education can conclude that a function *f* on the interval [*a*, *d*] contains four dots *a*, *b*, *c* and *d* sequentially with *f* can be integrated into [*a*, *d*], is $\int_a^d f(x)dx = \int_a^b f(x)dx + \int_b^c f(x)dx + \int_c^d f(x)dx.$.

Thus, the result of the mathematics education student's interview in understanding the definite integral concepts shows that it has used metacognition knowledge and metacognition skills in understand the concept of definite integrals in the explaining category.

The results above show that there are differences in metacognition profile between mathematics students and mathematics education students in understanding the concept of integral calculus, especially the integral concept of both the summarizing category and the explaining category. In the summarizing category, mathematics education students can review their understanding that the area of function $|x + 2|$ at interval [-2.2] is above the *x*-axis, and can explain the steps of calculating the area of the function $|x + 2|$ at the interval [-2,2]. While the mathematics student has not been able to review his understanding of the area of function $|x + 2|$ at interval [-2,2] and can not yet explain the steps to calculate the area of the function area. This is because math students are not yet accustomed to drawing graphs of functions of absolute values. Although the mathematics students are theoretically directed to think mathematics, because they are not familiar with the concept they are doing, the result is not good. This is in the opinion of Masami & Katagiri [11], there is no way a person could teach in such a way as to cultivate mathematical thinking without first understanding the kinds of mathematical thinking that exist. On the other hand, Wittrock [15], stated that some studies have found no effects of summarization and the conditions below are not well understood.

Then, in the explaining category, Mathematics education students can explain and write the integral form that if f is integral to $[a, d]$ then f can also be integrated into each of the intervals $[a, b]$, $[b, c]$, and $[c, d]$. While the mathematics student can write the integral form, but can not yet explain that if f is integral to $[a, d]$ then f can also be integrated into each of the intervals $[a, b]$, $[b, c]$, and $[c,d]$. This is because math students are not accustomed to voicing their thoughts. He is only skilled at completing a concept, but it is very difficult to reveal the results of his work to others.

Therefore, mathematics students need to learn about how to express their thoughts to others. This is as stated by Hattie et al, [9], teaching metacognitive strategies to the student can lead to marked improvement in their achievement. Then, Butler & Winn [2], suggests that Students can be learning to think their own thinking processes and apply specific learning strategies to think themselves through difficult task.

4. Conclusion

Based on the results of research and discussion above, it can be concluded that:

The summarizing category, mathematics students, and mathematics education students can use knowledge and metacognition skills in understanding the concept of indefinite integrals. While the definite integral concepts, mathematics students have not been able to use his/her knowledge and skills of metacognition. And mathematics education students only use skills metacognition.

The explaining category, mathematics students can use knowledge and skills of metacognition in understanding the concept of indefinite integrals, while the definite integral concepts only use metacognition skills. Meanwhile, mathematics education students can use knowledge and skills of metacognition in understanding the concept of both indefinite and definite integrals.

5. Reference

[1] Anderson, L. W., Krathwohl, D. R., Airasian, P. W., Cruikshank, K. A., Mayer, R. E., Pintrich, P. R., et al. 2001 *A Taxonomy for Learning, Teaching, and Assissing: A Revision of Bloom's Taxonomy of Educational Objectives* (New York: Longman)
[2] Butler, D.L., & Winne, P.H. 1995 *Review of educational Research,* **65** 3 p. 245
[3] Center for Curriculum Development (CCD) 2001 Ministry of Education Malaysia: http://myschoolnet. ppk.kpm.my/bcb8.pdf Accessed on 13 December 2016.
[4] Curriculum of Department of Mathematics and Mathematics Education Universitas Halu Oleo 2014 (Kendari: UHO)
[5] Dawson, TL. 2008 *Metacognition and Learning in Adulthood* (Paper presented in ODNI/CHCO/IC Leadership Development Office tanggal 23 Agustus 2008)
[6] Flavell, J.H. 1979 *Metacognition and Cognitive Monitoring a New Area of Cognitive—Developmental Inquiry* (California: American Psychologist)
[7] Gagne, R.M. 1983 *J. for Res. in Math. Edu.* **14** 1
[8] Goh, C. 2008 *RELC Journal* **39** 2 p.188
[9] Hattie J., Bibbs J. & Purdie, N. 1996 *Review of edu.Res.* **66** 2 p. 99
[10] Livingston, Jennifer A. 1997 *Metacognition: An Overview,* http://www.gse.buffalo.edu/fas/shuell/CEP564/Metacog.html Accessed on 27 November 2016.
[11] Masami Isoda & Shigeo Katagiri 2012 *Monograph on Lesson Study for Teaching Mathematics and Science-* **1** (World Scientific)
[12] Mayer, RE 2002 *Teaching for Meaningful Learning* (Upper Saddle River, NJ: Prentice-Hall)
[13] Miles, M.B. and Huberman, A.M. 2014 *Qualitative Data Analysis* (London: Sage Publications)
[14] Neisser, U. 1976 *Cognition and Reality* (S. Francisco, CA: Freeman)
[15] Wittrock, M.C. 1991 *Elementary school Journal* **9** 2 p.169

IOP Conf. Series: Journal of Physics: Conf. Series **974** (2018) 012023 doi:10.1088/1742-6596/974/1/012023

Portfolio return using Black-litterman single view model with ARMA-GARCH and Treynor Black model

Adri Arisena[1], Lienda Noviyanti[2], Achmad Zanbar S[3]

[1,2,3] Statistics Department, Faculty of Mathematics and Science, State University of Padjadjaran, Indonesia
[1] adriarisena@gmail.com
[2] lienda@unpad.ac.id
[3] zanbar1008@gmail.com

Abstrack. Establishing an optimal portfolio is a method that can help investors to minimize risk and optimize profits. Some models for optimal portfolio include Black-litterman Model and Treynor Black Model. The Black-litterman Model combines the elements of historical data and views of investors to form a new prediction of the portfolio as the basis for the preparation of weighted asset models. Predicted views in this study using time series ARMA-GARCH. The stock return data mostly have high volatility causing heteroscedasticity problems, so the GARCH model is chosen to overcome the problem. Treynor Black Model is active and passive portfolio. An active portfolio is that investors allocate their investment funds to individual securities in the capital market, while passive portfolios are investors allocating their investment funds to the stock market indices. In this research we will use Treynor-Black Model with active portfolio, so that investors can choose stock according to their allocation of funds. The purpose of this research is to form the weight of Black-litterman model portfolio with a single view investor using ARMA-GARCH and compare the profit result obtained from the formation of portfolio Black-litterman Model and Treynor-Black Model. The selected shares are PT. Bank BCA (BBCA) , PT. Gudang Garam (GGRM), and PT. Waskita Karya (WSKT).

1. Introduction

Stock according to Bodie (2009) is a securities that indicate the existence of a company and is. Based on Hiriyappa (2008) establishing an optimal portfolio is a method that can help investors to attract risk and optimize profits.

In optimal portfolio required a model that is Black-litterman Model and Treynor Black Model. Black-litterman model is a model used for input prediction portfolio optimization. This model repeats two types of historical data on equilibrium conditions with the views of investors so as to meet the results (Black & Litterman, 1990). In Black-litterman Model, estimates investors will use GARCH (General Autoregressive Conditional Heteroskedasticity) method. Stock return data mostly have high volatility so GARCH method selected to overcome the problem.

The next problem is how to determine the prior market parameters of the Bayesian example. The Bayesian framework on the Black-litterman model is the equilibrium return on the Capital Asset Pricing Model (CAPM) as an option prior to the investor's view so as to produce a new view of the return expectation (posterior distribution). To get the optimal benefit from the Black-litterman portfolio model, investors need to determine which combination to use from an investor's perspective. The combination of investor views can be one type of view (single-view) or some kind of combination of views.

Fortunately to find the optimal portfolio, then there is another model called the Treynor-Black Model. If the model portfolio of Black-litterman Model is viewed from view investor, then the Treynor-Black Model portfolio is divided into two, namely the active and passive portfolio. An active portfolio is the investor who allocates his investment funds to individual securities in the capital market, while the passive portfolio is the investor allocating his investment funds on the capital market index.

2. Black-litterman Model

Black-litterman Model is a model used to estimate inputs for portfolio optimization. This model combines two types of estimates, namely historical data on equilibrium conditions with the views of investors so as to update the estimation results (Black & Litterman, 1990).

The advantage of the Black-litterman Model is that investors can combine views in both absolute and relative terms with prior prior estimates, to generate new posterior estimates that include all view. The equation below is the Black-litterman equation and represents the expected return vector formed by the bayes rule of the equilibrium return vector (π) and the vector from the investor view (V).

$$E(R)=[(\tau\Sigma)^{-1}+P'\Omega^{-1}P]^{-1}[(\tau\Sigma)^{-1}\Pi+P'\Omega^{-1}Q]=\mu_{bl} \quad (1)$$

μ_{bl} : the average vector estimation of the Black-litterman model is sized p x 1
τ : the level of investor confidence in his views (constants)
Σ : variance-covariance matrix of return-sized p x p
P : vector / matrix from the investor's view of size q x p
Ω : the diagonal matrix of covariance from the investor's view with certain sized q x q
Π : equilibrium return vector of CAPM sized p x 1
Q : vector return value expectations according to investor's view q x1

2.1 CAPM (Capital Asset Pricing Model)

There is a model that can be used to estimate the return of a securities stock that is Capital Asset Pricing Model (CAPM) pioneered by Sharpe, Lintner, and Mossin in 1964-1966. According to Bodie et al (2014), the CAPM model is an important part of the financial field used to predict the expected return balance and the risk of an asset under equilibrium conditions. CAPM is an important part of finance that is used to predict the relationship between return and asset expectations of an asset. the equilibrium formation formula of each share according to CAPM is as follows:

$$E(R_i)_{CAPM}=R_f+\beta_i[E(R_m)-R_f] \quad ;i=1,2,...,p \quad (2)$$

$E(R_i)_{CAPM}$: return equilibrium stock
R_f : risk-free return value
β_i : estimation of stock beta coefficient value
$E(R_m)$: value of expectation of return from the market
p : the number of stocks studied

2.2 The views of investor

P is a vector / matrix of investor's view. Each matrix row represents an investor's view either certain or relative to a stock. While Q is a vector of expectation return value according to investor's view. According to He & Litterman (1999), the diagonal matrix is the covariance of the view with a certain degree of tau.

$$\Omega = P'\tau\sum P \tag{3}$$

τ	:	the level of investor confidence in his views (constants)
Σ	:	variance-covariance matrix of return-sized p x p
P	:	vector / matrix from the investor's view of size q x p
Ω	:	diagonal matrix of covariance of investor's view with a certain sized q x q
p	:	the number of stocks
q	:	the number of investors' views

2.2 Investor's view with time series GARCH method

Bollerslev (1986) refined the ARCH model to GARCH to facilitate the assessment of parameters at higher order. Bollerslev states that residual variance depends not only on the last period's residual but also the residual variance of the past period. Generally GARCH model can be written as follows:

$$\sigma_t^2 = \alpha_0 + \sum_{i=1}^{p} \alpha_i e_{t-1}^2 + \sum_{j=1}^{q} \beta_j h_{t-j}^2 \tag{4}$$

$\alpha_0 > 0, \alpha_i \geq 0,$ for $i = 1,..., p$ and $\beta_j \geq 0$ for $j = 1,..,q$ where

σ_t^2	: variance of the residual t-period
p	: ARCH order
q	: GARCH order
α_0	: constants
α_i	: parameters from ARCH
β_j	: parameters from GARCH
e_{t-1}^2	: variance of residual period to t-1
h_{t-j}^2	: squared from residual period to t-j

2.3 Portfolio of the Black-litterman Model

After forming an equilibrium return value with CAPM and estimating the views of investors, the next step is to establish the weight of the Black-litterman model portfolio. The stages of establishing a portfolio of Black-litterman models are as follows:

1. Calculating the return value vector return of the Black-litterman model

$$\mu_{bl} = \left[(\tau\Sigma)^{-1} + P'\Omega^{-1}P\right]^{-1}\left[(\tau\Sigma)^{-1} + P'\Omega^{-1}Q\right] \tag{5}$$

2. Determine the weight of assets / stocks on the Black-litterman model

$$W_{bl} = (\delta\Sigma)^{-1}\mu_{bl} \tag{6}$$

μ_{bl}	: the average vector estimate on the Black-litterman model is p x 1.
τ	: the level of investor confidence in his views (constants)
Σ	: variance-covariance matrix of return p x p size
P	: vector or matrix of investor sized view q x p
Ω	: diagonal matrix of covariance of investor's view with a certain sized q x q
Π	: vector return equilibrium of CAPM sized p x 1
Q	: vector return expectation value according to investor sized view q x 1
W_{bl}	: vector weight portfolio model Black-litterman p x 1 size

δ : value of stock coefficient
p : the number of stocks
q : the number of investors' views

2.4 Treynor Black Model

Treynor and Black deal with a scenario in which the mean-variance criterion (the Sharpe ratio) is used by investors; a specified market index is taken as the default efficient (passive) strategy, and the security analysts of a portfolio management firm cover a limited number of securities. Under these conditions, securities that are not analyzed are assumed to be efficiently priced, and a portfolio of only the covered securities cannot be efficient. The optimal portfolio must be a mix of the covered securities and the index portfolio. Treynor Black Model identify the portfolio of only the covered securities (the efficient Active Portfolio, A) that can be mixed with the index (Passive Portfolio, M) to obtain the optimal risky portfolio. Here are the Treynor-Black model optimization step.

1. Calculate the initial position of each of the securities in the active portfolio

$$W_i^0 = \frac{\alpha_i}{\sigma^2(e_i)};\ W_i = \frac{W_i^0}{\sum W_i^0} \tag{7}$$

W_i^o : the initial position of securities i on the active portfolio
α_i : alpha securities i
$\sigma^2(e_i)$: variance of i-th security
W_i : the weight scale of the securities i from the initial position

2. Calculates the active portfolio alpha and beta

$$\alpha_A = \sum W_i\alpha_i\ ;\ \beta_A = \sum W_i\beta_i \tag{8}$$

α_A : alpha active portfolio
W_i : the weight scale of the securities i from the initial position
α_i : alpha securities i
β_A : the sensitivity of the active portfolio to the market
β_i : the sensitivity of securities to the market

3. Calculates the starting position of active portfolio and optimal weight of active portfolio

$$W_A^0 = \left[\frac{\dfrac{\alpha_A}{\sigma^2(e_A)}}{\dfrac{E(R_m)}{\sigma_M^2}}\right]\ ;\ W_A^* = \frac{W_A^0}{1+(1-\beta_A)W_A^0} \tag{9}$$

3. Portfolio Analysis of Black-litterman Model

3.1 Establishment of Return Equilibrium with CAPM

The shares used in this research are PT. Bank BCA, PT. Gudang Garam and PT. Waskita Karya shares with daily period from January 2014 to May 2017. The result of CAPM equilibrium return from each stock are PT. Bank BCA 0.000842 percent, PT. Gudang Garam 0.001016 percent, and PT. Waskita Karya 0.002987 percent. The result will then be used to form the equilibrium return vector of CAPM as follows

$$\Pi = \begin{bmatrix} 0.000842 \\ 0.001016 \\ 0.002987 \end{bmatrix}$$

3.2 Investor's view with Time Series Approach

Before stepping on the establishment of the Black-litterman model, we must identify investor views with the time series GARCH. The assumption that must be stationary test, normality test, autocorrelation test and heteroscedasticity test. The three shares have met the assumption test so that the ARMA-GARCH order used is BBCA ARMA (1.2) -GARCH (1.1), GGRM ARMA (2.1)-GARCH (1.1) and WSKT ARMA (2.2)-GARCH (1.2) with result as follows

Table 1. ARMA-GARCH models.

Parameter	BBCA	GGRM	WSKT
ar1	p-value = 0.000150 estimates = 0.05455	p-value = < 2 x 10^{-16} estimates = 0.08160	p-value = < 2 x 10^{-16} estimates = 0.05942
ar2	-	p-value = 0.005547 estimates = -0.0102	p-value = < 2 x 10^{-16} estimates = -0.09359
ma1	p-value = 0.000197 estimates = -0.05506	p-value = < 2 x 10^{-16} estimates = -0.07944	p-value = < 2 x 10^{-16} estimates = 0.06111
ma2	p-value = 0.053192 estimates = -0.008245	-	p-value = < 2 x 10^{-16} estimates = 0.09080
omega	p-value = 0.001818 estimates = 0.000004405	p-value = 0.086378 estimates = 1.20 x 10^{-5}	p-value = 0.02341 estimates = 7.7 x 10^{-5}
alpha1	p-value = 0.00000373 estimates = 0.0149	p-value = 0.000613 estimates = 0.00632	p-value = 0.00180 estimates = 0.01558
beta1	p-value = 7.89 x 10^{-10} estimates = 0.06116	p-value = < 2 x 10^{-16} estimates = 0.09078	p-value = 0.00841 estimates = 0.03596
beta2	-	-	p-value = 0.01971 estimates = 0.03215
LM Arch	Statistic = 3.334952 p-Value = 0.9926816	Statistic = 9.356863 p-Value = 0.6721851	Statistic = 10.29474 p-Value = 0.5901208
AIC	-5.834237	-5.041847	-4.878264

Furthermore, investors' views on the three stock returns will be estimated using GARCH method with the following results:

Table 2. Return Estimation

Stock	Return Estimation
BBCA	0.000130
GGRM	0.00022
WSKT	0.00021

next is formed vector / matrix from investor's view (P) and vector return expectation value according to investor's view (Q). Whereas (Ω) is diagonal matrix the covariance of the view investor with the ($\tau = 1$). The variance-covariance matrix of return of each share are:

$$\sum = \begin{bmatrix} 4.018831e-04 & 9.071965e-05 & 1.067316e-04 \\ 9.071965e-05 & 1.880227e-04 & 8.243658e-05 \\ 1.067316e-04 & 8.243658e-05 & 4.713472e-04 \end{bmatrix}$$

The matrix (P), vector (Q) and matrix (Ω) values for each view investor are as follows

1. *Single View* 1

$$P = \begin{bmatrix} 1 & 0 & 0 \end{bmatrix} ; Q = \begin{bmatrix} 0.00013 \end{bmatrix} ; \Omega = \begin{bmatrix} 4.018831e-04 \end{bmatrix}$$

2. *Single View* 2

$$P = \begin{bmatrix} 0 & 1 & 0 \end{bmatrix} ; Q = \begin{bmatrix} 0.00022 \end{bmatrix} ; \Omega = \begin{bmatrix} 1.880227e-04 \end{bmatrix}$$

3. *Single View* 3

$$P = \begin{bmatrix} 0 & 0 & 1 \end{bmatrix} ; Q = \begin{bmatrix} 0.00021 \end{bmatrix} ; \Omega = \begin{bmatrix} 4.713472e-04 \end{bmatrix}$$

3.3 Portfolio Strategy with Single View Investor

The comparison between Treynor Black model portfolio weight as with the Black-litterman model generated from each investor's view is shown in **Table 3**. Black-litterman model with a Single View Investor and Treynor Black Model consistently allocates the biggest weight to the stock of PT. Waskita Karya

Table 3. Comparison of Portfolio Weight

Stock	Treynor Black Model	Black-litterman Model		
		View 1	View 2	View 3
PT. Bank BCA	0.2933	0.0676	0.0640	0.0677
PT. Gudang Garam	0.1716	0.0520	0.0522	0.0516
PT. Waskita Karya	0.5350	0.8804	0.8838	0.8807

4. Conclusion

ARMA-GARCH model can be used to determine investor's view on Black-litterman model with the assumption that time series assumption can be fulfilled. Black Litterman and Treynor Black Model models consistently allocate the greatest weight to the shares of PT. Waskita Karya, PT. Bank BCA and PT. Gudang Garam with different weight characteristics. However, weight generated by both models can be used as a reference for choosing an optimal portfolio.

5. References

[1] Alex Kane, T.-H. K. 2003. Active Portfolio Management: The Power of the Treynor-Black Model.
[2] Black, F., & Robert, L. 1992. Global Portfolio Optimization. *Financial Analysis Journal*.
[3] Bodie, K. M. 2014. *Investments*. McGraw-Hill Education.
[4] Bollerslev, T. 1986. Generelized Autoregressive Conditional Heteroskedasticity.
[5] Hiriyappa, B. 2008. *Investment Management*. New Age International.
[6] Idzroek, T. M. 2002. A Step by Step Guide To The Black Litterman Model.

Finite Volume Method for Pricing European Call Option with Regime-switching Volatility

Mey Lista Tauryawati, Chairul Imron, Endah RM Putri

Department of Mathematics, Institut Teknologi Sepuluh Nopember, Surabaya 60111, Indonesia

E-mail: meylistamey@yahoo.com

Abstract. In this paper, we present a finite volume method for pricing European call option using Black-Scholes equation with regime-switching volatility. In the first step, we formulate the Black-Scholes equations with regime-switching volatility. we use a finite volume method based on fitted finite volume with spatial discretization and an implicit time stepping technique for the case. We show that the regime-switching scheme can revert to the non-switching Black Scholes equation, both in theoretical evidence and numerical simulations.

1. Introduction

Finite volume method, also called Box method is used in the numerical solution of fluid flow problems. But the application of finite volume method is not limited in problems of fluid. This method can also be used in financial mathematics problems for example at issue option. In solving the problems of option is usually used Black Scholes model as a mathematical model. In this paper we use black scholes model with regime switching because the model with constant volatility can not represent the actual stock price. In 2001 finite volume method was successfully used by Zvan in numerical problems on an options case. In 2006 Julien Berton conducted a study of a finite volume method, it was used to determine an American option price. From the research it was found that the scheme of a finite volume method to be more accurate and efficient than the two methods compared a Dynamic programming method and an ADI method. In 2008 Zhang and Wang conducted research on a finite volume method to determine the price of European options. In determining the option price Zhang and Wang use the Black-Scholes model which has 1 partial diferential equation. From the results of his research, it was found that the scheme from a finite volume method in a space with an implicit scheme at a time is consistent and stable, so it is said to be convergent as the relevant financial solution. In this paper, we present a finite volume method in determining the price of European call option with regime-switching volatility as a measure of an uncertainty of stock movement. The option price is determined from Black-Scholes equations with regime-switching volatility. We shall show that the system matrix of discretization is an M-matrix.

2. European call option with regime-switching volatility

In this section, we will formulate a Black-Scholes model for European call option with regime-switching where the volatility moves randomly between two states. The Fluctuations of underlying assets are assumed to follow the stochastic process described in the following equation:

$$dS(t) = (\mu_{X_t} - D)S(t)dt + \sigma_{X_t} S(t)dW(t), \tag{1}$$

These equation is used to formulate a Black-Scholes model with regime-switching volatility. Regime-switching is a process of jumping on several states that occur on the Markov chain. Let X_t is a markov chain with continuous time that describes the state changes that occur, the state of the economy grows and the economy is weak. Because the economic state consists of two states, the change of the option price has a different value for each initial position of each state. When the initial position is state 1, the possible state change is fixed in state 1 or changed to state 2. So if the chance of change from state 1 to 2 is $\lambda_{12}dt$ then the chances of state 1 remain in state 1 is $1 - \lambda_{12}dt$. So we get the change of option price is:

$$dV = \begin{cases} V_1(S(t+dt), t+dt) - V_1(S(t), t), & p = 1 - \lambda_{12}dt \\ V_2(S(t+dt), t+dt) - V_1(S(t), t), & p = \lambda_{12}dt \end{cases}$$

where,

$$\begin{aligned} V_1(S(t+dt), t+dt) - V_1(S(t), t) &= dV_1 \\ V_2(S(t+dt), t+dt) - V_1(S(t), t) &= dV_2 + V_2 - V_1. \end{aligned}$$

The option price changes for initial state state 1 are:

$$dV = \begin{cases} dV_1 & , \quad p = 1 - \lambda_{12}dt \\ dV_2 + V_2 - V_1, & p = \lambda_{12}dt, \end{cases} \tag{2}$$

The possibility that occurs at state 2 is fixed at state 2 or changed to state 1. If the initial position of the economic state is state 2 then the probability state 2 turns to state 1 is $\lambda_{21}dt$ and the probability of remaining in state 2 is $1 - \lambda_{21}dt$. So we obtain the change of option price is as follows:

$$dV = \begin{cases} V_2(S(t+dt), t+dt) - V_2(S(t), t), & p = 1 - \lambda_{21}dt \\ V_1(S(t+dt), t+dt) - V_2(S(t), t), & p = \lambda_{21}dt. \end{cases}$$

where,

$$\begin{aligned} V_2(S(t+dt), t+dt) - V_2(S(t), t) &= dV_2 \\ V_1(S(t+dt), t+dt) - V_2(S(t), t) &= dV_1 + V_1 - V_2. \end{aligned}$$

The option price changes for initial state state 2 are:

$$dV = \begin{cases} dV_2 & , \quad p = 1 - \lambda_{21}dt \\ dV_1 + V_1 - V_2, & p = \lambda_{21}dt. \end{cases} \tag{3}$$

then we formulate a portfolio of options. The value of the portfolio is given by the equation:

$$\Pi = V - \Delta S \tag{4}$$

Over a period of time, portfolio value changes to:

$$d\Pi = dV - \Delta dS \tag{5}$$

So that the change of option value with dividend payment is:

$$d\Pi = dV - \Delta dS - \delta\Delta S dt \tag{6}$$

From equation (6) a portfolio change can be determined for each initial positions of different states. for the initial position of state 1, we obtain:

$$d\Pi \;=\; \begin{cases} dV_1 - \Delta dS - DS\Delta dt & , \quad p = 1 - \lambda_{12}dt \\ dV_2 + V_2 - V_1 - \Delta dS - DS\Delta dt, & p = \lambda_{12}dt, \end{cases}$$

Then we formulate an expected value of $d\Pi$ as follows:

$$E[d\Pi] = dV_1 - \Delta dS - \delta S\Delta dt + \lambda_{12}dt(dV_2 - dV_1 + V_2 - V_1).$$

The risk of regime-switching is risk-neutral. At risk-neutral, investors do not demand asset returns but profits from the portfolio as much as a change of the funds invested. Mathematically can be written as $E[d\Pi] = d\Pi$ so we get,

$$d\Pi = dV_1 - \Delta dS - \delta S\Delta dt + \lambda_{12}dt(dV_2 - dV_1 + V_2 - V_1). \tag{7}$$

Next we formulate dV_1 from Lemma ITO. And from equation (1) is obtained:

$$(dS)^2 \;=\; (\mu - \delta)^2 S^2 dt^2 + 2(\mu - \delta)S^2\sigma dt dW_t + \sigma^2 S^2 dW_t^2$$

because $dtdW_t = 0$ and $dt^2 = 0$ then:

$$(dS)^2 \;=\; \sigma^2 S^2 dt.$$

Lemma ITO

$$dV_i \;=\; \frac{\partial V_i}{\partial S}\sigma_i S dW_t + \left[\frac{\partial V_i}{\partial S}(\mu - \delta)S + \frac{\partial V_i}{\partial t} + \frac{1}{2}\frac{\partial^2 V_i}{\partial S^2}\sigma_i^2 S^2\right]dt \tag{8}$$

by substituting equation (1) and (8) into equation (7). Furthermore, it is assumed that $dW_t dt = 0$ and $dt^2 = 0$, then we get:

$$\begin{aligned} d\Pi \;=\; & \sigma_1 S\left(\frac{\partial V_1}{\partial S} - \Delta\right)dW_t + \left[\mu S\left(\frac{\partial V_1}{\partial S} - \Delta\right)\right. \\ & \left. -\delta S\left(\frac{\partial V_1}{\partial S} - \Delta\right) + \frac{1}{2}\sigma_1^2 S^2\frac{\partial^2 V_1}{\partial S^2} + \frac{\partial V_1}{\partial t} + \lambda_{12}(V_2 - V_1) - \delta\Delta S\right]dt. \end{aligned}$$

To eliminate random components in a random walk in order to get a riskless then selected

$$\Delta = \frac{\partial V_1}{\partial S}$$

so we get,

$$d\Pi = \left[\frac{1}{2}\sigma_1^2 S^2\frac{\partial^2 V_1}{\partial S^2} + \frac{\partial V_1}{\partial t} - \delta S\frac{\partial V_1}{\partial S} + \lambda_{12}(V_2 - V_1)\right]dt. \tag{9}$$

the same way as initial position state 1 we get the $d\Pi$ value at initial position state 2 as follows:

$$d\Pi = \left[\frac{1}{2}\sigma_2^2 S^2\frac{\partial^2 V_2}{\partial S^2} + \frac{\partial V_2}{\partial t} - \delta S\frac{\partial V_2}{\partial S} + \lambda_{21}(V_1 - V_2)\right]dt. \tag{10}$$

It is assumed that portfolio results are free arbitrage and obtain a risk free rate:

$$d\Pi = \Pi r dt \tag{11}$$

so we obtain the partial differential equation of Black-Scholes for initial position state 1 as follows:

$$\frac{\partial V_1}{\partial t} + \frac{1}{2}\sigma_1^2 S^2 \frac{\partial^2 V_1}{\partial S^2} + (r-\delta)S\frac{\partial V_1}{\partial S} - rV_1 = \lambda_{12}(V_1 - V_2). \tag{12}$$

Similarly, the partial differential equation of BlackScholes for initial position state 2 is

$$\frac{\partial V_2}{\partial t} + \frac{1}{2}\sigma_2^2 S^2 \frac{\partial^2 V_2}{\partial S^2} + (r-\delta)S\frac{\partial V_2}{\partial S} - rV_2 = \lambda_{21}(V_2 - V_1). \tag{13}$$

Furthermore, Equations are combined with boundary conditions to form a pair of partial differential equation (PDE) systems for an European call option with dividend and regime-switching as follows:
state 1

$$\frac{\partial V_1}{\partial t} + \frac{1}{2}\sigma_1^2 S^2 \frac{\partial^2 V_1}{\partial S^2} + (r-\delta)S\frac{\partial V_1}{\partial S} - rV_1 = \lambda_{12}(V_1 - V_2)$$

with boundary conditions:

$$\begin{aligned} V_1(0,t) &= 0 \\ \lim_{S\to\infty} V_1(S,t) &= S \\ V_1(S,T) &= \max\{S-E, 0\}. \end{aligned}$$

state 2

$$\frac{\partial V_2}{\partial t} + \frac{1}{2}\sigma_2^2 S^2 \frac{\partial^2 V_2}{\partial S^2} + (r-\delta)S\frac{\partial V_2}{\partial S} - rV_2 = \lambda_{21}(V_2 - V_1)$$

with boundary conditions:

$$\begin{aligned} V_2(0,t) &= 0 \\ \lim_{S\to\infty} V_2(S,t) &= S \\ V_2(S,T) &= \max\{S-E, 0\}. \end{aligned}$$

3. Finite volume method

In this section, we consider a numerical approximation of the problem using a finite volume method based on a fitted finite volume. We will determine the finite volume method for a Black-Scholes equation of the European call option with regime-switching volatility.

3.1. Spatial discretization

Divide I into two parts I_i and J_i. The first part consist of N sub-intervals $I_i = (S_i, S_{i+1})$, $i = 0, 1, 2, ...N-1$ with $0 = S_0 < S_1 < ... < S_N = S_{max}$. The second part J_i consist of $J_i = (S_{i-1/2}, S_{i+1/2})$ where $S_{i-1/2} = (S_{i-1} + S_i)/2$, $S_{i+1/2} = (S_i + S_{i+1})/2$, $S_0 = S_{-1/2}$ and $S_{i+1/2} = S_{N+1}$.

State 1

$$\frac{\partial V_1}{\partial \tau} = \frac{1}{2}\sigma_1^2 S^2 \frac{\partial^2 V_1}{\partial S^2} + (r-\delta)S\frac{\partial V_1}{\partial S} - rV_1 + \lambda_{12}(V_2 - V_1) \tag{14}$$

$$\frac{\partial V_1}{\partial \tau} = \frac{\partial}{\partial S}\left[aS^2\frac{\partial V_1}{\partial S} + bSV_1\right] + cV_1 + \lambda_{12}V_2 \tag{15}$$

where $a = \frac{1}{2}\sigma_1^2, b = r - \delta - \sigma_1^2, c = -b + r - \lambda_{12}$.
Integrating over J_i:

$$\int_{J_i}\frac{\partial V_1}{\partial \tau}dS = \int_{J_i}\frac{\partial V_1}{\partial S}\left[aS^2\frac{\partial V_1}{\partial S} + bSV_1\right]dS + \int_{J_i}cV_1 dS + \int_{J_i}\lambda_{12}V_2 dS \tag{16}$$

let $x_i = S_{i+1/2} - S_{i-1/2}$, we get:

$$\frac{\partial V_1}{\partial \tau}x_i = S_{i+1/2}\rho(V_1)\Big|_{S_{i+1/2}} - S_{i-1/2}\rho(V_1)\Big|_{S_{i-1/2}} + cV_1x_i + \lambda_{12}V_2x_i \tag{17}$$

where

$$\rho(V_1) = aS\frac{\partial V_1}{\partial S} + bV_1 \tag{18}$$

Then we need to derive an approximation of the continuous flux $\rho(V_1)$. This discussion is divided into two cases for i and $i = 0, 1$ and $i = 2, 3, ..., N-1$, respectively.
Case I. For $i = 2, 3, ...N - 1$
Approximation of $\rho(V_1)$ at the point $S_{i-1/2}$ and $S_{i+1/2}$ of the interval I_i. Let us consider the following two-point boundary value problem:
at the point $S_{i+1/2}$

$$(a_{1+1/2}SV_1' + b_{i+1/2}V_1)' = 0, S \in I \tag{19}$$

$$V_1(S_i) = V_1 i, V_1(S_{i+1}) = V_1 i + 1 \tag{20}$$

and at the point $S_{i-1/2}$

$$(a_{1-1/2}SV_1' + b_{i-1/2}V_1)' = 0, S \in I \tag{21}$$

$$V_1(S_i) = V_1 i, V_1(S_{i-1}) = V_1 i - 1 \tag{22}$$

Solving the problem analytically, we get: at the point $S_{i+1/2}$

$$\rho_i(V_1) = b_{i+1/2}\frac{V_iS_i^{k_i} - V_{i+1}S_{i+1}^{k_i}}{S_i^{k_i} - S_{i+1}^{k_i}} \tag{23}$$

and at the point $S_{i-1/2}$

$$\rho_i(V_1) = b_{i-1/2}\frac{V_iS_i^{k_i} - V_{i-1}S_{i-1}^{k_i}}{S_i^{k_i} - S_{i-1}^{k_i}} \tag{24}$$

where

$$k_i = \frac{b_{i+1/2}}{a_{i+1/2}} \tag{25}$$

Case II. For $i = 0, 1$ Note that the analysis in Case I does not apply to approximation of the flux on $(0, S_1)$ because the equation is degenerated for $S \to 0$. In this case we can consider in the following form

$$(a_{1/2}SV_1' + b_{1/2}V_1)' = c, S \in I \tag{26}$$

$$V_1(0) = V_10, V_1(S_1) = V_11 \tag{27}$$

Solving these equation, we get

$$\rho_i(V_1) = \frac{1}{2}((a_{1/2} + b_{1/2})V_1 - (a_{1/2} - b_{1/2})V_0) \tag{28}$$

$$V_1 = V_10 + \frac{S}{S1}(V_11 - V_10) \tag{29}$$

substituting (10), (11), (15) and (16), we obtain for $i = 1$

$$\frac{\partial V_1}{\partial \tau}x_i = S_{3/2}b_{3/2}\left[\frac{V_1S_1^{k1} - V_2S_2^{k1}}{S_1^{k1} - S_2^{k1}}\right] - S_{1/2}b_{1/2}\left[\frac{V_1S_1^{k1} - V_0S_0^{k1}}{S_1^{k1} - S_0^{k1}}\right] + cx_i + x_i\lambda_{12}V_2 \tag{30}$$

So we get

$$\frac{\partial V_11}{\partial \tau} = \xi_1 V_10 + \psi_1 V_11 + \phi_1 V_12 + \lambda_{12}V_21 \tag{31}$$

where

$$\xi_1 = \frac{S_1(a_{1/2} - b_{1/2})}{4\xi_1} \tag{32}$$

$$\psi_1 = \frac{-S_1(a_{1/2} + b_{1/2})}{4x_i} - \frac{S_{3/2}b_{3/2}S_1^{k1}}{x_i(S_2^{k1} - S_1^{k1})} - c \tag{33}$$

$$\phi_1 = \frac{S_{3/2}b_{3/2}S_2^{k1}}{x_i(S_2^{k1} - S_1^{k1})} \tag{34}$$

and for i=2,3,...N-1, we obtain,

$$\frac{\partial V1_i}{\partial \tau} = \xi_i V_{i-1} + \psi_i V_i + \phi_i V_{i+1} + \lambda_{12}V2_i \tag{35}$$

where

$$\xi_i = \frac{S_{i-1/2}b_{i-1/2}S_{i-1}^{ki}}{xi(S_i^{ki} - S_{i-1}^{ki})} \tag{36}$$

$$\psi_i = \frac{-S_{i-1/2}b_{i-1/2}S_i^{ki}}{xi(S_i^{ki} - S_{i-1}^{ki})} - \frac{S_{i+1/2}b_{i+1/2}S_i^{ki}}{xi(S_{i+1}^{ki} - S_i^{ki})} - c \tag{37}$$

$$\phi_i = \frac{S_{i+1/2}b_{i+1/2}S_{i+1}^{ki}}{xi(S_{i+1}^{ki} - S_i^{ki})} \tag{38}$$

3.2. Time discretization

Let τ denotes points from $[0, T]$ such that $0 = \tau_0 < \tau_1 < ... < \tau_M = T$ and $\Delta\tau_n = \tau_n - \tau_{n-1} > 0$, where $M > 1$ is a positive integer. Applying the fully implicit time discretization to (35) for simplicity, we get

$$\frac{V1_i^{n+1} - V1_i^n}{\Delta\tau_{n+1}} = \xi_i^{n+1}V1_{i-1}^{n+1} + \psi_i^{n+1}V1_i^{n+1} + \phi_i^{n+1}V1_{i+1}^{n+1} + \lambda_{12}V2_i^{n+1} \tag{39}$$

so that we obtain $V1_i^n$ i is the solution of S_i with time τ_n as follows:

$$V1^n = (1 - \Delta\tau_{n+1}M^{n+1})V1_i^{n+1} - \Delta\tau_{n+1}R^{n+1} \quad (40)$$

where

$$M^{n+1} = \begin{bmatrix} \psi_1^{n+1} & \phi_1^{n+1} & & & & \\ \xi_2^{n+1} & \psi_2^{n+1} & \phi_2^{n+1} & & & \\ & \cdot & \cdot & \cdot & & \\ & & \cdot & \cdot & \cdot & \\ & & & \cdot & \cdot & \cdot \\ & & & & \psi_{N-2}^{n+1} & \phi_{N-2}^{n+1} \\ & & & & \xi_{N-1}^{n+1} & \psi_{N-1}^{n+1} \end{bmatrix}_{(N-1)\times(N-1)}$$

$$V1^n = \begin{bmatrix} V1_1^n \\ V1_2^n \\ V1_3^n \\ \cdot \\ \cdot \\ \cdot \\ V1_{N-1}^n \end{bmatrix}, R_{n+1} = \begin{bmatrix} \xi_1^{n+1}V1_0^{n+1} + \lambda_{12}V2_0^{n+1} \\ \lambda_{12}V2_1^{n+1} \\ \lambda_{12}V2_2^n \\ \cdot \\ \cdot \\ \cdot \\ \phi_{N-1}^{N+1}V1_N^{n+1} + \lambda_{12}V2_{N-2}^{n+1} \end{bmatrix}$$

And in position state 2 we obtain,

$$V2^n = (1 - \Delta\tau_{n+1}M^{n+1})V2_i^{n+1} - \Delta\tau_{n+1}R^{n+1} \quad (41)$$

where:

$$M^{n+1} = \begin{bmatrix} \psi_1^{n+1} & \phi_1^{n+1} & & & & \\ \xi_2^{n+1} & \psi_2^{n+1} & \phi_2^{n+1} & & & \\ & \cdot & \cdot & \cdot & & \\ & & \cdot & \cdot & \cdot & \\ & & & \cdot & \cdot & \cdot \\ & & & & \psi_{N-2}^{n+1} & \phi_{N-2}^{n+1} \\ & & & & \xi_{N-1}^{n+1} & \psi_{N-1}^{n+1} \end{bmatrix}_{(N-1)\times(N-1)}$$

$$V2^n = \begin{bmatrix} V2_1^n \\ V2_2^n \\ V2_3^n \\ \cdot \\ \cdot \\ \cdot \\ V2_{N-1}^n \end{bmatrix}, R_{n+1} = \begin{bmatrix} \xi_1^{n+1}V2_0^{n+1} + \lambda_{21}V1_0^{n+1} \\ \lambda_{21}V1_2^{n+1} \\ \lambda_{21}V1_3^n \\ \cdot \\ \cdot \\ \cdot \\ \phi_{N-1}^{N+1}V1_N^{n+1} + \lambda_{21}V1_{N-2}^{n+1} \end{bmatrix}$$

4. Simulation and discussion

In this section, we conduct a computational process to examine our results. After we get two regime switching equations it will be done numerical simulation using MATLAB program. Let $V_1 = V_2$ to show the accuracy of models. So that the results are the same as conditions without regime switching and dividend. Then we will compare with the exact results of equations in the Black Scholes without regime-switching. We set up a parameter as follows: $T - t = 1$,$E = 100$, $r = 0.015$, $\sigma_1 = 0.2$,$\sigma_2 = 0.03$,$K = 100$, $S = 200$, $T = 0.25$, $\delta = 0.05$, $\lambda_{12} = 0.02$. It is seen that the settlement graph using finite volume method to European call option equations with regime-switching is almost same or near with exact solutions if $V_1 = V_2$ where V_1 is an option value of σ_1 and V_1 with σ_2.

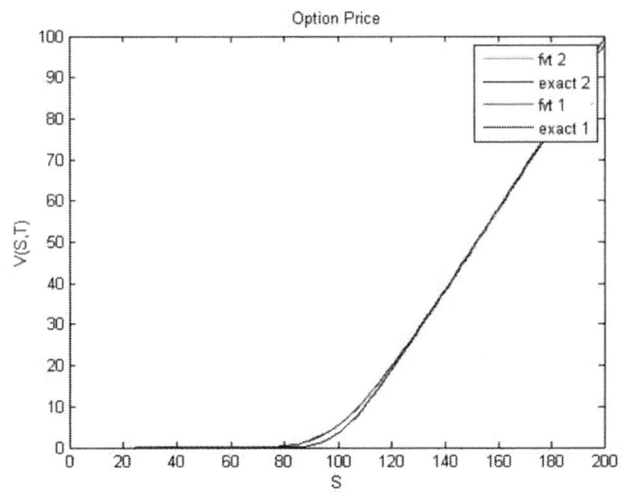

Figure 1. simulation results and exact solutions

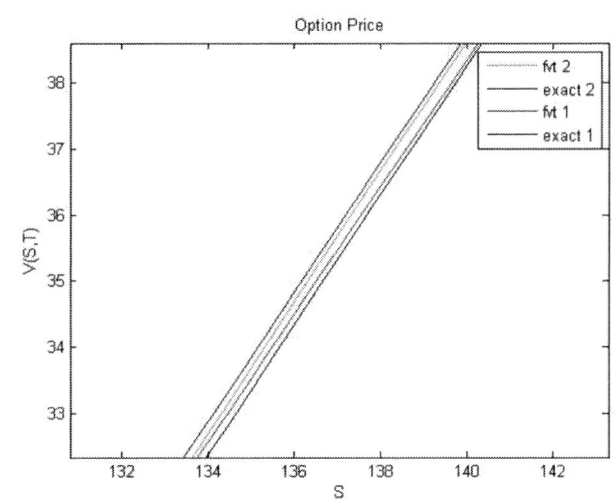

Figure 2. simulation results and exact solutions

5. Conclusion

In this paper, we presented a finite volume method based on fitted finite volume method for the numerical solution of 2 Black-Scholes equations governing European call option with regime-switching volatility. We have shown that the numerical sheme results with M-system matrixs. Numerical results were presented to demonstrate the usefulness of the method.

References

[1] McDonald, P.W., 1971,"The computation of transonic flow through two-dimensional gas turbine cascades", *American Society of Mechanical Engineers, Paper*, hal 71-89

[2] MacCormack, R.W., Paullay, A.J., 1972,"Computational Efficiency Achieved by Time Splitting of Finite Difference Operators", *American Institute of Aeronautics and Astronautics, Paper*, hal 72-154. American Institute of Aeronautics and Astronautics: San Diego, Calif.

[3] Zvan, R., Forsyth, P.A. and Vetzal, K.R.,2001, "A nite volume approach for contingent claims valuation", *IMA J. Numer. Anal. 21*, hal 703731.

[4] Julien, B., and Robert, E., 2006, "Finite volume methods for the valuation of american options", *ESAIM: Mathematical Modelling and Numerical Analysis Vol.40*, hal.311-330

[5] Zhang, K., Wang,S., 2008, "Pricing Options under Jump Difusion Processes with a Fitted Finite Volume Method", *Appl. Math. Comput 201*, hal.398-413

[6] Zhu, S., Badran, A., Lu, X., 2012, "A New Exact Solution for Pricing European Options in a two state regime-switching economy. Journal of Computers and Mathematics with Applications".*Journal of Computers and Mathematics with Applications 64*, hal. 2744 2755.

[7] Versteeg, H K., Malalasekera, W., 2007, "An Introduction to Computational Fluid Dynamics", *Appl. Math. Comput 201*, hal.115-129

[8] Cox, C. and Rubinstein, R., 1979, "Option pricing: a simplied approach", *Journal of Finance Economic 7*, hal.229-264

[9] Black, F., and Scholes, M.,1973, "The pricing of options and corporate liabilities". *The Journal of Political Economy 81*, hal. 637-654

[10] Willmot, P., Howison, S., Dewyne J.,1995), "The Mathematics of Financial Derivatives". New York: Press Syndicate of the Cambridge University.

Vehicle routing problem with time windows using natural inspired algorithms

A B Pratiwi[1], A Pratama[2], I Sa'diyah[3], and H Suprajitno[4]

[1,2,3,4]Department of Mathematics, Faculty of Science and Technology, Universitas Airlangga, Surabaya, Indonesia

[1]asri.bekti@fst.unair.ac.id

Abstract. Process of distribution of goods needs a strategy to make the total cost spent for operational activities minimized. But there are several constrains have to be satisfied which are the capacity of the vehicles and the service time of the customers. This Vehicle Routing Problem with Time Windows (VRPTW) gives complex constrains problem. This paper proposes natural inspired algorithms for dealing with constrains of VRPTW which involves Bat Algorithm and Cat Swarm Optimization. Bat Algorithm is being hybrid with Simulated Annealing, the worst solution of Bat Algorithm is replaced by the solution from Simulated Annealing. Algorithm which is based on behavior of cats, Cat Swarm Optimization, is improved using Crow Search Algorithm to make simplier and faster convergence. From the computational result, these algorithms give good performances in finding the minimized total distance. Higher number of population causes better computational performance. The improved Cat Swarm Optimization with Crow Search gives better performance than the hybridization of Bat Algorithm and Simulated Annealing in dealing with big data.

1. Introduction

Distribution is one of marketing activities in aim to facilitate the producers in sending goods to the customers. In the distribution process, a set of customers is served by a producer where located in an area. A set of crews operates a set of vehicles which is located in the producer place called a depot to fulfill the each customer demand. As a producer, for minimizing their total operational transportation cost of distribution, the depot have to find an appropriate road network for the vehicles. This problem is known as the Vehicle Routing Problem (VRP). The solution of VRP about determination of a set of routes where each performed by a vehicle that starts and ends on a depot, the demand of the customers are fulfilled, all the operational constrains are satisfied, and the total operational transportation cost is minimized [1]. One of the extension of VRP is VRP with Time Windows (VRPTW). In the VRPTW, each customer corresponds to service interval time which represents that the customer can only be served at the time interval given. A vehicle is allowed to serve the customer between known time intervals.

The road network is generally described as a graph. Given $G=(V,A)$ is a complete graph, where $V = \{0, 1, ..., L\}$ is the vertex set and A is the arc set. Vertex 0 is corresponds to the depot and vertices $i = 1, ..., L$ correspond to customers where L is number of customers. Each arc has a weighted value which represents the travel cost spent from vertex i to vertex j, c_{ij}. Each customer has a known nonnegative demand, d_i. Given a vertex set $S \subseteq V$, the total demand is described as $d(S) = \Sigma_{i \in S} d_i$. In a depot is available a set of K vehicles where each has capacity C. To ensure feasibility, assume that $d_i \leq C$ for each $i = 1, ..., L$. And each customer is served by a vehicle in a time interval, between earliest time and latest time $[e_i, l_i]$.

Many optimization methods for solving VRPTW are studied recently. Meta heuristic methods have been applied inspired by the behavior of natures. In solving VRPTW, Yaseen et. al proposes a Meta-Harmony Search [2]. Simulated Annealing is also an effective algorithm for solving VRPTW [3,4]. Bat algorithm is natural inspired algorithm based behavior of bats. Bat algorithm shows better perform than Particle Swarm Optimization, Genetic Algorithm and Harmony Search [5]. Improved Bat Algorithm has outperformed significantly all other alternatives, which are Evolutionary Simulated annealing, Genetic Algorithm and Firefly Algorithm, in most cases for symmetric and asymmetric routing problem, Traveling Salesman Problem [6]. In this paper, Bat Algorithm is studied to solve VRPTW. For increasing its performance, Bat Algorithm is combined with Simulated Annealing. This paper also presents VRPTW using Cat Swarm Optimization and Crow Search Algorithm. Cat Swarm Optimization Algorithm is adopted from behavior of cats where has two modes, seeking mode and tracing mode [7]. The complexity of Cat Swarm Optimization is helped by Crow Search Algorithm. Crow Search Algorithm has fewer parameter which based on the intelligent of crows [8]. For performance simulation, the algorithms given are tested and compared. This paper is organized as follows: the description of VRPTW is in the next section of this section. Next, the proposed natural inspired algorithms are explained. The comparison performances of each algorithm simulation is presented on the next section. Last, the result is concluded on last section.

2. Formulation of the Problems

The main objective of VRPTW is minimization of the transportation cost but still satisfy all the constraints. The routes on VRPTW can be represented as a graph theoretic problem [1]. Let $G = (V,A)$ be a complete graph where V is the vertex set and A is the arc set, $V = \{0, 1, ..., L\}$. A depot is represented as vertex 0 and customers are represented as vertices i, where $i = 1, ..., L$. Since G is a weighted graph then there is a travel cost from vertex i to vertex j which is denoted as c_{ij}. And K number of vehicles are available on depot. Each customer has time window $[e_i,l_i]$. When a vehicle comes before the upper bound, earliest time e_i, then the vehicle have to wait until time e_i. When a vehicle comes after the lower bound, latest time l_i, then the vehicle have to get back to the depot [9]. The VRPTW can be stated as follows:

$$Min\, Z, Z = \sum_{k=1}^{K}\sum_{i=1}^{L}\sum_{j=1}^{L} c_{ij}x_{ijk} \tag{1}$$

$$\sum_{i=1}^{L} q_i y_{ik} \le Q, \quad \forall k \tag{2}$$

$$\sum_{k=1}^{K} y_{ik} = 1, \quad \forall i \tag{3}$$

$$\sum_{i=1}^{L} x_{ijk} = y_{ik}, \quad \forall j,k \tag{4}$$

$$\sum_{j=1}^{L} x_{ijk} = y_{ik}, \quad \forall i,k \tag{5}$$

$$\sum_{i,j}\sum_{\in S\times S} x_{ijk} \le |S| - 1, \quad S \subset \{1,2,...,L\}, \quad \forall k \tag{6}$$

$$sl_i + t_{ij} \le se_j, \quad \forall i,j \tag{7}$$

$$e_i \le se_i \le l_i, \quad \forall i \tag{8}$$

where

$$x_{ijk}\begin{cases}1, \textit{vehicle k travels directly from customer i to customer j}\\ 0, \qquad \textit{otherwise}\end{cases}$$

$$y_{ik}\begin{cases}1, \textit{demand of customer i satiesfied by vehicle k}\\ 0, \qquad \textit{otherwise}\end{cases}$$

The c_{ij} is cost spent for travelling from customer i to customer j, Q is the vehicle capacity, demand quantity of customer i is denoted by q_i, time interval $[e_i,l_i]$ states that customer i can be served, se_i is the service time begin at customer i, sl_i is the time end at customer i, t_{ij} is time spent for travelling from customer i to customer j, L is number of customers, K is number of vehicles and S is the subset of customers [10]. Eq. (1) is the objective of VRPTW, Eq. (2) states that sum of demand of vehicle route is not exceed the capacity of the vehicle, and Eq. (3) states that each customer is visited only by one vehicle. Eq. (4)-(6) explains each route is complete, Eq. (7) explains that two customer which have closest distance are on the same route, and Eq. (8) explains the available time interval of services of each customer [9].

3. Natural Inspired Algorithms

Meta-heuristic algorithms which are based on the natural inspired algorithms are proposed to solve VRPTW. This section describes Hybrid Bat Algorithm and Simulated Annealing, and Improved Cat Swarm Optimization using Crow Search.

3.1. Hybrid Bat Algorithm and Simulated Annealing

A hybrid algorithm between bat algorithm and simulated annealing for solving VRPTW is begun with generating positions of bats population randomly as the initial solutions of VRPTW. For easier understanding of this hybrid algorithm, the overview of the bat algorithm and also simulated annealing is given.

3.1.1. Overview of Bat Algorithm. Bat is natural algorithm that inspired by micro-bats behavior when become predators, find locations and avoid obstacles when flies, using ecolocation [5]. Bats emit a very loud sound pulse where have a constant frequency. The loudness of the emitted pulse also varies. For each bat i th (i = 1,2,.., N), the procedure as follows:

1. Define pulse frequency (f_i), and loudness (A_i).
2. Initialize the bats population with position $x_{i,d}$, velocities (v_i), and pulse rates (r_i).
3. Evaluate the objective value and obtain the best solution so far.
4. Update the position of each bat by adjusting the new velocities and frequencies using Eq. (9)-(11) at time step t and D-dimensional space.
$$f_i = f_{\min} + (f_{\max} - f_{\min}) * rand \tag{9}$$
$$v_{i,d}^{t+1} = v_{i,d}^{t} + (x_{i,d}^{t} - x_{best,d}) * f_i \tag{10}$$
$$x_{i,d}^{t+1} = x_{i,d}^{t} + v_{i,d}^{t+1} \tag{11}$$
5. Generate randomly $rand$. Check the termination if $rand > r_i$ then select a solution among the best solution so far and generate a local solution using :
$$x_{i,d}^{t+1} = x_{i,d}^{t} + rand * A_i^t \tag{12}$$
6. If $rand < A_i$ and the objective value of position $x_{i,d}^{t+1}$ < the objective value of position $x_{i,d}^{t}$, then accept the new solutions, reduce A_i and increase r_i. Otherwise, accepting the $x_{i,d}^{t}$ as the new position.
7. Rank the bats and find the current best.

3.1.2. Overview of Simulated Annealing. Simulated Annealing is known as global search optimization inspired by annealing physical process of solids [11]. The idea of Simulated Annealing is that this algorithm not only accepts changes that improve the objective function, but also keeps changes that are not ideal. Annealing process needs temperature and cooling schedule controlling in aims to get the hard structures. The Simulated Annealing Algorithm procedure for time step t, D-dimensional spaces and i ($i = 1,2,..., N$) is described below [12] :

1. Define initial value for initial temperature, T_0 and generate randomly one initial solution $x_{i,d}^{0}$.
2. Set final temperature T_f and define cooling schedule using α where $0 < \alpha < 1$.
3. Evaluate the objective value of the initial solution.
4. Get modification on the solutions then obtain the new solutions $x_{i,d}^{t+1}$.
5. Evaluate the objective value of the new solutions $f(x_{i,d}^{t+1})$.
6. Calculate $\Delta f = f(x_{i,d}^{t+1}) - f(x_{i,d}^{t})$.
7. If the new solution $x_{i,d}^{t+1}$ is better than $x_{i,d}^{t}$, then accept the new solution and go to step 9. Otherwise calculate probability of acceptance solution p.

$$p = e^{-\frac{\Delta f}{T}} \tag{13}$$

8. Generate randomly *rand*. Check the acceptance condition if $p >$ *rand* then accept the new solutions.
9. Update the temperature $T_{t+1} = \alpha T_t$ where α is the reducing temperature factor.
10. Check the termination condition. If the condition unsatisfied then go to step 4.

3.1.3. Proposed Hybrid Bat Algorithm-Simulated Annealing. For increasing the performances of Bat Algorithm, this paper proposes a hybrid algorithm. The worst solution while using Bat Algorithm procedure is set as the initial solution for Simulated Annealing Algorithm. After obtain the best solution of Simulated Annealing, this solution will replace the solution before. The flowchart of the proposed hybrid is shown in figure 1.

3.2. Improved Cat Swarm Optimization with Crow Search

Cat Swarm Optimization Algorithm is a nature algorithm inspired by behavior of cats which have two modes, seeking mode and tracing mode [7]. Cat Swarm Optimization Algorithm has complex procedure which requires higher computational time. In this paper, Cat Swarm Optimization Algorithm performance is improved by the Crow Search Algorithm to help gaining faster convergences. Crow Search Algorithm has fewer parameter which easier to hybrid with other algorithm. The intelligence of crows inspiring this evolutionary algorithm [8]. Crows are known have good memorized and also have a good self-awareness. Crows will move their foods from hiding places for avoid being pilfered. The overview of Cat Swarm Optimization and Crow Search is given.

3.2.1. Overview of Cat Swarm Optimization. In Cat Swarm Optimization, there are two modes which are seeking mode and tracing mode [7]. Ratio number of cats distributed is determined by a mixture ratio (*mr*). Flag is given to help labelling whether a cat is in a seeking or tracing mode. For D-dimensional spaces, and N cats where $i = 1,2,...N$, the procedure of Cat Swarm Optimization Algorithm is explained as follows:

1. Initialize the position of each cat and the velocity randomly.
2. Based on *mr*, distributed cats randomly into seeking and tracing mode.
3. Evaluate the objective values of position $x_{i,d}^{t}$ for time step t.
4. Apply seeking mode if the cat is in seeking mode, otherwise apply tracing mode.
5. Evaluate the objectives values of posistion $x_{i,d}^{t+1}$.
6. If the objective value of position $x_{i,d}^{t+1}$ is better than $x_{i,d}^{t}$, then update the position.
7. Check the termination condition. If the condition unsatisfied then go to step 2.

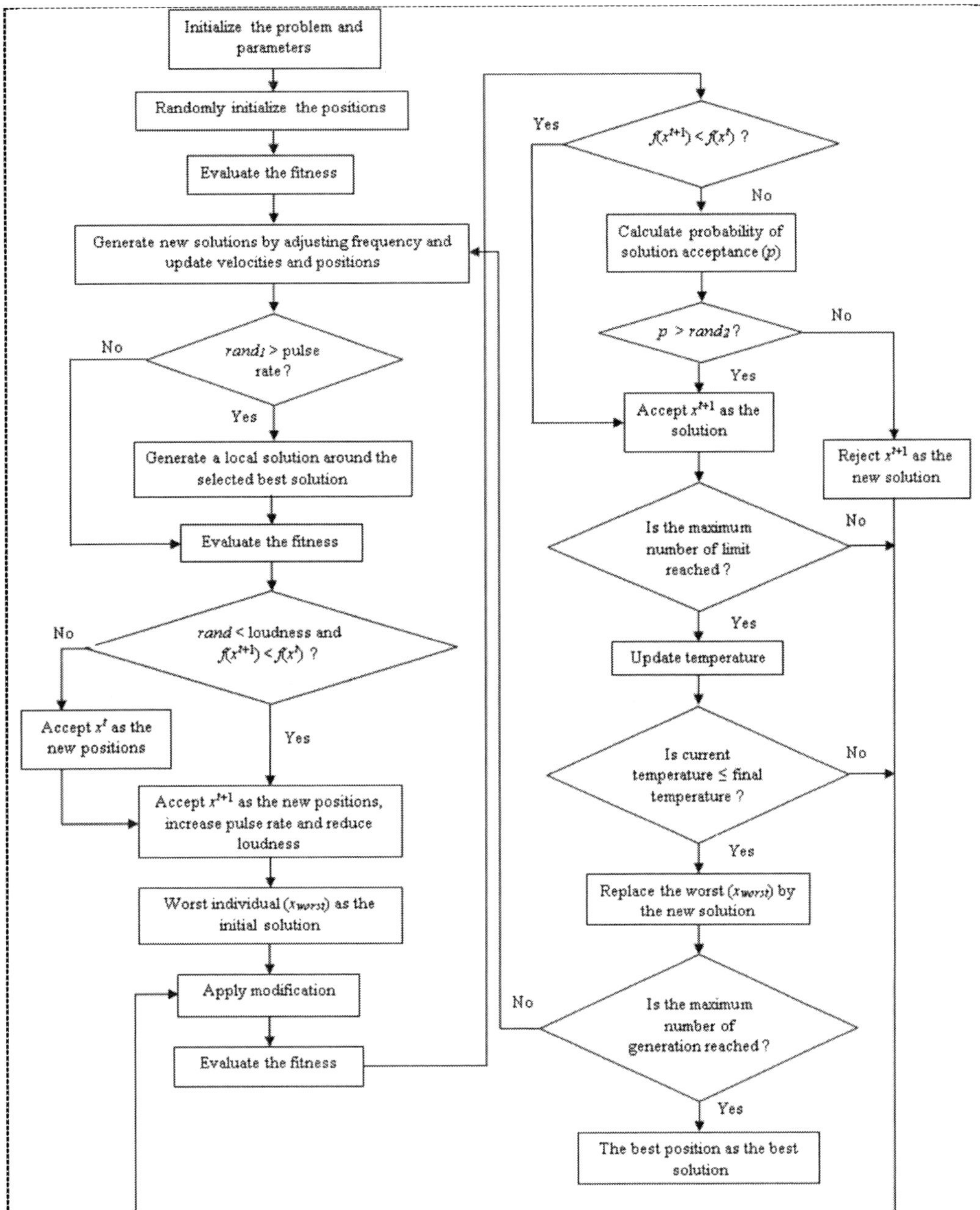

Figure 1. Flowchart of Hybrid Bat Algorithm and Simulated Annealing.

3.2.2. Overview of Crow Search. Crows are the most intelligent birds [8]. Crows have good memories, they can move their food from other thieves by flying and then hiding it on hiding places. This good self-awareness is one of distinctions of crows. The procedure of Crow Search Algorithm for N crows is described as follows:

1. Initialize the position of each i th bat where i =1,2,..,N and the memory randomly for D-dimensional spaces. In this step, the initial memory is assumed by the current position.
2. Evaluate the objective values of the position $x_{i,d}^{t}$ where t is the time step.
3. Get modification on the position.
4. Evaluate the objective values of the new positions, $x_{i,d}^{t+1}$.

5. Update the memory if the objective value of position $x_{i,d}^{t+1}$ is better than $x_{i,d}^{t}$.
6. Check the termination condition. If the condition unsatisfied then go to step 3.

3.2.3. Proposed Improved Cat Swarm Optimization with Crow Search. For time step t and D-dimensional spaces, the procedure of Improved Cat Swarm Optimization with Crow Search is explained by the flowchart on figure 2 and more clearly is described as follows:

1. Initialize parameters: number of individuals (N), seeking memory pool (smp), count dimension to change (cdc), seeking range dimension (srd), acceleration coefficient (c), mixing ratio (mr), flight length (fl) and awareness probability (AP).
2. Initialize the position ($x_{i,d}^{t}$), velocity ($v_{i,d}^{t}$), and memory ($m_{i,d}^{t}$) of the individual randomly. It is assumed that the initial position is set as initial memory.
3. Evaluate the fitness of each individual.
4. Individual which as the best solution ($x_{best,d}$) is labeled the self-position consideration (SPC) = 1, otherwise SPC = 0.
5. Randomly picked individuals from population based on mr and flagged in seeking mode. The flag remains is set to tracing mode.
6. For each individual in seeking mode, apply :
   a) If spc of j th cat = 1 then generates (smp-1) copies and keeps the origin position. Otherwise, generates smp copies.
   b) According cdc, apply a mutation operation to modify the k th dimension of each position.
   $$x_{i,k}^{t+1} = x_{i,k}^{t} + (-1)^{k} * srd * x_{i,k}^{t} \tag{14}$$
   c) Evaluate the fitness.
   d) Store the position in archive.
   e) Apply Roulette Wheel selection for each copies cat [13].
7. For each individual in tracing mode, apply :
   a) Update the velocities.
   $$v_{i,k}^{t+1} = v_{i,k}^{t} + c * rand * \left(x_{best,d} - x_{i,d}^{t}\right) \tag{15}$$
   b) Update the position of i th cat.
   $$x_{i,d}^{t+1} = x_{i,d}^{t} + v_{i,d}^{t+1} \tag{16}$$
   c) Evaluate the fitness.
   d) Store the position in archive.
8. Collect the new individual positions from seeking and tracing mode process in the archive.
9. If the fitness of new position is better than the memorized position then update the memory.
10. Generate r_i randomly with uniform distribution (0,1) and the new position from the chosen copy is modified using :
$$x_{i,d}^{t+1} = \begin{cases} x_{i,d}^{t} + r_i * fl * \left(m_{j,d} - x_{i,d}\right) & r_i \geq AP_j \\ a\ random\ position, & otherwise \end{cases} \tag{17}$$
11. Evaluate the fitness.
12. Update the memory when the fitness value of the modified position is better than the memorized position.
13. Check the termination criterion, if the termination is satisfied then the best position is as the best solution. Otherwise repeat step (4)-(13).

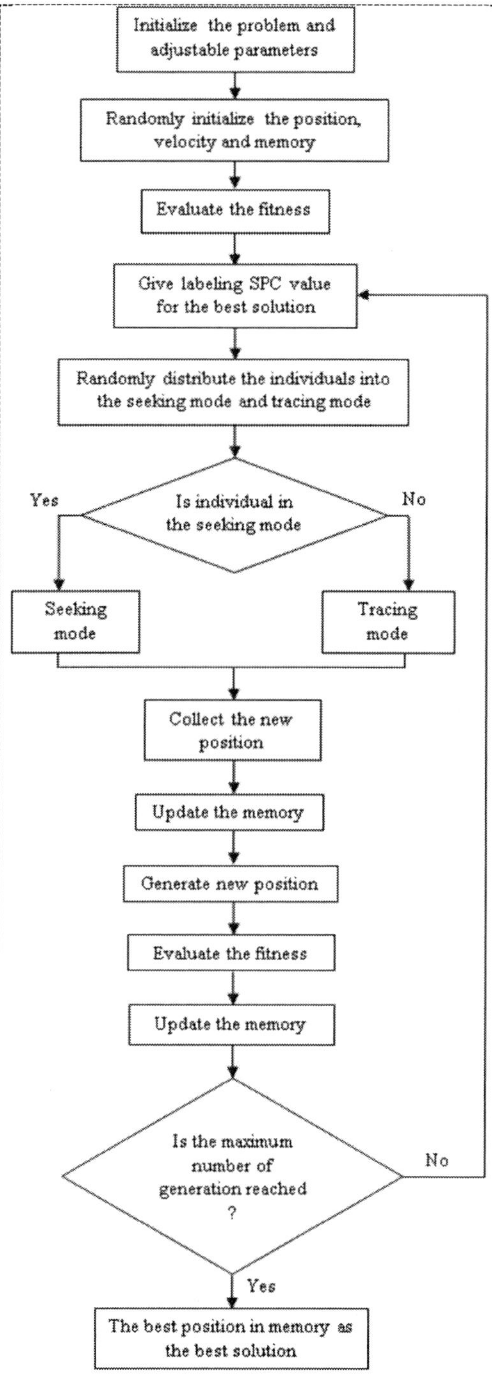

Figure 2. Flowchart of Improved Cat Swarm Optimization with Crow Search.

4. Computational Study

As an approximation for obtaining the optimal solution of VRPTW, the Bat Algorithm is being hybrid with Simulated Annealing and the Cat Swarm Optimization is being improved with Crow Search Algorithm. This proposed natural inspired algorithms are combined and gets to be compared in aims knowing which one is has a better performance for solving VRPTW. Both proposed algorithms have own technique for helps other weakness. Simulated Annealing is used to find better solution for replacing the worst solution of Bat Algorithm. The simplicity of Crow Search and how behavior of crows in memorizing hidden place helps the Cat Swarm Optimization to get its convergence faster.

VRPTW has capacities and time windows constrains. Because a route is only handled by one vehicle then the total customers demand in the same route is not exceed the capacity of the vehicle. And the service time of customers is according to the time windows of customer *i* th $[e_i,l_i]$. If a vehicle come after time l_i, then the vehicle have to back to the depot. And when a vehicle come before time e_i, then the vehicle should wait until time e_i. The proposed inspired algorithms are applied in test problems of VRPTW.

4.1. Test Problems

VRPTW will be more complex in a large number of customers. More customers in the problem, more computational time will be required. This paper uses data which is taken from http://neo.lcc.uma.es/vrp/vrp-instances/capacitated-vrp-with-time-windows-instances/. In the objective of performance comparison, this paper takes 25 customers, 50 customers and 100 customers. The data which available are the customers coordinate, demand of each customer, service duration of each customer, and time windows of each customer. The route of each vehicle is the solution of this problem. Each route must satisfy all the constraints and the solution can minimize the total travel cost spent of the vehicles.

4.2. Computational Results

In comparison the performance of the hybrid Bat Algorithm and Simulated Annealing, firstly shows the computational result using Bat Algorithm. Using vehicle capacity = 200, maximum number of vehicle = 25, number of bats (*N*) = (10, 50, 100), pulse rate = 0.9, loudness = 1, frequency_{min} = 0, and frequency_{max} = 100, the computation result of 1000 generations is shown in table 1. In table 1 also explain the computational result of hybrid Bat Algorithm and Simulated Annealing using the same parameters value with the Bat Algorithm, initial temperature = 1000000, final temperature = (905000, 369000, 43607), reducing temperature coefficient = 0.99, and maximum limits = 1000. From the comparison result, shows that the proposed hybrid Bat Algorithm and Simulated Annealing is performed better than the Bat Algorithm itself.

Table 1. Performance Comparison.

	Total Distance								
	25 customers			50 customers			100 customers		
	N = 10	*N* = *50*	*N* = 100	*N* = 10	*N* =50	*N* =100	*N* = 10	*N* =50	*N* =100
CSO	905,1	712,0	688,1	2128,2	2032,1	2010,0	4180,8	3954,5	3873,9
BA	623,8	568,1	599,0	1837,3	1581,1	1559,9	3764,3	3711,8	3757,4
BA-SA	616,4	597,8	442,9	1689,6	1666,3	1429,3	3631,5	3561,9	3648,1
CSO-CS	546,3	523,9	447,3	1647,3	1557,6	1447,9	3617,4	3495,9	3463,5

The Improved Cat Swarm Optimization is being improved with Crow Search Algorithm. Table 1 shows the computational result of Cat Swarm Optimization using *smp* = 5, *mr* = 0.4, *srd* = 0.5, *c* = 2, *cdc* = 0.2, with the same values of parameter VRPTW. Using *fl* = 2, *AP* = 0.5 for Crow Search parameters, table 1 shows the computational result of improved Cat Swarm Optimization with Crow Search. Comparison between the origin Cat Swarm Optimization and the improved Cat Swarm Optimization with Crow Search states that the improved algorithm gives better performance in minimizing the total distance of vehicle routes which is summary on table 1.

Overall computational results conclude that there are two algorithms which given good performance. A hybrid Bat Algorithm and Simulated Annealing and the improved Cat Swarm Optimization with Crow Search give a significant result. The number of population of the individuals (*N*) has high impact for the algorithm performances. Higher number of the population size gives higher performance of finding the solution which minimized the total distance. Hybrid Bat Algorithm and Simulated Annealing perform well in small data with the high population, but a higher number of data makes the algorithm gets slower convergences. The improved Cat Swarm Optimization performs better than hybrid of Bat Algorithm and Simulated Annealing in higher number of data.

5. Conclusions

Vehicle Routing Problem with Time Windows has complex constraints which needs meta-heuristic approximation for solving the problem. The objective is finding the optimal routes which minimized the total cost spent when some vehicles travel around to serve the customers. Simulated Annealing which has been applied in VRPTW is combined with Bat Algorithm to improve the performance of the Bat Algorithm. In computational result using variance of number of the customers, the hybrid Bat Algorithm and Simulated Annealing gives better performance comparing with the Bat Algorithm. But the improved algorithm, Cat Swarm Optimization with Crow Search has the better performance than the hybrid of Bat Algorithm and Simulated Annealing in dealing with large number of data which is the number of customers.

References

[1] Toth P and Vigo D 2001 The Vehicle Routing Problem (SIAM) pp 1-10
[2] Yaseen E T, Ayob M, Nazri M Z A and Sabar N R 2015 Meta-harmony search algorithm for the vehicle routing problem with time windows *Information Sciences* **325** pp 140-158
[3] Banos R, Ortega J, Gil C, Fernandez A and Toro F 2013 A Simulated Annealing-based parallel multi-objective approach to vehicle routing problems with time windows *Expert Systems with Applications* **40** pp 1696-1707
[4] Tavakkoli-Moghaddam R, Gazanfari M, Alinaghian M, Salamatbakhsh A and Norouzi N 2011 A new mathematical model for a competitive vehicle routing problem with time windows solved by simulated annealing *Journal of Manufacturing Systems* **30** 83-92
[5] Yang X S 2010 A New Metaheuristic Bat-Inspired Algorithm in *Nature Inspired Cooperatives Strategies for Optimization* (*NISCO 2010*) (Eds. J.R. Gonzalez et al.) *Studies in Computational Intelligence* Springer Berlin **284** pp 65-74
[6] Osaba E, Yang X S, Diaz F, Lopez-Garcia P and Carballedo R 2016 An improved discrete bat algorithm for symmetric and asymmetric *Engineering Applications of Artificial Intelligence* **48** pp 59-71
[7] Chu S C and Tsai P W 2007 Computational intelligence based on the behavior of cats *International journal of innovative computing* Taiwan
[8] Askarzadeh A 2016 A novel metaheuristic method for solving constrained engineering optimization problems : Crow Search algorithm *Computers and Structures* **169** pp 1-12
[9] Thangiah S R 1995 Vehicle routing with time windows using Genetic Algorithms *Applications Handbook of Genetic Algorithms : New Frontiers Vol II Lance Chambers*^{ed} (*CRC Press*) pp 253-277
[10] Pan F, Ye C, Wang K and Cao J 2013 Research on the vehicle routing problem with time windows using Firefly Algorithm *Journal of Computer vol 8 no 9* China : University Jiaoxuo
[11] Kirkpatrick S, Gelatt C D and Vecchi M P 1983 Optimization by Simulated Annealing *Science* **220**
[12] Yang X S 2012 Engineering Optimization : An introduction with metaheuristic application (John Wiley & Sons, Inc US)
[13] Deepa S N and Sivandanam S N 2007 Introduction to Genetic Algorithms (Springer) pp 47-48

Content knowledge of prospective elementary school teacher for fractional concepts

N Pattimukay[1,2,a], D Juniati[3,b] and M T Budiarto[3,c]

[1)]Faculty of Teacher Training and Education, Universitas Patimura
[2)]Student of Postgraduate Program, Universitas Negeri Surabaya)
[3)]Faculty of Mathematics and Science Universitas Negeri Surabaya

E-mail: *[a)]n355y_p@yahoo.com, [b)]dwijuniati@unesa.ac.id, [c)]megatbudiarto@unesa.ac.id*

Abstract. The aim of this study was to describe the content knowledge especially the concept of fraction of prospective elementary school teacher. The purpose of this study is to describe the content knowledge, especially the concept of fraction of prospective elementary school teacher. The subject of the study was one of prospective elementary school teacher of Pattimura University. This research is qualitative research. Data were collected through the provision of tests to explore the knowledge content of primary school teacher candidates about fractional concepts. Then continued with qualitative data analysis. The results of this study are as follows: that the prospective primary school teacher defines fractions as part of the whole if an object is divided into equal parts, so that the part that has been divided is part of the whole. Furthermore, the prospective elementary school teacher understood the fractions as division shown in two ways, namely the prospective elementary school teacher understood the fraction as a division operation, the primary school teacher candidate interpreted the fraction as a division when an object is divided be part of the same. Meanwhile, the fraction as a ratio is interpreted as the relationship between a pair of numbers. Then, the denominations are interpreted as a ratio between the numerator and the denominator of the same value. The prospective elementary school teacher also understands fractions of value when simplifying fractions. Primary school teacher candidates understand the concept of fractional operations.

1. Introduction

Primary school teacher education students are specifically prepared to become professional teachers who are expected to be able to teach almost all subjects in elementary school, one of which is mathematics. Mathematics is one of the subjects taught to students ranging from elementary to university level. At the elementary school level, the correct cultivation of mathematical concepts to students is essential, since the mathematical concepts will continue and are closely related to other mathematical concepts that will be taught at the next level of education. Mastery of mathematics material is one of the categories that must be mastered by teachers and prospective teachers who by Shulman (1986) called Pedagogical Content Knowledge (PCK). Ball (1990) defines PCK based on students and content knowledge, content learning and knowledge, and content curriculum and knowledge. Furthermore, Ball (2000) argues that when preparing teacher candidates we need to identify the essential content knowledge needed in learning, how the knowledge needs to be understood, how knowledge is taught in the classroom. The combination of specific content knowledge and pedagogic knowledge is known as Pedagogical Content Knowledge (PCK). Mathematical knowledge and knowledge of mathematical representation are related to content knowledge, while student knowledge and learning knowledge are related to PCK. Meanwhile, according to An, Kulm and Wu (2004), there are three components of PCK namely subject matter knowledge, curriculum knowledge, learning knowledge. The elementary mathematics material consists of integers, fractions, geometry, and statistics. One of the mathematics materials in elementary school that is quite difficult and vulnerable to misconceptions in students is the concept of fractions. The difficulty in understanding the fractional concept is not only experienced by elementary students but also by most students. The difficulty appears in the errors that are often made by the student if given a matter related to the fractional material.

Based on the author's experience when teaching calculus courses, not a few students make mistakes in operating the fractions. In addition, the author's experience in teaching some courses in PGSD, when giving a problem related to fractions, many students who make mistakes in solving the given problem. Errors students make vary. One example is, when the author gives a matter of summing the fractions with a different denominator, there are some students who solve the problem without equating the denominator, but directly sums the numerator with the numerator and sums the denominator with the denominator. Fractional concept is the basic concept that has been taught at primary school level and prospective elementary school teachers are prospective teachers who are prepared to teach at the primary school level. In addition, a deep understanding of a particular subject matter in this case math is very important. Fractional concepts must be well understood by elementary students because they become the basis for understanding other mathematical concepts in higher education.

Starting from some of the above, prospective elementary school teachers who are prepared to become elementary school teachers should be really prepared both in terms of mastery of material and delivery of material to students to be accepted and understood by students and should be familiarized to explain the material that is understood to students so that when there is a mistake in the delivery, can be fixed. In addition, with the mastery of good material from prospective teachers or teachers, prospective teachers or teachers can explain and improve misconceptions experienced by students. Preparing teachers to teach math effectively is one of the most pressing issues facing people who want to improve student learning (Morris, Hiebert, & Spitzer, 2009). Knowledge of prospective teacher content is an important focus for properly preparing teachers (Darling Hammond, 2000; Stohlmann, Moore, & Cramer, 2013).

The authors made initial observations of prospective elementary school teachers by giving some questions about the addition and subtraction of the same denominations and different sections. Based on preliminary observations, it was found that many students have not understood the concept of fractions correctly. This is seen through the work of students in adding the same fractions, the students still make mistakes. Some students solve the same amount of sums by multiplying by cross-sum then summing the numerator with the numerator and denominator with the denominator. As for the different problems disguised, most students equate the denominator by multiplying the denominator and then proclaiming the numerator. From some mistakes that happened, it turns out that there are still many students who do not understand about the concept of fractions worth, LCM, and GCD. The same error occurs when the student is asked to explain the answer to his friends. From some of the above, it appears that students' understanding of the basic concepts of mathematics is very low, even though these concepts have been studied since elementary school and will be taught by prospective students when they will become elementary school teachers.

Based on the above background, the purpose of this study is to describe the content knowledge of prospective elementary school teacher for the fractional concept material. Therefore, the question formulation in this study is how content knowledge of prospective elementary school teacher for the fractional concept material?

2. Theoretical background

2.1. Mathematics teaching

Delivering information in the form of knowledge to someone, and then the person understanding what is delivered is one of the characteristics of good teaching. Hamalik (2001: 44-53) suggests that teaching can be interpreted as (1) conveying knowledge to students, (2) passing culture to young generation, (3) the effort to organize the environment so as to create learning conditions for the students, (4) provide guidance to students, (5) preparing students to become good citizens, (6) a process of helping students face the daily life of the community. Associated with mathematics, according to Diene (1971), learning mathematics means understanding the relationship between symbols and relationships among others to obtain concepts that can be used in real life.

Based on the opinion of the experts above, the authors can conclude that teaching mathematics is the process of communicating information knowledge in the form of ideas and concepts of mathematics to students by using various ways in accordance with mathematical topics delivered in the hope that what has been delivered can be accepted with good and can be understood by students and can be applied in dealing with various problems in everyday life related to mathematics.

2.2. Pedagogical content knowledges

Shulman (1986) is one who introduces PCK (Pedagogic Content Knowledge) which is a slice of certain content knowledge and pedagogic. Based on Shulman's (1987) idea for PCK, effective teachers can have an indepth knowledge of how to represent subject matter to students (Parker & Heywood, 2000). Shulman (1987) also stated that PCK should include students' knowledge and their characteristics, knowledge of educational contexts, knowledge of outcomes, educational objectives and values, philosophical and historical basis of education. PCK leads to the ability of teachers to transform the content of learning materials in learning with a variety of methods that match the background of students' abilities.

PCK describes how specific subject matter is transformed to improve effective communication for teachers and students. Content knowledge refers to the body of knowledge and information that teachers teach and students are expected to study a given subject or content, such as English art, math, science, or social studies. Content knowledge generally refers to facts, concepts, theories, and principles taught and learned in the academic field rather than related skills such as reading, writing, or researching. Grossman (1990) describes four specific components of learning knowledge: (1) knowledge of strategy and representation of a particular topic of learning, (2) understanding of concepts and misconceptions of students on learning topics, (3) teacher knowledge and beliefs about specific learning objectives for students of different levels, covering their conceptions related to basic lessons and what topics are important for students to learn (4) curriculum material knowledge in learning. In line with the above opinion, Fennema and Franke (1992) determine the mathematics teacher's knowledge component as: (1) mathematical knowledge which includes knowledge of content consisting of the nature of mathematics and the mental organization of teacher knowledge, (2) knowledge of mathematical representation , (3) student knowledge, (4) knowledge of teaching and decision making. The first item is about having a conceptual understanding of mathematics. Fennema and Franke (1992) argue that if a teacher has a conceptual understanding of mathematics, it affects classroom teaching in a positive way; Therefore, it is important to have mathematical knowledge for teachers. Teachers and prospective teachers should be able to see the interrelationships between concepts taught because it is very important and is a procedural knowledge. Furthermore, Fennema and Franke emphasize the importance of knowledge about mathematical representation, since mathematics is seen as a composition of a very related set of abstractions. Fennema and Franke (1992: 153) state that if teachers do not know how to translate the abstraction into a form that allows learners

IOP Conf. Series: Journal of Physics: Conf. Series **974** (2018) 012026 doi:10.1088/1742-6596/974/1/012026

to relate mathematics to what they already know, they will not learn with understanding. Knowledge of mathematics and knowledge of mathematical representations related to content knowledge, while knowledge of students and knowledge of teaching is related to pedagogical content knowledge.

Based on the opinion of the experts above, it can be concluded that PCK includes the conceptual knowledge and procedural knowledge that students bring in the learning of a particular topic, misconceptions that may have developed in students, and understanding stages that students may skip from a little understanding of a particular topic to master the topic the. PCK also includes knowledge of techniques for assessing students' understanding and diagnosing their misconceptions, knowledge of teaching strategies that can be used to link what students learn with the knowledge they already possess, and knowledge of teaching strategies to eliminate misconceptions that have developed in student self. In this writing will be focused on the components of the PCK is the knowledge of mathematical content in this case the knowledge of prospective elementary school teachers in this case PGSD students about the fractional concept material.

3. Method research

This study aims to reveal the content knowledge of prospective elementary school teacher. This research is a qualitative research. The subject of this study is one student of prospective elementary school teacher Patimura University. A group of students will be given a math test. The results of the test were processed and categorized into three groups of highly skilled students (score ≥ 80), moderate math skills ($65 \leq$ score < 80), and low math skills ($0 \leq x < 65$) ranging from 0 to 100. Subjects in the study were subjects with high mathematics ability category (S1). Data were collected through the provision of tests to explore the content of primary school teacher candidates about fractional concepts. Then continued with qualitative data analysis.

4. Result and discussion

This research was conducted at Pattimura University of prospective elementary school teacher study program. The subject of this study is one students of prospective elementary school teacher Patimura University. Prior to conducting the research, researchers first gave a mathematics ability test to prospective elementary school teacher Patimura University amounting to 60 students. The test of mathematical ability given in the form of essay which amounted to 9 questions. Based on the test results obtained one student who has a score of 85. The score belongs to the category of high ability. Thus, obtained the subject of research and given the code S1. After that, to explore the content knowledge of the students of primary school teachers, the researchers designed the test questions consisting of 3 questions about the meaning of fractions, fractions of value and simple denominations, and understanding in solving various problems related to the concept of fractions. Problem test is given to S1 to be completed.

Understanding S1 about the subject matter especially the fractional concept can be seen based on the answers given by S1 in answering 3 items of questions designed by researchers to explore the understanding of prospective elementary school teachers about the concept of fractions. The following is the details of the questions and answers S1.

Problem number 1. Explain the meaning of fractions!

S1 answer for number 1.

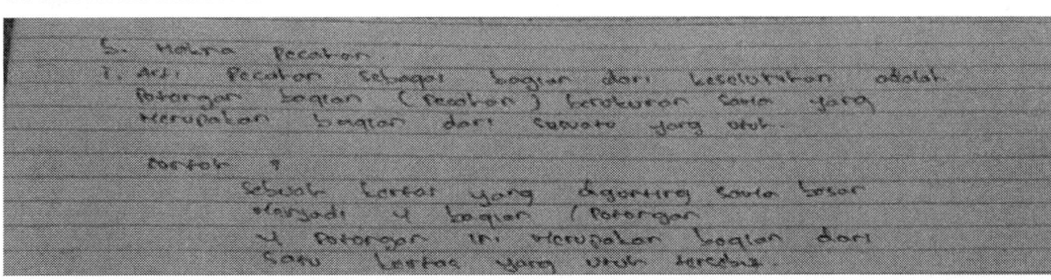
5. Makna Pecahan.
1. Arti Pecahan sebagai bagian dari keseluruhan adalah potongan bagian (pecahan) berukuran sama yang merupakan bagian dari sesuatu yang utuh.

Contoh :
Sebuah kertas yang digunting sama besar menjadi 4 bagian / potongan
4 potongan ini merupakan bagian dari satu kertas yang utuh tersebut.

Figure 1. S1's answer related with the meaning of fractions as part of the whole

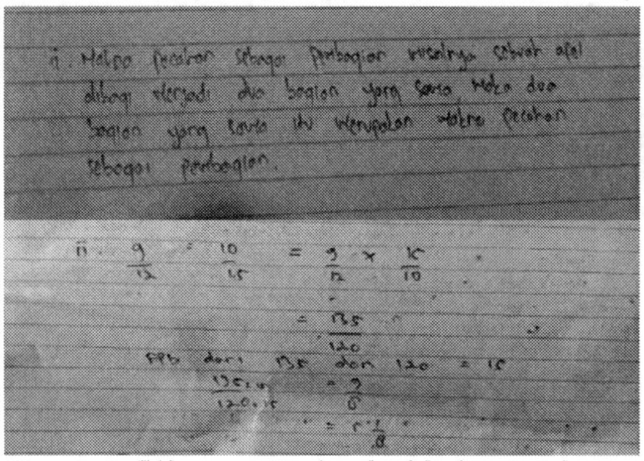

Figure 2. S1's answer related with the meaning of fractions as division

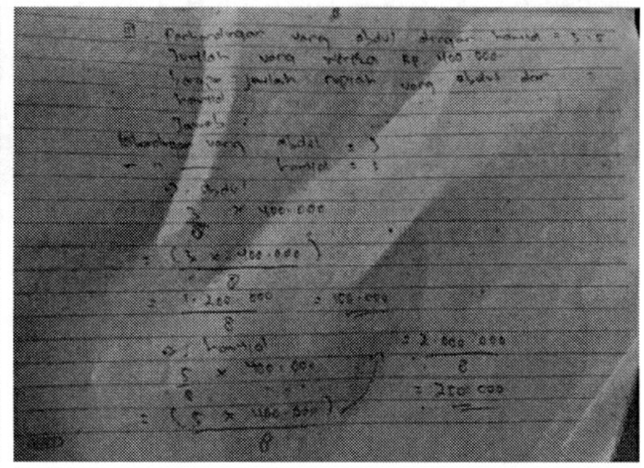

Figure 3. S1's answer related with the meaning of fractions as ratio

Based on the above data, obtained that S1 describes the meaning of fractions as part of the whole is pieces or parts of which each part has the same size that part or piece is part of something whole or whole. S1 gives a paper sample. The paper is cut into equal parts. S1 cut paper into 4 pieces of the same. according to S1, the piece of paper the piece is part of the whole paper. This shows that S1 has an understanding of fractions as part of the whole.

Furthermore, S1 defines fractions as division by giving an example of an object divided into equal parts. The object is an apple, an apple divided into two equal parts, so these two equal parts are the product of the division of an apple. Then, S1 exemplifies fractions as division: $\frac{9}{12} \div \frac{10}{15} = \frac{9}{12} \times \frac{15}{10} = \frac{135}{120}$. GCD of 135 and 120 are 15 so $\frac{135 \div 15}{120 \div 15} = \frac{9}{8} = 1\frac{1}{8}$. S1 defines fractions as divisions by dividing two common fractions and applying the fractional operating properties of fractions. After obtaining the result of operation, the fraction is then simplified by finding GCD from the numerator and denominator. After obtaining the GCD, the numerator and denominator are divided by the GCD and get the simplest value. This means that S1 defines fractions as a division operation. Based on the explanation S1, it can be concluded that S1 views fractions as a division that lies in the division operation. Thus, S1 denotes a fraction as an object divided into equal parts and fractions as a division operation.

The meaning of fractions as a ratio, S1 gives an example of a ratio of money owned by 2 children. The question of the example given S1 is about the amount of money held by two children if it is known the ratio of the money of both children and the amount of their money. Furthermore, S1 determines the amount of money each child has by applying a comparable ratio. This shows that S1 interpreted fractions as a ratio seen from the ratio of the amount of money owned by the two children against the amount of their money. This shows that S1 interpreted fractions as ratio by looking at the relationship between a pair of numbers so it can be concluded that S1 understands fractions as a ratio.

Based on the data analysis of the number 1, obtained that S1 understands the meaning of fractions as part of the whole, fractions as a division and fractions as a ratio and it relates to factual knowledge and conceptual knowledge. So it can be said that S1 has an understanding of factual and conceptual knowledge related to the concept of fractions.

Problem number 2.

a) Explain the meaning of a equivalent fraction!, b) Explain the meaning of fractions $\frac{1}{2}$, $\frac{2}{4}$, $\frac{4}{8}$!, c) Specify a equivalent fraction of $\frac{3}{5}$, Explain how you got a equivalent fraction!

S1 answer for number 2.

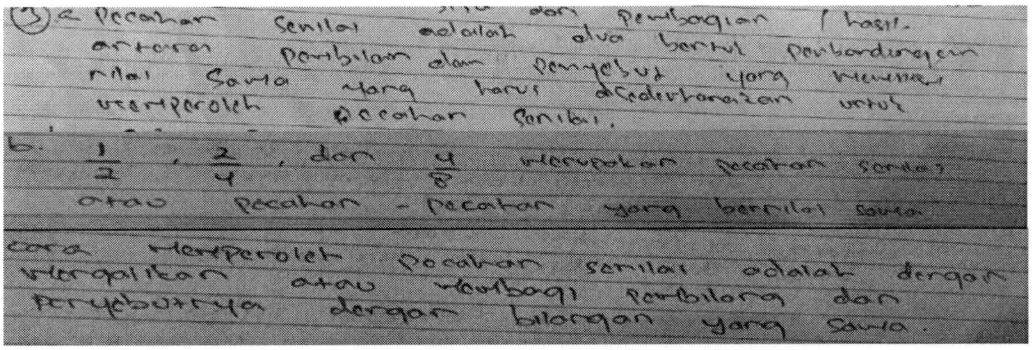

Figure 4. S1's answer related with equivalent fraction

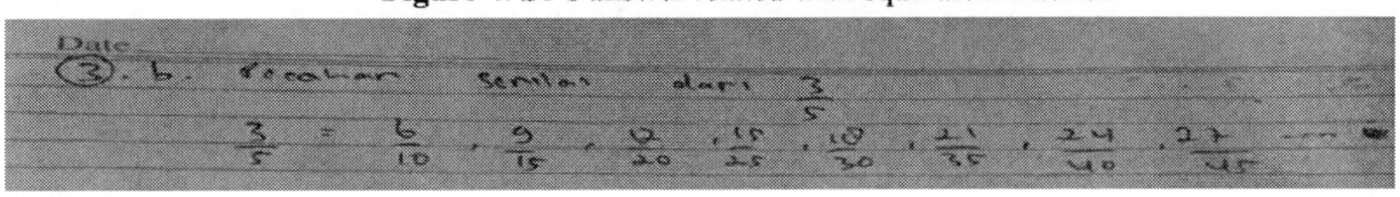

Figure 5. S1's answer related with equivalent fraction of $\frac{3}{5}$

Based on the above data, obtained that S1 defines equivalent fractions is two forms of comparison between the numerator and denominator who have the same value. According to S1, fractions of $\frac{1}{2}, \frac{2}{4}, \frac{4}{8}$ are equivalent fractions. According to S1, a equivalent fraction can be obtained by multiplying or dividing the numerator and denominator with the same number. S1 explains that a equivalent fraction is a fraction that has the same value. To obtain a equivalent fraction, the fraction must be multiplied or divided by the same number. S1 gives an example, to determaine equivalent fractions of $\frac{1}{2}$, then the $\frac{1}{2}$ fraction must be multiplied successively by 2, 3, 4 and so on and obtained fractions $\frac{1}{2}$ of value with fractions $\frac{2}{4}, \frac{3}{6}, \frac{4}{8}$. Equivalent fractions of $\frac{6}{10}$ are $\frac{9}{15}, \frac{12}{20}$, and $\frac{3}{5}$. This indicates that S1 understands the meaning of equivalent fractions, as well as how to determine a equivalent fraction. This also means that S1 understands how to simplify the fraction of searching for a fraction of value with the fraction in question in the simplest form of fraction.

Based on the data analysis of the number 2, obtained that S1 understands the meaning of equivalent fractions and it relates to conceptual knowledge. So it can be said that S1 has an understanding conceptual knowledge related to the equivalent fractions.

Problem number 3

If each child needs $\frac{3}{4}$ meters of ribbon, then 6 children need how many meters of ribbon section. Explain your answer!

S1 answer for number 3.

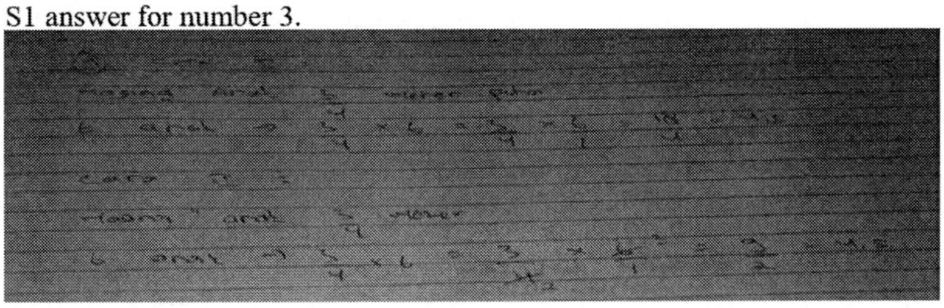

Figure 6. S1's answer related with problem solving

Based on the above data, obtained that, S1 solve the problem in two ways. The first way, S1 writes the information known in the problem ie each child needs $\frac{3}{4}$ meters ribbon. 6 children then $\frac{3}{4} \times 6 = \frac{3}{4} \times \frac{6}{1} = \frac{18}{4} = 4{,}5$. While the second way S1 write down the information known in the problem that is each child needs $\frac{3}{4}$ meters ribbon. 6 children then $\frac{3}{4} \times 6 = \frac{3}{4} \times \frac{6}{1} = \frac{9}{2} = 4{,}5$. This shows that S1 understands the meaning contained in the matter of the story so that S1 can solve the problem well and can apply the properties of fractional multiplication operations.

Based on the data analysis of the number 3, obtained that S1 understands problems solving related to the concept of fractions and it relates to conceptual and prosedural knowledge. So it can be said that S1 has an understanding of procedural knowledge that is problem solving related to the concept of fractions. It shows that, (1) S1 understand the meaning contained in the story so that S1 can solve problems related to the concept of fractions with well, (2) S1 understands fractional operations so as to apply the fractional operating properties.

5. Conclusion

The results above show that the prospective elementary school teachers interpret the fractions as part of the whole that is if an object is divided into equal parts, so that the part that has been divided is part of the whole. Furthermore, the prospective elementary school teacher understood the fractions as division shown in two ways, namely (1) the prospective elementary school teacher understood the fraction as a division operation; (2) the primary school teacher candidate interpreted the fraction as a division when an object is divided be part of the same. Meanwhile, the fraction as a comparison is interpreted as the relationship between a pair of numbers. The denominations are interpreted as the ratio between the numerator and denominator of the same value. The prospective elementary school teacher also understands fractions of value when simplifying fractions. Primary school teacher candidates understand the concept of fractional operations. This is shown when students can solve the story problem. In addition, prospective students have understood the meaning contained in the story and the prospective primary school teachers can solve the story well.

6. References

[1] Shulman L S 1986 Those who understand: Knowledge growth in teaching *Educational Researcher* 15(2) 4-14

[2] Ball D L 2000 Bridging Practices: Intertwining content and pedagogy in teaching and learning to teach *Journal of Teacher Education* 51(3) 241-247

[3] An S and Kulm G and Wu Z 2004 The Pedagogical content knowledge of middle school mathematics teachers in China and US *Journal of Mathematics Teacher Education* 7: 145-172

[4] Morris A K and Hiebert J and Spitzer S M 2009 Mathematical knowledge for teaching in planning and evaluating instruction: What can pre-service teachers learn? *Journal for Research in Mathematics Education 40*(5) 491-529

[5] Darling-Hammond L & Bransford J 2005 *Preparing teachers for a changing world* San Fransisco: Jossey Bass

[6] Park S H and Oliver J S 2008 Reconceptualization Of Pedagogical Content Knowledge (PCK): PCK as a conceptual tool to understand teachers as professionals *Research in Science Education* 38 261-284

[7] Grossman PL 1990 *The making of a teacher: teacher knowledge and teacher education* New York: Teacher College Press

[8] Fennema E & Franke M 1992 *Handbook of Research on Mathematics Teaching and Learning* Teachers' knowledge and its impact in: D A Grouws (Ed) (New York: Macmillan Publishing)
[9] Kula S Berna C G and Guzel B E and Ozgur Z and Elci N A 2013 Scala development for pre-service mathematics teachers' perceptions related to their pedagogical content knowledge *South African Journal of Education* 2013 33(2)

Classification of tumor based on magnetic resonance (MR) brain images using wavelet energy feature and neuro-fuzzy model

A Damayanti[1], and I Werdiningsih[2]

[1,2]Department of Mathematics, Faculty of Science and Technology, Universitas Airlangga, Surabaya, Indonesia

Email : [1]aulid@fst.unair.ac.id

Abstract. The brain is the organ that coordinates all the activities that occur in our bodies. Small abnormalities in the brain will affect body activity. Tumor of the brain is a mass formed a result of cell growth not normal and unbridled in the brain. MRI is a non-invasive medical test that is useful for doctors in diagnosing and treating medical conditions. The process of classification of brain tumor can provide the right decision and correct treatment and right on the process of treatment of brain tumor. In this study, the classification process performed to determine the type of brain tumor disease, namely Alzheimer's, Glioma, Carcinoma and normal, using energy coefficient and ANFIS. Process stages in the classification of images of MR brain are the extraction of a feature, reduction of a feature, and process of classification. The result of feature extraction is a vector approximation of each wavelet decomposition level. The feature reduction is a process of reducing the feature by using the energy coefficients of the vector approximation. The feature reduction result for energy coefficient of 100 per feature is 1 x 52 pixels. This vector will be the input on the classification using ANFIS with Fuzzy C-Means and FLVQ clustering process and LM back-propagation. Percentage of success rate of MR brain images recognition using ANFIS-FLVQ, ANFIS, and LM back-propagation was obtained at 100% .

1. Introduction

The brain is the organ that coordinates all the activities that occur in our bodies. Small abnormalities in the brain will affect body activity. The human brain weighs 2% of the adult's body weight which receives 20% of cardiac output and requires 20% of the body's oxygen consumption and about 400 kcal of energy per day. Therefore, brain tissue is very susceptible to changes in oxygen and blood glucose, stop blood flow 10 seconds alone can eliminate human consciousness and can damage the brain permanently if blood flow stops within minutes [1]. One of the brain disorders is a brain tumor. Tumor of the brain is one of the deadly illness at the world, namely the growth of abnormal cells in the brain structure [2]. Tumor of the brain is a lump that results from the growth of morbid and unbridled cells in the brain. Tumor of the brain can become malignant or benign. The benign brain tumor is a non-cancerous type and is not harmful to surrounding tissue, known as the non-metastatic tumor. While brain cancer is a cell that develops very quickly and destroys the surrounding tissue [3]. In Indonesia, within a decade of brain tumors can occur in children, are adults at the age of 30-70 with shoulders aged 40-65 years. occurred brain tumor incidents [4]. In America, there are 35,000 new cases of brain tumors each year, and central nervous system primary tumors are found in 10% of all neurological diseases found in General Hospital [5]. Decision-making becomes very complicated with many types of tumor of brain.

Therefore the classification of MRI images of the brain is very important to classify the types of brain tumors, which can affect people's lives. Magnetic resonance imaging (MRI) is a device of medicine on radiological diagnostic examination which results in high-quality image recording of the structure of anatomies body of human, particularly the brain and can be used for soft tissue imaging [6]. MRI is a noninvasive medical test that is beneficial to physicians in diagnosing and treating medical conditions. The process of classification of brain tumors can provide the right decision and correct treatment and appropriate in the process of treatment of tumor of brain. Classification of brain tumors can be done by using supervised classification and unsupervised classification, one of which is support vector machine (SVM) [7] and knearest neighbors (kNN) [8], self-organization map (SOM) [7] and fuzzy c-means [9]. All the classification methods that have been done, achieved good results, and the supervised classification of the success rate (accuracy) of the classification is better than the unsupervised classification. The method of classification used in this research is the Neuro-Fuzzy systems method, is a combination of inference fuzzy system mechanism described in the neural network architecture and a special approach to the development of Neuro-Fuzzy, Adaptive Neuro-Fuzzy Inference Systems (ANFIS). ANFIS has denoted prominent results in modeling functions of nonlinear. Anfis learning process through features of many data and based on the value of parameters and errors given [10]. In biomedical engineering, the most successful implementation of ANFIS is the process of classification [11].

The transform of the wavelet is an effective technique for extraction of a feature because wavelet transforms can perform various image analysis with various resolutions and can perform MRI data extraction process [6]. Since the results of wavelet transform require large memory storage, a method is needed to reduce the dimensions of the data. Energy coefficient of wavelet decomposition can be used to reduce dimensions and improve classification performance [12].

In this study, will use the coefficient of the energy of transformation of wavelet for dimension reduction in MR Brain Images data and be using ANFIS for MR Brain Images data classification process according to the type of diseases, such as Glioma, Carcinoma, and Alzheimer's. In previous studies, Damayanti (2014) have done classifications MR brain images using gradient descent backpropagation with the level of success of recognition pattern MR brain images of 95%. In this study classifications MR brain images using anfis , expected percentage this classifications success rate better than neural-networks

2. Methodology

Stages in this study there are three, namely: feature extraction, feature reduction, and classification process. Feature extraction stage is the stage to get the feature of MR brain image that is approximation of vector using wavelet decomposition. The feature reduction stage is a process step to reduce the feature by using energy coefficient of the approximation vector. The last stage is the MRI brain image classification stage using ANFIS. Classification results are divided into four kinds, namely normal, Alzheimer, Glioma, and Carcinoma. The process stages in this study can be seen in figure1.

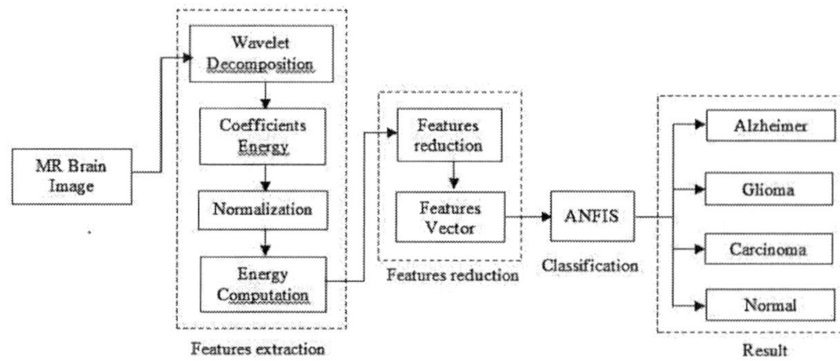

Figure 1. The system block diagram

2.1. Feature Extraction

Feature extraction is the process of capturing features of an image. At this study, feature extraction of MR brain image using wavelet decomposition, is decomposition of image process making use wavelet transform. The type of wavelet used is Haar. Wavelet haar is one of the simplest types of discrete wavelet transforms because the wavelet haar uses a square wave period [13]. The result of the feature extraction process is a vector feature. This feature vector is the output of the wavelet decomposition of the decomposition vector C and the matrix S [C, S]. Wavelet decomposition used up to level 3.

Feature Extraction is a process that takes the features of an image. In this study, feature extraction of MR brain image using wavelet decomposition is the process of image decomposition using wavelet transformation. The type of wavelet used is Haar. Wavelet Haar is one of the simplest types of discrete wavelet transforms because the wavelet Haar uses a square wave period [13]. The result of the feature extraction process is a vector feature. This feature vector is the output of the wavelet decomposition of the decomposition vector C and the matrix S [C, S].

Wavelet transform is a windowing technique using variable sizes to adjust the information and time of the signal. The advantage of wavelet transformation is taking a scale that is a time-scale rather than a time-frequency of a signal. [14]. Transformation of the Discrete wavelet is the implementation of wavelet which uses a scale and position. The basic of the wavelet of discrete is, suppose x(t) is a square-integrated function, then the continuous x(t) relative to w(t) is formulated as Eq.(1)-(2)

$$W_{\psi}(a,b) = \int_{-\infty}^{\infty} x(t)\psi_{a,b}(t)dt \tag{1}$$

where

$$\psi_{a,b}(t) = \frac{1}{\sqrt{a}}\psi\left(\frac{t-a}{b}\right) \tag{2}$$

The wavelet calculation is derived from the primary wavelet value with the translation and dilation process, parameter of a is the factor of the dilation and the parameter of b is the parameter of the translation which both must be real positive. on the development of wavelet analysis, The most commonly used wavelet is the Harr wavelet, which is the simplest and most preferred wavelet in many applications [15].

The discrete wavelet transform, which can be expressed as follows.

$$ca_{j,k}(n) = DS\left[\sum_n x(n)g_j^*(n-2^j k)\right], \tag{3}$$

$$cd_{j,k}(n) = DS\left[\sum_n x(n)h_j^*(n-2^j k)\right] \tag{4}$$

$ca_{j,k}$ and $cd_{j,k}$ is approximation components coefficients. g (n) and h (n) are the low-pass filter and high-pass filter. Index of j and k is scale of the wavelet and factors of translation. operator of DS represent down sampling [16]. Schematic diagram of 2D DWT can be seen at figure 2 with images of four sub-bands (LL, LH, HH, HL) at every scale.

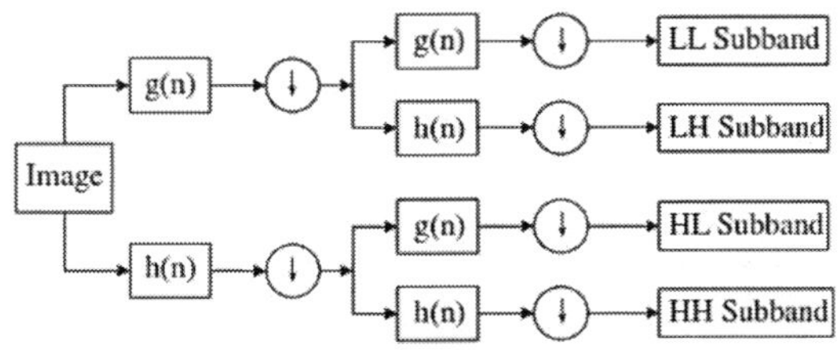

Figure 2. Schematic diagram of 2D DWT

2.2. Feature Reduction

The feature reduction is a process for reducing the feature by using the energy coefficient of the vector approximation. In this study, the feature reduction was performed by summing 100 energy per feature of the coefficient energy value obtained by squaring each vector approximation. The vector approximation is the result of the extraction of the MR brain images coefficients.

2.3. Adaptive Neuro-Fuzzy Inference Systems (ANFIS)

The ANFIS method was first introduced by Jang with embedded Fuzzy Inference System (FIS) into an adaptive networking framework. The network of adaptive is a network structure with a number of interconnected nodes. The output of this adaptive node depends on the modifications associated with this node. The learning rule determines how this parameter should be updated tminimize errors. Architecture of ANFIS can be seen in figure 3, it appears that the neuro-fuzzy system consists of five layers with different functions for each layer. Each layer consists of several vertices denoted by a box or circle. The square symbol represents an adaptive node meaning that its parameter value may change with learning and the circle symbol represents a non-adaptive node whose value is fixed.

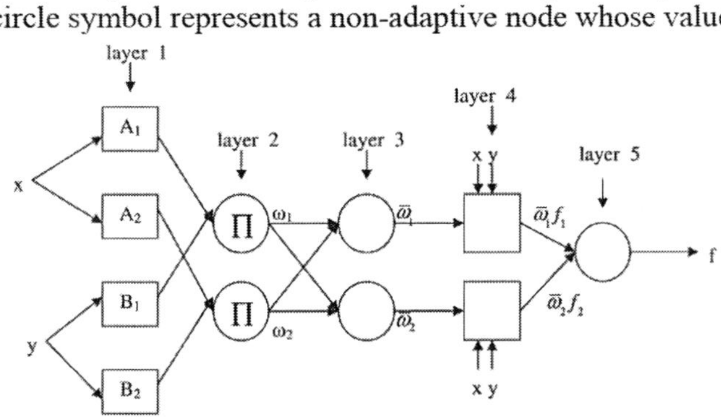

Figure 3. Adaptive neuro-fuzzy inference systems structure

The layers in the ANFIS structure can be explained as follows:

Layer 1. In this layer, all nodes are adaptive nodes (parameters may change) with function of node :

$$O_{1,i} = \mu_{Ai}(x) \tag{5}$$

$$O_{1,i} = \mu_{B_{i-2}}(y) \tag{6}$$

With x and y being inputs at the i node, Ai (or B_{i-2}) is the membership function of each node. O_1 node I serves to declare the membership degree of each input to the fuzzy set A and B. The membership function used is the generalized bell type. The parameters a, b, c, in the generalized bell membership function are called adaptive premise parameters.

Layer 2. In this layer all nodes are non-adaptive (fixed parameters). The function of this node is to multiply every incoming input signal. Node function :

$$O_{2,i} = w_i = \mu_{Ai}(x).\mu_{Bi}(y), \quad for \quad i = 1,2 \tag{7}$$

Each node output is the firing strength of each fuzzy rule. This function can be expanded if the premise section has more than two fuzzy sets. The number of vertices in this layer shows the number of rules that are formed. The multiplication function used is the interpretation of the hyphen and by using the t-norm operator.

Layer 3. Each node in this layer is a non-adaptive node that displays function of firing strength has normalized that is the output ratio of the i node of the previous layer to the entire output of the previous layer, with the shape of the node function:

$$O_{3,i} = \overline{w}_i = \frac{w_i}{w_1 + w_2}, \; i = 1,2 \tag{8}$$

When more than two rules are established, the function can be extended by dividing wi by the total number w for all rules.

Layer 4. Each node in this layer is an adaptive node with a node function:

$$O_{4,i} = \overline{w}_i f_i = \overline{w}_i (p_i x + q_i y + r_i) \tag{9}$$

by is the normalized degree of activation of layer 3 and the parameters p, q, r denotes the adaptive consequent parameters.

Layer 5. In this layer, there is only one fixed node whose function is to sum all the entries. Node function

$$O_{5,i} = \sum_i \overline{w}_i f_i = \frac{\sum w_i f_i}{\sum w_i} \tag{10}$$

The adaptive network with these five layers is equivalent to the TSK fuzzy inference system [10].

In the ANFIS structure of Fig.3, the adaptive nodes are in the first and fourth layers. The node in the first layer contains the nonlinear premise parameter while in the fourth layer contains the linear consequent parameter. Updates of ANFIS parameters are obtained from the learning process of the ANFIS neural network. Learning on ANFIS uses hybrid learning algorithm, that is forward pass using Recursive Least Square Estimator method and backward pass using gradient descent method.[10]

In the forward direction, the premise parameter is fixed. Using the Recursive Least Square Estimator (RLSE) method, the consequent parameters are corrected based on the input-output data pair. The RLSE method can be applied because the consequent parameters that are fixed are linear parameters. The RLSE method will speed up the hybrid learning process. Then after the consequent parameters are obtained, the input data is passed through the adaptive network again and the output of this adaptive network is compared with the actual output.[10]

In the backward direction, the consequent parameters are fixed. The error that occurs between the adaptive network output and the actual output is propagated back by using a gradient descent to correct the premise parameter [10]. This learning is known as back-propagation-error Algorithm. Stages of the learning process and ANFIS testing process can be seen in figure 4.

2.4. Fuzzy Learning Vector Quantization (FLVQ)

Fuzzy learning vector quantization (FLVQ) is a method that combines learning method of Learning Vector Quantization (LVQ) neural network and data grouping method of fuzzy C-Means (FCM). This method was first introduced by Tsao (1994). Stages of Fuzzy learning vector quantization training are as follows:

1. Initialization parameters include the total of clusters (C), values m_i and m_f , maximum iterations and error limits.
2. Determination of the initial value of each cluster (vo)
3. Calculation of the weight value of each iteration

$$m = m_i + k\left(\frac{m_f - m_i}{N}\right) \tag{11}$$

4. Calculation of the partition matrix of each cluster, using the FCM equation as follows

$$\alpha_{ij,k} = \left[\sum_{l=1}^{C}\left(\frac{\|x_i - v_j\|^2}{\|x_i - v_l\|^2}\right)^{1/(m-1)}\right]^{-m} \quad dengan\ 1 \le i \le M\ dan\ 1 \le j \le C \tag{12}$$

5. Calculation of cluster learning rate (η).

$$\eta_{j,k} = \left(\sum_{i=1}^{M}\alpha_{ij,k}\right); \quad 1 \le j \le C \tag{13}$$

6. Determination of new cluster center

$$v_{j,k} = v_{j,k-1} + \eta_{j,k}\left(\sum_{i=1}^{M}\alpha_{ij,k}\right)(x_i - v_{j,k-1}); \quad 1 \le j \le C \tag{14}$$

7. Calculation of the error value.

$$E_k = \sum_{j=1}^{C}\|v_{j,k} - v_{j,k-1}\|^2 \tag{15}$$

The iteration process continues until the target error is reached or the maximum iteration is reached.

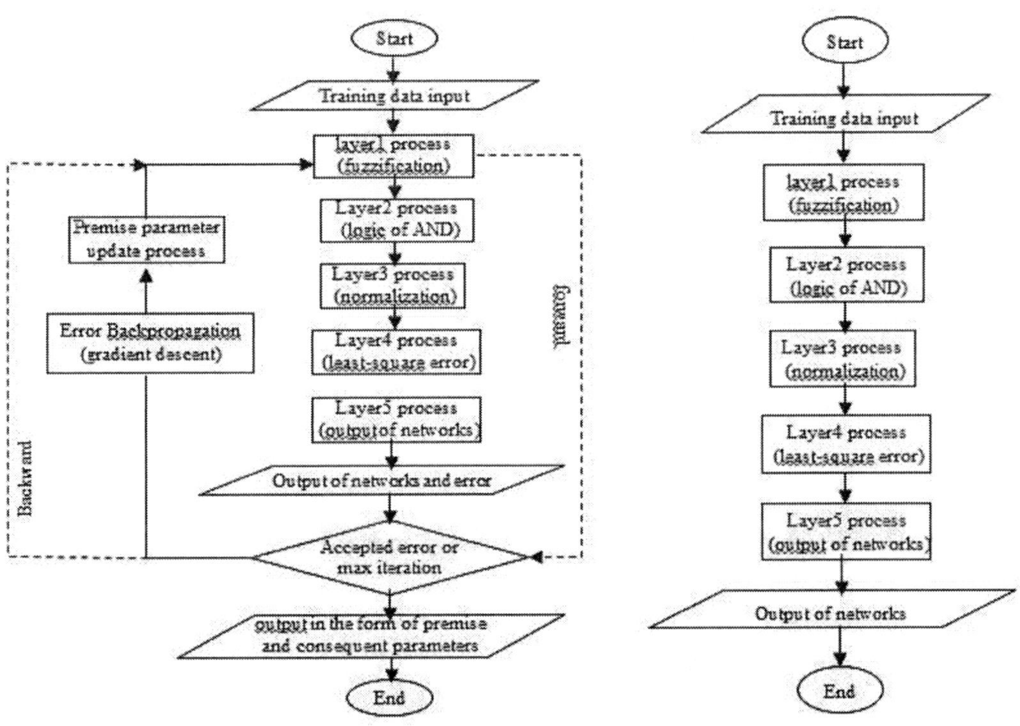

FIGURE 4. (a) Flowchart of ANFIS training process and (b) Flowchart of ANFIS testing process

2.5. Algorithm of Learning Back-propagation Neural Network

2.5.1. Gradient Descent Back-propagation
In the learning process using gradient descent backpropagation consists of 3 stages, namely the forward bait, propagation error, and renewal of weight. In the forward feed process, each input unit will spread the signal to each hidden unit, then each hidden unit will calculate its activation. From the hidden unit will send a signal to each unit of output. Then, each unit of output will be counted its activation function which is the result of a response to the input given the network. The process of propagation error is the process of comparing the activation function of the output with the target value. If the results obtained are different then it will result in an error value. Based on the error, the error correction factor used to distribute the error and output back to the previous layer. Then we perform the process of updating the weights on all layers [17].

2.5.2. Levenberg-Marquadt Algorithm
According to [18], Levenberg Marquardt (LM) algorithm is the development of backpropagation algorithm with the determination of changes of weight and bias using hessian matrix approach (H) which can be calculated with equation 16. and the gradient can be calculated by equation 17. The changes of weight on the network are calculated by equation 18 and weight fixing is determined by equation 19 .

$$H = J^T e \tag{16}$$

$$g = J^T J \tag{17}$$

$$\Delta X = [J^T J + \mu I]^- J^T e \tag{18}$$

$$X = X + \Delta X \tag{19}$$

where X = function of network weight and bias
e = vector that states all errors in the network output
μ = learning rate
I = identity matrix

3. The Dataset and Experimental Result
The MR Brain image dataset used for this research were downloaded from the Harvard Medical School website (http://www.med.harvard.edu/AANLIB/home.html). The MR Brain dataset image consists of 12 normal and 48 abnormal images. In the abnormal images consist of glioma, carcinoma, and Alzheimer's. Out of a total of 60, MR Brain image datasets are divided for 40 training data and 20 data testing. Brain data images used in this study are .gif type with pixel size 256 x 256. Image of MR brain can be shown in figure 5.

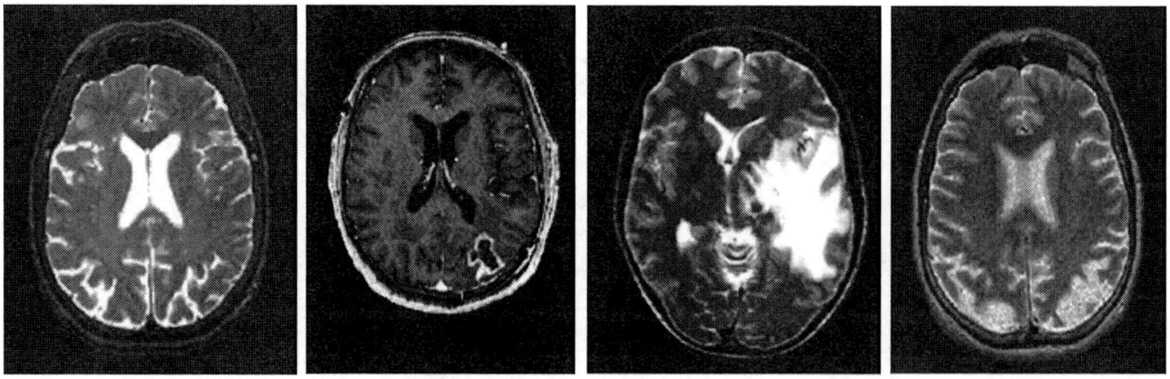

(a) (b) (c) (d)

Figure 5. MR Brain images (a). Alzheimer (b). Glioma (c). Carcinoma (d). Normal

Stages in this study are feature extraction, reduction of feature, and process of classification. In feature extraction process consists of four stages, namely: wavelet decomposition, coefficient extraction, normalization, and computational energy. The result of feature extraction is vector approximation of each level and ignore level 1. This is because at the decomposition of wavelet level 1 has many frequencies and noise detail. The size of the vector approximation at level 2 is 64 x 64 pixels and at level 3 is 32 x 32 pixels. Since the size of the approximation vector is too large, a feature reduction is performed. The result of feature reduction for energy coefficient of 100 per feature is 1 x 52 pixels. This is the vector that will be entered into the classification process. The feature reduction uses the energy coefficients of the vector approximation. Research stage after feature reduction is classification. And the classification method used in the next process is ANFIS.

The classification step using ANFIS is the first step is the determination of input on ANFIS. The input data used in the training process are 40 images consisting of 32 abnormal images of 12 Alzheimer, 8 Carcinoma, 12 Glioma and 8 normal images while the data testing as much as 20 images consisting of 16 abnormal images of 6 Alzheimer, 4 Carcinoma, 8 Glioma and 4 normal. The second stage is the determination of membership function. Membership functions are used generalized-bell because generalized-bell has the smoothness and simplicity of notation with much information and degrees of freedom to improve the steepness at the crossover point is the point where membership value is 0.5 [10]. The third stage is the determination of membership function number. The fourth stage is the initialization of the premise parameter. The premise parameter to be initialized with Bell's membership function is $\{a, b, c\}$ for each membership function. The value of parameter b is one, but for parameter a and c are initialized randomly. The fifth stage is the ANFIS learning process. In the process of learning ANFIS used 2 methods for the clustering process of Fuzzy C-Means (FCM) and fuzzy learning vector quantification (FLVQ). The last stage is the process of testing the MR brain images.

In this study, ANFIS classification results will be compared with the results of the classification process LM learning backpropagation and the result of Damayanti (2014). In ANFIS process using variation of learning rate parameter value (lr) and momentum (mc), that is 0.1, 0.4, and 0.9. In ANFIS classification using FCM for clustering process, 100% accuracy percentage is obtained for parameter variation lr = 0.1, mc = 0.1; lr = 0.1, mc = 0.4; lr = 0.4, mc = 0.9; and lr = 0.9, mc = 0.4, the results can be seen in figure 6. But the variation of the parameters resulting in the smallest error and with minimal iteration is at lr = 0.1 and mc = 0.4 with error of 0.0000664 with iteration as much 4. While the ANFIS classification with FLVQ for clustering process, obtained 100% accuracy percentage for all variations of learning rate parameter and momentum. The results of classification of MR brain images using ANFIS can be seen in table 1.

In classification LM learning back-propagation using a variation of the number of hidden layer node (hl) and learning rate (μ). The number of nodes in the hidden layers used are 26, 52 and 78. While the variation of learning rate value used is 0.1. 0.4 and 0.9. The result of the best accuracy percentage of 100% is at the value of hl = 26 and $\mu = 0.1$, $\mu = 0.9$, the result can be seen in figure 8. The results of the comparison of the process of classification of MR brain images using ANFIS, LM learning and result of Damayanti (2014) can be seen in table 2.

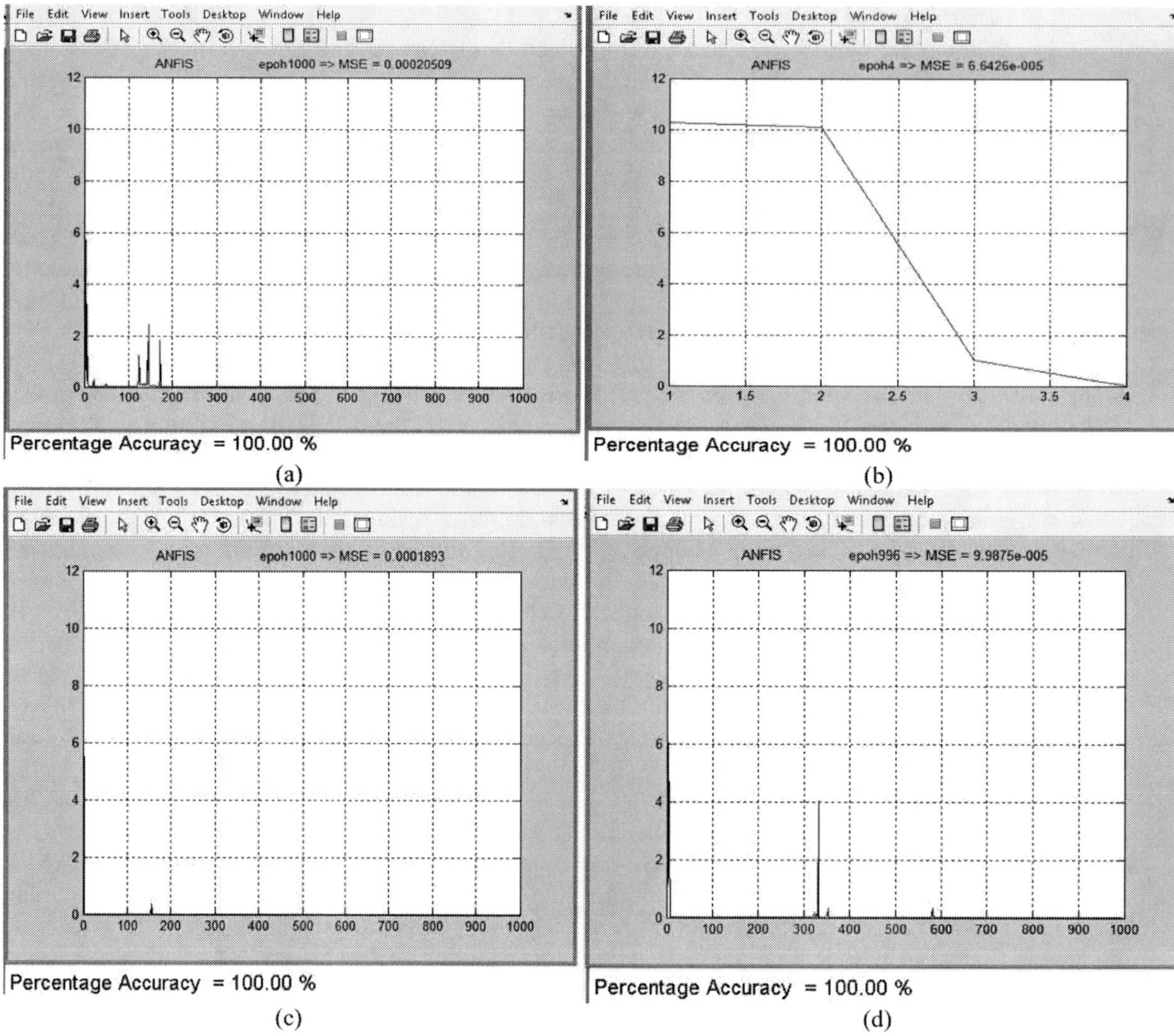

Figure 6. The results of the ANFIS classification process in MR brain images using FCM where (a). lr = 0.1, mc = 0.1, (b). lr = 0.1, mc = 0.4, (c). lr = 0.4, mc = 0.9, and (d). lr = 0.9, mc = 0.4

Table 1. The results of classification of MR brain images using ANFIS

Learning Rate	Momentum	Percentage Accuracy	
		ANFIS	ANFIS-FLVQ
0.1	0.1	100	100
0.1	0.4	100	100
0.1	0.9	85	100
0.4	0.1	85	100
0.4	0.4	90	100
0.4	0.9	100	100
0.9	0.1	95	100
0.9	0.4	100	100
0.9	0.9	95	100

Table 1 shows that the classification using ANFIS with FLVQ is the best classification result compared to ANFIS with FCM.

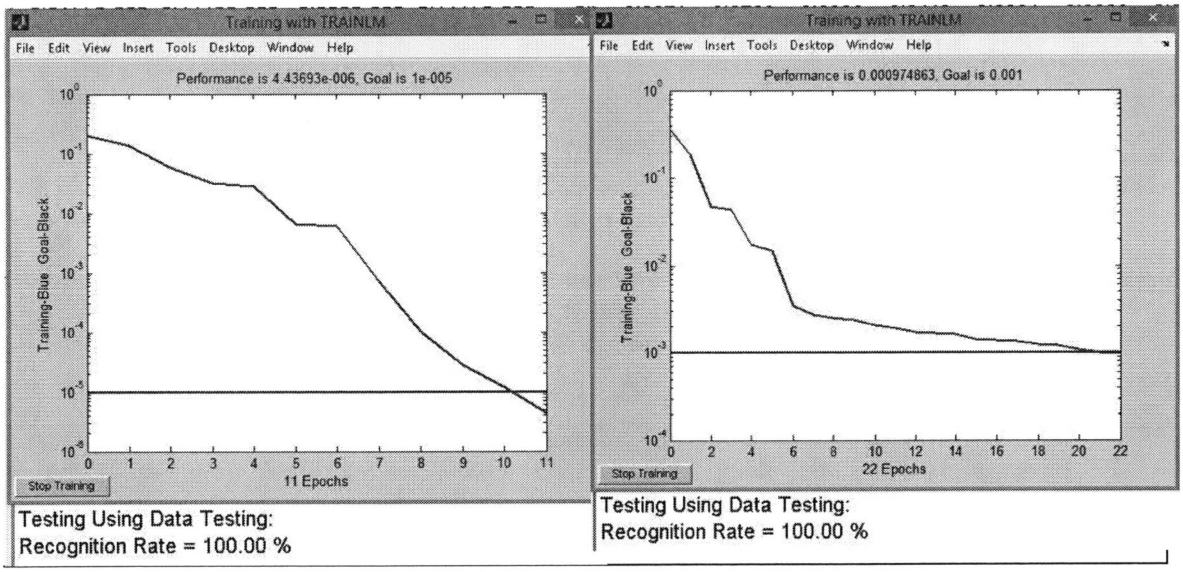

(a) (b)

Figure 7 The result of classification LM learning where (a). hl = 26 and $\mu = 0.1$, (b) hl = 26 and $\mu = 0.9$

Table 2. The Result of Classification MR brain images

Methods	Percentage Accuracy
GD Back-propagation (Damayanti,2014)	95%
LM Back-propagation	100 %
ANFIS-FLVQ	100%
ANFIS	100 %

4. Conclusion

The classification of MR brain images by type of disease, among others Glioma, carcinoma, and Alzheimer's using ANFIS is the aim of this study. Stages of the classification of MR brain images are feature extraction, reduction of feature, and process of classification. The resulting feature of the feature reduction process for 100 energy coefficient is 52 features, which is the input of the classification process. The result of classification MR brain image using ANFIS-FLVQ, ANFIS, and LM back-propagation by 100%,

References

[1] Wilson P 2006 Anatomi dan Fisiologi Sistem Syaraf *EGC Publisher* Jakarta
[2] Nagori M, Mutkule S and Sonarkar P 2013 Detection of Brain tumor by mining MRI Images Volume 2 Issue 4 and ISSN: 2278-1021
[3] John P 2012 Brain Tumor Classification Using Wavelet and Texture Based Neural Network *International Journal of Scientific & Engineering Research Vol 3* Issue 10 pp 1-7
[4] Jardi I PenyakitAlzheimer 24 Februari 2012 http://www.repository.usu.ac.id/bitstream/

12345678910/1996/1/bedah-skandarjapardi28.pdf

[5] Berttolone 1982 Tumor of The Central Nervous System Concepts in Cancer Medicine pp 649-659

[6] El-Dahshan E A, Hosny T and Salem A M 2010 Hybrid intelligent techniques for MRI *Brain Images* classification Cairo pp 433–441

[7] Chaplot S, Patnaik L M and Jagannathan N R 2006 Classification of magneticresonance brain images using wavelets as input to support vector machine and neural network. *Biomedical Signal Processing and Control* 1 pp 86–92

[8] Fletcher-Heath L M, Hall L O, Goldgof D B et al 2001 Automatic segmentation of non-enhancing brain tumors in magnetic resonance images *Artificial Intelligence in Medicine* 21 pp 43–63

[9] Maitra M and Chatterjee A 2008 Hybrid multi-resolution slantlet transform and fuzzy c-means clustering approach for normal pathological brain MR image segregation *Medical Engineering and Physics* 30(5) pp 615–623

[10] Jang J-SR 1997 ANFIS : Adaptive Neuro-Fuzzy based Inference Systems IEEI Trans On System Man and Cybernetics

[11] Belal S Y, Taktak A F G, Nevill A J, Spencer S A, Roden D and Bevan S 2002 Automatic detection of distorted plethysmogram pulses in neonates and paediatric patients using an adaptivenetwork-based fuzzy inference system *Artificial Intelligence in Medicine* 24 pp 149–165

[12] Mousa R, Munib Q and Mousa A 2005 Breast Cancer Diagnosis System Based on Wavelet Analysis and Fuzzy Neural Jordan pp 713–723

[13] Roy K and Bhattacharya P 2008 Optimal features subset selection and classification for Iris recognition *Journal Image Video Process* doi:10.1155/2008/743103

[14] Schmeelk J 2002 Wavelet transforms on two-dimensional images *Mathematical and Computer Modelling* 36(7) pp 939–948

[15] Kaur L, Gupta S, Chauhan R C and Saxena S C 2007 Medical ultrasound image compression using joint optimization of thresholding quantization and bestbasis selection of wavelet packets *Digital Signal Processing* 17(1) pp 189–198

[16] Nanni L and Lumini A 2008 Wavelet decomposition tree selection for palm and face authentication *Pattern Recognition Letters* 29(3) pp 343–353

[17] Fausett L 2003 Fundamental of Neural Network Architecture, Algorithm, and Application Printice-Hall Inc London

[18] Hagan M and Menhaj M B 1994 Training Feedforward Networks With The Marquard Algorithm *IEEE Transactions On Neural Networks* 5(6) pp 989-993

Implementation of digital image encryption algorithm using logistic function and DNA encoding

Suryadi MT[1], Yudi Satria[2], Muhammad Fauzi[3]

[1,2,3] Department of Mathematics, Universitas Indonesia, Depok, 16424, Indonesia

E-mail: muhammad.fauzi22@sci.ui.ac.id

Abstract. Cryptography is a method to secure information that might be in form of digital image. Based on past research, in order to increase security level of chaos based encryption algorithm and DNA based encryption algorithm, encryption algorithm using logistic function and DNA encoding was proposed. Digital image encryption algorithm using logistic function and DNA encoding use DNA encoding to scramble the pixel values into DNA base and scramble it in DNA addition, DNA complement, and XOR operation. The logistic function in this algorithm used as random number generator needed in DNA complement and XOR operation. The result of the test show that the PSNR values of cipher images are 7.98-7.99 *bits*, the entropy values are close to 8, the histogram of cipher images are uniformly distributed and the correlation coefficient of cipher images are near 0. Thus, the cipher image can be decrypted perfectly and the encryption algorithm has good resistance to entropy attack and statistical attack.

1. Introduction

With the vast development of computer technology, digital image usage as image information storage used in the society tend to increase. Sometimes, the digital image need to be concealed because it could inflict detriments to the owner if somebody know or could manipulate those digital image. Thus, digital image cryptography emerges as one of the method to overcome this problem.

Chaos is a random or uncertain movement in a particular system. Chaos system has certain properties including aperiodic, non-convergent, and initial value sensitive [1-2]. Therefore, chaos system could be used in the cryptography field. Research on chaos based cryptography already performed [2-11] before on digital image. But, according to their research [8-10] a lot of encryption algorithms [12-14] which use a single chaos function are vulnerable to be interpreted.

Deoxyribonukleat acid (DNA) cryptography is a study which use DNA as information carrier and uses biological technology to perform encryption where it's commonly uses common dogma in molecular biology to implement encryption. But, DNA encryption has some demerit points, it needs expensive equipment, complex operation, and difficult to grasp its biotechnology [2].

Based on the reasons that DNA encryption is hard to be implemented and the usage of single chaos function is vulnerable to be interpreted, Liu, Zhang & Wei [2] proposed encryption algorithm based on chaos logistic function and DNA encoding. This algorithm split the channel contained in the RGB digital image, and transformed its pixel values into binary values in order to change it into DNA components according to the DNA map rule already defined. Then, DNA addition operation and DNA complement operation are performed with the help of chaos logistic function to scramble the DNA information. Finally, it changes the DNA information into binary values to be used in the XOR operation against values created with the chaos logistic function, the output values of the XOR operation will represent the cipher image's pixels.

2. Experimental Methods

Logistic function is a recursive function which used to be referenced as an example of chaos properties complexity which may occur in the simple dynamic non-linear equation. This function is defined as follows:

$$x_{n+1} = rx_n(1-x_n) \quad (1)$$

where $r \in (0,4]$, $x_n \in (0,1)$, $n = 0,1,2,\ldots$, both r and x_0 are the *initial value* of the logistic function and x_n is the logistic function value with n element of the whole number [1].

The *chaos* logistic function values generated in this algorithm will be transformed into binary value with threshold function $f(x)$ in the equation (2). Special function $g(x)$ shown in equation (3) will be used to transform the *chaos* logistic function values into *pixel* values.

$$f(x) = \begin{cases} 0, & 0 < x_n \leq 0.5 \\ 1, & 0.5 < x_n < 1 \end{cases} \quad (2)$$

$$g(x) = \lfloor 100000 \times x \rfloor \bmod 256 \quad (3)$$

Adenin (A), *Cytosine* (C), *Guanine* (G), and *Thymine* (T) are DNA components which create a single strand of DNA, where A and T are complement to each other, so are C and G. In binary system theory, 0 and 1 are complement to each other, so 00 and 01, 10 and 11 are also complement to each other. DNA encoding in this algorithm transform binary value into 4 DNA components with 2 *bit* binary value to represent a DNA component, and DNA decoding transform DNA components into binary value. If 00, 01, 10, and 11 are used to represent 4 DNA components, there will be 24 possible DNA map combinations. But because there is a complement relation between DNA component and binary value, there is simply 8 kind of DNA map which satisfy complement relation from 24 possible DNA combinations [2]. This can be seen at Table 1.

Table 1. DNA Map Rules.

1	2	3	4	5	6	7	8
00-A	00-A	00-C	00-C	00-G	00-G	00-T	00-T
01-C	01-G	01-A	01-T	01-A	01-T	01-C	01-G
10-G	10-C	10-T	10-A	10-T	10-A	10-G	10-C
11-T	11-T	11-G	11-G	11-C	11-C	11-A	11-A

Also, this algorithm using DNA addition and subtraction rules as shown at Table 2.

Table 2. DNA Addition and Subtraction Rule.

±	A	T	C	G
A	G	C	T	A
T	C	G	A	T
C	T	A	G	C
G	A	T	C	G

IOP Conf. Series: Journal of Physics: Conf. Series **974** (2018) 012028 doi:10.1088/1742-6596/974/1/012028

3. Results and Analysis

The experiment will be performed in a computer with processor specification Intel(R) Core(TM)2 Duo CPU T6500 ~ 2.1 Ghz, 2.00 GB, Windows 7 Ultimate 32-bit Operating System. Digital images used in this experiment are *8-bit* Baboon.bmp and Lena.bmp shown in Figure 1 (a) and Figure 2 (a) with 3 sizes in *pixel* shown in Table 3 and Table 4, and the *cipher image* shown in Figure 1 (b) and Figure 2 (b). The encryption and decryption keys used through all the experiment are $x_0^{(1)} = 0.25$, $r^{(1)} = 3.94$, $x_0^{(2)} = 0.52$, $r^{(2)} = 3.78$, $key1 = 7$, $key2 = 4$.

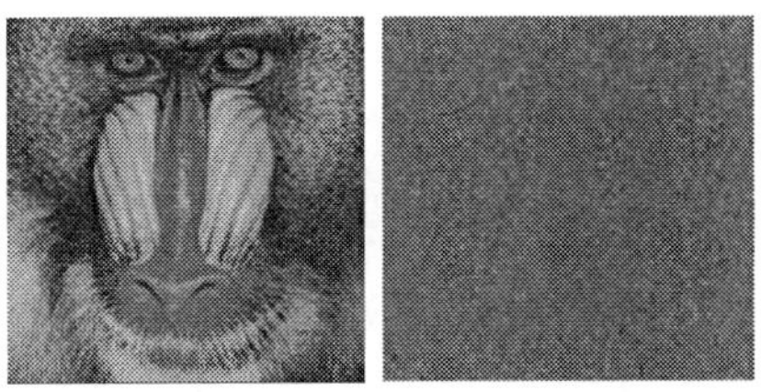

(a) (b)

Figure 1. Display of Baboon.bmp (a) *Plain Image* (b) *Cipher Image*

(a) (b)

Figure 2. Display of Lena.bmp (a) *Plain Image* (b) *Cipher Image*

Table 3. Baboon.bmp Data Sample Sizes.

Data Sample	Size (*pixel*)	Size (*bytes*)
1	128 × 128	49.206
2	256 × 256	196.664
3	512 × 512	786.488

Table 4. Lena.bmp Data Sample Sizes.

Data Sample	Size (*pixel*)	Size (*bytes*)
1	128 × 128	49.206
2	256 × 256	196.662
3	512 × 512	786.486

3.1. *Histogram and Goodness of Fit Analysis*

Table 5. Histograms of Baboon.bmp.

Data Sample	R Channel (*plain*)	G Channel (*plain*)	B Channel (*plain*)	R Channel (*cipher*)	G Channel (*cipher*)	B Channel (*cipher*)
1	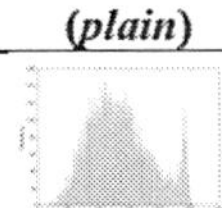	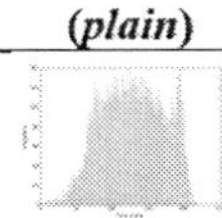	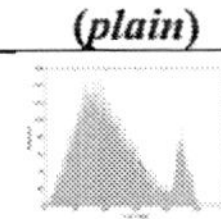	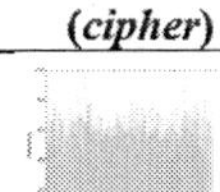	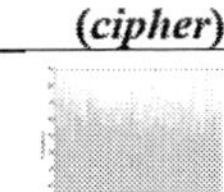	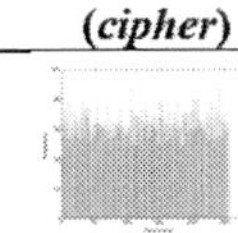
2	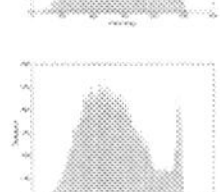	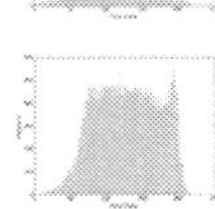	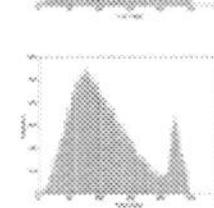	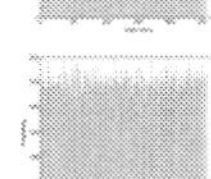	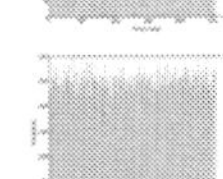	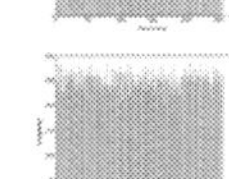
3	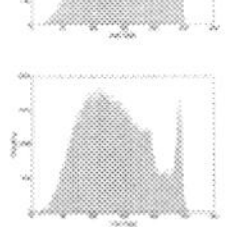	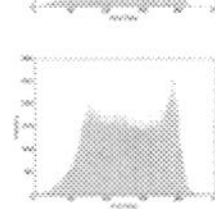	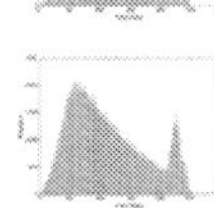	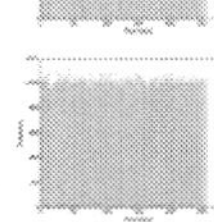	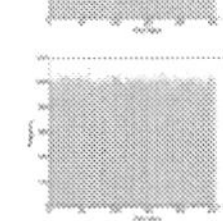	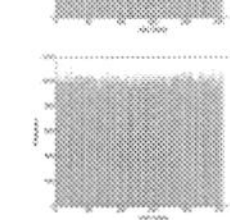

Table 6. Histograms of Lena.bmp.

Data Sample	R Channel (*plain*)	G Channel (*plain*)	B Channel (*plain*)	R Channel (*cipher*)	G Channel (*cipher*)	B Channel (*cipher*)
1	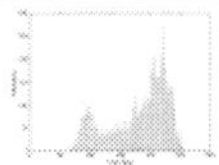	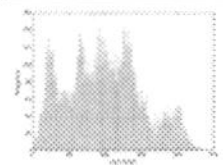	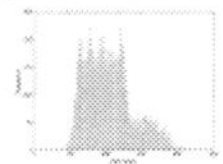	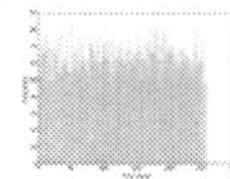	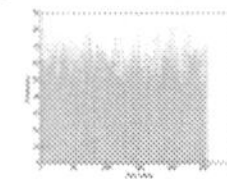	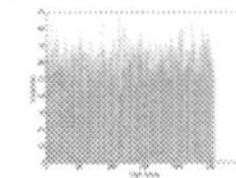
2	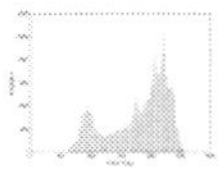	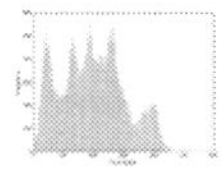	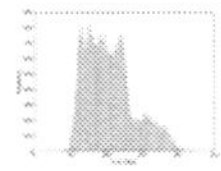	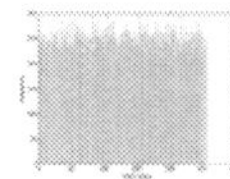	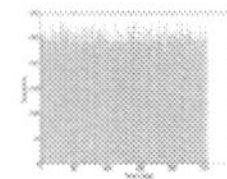	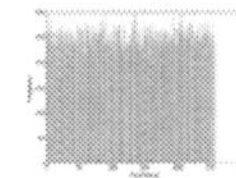
3	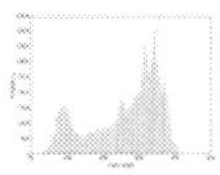	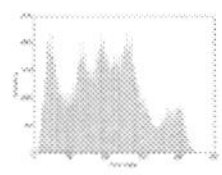	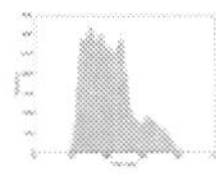	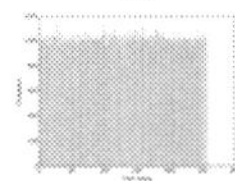	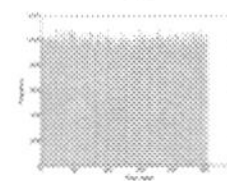	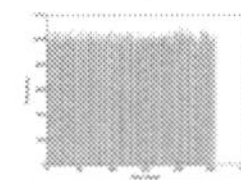

Table 5 and Table 6 shows the histogram of the data sample in Table 3 and Table 4. It can be seen that the histograms of the plain image data samples tend to fluctuate but the histograms of the cipher image data samples tend to have a flat shape. In order to prove the uniformity of the cipher image data samples histograms, *Goodness of Fit* test are performed.

Table 7. Statistical Value of Baboon.bmp *Cipher Images*.

Data Sample	R Channel	G Channel	B Channel
1	274.68750	234.40625	254.15625
2	238.55468	273.79687	241.14062
3	274.26757	257.26757	256.21093

Table 8. Statistical Value of Lena.bmp *Cipher Images*.

Data Sample	R Channel	G Channel	B Channel
1	244.93750	225.12500	246.84375
2	268.67968	218.30468	245.10937
3	244.36914	305.78710	231.98046

Table 7 and Table 8 shows the statistical test value. Because the degree of freedom is 255 and with 1% level of significance, the critical value in this *Goodness of Fit* test are 310.4574. It can be seen from Table 7 and Table 8 that all of the data sample statistical test values are smaller than 310.4574 which can be implied that the distribution of the cipher image pixel values are uniformly distributed. Therefore, it shows that the encryption algorithm has a good resistance against statistical attack.

3.2. Coefficient Correlation Analysis

Correlation coefficient test are also performed by using 2000 random pairs of adjacent pixels horizontally, vertically, and diagonally from the plain image and the cipher image to compute its correlation coefficient value.

Table 9. Horizontal Correlation Coefficient Values of Baboon.bmp.

Data Sample	R Channel (*plain*)	G Channel (*plain*)	B Channel (*plain*)	R Channel (*cipher*)	G Channel (*cipher*)	B Channel (*cipher*)
1	0.88327	0.74551	0.88117	0.04630	0.04001	0.01118
2	0.90367	0.81081	0.90927	-0.00910	0.00831	0.01123
3	0.88639	0.75959	0.89902	-0.00595	0.02003	-0.00521

Table 10. Vertical Correlation Coefficient Values of Baboon.bmp.

Data Sample	R Channel *(plain)*	G Channel *(plain)*	B Channel *(plain)*	R Channel *(cipher)*	G Channel *(cipher)*	B Channel *(cipher)*
1	0.89311	0.73341	0.82691	-0.01210	0.02432	-0.01319
2	0.93051	0.89391	0.90991	0.01190	0.03859	0.03328
3	0.91277	0.88798	0.89985	0.00905	-0.01137	-0.00139

Table 11. Diagonal Correlation Coefficient Values of Baboon.bmp.

Data Sample	R Channel *(plain)*	G Channel *(plain)*	B Channel *(plain)*	R Channel *(cipher)*	G Channel *(cipher)*	B Channel *(cipher)*
1	0.83893	0.69898	0.70857	-0.01137	0.02120	0.00531
2	0.92393	0.79898	0.80373	-0.02636	-0.04144	0.04011
3	0.85858	0.72212	0.88656	0.01955	0.01342	0.00910

Table 12. Horizontal Correlation Coefficient Values of Lena.bmp.

Data Sample	R Channel *(plain)*	G Channel *(plain)*	B Channel *(plain)*	R Channel *(cipher)*	G Channel *(cipher)*	B Channel *(cipher)*
1	0.92771	0.91070	0.88345	-0.01038	0.00890	0.04182
2	0.95805	0.93792	0.89743	-0.02696	-0.03394	-0.02621
3	0.98885	0.98316	0.96021	-0.02703	-0.02576	-0.02131

Table 13. Vertical Correlation Coefficient Values of Lena.bmp.

Data Sample	R Channel *(plain)*	G Channel *(plain)*	B Channel *(plain)*	R Channel *(cipher)*	G Channel *(cipher)*	B Channel *(cipher)*
1	0.85383	0.82398	0.80136	-0.01818	-0.00568	-0.03831
2	0.92341	0.88146	0.83511	0.02614	0.00641	-0.01494
3	0.97940	0.97123	0.93368	-0.01001	-0.01758	-0.02634

Table 14. Diagonal Correlation Coefficient Values of Lena.bmp.

Data Sample	R Channel *(plain)*	G Channel *(plain)*	B Channel *(plain)*	R Channel *(cipher)*	G Channel *(cipher)*	B Channel *(cipher)*
1	0.81588	0.80449	0.78441	0.02290	0.02634	0.00999
2	0.89650	0.86048	0.82421	-0.00847	0.00287	0.00020
3	0.96751	0.95926	0.92277	-0.00770	-0.00062	0.00126

Table 9, Table 10, Table 11, Table 12, Table 13, Table 14 show the plain image and the cipher image horizontal, vertical, and diagonal correlation coefficient values of Baboon.bmp and Lena.bmp. The horizontal, vertical, and diagonal correlation coefficient values of the plain image are close to 1 which implied that the plain image pixels have a strong linear correlation, but the horizontal, vertical and diagonal correlation coefficient values of the cipher image are close to 0 which implied that the linear correlation in their corresponding plain image were lost. Thus, it can be said that the encryption algorithm has a high security level against pixel correlation statistical attack.

3.3. PSNR and Entropy Analysis

Table 15 and Table 17 shows the PSNR values between the plain image and the cipher image of Baboon.bmp and Lena.bmp data samples.

Table 15. PSNR Value of Baboon.bmp *Cipher-Plain*.

Data Sample	R Channel	G Channel	B Channel
1	8.93554	9.41361	8.45741
2	8.88767	9.43164	8.52543
3	8.77369	9.23965	8.34853

Table 16. PSNR Value of Baboon.bmp *Plain-Decrypted*.

Data Sample	R Channel	G Channel	B Channel
1	∞	∞	∞
2	∞	∞	∞
3	∞	∞	∞

Table 17. PSNR Value of Lena.bmp *Cipher-Plain*.

Data Sample	R Channel	G Channel	B Channel
1	7.86522	8.49217	9.54076
2	7.87227	8.52729	9.55073
3	7.86159	8.54053	9.59230

Table 18. PSNR Value of Lena.bmp *Plain-Decrypted*.

Data Sample	R Channel	G Channel	B Channel
1	∞	∞	∞
2	∞	∞	∞
3	∞	∞	∞

It can be seen that the PSNR values are quite small which ranges from 7.86-9.59 dB. PSNR values ≤ 30 dB implied that the encryption algorithm gives good quality cipher image [15]. Table 16 and Table 18 shows PSNR values between the plain image and the decrypted image. It can be seen that all the values are infinite, thus it proves that there are not any pixel differences between the plain image and the decrypted image and the decryption algorithm returns the decrypted image perfectly.

Table 19. Entropy Value of Baboon.bmp *Cipher Image*.

Data Sample	R Channel	G Channel	B Channel
1	7.98793	7.98972	7.98891
2	7.99736	7.99697	7.99732
3	7.99924	7.99929	7.99929

Table 20. Entropy Value of Lena.bmp *Cipher Image*.

Data Sample	R Channel	G Channel	B Channel
1	7.98913	7.99004	7.98908
2	7.99702	7.99759	7.99729
3	7.99932	7.99910	7.99936

Table 19 and Table 20 shows the entropy values between the plain image and the cipher image of Baboon.bmp and Lena.bmp data samples. It shows that its entropy values ranges from 7.98-7.99 *bits*, close to 8 *bits*. Thus, it can be implied that the encryption algorithm has a good resistance against entropy attack[2].

3.4. Encryption and Decryption Time Analysis

Table 21. Average Encryption and Decryption Time of Baboon.bmp.

Data Sample	Encryption (second)	Decryption (second)
1	1.83520	1.89979
2	7.31240	7.52479
3	29.39357	30.50018

Table 22. Average Encryption and Decryption Time of Lena.bmp.

Data Sample	Encryption (second)	Decryption (second)
1	1.81410	1.86235
2	7.14926	7.34866
3	29.11883	30.48092

Table 21 and Table 22 shows the average of encryption and decryption time measured in seconds. This was taken from the average of 5 test with the same key. Table 21 and Table 22 shows that the encryption and decryption time tend to increase as the image size increased.

4. Conclusion

According to the program implementation and results analysis on the data samples in Table 4 and Table 5 we can see that the cipher image's histograms and pixels are uniformly distributed hence the encryption algorithm can resist statistical attack, all the data sample cipher image's coefficient correlation values close to 0 hence the encryption algorithm can resist pixel correlation statistical attack, cipher-plain image's PSNR values range around 7.8-9.5 dB under 30 dB hence the encryption algorithm has a good quality, plain-decrypted image's PSNR values are infinite hence the decrypted images are equal plain images, cipher image's entropy values range around 7.98-7.99 *bits* which is close to 8 *bits* hence the encryption algorithm can resist entropy attack, and the bigger the data sample image size the bigger the encryption and decryption time of the data sample.

Acknowledgment

The authors would like to thank Kemenristekdikti RI and Directorate of Research and Community Service (DRPM) Universitas Indonesia for the grant PTUPT 2017 who support us in this research.

References

[1] Kocarev L and Lian S 2011 *Chaos-Based Cryptography: Theory, Algorithm and Applications*, (Berlin: Springer-Verlag)
[2] Liu L, Zhang Q and Wei X 2012 A RGB image encryption algorithm based on DNA encoding and Chaos Map *Computers and Electrical Engineering design* **38** 1240-1248
[3] Eva N and Suryadi MT 2013 Chaos-Based Encryption Algorithm for Digital Image *Proceedings IICMA* 169-177
[4] Suryadi MT, Maria YTI, and Satria Y 2015 Encryption Algorithm using New Modified map for digital image *Proceedings of IICMA* 71-78
[5] Suryanto Y, Suryadi MT, and Ramli K 2016 A Secure and Robust Image Encryption Based on Chaotic Permutation Multiple Circular Shrinking and Expanding *Journal of Information Hiding and Multimedia Signal Processing* **7** 697-713
[6] Suryanto Y, Suryadi MT, and Ramli K 2017 A New Image Encryption using color scrambling based on chaotic permutation multiple circuler shrinking and Expanding *Multimedia Tools and Applications* **76** 16831-16854
[7] Ahmad J and Ahmed F 2012 Efficiency Analysis and Security Evaluation of Image Encryption

Schemes *International Journal of Video & Image Processing and Network Security* **12** 18-31
[8] Laptyeva TV, Flach S and Kladko K 2011 The weak password problem: Chaos, criticality, and encrypted p-CAPTCHAs *EPL* **95** 981-981
[9] Shujun L and Xuan Z 2002 Cryptanalysis of a chaotic image encryption method *International Symposium on Circuits and System* **2** 87-91
[10] Qinan L 2011 Color image encryption algorithm and its decryption method protecting from shearing attack *Computer engineering and design* **32** 509-512
[11] Suryadi MT, Nurpeti E, Widya D 2014 Performance of Chaos-Based Encryption Algorithm for Digital Image *TELKOMNIKA* **12** 675-682
[12] Sun F, Liu S and Li Z 2008 A novel image encryption scheme based on spatial chaos map *Chaos Solitons Fractals* **38** 631-640
[13] Vinod P, Pareek NK, Purohita G and Suda KK 2011 A robust and secure chaotic standard map based pseudorandom permutation substitution scheme for image encryption *Optics Commun* **284** 4331-4339
[14] Yunpeng Z, Fei Z, Zhengjun Z and Cai X 2008 A new image encryption Algorithm based on Multiple Chaos System *International Symposium on Electronic Commerce and Security* **142** 347-350
[15] Pareek NK 2012 Design and Analysis of a Novel Digital Image Encryption Scheme *International Journal of Network Security & Its Applications* **4** 95-108

IOP Conf. Series: Journal of Physics: Conf. Series **974** (2018) 012029 doi:10.1088/1742-6596/974/1/012029

Value-at-Risk analysis using ARMAX GARCHX approach for estimating risk of banking subsector stock return's

Iis Dewi Ratih[1], **Brodjol Sutijo Supri Ulama**[2] , **Mike Prastuti**[3]
[1,2,3] Department of Business Statistics, Faculty of Vocation, Institut Teknologi Sepuluh Nopember, Surabaya, Indonesia

E-mail: iisdewiratih@gmail.com

Abstract. Value at Risk (VaR) is one of the statistical methods used to measure market risk by estimating the worst losses in a given time period and level of confidence. The accuracy of this measuring tool is very important in determining the amount of capital that must be provided by the company to cope with possible losses. Because there is a greater losses to be faced with a certain degree of probability by the greater risk. Based on this, VaR calculation analysis is of particular concern to researchers and practitioners of the stock market to be developed, thus getting more accurate measurement estimates. In this research, risk analysis of stocks in four banking sub-sector, Bank Rakyat Indonesia, Bank Mandiri, Bank Central Asia and Bank Negara Indonesia will be done. Stock returns are expected to be influenced by exogenous variables, namely ICI and exchange rate. Therefore, in this research, stock risk estimation are done by using VaR ARMAX-GARCHX method. Calculating the VaR value with the ARMAX-GARCHX approach using window 500 gives more accurate results. Overall, Bank Central Asia is the only bank had the estimated maximum loss in the 5% quantile.

1. Introduction

Value at Risk (VaR) is one of the statistical methods used to measure market risk by estimating the worst losses over time and certain levels of confidence (Messaoud and Aloui, 2015). The accuracy of this measuring tool is very important in determining the amount of capital that must be provided by the company to cope with possible losses. As for users / investors, it is important to determine trust level that reflects the risks to be considered in investment making process. This is because the greater the risk, the greater the losses to be faced with a certain degree of probability (Dharmawan, 2014). Based on this, the VaR calculation analysis is of particular concern to researchers and stock market practitioners to be developed, thus getting more accurate measurement estimates.

VaR estimates on the stock market is important to know the risks that will occur in the stock market at the next period. The condition of the stock that fluctuates every day, caused investors need to pay attention and research the past data of a company that will be selected to invest. This is very important for investors to know the future prospects of stock prices in a company. Therefore, stock return forecast for several future period is needed as one step in estimating VaR.

ARIMA is one of many forecasting method, it has the residual assumption of a homogeneous model. However, Donaldson (2015) shows that stock return data is stochastic and stationary also has a volatile nature of non-homogeneous data (heteroscedasticity). So the ARIMA method is less appropriate to use. One method that can be used to analyze non-homogenous data is ARCH, Autoregressive Conditional Heteroscedasticity (Wei, 2006). In 1986, Bollerslev and Taylor developed the ARCH model into a GARCH (Generalized Autoregressive Conditional Heteroscedasticity) model. GARCH model is used to avoid large order in ARCH model.

ARMAX (Autoregressive Moving Average with Exogenous Variable) is the development of ARMA methods used to model and forecast by including exogenous variables therein. In generating better and accurate models and also forecast of non-homogenous stock returns by including exogenous variables, GARCHX (Generalized Autoregressive Conditional Heteroscedasticity with Exogenous Variable) can be used as one of the forecasting method. This research used ARMAX-GARCHX approach, assuming that stock price of a company was not only influenced by stock price in the previous period, but also influenced by exogenous variables namely Indonesia Composite Index (ICI) and exchange rate IDR/USD.

On May 25, 2016, Forbes 2000 The World Biggest Companies listed 2000 lists of the world's largest public companies. There are 6 Indonesian public companies listed on the Forbes 2000 list: Bank Rakyat Indonesia, Bank Mandiri, Bank Central Asia, Telkom Indonesia, Bank Negara Indonesia, and Gudang Garam. It is seen that 4 from 6 entering companies are Banking companies (SahamOK, 2016). Stock return analysis on four banks, namely Bank Rakyat Indonesia, Bank Mandiri, Bank Central Asia, and Bank Negara Indonesia, will be done in this research because these companies were included in Forbes 2000 as the largest company in the world for several times. In addition, those banks are the companies incorporated in the LQ45 index over the last period.

Based on these reviews, VaR analysis to estimate the risk of stock returns in banking sub-sector using Copula ARMAX-GARCHX approach will be done in this reesearch. The ARMAX method is used to estimate the μ parameter. While GARCHX is used to estimate parameter σ_t^2 (because stock data is non-homogenous/heteroscedasticity). So hopefully this research can yield more accurate VaR estimation.

2. Material and Methods

2.1. Material

The data used in this research is secondary data consist of closing price daily, ICI and exchange rate IDR/USD. The closing price and ICI data are sourced from www.finance.yahoo.com. Exchange rate IDR/USD data is sourced from bi.go.id. Period of closing price data of Bank Rakyat Indonesia (Persero) Tbk., Bank Mandiri (Persero) Tbk., Bank Central Asia Tbk., and Bank Negara Indonesia (Persero) Tbk., used is January 1, 2010 until February 28, 2017. Variables used in this research consist of response variables and predictor variables. Response variables are stock return variable at Bank Rakyat Indonesia (Persero) Tbk., Bank Mandiri (Persero) Tbk., Bank Central Asia Tbk., and Bank Negara Indonesia (Persero) Tbk. While the ICI and exchange IDR/USD are predictor variables. The data structure used in this research is presented in Table 1.

Table 1. Data Structure

t	Month	Year	*Closing Price*				ICI ($X_{1,t}$)	Exchange rate IDR-USD ($X_{2,t}$)
			BBRI ($Y_{1,t}$)	BMRI ($Y_{2,t}$)	BBCA ($Y_{3,t}$)	BBNI ($Y_{4,t}$)		
1			$Y_{1,1}$	$Y_{2,1}$	$Y_{3,1}$	$Y_{4,1}$	$X_{1,1}$	$X_{2,1}$
2			$Y_{1,2}$	$Y_{2,2}$	$Y_{3,2}$	$Y_{4,2}$	$X_{1,2}$	$X_{2,2}$
3	January	2010	$Y_{1,3}$	$Y_{2,3}$	$Y_{3,3}$	$Y_{4,3}$	$X_{1,3}$	$X_{2,3}$
⋮			⋮	⋮	⋮	⋮	⋮	⋮
28			$Y_{1,28}$	$Y_{2,28}$	$Y_{3,28}$	$Y_{4,28}$	$X_{1,28}$	$X_{2,28}$
⋮	⋮	⋮	⋮	⋮	⋮	⋮	⋮	⋮
1770			$Y_{1,1770}$	$Y_{2,1770}$	$Y_{3,1770}$	$Y_{4,1770}$	$X_{1,1770}$	$X_{2,1770}$
1771			$Y_{1,1771}$	$Y_{2,1771}$	$Y_{3,1771}$	$Y_{4,1771}$	$X_{1,1771}$	$X_{2,1771}$
1772	February	2017	$Y_{1,1772}$	$Y_{2,1772}$	$Y_{3,1772}$	$Y_{4,1772}$	$X_{1,1772}$	$X_{2,1772}$
⋮			⋮	⋮	⋮	⋮	⋮	⋮
n			$Y_{1,n}$	$Y_{2,n}$	$Y_{3,n}$	$Y_{4,n}$	$X_{1,n}$	$X_{2,n}$

Stages of the data analysis are described as follows:

a. Identify ARMAX model on stock return BBRI, BMRI, BBCA, and BBNI as a whole.
b. Estimating parameters on the ARMAX model where the predictor variables are IHSG and rupiah exchange rate return against dollar.
c. Testing the significance of ARMAX model parameters.
d. Perform assumption testing on residual which includes white noise assumption and normal distribution on residual data.
e. Choose the best ARMAX model using AIC criteria.
f. Identify ARCH-GARCH effect on error value of ARMAX model formed by using Ljung Box and Lagrange Multiplier test.
g. Identify the GARCHX model.
h. Estimating parameters on the GARCHX model.
i. Conducting a test of significance to GARCHX model parameters.
j. Getting the appropriate ARMAX-GARCHX model.
k. Calculates the VaR value based on ARMAX-GARCHX model parameter estimation

2.2. Autoregressive Moving Average with Exogenous Variables (ARMAX)

ARMAX or Autoregressive Moving Average with Exogenous Variables is the development of ARIMA model by including the exogenous variables into the equation. ARMA model is not only influenced by data of previous period (historical), but also influenced by exogenous variable. The following is a general equation of the ARMAX model as follows (Hyndman, 2010).

$$Y_t = \varpi_1 X_{1,t} + \ldots + \varpi_U X_{U,t} + \phi_1 Y_{t-1} + \ldots + \phi_p Y_{t-p} - \theta_1 a_{t-1} - \ldots - \theta_q a_{t-q} + a_t$$

$$Y_t = \sum_{u=1}^{U} \varpi_u X_{u,t} + \sum_{j=1}^{p} \phi_j Y_{t-j} - \sum_{j=1}^{q} \theta_j a_{t-j} + a_t,$$

Y_t is the response variable at time *t*, X_t is the exogenous variable at time *t*, a_t is white noise error, and (p, q, U) indicates the order of the ARMAX model. Based on ARMAX model it can be seen that ARMAX modeling concept is similar as transfer function model, but with condition where *s* and *b* value equal to 0 and *r* value equal to *p*. So the steps of ARMAX model identification are as follows.

a. Setting order (*p,q*), order value (*p,q*) on ARMAX model is obtained from ACF and PACF response variable data or Y_t .
b. Assign (*r, s, b*), where in the ARMAX model the value of $r = p$, $s = 0$, and $b = 0$.
c. Estimate the ARMAX model that has been formed.

2.3. Generalized Autoregressive Conditional Heteroscedasticity with Exogenous Variable (GARCHX)

In generating better and accurate models and forecast of volatility in the time series of financial and economic data, researchers and practitioners often incorporate exogenous predictor variables into data volatility. One method that can be used to overcome this problem is GARCHX or Generalized Autoregressive Conditional Heteroscedasticity with Exogenous Variable method. GARCHX was a GARCH model developed by Bollerslev in 1986, by adding exogenous variables into model equations. Where the value of Z_t in equation is a random variable that identical, independent, and normally distibuted (0, 1). Whereas σ_t^2 is a process of volatility that is influenced by exogenous variables i.e.

$$\sigma_t^2 = \omega + \sum_{j=1}^{r} \varphi_j \varepsilon_{t-j}^2 + \sum_{j=1}^{s} \beta_j \sigma_{t-j}^2 + \sum_{u=1}^{U} \pi_u X_{u,t}^2$$

Where $\omega > 0$, $\varphi > 0$, $\beta > 0$, and $\pi > 0$. The value X_t represents the exogenous variable, the X_t value squared to guarantee that $\sigma_t^2 > 0$. The addition parameter of X_t regressor is often used to explain the volatility of stock return, exchange rate returns, or interest rates, and tend to lead to better in-sample and out-sample in forecasting performance. In this case, the exogenous variable is allowed in a

IOP Conf. Series: Journal of Physics: Conf. Series **974** (2018) 012029 doi:10.1088/1742-6596/974/1/012029

stationary and non-stationary state. However, the intercept parameter ω is not identifiable when the exogenous variable is not stationary. (Han and Kristensen, 2013).

2.4. *Value-at-Risk*

The Value-at-Risk (VaR) method is a risk assessment method that summarizes the possible minimum losses for an investor in a portfolio at a certain level of confidence. VaR aims to measure the rate of gain or loss of a stock investment. The amount of profit earned is proportional to the risk obtained. Here is a VaR model based on the return value.

$$P(R_t < -VaR) = (1 - CI)\% = \tau$$

where the value of R_t denotes the stock return value of the period to-*t*, *CI* is the Confident Interval value, and τ is the quantity used to measure the risk level. The greater the VaR the higher the portfolio risk should be considered. The advantage in VaR allows the user to determine trust level that reflects the risk-averseness of the individual. In general, the method for measuring VaR often assumes a portfolio return value following a normal distribution. VaR calculations are generally shown in the following equation.

$$\hat{VaR}_t(\tau) = \hat{\mu}_t + \hat{\sigma}_t F^{-1}(\tau)$$

3. Result

Descriptive statistics for stock prices of BRI (BBRI), BCA (BBCA), Bank Mandiri (BMRI), BNI (BBNI), ICI, and exchange rate are presented in Table 2.

Table 2. Data Charasteristic

Variable	*Mean*	*Variance*	*Min*	*Max*
BBRI	11,599	1,528,194	8,300	15,300
BBCA	13,324	2,283,945	11,300	18,250
BMRI	10,485	1,418,167	7,525	12,900
BBNI	5,612.9	603,904.6	3,940	7,275
ICI	5,081	163,731.1	3,120.5	5,829.7
Exchange Rate IDR/USD	13,413	142,768	12,506	13,802

Bank with the highest average stock price is BCA and the lowest is BNI. The highest stock price in a 2.5 years is Rp7,275 and the lowest stock price is Rp3,940.00, the stock price has the lowest price compared to other banks. For ICI in January 2015 to June 2016 has an average of 5,081 with the highest ICI score is 5,829.7 and 3,120.5 is the lowest ICI score. The rupiah-denominated selling rate against the dollar shows an improvement although it is still very volatile. It is known that the average selling rate of rupiah against the dollar in January 2015 to June 2017 was 13,413 with the highest selling rate at 13,802 and the lowest selling rate at 12,506. For the diversity of data in each bank, ICI, and exchange rate IDR/USD can be known very high, it can be known from the high value of variance. The characteristics of stock return can also be seen through the time series plot.

Figure 1 Time Series Plot of BBRI, BBCA, BMRI and BBNI

Figure 1 shows that Bank BCA's stock price from 2015 to June 2017 is higher than other banking stocks, while BNI is the lowest among the three other banks. The four banks have a fluctuating stock value. Competitiveness of stock prices appears in Bank Mandiri and Bank BRI, but Bank BRI is superior in stock price in a longer time than Bank Mandiri..

Characteristics for the exogenous variables of the rupiah exchange rate against the dollar and ICI can be seen in Figure 2.

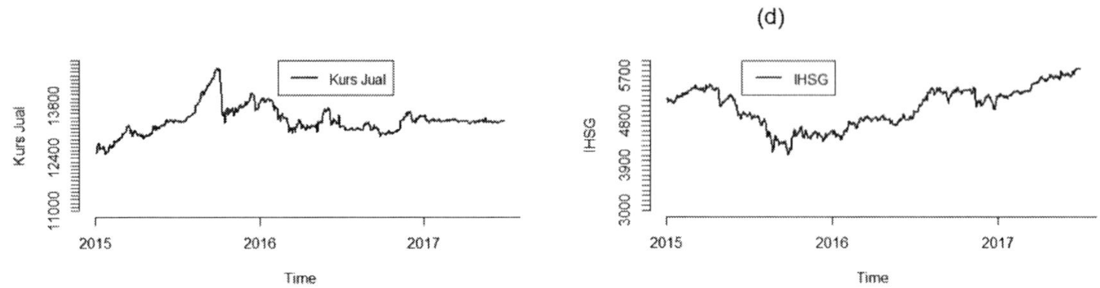

(a) (b)

Figure 2 Time Series Plot (a) Exchange rate and (b) ICI

The movement pattern of the IDR / USD exchange rate from day to day tend to increase, but also decreased in mid 2011 which reached the lowest value of Rp 8,502.00. But since then, the IDR / USD exchange rate has always increased. The drastic increase occurred in mid-2013 to Rp 12.331,00 and the highest IDR / USD rate occurred in mid-2015 reached Rp14.802,00. This phenomenon also coincided with the increase in BI rate to 7.5% which also affected the weakening of ICI at that time, as shown in Figure 2 (b), it appears that when the rupiah weakened in mid-2013 and 2015, the rate of IHSG precisely decrease. The decline of IHSG is affecting the declining value of stocks from several sectors such as property sector. So that the movement rate of IDR / USD and IHSG exchange rate is able to influence the increase and decrease the stock price of a company.

3.1. ARMAX GARCHX Modeling

Analysis of Capital Asset Pricing Model (CAPM) aims to prove the theory about ICI influence. Where the window used 250 days of transactions and generate parameters $\beta_{\alpha.}$ In CAPM analysis between IHSG return with stock return of the four companies, the variable that used as risk free rate is the interest rate. The significance test of β_α parameter can be seen in Table 3.

Table 3. Significant Test of Parameter $\boldsymbol{\beta_\alpha}$

Return Stock	*Window*	β_α	*Pvalue*	Inf.
BBRI	1	1.613	1.63E-30	Significant
BBCA		1.518	1.22E-10	Significant
BMRI		0.832	7.04E-11	Significant
BBNI		1.416	1.50E-10	Significant
...	...	...	...	...
BBRI	1323	1.196	1.32E-14	Significant
BBCA		0.927	7.49E-17	Significant
BMRI		0.733	7.01E-09	Significant
BBNI		1.125	0.000186	Significant

Known from Table 3, at α = 5% the result of *Pvalue* β_α the five firms in each window have value less than α, it can be concluded that the return of IHSG influence significantly to stock return of four company. Thus, the results of this CAPM analysis support the theory that ICI affects individual stocks.

Furthermore, modeling stock prices with ARMAX-GARCHX with exogenous variables are ICI (X_1) and exchange rate IDR / USD. Table 4 is the estimation result of BBRI stock return model.

Table 4. BBRI Stock Return Model

Model	Par.	Estimation	*Pvalue*
GARCHX (0,1,1)	ω	0.000003	0.00
	β_1	0.876576	0.00
	$\pi_1\ (X_2)$	0.000238	0.00

Table 4 show that model parameters are significant because each model parameter has | *t value* | > value | t_{table} | amounted to 1.96, so the model formed on stock BBRI is GARCHX (0,1,1) with exogenous variable that is rate of IDR / USD. BBRI stock return model is as follows.

$$\hat{\sigma}_t^2 = 0.000003 + 0.876576\hat{\sigma}_{t-1}^2 + 0.000238X_{2,t}^2$$

Table 5. BMRI Stock Return Model

Model	Par.	Estimation	*Pvalue*
GARCHX (0,1,1)	ω	0.000001	0.00
	β_1	0.995643	0.00
	$\pi_1\ (X_2)$	0.000358	0.00

Table 5 show that model parameters are significant because each model parameter has | *t value* | > value | t_{table} | amounted to 1.96, so the model formed on stock BMRI is GARCHX (0,1,1) with exogenous variable that is rate of IDR / USD. BMRI stock return model is as follows.

$$\hat{\sigma}_t^2 = 0.000001 + 0.995643\hat{\sigma}_{t-1}^2 + 0.000358X_{2,t}^2$$

Table 6 is the result of estimation of BBCA return model.

Table 6. BBCA Stock Return Model

Model	Par.	Estimation	*Pvalue*
ARMAX (1,0,2) GARCHX (1,1,1)	ϕ_1	0.089	0.0002
	$\varpi_1\ (X_1)$	0.987	0.0000
	$\varpi_2\ (X_2)$	-0.611	0.0009
	ω	0.000002	0.0305
	φ_1	0.020	0.0000
	β_1	0.887	0.0000
	$\pi_1\ (X_1)$	0.0005	0.0306

The corresponding model in the stock return of BBCA is ARMAX (1,0,2) -GARCHX (1,1,1), where exogenous variable is significant in ARMAX model that is ICI and IDR / USD rate while in GARCHX model that is IHSG. The LPKR stock return model is as follows.

ARMAX(1,0,2): $\hat{R}_t = 0{,}089R_{t-1} + 0{,}987X_{1,t} - 0{,}611X_{2,t}$

GARCHX(1,1,1): $\hat{\sigma}_t^2 = 0{,}000002 + 0{,}020\hat{\sigma}_{a_{t-1}}^2 + 0{,}887\hat{\sigma}_{t-1}^2 + 0{,}0005X_{1,t}^2$.

Table 7. Model *Return* Saham BBNI

Model	Par.	Estimation	*Pvalue*
GARCHX (0,1,1)	ω	0.000009	0.00
	β_1	0.888899	0.00
	$\pi_1\ (X_2)$	0.000913	0.00

In Table 7, we can obtain the return model of BBNI company's stock is GARCHX (0,1,1) where the IDR / USD rate as its exogenous variable, with equation that is

$$\hat{\sigma}_t^2 = 0.000009 + 0.888899\hat{\sigma}_{t-1}^2 + 0.000913X_{2,t}^2 .$$

3.2. Value-at Risk

The estimated risk of stock return of the five companies using ARMAX-GARCHX approach. the mean parameter in the VaR model is approximated by the ARMAX model, whereas the variance parameter is approximated by the GARCHX model. VaR calculations are performed on each window with a total of 250, 375, and 500 days transactions, and 5% quantiles are used.

Table 8. Risk Estimation with VaR

	Risk		
	250	375	500
BMRI	-0.043	-0.043	-0.043
BBCA	-0.050	-0.050	-0.051
BMRI	-0.032	-0.031	-0.032
BBNI	-0.043	-0.044	-0.046

Based on Table 8 it is known that at the 95% confidence level, the biggest maximum loss value experienced by an investor when investing Rp 1 Billion in BBRI company is Rp 43.000.000,- , BBCA company Rp 51.000.000,- , BMRI Rp 32.000.000,- , BBNI company Rp 46.000.000,- , Also note that the companies that provide the highest and lowest loss rates on the windows 250, 375 and 500 are BBCA. This VaR value estimate applies when the economy is in a normal.

4. Conclusion

Banks with the highest average stock price is BCA and the lowest is BBNI. BBRI and BBCA stocks tend to be very volatile but there is an increase. The movement pattern of the exchange rates IDR / USD and ICI is always increasing every year. In CAPM analysis, it is proven that ICI has significant effect on all four stocks. In the VaR estimation, the appropriate model for BBRI is GARCHX (0,1,1) with exogenous variable ie IDR / USD rate. The BMRI model is GARCHX (1,0,1) where the exogenous variable is the IDR / USD rate. BBCA model is ARMAX (1,0,2) -GARCHX (1,1,1) with exogenous variable of ARMAX model that is ICI and IDR / USD rate while at GARCHX that is IHSG. The BBNI model is GARCHX (0,1,1) where the IDR / USD rate is the exogenous variable. It is also known that the company that gives the highest and lowest loss rate is BCA. The VaR estimate using the GARCHX approach yields a more accurate VaR score than ARMAX-GARCHX, this is because the ARMAX model is redundant

5. Acknowledgments

Thank you for Business Statistics Department who has provided financial support.

References

[1] Messaoud, S.B dan Aloui, C, (2015) Measuring Risk of Portfolio : GARCH-Copula Model, Journal of Economic Integration, Vol.30 No.1,, 172~205

[2] Dharmawan, K. (2014). *Estimasi Nilai Value at Risk Portofolio Menggunakan Metode t-Copula.*, Jurnal Matematika, Saint, dan Teknologi, Volume 15, Nomor 1, Maret 2014, 1-11.

[3] Donaldson, J. B. (2015). *Intermediate Financial Theory, Third Edition.* New York: Elsevier Inc

[4] Wei, W. W. (2006). *Time Series Analysis Univariate and Multivariate Methods.* New York: Pearson Education.

[5] Hyndman, R. J. (2010). *The ARIMAX Model Muddle.* Tersedia pada situs: http://robjhyndman.com/hyndsight/arimax/. Diakses 8 November 2016

[6] Han, H., & Kristensen, D. (2013). *Asymptotic Theory for the QMLE in GARCH-X Models with Stationary and Non-Stationary Covariates.* Journal of Financial Econometrics. 18/13.

Forecasting the mortality rates of Indonesian population by using neural network

Lutfiani Safitri[1], Sri Mardiyati[1], and Hendrisman Rahim[2]

[1] Department of Mathematics, FMIPA Universitas Indonesia, Kampus UI Depok, Depok 16424, Indonesia
[2] PT. Asuransi Jiwasraya, Indonesia

Abstract. A model that can represent a problem is required in conducting a forecasting. One of the models that has been acknowledged by the actuary community in forecasting mortality rate is the Lee-Certer model. Lee Carter model supported by Neural Network will be used to calculate mortality forecasting in Indonesia. The type of Neural Network used is feedforward neural network aligned with backpropagation algorithm in python programming language. And the final result of this study is mortality rate in forecasting Indonesia for the next few years

1. Introduction

Mortality is the number of death sizes of a population. Mortality is one component that influences the change of population structure [3]. In general, the mortality of an area is expressed in a mortality rate, which describes how large the rate of death in an area per year and at a certain age. Mortality rate is beneficial in several areas, such as economics, health and life insurance [13]. For example in life insurance, mortality rates are typically used to determine the amount of a premium a person pays to a life insurance company [7].

Due to the importance of this mortality rate, it is necessary to forecast mortality rates for the foreseeable future. In order to predict future mortality rates, a model can be used to describe the mortality rate of a region. There are two approaches in a modeling, namely a deterministic approach and a stochastic approach. In the deterministic model, it will produce a fixed output value, and the output is fully influenced by the value of the parameter and the initial value. While the output value of the stochastic model is probability, and the output value is not fully dependent on the value of the parameter and the initial value [6]. One model of mortality rate using a deterministic model is the model proposed by Helligman and Pollard in 1980, they used a deterministic approach based on their age [5]. one model of mortality rate with a stochastic approach is the Lee-Carter model. In 1992, Lee and Carter introduced a stochastic model to predict mortality rates in the United States of America. The Lee-carter model has been successfully applied for mortality rates forecasting in several countries and at different time periods, including the United States [8], Chile (1994) [9], Singapore (2004) [10], Hungary (2007) [1], and Malaysia (2016) [11].

From Lee Carter model, The parameters will be estimate using least square method and Singular Value Decomposition (SVD) [11]. After the parameter value obtained then forecasting from stochastic process will be conducted. This paper will present forecasting mortality rate by using Neural Network. Neural network (NN) is a tool that used to complete a calculation process, where the NN work system is following the learning process in the human brain that is implemented in computing systems. NN itself is divided into 2 types, namely feedforward NN and recurrent NN, This Research will use Feedforward Neural Network type [2].

2. Methodology

2.1 Feedforward NeuralNetwork

Before going to deeper understanding of Feedforward Neural Network, we will explain Lee Carter model. Lee and Carter introduced a stochastic model that was used to predict a region's mortality rate, in 1992. Its first study predicted mortality rates in the United States. The study was quite successful because it adequately describes the mortality rate in the American population. Then several years after that, many countries use the model to forecast mortality rates in their countries and get pretty good results. Here is a model from Lee Carter

$$\ln(m_{x,t}) = a_x + b_x \cdot k_t + \varepsilon_{x,t} \tag{1}$$

with constraints:

$$\sum_{t=1}^{T} k_t = 0 \text{ and } \sum_{x=1}^{N} {b_x}^2 = 1$$

where $m_{x,t}$ is the central death rate at age x in year t, with x = 1,2, ..., N, and t = 1,2, ..., T. Then a_x and b_x are age-dependent parameters, and k_t is a stochastic process that depends on the observed year and $\varepsilon_{x,t}$ is independent error, with mean 0 and its variance σ^2 [11].

The early step is to estimate parameter value of a_x by using Least Square method, so the equation is obtained as follows:

$$\hat{a}_x = \frac{\sum_{t=1}^{T} \ln(m_{x,t})}{T} \tag{2}$$

then parameter value of b_x and k_t will be estimated by using Singular Value Decomposition, so the equation is obtained as follows:

$$\hat{b}_x = \frac{1}{\sum_x u_{x,1}} (u_{1,1}, u_{2,1}, \cdots, u_{x,1})^T \tag{3}$$

$$\hat{k}_t = \sum_x u_{x,1} \times \sigma_1 \times (v_{1,1}, v_{2,1}, \cdots, v_{t,1}) \tag{4}$$

then the value of k_t for t = T + 1, T + 2 will be forecasted and so on by using the neural network, since the main discussion of this paper is forecasting using Neural Network [11].

According to the encyclopedia of Machine learning and data mining, Neural network is a learning algorithm whose work system follows the brain's working system in humans. The learning is obtained by adjusting the value of weights between nodes, analogous to synapses and neurons [12]. In building a neural network framework that will be used, it requires 3 things, namely:

1. Pattern of relationship between neurons, or its architecture.
   Based on the architecture, there are 2 types of single-layer net and multi-layer net in the feedforward neural network. In single-layer net there is only one layer with the connected weights. This network only accepts input and is directly processed into output value through its weights, without going through the hidden layer. While in a multi-layer net network there is one layer or more located between the input layer and the output layer, called the hidden layer.
2. The method used to determine the weight of the correction (learning algorithm)
   there are many of Algorithms that can be used in the learning process such as, Backpropagation, Delta Learning Rule, Forward Propagation, Hebb Learning Rule, and Simulated Annealing.
3. The activation function used
   Some commonly used activation functions are the Identity function, Binary step function, Binary sigmoid, and Sigmoid Bipolar [4].

Feedforward neural network with multi-layer net will be used in this paper, which there is one layer in hidden layer. Then the training algorithm used is backpropagation algorithm, and the activation function used is sigmoid function. Here is an overview of the neural network structure to be applied in this paper.

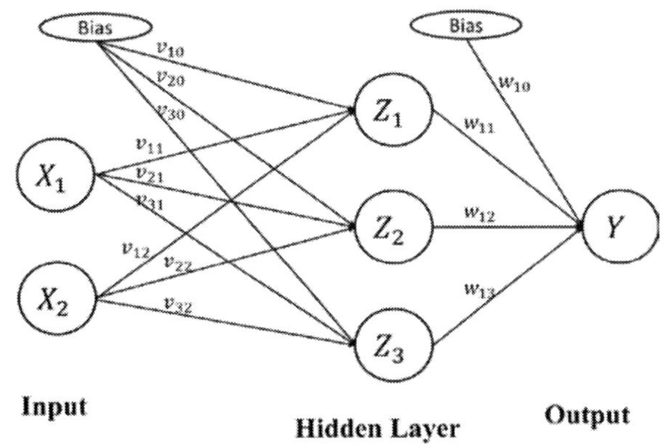

Figure 1. Multilayer Structure of Feedforward Neural Network

The steps in the backpropagation algorithm can be explained as follow:

1. The number of input nodes will be determined by looking at the relationship proximity between variables, by calculating the correlation value between variables, or by describing the correlogram by using the device R
2. Data will be divided into two parts, namely training data and testing data. . This training data will be used for learning while the testing data will be used to validate the model obtained, whether the model obtained is sufficient to describe the existing data or not. After that, data normalization will be conducted. Then, early initialization will be conducted on all weights and bias values, with random number range (0,1) and determine the activation function to be used.
3. Then, enter the input value and the input will be forwarded to the next layer that is the hidden layer, by multiplying each input by its respective weight and adding it, like the following formula

$$Z\_net_j = v_{j0} + \sum_{i=1}^{n} x_i v_{ji} \tag{5}$$

   once the Z_net_j value obtained on each node in the hidden layer, calculate the output signal from the hidden layer as follows $Z_j = f(Z\_net_j)$ where f is the predefined activation function
4. Perform step 3 as much as the number of hidden layers used.
5. Then the output value from the hidden layer will be obtained. The output of the hidden layer will be passed to the output layer, through this following formula

$$Y\_net_k = w_{k0} + \sum_{j=1}^{p} z_j w_{kj} \tag{6}$$

   the value of the activation function of $Y_k = f(Y\_net_k)$, will be determined and the output value will be obtained.
6. The output obtained and the expected output then will be checked. If not appropriate or the error is significant, weight correction on each layer will be done. The weight correction can be conducted in two ways. It can starts from the weight between hidden layer with output layer, or

done from behind. Here is a formula for updating the weights and biases between the hidden layer and the output layer

$$w_{kj}(new) = w_{kj}(existing) + \Delta w_{kj} \tag{7}$$

$\Delta w_{kj} = \alpha\delta_k z_j, j \neq 0$ is correction formula for weight value and $\Delta w_{k0} = \alpha\delta_k$ is correction formula for bias value, where δ_k was obtained from formula presented below:

$$\delta_k = (t_k - Y_k)f'(Y\_net_k)$$

7. Then value of weight and bias between input layer and hidden layer will be renewed with formula as follows:

$$v_{ji}(new) = v_{ji}(existing) + \Delta v_{ji} \tag{8}$$

$\Delta v_{ji} = \alpha\delta_j x_i, i \neq 0$ is correction formula for weight value, and $\Delta v_{j0} = \alpha\delta_j$. is correction formula for bias value, where δ_j was obtained from formula presented below:

$\delta_j = \delta\_net_j f'(Z_{net\,j})$

$\delta\_net_j = \sum_{k=1}^{m} \delta_k w_{kj}$

this process is repeated until the smallest error value is obtained. Then we will get the model with the best estimation weight, which will then be done for future forecasting [4].

3. Main Result

The data used in this study is the data of Indonesia's state mortality rate from 1950-2015, with the data range of five years. The death rate data is taken from https://www.un.org. Available data from age 0 to 90 years, with age range is 0, 1-5, 6-10, and so on until age 86-90. Here is an overview of the data used in this study

Table 1. Indonesia Mortality rate

		1	2	3	4	5	6	7	8	9	10	11	12	13
	Age Group	**1950-1955**	**1955-1960**	**1960-1965**	**1965-1970**	**1970-1975**	**1975-1980**	**1980-1985**	**1985-1990**	**1990-1995**	**1995-2000**	**2000-2005**	**2005-2010**	**2011-2015**
1	**0**	0.22061	0.18733	0.15908	0.13508	0.11285	0.09782	0.08490	0.07203	0.05879	0.04701	0.03766	0.03047	0.02553
2	**1-5**	0.03605	0.02804	0.02182	0.01697	0.01288	0.01012	0.00805	0.00621	0.00454	0.00326	0.00237	0.00178	0.00136
3	**6-10**	0.00553	0.00472	0.00403	0.00344	0.00289	0.00231	0.00187	0.00157	0.00129	0.00106	0.00092	0.00079	0.00066
	...	...	...	...	...	...	...	...	...	...	...	...	...	...
18	**81-85**	0.20386	0.19746	0.19130	0.18537	0.17906	0.17546	0.17179	0.16817	0.16336	0.15741	0.15194	0.14671	0.14275
19	**86-90**	0.30937	0.30388	0.29847	0.29327	0.28777	0.28486	0.28175	0.27812	0.27327	0.26748	0.26106	0.25456	0.25093

From the data of Indonesia mortality rate, the estimation of parameters value of a_x, b_x and k_t can be defined by using Least Square (2) and Singular Value Decomposition method (3), (4), and the estimation result which will be used in the calculation of mortality forecasting in Indonesia are presented in **Table 2** and **Table 3.**

Later, the value of k_t will be forecasted by using the neural network that is implemented in python programming language, with the steps described as follows:

1. Determine the number of input nodes, in this paper 2 nodes are required.
2. Data shall be divided into training data and testing data. Since the data obtained is in small amount, then all data will be used for training data. The testing data will use random data picked from training data. Then, data normalization and initialization of weight and value of bias will be conducted.
3. Calculate the output from hidden layer according to formula (5)
4. Calculate the output value of output layer according to formula (6)
5. Update the weight of each edge and bias by using formulas (7) and (8).

6. After weight and bias value being updated, do step 3 repeatedly until minimum error value is achieved.
7. Then will be obtained neural network model for forecasting,

Table 2. The estimation results of parameter a_x, b_x

X	Age Group	a_x	b_x
1	0	-2.535305571	0.113121044
2	1-5	-4.903325525	0.114527372
3	6-10	-6.257768336	-0.288364691
4	11-15	-6.642158292	-0.361029496
5	16-20	-6.227792463	-0.207156019
6	21-25	-5.985092803	-0.143706519
7	26-30	-5.931403207	-0.064089919
8	31-35	-5.789418083	-0.010321486
9	36-40	-5.557747938	0.03611584
10	41-45	-5.280767417	0.100378846
11	46-50	-4.961430283	0.131166975
12	51-55	-4.580377048	0.212028902
13	56-60	-4.183308393	0.294081098
14	61-65	-3.621794699	0.392838978
15	66-70	-3.167211589	0.386845962
16	71-75	-2.676287374	0.345871638
17	76-80	-2.204364696	0.266083438
18	81-85	-1.766726711	0.175920359
19	86-90	-1.273669262	0.086247194

Table 3. The estimation results of parameter k_t

t	Year Group	k_t
1	1950-1955	-0.036811276
2	1955-1960	0.016137596
3	1960-1965	-0.00393337
4	1965-1970	0.003227391
5	1970-1975	0.011400005
6	1975-1980	-0.069699439
7	1980-1985	0.010435503
8	1985-1990	0.013913731
9	1990-1995	0.005765721
10	1995-2000	-0.029592153
11	2000-2005	0.040735604
12	2005-2010	0.040735604
13	2010-2015	0.063964828

There are the results of testing the model that has been obtained before

Table 4. Data testing

t	Year Group	k_t	$\widehat{k_t}$	Error
3	1960-1965	-0.00393337	0.08050727200523929	0.026025367622659307
4	1965-1970	0.003227391	0.0836324332623512	0.000007652517448858718
5	1970-1975	0.011400005	0.08261727004489795	0.0004112266640321528
6	1975-1980	-0.069699439	0.07912088530554803	0.00054217787477231
7	1980-1985	0.010435503	0.10803265969791215	0.0005431528543469121
8	1985-1990	0.013913731	0.08518219604717701	0.0005165814968914099
9	1990-1995	0.005765721	0.07766817950755707	0.00048689041110894194
10	1995-2000	-0.029592153	0.0803209450877136	0.0004600989038951871
11	2000-2005	0.040735604	0.09375274425816599	0.00043739012275880324
12	2005-2010	0.040735604	0.07108085029955004	0.00041861177683834
13	2010-2015	0.063964828	0.06590842496977312	0.0004032820078852832

The forecast value obtained of k_t for 5 years ahead is:

Table 5. Forecasting value of k_t

t	Year Group	k_t
14	2015-2020	−0.03203691
15	2020-2025	0.06385799
16	2025-2030	0.06269202
17	2030-2035	−0.06117693
18	20035-2040	0.06393609

From Table 5 will be calculate the mortality rate with the parameters obtained in table 2 using equation (1). The result of the mortality rate forecasting will be presented in table 6

Table 6. The result of the mortality rate forecasting in Indonesia

	Year Group	2015-2020	2020-2025	2025-2030	2030-2035	2035-2040
Age Group		1	2	3	4	5
0	0	0.078950862	0.079811962	0.079801436	0.078691041	0.079812667
1-5	1	0.007394679	0.007476339	0.007475341	0.007370041	0.007476406
6-10	2	0.001933294	0.001880566	0.001881198	0.001949608	0.001880523
11-15	3	0.001319382	0.001274485	0.001275022	0.001333335	0.001274449
16-20	4	0.001986947	0.001947866	0.001948336	0.001998978	0.001947834
21-25	5	0.00252759	0.002492997	0.002493415	0.002538197	0.002492969
26-30	6	0.002660211	0.002643911	0.002644109	0.002665183	0.002643898
31-35	7	0.003060774	0.003057746	0.003057783	0.003061695	0.003057744
36-40	8	0.003852993	0.003866361	0.003866198	0.00384894	0.003866371
41-45	9	0.005072187	0.005121246	0.005120647	0.005057372	0.005121287
46-50	10	0.006973539	0.007061808	0.007060728	0.006946935	0.00706188
51-55	11	0.010181634	0.010390771	0.010388202	0.01011892	0.010390943
56-60	12	0.015104994	0.015537032	0.015531705	0.014976105	0.015537389
61-65	13	0.026400296	0.027413799	0.027401245	0.026099806	0.02741464
66-70	14	0.041602087	0.043174364	0.043154895	0.041135753	0.043175669
71-75	15	0.068059835	0.070355051	0.070326684	0.067377325	0.070356951
76-80	16	0.10938416	0.112211124	0.112176317	0.10853931	0.112213456
81-85	17	0.169931026	0.172822056	0.172786611	0.169062133	0.172824431
86-90	18	0.27903101	0.281348352	0.281320061	0.278330617	0.281350247

4. Conclusion

From table 4 it can be seen that the average error for each value of k_t is small enough, and the forecasting result can present the original value, so the Neural Network Method can be used in forecasting mortality rate in Indonesia.

References

[1] Baran, S., Gall, J., Ispany, M., & Pap, M. (2007). Forecasting Hungarian Mortality Rates Using The Lee-Carter Method. *Acta Oeconomica, 57*, 21-34.
[2] Bishop, C. M. (2006). Neural Network. In *Pattern Recognition and Machine Learning* (pp. 225-290). Cambridge: springer.
[3] Cox, P. R. (1976). *Demography fifth edition.* New York: Cambridge University Press.
[4] Fausett, L. (1994). Backpropagation Neural Net. In *Fundamentals of Neural Network: architectures, algorithms, and applications* (pp. 289-328). USA: rentice-Hall, Inc. Upper Saddle River, NJ.
[5] Heligman, L., & Pollard, J. H. (1980). The Age Pattern of Mortality. *Journal Institue of Actuaries*, 49-80.
[6] Hinderer, K., Rieder, U., & Stieglitz, M. (2016). *Dynamic Optimization: Deterministic and Stochastic Models.* Switzerland: Springer Nature.

[7] Khamladze, E. V. (2013). *Statistical Methods With Application to Demography and Life Insurance.* New York: Taylor & Franciss Group.

[8] Lee, R. D., & Carter, L. R. (1992). Modeling and Forecasting U.S Mortality. *Journal of the American Statistical Association, 87*(419), 659-671.

[9] Lee, R. D., & Rofman, R. (1994). Modeling and Forecasting Mortality in Chile. *Notas, 22*(59), 182-213.

[10] Li, S. H., & Chan, W. S. (2004). Estimation of Complete period life tables for singaporeans. *Journal of Actuarial Practice, 11*, 129-146.

[11] Ngataman, N., Ibrahim, R. I., & Yusuf, M. M. (2016). Forecasting the Mortlaity Rates of Malaysian Population Using Lee-Carter Method. *American Institue of Physics.*

[12] Sammuel, C., & Webb, G. I. (2017). *Encyclopedia of Machine Learning and Data Mining.* New York: Springer Nature.

[13] Severine, G. (2012). Forecasting mortality: when academia meets practice. *European Actuarial Journal*, Volume 2 (1), 49-76.

[14] Zhang, F. (2010). *Matrix Theory, Basic Results and Techniques.* New York: Springer.

Rainfall estimation with TFR model using Ensemble Kalman filter

Nabila Asyiqotur Rohmah[1] **and Erna Apriliani**[2]
Department of Mathematics, Institut Teknologi Sepuluh Nopember, Surabaya 60111, Indonesia

E-mail: [1] nabila.asya@gmail.com, [2] aprilmath12@gmail.com

Abstract. Rainfall fluctuation can affect condition of other environment, correlated with economic activity and public health. The increasing of global average temperature is influenced by the increasing of CO2 in the atmosphere, which caused climate change. Meanwhile, the forests as carbon sinks that help keep the carbon cycle and climate change mitigation. Climate change caused by rainfall intensity deviations can affect the economy of a region, and even countries. It encourages research on rainfall associated with an area of forest. In this study, the mathematics model that used is a model which describes the global temperatures, forest cover, and seasonal rainfall called the TFR (temperature, forest cover, and rainfall) model. The model will be discretized first, and then it will be estimated by the method of Ensemble Kalman Filter (EnKF). The result shows that the more ensembles used in estimation, the better the result is. Also, the accurateness of simulation result is influenced by measurement variable. If a variable is measurement data, the result of simulation is better.

1. Introduction

Rainfall is one of the most important elements in weather in Indonesia. Because of its location on the equator, temperature in Indonesia does not have much fluctuation. In certain years, the duration of the rainy season and the intensity of drought during the dry season vary greatly in Indonesia. Certain trajectories of monsoons vary from one year to the next, leading to widespread variation in rainfall across the archipelago from year to year [1].

The climate in Indonesia consists of one rainy season and one dry season throughout each year. The characteristics of the rainy season are at least 200 mm of rainfall's amount per month. This provision is based on the minimum required for rice production and calculates account evaporation and seepage through the soil [2]. The rainy season and the dry season are in varying months throughout the provinces of Indonesia.

The average deviations of rainfall at the district level have positive impact to the average deviation of rice production at the district level in Indonesia in the 1990s. (The year in which very high rainfall produces very high rice production, and the year in which very low rainfall produces very low rice production) [3]. The secondary report also shows that the higher rainfall was, the more agricultural productivity earned in Indonesia [2]. Production of rice in the rainy season and dry season depends on the time of the monsoon wind. Unexpectedly, it is clear that drought in the dry season is driven by the El Nino weather phenomenon. For example, there were ten long or short droughts (large-scale crop failures) between 1921 and 1954. The planting time during the rainy season was based on the threshold of accumulated rainfall. Generally, when the planting time is delayed, the harvest season is reduced. The delaying in the start of the monsoon winds in the rainy season, and the ongoing drought during the season, leads to a reduction in the rainy season's harvest. Secondary crop yields may decrease during the following dry season due to harvest delays in the previous monsoon season (eg, El Nino's 1997-1998 perforation). On the other hand, in La Niña year (when rainfall is very high), the planting season can start earlier and produce above average crop.

Due to the importance of the rainy season cycle, food insecurity also tends to differ each season. Food insecurity tends to be highest at the end of the dry season and the beginning of the next rainy season, when food supply from the previous rainy season is low and demand is high with initiation of planting [4]. Thus, the drought makes the food scarcity worse.

The rainfall is not only impacting in agricultural production and human income, but also child growth and health. Unusual rainfall can affect the growth of disease in the environment, and diseases directly affect the absorption of nutrients, especially in the early years of children growth, when kids are weak from diseases. Fluctuations in rainfall can affect other environmental conditions that connected with economic and public health activities, such as increasing rate of forest wildfires, rate of floods, or rate of landslides, the availability of clean water, and agricultural pest control [5]. Some of these relationships imply the negative effects of rainfall that offset the positive effects through increased yields.

Most of the increase in global average temperature is caused by a certain gas increase in the atmosphere. The atmosphere is made up of a wide variety of naturally occurring gases. Gas is also produced by human activities. If some of the gas produced is excessive, it can cause changes in natural processes that eventually lead to climate change [6].

Some of the atmospheric gases are capable of extracting or absorbing heat from the sun and the earth and keeping it in the lowest atmospheric parts closest to the earth. There are a lot of greenhouse gases in the atmosphere, but the most important greenhouse gas is carbon dioxide (CO2). This gas is produced when carbon substances join the oxygen in the air. The increase in atmospheric CO2 is the biggest cause of climate change.

Carbon can be found in three ways. The carbon can be 1) absorbed from the air as part of carbon dioxide by plants and trees, then used as energy and food for growth; 2) released back into the air as part of CO2 by plants, trees, animals and humans by breathing; and 3) stored in tree trunks, animal bodies, human bodies, and rocks and other inanimate objects.

Different types of areas cause different amounts of carbon storage. Forests with many trees can store large amounts of carbon, while grasslands or farms store less carbon.

The main reason the climate is changing is that human activities interfere with Earth processes and cycles that control the Earth's climate, such as the greenhouse effect and carbon cycle. The more CO2 emissions from human activities that alter the balance of the Earth's natural processes, this leads to global warming and climate change.

Forests and natural areas play a very important role in maintaining natural processes. Forests are one of the largest carbon sinks to help keep the carbon cycle and other natural processes running well and help mitigate climate change. However, forests can also be one of the largest sources of CO2 emissions. Because forests and other plants also absorb CO2 out of the atmosphere, this dual role makes forests more and more important. The earlier investigations say that destruction of forest caused by human is delivering carbon in atmosphere is between 12-17% of all CO2 around the world [7].

It can be concluded that climate change that causes deviations of rain intensity can affect the economy of a region, even the country. Climate change is affected by forests and crop areas. Result obtained through a mathematical model for global temperature, forest area, and seasonal rainfall show that the higher the forest cover, the less fluctuation there is between rainy-season and summer rainfalls. Moreover, growth in forest cover also correlates with an increase in summer rainfalls [8].

Therefore, it is important to estimate the behavior of rainfall in Indonesia. Some researches about rainfall prediction are done with Kalman Filter method. Kalman Filter is an algorithm to estimate the state variable of the stochastic dynamical linear system. The work are about implementing Kalman Filter analysis for modeling and forecasting rainfall in Semarang city with annual rainfall data in

Semarang form 2005-2012 [9] and the designed model from the relationship between rainfall and climate anomaly indicator to forecast rainfall condition in the future to support agriculture planning [10].

There are many modification of Kalman Filter algorithm. The modification has been done because we need faster computational time, lesser error, or better algorithm. One of modified algorithm in Kalman Filter is the Ensemble Kalman Filter. Ensemble Kalman Filter algorithm has been implemented in some researches. One of it is the groundwater pollution estimation in Surabaya, where the groundwater quality is influenced by industrial pollution [11]. Also, the Ensemble Kalman Filter has been used to estimate position of an autonomous underwater vehicle based on dynamical system of AUV motion [12].

In this work we use Ensemble Kalman Filter to estimate the rainfall behavior using a dynamical model that shows the relationship between temperature, forest area, and amount of rainfall. The paper is organized as follows: mathematics model of seasonal rainfall is studied in Section 2, discretization of the mathematics model is described in Section 3, implementation of Ensemble Kalman Filter for the model in Section 4, simulation is described in section 5, and finally, Section 6 concludes the result of the study.

2. Seasonal rainfall model

The change in global temperature, forest area, and seasonal rainfall is related. Here, we use $T = T(t)$ that represents global temperature at time t, $F = F(t)$ that represents forest cover area at time t, $R = R(t)$ that represents amount of rainfall at time t, and $S = S(t)$ that represents rate of change of rainfall. Our work is studying the mathematical model that represents a relationship about global temperature, forest area, and amount of seasonal rainfall.

The global temperature and forest cover are represented by logistic equation, not exponential equation. The global temperature is increased exponentially, while forest area is decreased exponentially [13]. However, they cannot increase or decrease forever. While exponential growth is unbounded, logistic growth is bounded. Therefore we assume logistic equations of global temperature and forest area are as follow:

$$\frac{dT}{dt} = a_T\left(1 - \frac{T - m_T}{b_T}\right)(T - m_T) \tag{1}$$

$$\frac{dF}{dt} = -a_F\left(1 - \frac{F - m_F}{b_F}\right)(F - m_F) \tag{2}$$

With a_T and a_F are increasing rate of global temperature and decreasing rate of forest area respectively, m_T and m_F area minimal global temperature and minimal forest area, respectively, where b_T and b_F are gaps between two equilibrium points.

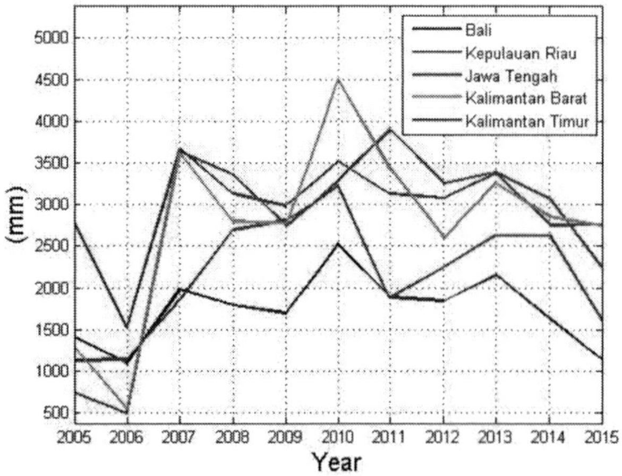

Figure 1: Annual rainfall amount in some area in Indonesia from 2005-2015

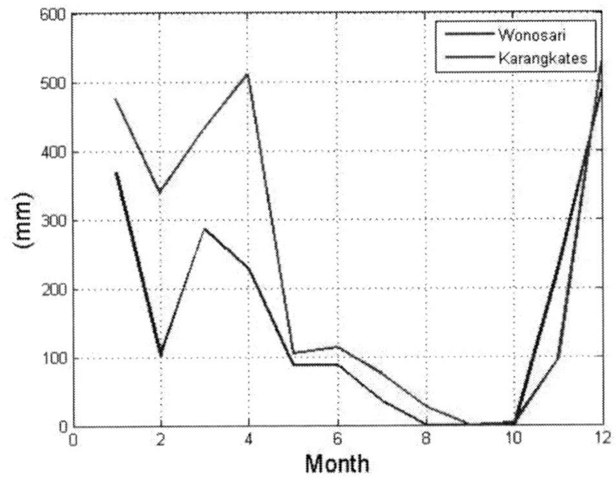

Figure 2: Monthly rainfall amounts in two spots in Malang (2014)

The amount of rainfall in Indonesia is seasonal and periodical (see Figure 1 and Figure 2) according to the data from Badan Pusat Statistik for the year 2000-2010[14], year 2011-2015[15] and Badan Meteorologi, Klimatologi, dan Geofisika [16]. Therefore the solution of second order differential equation should be periodic. Because the rest variables state can be represented by first order differential equations, the second order differential order is reduced to two first order differential equations. With S as the rate of change of rainfall, the periodic rainfall's amount is represented by the dynamical system as follows

$$\frac{dR}{dt} = S \tag{3}$$

$$\frac{dS}{dt} = -e_S^2(R - c) \tag{4}$$

With e_s and c are periodic of predicted solution and the coefficient of predicted solution shifting.

However, global temperature and amount of seasonal rainfall have correlation to other variables. Therefore, the model should show that the rate of change of each variable is affected by other variable. By that assumption, we add prey-predator term to the model to show the relationship. Here the new model is:

$$\frac{dT}{dt} = a_T\left(1 - \frac{T - m_T}{b_T}\right)(T - m_T) \tag{5}$$

$$\frac{dF}{dt} = -a_F\left(1 - \frac{F - m_F}{b_F}\right)(F - m_F) + e_F R + \alpha_F FR + \beta_F TF - \pi_F TR \tag{6}$$

$$\frac{dR}{dt} = f_R S - \alpha_R FR - \pi_R TR + \gamma_R FS + \mu_R TS - \delta_R RS \tag{7}$$

$$\frac{dS}{dt} = -e_S^2(R - c) + \beta_S TF + \pi_S TR - \gamma_S FS - \mu_S TS - \delta_S RS \tag{8}$$

The parameter e_F and f_R represent the rate of change of forest area caused by amount of rainfall absorption and increasing rate of rainfall's amount caused by rest of water in forest, respectively.

The term FR represents proportion of the amount of water in forest. The coefficients in term FR, α_F and α_R, indicate the growth rate of forest area and decreasing rate of rainfall caused by proportion of the amount of water in forest, respectively. While the term FS represents proportion of the rainfall difference change in the forest, we use the coefficients γ_R and γ_S both to show the difference rates of the amount of rainfall affected by the change of rainfall pattern in the forest.

IOP Conf. Series: Journal of Physics: Conf. Series **974** (2018) 012031 doi:10.1088/1742-6596/974/1/012031

The term TF represents proportion of the forest area at each temperature level. The coefficients for TF term, β_F and β_S indicate the difference rate of forest area and rainfall at temperature level, respectively. The term TR represents proportion of the rainfall at each temperature level, that's why the coefficients π_F and π_R both show the decreasing or increasing of rainfall's amount caused by proportion of the rainfall at each temperature level. While the term TS represents the proportion of rate of change of rainfall at each temperature level, the coefficients μ_R and μ_S both indicate the difference rates of the amount of rainfall affected by the change of rainfall pattern at each temperature level

The last term, RS, represents the correlation between amount of rainfall and rate of rainfall change. We use δ_R and δ_S to show the difference rate of the amount of rainfall and the rainfall difference rate affected by the relationship between the change in the pattern and the amount of rainfall, respectively.

3. Dicretization of model

Ensemble Kalman Filter algorithm can be implemented only for discrete system. First, the continuous model (5)-(8) must be discretized. We can use finite difference method to discretized the model. After the model discretizate with the finite forward difference method, the model can be written as follows

$$\begin{bmatrix} \dot{T} \\ \dot{F} \\ \dot{R} \\ \dot{S} \end{bmatrix}_{k+1} = \begin{bmatrix} a_T\left(1-\dfrac{T-m_T}{b_T}\right)(T-m_T)\Delta t + T \\ \left(-a_F\left(1-\dfrac{F-m_F}{b_F}\right)(F-m_F) + e_F R + \alpha_F FR + \beta_F TF - \pi_F TR\right)\Delta t + F \\ (f_R S - \alpha_R FR - \pi_R TF + \gamma_R FS + \mu_R TS - \delta_R RS)\Delta t + R \\ (-e_S^2(R-c) + \beta_S TF + \pi_S TR - \gamma_S FS - \mu_S TS - \delta_S RS)\Delta t + S \end{bmatrix}_k \quad (9)$$

The model in equations (5)-(8) used to understand the effect of global temperature and forest cover on seasonal rainfall. In fact, there is temperature, forest cover, and amount of seasonal rainfall that didn't match with the model, which was called by noise. Noise caused deterministic model turn into stochastic model. The equation (9) can be written in the form

$$x_{k+1} = f(x_k, u_k) + w_k \quad (10)$$
$$z_k = Hx_k + \varepsilon_k \quad (11)$$

With $f(x_k, u_k) + w_k$ is a nonlinear function defined in equation (9).

The system noise, w_k, and noise measurements, ε_k, are generated through the computer and are generally taken normally distributed and have mean value zero, the system noise variation is expressed by Q_k and the measurement noise variation is expressed by R_k. Both depend on time and the value assumed to be constant.

The matrix of square-shaped system noise variations corresponds to the size of the error covariance of the estimates. While the variance matrix size of the measurement noise is square-shaped in accordance with the number of rows from the observed H -vector.

The measurement vector H is determined from the state variable that is used as the measurement variable. In this case, it is assessed by the measurement variable that there are state variables T and R so that the measurement vector (H) is

$$H = \begin{bmatrix} 1 & 0 & 0 & 0 \\ 0 & 0 & 1 & 0 \end{bmatrix} \quad (12)$$

4. Assimilation method

The assimilation method is varied. In this work, we use Ensemble Kalman filter method. Ensemble Kalman filter is estimation method for nonlinear dynamical stochastic system. This method used ensemble value as initial estimation for state variable and measurement data based on real data. This is EnKF steps shown as follow [17]:

First Step: Initialization

First generation of ensembles is generated in accordance with the initial value for each state by providing system noise.

$$x_{0,i} = [x_{0,1} \quad x_{0,2} \quad x_{0,3} \quad \cdots \quad x_{0,N}]$$

Next is to find the average value of each state of the ensemble generation.

$$\hat{x}_0 = \frac{1}{N}\sum_{i=0}^{N} x_{0,i}$$

Second Step: Prediction

We get prediction value from $\hat{x}_k$ from the previous step by adding noise system w_k

$$\hat{x}_{k,i}^- = f(\hat{X}_{k-1}, u_{k-1}) + w_{k,i} \text{ for } i = 1,2,\ldots,N$$

Mean value from the estimation is

$$\hat{x}_k^- = \frac{1}{N}\sum_{i=0}^{N} \hat{X}_{k,i}^-$$

With error covariance

$$P_k^- = \frac{1}{N-1}\sum_{i=1}^{N} (\hat{x}_{k,i}^- - \hat{x}_k^-)(\hat{x}_{k,i}^- - \hat{x}_k^-)^T$$

Third Step: Correction

Ensemble from measurement data as follow:

With $v_{k,i} \sim N(0, R_k)$ is ensemble from measurement noise.

Kalman Gain:

$$K_k = P_k^- H^T (H P_k^- H^T + R_k)^{-1}$$

Ensemble estimation:

$$\hat{x}_{k,i} = \hat{x}_{k,i}^- + K_k(z_{k,i} - H\hat{x}_{k,i}^-)$$

Estimation mean:

$$\hat{x}_k = \frac{1}{N}\sum_{i=1}^{N} \hat{x}_{k,i}$$

Error covariance:

$$P_k = [1 - K_k H]P_k^-$$

5. Simulation of the model

In this section, the model is simulated with given noise value in state space (Q) and noise measurement. Noise in the state space for variable T is 0.001. For state variable F, given noise is 0.01. While the state variables R and S, the given noises are both 0.01. Generally, the noise on the system state and noise on the measurement are as follows:

$$Q = \begin{bmatrix} 0.001 & 0 & 0 & 0 \\ 0 & 0.01 & 0 & 0 \\ 0 & 0 & 0.01 & 0 \\ 0 & 0 & 0 & 0.01 \end{bmatrix}, R = \begin{bmatrix} 0.001 & 0 \\ 0 & 0.01 \end{bmatrix}$$

The parameter values used are as follows:

$a_T = 0.021294$; $e_F = 0.000104$; $\gamma_R = 0.0024442$; $\pi_F = 0.000998$;
$a_F = 0.107198$; $e_S = 6.284240$; $\gamma_S = 0.000087$; $\pi_R = 0.000099$;
$b_T = 0.609899$; $\alpha_F = 0.012701$; $\mu_R = 0.001440$; $\pi_S = 0.357472$;
$b_F = 0.591300$; $\alpha_R = 0.000998$; $\mu_S = 0.000193$; $f_R = 1.003292$;
$m_T = 0.303309$; $\beta_F = 0.001001$; $\delta_R = 0.002698$; $c = 0.421764$;
$m_F = 0.314800$; $\beta_S = 0.009999$; $\delta_S = 0.000090$;

The original unit of global temperature is degree Celcius with the maximum value of 45°C, the maximum body temperature that normal human can tolerate [18]. The unit of the forest cover is squared kilometers and maximum is set to region's total area, Kabupaten Malang, 3535 km^2 [19]. The maximum possible monthly rainfall in the region is 700 mm making millimeter the unit of the amount of rainfall according to the data in 2014 from Badan Meteorologi, Klimatologi, and Geofisika Malang (see Figure. 2) [16].

For this simulation, the initial value when the initial time is January 2014. We use monthly temperature and rainfall's amount data from Karangkates, Malang. The temperature of Karangkates is 25.5 °C, the forest area of Malang is 110,494.66 ha or 1104,946.6 km^2 according to data from Dinas Kehutanan Provinsi Jawa Timur [20], and amount of rainfall is 369 mm. The time change value (Δt) used is $\Delta t = 0.0000001$. Estimated rainfall is done as much as 100 iterations. Rainfall estimates use different number of ensembles, 100, 200, and 500 ensembles.

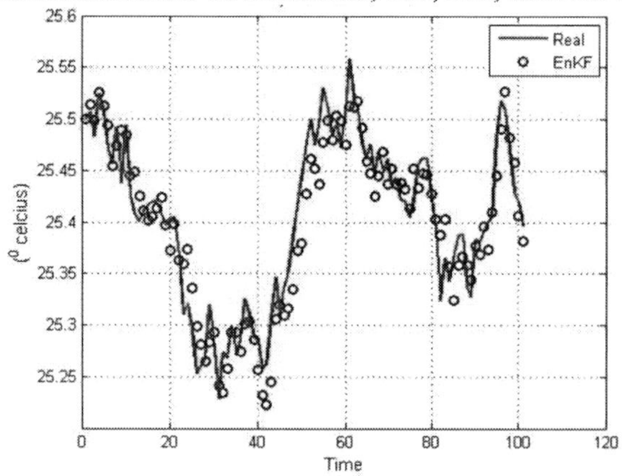

Figure 3. Temperature estimation

Figure 3 shows real and estimation graphic of global temperature in Kabupaten Malang until 100th time with Ensemble Kalman filter method. It shows the fluctuation of temperature in the range of 25.2°C to 25.6°C. The graphic shows that distance between real value and estimation value is small, so the RMSE value is small too.

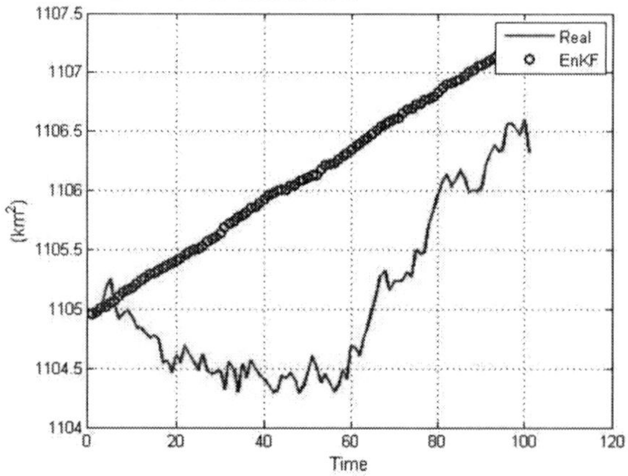

Figure 4. Forest cover estimation

Figure 4 shows real and estimation graphic of forest cover in Kabupaten Malang until 100th time with Ensemble Kalman filter method. It shows that the forest area will be wider from 1106,2 km^2 to 1107.2 km^2. The distance between real and EnKF is far enough (about 1 km^2) so we can see that RMSE value is big.

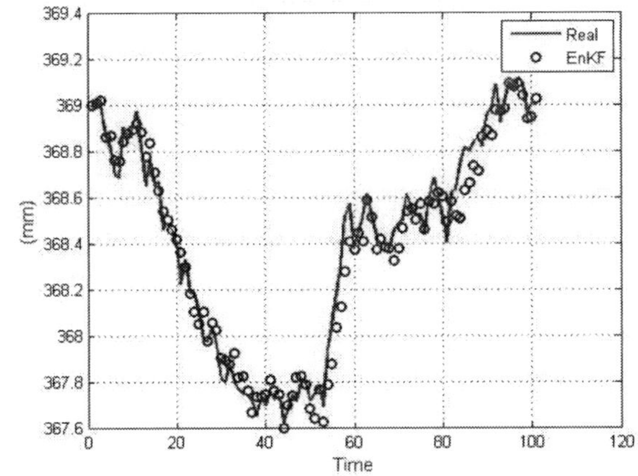

Figure 5. Seasonal rainfall estimation

Figure 5 shows real and estimation graphic of seasonal rainfall in Kabupaten Malang until 100th time with Ensemble Kalman filter method. It shows the fluctuation of seasonal rainfall. The graphic shows that distance between real value and estimation value is small, so the RMSE value is small too.

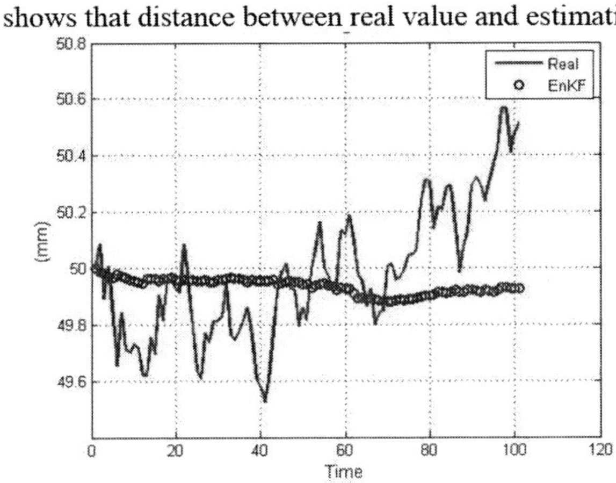

Figure 6. Rainfall change estimation

Figure 6 shows real and estimation graphic of rainfall change in Kabupaten Malang until 100th time with Ensemble Kalman filter method. It shows that the rainfall change is fluctuated. The distance between real and EnKF is far enough so the RMSE value is big.

Table 1. RMSE value with T and R as measurements variable

State	T and R as measurements data		
	100 ensembles	200 ensembles	500 ensembles
T	0.029403	0.029012	0.026497
F	0.81841	0.61396	0.60477
R	0.084367	0.083865	0.080612
S	0.40064	0.64698	0.33419

For every state variable using $N = 500$, RMSE value in the simulation is lower than RMSE value using $N = 100$ and 200. We can conclude that estimation with 500 ensembles is better than 100 and 200 ensembles. In other way, RMSE value in state variables F and S is higher than RMSE value in state variable *R* although the given noise value is the same. This is because in this simulation, variables *F* and *S* are not used as measurement data.

6. Conclusion and discussion

In this work, TFR model is estimated with Ensemble Kalman filter. The investigation brings us to the conclusion as follow:

1. The result of simulation with more ensembles gives better result than few one.
2. The accurateness of simulation result is influenced by measurement variable. If a variable is measurement data, the result of simulation is better. This is shown from short distance of real and estimation value in graphic or low RMSE value.

References

[1] Library of Congress, (2004). Indonesia. Washington, DC: Federal Research Division. https://www.loc.gov/rr/frd/cs/profiles/Indonesia.pdf.

[2] Kishore, K., Subbiah, A.R., Sribimawati, T., Diharto, I.S., Alimoeso, S., Rogers, P., Setiana, A., (2000). *Indonesia Country Study*. http://www.unu.edu/env/govern/ElNIno/CountryReports/pdf/indonesia.pdf.

[3] Levine, D.I., Yang, D., (2006). *A Note on the Impact of Local Rainfall on Rice Output in Indonesian Districts*. <https://www.researchgate.net/profile/David_Levine6/publication/228813081_A_Note_on_the_Impact_of_Local_Rainfall_on_Rice_Output_in_Indonesian_Districts/links/53eeca840cf2711e0c42de90/A-Note-on-the-Impact-of-Local-Rainfall-on-Rice-Output-in-Indonesian-Districts.pdf> (accessed 25.10.17).

[4] Herdt, R.W., (1989). *The Impact of Technology and Policy on Seasonal Household Food Security in Asia. in seasonal Variability in Third World Agriculture: The consequences for food security*, ed. D. E. Sahn, 235–45. Baltimore: Johns Hopkins University Press for the International Food Policy Research Institute.

[5] Maccini, S., Yang, D., (2009). *Under the Weather: Health, Schooling, and Economic Consequences of Early-Life Rainfall*. American Economic Review 99:3, 1006–1026

[6] Stone, S., Leon, M.C., Fredericks, P., (2010). *Perubahan Iklim dan Peran Hutan: Manual Komunitas*. Conservation International. ISBN: 978-1-934-151-41-9.

[7] Van der Werf, G.R., Morton, D.C., DeFries, R.S., Olivier, J. G. J., Kasibhatla, P.S., Jackson, R.B., Collatz, G. J. and Randerson, J. T., (2009). *CO_2 Emissions from Forest Loss*. Nature Geoscience. Vol. 2, November 2009.

[8] Duangdai, E., Likasiri, C., (2017). *Rainfall model investigation and scenario analyses of the effect of government reforestation policy on seasonal rainfall: A case study from Northern Thailand*. ScienceDirect Journal, Vol 185, Page 1-12.

[9] Mirawati, T.D., Yasin, H., Rusgiyono, A., (2013). *Prediksi Curah Hujan di Kota Semarang dengan Metode Kalman Filter*. Prosiding Seminar Nasional Statistika. Universitas Diponegoro. ISBN: 978-602-14387-0-1.

[10] Estiningtyas, W., dan Amien, L.I., (2006). *Pengembangan Model Prediksi Hujan dengan Metode Filter Kalman untuk Menyusun Skenario Masa Tanam. Badan Penelitian dan Pengembangan Pertanian*, Departemen Pertanian, ISSN 1907-0799. Vol 1 No 2, Page 1-12.

[11] Apriliani, E., Sanjoyo, B.A., Adzkiya, D., (2011). *The groundwater pollution estimation by the Ensemble Kalman Filter*. Canadian Journal on Science and Engineering Mathematics. Vol. 2 No. 2, June 2011.

[12] Ngatini, Apriliani, E., Nurhadi, H., (2017). *Ensemble and Fuzzy Kalman Filter for position estimation of an autonomous underwater vehicle based on dynamical system of AUV motion*. Expert Systems with Applications: An international Journal. Vol. 68 Issue C, February 2017. Page 29-35.

[13] Likasiri, C., Duangdai, E., Pongvuthithum, R., (2014). *Mathematical model on the effect of*

global climate change and decreasing forest cover on seasonal rainfall in Northern Thailand. Ecol, Model. 272, 388-393.

[14] Badan Pusat Statistik. (2015). *Jumlah Hujan dan Jumlah Pengamatan Hari Hujan di Stasiun Pengamatan BMKG, 2000-2010*. <https://www.bps.go.id/linkTabelStatis/view/id/1349> (accessed 25.10.17).

[15] Badan Pusat Statistik. (2015). *Jumlah Hujan dan Jumlah Pengamatan Hari Hujan di Stasiun Pengamatan BMKG, 2011-2015*. < https://www.bps.go.id/linkTabelStatis/view/id/1959> (accessed 25.10.17)

[16] Badan Meteorologi, Klimatologi, dan Geofisika. (2017). *Data Iklim Tahun 2014. Badan Meteorologi, Klimatologi, dan Geofisika Malang: Stasiun Perkebunan Wonosari dan Karangkates*.

[17] Evensen, G., (2009). *Data Assimilation: The Ensemble Kalman Filter 2nd Edition*, Springer-Verlag Berlin Heidelberg, New York.

[18] Sherwood, S.C., Huber, M., (2010). *An adaptability limit to change due to heat stress*, PNAS 107, 9552-9555.

[19] Kementerian Dalam Negeri, (2017). <http://www.kemendagri.go.id/pages/profil-daerah/kabupaten/id/35/name/jawa-timur/detail/3507/malang> (accessed 25.10.17)

[20] Dinas Kehutanan Provinsi Jawa Timur. (2015). *Statistik Dinas Kehutanan Provinsi Jawa Timur Tahun 2011– 2015*. <http://dishut.jatimprov.go.id/gambar/statistika/stat_2015_bab_1.pdf> (accessed 25.10.17)

The modification of hybrid method of ant colony optimization, particle swarm optimization and 3-OPT algorithm in traveling salesman problem

Hertono. G.F.[1], Ubadah[1], Handari, B.D.[1]

[1] Department of Mathematics, Universitas Indonesia, Depok, 16424, Indonesia

gatot-f1@sci.ui.ac.id, ubadah@sci.ui.ac.id, bevina@sci.ui.ac.id

Abstract. The traveling salesman problem (TSP) is a famous problem in finding the shortest tour to visit every vertex exactly once, except the first vertex, given a set of vertices. This paper discusses three modification methods to solve TSP by combining Ant Colony Optimization (ACO), Particle Swarm Optimization (PSO) and 3-Opt Algorithm. The ACO is used to find the solution of TSP, in which the PSO is implemented to find the best value of parameters α and β that are used in ACO.In order to reduce the total of tour length from the feasible solution obtained by ACO, then the 3-Opt will be used. In the first modification, the 3-Opt is used to reduce the total tour length from the feasible solutions obtained at each iteration, meanwhile, as the second modification, 3-Opt is used to reduce the total tour length from the entire solution obtained at every iteration. In the third modification, 3-Opt is used to reduce the total tour length from different solutions obtained at each iteration. Results are tested using 6 benchmark problems taken from TSPLIB by calculating the relative error to the best known solution as well as the running time. Among those modifications, only the second and third modification give satisfactory results except the second one needs more execution time compare to the third modifications.

1. Introduction

Traveling salesman problem (TSP) is known as a problem to determine the salesman's route visiting some places (cities) that start and end in a certain city. Furthermore, a condition is applied that every city must be visited once (except the initial city). The chosen route must have the minimal distance (Gupta, 2013).

Some methods have been developed to solve the TSP, and one of them is the hybrid method proposed by Mahi et al. (2015), that is based on Ant Colony Optimization (ACO) (Dorigo, 1996, Deneubourg and Goss, 1989), Particle Swarm Optimization (PSO) (Kennedy, 1995, Reynold, 1987, Heppner and Grenander, 1990) and 3-Opt Algorithm.

In this paper, we modified the method proposed by Mahi et al with three different strategies. The difference among those modifications is lay on the usage of 3-Opt algorithm during the process. In the first modification method, the 3-Opt algorithm is used to reduce the total distance of the feasible solution that is obtained from each iteration. Meanwhile, in the second modification, the 3-Opt algorithm is used to reduce the total distance of all solutions that are obtained in each iteration. In the third modification, the 3-Opt algorithm is used to reduce the total distance of different solutions that are obtained in each iteration. In order to see the performance of those modifications, 6 benchmark problems from TSPLIB are used. The relative error of the mean solution to the best known solution will be compared as well as the related running time.

2. Preliminaries

In [6], Mahi's et al proposed the hybrid method that is described in the scheme as shown in Figure 1.

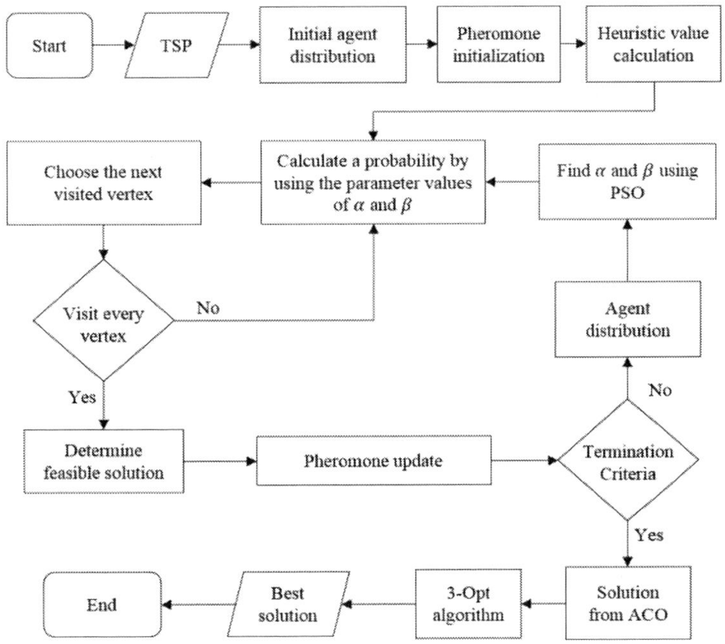

Figure 1. Mahi's Hybrid Method.

The hybrid method starts by implementing the ACO steps, which consist of: the initial agent distribution process, pheromone initialization and heuristic value calculation. In order to create solutions, the second step of ACO will calculate a probability by using the parameter values of α and β. An agent will visit the next vertex based on this probability value. This step will be repeated until every agent has visited every vertex. The feasible solution, which is the route with the minimum total distance will be determined from all agent's route. The third step is used to update the pheromone of the feasible solution. The ACO process will be ended by termination criteria. If the criteria do not meet the termination condition, the process will be continued by redistributing agents to the vertices. In this phase, the PSO will be used to determine the new parameter values α and β that will be used in the next ACO processes. The whole process will be repeated until the termination criteria is met and the candidate solution can be obtained from all feasible solutions. This candidate solutions will be optimized by 3-Opt algorithm to obtain the best solution.

3. Discussion

In Mahi's, 3-Opt algorithm is applied to the candidate solution. In our first modification method, 3-Opt algorithm is applied to the feasible solution, as described in Figure 2. The differences can be seen after every agent has visited every vertex (visitation complete). The feasible solution will be chosen from all solutions generated by all agents. The 3-Opt algorithm will be applied to these feasible solutions. The next step is the third ACO process that updates the pheromone and followed by the same steps as Mahi's until the process satisfies the terminal criteria. When the criteria is fulfilled, the best of the feasible solution will be the best solution of the first modification hybrid method.

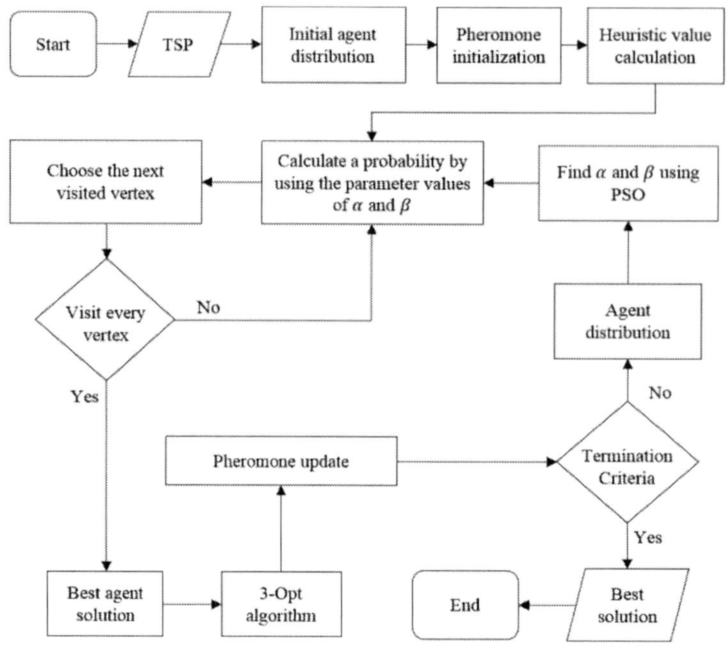

Figure 2. First Modification of Hybrid Method.

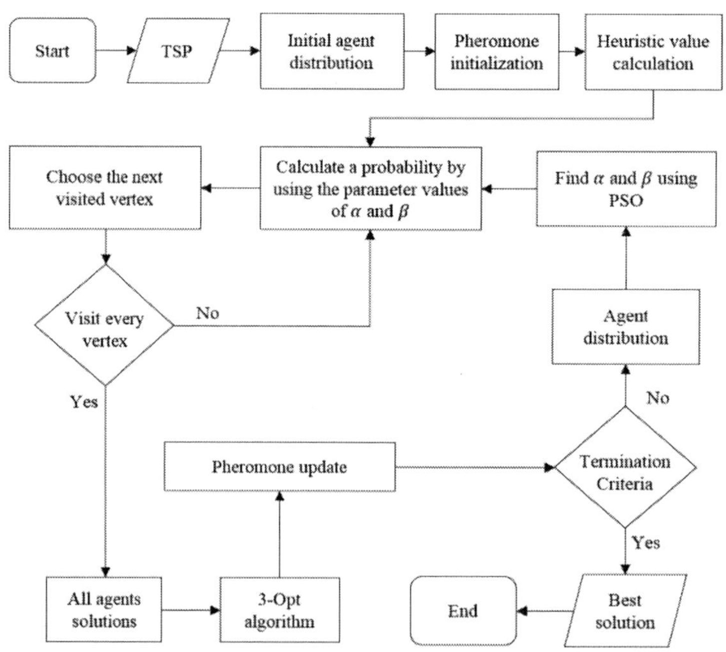

Figure 3. Second Modification of Hybrid Method.

In the second hybrid modification method, we implement the 3-Opt algorithm to all agent's solution in each iteration as shown in Figure 3. In the first modification method, the 3-Opt algorithm is only applied to the feasible solution in each ACO iteration, while in the second modification method, the 3-Opt algorithm is applied to all agent's solution in each iteration. After all solutions have been optimized by 3-Opt, the solution with the minimum total distance is selected as a feasible solution.

The next step is the third step of ACO (update the pheromone) and testing criteria. The process will continue until the process satisfies the terminal criteria and the best solution is obtained.

The third hybrid modification method is described as shown in Figure 4:

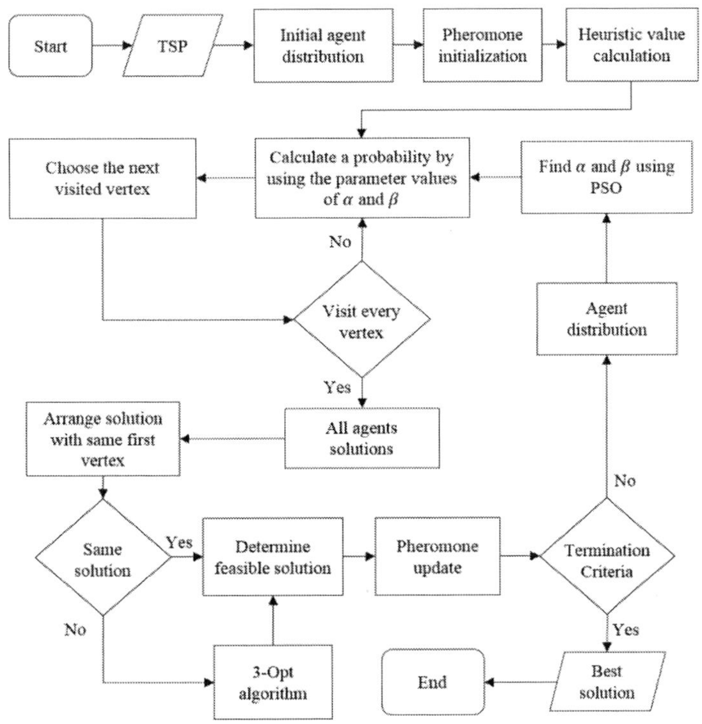

Figure 4. Third Modification of Hybrid Method.

In this modification, we only implement the 3-Opt algorithm to solutions which are different from solutions obtained in the previous iterations. The idea of this strategy is since some agents may produce the same solutions which represent the same route even though they have different vertex order. The checking step can be done by setting one vertex as the initial vertex and put the other vertices in a corresponding order. Hence, the route of each agent will be obtained with the same initial vertex. When one route has been applied the 3-Opt algorithm, and if there is other route with the same vertex's order exists then the corresponding route will be skipped from the 3-Opt algorithm. In other words, the 3-Opt algorithm only applied to the routes with the different vertex's order. Hence, this strategy will reduce the running time. After all different routes have been optimized, the solution is chosen to be the feasible solution and the process continue with the third ACO process. The whole process will be repeated until the termination criteria is held and the best solution is obtained.

The implementation uses 6 benchmarks problems from TSPLIB (Reinelt, 1991): Eil51, Berlin52, St70, Eil76, Rat99, and kroA200. Every problem is tested in 10 runs with 50 iterations each. Mean solution (MS) is calculated by using the following equation:

$$MS = \frac{\sum_{i=1}^{n} D_i}{n}, \quad (1)$$

where D_i is the solution of problem i, $i = 1,2,\ldots,n$. The percentage of error relative (ER) is used to determine how good the method solving the TSP and calculated by equation:

$$ER = \frac{MS-BKS}{BKS} \times 100, \quad (2)$$

where BKS is the best known solution.

In the implementation, we also use the parameters value as suggested in the previous papers, as follows: $\rho = 0.1$, $c_1 = 2$, $c_2 = 2$ (Mahi, 2015) and $Q = 100$ as the best constant value for the ACO method (Dorigo, 1996). We also use $0 \leq \alpha \leq 2$ and $0 \leq \beta \leq 2$ as suggested by Mahi (2015) to get the best result for the ACO method. Mahi (2015) also suggests using 10 agents when we use the hybrid method using ACO, PSO and 3-Opt algorithm.

Table 1. The results of the first hybrid modification method.

Problems	BKS	Best Solution	Worst Solution	Mean Solution	ER (%)	Time (second)
Eil51	426	437	514	480.8	12.86	325.55
Berlin52	7542	7930	8832	8351.6	10.73	328.27
St70	675	683	777	732.8	8.56	427.28
Eil76	538	605	663	634.4	17.92	488.64
Rat99	1211	1311	1499	1439.6	18.88	818.51
KroA200	29368	29957	36761	34039.4	15.91	1042.13

Table 1 shows that the first hybrid modification method does not give good results for all problems. It shows the corresponding best solution is larger than the best known solution (BKS) in Mahi (2015). Furthermore, the corresponding ER is also greater than 8%. The results also show high running time.

Table 2. The results of the second hybrid modification method.

Problems	BKS	Best Solution	Worst Solution	Mean Solution	ER(%)	Time (second)
Eil51	426	426	426	426	0.00	658.43
Berlin52	7542	7542	7542	7542	0.00	547.11
St70	675	675	675	675	0.00	1138.9
Eil76	538	538	538	538	0.00	1413.8
Rat99	1211	1211	1211	1211	0.00	3941.3
KroA200	29368	29368	29368	29368	0.00	7261.4

The second hybrid modification method gives better results for every problem used. Table 2 shows that the worst solutions exactly match the BKS for every problem. However, the running time that represents the mean time needed to solve problem in one trial is greater than the one of the corresponding problems in the first hybrid modification method. The hypothesis is that it causes too many executions of 3-Opt algorithm in the method. This 3-Opt algorithm works for every solution generated by every agent in each iteration. Since there are 10 agents and each runs in 50 iterations there are 500 executions 3-Opt algorithm.

In Table 3, the third modification method relatively gives a better result than the previous methods. The difference of the relative error is not more than 0.1% compared to the second one. Fortunately, the mean time needed to execute one trial is also less than the corresponding time that needed to solve one trial of the second modification problems, eq. problem Rat99 needs 3941.3 seconds in second hybrid modification and it needs 1103.3 seconds in the third modification.

Table 3. The result of the third hybrid modification method.

Problems	BKS	Best Solution	Worst Solution	Mean Solution	ER(%)	Time (second)
Eil51	426	426	428	426.4	0.09	408.82
Berlin52	7542	7542	7542	7542	0.00	450.72
St70	675	675	676	675.2	0.03	526.51
Eil76	538	538	539	538.2	0.04	570.46
Rat99	1211	1211	1212	1211.8	0.07	1103.3
KroA200	29368	29368	29648	29424	0.19	1405.71

In Table 4, we try with a larger number of iterations with two TSPLIB problems, Ei51 (problem with 51 vertices) and KroA200 (problem with 200 vertices). With 5 runs and 100 iterations each, the first modification of hybrid method has a significant reduction in percentage of relative error by increasing the number of iterations.

Table 4. The Comparison Results of Three Methods on Ei51 and KroA200 with 50 and 100 iterations.

Problems	Iterations	ER(%)			Time(Second)		
		1	2	3	1	2	3
Eil51	50	12,28	0,00	0,13	304.04	655,01	408,00
	100	7,04	0,00	0,00	533,54	1004,98	690,95
KroA200	50	20,17	0,00	0,22	1338,13	7371,02	1801,71
	100	12,01	0,00	0,09	1985,14	13292,01	2915,16

4. Conclusion

Based on six TSPLIB problems used with 10 agents, with 10 runs in 50 iterations each, we can see that the third hybrid modification gives the best result in terms of relative error and running time. The second modification gives a good relative error but needs more time for execution. The performance of the first hybrid modification method can be improved by increasing the number of iterations.

Acknowledgments
This research is partially funded by Cluster research UI 2015 Grant.

References

[1] Deneubourg J L and Goss S 1989 Collective patterns and decision-making *Ethology Ecology & Evolution* **1** pp 295-311.

[2] Dorigo M, Maniezzo V and Colorni A 1996 The Ant System: Optimization by a colony of cooperating agents. *IEEE Transactions on Systems, Man, and Cybernetics–Part B* 26 **1** pp 1-13.

[3] Gupta S and Panwar P 2013 Solving Travelling Salesman Problem Using Genetic Algorithm *International Journal of Advanced Research in Computer Science and Software Engineering* 3 **6** pp 376-380.

[4] Heppner F and Grenander U 1990 A stochastic nonlinear model for coordinated bird flocks *The Ubiquity of Chaos* (Washington DC: AAAS Publications).

[5] Kennedy J and Eberhart R 1995 Particle swarm optimization *1995 IEEE International Conference on Neural Networks Proceedings* **4** pp 1942–1948.

[6] Mahi M, Baykan Ö K and Kodaz H 2015 A new hybrid method based on Particle Swarm Optimization, Ant Colony Optimization and 3-Opt algorithms for Traveling Salesman Problem *Applied Soft Computing* **30** pp 484–490.
[7] Reinelt G 1991 TSPLIB—A Traveling Salesman Problem *ORSA Journal on Computing* 3 **4** pp 376-384.

IOP Conf. Series: Journal of Physics: Conf. Series **974** (2018) 012033 doi:10.1088/1742-6596/974/1/012033

K-means cluster analysis of tourist destination in special region of Yogyakarta using spatial approach and social network analysis (a case study: post of @explorejogja instagram account in 2016)

N Iswandhani[1], M Muhajir[2]

[1,2]Department of Statistics Islamic University of Indonesia, Sleman, Yogyakarta, Indonesia, 55584.

E-mail: [1]13611066@students.uii.ac.id , [2]mmuhajir@uii.ac.id

Abstract. This research was conducted in Department of Statistics Islamic University of Indonesia. The data used are primary data obtained by post @explorejogja instagram account from January until December 2016. In the @explorejogja instagram account found many tourist destinations that can be visited by tourists both in the country and abroad, Therefore it is necessary to form a cluster of existing tourist destinations based on the number of likes from user instagram assumed as the most popular. The purpose of this research is to know the most popular distribution of tourist spot, the cluster formation of tourist destinations, and central popularity of tourist destinations based on @explorejogja instagram account in 2016.Statistical analysis used is descriptive statistics, k-means clustering, and social network analysis. The results of this research were obtained the top 10 most popular destinations in Yogyakarta, map of html-based tourist destination distribution consisting of 121 tourist destination points, formed 3 clusters each consisting of cluster 1 with 52 destinations, cluster 2 with 9 destinations and cluster 3 with 60 destinations, and Central popularity of tourist destinations in the special region of Yogyakarta by district.

1. Introduction

The development of internet technology continues to progress very rapidly. With the advent of various kinds of facilities offered the internet makes users more access to information. The development of the internet that is rife among the public is social media.

One of the most popular social media is instagram. Instagram is a social media containing photo content, information media and references. Instagram users are now more than 50 million inhabitants thus making Instagram occupy the 4th position of most users for social media category [1].

Instagram contains about a variety of account content, ranging from online shop accounts that sell goods and services, campus organization accounts and also instagram accounts that promote tourism in a country or city.

Tourism is a journey from one place to another, temporary, done by individuals or groups, in an attempt to find balance or harmony and happiness with the environment in the social, cultural, natural and scientific dimension [2].

Not a few tourists who seek ideas through Facebook, Twitter, and other social networking. 65% of travelers are looking for travel ideas through social search. 52% Facebook users are heavily influenced by the photos of friends in his Facebook network to determine the sights. 33% of travelers changed their original plans after viewing the photos [3]. @explorejogja Instagram account serves as a tourism promotion tool for Yogyakarta area, with Follower 510k (Five hundred ten thousand) they are able to promote or introduce new and old tourist spot to people outside Yogyakarta and Yogyakarta society itself.

Popularity comes from the popular word, meaning known and liked by the crowd [4]. In the Indonesian General Dictionary of popularity means the fame of a person [4].

Yogyakarta is a province with a variety of tourist destinations. Ranging from mountains, waterfalls, zoos to complete beaches are available in the region. Newer attractions will soon become booming as many social media users share when visiting the place [5]. With the increasing number of new places, there is a need for facilities that can summarize information from those places..

The tourism sector is a potential sector to be developed as one source of local revenue. For that in need of way for tourist destinations in the province of Yogyakarta can be known by the public, one of them with the media publications and information.

One solution to be able to display information in the right format is a dynamic map that developed a web-based application that is able to manage and display tourist destination information. This app is expected to address some of the limitations of the manual map. Maps used with features owned by Google Map. With the development of the dynamic map is then realized a geographic information system with the theme of the spread of tourist destinations in yogyakarta [6].

. In the @explorejogja instagram account found many tourist destinations that can be visited by tourists both in the country and abroad, Therefore it is necessary to form a cluster of existing tourist destinations based on the number of likes from user instagram assumed as the most popular.

Cluster analysis is one type of problem in data mining. Cluster analysis in data mining (also known as clustering) is a method used to divide the data set into clusters based on similarities that have been determined [7]. The object of this research is the number of likes of post account of Instagram of Explorejogja relating to tourist destinations in Yogyakarta province in 2016, to categorize the object of the researcher using K-Means method.

Based on post account instagram @explorejogja get the data name of tourist destinations and also the location of tourist destinations (in this case the city), then will be searched the popularity of tourist destinations that most many get like from instagram users who see posts from account instagram @explorejogja based on city location tourist destination . The existence of this popularity center will be determined through graphs on Social Network Analysis (SNA). SNA is one of the analysis in data mining that connects several interrelated objects through graph [8].

2. Basic theory

2.1. Spatial analysis

Spatial analysis is a visual inference to a map that is a combination of spatial data and attribute data. Spatial data refers to a location or position on the surface of the earth, which is the coordinates, raster or administrative boundaries of the region.

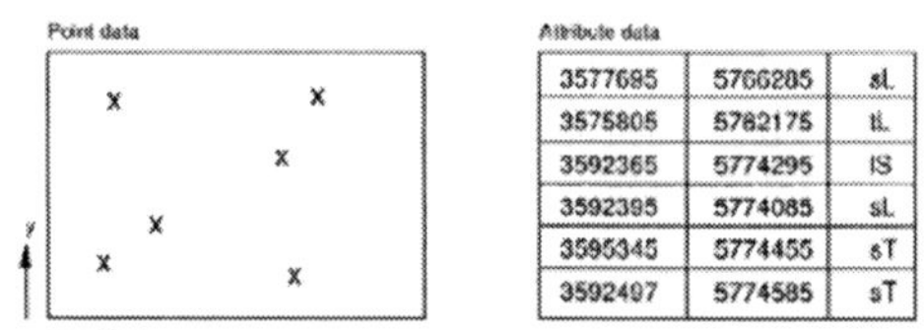

Figure 1. Differences of spatial data and attribute data [9].

2.2. Data mining

Data mining is the mining or discovery of new information by searching for a particular pattern or rule from a large amount of data [10]. Data mining is also called a series of processes to explore the added value of knowledge that so far can not be known manually from a data collection [11]. Data mining has stages like in figure 2.

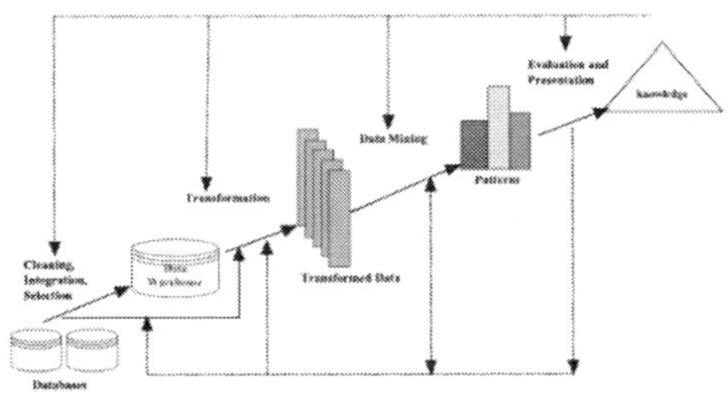

Figure 2. Data mining process [12].

2.3 K-means

Clustering is used to create a group (cluster) of the data so that it can easily find the necessary data. Clustering is a classification of similar objects into several different groups, it is usually applied in the analysis of statistical data which can be utilized in various fields, for example, machine learning, data mining, pattern recognition, image analysis and bioinformatics [13].

Clustering including supervised learning types. There are four types of clustering algorithms that have been compared based on performance, such as K-Means, hierarchical clustering, self-organization map (SOM) and expectation maximization (EM Clustering). Based on these test results can be concluded that the k-means algorithm performance and EM better than a hierarchical clustering algorithm. In general, partitioning algorithms such as K-Means and EM highly recommended for use in large-size data. This is different from a hierarchical clustering algorithm that has good performance when they are used in small size data [14].

The method of K-means algorithm as follows [15] :

1) Determine the number of clusters k as in shape. To determine the number of clusters K was done with some consideration as theoretical and conceptual considerations that may be proposed to determine how many clusters.
2) Generate K centroid (the center point of the cluster) beginning at random. Determination of initial centroid done at random from objects provided as K cluster, then to calculate the i cluster centroid next, use the following formula:

$$v = \frac{\sum_{i=1}^{n} x_i \,; i=1,2,\ldots,n}{n} \quad (1)$$

v : cluster centroid
n : the number of objects to members of the cluster
xi: the object to-i

3) Calculate the distance of each object to each centroid of each cluster. To calculate the distance between the object with the centroid author using Euclidian Distance.

$$d(x,y) = \|x-y\| = \sqrt{\sum_{j=1}^{n}(x_i - y_i)^2} \quad (2)$$

xi: object x to-i
yi: object y to-i
n : the number of object

4) Allocate each object into the nearest centroid. To perform the allocation of objects into each cluster during the iteration can generally be done in two ways, with a hard K-means, where it is explicitly every object is declared as a member of the cluster by measuring the distance of the proximity of nature towards the center point of the cluster, another way to do with fuzzy C-Means.
5) Do iteration, then specify a new centroid position using equation (1).

6) Repeat step 3 if the new centroid position is not the same.

2.4. Social Network Analysis

Social Network Analysis (SNA) is one of the analysis in data mining that connects several interrelated objects through graph. Objects in SNA called actor terms are the main focus in this analysis.

There are two types of relationships that can be explained in SNA, namely [16]:

1) *Directional Relations*: the type of relationship "self choices" that the relationship that occurs between actors is the choice of each actor and does not apply to each other in opposite, eg friendship relationship between A and B. If A recognizes B as a friend then not B will recognize A as his friend.
X: friendship sosiomatriks A, B and C. If it is known that A is friends with B (A B), B be friends with C (B → C), C be friends with A (C → A), dan C be friends with B (C → B), If a relationship is dichotomous then the element of the matrix **X** ($\boldsymbol{X_{ij}}$), with i= A,B,C and j= A,B,C are:

$$\boldsymbol{X} = \begin{matrix} - & \mathbf{1} & \mathbf{0} \\ \mathbf{0} & - & \mathbf{1} \\ \mathbf{1} & \mathbf{1} & - \end{matrix} \tag{3}$$

2) Nondirectional Relations: the type of relationship between actors symmetrical each other. Example of this relationship is the neighboring neighborhood. If A-neighbor is next door with B then it is definitely B after the house again with A. This type of relationship will be denoted by a line (without arrows) on the sosiogram. In the form of matrix notation, this relationship can be described as follows:
X: sosiomatriks next door neighbor A, B and C. If it is known that A next door neighbor with B (A-B and BA), adjacent neighbor B with C (B-C and C-B). If a relationship is dichotomous then the element of the X matrix ($\boldsymbol{X_{ij}}$), with i=A,B,C and j= A,B,C are:

$$\boldsymbol{X} = \begin{matrix} - & \mathbf{1} & \mathbf{0} \\ \mathbf{1} & - & \mathbf{1} \\ \mathbf{0} & \mathbf{1} & - \end{matrix} \tag{4}$$

In SNA, the relationships between actors can be valuable dichotomes and have value. The relationship is dichotomous if the relation exists then it is worth 1 and if there is no relationship will be worth 0. The relation can also be valuable so that each relationship between actors has different values, it can be worth the strength of the relationship between the actor, the intensity of the relationship or the frequency of the relationship.

3. Research Methods

The first step in conducting this research is data collection, based on previous research conducted by researchers to update data so that get 313 pots of tourism destinations from @explorejogja then get 121 points of tourist destination location by collecting the point of coordinat based on lattitude and longitude of each tourist destinations, descriptive statistical analysis, create a map of web-based tourism destination distribution, Cluster k-means analysis, and create Social network analysis graphs.

Data analysis method used in this research is descriptive analysis, tourism destination distribution in the form of web based map, non hierarchy cluster analysis that is K-means method and Social network analysis. The tools used in data analysis in this study are Ms. Excel, QGIS 2.10, R, and gephi.

4. Result

4.1. Distribution of tourist destination

Instagram accounts explorejogja have a fairly large existence as an account that promotes tourist destinations contained in the province of Yogyakarta. these accounts also get a fairly positive response from instagram users, especially followers (followers) of the explorejogja account.

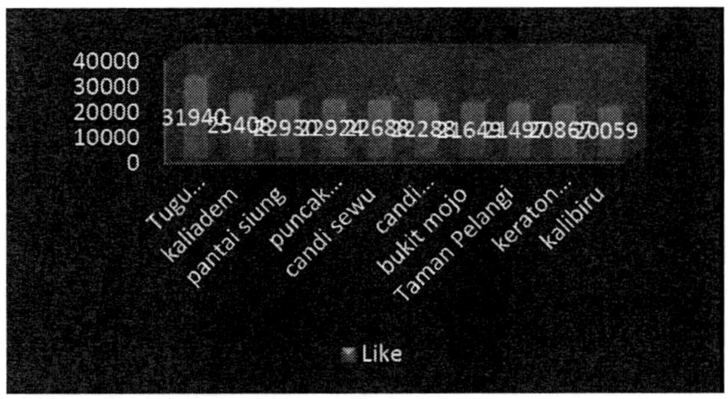

Figure 3. Chart the number of posts explorejogja

From the picture above in figure 3, it can be seen the top 10 of most popular tourist destination from January until December 2016. Namely is Tugu, Kaliadem, Siung beach, Suroloyo peak, Sewu temple, Prambanan temple, Mojo Hill, Taman Pelangi, keraton of Yogyakarta and Kalibiru.

Based on 121 tourist destinations obtained, then in the form of map web-based tourist destination with 121 point coordinates in figure 4.

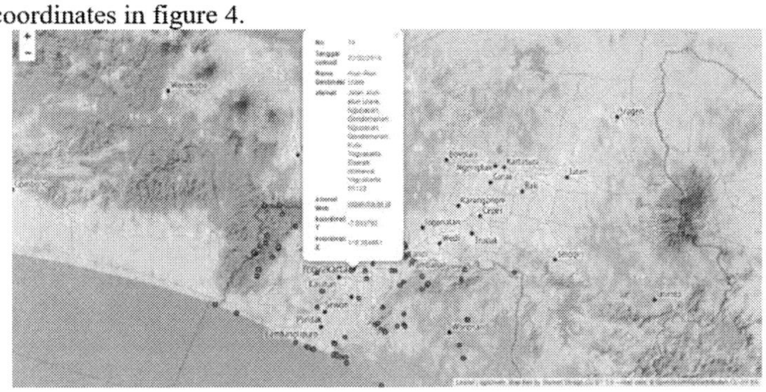

Figure 4. The map of Tourist Destination Destinations in special region of Yogyakarta

As for the explanation of tourist destinations in accordance with the coordinate point as follows:

Table 1. A list of tourist destinations along with point coordinates

No	Destination Names	Coordinate point	
		Latitude	Longitude
1	Kedung kandang waterfall	-7.855807	110.535138
2	Kedung Pedut waterfall	-7.766689	110.120270

3	Randusari Waterfall	7.917892	110.490050
....			
121	Sendangsari pine forest	-7.871076	110.459740

4.2. K-means clustering

The K-means method processes all objects simultaneously where k is the number of groups. So in the K-Means method also determined the number of groups formed, in this study in the form of three groups.

Table 2. A list of tourist destinations along with point coordinates

	Cluster		
	1	2	3
Number of tourist destination	9	60	52

In table 2 above can be seen that the grouping by using K-means method of popularity of tourist destinations based on the number of like from account instagram explorejogja consists of 121 tourist destinations that have been in execution as much as 50 times by using statistical software. There are 3 clusters each consisting of cluster 1 with 9 tourist destinations, cluster 2 consists with 60 tourist destinations and cluster 3 with 52 tourist destinations. Of the three clusters are then performed profiling cluster that obtained the following results.

Table 3. A list of tourist destinations along with point coordinates

Cluster Groups	Average amount like
Cluster 1 (hig)	19683.78
Cluster 2 (medium)	14213.23
Cluster 3 (low)	10538.98

In table 3 can be seen average in each cluster. In cluster 1 it can be seen that it has an average of 19683.78 like number of 9 tourist destinations contained in cluister 1 members. so it is concluded that cluster 1 is categorized as a group with like "high" with range of like numbers between 17153 to 25539 like.

In cluster 2 it can be seen that it has an average of 14213.23 like number of 60 destinations within cluister member 2. It is concluded that cluster 2 is categorized as a "moderate" group with a range of like between 12324 to 16555 like.

In cluster 3 it can be seen that it has an average of 10538.98 like numbers of 52 destinations within cluister 3. so it can be concluded that cluster 3 can be categorized as a group with like "low" with a range of like numbers between 7968 and 12208 like

4.3. Social netwrok analysis

Social Netwrok Analysis (SNA) is formed between tourist destinations based on tourist sites (sub-district) seen there are some locations that become a big role as the location that has the most tourist destinations.

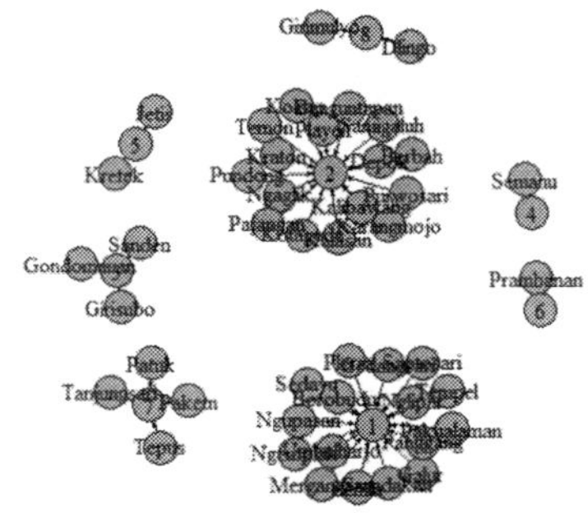

Figure 5. SNA graphs between locations (sub-district)

In the SNA Graph above in figure 5, it can be concluded that the sections 8.7 and 6 are expressed as the center of popularity based on the location (sub-district) tourist destination. So to be more clearly done the analysis by looking at the percentage of the main popularity by using the gephi program as in the following figure.

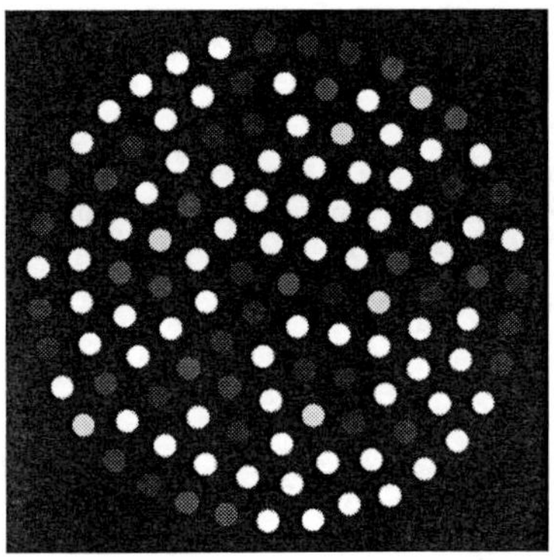

Figure 6. SNA graphs between locations (sub-districts)

As in the picture above in figure 6, there are 16 red dots, each of which is 8 tourist destinations located in Dlingo district and 8 other tourist destinations located in Girimulyo district with each percentage of 6.61%. then at the blue point there are 28 points, each of which is 7 tourist destinations located in Pakem sub-district, Tepus, Patuk and Tanjungsari with each percentage percentage of 5.79%. And lastly on the yellow point are 6 tourist destinations located in the district of Prambanan.

5. Conclusion

The Distribution of tourist destinations in Yogyakarta province is displayed in the form of a web-based dynamic map in which there are 121 points coordinates that symbolize each tourist destination. Of the 121 destinations are then obtained top 10 most popular destinations based on the number of like. Namely, Tugu yogyakarta, Kaliadem, Siung beach, Suroloyo peak, Sewu temple, Prambanan temple, Mojo hill, Taman Pelangi, and Kalibiru.

Grouping by means of K-means method to classify the popularity of tourist destination based on the number of instagram accounts from explorejogja consists of 121 tourist destinations that have been in execution 50 times using statistical software. There are 3 groups, each of which consists of cluster 1 of 9 tourist destinations, cluster 2 consists of 60 tourist destinations and for cluster 3 as many as 52 tourist destinations.

In the SNA graphic it is found that among the 8 sections present, sections 8,7 and 6 are expressed as the popularity centers based on the location (sub-district) of the tourist destination. central popularity of tourist destinations in the province of Yogyakarta based on the sub-district is located in 7 districts namely Dlingo, Girimulyo, Pakem, Tepus, Patuk, Tanjungsari and Prambanan.

Based on the conclusions obtained by the researcher wants the results of this study can be used by the Yogyakarta government, especially in the field of tourism in order to be used as a reference to improve quality and publications related tourism that has been considered popular. This is certainly in order to bring in more tourists-both from within the country and from abroad.

References

[1] Huraerah A 2016 *Communication behavior of explorejogja instagram account in Yogyakarta tourism journey* (Yogyakarta: Respositary UMY)
[2] Kodyat H 1998 *history of tourism and its development of Indonesia* (Jakarta: Grasindo)
[3] Nevita S 2016 *optimize social media for tourism promotion* http://m.radarbangka.co.id/rubrik/detail/perspektif/13837/mengoptimalkan-medsos-untuk-promosi-pariwisata.html
[4] Poerwadarminta 2006 *Indonesia dictionary Third edition* (Jakarta: Balai pustaka)
[5] Nita V 2015 *#exploreJogja As Reference Places in Yogyakarta* https://citraceritajogja.wordpress.com/2015/01/12/explorejogja-sebagai-referensi-tempat-wisata-di-yogyakarta/
[6] Dewanto 2012 *Design of Geographic Information System of Web-based Culinary Tour with Google API.*(Siliwangi: Siliwangi University)
[7] Gorunescu F 2011 *Data mining: Concepts, Models and Techniques* (Verlag Berlin Heidel: Springer)
[8] Han J, Kamber M 2006 *Data Mining Concepts and Techniques Second Edition* (San Francisco: Morgan Kaufmann)
[9] Neteler M and Mitasova H 2004 *Open Source GIS: A GRASS GIS Approach* (Boston: Kluwer Academic Publishers)
[10] Davies and Beynon P 2004 *Database Systems Third Edition* (Palgrave Macmillan: New York)
[11] Pramudiono 2007 *Introduction to Data Mining: Mining Gems of Knowledge in Mount Data* http://www.ilmukomputer.org/wpcontent/uploads/2006/08/iko-datamining.zip
[12] Lei Xu C J, Wang J, Yuan J, and Ren Y 2014 *Information Security in Big Data : Privacy and Data Mining* pp. 1149–1176
[13] Chauhan A, Mishra G, and Kumar G 2011 *Survey on Data Mining Techniques in Intrusion Detection* vol. 2, no. 7, pp. 2–5
[14] Riadi I, Istiyanto J E, and Saleh S 2013 *Internet Forensics Framework Based-on Internet Forensics Framework Based-on Clustering*
[15] Eiyanto N S, Mara M N 2013 *Characteristics classification by method k-means cluster analysis* Bul. Llm., vol.02.,no.2, pp.133-136,2013.
[16] Wasserman S, and Faust K. 1994. *Social Network Analysis: Methods and Application* (New York: Cambridge University Press)

IOP Conf. Series: Journal of Physics: Conf. Series **974** (2018) 012034 doi:10.1088/1742-6596/974/1/012034

Application of selection and estimation regular vine copula on go public company share

Hasna Afifah R[1], **Lienda Noviyanti**[2], **Achmad Bachrudin**[3]

[1,2,3] Statistics Department, Faculty of Mathematics and Science, State University of Padjadjaran, Indonesia
[1] hasnaafifahrusyda@gmail.com
[2] lienda@unpad.ac.id
[3] achmad.bachrudin@unpad.ac.id

Abstract. The accuracy of financial risk management involving a large number of assets is needed, but information about dependencies among assets cannot be adequately analyzed. To analyze dependencies on a number of assets, several tools have been added to standard multivariate copula. However, these tools have not been adequately used in apps with higher dimensions. The bivariate parametric copula families can be used to solve it. The multivariate copula can be built from the bivariate parametric copula which is connected by a graphical representation to become Pair Copula Constructions (PCCs) or vine copula. The application of C-vine and D-vine copula have been used in some researches, but the use of C-vine and D-vine copula is more limited than R-vine copula. Therefore, this study used R-vine copula to provide flexibility for modeling complex dependencies on a high dimension. Since copula is a static model, while stock values change over time, then copula should be combined with the ARMA-GARCH model for modeling the movement of shares (volatility). The objective of this paper is to select and estimate R-vine copula which is used to analyze PT Jasa Marga (Persero) Tbk (JSMR), PT Waskita Karya (Persero) Tbk (WSKT), and PT Bank Mandiri (Persero) Tbk (BMRI) from august 31, 2014 to august 31, 2017. From the method it is obtained that the selected copulas for 2 edges at the first tree are survival Gumbel and the copula for edge at the second tree is Gaussian.

1. Introduction

The accurate financial risk management involving a large number of assets is needed. Especially the interdependence of assets has not been adequately estimated. Although some advantages were added in multivariate standard, they are inadequate in higher dimensional applications.

Unfortunately, the choice of multivariate copula-copula is quite limited, as opposed to bivariate cases, whereas different types of copula exhibit flexible and complex dependence patterns. Salmon (2009) describes copula Gaussian has a weakness to see the stock dependency structure during a financial crisis. In addition, Student's copula only can capture symmetric dependencies using correlation matrices and Archimedean copula can be used only one or two parameters to model dependencies between possible dozens of variables. For high-dimensional distributions, there is a large number of possible copula pairs. For example, there are 240 constructions for five-dimensional density. To help organize it, Joe (1996) forms a copula model and further explored by Bedford and

Cooke (2001, 2002) and Kurowicka and Cooke (2006). The copula is Regular Vine (R-vine Copula). R-Vine Copula is an easy graphical model for classifying pairs and a hierarchical copula construction. Each degree only involves the specification of the bivariate copula that changes as the built blocks. Therefore it allows for a very flexible model and effectively shows each possible dependency structure.

Since copula is a static model, they must be combined with time series models for modeling movements of stock value (Cherubini et al, 2004). In the financial market, the economic cycle can be observed, where the value of stock prices, indexes, and currencies can sustain value movements, both increase in value (profit), and impairment (loss). The movement of the value of the stock is called volatility. Asset volatility is estimated in order to know how asset movements are. Greater value of volatility return can be concluded that the change of asset value is very large. Historical return data can be used to estimate volatility, i.e. current volatility and future volatility.

The volatility model used can accommodate the characteristics of stock return data. According to Bollerslev et al. 1994, the time series data in finance has two important properties: the probability distribution returns of stock is fat tails and contain of volatility clustering. The volatility model used is Generalized Autoregressive Conditionally Heteroskedasticity (GARCH).

2. Copula

Copula is a Latin which means relationship or bond. The word copula was first used in 1959 for Theorem of Sklar. This theorem is used to describe a function that can combine several distribution functions from several random variables to a joint distribution function. The joint distribution function of some random variables can be determined using copula. These distribution functions can come from several different distributions.

The distribution function of randomized variables that have been combined to form a joint distribution function in the copula form is called the marginal function of the copula. This marginal function spreads uniformly (0,1) (Nelsen, 2006).

For each continuous random variable, the density function of copula is related to the joint density function of random variables in the copula, ie $(x_{1,2},\cdots,x_d)$. The relationship of the two functions can be described canonically as follows:

$$f(x_1,x_2,\cdots,x_d)=c(F_1(x_1),F_2(x_2),\cdots,F_n(x_d))\prod_{i=1}^{d} f_i(x_i) \quad (1)$$

2.1 Sklar's theorem

If H is a distribution function having d dimension with a cumulative distribution function F_1, F_2, $\cdots$, F_d, then will appear a copula for all $\boldsymbol{x}$ in $\overline{\boldsymbol{R}}^d$, the copula associated with F is a function distribution $C: [0,1]^d \rightarrow [0,1]$ with margin $U(0,1)$, as follows:

$$H(x_1,x_2,\cdots,x_d)=C(F_1(x_1),F_2(x_2),\cdots,F_d(x_d)) \quad (2)$$

- If F is a continuous d-variate distribution function with univariate margins F_1, ..., F_d, and quantile functions F_1^{-1}, ..., F_d^{-1}, then

$$C(u) = F\left(F_1^{-1}(u_1), \dots, F_d^{-1}(u_d)\right), \quad u \in [0,1]^d, \quad (3)$$

- If F is the d-variate distribution of discrete random variables (in general, some are continuous and partly discrete), then copula is the only unique part of that collection

$$Range(F_1) \times \dots \times Range(F_d).$$

In Sklar's theorem, $RanF_1 \times RanF_2 \times \cdots \times RanF_d$ are combination of all possibility margin value F_i.

- If F is the d-variate distribution of some or all of the discrete random variables, then the copula associated with F is not unique. If H is non-continuous or discrete univariate cdf and $Y \sim H$, then $H(Y)$ doesn't have a $U(0,1)$ distribution. The copula of equation (2) becomes un-unique. Copula will only be unique in $Range(F_1) \times \ldots \times Range(F_d)$, since C on equation (2) will only be needed to be defined in this set. Such a C must satisfy $C(1,\ldots,1,u_j,1,\ldots 1) = u_j$ where $u_j \in Range(F_j)$ for $j = 1, \ldots, d$; and C can be extended to multivariate distribution with $U(0,1)$ margin.

Based on the Sklar's theorem, it can be seen that copula has the following properties:

- $Dom\ C = I^n = [0,1]^n = [0,1] \times [0,1] \ldots \times [0,1]$
- $C(u_1, \ldots, u_n)$ is non-decreasing function for all component u_i
- $C(0,\ldots,0,u_i,0,\ldots,0) = 0$ for each $i \in \{1,\ldots,n\}$, $u_i \in [0,1]$
- $C(1,\ldots,1,u_i,1,\ldots,1) = u_i$ for each $i \in \{1,\ldots,n\}$, $u_i \in [0,1]$
- For all $(\alpha_1,\ldots,\alpha_n),(b_1,\ldots,b_n) \in [0,1]^n$ with $\alpha_i \le b_i$ be valid with $\sum_{i_1=1}^{2},\ldots,\sum_{i_n=1}^{2}(-1)^{i_1+\cdots+i_n} C(u_{1i_1},\ldots,u_{ni_n}) \ge 0$ where $u_{j1} = \alpha_j$ dan $u_{j2} = b_j$, $j \in \{1,2,\ldots,n\}$.

3. Build the volatility model

Before solving the problem of heteroscedasticity in data through modeling its volatility using GARCH, data must be modelled by Autoregressive Integrated Moving Average (ARIMA). This model is a Box-Jenkins variant of the ARMA model where the model requires analysis of stationarity. In general, regular ARIMA model (without seasonal patterns) have the following mathematical models.

$$\phi_p(B)(1-B)^d Z_t = \theta_0 + \theta_q(B) a_t \tag{4}$$

where $\phi_p(B) = (1 - \phi_1 B - \cdots - \phi_p B^p)$ and $\theta_q(B) = (1 - \theta_1 B - \cdots - \phi_p B^q)$ and $(1-B)^d$ is a non seasonal differencing in order d. This equation can also be written as ARIMA (p, d, q). When the data is stationary without going through the process of differentiation then the used model is ARMA (p, q). ARMA model have the following mathematical models.

$$X_t - a_1 X_{t-1} - \cdots - a_p X_{t-p} = \varepsilon_t + b_1 \varepsilon_{t-1} + \cdots + b_q \varepsilon_{t-q} \tag{5}$$

Where $a_1, a_2, \ldots, a_p, b_1, b_2, \ldots, b_p \in \mathbb{R}, \varepsilon_t \sim WN(0,\sigma^2)$

After obtaining a suitable ARIMA or ARMA model, the next step is build a GARCH model from square residual ARIMA or ARMA model. Bollerslev introduces the GARCH model. In the GARCH model, it is assumed that the model mean follows the ARIMA or ARMA model. Example $\varepsilon_t = r_t - \mu_t$ and ε_t follow the GARCH form (m, s) if

$$\varepsilon_t = \sigma_t v_t,\quad \sigma_t^2 = \alpha_0 + \sum_{i=1}^{m} \alpha_i \varepsilon_{t-i}^2 + \sum_{j=1}^{s} \beta_j \sigma_{t-j}^2 \tag{6}$$

Where v_t is independent random variabel and identic with mean = 0 and varians = 1, $\alpha_0 > 0, \alpha_i \ge 0, i = 1,2,\ldots,m, \beta_j \ge 0$, $j = 1,2,\ldots,m$ with $\sum_{i=1}^{\max(m,s)}(\alpha_i + \beta_j) < 1$. Assume that $E(\varepsilon_t|\mathcal{F}_{t-1}) = 0$, $Var(\mathcal{E}_t|\mathfrak{I}_{t-1}) = E(\mathcal{E}_t^2|\mathcal{F}_{t-1}) = \sigma_t^2$.

4. Vine copula

Vine copula is another name of the Pair Copula Constructions (PCCs) introduced by Aas, et al (2009). Vine copula is a graphical representation to determine PCC. This is explained with more detail by Kurowicka and Joe (2011). Regular statistical inference vines have been reinforced by Dißmann, et al (2013).

IOP Conf. Series: Journal of Physics: Conf. Series **974** (2018) 012034 doi:10.1088/1742-6596/974/1/012034

A PCC can be illustrated by three dimensions. Example $\boldsymbol{x} = (X_1, X_2, X_3)' \sim F$ and assumed that all densities needed exist, then density function of $\boldsymbol{x}$ is

$$f(x_1, x_2, x_3) = f_1(x_1) f(x_2 \mid x_1) f(x_3 \mid x_1, x_2) \tag{7}$$

With Sklar's theorem, joint density function of three dimmensions can be described in bivariat copula as follow

$$f(x_1, x_2, x_3) = f_1(x_1) f_2(x_2) f_3(x_3) c_{1,2}(F_1(x_1), F_2(x_2)) c_{1,3}(F_1(x_1), F_3(x_3)) \\ c_{2,3|1}(F(x_2 \mid x_1), F(x_3 \mid x_1)) \tag{8}$$

Bivariate $C_{1,2}, C_{1,3},$ and $C_{2,3|1}$ can be independently selected from each other, so that different dependency structures can be modeled using PCCs, as well as constructing high dimension. For ease of inference, copula $C_{2,3|1}$ is assumed to depend only on x_1 through arguments $F(x_2 \mid x_1)$ and $F(x_3 \mid x_1)$. This is called simplification of the principle which has been investigated by Hobaek Haff, et al (2010).

5. Regular vine copula

The R-vine structure is graphically used for determining the copula required for the construction of a series of copula pairs, where the copula is a multivariate distribution of hypercube units $[0,1]^n$ with a marginal uniform distribution (Nelsen (2006)).

Referring to Kurowicka and Cooke (2006), a regular vine (R-vine) in the d variable consists of a series of first related trees (related to acylic graphs). $T_1, \ldots, T_{d-1}$ with nodes N_i and edges E_i for $i=1,\ldots,d-1$, where T_1 has *nodes* $N_1 = \{1,\ldots,d\}$ and *edge* E_1 and for $i = 2,\ldots,d-1$ tree T_i has *nodes* $N_i = E_{i-1}$. Besides that, *proximity condition* needs 2 *edges* in tree T_i. It is combined with tree T_{i+1} if only *edge* shares the same *nodes* in tree T_i.

Bedford and Cooke (2001) and Kurowicka and Cooke (2006) have shown that an edges of an R-vine tree can be uniquely identified with 2 nodes: a conditioned nodes and series of conditioning nodes. It is denoted by $e=j(e),k(e)|D(e)$ where $D(e)$ is a conditioning set.

The multivariate copula connecting $T_1,\ldots,T_{d-1}$ is then constructed by connecting each edge $e=j(e),k(e)|D(e)$ to E_i with a bivariate copula density $C_{j(e),k(e)|D(e)}$. Referring to the theorem in Kuroicka and Cooke (2006), the density of R-vine copula is:

$$c(F_1(x_1), \ldots, F_d(x_d)) = \prod_{i=1}^{d-1} \prod_{e \in E_i} c_{j(e),k(e)|D(e)}(F(x_{j(e)} \mid \mathbf{x}_{D(e)}), F(x_{k(e)} \mid \mathbf{x}_{D(e)})), \tag{9}$$

where $\boldsymbol{x}_{D(e)}$ denotes subvector from $\boldsymbol{x}=(x_1,\ldots,x_d)'$ showing with contained index in $D(e)$.

Two special types from *reguler vine* (*R-vine*) copula are *canonical vine* (*C-vine*) and *drawable vine* (*D-vine*). *R-vine* is called a *C-vine* if each tree (T_i) have unique *nodes* with degree $d-i$ (maximum degree) for $i = 1,\ldots,d-2$. Therefore, a *root node* or *R-vine* is contained in a *node* with maximum degree in each tree (*star structure*). And *R-vine* calls *D-vine* if all *nodes* in the first tree T_1 have degree no more than two (*path structures*).

6. Modelling dependency among daily returns of stock

Data used in this research is daily stock return from go public company such as PT Jasa Marga (Persero) Tbk (JSMR), PT Waskita Karya (Persero) Tbk (WSKT), and PT Bank Mandiri (Persero)

Tbk (BMRI) from august 31, 2014 to august 31, 2017 (734 data). The extracted data is the daily closing price of each stock. R software is used in data processing with 'VineCopula' package.

Value of stock return is calculated as follows.

$$r_t = \ln\left(\frac{p_t}{p_{t-1}}\right), \tag{10}$$

where:

r_t = return of stock,
P_t = closing price stock in t period ,
P_{t-1} = closing price stock in $t-1$ period.

Descriptive of return JSMR, WSKT, and BMRI is as follow

Table 1. Summary statistics of JSMR, WSKT, and BMRI returns.

Parameter	JSMR	WSKT	BMRI
Min.	-0.0819166	-0.060625	-0.0762274
1st Qu.	-0.0093896	-0.008801	-0.0082475
Median	0.000000	0.000000	0.000000
Mean	-0.0001144	0.001254	0.0003014
3rd Qu.	0.0086918	0.011116	0.0099503
Max.	0.0909715	0.07509	0.0819171
Kurtosis	3.664556	1.160076	2.512133
Skewness	0.2680271	0.3399456	0.08370384
Jarque Bera Test	p-value < 2.2e-16	p-value = 9.834e-13	p-value < 2.2e-16

Since the return data contain volatility clustering, first of all, GARCH model is applied on each return before it is modeled by copula. The result is as follow.

Table 2. Estimates from univariate ARMA-GARCH models.

Parameters	JSMR (P-value)	WSKT (P-value)	BMRI (P-value)
ar1	0.5403 (0.008001)	-0.626805 (< 2e-16)	0.5483 (0.000385)
ar2	-	-0.8653877 (< 2e-16)	-
ma1	-0.5681 (0.004477)	0.6531046 (8.64e-14)	-0.5255 (0.000674)
ma2	-0.1064 (0.020392)	0.8371652 (< 2e-16)	-0.1239 (0.002816)
omega	0.00002097 (0.031505)	0.0001008 (0.019084)	0.00007641 (0.007782)
alpha1	0.2089 (0.000614)	0.1602536 (0.002706)	0.1645 (0.000993)
beta1	0.2049 (0.012611)	0.5739191 (0.000125)	0.5946 (0.00000192)
beta2	0.5461 (0.000000000347)		

LM Arch	Statistic = 11.73823 p-Value = 0.4669249	Statistic = 7.944844 p-Value = 0.7894257	Statistic = 7.852354 p-Value = 0.7965593
AIC	-5.253642	-5.085729	-5.288613

Descriptive standardize residual from GARCH model such as histogram, dependency structures, and corelation value are described as follow:

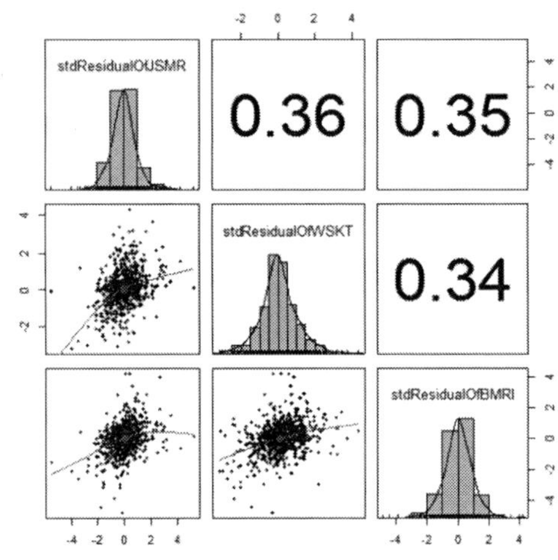

Figure 1. Pair-plots and Kendall's taus for representatives of each index group.

From GARCH model will be generated standardized residual. The dependencies among these standardized residuals are then modeled with copula after being transformed into marginal uniform data using an integral transformation of empirical or parametric probabilities. To model R-vine copula, the following steps are should be taken:

- Plot any possible pairs of variables on the data and estimate values of Kendall's tau.
- Determine the class of R-vine along with the choice of the copula family according to the dataset of each variable pairs. The pair-copula can be individually selected from 7 types of bivariate copulas, those are: Gauss, Student-t, Gumbel, Gumbel survival, rotated gumbel 90 degrees, rotated gumbel 270 degrees, and Frank. To select one possible R-vine for a given dataset, pairs must be choiced to determine used family copula.
  For the selection process of the R-vine copula specification based on Kendall's tau, the first stage is to define the first tree $T_1 = (N_1, E_1)$ for R-vine, followed by the second tree, and so on. Trees are chosen in such a way as to produce the most robust dependency model choices. Maximum Spanning Tree (MST) algorithm is used to select the tree. It can maximize the number of absolute *absolute Kendall's tau*. A Maximum Spanning Tree (MST) algorithm is clearly described on Dißman (2013).
- Select the suitable pair-copula with the data through AIC criteria.
- Estimated parameters for pair-copula through maximum likelihood estimation.

Since the possibility of R-vine on n variables increases with increasing n by the number of n n ($n!/2 \ x \ 2^{\binom{n-2}{2}}$), algorithms are needed to facilitate that peocess. Dißman (2013) developed a sequential heuristic method for selecting tree structures from R-vine.

After going through the process described previously then obtains the suitable R-vine structure with data as follow.

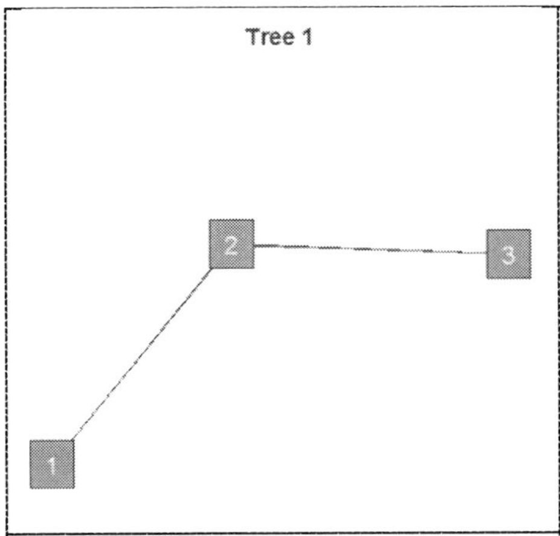

Figure 2. Tree 1 of R-vine copula

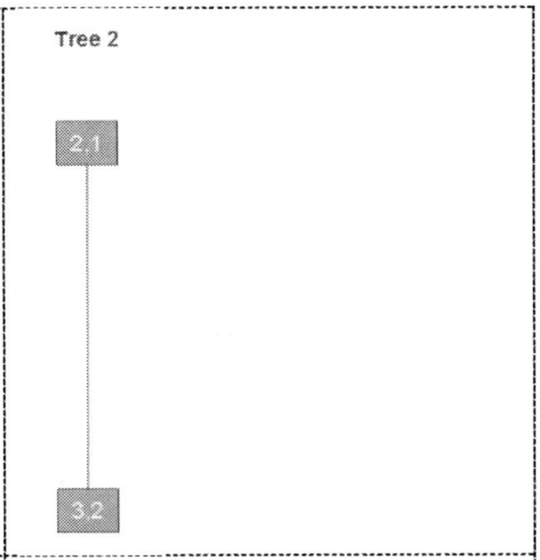

Figure 3. Tree 2 of R-vine copula

and the parameters for pair-copula is as follow.

Table 3. Estimation results of R-vine copula

Edge	Bivariate Copula	par	par2	tau	utd	ltd
	Tree 1					
WSKT, JSMR	Survival Gumbel / SG	1.33	0	0.25	-	0.31
BMRI, WSKT	Survival Gumbel / SG	1.29	0	0.23	-	0.29
	Tree 2					
BMRI, JSMR; WSKT	Gaussian / N	0.25	0	0.16	-	-

Result from the selection process of the R-vine copula specification based on Kendall's tau is 2 trees. The first tree's bivariate copula family from edges $C_{2,1}$, $C_{3,2}$ are survival Gumbel and the second tree's bivariate copula family from edge $C_{1,3|2}$ is Gaussian. Those copulas are chosen in such a way as to produce the most robust dependency model choices.

7. Conclusion

R-vine copula has different choices for individual copula types for each pair-copula, these choices are too many to be investigated. With R-vine copula, multivariate copula modeling become more flexible.

In the formation of R-vine copula, a general selection approach have been used sequentially to select the representation of trees together. Selection is done by selecting the copula type from the bivariate copula family and corresponding parameter prediction. Algorithm for finding the maximum spanning tree is used in the selection of each theoretical graphics. Absolute Kendall empirical values are used as weights, but other weights are possible. In financial data, tail behavior may be useful to be investigated.

The output of this selection procedure gives the best result of R-vine tree, the corresponding pair-copula types, and corresponding parameters. The results of this selection process can be used as a starting option for determining maximum likelihood estimates.

The best result of R-vine trees from the data is that contained edges $C_{2,1}$, $C_{3,2}$ at the first tree and $C_{1,3|2}$ at the second tree and the most corresponding pair-copula types from edges $C_{2,1}$, $C_{3,2}$ is survival Gumbel, and from edge $C_{1,3|2}$ is Gaussian. Result of R vine copula formation can be used to measure risk of investment portfolio.

IOP Conf. Series: Journal of Physics: Conf. Series **974** (2018) 012034 doi:10.1088/1742-6596/974/1/012034

References

[1] Aas K, Czado C, Frigessi A and Bakken H 2009 Pair-copula constructions of multiple dependence. *Insurance: Mathematics and Economics* **44** (**2**) pp 182– 198.

[2] Bedford T and R M Cooke 2002 Vines - a new graphical model for dependent random variables *Annals of Statistics* **30** pp 1031-1068.

[3] Cherubini U, E Luciano and W Vecchiato 2004 *Copula Methods in Finance* (Chichester: John Wiley & Sons)

[4] Dißmann J, E C Brechmann, C Czado, and D Kurowicka 2013 Selecting and estimating regular vine copulae and application to financial returns *Journal of Computational Statistics & Data Analysis* **59** pp 52-69.

[5] Joe H and J J Xu 1996 The estimation method of inference functions for margins for multivariate models *Technical Report* **166** Department of Statistics, University of British Columbia

[6] Kurowicka D and R M Cooke 2006 *Uncertainty Analysis with High Dimensional Dependence Modelling* (Chichester: John Wiley)

[7] Kurowicka D 2011. Optimal truncation of vines. In: Kurowicka, D., Joe H. (Eds), Dependence Modeling *Handbook on vine copulae* (Singapore : World Scientific Publishing Co)

[8] Nelsen R B 2006 *An Introduction to Copulas 2nd Edition*. (New York : Springer)

[9] Salmon, F 2009 Recipe for disaster: The formula that killed Wall Street *Wired Magazine* **17** (**3**)

[10] Yahoo! Finance. 1997. *Pt Jasa Marga (Persero) Tbk (JSMR.JK)*. https://finance.yahoo.com/quote/JSMR.JK?p=JSMR.JK. September 16th 2017.

[11] Yahoo! Finance. 1997. *PT Wijaya Karya (Persero) Tbk (WIKA.JK)*. https://finance.yahoo.com/quote/WIKA.JK?p=WIKA.JK. September 16th 2017.

[12] Yahoo! Finance. 1997. *PT Bank Mandiri (Persero) Tbk (BMRI.JK)*. https://finance.yahoo.com/quote/BMRI.JK?p=BMRI.JK. September 16th 2017.

IOP Conf. Series: Journal of Physics: Conf. Series **974** (2018) 012035 doi:10.1088/1742-6596/974/1/012035

Mathematical model for HIV spreads control program with ART treatment

Maimunah, Dipo Aldila*

Department of Mathematics, Universitas Indonesia, Depok 16424, Indonesia

E-mail: *aldiladipo@sci.ui.ac.id

Abstract. In this article, using a deterministic approach in a seven-dimensional nonlinear ordinary differential equation, we establish a mathematical model for the spread of HIV with an ART treatment intervention. In a simplified model, when no ART treatment is implemented, disease-free and the endemic equilibrium points were established analytically along with the basic reproduction number. The local stability criteria of disease-free equilibrium and the existing criteria of endemic equilibrium were analyzed. We find that endemic equilibrium exists when the basic reproduction number is larger than one. From the sensitivity analysis of the basic reproduction number of the complete model (with ART treatment), we find that the increased number of infected humans who follow the ART treatment program will reduce the basic reproduction number. We simulate this result also in the numerical experiment of the autonomous system to show how treatment intervention impacts the reduction of the infected population during the intervention time period.

1. Introduction

Human Immunodeficiency Virus (HIV) is a virus that could infect limfosit in a human body and will result in the decline of human immune system, which well known as *Acquired Immune Deficiency Syndrome* (AIDS). HIV (*Human Immunodeficiency Virus*) contained in the human fluids body who are living with HIV. Even though this virus contains on saliva, tear, and urine, but it does not shown to be at risk of transmitting the infection among human population since the virus levels on these liquids are very low. The spread of this HIV normally spread from unhealthy sex behavior, the use of unsterile syringes, and human blood transfusions who have been infected by HIV to healthy humans [1].

According to *World Health Organization* (WHO)[2] and *United Nations Programme on HIV/AIDS* (UNAIDS) [3], in 2016, it is estimated that there are 36,7 million of humans who get infected by HIV/AIDS in all over the world. Most of the HIV-infected human population comes from developing countries, such as South Africa, India, Kenya, Mozambique, Uganda, China, Zimbabwe, Zambia, Malawi, Indonesia, and Thailand. The highest case of HIV spread occurs in the South African country as many as 7 million people are infected with HIV / AIDS [4].

Since 1999, there has been an increase in the number of people living with HIV in a group of people with high-risk behaviors, including commercial sex workers and injecting drug users in several provinces such as DKI Jakarta, Riau, Bali, West Java and East Java. With this reason, therefore these provinces are classified as concentrated level of epidemic areas. Papua province is categorized as a province in a generalized epidemic area. The estimation result of 2009, there are 186.000 people with HIV positive in Indonesia. Humans who get infected with HIV / AIDS

IOP Conf. Series: Journal of Physics: Conf. Series **974** (2018) 012035 doi:10.1088/1742-6596/974/1/012035

cannot be cured. The treatment that can be done is only to inhibit the development of viruses in the body through the process of emphasis *Anti-Retroviral Therapy*(ART). From the Report on HIV Progress Situation in Indonesia until September 2011, it was recorded that the number of PLHA who get *Anti Retroviral Therapy (ART)* are 22.843 from 33 provinces and 300 districts cities, with male and female ratio, is 3: 1, and the highest percentage in the age group 20-29 years (Ministry of Health, 2011)[5].

Many mathematical models have been developed to understand how HIV infection might spread among the human population, such as a mathematical model that discussed how HIV might spread among the community of injecting drug users (I Mardhiyah, 2012). The authors use a basic reproduction number analysis and equilibrium analysis about the existence and their global stability criterion and also Bendixon-Dulac criterion for the local stability check. They found that HIV will spread among the injecting drug users if and only if the basic reproduction number is larger than one, and the disease will disappear if and only if the basic reproduction number is smaller than one[6]. The other authors that also discuss the spread of HIV are (Lindsay Simpson and Abba B. Gumel, 2016) that include Anti-retroviral intervention among the homosexual community. They also using basic reproduction number and the local stability and existence criteria about the equilibrium points to understand how HIV might spread. They found that an infection among human in the homosexual category might suppress with the Anti-retroviral intervention[7].

Some authors explained about benefits and risks treatment with ART for pregnant women (M.G. Fowler, M. Qin, S.A. Fiscus, J.S. Currier, 2016) with randomized-trial data on the risks and benefits of antiretroviral therapy (ART) as compared with zidovudine and single-dose nevirapine to prevent transmission of the human immunodeficiency virus (HIV) in HIV-infected pregnant women with high CD4 counts are lacking, then We randomly assigned HIV-infected women at 14 or more weeks of gestation with CD4 counts of at least 350 cells per cubic millimeter to zidovudine and single-dose nevirapine plus a 1-to-2-week postpartum "tail" of tenofovir and emtricitabine (zidovudine alone); zidovudine, lamivudine, and lopinavir-ritonavir (zidovudine-based ART); or tenofovir, emtricitabine, and lopinavir-ritonavir (tenofovir-based ART). The primary outcomes were HIV transmission at 1 week of age in the infant and maternal and infant safety. Then the conclusion is antenatal ART resulted in significantly lower rates of early HIV transmission than zidovudine alone but a higher risk of adverse maternal and neonatal outcomes[8]. The other authors that also discuss enhanced prophylaxis plus antiretroviral therapy for advanced HIV infection in Africa. In sub-Saharan Africa, among patients with advanced human immunodeficiency virus (HIV) infection, the rate of death from infection (including Tuberculosis and Cryptococcus) shortly after the initiation of antiretroviral therapy (ART) is approximately 10%, in this factorial open-label trial conducted in Uganda, Zimbabwe, Malawi, and Kenya, we enrolled HIV-infected adults and children 5 years of age or older who had not received previous ART and were starting ART with a CD4+ count of fewer than 100 cells per cubic millimeter, and the conclusion is among HIV-infected patients with advanced immunosuppression, enhanced antimicrobial prophylaxis combined with ART resulted in reduced rates of death at both 24 weeks and 48 weeks without compromising viral suppression or increasing toxic effects[9].

Different from above model, the spread of HIV among the human population with the ART intervention will be included into the model. We also will include the human behavior about their awareness in following the treatment process. The ART intervention is only given to an infected human in a chronic infected category. The article will be constructed as follows. In the next section, mathematical model construction along with the assumptions descriptions will be discussed. The analysis of the model about the existence and local stability criteria of the model along with the construction of the next-generation matrix will be discussed in section3. Finally, some numerical simulations and conclusions will be given in section 4 and 5, respectively.

2. Mathematical model construction

Before constructing the model, let us assume that the human population might be separated into seven classes, i.e :

(i) Susceptible human compartment ($x_1(t)$), described as a group of susceptible humans, who might be infected by HIV. They might be infected because of direct contact (free sex), blood transfusion from an infected human, or from the use of a syringe from an infected human.

(ii) Infected human of acute level ($x_2(t)$), described as a group of humans who are already infected by HIV, but still in a non-harmful stage. Humans in this category still might perform daily activities like susceptible humans. We assume that this infected compartment does not receive ART treatment since they do not realize that they are already infected with HIV.

(iii) Infected humans of chronic level ($x_3(t)$), which includes all infected humans who have shown the symptoms of HIV infection. This group of humans cannot perform daily activities like susceptible humans anymore. Therefore, the ART treatment is implemented in this group of infected humans.

(iv) Infected humans who receive the ART treatment intervention ($x_4(t)$), are a group of infected humans at the chronic stage and follow the therapy process with a high level of awareness.

(v) Infected humans who receive ART treatment but with a low level of awareness ($x_5(t)$), described as a group of infected humans at the chronic level, following the ART treatment procedure, but with a low awareness.

(vi) Infected humans who have failed in the ART treatment procedure ($x_6(t)$). This compartment only increases from the transition of infected humans with the ART treatment intervention with a low level of awareness.

(vii) Infected humans with HIV/AIDS complication ($x_7(t)$). This group of infected humans already have AIDS, and their condition might be complicated by other diseases due to the decline of immunity in their body.

The transmission process that involves the seven compartments above is given in Figure 1. The process of interaction between each compartment is given as follows. The recruitment rate only occurs in susceptible human x_1 with a constant rate A. Each compartment is decreasing due to the natural death rate (μ), except for the AIDS human compartment (x_7) where mortality caused by AIDS has a rate of δ. Please note that $\delta > \mu$. Susceptible humans might be infected by direct contact with infected humans with the probability of infection β_a if they are infected by x_2 compartment, β_c if they are infected by x_3, x_4, x_5 and x_6, and β_s if they are infected by an individual in the (x_7) compartment. Since we assume that more infected human, than it will limit their capability to do an activity in daily life; therefore, we assume that $\beta_a > \beta_c > \beta_s$. After an incubation period of approximately 3-5 years, individual in compartment x_2 will be transferred into compartment x_3. The ART treatment intervention is given to chronically infected humans at a constant rate u, which will separate them into a specific compartment, (x_4). If individual in the (x_4) compartment following the treatment process with high-level awareness, it will prevent them becoming an AIDS-infected human. On the other hand, if they do not follow the treatment process obediently, they will transfer into the (x_5) compartment at a constant rate ξ_4. If they come back following the treatment process with a high level of awareness, they will come back into the (x_4) compartment. If individuals in the (x_5) compartment finally stop following the treatment, they will transfer into the (x_6) compartment at a constant rate ρ_c. There is a possibility that infected humans who have failed in the treatment process ((x_6)) may or may not follow the ART treatment again, with the probability of $1 - r_c$ and r_c, respectively. There is also a possibility that if the infected humans in the (x_3) compartment do not receive an ART treatment, they will be transferred directly into AIDS compartment x_7.

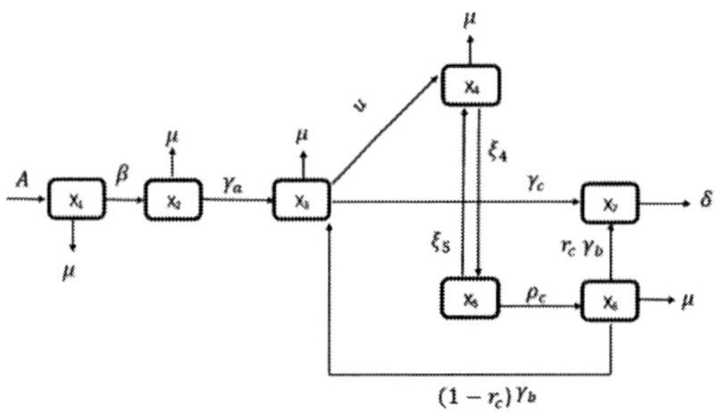

Figure 1. Transmission diagram of HIV spread with ART treatment.

Based on transmission diagram in Figure 1, the mathematical model of HIV spread with ART treatment now is given by 7-dimensional non linear ordinary differential equation, i.e :

$$\begin{aligned}
\frac{dx_1}{dt} &= A - \frac{\beta_a x_1 x_2}{N_x} - \frac{\beta_c x_1 (\theta_h x_4 + \theta_l x_5 + x_3 + x_6)}{N_x} - \frac{\beta_s x_1 x_7}{N_x} - \mu x_1 \\
\frac{dx_2}{dt} &= \frac{\beta_a x_1 x_2}{N_x} + \frac{\beta_c x_1 (\theta_h x_4 + \theta_l x_5 + x_3 + x_6)}{N_x} + \frac{\beta_s x_1 x_7}{N_x} - \gamma_a x_2 - \mu x_2 \\
\frac{dx_3}{dt} &= \gamma_a x_2 + (1 - r_c)\gamma_b x_6 - u x_3 - \gamma_c x_3 - \mu x_3 \\
\frac{dx_4}{dt} &= -\mu x_4 + u x_3 - x_4 \xi_4 + x_5 \xi_5 \qquad (1) \\
\frac{dx_5}{dt} &= -\mu x_5 - \rho_c x_5 + x_4 \xi_4 - x_5 \xi_5 \\
\frac{dx_6}{dt} &= \rho_c x_5 - (1 - r_c)\gamma_b x_6 - r_c \gamma_b x_6 - \mu x_6 \\
\frac{dx_7}{dt} &= r_c \gamma_b x_6 - \delta x_7 - \mu x_7 + \gamma_c x_3,
\end{aligned}$$

which supplemented with initial condition $x_i(t = 0) = x_{i0}$ for $i = 1, 2, \ldots, 7$. The parameters interpretation of system 1 is given in Table 1.

Table 1. Parameters description.

Par.	Description	Unit
A	recruitment rate of human	$\frac{human}{time}$.
β_a	rate of success infection of infected human in acute stage	$\frac{1}{time}$.
β_c	rate of success infection of infected human in chronic stage	$\frac{1}{time}$.
β_s	rate of success infection of infected human in AIDS stage	$\frac{1}{time}$.
θ_h	reduction of β_a for infected human who undergo with treatment (x_4)	-
θ_l	reduction of β_a for infected human who undergo with treatment (x_5)	-
γ_a	Transition rate from x_2 into x_3	$\frac{1}{time}$.
γ_b	Rate of transition for the failure of Treatment (x_4 to x_5)	$\frac{1}{time}$.
γ_c	Rate of transition from chronic to AIDS infected human	$\frac{1}{time}$.
r_c	Proportion of human who failed and not repeat the ART treatment	-
ρ_c	Rate of infected human x_5 who finally discontinue using ART treatment	$\frac{1}{time}$.
u	ART treatment rate	$\frac{1}{time}$.
ξ_4	Transition due to decreased awareness for the use of treatment	$\frac{1}{time}$.
ξ_5	transition due to increased awareness in the use of treatment	$\frac{1}{time}$.
μ	Natural death rate	$\frac{1}{time}$.
δ	Death rate caused by AIDS	$\frac{1}{time}$

3. Analysis of mathematical model

In this section , mathematical analysis of the system 1 will be given to discuss about the existence and local stability criteria of the equilibrium point (numerically and analytically), and also the basic reproduction number of system 1. We begin with a simpler model in the next subsection, when there is no ART treatment implemented in to the model.

3.1. Mathematical model without ART treatment

In this subsection, a mathematical model analysis for a simpler model of system 1 will be discuss. A simplification is taken in the form of when $u = 0$, which means that no more compartment of infected human who undergo with an ART treatment (x_4, x_5, x_6). With this simplification, the model in system 1 is simplified into :

$$\begin{aligned}
\frac{dx_1}{dt} &= A - \frac{\beta_a x_1 x_2}{N_x} - \frac{\beta_c x_1 x_3}{N_x} - \frac{\beta_s x_1 x_7}{N_x} - \mu x_1 \\
\frac{dx_2}{dt} &= \frac{\beta_a x_1 x_2}{N_x} + \frac{\beta_c x_1 x_3}{N_x} + \frac{\beta_s x_1 x_7}{N_x} - \gamma_a x_2 - \mu x_2 \\
\frac{dx_3}{dt} &= -\mu\, x_3 + \gamma_a x_2 - \gamma_c x_3 \\
\frac{dx_7}{dt} &= -\mu\, x_7 - \delta\, x_7 + \gamma_c x_3,
\end{aligned} \tag{2}$$

with initial condition x_{10}, x_{20}, x_{30} and x_{70} are given.

There are two equilibrium point of system 2, i.e the disease free equilibrium when there are no infected human in the system, which given by :

$$DFE = \left\{ x_1 = \frac{A}{\mu}, x_2 = 0, x_3 = 0, x_7 = 0 \right\}. \tag{3}$$

IOP Conf. Series: Journal of Physics: Conf. Series **974** (2018) 012035 doi:10.1088/1742-6596/974/1/012035

The other equilibrium is the endemic equilibrium, which represented as a condition when all compartments in system 2 are positive, which given by :

$$EE = \{x_1 = \bar{x_1}, x_2 = \bar{x_2}, x_3 = \bar{x_3}, x_7 = \bar{x_7}\}. \tag{4}$$

where:

$\bar{x_1} = \frac{A}{R_0\mu}$

$\bar{x_2} = \frac{\mu N_x(\mu+\gamma_c)(\mu+\delta)(R_0-1)}{k_1}$

$\bar{x_3} = \frac{\gamma_a,\mu N_x(\mu+\delta)(R_0-1)}{k_2}$

$\bar{x_7} = \frac{\gamma_a\gamma_c\mu N_x(R_0-1)}{k_3}$

$k_1 = \delta\,\mu\,\beta_a + \delta\,\beta_a\gamma_c + \delta\,\beta_c\gamma_a + \mu^2\beta_a + \mu\,\beta_a\gamma_c + \mu\,\beta_c\gamma_a + \beta_s\gamma_a\gamma_c$

$k_2 = \delta\,\mu\,\beta_a + \delta\,\beta_a\gamma_c + \delta\,\beta_c\gamma_a + \mu^2\beta_a + \mu\,\beta_a\gamma_c + \mu\,\beta_c\gamma_a + \beta_s\gamma_a\gamma_c$

$k_3 = \delta\,\mu\,\beta_a + \delta\,\beta_a\gamma_c + \delta\,\beta_c\gamma_a + \mu^2\beta_a + \mu\,\beta_a\gamma_c + \mu\,\beta_c\gamma_a + \beta_s\gamma_a\gamma_c$

and

$\mathcal{R}_0^* = \frac{A\left(\delta\,\mu\,\beta_a+\delta\,\beta_a\gamma_c+\delta\,\beta_c\gamma_a+\mu^2\beta_a+\mu\,\beta_a\gamma_c+\mu\,\beta_c\gamma_a+\beta_s\gamma_a\gamma_c\right)}{\mu\,N_x(\mu+\gamma_a)(\mu+\gamma_c)(\delta+\mu)}.$

Since number of human are should be positive, we have that EE will have a biological interpretation ($x_i > 0$) if and only if

$$\mathcal{R}_0^* = \frac{A\left(\delta\,\mu\,\beta_a + \delta\,\beta_a\gamma_c + \delta\,\beta_c\gamma_a + \mu^2\beta_a + \mu\,\beta_a\gamma_c + \mu\,\beta_c\gamma_a + \beta_s\gamma_a\gamma_c\right)}{\mu\,N_x\,(\mu+\gamma_a)\,(\mu+\gamma_c)\,(\delta+\mu)} > 1. \tag{5}$$

The local stability of the DFE is analyzed using the Jacobian matrix of system 2 which evaluated in DFE. The Jacobian matrix is given by :

$$J = \begin{bmatrix} -\mu & -\frac{\beta_a A}{\mu N_x} & -\frac{\beta_c A}{\mu N_x} & -\frac{\beta_s A}{\mu N_x} \\ 0 & \frac{\beta_a A}{\mu N_x} - \gamma_a - \mu & \frac{\beta_c A}{\mu N_x} & \frac{\beta_s A}{\mu N_x} \\ 0 & \gamma_a & -\mu-\gamma_c & 0 \\ 0 & 0 & \gamma_c & -\delta-\mu \end{bmatrix}. \tag{6}$$

The characteristic polynomial of matrix J is given by :

$$\lambda^4 h_0 + \lambda^3 h_1 + \lambda^2 h_2 + \lambda\, h_3 + h_4 = 0. \tag{7}$$

where

$$h_0 = 1$$

$$h_1 = -\frac{-\delta\mu N_x - 4\mu^2 N_x - \mu N_x\gamma_a - \mu N_x\gamma_c + A\beta_a}{\mu N_x}$$

$$
\begin{aligned}
h_2 &= -\frac{1}{\mu N_x}\left(-3\delta\mu^2 N_x - \delta\mu N_x\gamma_a - \delta\mu N_x\gamma_c - 6\mu^3 N_x - 3\mu^2 N_x\gamma_a - 3\mu^2 N_x\gamma_c\right. \\
&\quad \left.-\mu N_x\gamma_a\gamma_c + A\delta\beta_a + 3A\mu\beta_a + A\beta_a\gamma_c + A\beta_c\gamma_a\right) \\
h_3 &= \frac{1}{\mu N_x}\left(4\mu^4 N_x + 3N_x\left(\delta+\gamma_a+\gamma_c\right)\mu^3 + \left(\left(\left(2\delta+2\gamma_c\right)\gamma_a + 2\delta\gamma_c\right)N_x - 3A\beta_a\right)\mu^2\right. \\
&\quad \left.+\left(\delta N_x\gamma_a\gamma_c - 2A\left(\beta_c\gamma_a + \beta_a\left(\delta+\gamma_c\right)\right)\right)\mu - A\left(\left(\delta\beta_c + \beta_s\gamma_c\right)\gamma_a + \delta\beta_a\gamma_c\right)\right) \\
h_4 &= \frac{1}{N_x}\left(\mu^4 N_x + N_x\left(\delta+\gamma_a+\gamma_c\right)\mu^3 + \left(\left(\left(\delta+\gamma_c\right)\gamma_a + \delta\gamma_c\right)N_x - A\beta_a\right)\mu^2\right. \\
&\quad \left.+\left(\delta N_x\gamma_a\gamma_c - A\left(\beta_c\gamma_a + \beta_a\left(\delta+\gamma_c\right)\right)\right)\mu - \left(\left(\delta\beta_c + \beta_s\gamma_c\right)\gamma_a + \delta\beta_a\gamma_c\right)A\right).
\end{aligned}
$$

Using the Routh-Hurwitz criterion [12], all eigenvalues of J will be negative if and only if :

$$
\begin{cases} h_1 > 0, \;\; h_3 > 0, \;\; h_4 > 0 \\ h_1 h_2 h_3 > h_4 \end{cases} \tag{8}
$$

Next, the basic reproduction number of system 2 will be analyzed. Basic reproduction number is define as the expected number of secondary cases caused by one primary case in a closed population during one infection period [10]. The illustration of basic reproduction number $\mathcal{R}_0$ when it is larger than one, smaller than one, or equal to one is given in Figure 2. There are some

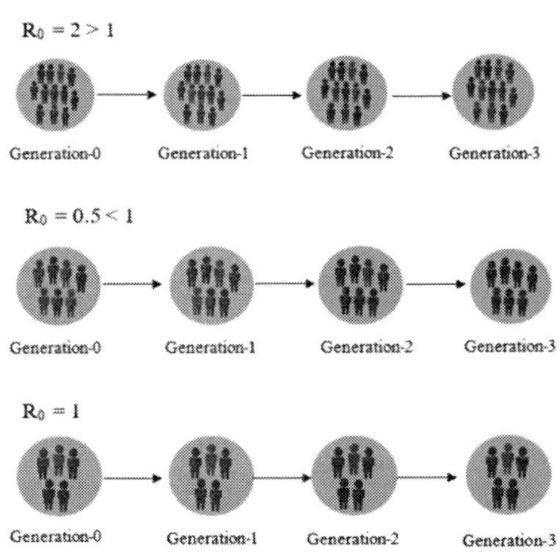

Figure 2. Interpretation of disease spread using $\mathcal{R}_0$.

method to calculate the basic reproduction number, such as using next-generation matrix [10], graph theory [11], etc. In this article, next-generation matrix approach will be implemented to calculate the basic reproduction number of system 2. Using the next-generation matrix approach implemented to system 2, the basic reproduction number as the spectral radius of the respected next-generation matrix is given by :

$$
\mathcal{R}_0 = \frac{A\left(\delta\,\mu\,\beta_a + \delta\,\beta_a\gamma_c + \delta\,\beta_c\gamma_a + \mu^2\beta_a + \mu\,\beta_a\gamma_c + \mu\,\beta_c\gamma_a + \beta_s\gamma_a\gamma_c\right)}{\mu\,N_x\left(\mu+\gamma_a\right)\left(\mu+\gamma_c\right)\left(\delta+\mu\right)}. \tag{9}
$$

Let us define the parameters value for system 2 as follows : $A = \frac{1000}{65\times365}, \mu = \frac{1}{65\times365}, N_x = 1000, \beta_a = \frac{0.0025}{1000}, \beta_c = 0.75\beta_a, \beta_s = 0.5\beta_a, \gamma_a = \frac{1}{5\times365}, \gamma_c = \frac{1}{3\times365}$. With this parameters value,

IOP Conf. Series: Journal of Physics: Conf. Series **974** (2018) 012035 doi:10.1088/1742-6596/974/1/012035

the basic reproduction number $\mathcal{R}_0$ reach 4.5781 which leads in to the existence of the endemic equilibrium, while the disease free equilibrium became unstable. Also with this parameters, the endemic equilibrium is given by

$$\bar{x}_1 = 164.6141532, \bar{x}_2 = 59.67041757, \bar{x}_3 = 34.22273950, \bar{x}_7 = 0.4462119387. \tag{10}$$

Now, we want to see how the parameters sensitivities could determine the magnitude of each compartment in the endemic equilibrium. This will be perform with differentiating the variables in endemic equilibrium respect to the parameters, respectively for $\beta_a, \beta_c, \beta_s, \gamma_a, \gamma_c$ and δ. Do the same thing for variables x_2, x_3 and x_7. This results is given in Jacobian.

$$J = \begin{bmatrix} -45921.12618 & -26337.11649 & -343.3955311 & 188056.5074 & 51667.0652 & 6.128373 \\ 3280.08045 & 1881.222623 & 24.5282524 & -114552.6545 & -3690.50458 & -0.4377402 \\ 1881.222608 & 1078.936498 & 14.06767408 & -3242.81700 & -37937.25243 & -0.2510568 \\ 24.5282523 & 14.06767406 & 0.1834208552 & -42.2813513 & -6.0414070 & -6.373893674 \end{bmatrix}. \tag{11}$$

Based on the matrix J, it can be seen that γ_a is the most influential parameter that can determine the magnitude of x_1 in the endemic equilibrium. Since the sign is positive, it means that larger the transition rate from acutely infected into chronic infected will enlarge total of susceptible human. Using same approach, the order of the most influential parameters in determining the magnitude of the variable x_1 at the endemic equilibrium point is $\gamma_a, \gamma_c, \beta_a, \beta_c, \beta_s$ and δ. Using the same approach applied to all row of matrix J, we find that γ_a is the most influent parameter in determining the size of each variable in the endemic equilibrium point.

3.2. Mathematical model with ART treatment

In this section, we analyzed the complete model of HIV spread with ART treatment, which given in system 1. Using the same approach with in the previous section, the disease free equilibrium point of system 1 is given by :

$$DFE = \left\{ x_1 = \frac{A}{\mu}, x_2 = 0, x_3 = 0, x_4 = 0, x_5 = 0, x_6 = 0, x_7 = 0 \right\}. \tag{12}$$

Using the Jacobian matrix approach, the local stability of system 1 is determined by roots of 7 degree polynomial which given by :

$$\lambda^7 a_0 + \lambda^6 a_1 + \lambda^5 a_2 + \lambda^4 a_3 + \lambda^3 a_4 + \lambda^2 a_5 + \lambda a_6 + a_7 = 0, \tag{13}$$

where :

$$a_0 = 1, a_1 = -\frac{Q_1}{\mu N_x}, a_2 = -\frac{Q_2}{\mu N_x}, a_3 = -\frac{Q_3}{\mu N_x}, a_4 = -\frac{Q_4}{\mu N_x},$$

$$a_5 = -\frac{Q_5}{\mu N_x}, a_6 = -\frac{Q_6}{\mu N_x}, a_7 = -\frac{Q_7}{N_x}$$

$$Q_1 = -\mu\left(\gamma_a + \gamma_b + \gamma_c + u + 7\mu + \delta + \xi_4 + \xi_5 + \rho_c\right) N_x + A\beta_a$$

$$\begin{aligned} Q_2 = & -21 N_x \mu^3 - 6 N_x \left(\gamma_a + \gamma_b + \gamma_c + u + \delta + \xi_4 + \xi_5 + \rho_c\right) \mu^2 \\ & + \begin{pmatrix} ((-\gamma_b - \gamma_c - u - \delta - \xi_4 - \xi_5 - \rho_c)\gamma_a \\ + (-\gamma_c - u - \delta - \xi_4 - \xi_5 - \rho_c)\gamma_b \\ + (-\gamma_c - u - \xi_4 - \xi_5 - \rho_c)\delta \\ + (-\xi_4 - \xi_5 - \rho_c)\gamma_c \\ + (-u - \xi_4)\rho_c - u(\xi_4 + \xi_5)) N_x + 6A\beta_a \end{pmatrix} \mu \\ & + A\left(\beta_c \gamma_a + \beta_a\left(\gamma_b + \gamma_c + u + \delta + \xi_4 + \xi_5 + \rho_c\right)\right) \end{aligned}$$

$$\begin{aligned} Q_3 = & -35 N_x \mu^5 - 20 N_x \left(\gamma_a + \gamma_b + \gamma_c + u + \delta + \xi_4 + \xi_5 + \rho_c\right) \mu^4 \\ & + \begin{pmatrix} ((-10\gamma_a - 10\gamma_c - 10u - 10\delta - 10\xi_4 - 10\xi_5 - 10\rho_c)\gamma_b \\ + (-10\gamma_c - 10u - 10\delta - 10\xi_4 - 10\xi_5 - 10\rho_c)\gamma_a \\ + (-10\gamma_c - 10u - 10\xi_4 - 10\xi_5 - 10\rho_c)\delta \\ + (-10\gamma_c - 10u - 10\xi_4)\rho_c - 10(\xi_4 + \xi_5)(\gamma_c + u)) N_x + 20A\beta_a \end{pmatrix} \mu^3 \end{aligned}$$

$$
+\left(\begin{array}{c}
(((-4\gamma_c-4u-4\delta-4\xi_4-4\xi_5-4\rho_c)\,\gamma_a+(-4\gamma_c-4u-4\xi_4-4\xi_5-4\rho_c)\,\delta\\
+\ (-4\gamma_c-4u-4\xi_4)\,\rho_c-4\,(\xi_4+\xi_5)\,(\gamma_c+u))\,\gamma_b+((-4\gamma_c-4u-4\xi_4-4\xi_5-4\rho_c)\,\delta\\
+\ (-4\gamma_c-4u-4\xi_4)\,\rho_c-4\,(\xi_4+\xi_5)\,(\gamma_c+u))\,\gamma_a\\
+\ ((-4\gamma_c-4u-4\xi_4)\,\rho_c-4\,(\xi_4+\xi_5)\,(\gamma_c+u))\,\delta-4\xi_4\rho_c\,(\gamma_c+u))\,N_x\\
+10\,(\beta_a\gamma_b+\beta_c\gamma_a+\beta_a\,(\gamma_c+u+\delta+\xi_4+\xi_5+\rho_c))\,A
\end{array}\right)\mu^2
$$

$$
+\left(\begin{array}{c}
((((-\gamma_c-u-\xi_4-\xi_5-\rho_c)\,\delta+(-\gamma_c-u-\xi_4)\,\rho_c-(\xi_4+\xi_5)\,(\gamma_c+u))\,\gamma_a\\
+\ ((-\gamma_c-u-\xi_4)\,\rho_c-(\xi_4+\xi_5)\,(\gamma_c+u))\,\delta-\xi_4\rho_c\,(ur_c+\gamma_c))\,\gamma_b\\
+\ (((-\gamma_c-u-\xi_4)\,\rho_c-(\xi_4+\xi_5)\,(\gamma_c+u))\,\delta-\xi_4\rho_c\,(\gamma_c+u))\,\gamma_a-\delta\xi_4\rho_c\,(\gamma_c+u))\,N_x\\
+4\,((\beta_c\gamma_a+\beta_a\,(\gamma_c+u+\delta+\xi_4+\xi_5+\rho_c))\,\gamma_b+(u\beta_c\theta_h+\delta\beta_c+\beta_c\rho_c+\beta_c\xi_4+\beta_c\xi_5+\beta_s\gamma_c)\,\gamma_a\\
+\ ((\gamma_c+u+\xi_4+\xi_5+\rho_c)\,\delta+(\gamma_c+u+\xi_4)\,\rho_c+(\xi_4+\xi_5)\,(\gamma_c+u))\,\beta_a)\,A
\end{array}\right)\mu
$$

$$
+A\left(\begin{array}{c}
((u\beta_c\theta_h+\delta\beta_c+\beta_c\rho_c+\beta_c\xi_4+\beta_c\xi_5+\beta_s\gamma_c)\,\gamma_a\\
+\ ((\gamma_c+u+\xi_4+\xi_5+\rho_c)\,\delta+(\gamma_c+u+\xi_4)\,\rho_c\\
+\ (\xi_4+\xi_5)\,(\gamma_c+u))\,\beta_a)\,\gamma_b+(\beta_c\,(u\theta_h+\rho_c+\xi_4+\xi_5)\,\delta\\
+\ (u\beta_c\theta_h+\beta_c\xi_4+\beta_s\gamma_c)\,\rho_c+(u\beta_c\theta_l+\beta_s\gamma_c)\,\xi_4+\xi_5\,(u\beta_c\theta_h+\beta_s\gamma_c))\,\gamma_a\\
+\beta_a\,(((\gamma_c+u+\xi_4)\,\rho_c+(\xi_4+\xi_5)\,(\gamma_c+u))\,\delta+\xi_4\rho_c\,(\gamma_c+u))
\end{array}\right)
$$

$$
Q_4=-35\,\mu^5N_x-20\,N_x\,(\gamma_a+\gamma_b+\gamma_c+u+\delta+\xi_4+\xi_5+\rho_c)\,\mu^4
$$

$$
+\left(\begin{array}{c}
(-10\,\delta\,uN_x-10\,\delta\,N_x\gamma_a-10\,\delta\,N_x\gamma_b-10\,\delta\,N_x\gamma_c\\
-\ 10\,\delta\,N_x\rho_c-10\,\delta\,N_x\xi_4-10\,\delta\,N_x\xi_5\\
-\ 10\,uN_x\gamma_a-10\,uN_x\gamma_b-10\,uN_x\rho_c-10\,uN_x\xi_4\\
-\ 10\,uN_x\xi_5-10\,N_x\gamma_a\gamma_b-10\,N_x\gamma_a\gamma_c-10\,N_x\gamma_a\rho_c-10\,N_x\gamma_a\xi_4\\
-\ 10\,N_x\gamma_a\xi_5-10\,N_x\gamma_b\gamma_c-10\,N_x\gamma_b\rho_c-10\,N_x\gamma_b\xi_4-10\,N_x\gamma_b\xi_5-10\,N_x\gamma_c\rho_c\\
-\ 10\,N_x\gamma_c\xi_4-10\,N_x\gamma_c\xi_5-10\,N_x\rho_c\xi_4+20\,A\beta_a)
\end{array}\right)\mu^3
$$

$$
+\left(\begin{array}{c}
(-4\,\delta\,uN_x\gamma_a-4\,\delta\,uN_x\gamma_b-4\,\delta\,uN_x\rho_c-4\,\delta\,uN_x\xi_4\\
-\ 4\,\delta\,uN_x\xi_5-4\,\delta\,N_x\gamma_a\gamma_b-4\,\delta\,N_x\gamma_a\gamma_c-4\,\delta\,N_x\gamma_a\rho_c\\
-\ 4\,\delta\,N_x\gamma_a\xi_4-4\,\delta\,N_x\gamma_a\xi_5-4\,\delta\,N_x\gamma_b\gamma_c-4\,\delta\,N_x\gamma_b\rho_c\\
-\ 4\,\delta\,N_x\gamma_b\xi_4-4\,\delta\,N_x\gamma_b\xi_5-4\,\delta\,N_x\gamma_c\rho_c-4\,\delta\,N_x\gamma_c\xi_4-4\,\delta\,N_x\gamma_c\xi_5\\
-\ 4\,\delta\,N_x\rho_c\xi_4-4\,uN_x\gamma_a\gamma_b-4\,uN_x\gamma_a\rho_c-4\,uN_x\gamma_a\xi_4-4\,uN_x\gamma_a\xi_5-4\,uN_x\gamma_b\rho_c\\
-\ 4\,uN_x\gamma_b\xi_4-4\,uN_x\gamma_b\xi_5-4\,uN_x\rho_c\xi_4-4\,N_x\gamma_a\gamma_b\gamma_c-4\,N_x\gamma_a\gamma_b\rho_c-4\,N_x\gamma_a\gamma_b\xi_4-4\,N_x\gamma_a\gamma_b\xi_5\\
-\ 4\,N_x\gamma_a\gamma_c\rho_c-4\,N_x\gamma_a\gamma_c\xi_4-4\,N_x\gamma_a\gamma_c\xi_5-4\,N_x\gamma_a\rho_c\xi_4-4\,N_x\gamma_b\gamma_c\rho_c-4\,N_x\gamma_b\gamma_c\xi_4\\
-\ 4\,N_x\gamma_b\gamma_c\xi_5-4\,N_x\gamma_b\rho_c\xi_4-4\,N_x\gamma_c\rho_c\xi_4+10\,A\delta\,\beta_a+10\,Au\beta_a\\
+\ 10\,A\beta_a\gamma_b+10\,A\beta_a\gamma_c+10\,A\beta_a\rho_c+10\,A\beta_a\xi_4+10\,A\beta_a\xi_5\\
+\ 10\,A\beta_c\gamma_a)
\end{array}\right)\mu^2
$$

$$
+\left(\begin{array}{c}
(-uN_x\gamma_br_c\rho_c\xi_4+4\,Au\beta_c\gamma_a\theta_h-\delta\,uN_x\gamma_a\gamma_b-\delta\,uN_x\gamma_a\rho_c-\delta\,uN_x\gamma_a\xi_4-\delta\,uN_x\gamma_a\xi_5\\
-\ \delta\,uN_x\gamma_b\rho_c-\delta\,uN_x\gamma_b\xi_4-\delta\,uN_x\gamma_b\xi_5-\delta\,uN_x\rho_c\xi_4-\delta\,N_x\gamma_a\gamma_b\gamma_c\\
-\ \delta\,N_x\gamma_a\gamma_b\rho_c-\delta\,N_x\gamma_a\gamma_b\xi_4-\delta\,N_x\gamma_a\gamma_b\xi_5-\delta\,N_x\gamma_a\gamma_c\rho_c\\
-\ \delta\,N_x\gamma_a\gamma_c\xi_4-\delta\,N_x\gamma_a\gamma_c\xi_5-\delta\,N_x\gamma_a\rho_c\xi_4-\delta\,N_x\gamma_b\gamma_c\rho_c\\
-\ \delta\,N_x\gamma_b\gamma_c\xi_4-\delta\,N_x\gamma_b\gamma_c\xi_5-\delta\,N_x\gamma_b\rho_c\xi_4\\
-\ \delta\,N_x\gamma_c\rho_c\xi_4-uN_x\gamma_a\gamma_b\rho_c-uN_x\gamma_a\gamma_b\xi_4-uN_x\gamma_a\gamma_b\xi_5-uN_x\gamma_a\rho_c\xi_4-N_x\gamma_a\gamma_b\gamma_c\rho_c\\
-\ N_x\gamma_a\gamma_b\gamma_c\xi_4-N_x\gamma_a\gamma_b\gamma_c\xi_5-N_x\gamma_a\gamma_b\rho_c\xi_4\\
-\ N_x\gamma_a\gamma_c\rho_c\xi_4-N_x\gamma_b\gamma_c\rho_c\xi_4+4\,A\delta\,u\beta_a+4\,A\delta\,\beta_a\gamma_b+4\,A\delta\,\beta_a\gamma_c+4\,A\delta\,\beta_a\rho_c\\
+\ 4\,A\delta\,\beta_a\xi_4+4\,A\delta\,\beta_a\xi_5+4\,A\delta\,\beta_c\gamma_a+4\,Au\beta_a\gamma_b+4\,Au\beta_a\rho_c+4\,Au\beta_a\xi_4\\
+\ 4\,Au\beta_a\xi_5+4\,A\beta_a\gamma_b\gamma_c+4\,A\beta_a\gamma_b\rho_c+4\,A\beta_a\gamma_b\xi_4\\
+\ 4\,A\beta_a\gamma_b\xi_5+4\,A\beta_a\gamma_c\rho_c+4\,A\beta_a\gamma_c\xi_4+4\,A\beta_a\gamma_c\xi_5+4\,A\beta_a\rho_c\xi_4+4\,A\beta_c\gamma_a\gamma_b\\
+\ 4\,A\beta_c\gamma_a\rho_c+4\,A\beta_c\gamma_a\xi_4+4\,A\beta_c\gamma_a\xi_5+4\,A\beta_s\gamma_a\gamma_c)
\end{array}\right)\mu
$$

$$
+\left(\begin{array}{c}
(\delta\,u\beta_c\gamma_a\theta_h+u\beta_c\gamma_a\gamma_b\theta_h+u\beta_c\gamma_a\rho_c\theta_h\\
+\ u\beta_c\gamma_a\theta_h\xi_5+u\beta_c\gamma_a\theta_l\xi_4+\delta\,u\beta_a\gamma_b\\
+\ \delta\,u\beta_a\rho_c+\delta\,u\beta_a\xi_4+\delta\,u\beta_a\xi_5+\delta\,\beta_a\gamma_b\gamma_c+\delta\,\beta_a\gamma_b\rho_c+\delta\,\beta_a\gamma_b\xi_4+\delta\,\beta_a\gamma_b\xi_5+\delta\,\beta_a\gamma_c\rho_c+\delta\,\beta_a\gamma_c\xi_4\\
+\ \delta\,\beta_a\gamma_c\xi_5+\delta\,\beta_a\rho_c\xi_4+\delta\,\beta_c\gamma_a\gamma_b\\
+\ \delta\,\beta_c\gamma_a\rho_c+\delta\,\beta_c\gamma_a\xi_4+\delta\,\beta_c\gamma_a\xi_5+u\beta_a\gamma_b\rho_c+u\beta_a\gamma_b\xi_4+u\beta_a\gamma_b\xi_5+u\beta_a\rho_c\xi_4\\
+\ \beta_a\gamma_b\gamma_c\rho_c+\beta_a\gamma_b\gamma_c\xi_4+\beta_a\gamma_b\gamma_c\xi_5+\beta_a\gamma_b\rho_c\xi_4+\beta_a\gamma_c\rho_c\xi_4+\beta_c\gamma_a\gamma_b\rho_c\\
+\ \beta_c\gamma_a\gamma_b\xi_4+\beta_c\gamma_a\gamma_b\xi_5+\beta_c\gamma_a\rho_c\xi_4+\beta_s\gamma_a\gamma_b\gamma_c+\beta_s\gamma_a\gamma_c\rho_c+\beta_s\gamma_a\gamma_c\xi_4+\beta_s\gamma_a\gamma_c\xi_5)
\end{array}\right)A
$$

$$
Q_5=-21\mu^6N_x-15N_x\,(\gamma_a+\gamma_b+\gamma_c+u+\delta+\xi_4+\xi_5+\rho_c)\,\mu^5
$$

$$
+\left(\begin{array}{c}
((-10\gamma_a-10\gamma_c-10u-10\delta-10\xi_4-10\xi_5-10\rho_c)\,\gamma_b\\
+\ (-10\gamma_c-10u-10\delta-10\xi_4-10\xi_5-10\rho_c)\,\gamma_a\\
+\ (-10\gamma_c-10u-10\xi_4-10\xi_5-10\rho_c)\,\delta\\
+\ (-10\gamma_c-10u-10\xi_4)\,\rho_c-10\,(\xi_4+\xi_5)\,(\gamma_c+u))\,N_x+15A\beta_a
\end{array}\right)\mu^4
$$

$$
+\left(\begin{array}{c}
(((-6\gamma_c-6u-6\delta-6\xi_4-6\xi_5-6\rho_c)\,\gamma_a+(-6\gamma_c-6u-6\xi_4-6\xi_5-6\rho_c)\,\delta\\
+\ (-6\gamma_c-6u-6\xi_4)\,\rho_c-6\,(\xi_4+\xi_5)\,(\gamma_c+u))\,\gamma_b\\
+\ ((-6\gamma_c-6u-6\xi_4-6\xi_5-6\rho_c)\,\delta+(-6\gamma_c-6u-6\xi_4)\,\rho_c\\
-6\ (\xi_4+\xi_5)\,(\gamma_c+u))\,\gamma_a+((-6\gamma_c-6u-6\xi_4)\,\rho_c\\
-\ 6\,(\xi_4+\xi_5)\,(\gamma_c+u))\,\delta-6\xi_4\rho_c\,(\gamma_c+u))\,N_x\\
+10\,(\beta_a\gamma_b+\beta_c\gamma_a+\beta_a\,(\gamma_c+u+\delta+\xi_4+\xi_5+\rho_c))\,A
\end{array}\right)\mu^3
$$

$$
+\left(\begin{array}{c}
((((-3\gamma_c-3u-3\xi_4-3\xi_5-3\rho_c)\,\delta+(-3\gamma_c-3u-3\xi_4)\,\rho_c-3\,(\xi_4+\xi_5)\,(\gamma_c+u))\,\gamma_a\\
+\ ((-3\gamma_c-3u-3\xi_4)\,\rho_c-3\,(\xi_4+\xi_5)\,(\gamma_c+u))\,\delta-3\xi_4\rho_c\,(ur_c+\gamma_c))\,\gamma_b\\
+\ (((-3\gamma_c-3u-3\xi_4)\,\rho_c-3\,(\xi_4+\xi_5)\,(\gamma_c+u))\,\delta-3\xi_4\rho_c\,(\gamma_c+u))\,\gamma_a-3\delta\xi_4\rho_c\,(\gamma_c+u))\,N_x\\
+6\,((\beta_c\gamma_a+\beta_a\,(\gamma_c+u+\delta+\xi_4+\xi_5+\rho_c))\,\gamma_b+(u\beta_c\theta_h+\delta\beta_c+\beta_c\rho_c+\beta_c\xi_4+\beta_c\xi_5+\beta_s\gamma_c)\,\gamma_a\\
+\ ((\gamma_c+u+\xi_4+\xi_5+\rho_c)\,\delta+(\gamma_c+u+\xi_4)\,\rho_c+(\xi_4+\xi_5)\,(\gamma_c+u))\,\beta_a)\,A
\end{array}\right)\mu^2
$$

$$
+\left(\begin{array}{c}
(((((-\gamma_c-u-\xi_4)\,\rho_c-(\xi_4+\xi_5)\,(\gamma_c+u))\,\delta-\xi_4\rho_c\,(ur_c+\gamma_c))\,\gamma_a-\delta\xi_4\rho_c\,(ur_c+\gamma_c))\,\gamma_b\\
-\ \gamma_a\delta\xi_4\rho_c\,(\gamma_c+u))\,N_x+3\,(((u\beta_c\theta_h+\delta\beta_c+\beta_c\rho_c+\beta_c\xi_4+\beta_c\xi_5+\beta_s\gamma_c)\,\gamma_a\\
+\ ((\gamma_c+u+\xi_4+\xi_5+\rho_c)\,\delta+(\gamma_c+u+\xi_4)\,\rho_c+(\xi_4+\xi_5)\,(\gamma_c+u))\,\beta_a)\,\gamma_b\\
+\ (\beta_c\,(u\theta_h+\rho_c+\xi_4+\xi_5)\,\delta+(u\beta_c\theta_h+\beta_c\xi_4+\beta_s\gamma_c)\,\rho_c+(u\beta_c\theta_l+\beta_s\gamma_c)\,\xi_4\\
+\ \xi_5\,(u\beta_c\theta_h+\beta_s\gamma_c))\,\gamma_a+\beta_a\,(((\gamma_c+u+\xi_4)\,\rho_c+(\xi_4+\xi_5)\,(\gamma_c+u))\,\delta+\xi_4\rho_c\,(\gamma_c+u)))\,A
\end{array}\right)\mu
$$

$$
+A\left(\begin{array}{c}
(((\beta_c\,(u\theta_h+\rho_c+\xi_4+\xi_5)\,\delta+(u\beta_c\theta_h+\beta_c\xi_4+\beta_s\gamma_c)\,\rho_c+(u\beta_c\theta_l+\beta_s\gamma_c)\,\xi_4\\
+\ \xi_5\,(u\beta_c\theta_h+\beta_s\gamma_c))\,\gamma_a+\beta_a\,(((\gamma_c+u+\xi_4)\,\rho_c+(\xi_4+\xi_5)\,(\gamma_c+u))\,\delta+\xi_4\rho_c\,(ur_c+\gamma_c)))\,\gamma_b\\
+\ (((u\theta_h+\xi_4)\,\rho_c+u\,(\theta_h\xi_5+\theta_l\xi_4))\,\beta_c\delta+\xi_4\rho_c\,(u\beta_c+\beta_s\gamma_c))\,\gamma_a+\beta_a\delta\xi_4\rho_c\,(\gamma_c+u))
\end{array}\right)
$$

$$
Q_6=-7\mu^7N_x-6N_x\,(\gamma_a+\gamma_b+\gamma_c+u+\delta+\xi_4+\xi_5+\rho_c)\,\mu^6
$$

$$
+\left(\begin{array}{c}
((-5\gamma_a-5\gamma_c-5u-5\delta-5\xi_4-5\xi_5-5\rho_c)\,\gamma_b\\
+\ (-5\gamma_c-5u-5\delta-5\xi_4-5\xi_5-5\rho_c)\,\gamma_a\\
+\ (-5\gamma_c-5u-5\xi_4-5\xi_5-5\rho_c)\,\delta\\
+\ (-5\gamma_c-5u-5\xi_4)\,\rho_c-5\,(\xi_4+\xi_5)\,(\gamma_c+u))\,N_x+6A\beta_a
\end{array}\right)\mu^5
$$

$$
\begin{aligned}
&+\left(\begin{array}{c}
(((-4\gamma_c-4u-4\delta-4\xi_4-4\xi_5-4\rho_c)\,\gamma_a+(-4\gamma_c-4u-4\xi_4-4\xi_5-4\rho_c)\,\delta\\
+\,(-4\gamma_c-4u-4\xi_4)\,\rho_c-4\,(\xi_4+\xi_5)\,(\gamma_c+u))\,\gamma_b+((-4\gamma_c-4u-4\xi_4-4\xi_5-4\rho_c)\,\delta\\
+\,(4\gamma_c-4u-4\xi_4)\,\rho_c-4\,(\xi_4+\xi_5)\,(\gamma_c+u))\,\gamma_a+((-4\gamma_c-4u-4\xi_4)\,\rho_c\\
-4\,(\xi_4+\xi_5)\,(\gamma_c+u))\,\delta-4\xi_4\rho_c\,(\gamma_c+u))\,N_x+5\,(\beta_a\gamma_b+\beta_c\gamma_a\\
+\,\beta_a\,(\gamma_c+u+\delta+\xi_4+\xi_5+\rho_c))\,A
\end{array}\right)\mu^4\\
&+\left(\begin{array}{c}
((((-3\gamma_c-3u-3\xi_4-3\xi_5-3\rho_c)\,\delta+(-3\gamma_c-3u-3\xi_4)\,\rho_c-3\,(\xi_4+\xi_5)\,(\gamma_c+u))\,\gamma_a\\
+\,((-3\gamma_c-3u-3\xi_4)\,\rho_c-3\,(\xi_4+\xi_5)\,(\gamma_c+u))\,\delta-3\xi_4\rho_c\,(ur_c+\gamma_c))\,\gamma_b\\
+\,(((-3\gamma_c-3u-3\xi_4)\,\rho_c-3\,(\xi_4+\xi_5)\,(\gamma_c+u))\,\delta-3\xi_4\rho_c\,(\gamma_c+u))\,\gamma_a\\
-\,3\delta\xi_4\rho_c\,(\gamma_c+u))\,N_x+4\,((\beta_c\gamma_a+\beta_a\,(\gamma_c+u+\delta+\xi_4+\xi_5+\rho_c))\,\gamma_b\\
+\,(u\beta_c\theta_h+\delta\beta_c+\beta_c\rho_c+\beta_c\xi_4+\beta_c\xi_5+\beta_s\gamma_c)\,\gamma_a+((\gamma_c+u+\xi_4+\xi_5+\rho_c)\,\delta\\
+\,(\gamma_c+u+\xi_4)\,\rho_c+(\xi_4+\xi_5)\,(\gamma_c+u))\,\beta_a)\,A
\end{array}\right)\mu^3\\
&+\left(\begin{array}{c}
(((((-2\gamma_c-2u-2\xi_4)\,\rho_c-2\,(\xi_4+\xi_5)\,(\gamma_c+u))\,\delta-2\xi_4\rho_c\,(ur_c+\gamma_c))\,\gamma_a\\
-\,2\delta\xi_4\rho_c\,(ur_c+\gamma_c))\,\gamma_b-2\gamma_a\delta\xi_4\rho_c\,(\gamma_c+u))\,N_x\\
+3\,(((u\beta_c\theta_h+\delta\beta_c+\beta_c\rho_c+\beta_c\xi_4+\beta_c\xi_5+\beta_s\gamma_c)\,\gamma_a+((\gamma_c+u+\xi_4+\xi_5+\rho_c)\,\delta\\
+\,(\gamma_c+u+\xi_4)\,\rho_c+(\xi_4+\xi_5)\,(\gamma_c+u))\,\beta_a)\,\gamma_b\\
+\,(\beta_c\,(u\theta_h+\rho_c+\xi_4+\xi_5)\,\delta+(u\beta_c\theta_h+\beta_c\xi_4+\beta_s\gamma_c)\,\rho_c\\
+\,(u\beta_c\theta_l+\beta_s\gamma_c)\,\xi_4+\xi_5\,(u\beta_c\theta_h+\beta_s\gamma_c))\,\gamma_a+\beta_a\,(((\gamma_c+u+\xi_4)\,\rho_c\\
+\,(\xi_4+\xi_5)\,(\gamma_c+u))\,\delta+\xi_4\rho_c\,(\gamma_c+u)))\,A
\end{array}\right)\mu^2\\
&+\left(\begin{array}{c}
-\gamma_a\gamma_b\delta\xi_4\rho_c\,(ur_c+\gamma_c)\,N_x+2A\,(((\beta_c\,(u\theta_h+\rho_c+\xi_4+\xi_5)\,\delta\\
+\,(u\beta_c\theta_h+\beta_c\xi_4+\beta_s\gamma_c)\,\rho_c+(u\beta_c\theta_l+\beta_s\gamma_c)\,\xi_4+\xi_5\,(u\beta_c\theta_h+\beta_s\gamma_c))\,\gamma_a\\
+\,\beta_a\,(((\gamma_c+u+\xi_4)\,\rho_c+(\xi_4+\xi_5)\,(\gamma_c+u))\,\delta+\xi_4\rho_c\,(ur_c+\gamma_c)))\,\gamma_b\\
+\,(((u\theta_h+\xi_4)\,\rho_c+u\,(\theta_h\xi_5+\theta_l\xi_4))\,\beta_c\delta+\xi_4\rho_c\,(u\beta_c+\beta_s\gamma_c))\,\gamma_a+\beta_a\delta\xi_4\rho_c\,(\gamma_c+u))
\end{array}\right)\mu\\
&+A\left(\begin{array}{c}
((((u\theta_h+\xi_4)\,\rho_c+u\,(\theta_h\xi_5+\theta_l\xi_4))\,\beta_c\delta+\beta_s\xi_4\rho_c\,(ur_c+\gamma_c))\,\gamma_a\\
+\,\beta_a\delta\xi_4\rho_c\,(ur_c+\gamma_c))\,\gamma_b+\beta_c\gamma_a u\delta\xi_4\rho_c
\end{array}\right)
\end{aligned}
$$

$$
\begin{aligned}
Q_7 = &-N_x\mu^7-N_x\,(\gamma_a+\gamma_b+\gamma_c+u+\delta+\xi_4+\xi_5+\rho_c)\,\mu^6\\
&+\left(\begin{array}{c}
((-\gamma_a-\gamma_c-u-\delta-\xi_4-\xi_5-\rho_c)\,\gamma_b+(-\gamma_c-u-\delta-\xi_4-\xi_5-\rho_c)\,\gamma_a\\
+\,(-\gamma_c-u-\xi_4-\xi_5-\rho_c)\,\delta+(-\gamma_c-u-\xi_4)\,\rho_c-(\xi_4+\xi_5)\,(\gamma_c+u))\,N_x+A\beta_a
\end{array}\right)\mu^5\\
&+\left(\begin{array}{c}
(((-\gamma_c-u-\delta-\xi_4-\xi_5-\rho_c)\,\gamma_a+(-\gamma_c-u-\xi_4-\xi_5-\rho_c)\,\delta\\
+\,(-\gamma_c-u-\xi_4)\,\rho_c-(\xi_4+\xi_5)\,(\gamma_c+u))\,\gamma_b\\
+\,((-\gamma_c-u-\xi_4-\xi_5-\rho_c)\,\delta+(-\gamma_c-u-\xi_4)\,\rho_c-(\xi_4+\xi_5)\,(\gamma_c+u))\,\gamma_a\\
+\,((-\gamma_c-u-\xi_4)\,\rho_c-(\xi_4+\xi_5)\,(\gamma_c+u))\,\delta-\xi_4\rho_c\,(\gamma_c+u))\,N_x\\
+\,(\beta_a\gamma_b+\beta_c\gamma_a+\beta_a\,(\gamma_c+u+\delta+\xi_4+\xi_5+\rho_c))\,A
\end{array}\right)\mu^4\\
&+\left(\begin{array}{c}
((((-\gamma_c-u-\xi_4-\xi_5-\rho_c)\,\delta+(-\gamma_c-u-\xi_4)\,\rho_c-(\xi_4+\xi_5)\,(\gamma_c+u))\,\gamma_a\\
+\,((-\gamma_c-u-\xi_4)\,\rho_c-(\xi_4+\xi_5)\,(\gamma_c+u))\,\delta-\xi_4\rho_c\,(ur_c+\gamma_c))\,\gamma_b\\
+\,(((-\gamma_c-u-\xi_4)\,\rho_c-(\xi_4+\xi_5)\,(\gamma_c+u))\,\delta-\xi_4\rho_c\,(\gamma_c+u))\,\gamma_a\\
-\,\delta\xi_4\rho_c\,(\gamma_c+u))\,N_x+((\beta_c\gamma_a+\beta_a\,(\gamma_c+u+\delta+\xi_4+\xi_5+\rho_c))\,\gamma_b\\
+\,(u\beta_c\theta_h+\delta\beta_c+\beta_c\rho_c+\beta_c\xi_4+\beta_c\xi_5+\beta_s\gamma_c)\,\gamma_a+((\gamma_c+u+\xi_4+\xi_5+\rho_c)\,\delta\\
+\,(\gamma_c+u+\xi_4)\,\rho_c+(\xi_4+\xi_5)\,(\gamma_c+u))\,\beta_a)\,A
\end{array}\right)\mu^3\\
&+\left(\begin{array}{c}
(((((-\gamma_c-u-\xi_4)\,\rho_c-(\xi_4+\xi_5)\,(\gamma_c+u))\,\delta-\xi_4\rho_c\,(ur_c+\gamma_c))\,\gamma_a\\
-\,\delta\xi_4\rho_c\,(ur_c+\gamma_c))\,\gamma_b-\gamma_a\delta\xi_4\rho_c\,(\gamma_c+u))\,N_x+(((u\beta_c\theta_h+\delta\beta_c+\beta_c\rho_c+\beta_c\xi_4+\beta_c\xi_5+\beta_s\gamma_c)\,\gamma_a\\
+\,((\gamma_c+u+\xi_4+\xi_5+\rho_c)\,\delta+(\gamma_c+u+\xi_4)\,\rho_c+(\xi_4+\xi_5)\,(\gamma_c+u))\,\beta_a)\,\gamma_b\\
+\,(\beta_c\,(u\theta_h+\rho_c+\xi_4+\xi_5)\,\delta+(u\beta_c\theta_h+\beta_c\xi_4+\beta_s\gamma_c)\,\rho_c+(u\beta_c\theta_l+\beta_s\gamma_c)\,\xi_4\\
+\,\xi_5\,(u\beta_c\theta_h+\beta_s\gamma_c))\,\gamma_a+\beta_a\,(((\gamma_c+u+\xi_4)\,\rho_c+(\xi_4+\xi_5)\,(\gamma_c+u))\,\delta+\xi_4\rho_c\,(\gamma_c+u)))\,A
\end{array}\right)\mu^2\\
&+\left(\begin{array}{c}
-\gamma_a\gamma_b\delta\xi_4\rho_c\,(ur_c+\gamma_c)\,N_x+A\,(((\beta_c\,(u\theta_h+\rho_c+\xi_4+\xi_5)\,\delta+(u\beta_c\theta_h+\beta_c\xi_4+\beta_s\gamma_c)\,\rho_c\\
+\,(u\beta_c\theta_l+\beta_s\gamma_c)\,\xi_4+\xi_5\,(u\beta_c\theta_h+\beta_s\gamma_c))\,\gamma_a+\beta_a\,(((\gamma_c+u+\xi_4)\,\rho_c\\
+\,(\xi_4+\xi_5)\,(\gamma_c+u))\,\delta+\xi_4\rho_c\,(ur_c+\gamma_c)))\,\gamma_b\\
+\,(((u\theta_h+\xi_4)\,\rho_c+u\,(\theta_h\xi_5+\theta_l\xi_4))\,\beta_c\delta+\xi_4\rho_c\,(u\beta_c+\beta_s\gamma_c))\,\gamma_a+\beta_a\delta\xi_4\rho_c\,(\gamma_c+u))
\end{array}\right)\mu\\
&+A\left(\begin{array}{c}
((((u\theta_h+\xi_4)\,\rho_c+u\,(\theta_h\xi_5+\theta_l\xi_4))\,\beta_c\delta+\beta_s\xi_4\rho_c\,(ur_c+\gamma_c))\,\gamma_a\\
+\,\beta_a\delta\xi_4\rho_c\,(ur_c+\gamma_c))\,\gamma_b+\beta_c\gamma_a u\delta\xi_4\rho_c
\end{array}\right).
\end{aligned}
$$

Using the Routh-Hurwitz stability criterion for 7 degree polynomial [12], the system 1 will locally stable if and only if :

Table 2. Local stability criteria of polynomial 13.

$a_0, a_1, a_2, a_3, a_4, a_5, a_6, a_7 > 0,$
$b_1 = \frac{a_1a_2 - a_0a_3}{a_1} > 0, b_2 = \frac{a_1a_4 - a_0a_5}{a_1} > 0, b_3 = a_6,$
$c_1 = \frac{b_1a_3 - a_1b_2}{b_1} > 0, c_2 = \frac{b_1a_5 - a_1b_3}{b_1} > 0,$
$d_1 = \frac{c_1b_2 - b_1c_2}{c_1} > 0, d_2 = a_6 > 0,$
$e_1 = \frac{d_1c_2 - c_1d_2}{d_1} > 0$
$f_1 = a_6 > 0.$

IOP Conf. Series: Journal of Physics: Conf. Series **974** (2018) 012035 doi:10.1088/1742-6596/974/1/012035

Basic reproduction number $(\mathcal{R}_0)$

Next, we want to analyzed the basic reproduction number of the complete system 1 where the ART treatment is involved into the model. We will use the next-generation matrix approach to determine the magnitude of basic reproduction number. First we define the transition matrix of system 1 which given by

$$H = \begin{bmatrix} M_1 & 0 & 0 & 0 & 0 & 0 \\ \gamma_a & M_2 & 0 & 0 & (1-r_c)\gamma_b & 0 \\ 0 & u & M_3 & \xi_5 & 0 & 0 \\ 0 & 0 & \xi_4 & M_4 & 0 & 0 \\ 0 & 0 & 0 & \rho_c & M_5 & 0 \\ 0 & \gamma_c & 0 & 0 & r_c\gamma_b & M_6 \end{bmatrix},$$

where

$$M_1 = -\gamma_a - \mu$$

$$M_2 = -u - \gamma_c - \mu$$

$$M_3 = -\mu - \xi_4$$

$$M_4 = -\mu - \rho_c - \xi_5$$

$$M_5 = -(1-r_c)\gamma_b - r_c\gamma_b - \mu$$

$$M_6 = -\delta - \mu$$

and the transmission matrix is given by

$$V = \begin{bmatrix} \frac{\beta_a x_1}{N_x} & \frac{\beta_c x_1}{N_x} & \frac{\beta_c x_1 \theta_h}{N_x} & \frac{\beta_c x_1 \theta_l}{N_x} & \frac{\beta_c x_1}{N_x} & \frac{\beta_s x_1}{N_x} \\ 0 & 0 & 0 & 0 & 0 & 0 \\ 0 & 0 & 0 & 0 & 0 & 0 \\ 0 & 0 & 0 & 0 & 0 & 0 \\ 0 & 0 & 0 & 0 & 0 & 0 \\ 0 & 0 & 0 & 0 & 0 & 0 \end{bmatrix}.$$

We find that the basic reproduction number as the spectral radius of the next generation matrix $\mathbf{K} = -\mathbf{E}^T\mathbf{V}\mathbf{H}^{-1}\mathbf{E}$ which given by:

$$R_0 = \frac{A(P_1 + P_2 + P_3 + P_4)}{\mu M_1 M_6 N_x P_5}, \tag{14}$$

where :

$$P_1 = -uM_4M_5M_6\beta_c\gamma_a\theta_h + uM_5M_6\beta_c\gamma_a\theta_l\xi_4 - uM_6\beta_a\gamma_b r_c\rho_c\xi_4$$

$$P_2 = u\beta_s\gamma_a\gamma_b r_c\rho_c\xi_4 + uM_6\beta_a\gamma_b\rho_c\xi_4 - uM_6\beta_c\gamma_a\rho_c\xi_4$$

$$P_3 = -M_2M_3M_4M_5M_6\beta_a + M_2M_5M_6\beta_a\xi_4\xi_5 + M_3M_4M_5M_6\beta_c\gamma_a$$

$$P_4 = -M_3M_4M_5\beta_s\gamma_a\gamma_c - M_5M_6\beta_c\gamma_a\xi_4\xi_5 + M_5\beta_s\gamma_a\gamma_c\xi_4\xi_5$$

$$P_5 = -u\gamma_b r_c\rho_c\xi_4 + u\gamma_b\rho_c\xi_4 + M_2M_3M_4M_5 + M_2M_5\xi_5\xi_4$$

and $\mathbf{E} = \begin{bmatrix}1 \ 0 \ 0 \ 0 \ 0 \ 0\end{bmatrix}^T$.

4. Numerical experiment

In this section, we will see how the ART treatment intervention will impact the magnitude of $\mathcal{R}_0$ and also impact the dynamic of infected human. Firstly, we will see how the basic reproduction number will change respect to u and r_c. Using the set of parameter given by : $A = 1000/(65\times365)$, $\beta_a = 0.000025$, $\beta_c = 0.008\times0.000025$, $\beta_s = 1.5\times0.000025$, $\gamma_a = 1/(5\times365)$, $\gamma_c = 1/(3\times365)$, $\delta = 0$, $\mu = 1/(65\times365)$, $N_x = 1000$, $\xi_4 = 0.75/365$, $\xi_5 = 0.25/365$, $\theta_h = 1$, $\theta_l = 2$, $\rho_c = 0.1/365$, $\gamma_b = 0.5/365$,

and substitute it into $\mathcal{R}_0$ in 14, we find the function of $\mathcal{R}_0$ now as a function of two variables $(f(u, r_c))$. Plot it in the $u-r_c$ plane with various value of $\mathcal{R}_0$, the figure is given in Figure 3.

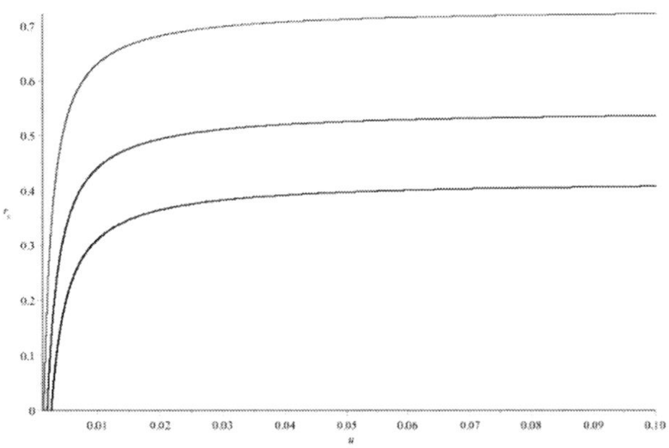

Figure 3. Sensitivity analysis of $\mathcal{R}_0$ with respect to u and r_c, with black, blue and red represent $\mathcal{R}_0$ are $5.5, 6$ and 6.5 respectively.

It can be seen that increasing the ART treatment rate (u) will reduce basic reproduction number $\mathcal{R}_0$, which will lead us into extinction of HIV-AIDS. In the other hand, increasing of transition rate into AIDS compartment (r_c) will increase basic reproduction number $\mathcal{R}_0$. This means that if the individual who have failed in ART treatment do not re-following the ART treatment will make the HIV-AIDS exist in the field.

The numerical simulation to see how the dynamic of infected population will change respect to the various value of ART treatment (u) will be performed in the next simulation. To do this simulation, we use the set of parameters :

$A = 1000/(65*365)$, $\beta_a = 0.000025$, $\beta_c = 0.008*0.000025$, $\beta_s = 1.5*0.000025$, $\gamma_a = 1/(5*365)$, $\gamma_c = 1/(3*365)$, $\delta = 0$, $\mu = 1/(65*365)$, $N_x = 1000$, $\xi_4 = 0.75/365$, $\xi_5 = 0.25/365$, $\theta_h = 1$, $\theta_l = 2$, $r_c = 0$, $\rho_c = 0.1/365$, $\gamma_b = 0.5/365$
With this set of parameters, we find the basic reproduction number when $u = 0$ is 1.664503135.

Please note that with this set of parameter, it can be seen that this situation represent a situation where no ART treatment is implemented $u = 0$. It can be seen in Figure 4 that the gradation value of ART treatment rate do not impact the change of dynamic on susceptible human, infected human in acute level, and infected human in HIV/AIDS level significantly. Different phenomenon appear in the other four compartment, infected human in chronic, humans in ART treatment and human who failed in ART treatment, when treatment rate success to reduce number of infected human significantly.

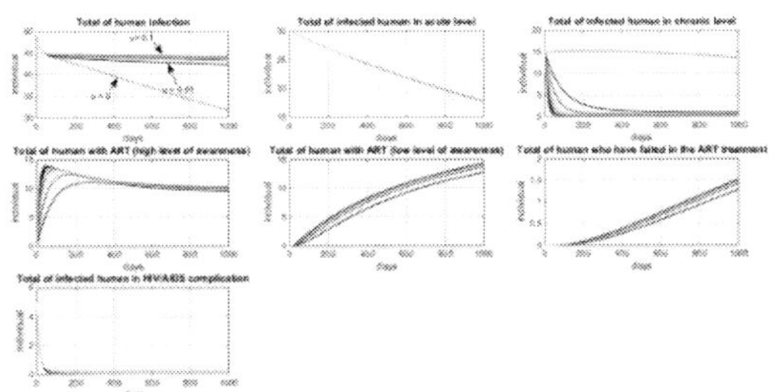

Figure 4. Dynamic of all compartment for various value of u.

5. Conclusions

HIV, or Human Immunodeficiency Virus, attacks the immune system of the human body and will weaken it to attack the disease or infection. There is no medicine to cure people if the infected human already has HIV status. However, early diagnosis and effective handling of the infective people in the early stage with ART treatment will prevent the infected people progressing to the next stage, the AIDS stage.

In this article, a mathematical model to understand ART treatment as an alternative way to control the spread of HIV has been developed. The mathematical model analysis when the ART treatment is not implemented in the model has shown that the transition rate between infected compartment should be reduced to control the spread of HIV. Therefore, implementation of ART treatment is the best way to control the spread of HIV.

Disease-free equilibrium and the form of basic reproduction number as the endemic indicator for the model have been shown analytically. Mathematical analysis to show the sensitivity of ART treatment to reduce the basic reproduction number has been shown, along with a numerical simulation to see how the dynamics of each compartment respond to the ART treatment intervention. We find that ART treatment does not increase the number of susceptible humans, but could reduce the number of infected humans who progress to AIDS status.

Acknowledgment

We are thank to all referee for all his or her helpful comments into the earlier version of this article. This research is funded by Universitas Indonesia with PITTA 2017 research grant (Project ID : 644/UN2.R3.1/HKP.05.00/2017).

IOP Conf. Series: Journal of Physics: Conf. Series **974** (2018) 012035 doi:10.1088/1742-6596/974/1/012035

References

[1] IOP Publishing is to grateful Mark A Caprio, Center for Theoretical Physics, Yale University, for permission to include the `iopart-num` BIBTEXpackage (version 2.0, December 21, 2006) with this documentation. Updates and new releases of `iopart-num` can be found on `www.ctan.org` (CTAN)

[2] *People living with HIV* [accessed on July 2017, available from: `http://www.who.int/hiv/en/`]

[3] *Number of people living with HIV* [accessed on August 2017, available from:`http://aidsinfo.unaids.org/`]

[4] H D Gayle, G L Hill 2001 Global impact of human immunodeficiency virus and AIDS **14** pp 327–35

[5] Kementrian Kesehatan Republik Indonesia Direktorat Jendral Pengendalian Pengakit dan Penyehatan Lingkungan 2011 Pedoman nasional tatalaksana klinis infeksi HIV dan terapi antiretroviral pada orang dewasa

[6] I Mardhiyah 2012 Model matematika penyebaran infeksi HIV pada komunitas IDU

[7] L Simpson, A B Gumel 2016 Mathematical assessment of the role of pre-exposure prophylaxis on HIV transmission dynamics pp 168–93

[8] J Hakim, V Musiime, A J Szubert, J Mallewa, A Siika et al 2017 Enhanced prophylaxis plus antiretroviral therapy for advanced HIV infection in Africa *N Engl J Med* **377** pp 233-45

[9] M G Fowler, M Qin, S A Fiscus, J S Currier et al 2016 Benefits and risks of antiretroviral therapy for perinatal HIV prevention *N Engl J Med* **375** pp 1726-37

[10] O Diekmann, J A P Heesterbeek, M G Roberts 2010 The construction of next-generation matrices for compartmental epidemic models *J. R. Soc. Interface* **7** pp 873-85

[11] T de-Camino-Beck et al 2009 A graph-theoretic method for the basic reproduction number in continuous time epidemiological models *J Math Biol* **59** pp 503–16

[12] K Ogata 1990 *Modern Control Engineering*

Optimal blood glucose level control using dynamic programming based on minimal Bergman model

Maria Rettian Anggita Sari[1] and Hartono[2]

Sanata Dharma University, Paingan, Maguwoharjo, Depok, Sleman, Yogyakarta, Indonesia 55282

E-mail : [1]rettian.anggitasari@gmail.com
[2]yghartono@usd.ac.id

Abstract. The purpose of this article is to simulate the glucose dynamic and the insulin kinetic of diabetic patient. The model used in this research is a non-linear Minimal Bergman model. Optimal control theory is then applied to formulate the problem in order to determine the optimal dose of insulin in the treatment of diabetes mellitus such that the glucose level is in the normal range for some specific time range. The optimization problem is solved using dynamic programming. The result shows that dynamic programming is quite reliable to represent the interaction between glucose and insulin levels in diabetes mellitus patient.

1. Introduction

Diabetes mellitus (DM) is a metabolic disorders disease, i.e. carbohydrate metabolism, caused by reduced or unavailability of insulin hormone from beta cells in the pancreas. It may also be caused by or disorders insulin function. In healthy people, normal glucose level for adults after fasting (not eating approximately 8 hours) is 70-110 mg/dl [3]. Thus, diabetes mellitus patients are those who have glucose level outside that normal range.

Patients with DM have too much glucose in the blood and urine, but not enough in the cells. On the other hand, these cells still need glucose, either as food or as a material to produce energy. Insulin produced by beta cells, serves to keep blood glucose in normal range by turning glucose into glycogen and storing it in muscles or tissues as reserves. This hormone works to speed up the transport of glucose from the blood into the cells. Insufficient insulin leads to high glucose level (hyperglycemia) due to the reduction of the transport of glucose into cells. Conversely, insufficient glucagon may lead to low glucose level (hypoglycemia) below the normal range.

Treatment of diabetes mellitus is usually done using oral anti-diabetic drugs (OAD) or insulin injection. However, the use of these drugs may cause side effects and insulin injection is relatively expensive. Therefore, it is necessary to have a good way to control blood glucose so that treatment can be optimal. In this article, the formulated optimal control problem will be solved with dynamic programming method.

2. Minimal Bergman Model

Minimal Bergman model consists of two parts i.e. glucose dynamic and insulin kinetic. The non-linear Minimal Bergman model in Chee and Fernando [1] is as follows

$$\frac{dG(t)}{dt} = -(p_1 + X(t))G(t) + p_1 G_B \quad (1)$$
$$\frac{dX(t)}{dt} = -p_2 X(t) + p_3[I(t) - I_B] \quad (2)$$
$$\frac{dI(t)}{dt} = -n[I(t) - I_B] + u(t) \quad (3)$$

where

$G(t)$ = blood glucose level in plasma [mg/dl] at time t

$X(t)$ = a function that is proportional to the insulin concentration in a remote compartment or insulin level in interstitial [min^{-1}] at time t

$I(t)$ = insulin level in plasma [μU/ml] at time t

G_B = basal glucose level in plasma [mg/dl]

I_B = basal insulin level [μU/ml]

n = fractional disappearance rate constant for endogenous insulin

u = rate of exogenous insulin injection

p_1 = glucose clearance rate independent of insulin

p_2 = rate of clearance of active insulin

p_3 = rate of change of insulin level in interstitial caused by insulin in plasma

From equation (1), it can be seen that glucose will leave or enter plasma compartment when there is a difference between plasma glucose level $G(t)$ and basal glucose level G_B. If plasma glucose level is below basal glucose level, then glucose enters plasma. On the other hand, when plasma glucose level is above basal glucose level, glucose will leave plasma compartment. Glucose also leaves the plasma compartment at a rate proportional to insulin in interstitial tissue $X(t)$. Analogously, from equation (2) and (3), the difference between plasma insulin $I(t)$ and basal insulin I_B will affect the amount of $X(t)$ and $I(t)$.

Parameters of diabetic patient used in this article are for glucose-resistant patient in minimal Bergman model as shown in Chee and Fernando [1] i.e. $p_1 = 0$, $p_2 = 0.025$, $p_3 = 0.000013$, and n=5/54 [2]. Glucose-resistant patients are also called type I diabetic patient. The destruction of pancreas cells that producing insulin in type I diabetic patient leads to insulin dependence. Thus, type I diabetic patient requires more insulin injections than type II. In this case, an optimal insulin injection is really necessary to obtain.

3. Dynamic Programming

To solve the problem, optimal control theory is used to model it. The general problem of optimal control is to

$$\text{minimize } I(x(0), t_f) = \int_0^{t_f} \psi(x, u, t)\, dt + \phi(x(t_f))$$
$$\text{subject to } \frac{dx}{dt} = f(x, u, t), x(0) = x_0$$
$$u_a \le u_i(t) \le u_b, \quad i = 1,2,3 \ldots m$$

where $\boldsymbol{I}: \boldsymbol{R^n} \times \boldsymbol{R^m} \times \boldsymbol{R} \to \boldsymbol{R}$, $\boldsymbol{f}: \boldsymbol{R^n} \times \boldsymbol{R^m} \times \boldsymbol{R} \to \boldsymbol{R^n}$, $\boldsymbol{x} = (\boldsymbol{x_1}, \cdots, \boldsymbol{x_n}) \in \boldsymbol{R^n}$, $\boldsymbol{u} = (\boldsymbol{u_1}, \cdots, \boldsymbol{u_m}) \in \boldsymbol{R^m}$.

Then, dynamic programming method originally developed by Richard Bellman is applied to the model. Rein Luus in [4] proposes a kind of dynamic programming called the iterative dynamic programming method to solve the general optimal control problem. This method separates the problem with $[0,t_f]$ time interval into several time stages with initial condition placed at stage 1. The first step is the construction of the P time stages with length $L = t_f / P$. The performance index is approximated by

$$I(x(0), P) = \psi(x(t_f)) + \sum_{k=1}^{P} \int_{t_{k-1}}^{t_k} \phi(x(t), u(k-1), t)\, dt \quad (4)$$

Next step is the construction of grid for state x. At each stage component x_i of state vector x is allowed to take N values except for the first stage. Each control component u_j of control vector u is allowed to take M values. Thus, there are N^n grid points at stages 2, 3, …, P and at each stage M^m values of the control vector u are taken.

The calculation are started from stage P corresponding with time interval $t_f - L \leq t \leq t_f$. For each x-grid point, evaluate M^m values for control u based on the following performance index

$$I(x(t_f - L), 1) = \psi\left(x(t_f)\right) + \int_{t_f-L}^{t_f} \phi(x(t), u(P-1))\ dt \tag{5}$$

where M^m control values used for $u(P-1)$. Then compare M^m value of the performance index (5) to determine the control value u giving the minimum value.

The next step is to go backwards to stage $P-1$ associating with time $t_f - 2L \leq t \leq t_f - L$. For each x-grid point, evaluate again M^m values for control u based on the performance index and determine the control value u which gives the minimum value. Then, integration until the final time t_f is done using the control u corresponding to the grid point that is closest to state $x(t_f - L)$ if $x(t_f - L)$ is not one of grid points at state P.

This procedure should be repeated for stages $P-2$, $P-3$, and so on until stage 1 which consists of the initial condition $x(0)$ as a single grid point corresponding to the time interval $0 \leq t \leq L$. The process is terminated by comparing the M^m value of the performance index to determine the control u which gives the minimum value.

The process ends with storing the state trajectory. The selected trajectory is traced back by integrating an initial point of state with control u which gives the minimum value at the corresponding time. Finally, forward integration is done until the final time reached.

4. Computer Simulation

The performance index of the problem is to minimize glucose level and insulin injections in the Lagrange form as follows

$$J = \int_0^{t_f} (\rho|u| + (1-\rho)|G - G_T|)\ 10^5 dt \tag{6}$$

with ρ is control weight factor and G_T is blood glucose level target.

Simulation is done using MATLAB over $P = 30$ stages where final time $t_f = 140$ with $L = 140/30$. The number of grid points are $N = 30$. Each stage takes $M = 15$ values of the control vector u with the lower bound of control $u_a = 0$ and the upper bound of control $u_f = 6$. So, the control vector is bounded by $0 \leq u \leq 6$. The initial value of glucose level, the initial value of the interstitial insulin and the initial value of insulin level used in this simulation are $G(0) = 500$, $X(0) = 0$, $I(0) = 30$ respectively. The basal insulin level is set to $I_B = 7$ and basal glucose level is $G_B = 81$.

Therefore, the Minimal Bergman model become as follows

$$\frac{dG(t)}{dt} = -X(t)G(t) \tag{7}$$

$$\frac{dX(t)}{dt} = -0.025X(t) + 0.000013[I(t) - 7] \tag{8}$$

$$\frac{dI(t)}{dt} = -\frac{5}{54}[I(t) - 7] + u(t) \tag{9}$$

In addition to minimize the use of insulin injection, glucose level of diabetic patients is also maintained to approach as close as possible to target glucose levels i.e. $G_T = 90$ with $\rho = 0.25$. Therefore based on equation (6), the performance index of the problem become as follows

$$J = \int_0^{t_f} (0.25|u| + 0.75|G - 90|)\ 10^5 dt \tag{10}$$

The value of the coefficient $\rho = 0.25$ means that keeping the blood glucose level in the normal range is considered more important than the control of insulin injection. The simulation results are shown in Figure 1, Figure 2, Figure 3 and Figure 4.

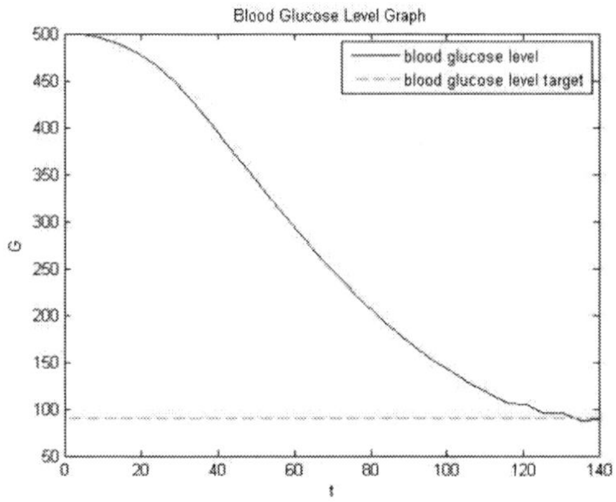

Figure 1. Glucose level graph

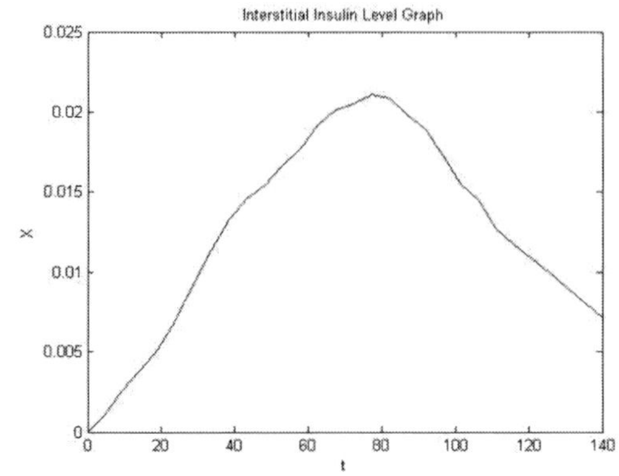

Figure 2. Interstitial insulin level graph

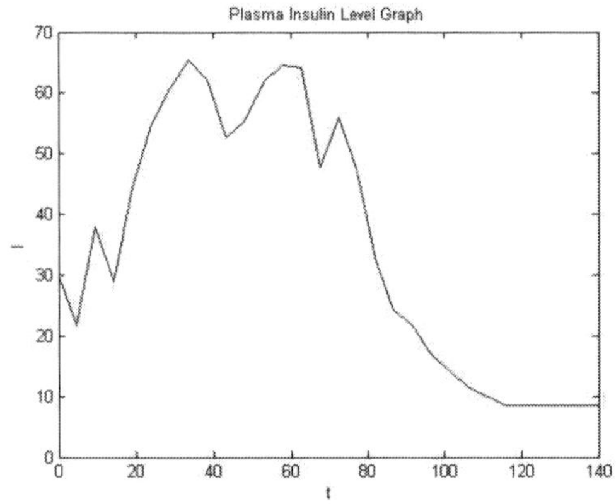

Figure 3. Plasma insulin level graph

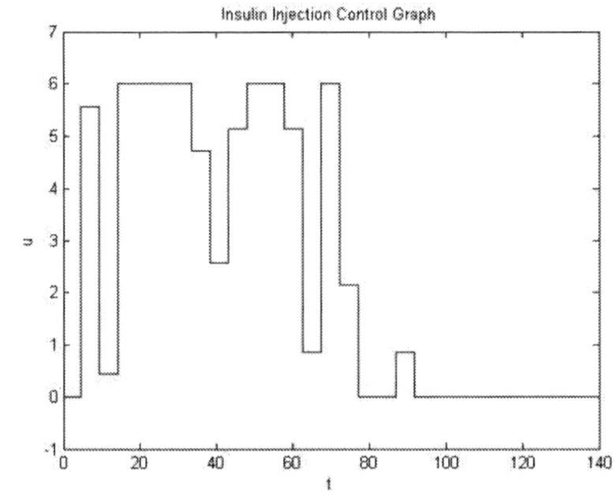

Figure 4. Insulin injection control graph

From the graphs, it is seen that at $t < 80$ the glucose level decreases rapidly due to the increase of injected insulin. Consequently, this increases the insulin either in plasma or interstitial cells as well. On the other hand, when the injected insulin becomes zero ($t > 80$) the plasma and interstitial insulin start to decrease and cause the decrease of the blood glucose level more slowly than before. Interstitial insulin level increases and reaches the highest point at time t = 80. At the final time the glucose level of diabetic patient can decrease until $G(t_f) = 88.2005$, whereas insulin interstitial is $X(t_f) = 0.0072$ and insulin level is $I(t_f) = 8.4407$. The minimum performance index (6) achieved is 1.69251. Thus, the glucose change is 411.7995. Glucose level at final time reaches the desired glucose level located around the target glucose level and the required insulin injection is 69.4286.

5. Conclusion

Based on the results of the simulation program it can be concluded that Minimal Bergman model could represent well the glucose dynamic and the insulin kinetic in people with diabetes mellitus. Optimal control problem with Minimal Bergman model could be solved by applying dynamic programming method. Glucose level of diabetic patients can reach the desired target and the dynamic of glucose is around the target glucose level by minimizing the performance index in Lagrange form. Hence, dynamic programming can be used to control glucose level in diabetic patients with an insulin infusion therapy based on the resulted insulin injection control behaviour.

IOP Conf. Series: Journal of Physics: Conf. Series **974** (2018) 012036 doi:10.1088/1742-6596/974/1/012036

References

[1] Chee F and Fernando T 2007 *Closed-Loop Control of Blood Glucose* (Springer)

[2] Chee F, Fernando T and Heerden V 2002 Simulation Study on Automatic Blood Glucose Control *Proc. of the 7th Australian and New Zealand Intelligent System Conf.* (IEEE) pp. 423–427

[3] González A A, Voos H and Darouach M 2015 Glucose-Insulin System based on Minimal Model: Realistic Approach *Proc. of the 17th UKSIM-AMSS Int. Conf. on Modelling and Simulation* (IEEE) pp. 55–60

[4] Luus R 2000 *Iterative Dynamic Programming* (Boca Raton: Chapman & Hall/CRC)

Solving large scale unit dilemma in electricity system by applying commutative law

Supriadi Legino, Rakhmat Arianto

Electricity Department
Sekolah Tinggi Teknik PLN, Jakarta - Indonesia

Email: supriadi@sttpln.ac.id; arianto@sttpln.ac.id

Abstract. The conventional system, pooling resources with large centralized power plant interconnected as a network. provides a lot of advantages compare to the isolated one include optimizing efficiency and reliability. However, such a large plant need a huge capital. In addition, more problems emerged to hinder the construction of big power plant as well as its associated transmission lines. By applying commutative law of math, ab = ba, for all a,b €-R, the problem associated with conventional system as depicted above, can be reduced. The idea of having small unit but many power plants, namely "Listrik Kerakyatan," abbreviated as LK provides both social and environmental benefit that could be capitalized by using proper assumption. This study compares the cost and benefit of LK to those of conventional system, using simulation method to prove that LK offers alternative solution to answer many problems associated with the large system. Commutative Law of Algebra can be used as a simple mathematical model to analyze whether the LK system as an eco-friendly distributed generation can be applied to solve various problems associated with a large scale conventional system. The result of simulation shows that LK provides more value if its plants operate in less than 11 hours as peaker power plant or load follower power plant to improve load curve balance of the power system. The result of simulation indicates that the investment cost of LK plant should be optimized in order to minimize the plant investment cost. This study indicates that the benefit of economies of scale principle does not always apply to every condition, particularly if the portion of intangible cost and benefit is relatively high.

1. Introduction

Many power system in the world still applies conventional system, pooling resources with large centralized power plant interconnected one into others as a network. From this network, power then delivers to the customers using transmission and distribution lines. Interconnected system provides a lot of advantages compare to the isolated one include intermingling power, variety of loads and access to a variety of power resources. Until recently, we are still relying on conventional system, which aimed at optimizing efficiency and reliability. However, such a large plant need a huge capital that should be borrowed from financial institution with complicated lender guidelines. In addition, more problems emerged to hinder the construction of big power plant as well as its associated transmission lines. Those problems include land acquisition, right of way permit, environmental as well as social obligation. The above challenges have resulted in the slippage of many big power plants and transmission lines projects in Indonesia. The fact described above becomes the prognosis of this paper:

The overall cost due to the problems risen in constructing large power plant under the conventional system may offset its benefit coming from its economies of scale and interconnection system.

Mathematical approach has potential benefit for problem solving if we could use it beyond the routine acquisition of isolated techniques and procedures [1]–[4]. Furthermore, mathematical can be considered as a social process [5] as reported by some studies. For instance, Carraher et al. [6] studied the relations between mathematic and cultural activities in Brazil and Cobb [7] studied the sociocultural perspectives on mathematical development.

Under the micro scientific view, electricity concept cannot be separated from mathematical theory, from complex number to analyze electric circuit to Fourier and Laplace Transform to analyze magnetic field and wave propagation [8], [9]. However, for the macro view in the real world, there is very rare studies that use mathematical approach to solve the real world problem, such as a large power system dilemma, a focal issue in this study. This paper will show that such problem can be analyzed and solved by applying simple mathematical principle, known as commutative law of math, ab = ba, for all a,b [10]. Commutative algebra has been known for a long time [11] and some author call it as commutative property [12].

By using commutative law, the problem associated with conventional system as depicted above, can be reduced as follows. Such principle can be illustrated as an algebraic expression that 1000 x 1 = 1x 1000, which means that "to build one large unit plant of 1000 MW equals to building 1000 small units of 1 MW. The idea of having small unit but many power plants in this paper is relevant to the recent initiative of Sekolah Tinggi Teknik PLN, Jakarta, namely "Listrik Kerakyatan" abbreviated as LK. This initiative offers an option to avoid problems that challenge conventional system by adopting the concept of small scale distributed generation, which plants are built scattered closed to the customer side. Distributed generation provides benefits including voltage improvement and stability and could reduce power losses [13] . For example, Arunachalam et.al. reported the implementation of Decentralization Distributed Generation in India [14] and [15] reported that distributed generation may increase energy security and reduce carbon pollution. In addition, LK creates opportunity for local people to own and manage electricity their selves using available and easy renewable energy resources.

This study conduct cost and benefit analysis to compare the tangible and intangible value of large scale centralized conventional system with those of Listrik Kerakyatan by using simulation.

2. The conceptual framework of LK

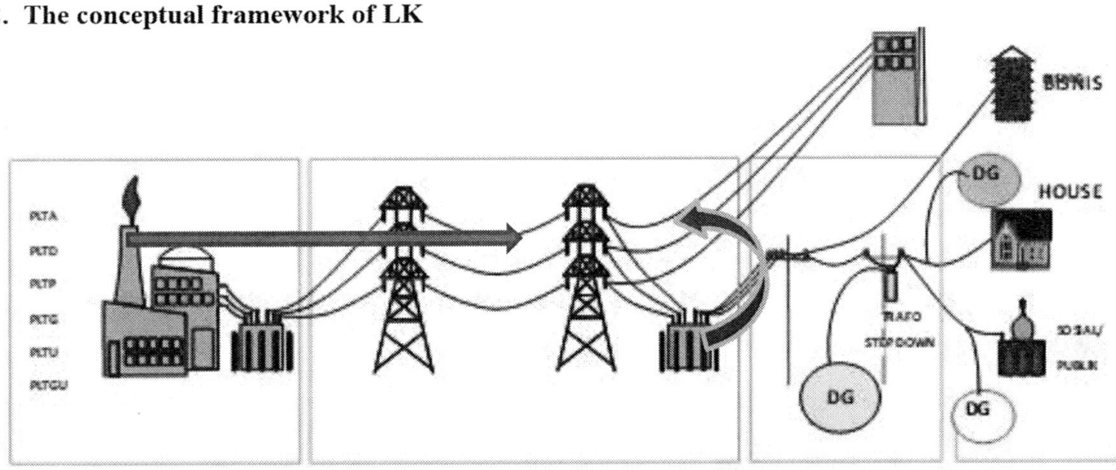

Figure 1. Conventional system vs distributed generation

As illustrated in Figure 2, it takes around five years serial process to build a large scale power plants such as for 1000 MW Coal Fired Plant, much longer than building a small LK unit, which takes less than six months and can be constructed simultaneously as many numbers we want.

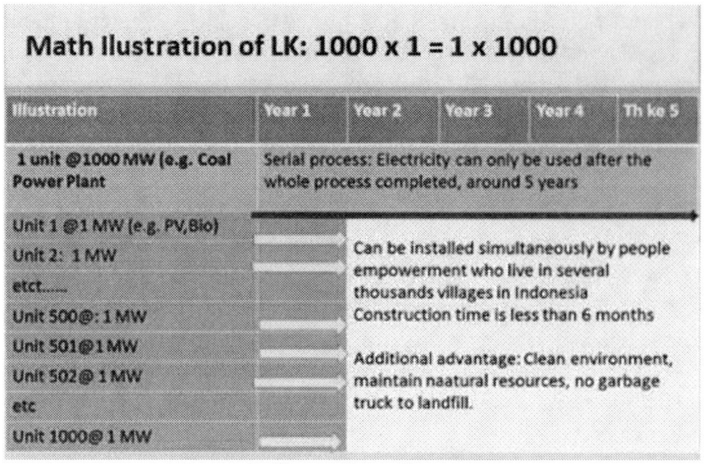

Figure 2. Illustration of comparison

3. The benefit of LK

The small size of LK, provides opportunity for local business players to own small scale power plant or waste to energy plant because the investment cost per package of LK is relatively low, less than one billion rupiah for the unit size of 30 kW to 50 kW. Such small size plant also needs only small area (200-1000 m2) and flexible to be located anywhere around school, housing, market, mall, hotel, and office. For remote areas that have no electricity, LK can be managed locally as a hybrid system. Technically, LK will reduce the transmission losses and the problem of excessive transformer overload and drop voltage. From the energy balance aspect, LK offers the additional renewable energy capacity to the system without any transmission lines construction.

On top of the above advantages, LK will promote to maximize the use of local material and man power especially for biomass power plant. LK may create energy security because very little dependency to the import material and expertise.

If LK can be risen up to the national scale, the government target to have 100 percent electrification ratio and the target of national energy policy to have 23 percent renewable energy by 2025 can be achieved in less than three years. Unlike, the giant project that needs foreign funding, LK can be funded by local people equity and national bank with special interest for small and medium business scheme.

For the big city, the biomass plant of LK offers the option to apply zero waste concept without any truck go to landfill. For the remote areas that are still lack of electricity and cannot be reached by the large size project, LK offers the option to have their own power plant to manage and serve their own community using hybrid of Solar and Biomass plant.

4. Power plant type of LK

LK is a hybrid power plants that can be mixed of Solar PV, Biogas, Biomas, or Wind as main sources. If we want to improve reliability, battery backup or diesel engine may be added based on the requirement. Micro hydro may be also included if the resource is available near the area. STT PLN has conducted trial for LK model at two locations, LK 1 at Pondok Kopi district East Jakarta and LK2 in STT PLN Campus Duri Kosambi West Jakarta. The hybrid for trial consists of solar PV and biomass plant using municipal waste.

5. From waste to energy

Municipal waste can be converted to thermal and electrical energy by using dual processes, digestation and bricketting.

5.1. Anaerobic digesting
The food waste and any degradable organic waste is putting into the digester and convert into methan gas and leachate through Anaerobic digestation process. This is the first stage of zero waste process that can eliminate the bad smell of waste. This process can eliminate the waste odor and produce leachate that can be used as an alternative enzyme required for "peuyeumisasi" (composting process for the rest of mixed waste) This process also produce methan gas that can be used for cooking or generate electricity.

5.2. Bricketting peuyeumisasi
After digesting process, there are still more than 50 percent of waste remaining, not taking into account about 30%-40% of usable waste such as plastic bottle, can, and other valuable material. This remaining waste can be converted into bricket through the process similar to composting, called "peuyeumisasi.". This process will dry the waste and reduce the waste volume and eliminate the waste odor. It takes 10 days to proceed peuyeumisasi, that will be converted further into bricket or pellet form using shredder machine.

Fresh domestic or municipal waste typically has low caloric value around 1000 kcal/kg waste. The process of peuyeumisasi may increase the caloric value to the range of 2400 kcal/kg to 3000 kcal/kg. The gasifier unit is used to converted waste "pellet" or waste bricket into syngas as fuel to run the generator set engine.

5.3. The drawback of LK
There are some disadvantages of LK or distributed generation as a consequence of a huge number and scattered plans that should be supported by a proper control and monitoring system, which means additional capital for control and information technology support facilities. The second drawback is the lack of supporting regulation or incentive for small scale, people owned generation, kind of small scale IPP. Although LK provides a lot of benefit for clean environment and additional income for local people, there is no legal base to provide compensation for such social and environmental benefit. The minister decree for Renewable Energy Feed in Tariff does not count such intangible benefit for the nation and this rule is applied for high level IPP investor, not suitable for local people. Another drawback of LK is the lack of skilled labour, particularly for renewable energy power plants. Moreover, there is additional cost related to improve reliability and stability of LK by putting battery for stabilizer, which price is still very expensive.

6. Pilot projects
STT PLN has undertaken pilot projects of LK in Duri Kosambi Campus (LK2) and in public space at Pondok Kopi Village (LK 1), aimed at processing one ton of municipal waste using dual process of digesting and bricketing. The waste volume in each location is 1 (one) ton of mixed waste per day and the PV total capacity of the two locations is 13 kW. The actual expenses spent for LK 1 and LK 2 will be used for further calculation in this paper. The capital expenditure (capex) and operational expenditure (opex) of the two locations are shown below:

- Building and facilities : IDR 198,353,540
- Anaerobic digestion facilities : IDR 56,965,000
- Bricketing organic waste facilities (Peuyeumisasi) : IDR 482,250,000
- Gasification and genset facilities : IDR 152,483,100
- PV and installation facilities : IDR 246,613,400
- Total Capex of LK2 at STT PLN Campus : IDR 1.136.715.040; Eq. US$ 87,500
- The annual opex of LK1 Pondok Kopi : IDR 148.800.000; Eq. US$: 11,500
- The annual opex of LK2 at STT PLN Campus : IDR 74.400.000 ;Eq. US$: 5,700

IOP Conf. Series: Journal of Physics: Conf. Series **974** (2018) 012037 doi:10.1088/1742-6596/974/1/012037

7. Products and revenue projection from LK1 and LK2

After a year trial at two locations has resulted in various products that may potentially make money including waste treatment or tipping fee, valuable waste material (e.g. bottle, can, wood, metal), bricket for fuel, organic fertilizer, and income from selling electrical energy, as shown in Table 1. In addition, LK will provide social benefit as more local people involved in LK related business. LK will also render environmental benefits because LK applies a zero waste concept that can reduce huge number of waste transportation from communities to landfill area. More over, LK will save fossil fuel as natural resources and contribute to the global green environment program.

7.1. Perspective Revenue of LK

LK will be sustained if it is treated as a proper business model that should have secured revenue that are potentially can be coming from the following products and services:

a) Waste treatment or tipping fee, which due should be paid by local communities or by government or combination of both ways. For the purpose of simulation in this paper, the tipping fee is IDR 200,000 per ton of waste, which number is realistic because it is much lower than the current provincial waste treatment spending in Jakarta (IDR 325,000/ton). The local government may get budget saving and environmental bonus as LK will make less and less truck transporting waste to the Landfill area.
b) Income from selling usable and valuable material (bottle, can, cartoon, wood, etc), which this kind of business has been done by ordinary people for decades. Based on observation during one year trial, the potential quantity of this material is 40% of total waste. We pick up a moderate number of IDR 1000 per ton of material, which number is the average actual value between the minimum IDR 600 and IDR 3600
c) Organic fertilizer as by product of anaerobic digesting process for digestable waste or composting process of organic waste. We pick a number IDR 5000 per litre from the market price of similar product.
d) Although energy sales can be a good potential income, but there is no supporting regulation available that provide suitable Feed in Tariff (FIT) for people owned small scale renewable in Indonesia. Currently, the existing rule is applied for big IPP (independent power producer) investors and the FIT is very low, only less than 10 cents US$, same for every kind of renewable energy. For the purpose of simulation, this paper will use the FIT for Java System, which is IDR 867,- equal to 7 cents US$.
e) Energy from biogas (200 kg degradable waste) is around 6 kWh per day @ IDR 867 /kWh
f) Energy from biomass (480 kg organic waste x 0,75 kWh) is around 435 kWh per day @ IDR 867/kWh

Table 1. LK 1 and LK 2 revenue estimation from 2 ton of waste

Revenue Projections of LK 1 & LK 2	Total per period	Number/year	Unit	Total per year
Estimated Tipping Fee: 2 ton @ IDR 200.000/ton	IDR 400,000	365	day	IDR 146,000,000
Valuable waste @IDR 1000 per kg x 400 kg	IDR 400,000	365	day	IDR 146,000,000
Organic fertilizer (@IDR 5000 x 2000 liter/month)	IDR 10.000.000	12	month	IDR 120,000,000
Biogas energy from degradable waste (6kWh/day) @IDR 867/kWh	IDR 4,482	365	day	IDR 1,635,930
Biomass energy from bricket gasifier plant (435 kWh/day @IDR 867/kWh)	IDR 537,840	365	day	IDR 196,311,600
PV Solar 13 kWp @4hr x IDR 867/kWh	IDR 60,690	365	day	IDR 22,151,850

Total potential Revenue per 2 ton	IDR 512,339,380

7.2. Specific cost and revenue of LK

Based on the result from pilot project of LK1 and LK2, we can calculate the specific cost per item of work as shown in Table 2. The calculation has considered some additional equipment required to improve the production and quality.

Table 2. Specific cost of LK

Item	Unit cost (IDR)	Unit	Eq.US$
Civil work	800,000	IDR/m2	62 US$/m2
Biogas unit	615,225	IDR/kg	47 US$/kg
Bricketing unit without gasifier	219,650,000	IDR/ton	16896 US$
Gasifier unit	10,516,076	IDR/kW	809 US$/kW
PV unit	21,800,303	IDR/kWp	1,677 US$/kWp
Operational Cost			
Supervisor	60,000,000	IDR/year	4,615US$/year
Operator	38,400,000	IDR/year	2,954 /year
Operation and maintenance material	15,000,000	IDR/year	
Over head	10,000,000	IDR/unit office	769 US$/unit

Revenue is estimated based on available data and market price of similar product and service as follow:

a) Valuable waste (bottle, can, paper, etc) estimate volume is 40% of the total waste and picking the smallest market price of similar products (IDR 500 /kg). If the volume of waste is 1 ton, it may generate revenue: 40%x1000x500=IDR 200,000/day
b) Tipping fee is assumed IDR 200.000/ton, much lower compare to the existing cost in Jakarta (IDR325,000 per ton)
c) Organic fertilizer IDR 1,000 /litre, much lower than market price of similar product.
d) Electric energy (kWh) is calculated using minister decree: For Jawa/Bali, the production cost (BPP) is IDR 867 /kWh. Energy from biogas organic 2 kWh per ton waste, and for bricket may produce: 480kg/ton x 0,75 kWh/kg=216 kWh/ton
e) If PV installed, additional revenue will be get from PLN: 4 hour time Capacity kWp).

Table 3. Specific revenue of LK per ton per day

Revenue item	Per Day (IDR)	Per Year (IDR)	Eq. US$/year*
Valuable waste IDR per ton/day	200,000	73,000,000	5615
Tipping fee /ton/day	100,000	36,500,000	2,808
Organic fertilizer /liter	1,000	365,000	28
kWh digester/ton	2	639,480	49
kWh Briket/ton	216	69,063,840	5,313
Income from waste/ton/year	181,486,760	181,486,760	13,961
Additional rev from PV per year		3,836,880	295
Total income for 1 ton waste		**191,078,960**	14,698

7.3. Specific cost per ton of waste

Specific cost per ton of waste is helpful to get a rough estimation of LK project investment as a preliminary figure before moving to further feasibility study. One of LK criteria depicts that the unit LK should be small enough in order to limit the investment requirement and local people affordability. Therefore, we propose the largest LK package capacity is 10 ton per day. Table 4 shows ten typical cost of LK from the lowest one ton to the largest 10 tons.

IOP Conf. Series: Journal of Physics: Conf. Series **974** (2018) 012037 doi:10.1088/1742-6596/974/1/012037

Table 4. Typical specific cost per ton of waste

Volume (ton/day)	Specific cost (USD)	Capex (USD)	Opex (USD)	kWh/day	Capacity (kW)	Op Hr/day
1	31,000	31,000	7,185	144	25	6
2	31,000	62,000	10,585	288	45	6
3	24,800	74,360	14,435	432	75	6
4	24,800	99,150	18,285	576	29	6
5	24,800	123,940	22,870	720	120	6
6	24,800	146730	27,750	864	150	6
7	24,800	173,520	30,860	1.008	175	6
8	24,800	198,310	36,927	1.152	200	6
9	24,800	223,100	41,515	1.296	225	6
10	22,320	223,156	47,546	1.440	250	6

8. LK packages and financial simulation

One of LK criteria depicts that the unit LK should be small enough in order to limit the investment requirement and local people affordability. Therefore, we propose six categories of small size of LK capacity from the smallest 1 (one) ton per day to the largest LK package of 10 ton per day, the size of which is based on waste volume in each location. Table 5 shows the simulation result of financial indicators for each category including payback period, Net Present Value (NPV), Internal Rate of Return, Electricity Product and Capacity.

$$NPV = -C_0 + \sum_{i=1}^{T} \frac{Ci}{(1+r)}i \quad (1)$$

Where IRR is the value of i if NPV = 0, Energy is electrical energy (E), which is power (P) or LK capacity multiplies by time E (kWh) = P (kW) x h (hours).

Table 5. NPV, IRR, and payback period of five module packages (10 years basis)

Pack. Modul	Volume (ton/day)	Payback Period	Net Present Value (USD)	Internal Rate of Return*	Energy kWh/day	Capacity (kW)**
I	1	5 years	8.352	5,2%	144	25
II	2	4 years	38.088	11,9%	288	50
III	3	3 years	83.919	21%	432	75
IV	4	3 years	117.325	22%	576	100
V	5	3 years	146.558	22%	720	120
VI	10	3 years	307.673	26%	1.440	250

9. Intangible benefit

In addition to the real revenue as written above, LK provides intangible benefit in the form of social benefit and environmental benefit. Social benefit can be earned from people opportunity to get income and wellbeing improvement under the zero waste milieu. Environmental benefit can be valued from fossil fuel cost saving and carbon reduction [16]–[18]. For the purpose of simulation in this paper, the social benefit that will be taken into consideration is potential additional income tax, both from LK as company and from thousands of individuals who get new jobs as LK employees.

9.1. Cost benefit comparison between conventional system and LK system

Compare to LK distributed generation system, the conventional system (CS) has two main merits. First CS is cheaper in price per unit capacity due to the principle of economies of scale [19]. Second, CS has much better reliability than LK because of interconnection system advantage. However, CS has to incur at least three additional indirect costs. First, CS plants construction time (around 5 years) is much longer than those of LK plants (6 months). Second, the high cost of transmission lines construction, around USD 114/kW to USD227 /kW [20]. Third, large scale fossil fuel power plant construction time is much longer than LK and even worse, most of the big size power plants construction are facing many problems that cause delay or project slippage. Delay means cost due to the loss of opportunity to get income from energy services.

On the other hand, LK offers three benefits that may answers problems of CS system. First, land acquisition problem is not a big issue in LK since LK capacity is small and the plants can be scattered wherever the land available. Second, LK construction time is less than a year, much shorter than CS that takes around five years. Third, LK power plants does not require a high cost transmission lines because it can be connected directly to the low voltage or medium voltage distribution lines. In addition to the three advantages described above, LK provides social and environmental benefits. Among many social benefits, for simple simulation in this paper, we will take the social benefit derived from the increasing start up businesses and job opportunities for local people. This intangible benefit is capitalized in term of additional tax revenue, both from company and individual taxes. The environmental benefit is derived from the less CO2 pollution that will be capitalized using equivalent carbon reduction (CDM) formula. Despite its benefits, there are two disadvantages that may need additional costs related to LK, higher per unit capital expenditure and cost for control and monitoring a huge numbers and scattered LK power plants. The cost-benefit comparison between LK and CS can be represented in simple mathematical expression namely Delta Value as follows:

$$\begin{aligned} \Delta V_{LK-CS} &= \Delta \mathrm{B}_{\mathrm{LK-CS}} - \Delta \mathrm{C}_{\mathrm{LK-CS}} \\ &= (B_{LK} - B_{CS}) - (C_{LK} - C_{CS}) \\ \Delta V_{LK-CS} &= \{(B_{LK1} + B_{LK2}) - (B_{CS1} + B_{CS2})\} - \{C_{LK} - (C_{CS1} + C_{CS2})\} \end{aligned} \tag{2}$$

Where, $\Delta V_{LK\text{-}CS}$ is a monetary value of LK (USD), benefit if the value is positive and cost if the value is negative. B_{LK} is a total benefits of LK consist of B_{LK1} (Social benefit) and B_{LK2} (Environmental benefit). B_{CS} is a total benefits of CS consist of B_{CS1} (Benefit of economies of scale) and B_{CS2} (Benefit of reliability). C_{LK} is additional cost for control and monitoring cost of LK. C_{CS} is a total costs of CS consist of C_{CS1} (transmission lines cost) and C_{CS2} (construction delay cost)

9.2. LK benefit – cost simulation

The purpose of this simulation in this paper, we will use the following numbers: Total LK capacity (P) = 1000 MW, which number is taken from the largest unit of power plant in Indonesia. Production time is annual based, h = 8760 hours. Period of simulation: 10 (ten) year basis. Feed in tariff, f = 0.01 USD/kWh, the price of which is taken from the current government regulated tariff for electrical energy.

As depicted previously, LK provides opportunity for local business players to own small scale power plant or waste to energy plant. It means that LK will give social benefit as a lot job opportunity to people to run thousands of LK plants around the country. The method that will be used to quantify such social benefit in this paper will be derived from the potential additional tax income for the nation, both from company taxes and individual income taxes. For simplification, other social benefit of LK including more comfortable neighbourhood, increasing buying power, reducing unemployment rate, and energy security improvement are not be included in this calculation. For simulation purpose, we will take the moderate package of 5 ton waste per day, which employs around 12 workers with regional minimum wage per year is IDR 40,000,000 or equivalent USD 3,000

IOP Conf. Series: Journal of Physics: Conf. Series **974** (2018) 012037 doi:10.1088/1742-6596/974/1/012037

Refer to Table 5, the package of 5 ton of waste per day is estimated to have Net Present Value of USD 146.558. The country will have potential income from added value and corporate income tax at least 20 % or USD29,312. Estimated capacity of LK with 5 ton waste per day is 120 kW, so the total potential corporate income tax from 1000 MW LK is 1000,000/120 x USD29,312 = USD 244.263.410.

The individual tax potential is calculated from the income of 12 workers who work in 5 ton package waste (120 kW capacity). The total income in this LK is 12 x USD3000 = USD36,000 and the potential tax (15%) from 1000 MW LK is 15% x 1,000,000/120 x USD 36,000 = USD 300,000,000.

Every kWh of electrical energy is similar to 0.00012 Ton coal equivalent and if the price of coal is USD40/ton, then any kWh used of renewable LK plant will save: 0.00012 x USD40 = USD 0.0048. Since the package of 5 ton waste may produces 720 kWh/day (Table 5), the total environmental benefit resulted from 1000 MW LK plants will be: 1,000,000/120 x 720 x 365 x USD 0.0048 = USD 10.512.000

A huge number of plants under distributed generation system, require rigorous control and monitoring that can be expressed as additional cost. For the purpose of this simulation, we will pick the moderate value of this cost as 5% of Capital Expenditure (Capex). For the package of 5 ton waste, the capex is USD123,940 and the total cost of 1,000 MW LK will be: 1,000,000/120 x 5% x USD123,940 = USD 51,641,667

The benefit of the conventional system is derived from the lower investment cost per unit of large power plant and the better system reliability compare to that of LK. We will use Coal Power Plant to represent the large CS power plant and the average capital cost of this kind in Indonesia is USD 1,218/kW (Wahid, 2016). While the capital cost for LK plant is calculated from the typical cost of 5 ton waste package LK (Table 4) with 20 hours daily operation. Capex for 5 ton is USD 123,940 and the capacity is 720 kWh/20 h = 36 kW. The cost per kW of LK= 123,940/36 = 3,443 USD/kW. So, the total cost of 1000 MW LK plants will be: 1000,000 x USD (3,443 – 1,218) = USD 2.224.777.778

Conventional system has better reliability since it has a lot of redundant power plants that interconnected one into another. Reliability can be represented by the delta reliability (less duration of interruption compare to the LK system), in hours per year. For the purpose of this calculation we assume that the maximum of shortage in LK plants is 10 days or 240 hours a year. So, we can pick three conditions of delta; 72 hours, 120 hours, days, and 240 hours outages per year. For instance, if the delta is 120 hours, the benefit will be: 1000,000 kW x 120 hours x 0,01 USD/kWh = USD1,200,000

Transmission line (T/L) cost is taken from Andrade, Juan et. Al (2016), which stipulated that the Grid connection cost for Generating Technology is within the range of 114 USD/kW for the small plant and 227USD/kW for the large plant that far from load. So, we can pick the number of 227USD/kW as transmission lines cost for this simulation. If the total capacity of plants is 1000 MW, the T/L cost will be: 1000,000 x USD 227 = USD 227,000,000

In the electricity industry, revenue is coming from energy sales in USD/kWh. It other words, any single hour delay means the potential loss of income is similar to the capacity of unit in kW times the tariff. For example: if capacity is 1000 MW or 1,000,000 kW and average tariff in Indonesia 0,07 USD, then the potential loss is USD 70,000 per hour or USD 1,680,000 per day of delay. This value will be used to calculate the total delay cost for several conditions. For simulation purpose, we will pick 3 (three) different conditions: one month delay, six month delay, and one year delay. For example, if the project slippage is one month, the cost will be: 30 x USD 1,680,000 = USD 50,400,000.

10. Result

The overall simulation of Cost-Benefit calculation is using the equation (2) and simulated for nine different cases by combining three different numbers for reliability and project delay. Case 1 : 72 hours outage; 1 month slippage, Case 2: 72 hours outage; 6 month slippage, Case 3: 72 hours outage; 1 2month slippage, Case 4: 120 hours outage; 1 month slippage, Case 5:

120 hours outage; 6 month slippage, Case 6: 120 hours outage; 12 month slippage, Case 7: 240 hours outage; 1 month slippage, Case 8: 240 hours outage; 6 month slippage, Case 9: 120 hours outage; 12 month slippage. The overall simulation result of Cost-Benefit calculation is put in the Table 6.

Table 6. Summary of cost-benefit comparison between LK and CS (in thousand USD)

Cases	B_{LK1}	B_{LK2}	B_{CS1}	B_{CS1}	C_{LK}	C_{CS1}	$C_{CS2)}$	$\Delta V_{LK\text{-}CS}$
1	290.417	10.512	503.389	720	51.642	227.000	50.400	22.579
2	290.417	10.512	503.389	720	51.642	227.000	302.400	274.579
3	290.417	10.512	503.389	720	51.642	227.000	604.800	576.979
4	290.417	10.512	503.389	1.200	51.642	227.000	50.400	22.099
5	290.417	10.512	503.389	1.200	51.642	227.000	302.400	274.099
6	290.417	10.512	503.389	1.200	51.642	227.000	604.800	576.499
7	290.417	10.512	503.389	2.400	51.642	227.000	50.400	20.899
8	290.417	10.512	503.389	2.400	51.642	227.000	302.400	272.899
9	290.417	10.512	503.389	2.400	51.642	227.000	604.800	575.299

Table 6 can be used as a template to calculate and compare Listrik Kerakyatan system to Conventional system by simulating different input variable that match the local condition. From table 6, we can observe that the most dominant variable is the economies of scale $B_{CS}1$that mainly depends on specific cost per kW of LK power plant. Result in Table 6 shows that in all cases, $\Delta V_{LK\text{-}CS}$ positives. It means that LK system may offer more benefit compare to the conventional system if LK operates only 6 hours per day. In this case, LK does not operate as a base load but as a peaker or load follower plant, which runs during the peak load condition.

However, the case will be different if the 5 ton LK power plants operates longer than 6 hours. For instance if LK operates for 11 hours a day, leads to the capacity of 65 kW (720 kWh/11 h). In this case the cost per kW of LK becomes 1,894 USD/kW and the result shows ΔVLK-CS negative for case 1, case 4, and case 7 as shown in Table 7.

Table 7. Summary of cost-benefit comparison between LK and CS (in USD)

Cases	BLK1	BLK2	BCS1	BCS2	CLK	CCS1	CCS2	$\Delta V_{LK\text{-}CS}$
1	290,417,179	10,512,000	675,527,778	720,000	51,641,667	227,000,000	50,400,000	-149,560,265
2	290,417,179	10,512,000	675,527,778	720,000	51,641,667	227,000,000	302,400,000	102.439.735
3	290,417,179	10,512,000	675,527,778	720,000	51,641,667	227,000,000	604,800,000	404.839.735
4	290,417,179	10,512,000	675,527,778	1,200,000	51,641,667	227,000,000	50,400,000	-150.040.265
5	290,417,179	10,512,000	675,527,778	1,200,000	51,641,667	227,000,000	302,400,000	101.959.735
6	290,417,179	10,512,000	675,527,778	1,200,000	51,641,667	227,000,000	604,800,000	404.359.735
7	290,417,179	10,512,000	675,527,778	2,400,000	51,641,667	227,000,000	50,400,000	-151.240.265
8	290,417,179	10,512,000	675,527,778	2,400,000	51,641,667	227,000,000	302,400,000	100.759.735
9	290,417,179	10,512,000	675,527,778	2,400,000	51,641,667	227,000,000	604,800,000	403.159.735

11. Conclusion

Commutative Law of Algebra can be used as a simple mathematical model to analyse whether the Listrik Kerakyatan model as an eco-friendly distributed generation can be applied to solve various problems associated with a large scale Conventional System. The result of simulation shows that LK provides more value if its plants operate in less than 11 hours as peaker power plant or load follower power plant to improve load curve balance of the power system. The result of simulation indicates that the investment cost of LK plant should be optimized in order to minimize the plant investment cost. This study indicates that the benefit of economies of scale principle does not always apply to every conditions, particularly if the portion of intangible cost and benefit is relatively high.

IOP Conf. Series: Journal of Physics: Conf. Series **974** (2018) 012037 doi:10.1088/1742-6596/974/1/012037

References

[1] J. Vincent, M. Huggan, and M. A. of Victoria, *Mathematics - Making Sense of Our World: The MAV's 44th Annual Conference, Thursday, 6th and Friday, 7th December 2007, La Trobe University, Bundoora*. Mathematical Association of Victoria, 2007.
[2] J. J. Kaput, D. W. Carraher, and M. L. Blanton, *Algebra in the early grades*. New York: Lawrence Erlbaum Associates : National Council of Teachers of Mathematics, 2008.
[3] D. Carraher and A. Schliemann, *Early algebra and algebraic reasoning*, vol. 2. 2007.
[4] W. Windsor, "How Problem Solving Can Develop an Algebraic Perspective of Mathematics.," *Aust. Prim. Math. Classr.*, vol. 16, no. 4, pp. 8–12, 2011.
[5] J. Voigt, "Negotiation of mathematical meaning and learning mathematics," *Educ. Stud. Math.*, vol. 26, no. 2, pp. 275–298, Mar. 1994.
[6] T. N. Carraher, D. W. Carraher, and A. D. Schliemann, "Mathematics in the streets and in schools," *Br. J. Dev. Psychol.*, vol. 3, no. 1, pp. 21–29, Mar. 1985.
[7] P. Cobb, "Where Is the Mind? Constructivist and Sociocultural Perspectives on Mathematical Development," *Educ. Res.*, vol. 23, no. 7, pp. 13–20, 1994.
[8] V. D. Sharma, K. Nagar, and L. B. Almeida, "Operation Transform Formulae of Fourier-Laplace Transform," *Int. J. Pure Appl. Sci. Technol.*, vol. 15, no. 2, pp. 62–67, 2013.
[9] W. L. Briggs and V. E. Henson, "The DFT an owner's manual for the discrete Fourier transform." Society for Industrial and Applied Mathematics, Philadelphia, PA, 1995.
[10] G. Kemper, *A Course in Commutative Algebra*, 256th ed. Springer-Verlag Berlin Heidelberg, 2011.
[11] O. Zariski and P. Samuel, *Commutative Algebra*, Volume II. Springer-Verlag Berlin Heidelberg, 1960.
[12] S. Hannabuss, "The Gale Encyclopedia of Science (4th edition)," *Ref. Rev.*, vol. 22, no. 7, pp. 31–33, Sep. 2008.
[13] A. Yadav and L. Srivastava, "Optimal placement of distributed generation: An overview and key issues," in *Power Signals Control and Computations (EPSCICON), 2014 International Conference on*, 2014, pp. 1–6.
[14] K. Arunachalam, V. S. Pedinti, and S. Goel, "Decentralized distributed generation in India: A review," *J. Renew. Sustain. Energy*, vol. 8, no. 2, 2016.
[15] Z. Abdmouleh, R. A. M. Alammari, and A. Gastli, "Recommendations on renewable energy policies for the GCC countries," in *Renewable and Sustainable Energy Reviews*, 2015, vol. 50, pp. 1181–1191.
[16] P. Pathak and P. Dattani, "Social return on investment: three technical challenges," *Soc. Enterp. J.*, vol. 10, no. 2, pp. 91–104, 2014.
[17] A. Tabi, "Using the stated preference method for the calculation of social discount rate," *Soc. Econ.*, vol. 36, no. 2, pp. 167–186, 2013.
[18] M. Gatto and G. A. De Leo, "Pricing biodiversity and ecosystem services: the never-ending story," *Bioscience*, vol. 50, no. 4, pp. 347–355, 2000.
[19] G. Patacconi and F. Russo, "Rethinking Economies of Scale Towards Network Economies," *Development*, vol. 58, no. 4, pp. 521–527, 2015.
[20] J. Andrade and R. Baldick, "Estimation of transmission costs for new generation," University of Texas, 2016.

IOP Conf. Series: Journal of Physics: Conf. Series **974** (2018) 012038 doi:10.1088/1742-6596/974/1/012038

Optimal blood glucose control in diabetes mellitus treatment using dynamic programming based on Ackerman's linear model

Paskalia Pradanti[1], Hartono[2]

Sanata Dharma University, Paingan, Maguwoharjo, Depok, Sleman, Yogyakarta, Indonesia 55282

E-mail:[1] paskaliapradanti@gmail.com
[2] yghartono@usd.ac.id

Abstract. Determination of insulin injection dose in diabetes mellitus treatment can be considered as an optimal control problem. This article is aimed to simulate optimal blood glucose control for patient with diabetes mellitus. The blood glucose regulation of diabetic patient is represented by Ackerman's Linear Model. This problem is then solved using dynamic programming method. The desired blood glucose level is obtained by minimizing the performance index in Lagrange form. The results show that dynamic programming based on Ackerman's Linear Model is quite good to solve the problem.

1. Introduction

Diabetes mellitus is a disease where the blood glucose level cannot be maintained in the normal range, i.e. 60 – 110 mg/dl as stated in [2]. Diabetes mellitus affects insulin production or the insulin response ability to the body. On the other hand, insulin hormone is responsible for the glucose absorption. Insulin injection is one of the most common diabetes mellitus treatments. In this research, the optimal insulin dose for diabetes mellitus treatment is determined based on the optimal control of blood glucose level.

The use of dynamic programming based on Ackerman's linear model to solve the optimal blood glucose control problem will be discussed in this paper. A MATLAB program is built based on the dynamic programming principle to solve the optimal control problem representing the problem. In this case, Ackerman's linear model is chosen to represent the regulation of glucose concentration in blood.

2. Ackerman's linear model

Ackerman's linear model is able to simulate glucose regulation system although this model is an abstraction of the reality as stated in [1]. Ackerman's linear model based on [1] and [3] is as follows.

$$\frac{dg}{dt} = -m_1 g - m_2 h + J \tag{1}$$

$$\frac{dh}{dt} = -m_3 h + m_4 g + K \tag{2}$$

It is defined that $g \equiv G - G_0$ where G is glucose concentration and G_0 is fasting glucose concentration. Furthermore, h is defined as $h \equiv H - H_0$ where H is blood hormone concentration (including insulin) and H_0 is fasting blood hormone concentration.

The parameters in (1) and (2) as given in [1] and [3] are:

m_1: constant rate for the removal of glucose above the initial (fasting) level due to its own excess above the initial level

m_2: constant rate for the removal of glucose above the initial level due to blood-hormone concentrations above the initial level

m_3: constant rate for the removal of hormone above the initial (fasting) level due to its own excess above the initial level

m_4: constant rate for the release of hormone above the initial level due blood-glucose concentrations above the initial level

In this paper, the exogenous glucose infusion rate J is not considered. J is set to 0 because it is assumed that there is no given exogenous glucose. Both h and K are assumed as insulin where K is the insulin infusion rate.

3. Dynamic programming

The general problem of optimal control is to

$$\text{minimize } I(x(0), t_f) = \int_0^{t_f} \psi(x, u, t)\, dt + \phi(x(t_f))$$

$$\text{subject to } \frac{dx}{dt} = f(x, u, t), x(0) = x_0$$

$$\alpha_i \le u_i(t) \le \beta_i, \quad i = 1,2,3 \dots m$$

where $I: R^n \times R^m \times R \to R$, $f: R^n \times R^m \times R \to R^n$, $x = (x_1, \cdots, x_n) \in R^n$ $u = (u_1, \cdots, u_m) \in R^m$.

Dynamic programming that is implemented in this research is based on iterative dynamic programming introduced by Luus in [4]. In the dynamic programming, the optimal control problem is set up into a sequence of stages. The optimal control problem with $[0,t_f]$ time interval is then approximated by a piecewise constant control u over P stages with length $L = t_f/P$. The performance index is estimated by

$$I(x(0), P) = \psi(x(t_f)) + \sum_{k=1}^{P} \int_{t_{k-1}}^{t_k} \phi(x(t), u(k-1), t)\, dt \tag{3}$$

Each state component x_i of x takes N values over a region at each time stage except for the first stage which is the given initial condition. Hence, stage 2 until P has each N^n grid points. Each control component u_j of u takes M values over a region at each stage. Thus, each state grid point is allowed to take M^m control vector values.

The calculations are started from stage P corresponding to time interval $t_f - L \le t < t_f$. For each grid point of state x, M^m values of the performance index are evaluated

$$I(x(t_f - L), 1) = \psi\left(x(t_f)\right) + \int_{t_f - L}^{t_f} \phi(x(t), u(P-1))\, dt \tag{4}$$

where M^m control values used for $u(P-1)$. These M^m values of the performance index are compared to determine the particular value of $u(P-1)$ which gives the minimum value. This particular value is the best control to use at that particular grid point of state x. By repeating the same procedure for all N^n grid points, the optimal control to use at this last stage can be determined.

Next, the integration is done backward to stage $P - 1$ corresponding to time interval $t_f - 2L \le t < t_f - L$. For each grid point, compare M^m control values to minimize the performance index. However, it is possible that when the integration from $t_f - 2L$ to $t_f - L$ is done, the state $x(t_f - L)$ is not one of the grid points at state P. In this case, the optimal control policy corresponding to the closest grid point to the state $x(t_f - L)$ is taken. The same calculation procedure is continued to stages $P-2$, $P-3$, and so on until reaching the first stage. Grid points at stage 1 corresponding to the time interval $0 \le t < L$ consists of the initial condition $x(0)$ as a single grid point. M^m performance index values are compared to determine the control policy giving the minimum value. See figure 1, for the case $n = 2$, $N = 5$, $m = 1$ and $M = 4$.

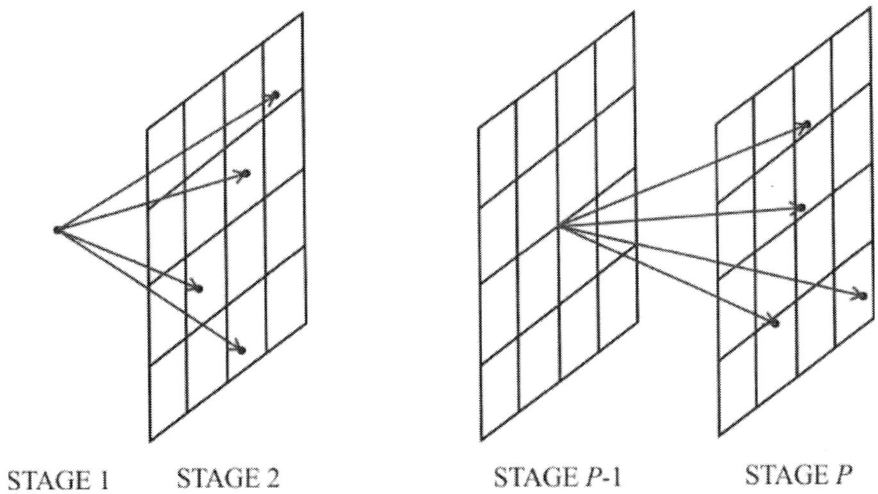

Figure 1. Illustration of control values assignment on each grid point.

Finally, the state values at the beginning of each stage are stored to generate state trajectories by doing a forward integration. The trajectories store states with controls giving the minimum value performance index.

4. Simulation results

Ackerman's model diabetic patient parameters as stated by Yipintsoi et al. in [5] is used in the program simulation, i.e. $m_1 = 0.0009$, $m_2 = 0.0031$, $m_3 = 0.0415$, and $m_4 = 0$. Hence, the Ackerman's linear model becomes

$$\frac{dg}{dt} = -0.0009g - 0.0031h \tag{5}$$

$$\frac{dh}{dt} = -0.0415h + K \tag{6}$$

The program is simulated over $P = 250$ with final time $t_f = 300$, number of state g and h grid points $N = 10$, and control discretization $M = 10$. The optimal control problem is solved by minimizing the performance index in Lagrange form, i.e.

$$J = \int_0^{t_f} \rho_1 |K| + \rho_2 |G - G_T| \, dt \tag{7}$$

where ρ_1 is control u weight factor, ρ_2 is state g weight factor, and G_T is desired glucose level. Weight factors $\rho_1 = \rho_2 = 1$ used in this simulation means that minimizing the dose of insulin injection is as important as minimizing glucose concentration. The initial glucose level is $g(0) = 300$ and the initial hormone level is $h(0) = 60$. The control vector is bounded by $0 \le u \le 10$ and the desired glucose level as the target is $G_T = 100$. The simulation results are shown in figure 2, figure 3, and figure 4.

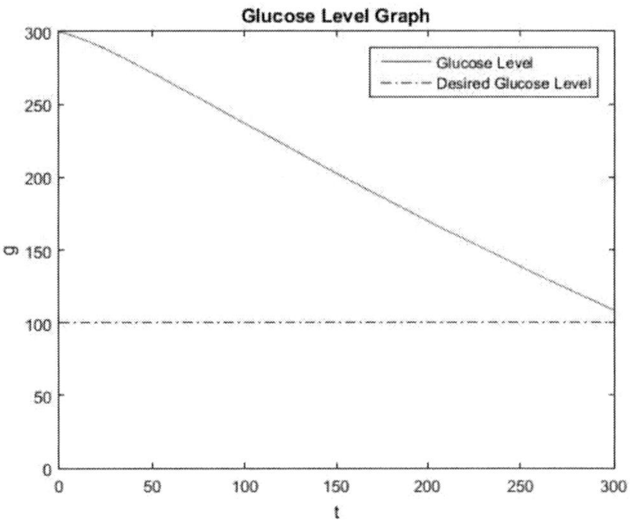

Figure 2. Glucose level graph.

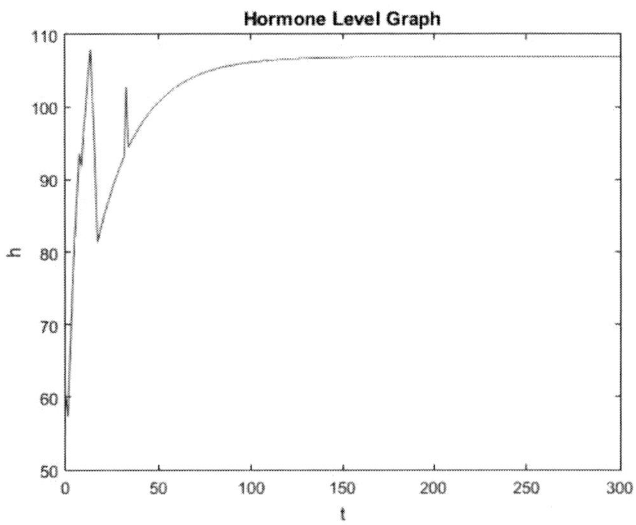

Figure 3. Hormone level graph.

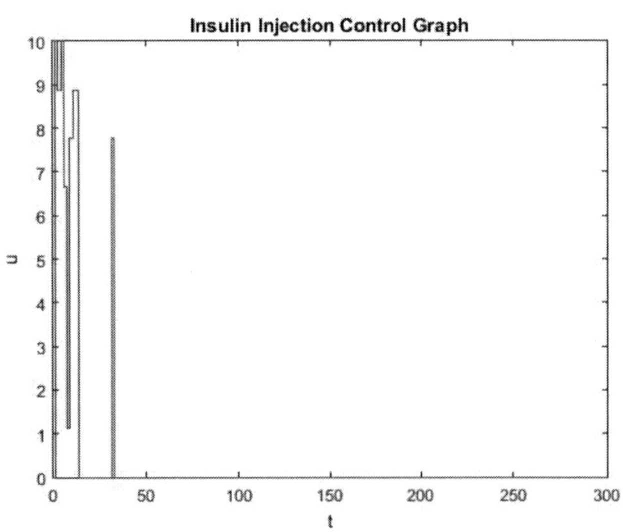

Figure 4. Insulin injection control graph.

From the graphs, it is clearly seen that the final state glucose level approaching the desired level could be reached with the obtained insulin injection control behaviour. When t is less than 20 the maximum value of control is used to decrease the blood glucose level rapidly. At the same time, this causes the increase of hormone level till reach the steady state condition around $t = 150$. On the other hand, the control chooses the minimum value after $t > 50$. The glucose concentration at the final stage is $g(t_f) = 108.5668$ and the final hormone level is $h(t_f) = 106.9235$. Therefore, the glucose concentration change is 191.4332.

5. Conclusions

It could be concluded from the result that Ackerman's linear model could represent the regulation of glucose concentration quite well. The results also show that the used method, i.e. dynamic programming, could be applied to solve the blood glucose optimal control problem discussed in this paper. The desired glucose level could be approached by the achieved final glucose level with the determined control policy by minimizing the performance index in Lagrange form. An insulin infusion therapy can be practiced based on the resulted insulin injection control behaviour in this paper. Nevertheless, several kinds of adjustments to the real problem are still needed to apply this result.

References

[1] Ackerman E, Gatewood L C, Rosevear J W and Molnar G D 1965 Model studies of blood-glucose regulation *Bulletin of Mathematical Biophysics* **27** pp 21-37

[2] Brand-Miller J, Foster-Powell K and Mendosa D 2006 *The New Glucose Revolution: What Makes My Blood Glucose Go Up...and Down?* (New York: Marlowe & Company)

[3] Chee F and Fernando T 2007 *Closed-Loop Control of Blood Glucose* (Springer)

[4] Luus R 2000 *Iterative Dynamic Programming* (Boca Raton: Chapman & Hall/CRC)

[5] Yipintsoi T, Gatewood L C, Ackerman E, Spivak P L, Molnar G D, Rosevear J W and Service J V 1973 Mathematical analysis of blood glucose and plasma insulin responses to insulin infusion in healthy and diabetic subjects *Computers in Biology and Medicine* **3** pp 71-78

IOP Conf. Series: Journal of Physics: Conf. Series **974** (2018) 012039 doi:10.1088/1742-6596/974/1/012039

Comparison between GSTAR and GSTAR-Kalman Filter models on inflation rate forecasting in East Java

Jessica Rahma Prillantika[1], Erna Apriliani[2], Nuri Wahyuningsih[3]

[1] Student at Mathematics Department, Sepuluh Nopember Institute of Technology, Indonesia
[2,3] Lectures at Mathematics Department, Sepuluh Nopember Institute of Technology, Indonesia

Abstract—Up to now, we often find data which have correlation between time and location. This data also known as spatial data. Inflation rate is one type of spatial data because it is not only related to the events of the previous time, but also has relevance to the other location or elsewhere. In this research, we do comparison between GSTAR model and GSTAR-Kalman Filter to get prediction which have small error rate. Kalman Filter is one estimator that estimates state changes due to noise from white noise. The final result shows that Kalman Filter is able to improve the GSTAR forecast result. This is shown through simulation results in the form of graphs and clarified with smaller RMSE values.

1. Introduction

Up to now, we often find data which have correlation between time and location. This data also known as spatial data. Inflation rate is one type of spatial data because it is not only related to the events of the previous time, but also has relevance to the other location or elsewhere. Inflation rate can not be destroyed clearly. We only can reduce and control it. Because the controlled inflation can increase economics matters [1]. This fact makes forecast analysis be the important things in inflation rate in Malang, Probolinggo, and Surabaya. Because these locations have an important role to oversee the economics stability in East Java.

One of the methods that can be used to forecast the inflation rate is *Generalized Space Time Autoregressive method* (GSTAR). This method is used to forecast data which have correlation between time and location. In forecasting method, we need some parameters from GSTAR model. To estimate the value of the GSTAR parameters, we need method that can give the best forecasting. In this research, *Ordinary Least Square* methods (OLS) and Kalman Filter Estimator are used to estimate the GSTAR parameters [2].

Filter Kalman method is one of the estimator that estimate toward the conversion of state because of white noise disruption. The advantage of Kalman Filter is in estimation process, recursive form is used to increase *Root Mean Square Error* (RMSE) value and error [3]. In the past researches, Kalman Filter had done to estimate the ARIMA and VAR parameters. The result showed that estimation using Kalman Filter was better than OLS. It was shown with RMSE value in estimation using Kalman Filter is smaller than OLS [4][5][6].

In this research, we've already comparison the forecasting result in the inflation rate month to monthly in Malang, Probolinggo, and Surabaya using GSTAR method and GSTAR-Kalman Filter method. The best method is the method that have smaller RMSE.

2. GSTAR (Generalized Space-Time Autoregressive) Model

The GSTAR model is specific form of VAR (Vector Autoregressive) model. It reveals linear dependencies of space and time. The main difference is on the spatial dependent, that in GSTAR model, it is expressed by weight matrix. Let $\{\boldsymbol{Z}(t): t = 0, \pm 1, \pm 2, \ldots\}$ be a multivariate time series of N components. In matrix notation, the GSTAR model of autoregressive order p and spatial orders $\lambda_1, \lambda_2, \ldots, \lambda_p$, GSTAR $(\lambda_1, \lambda_2, \ldots, \lambda_p)$ could be written as [7]:

IOP Conf. Series: Journal of Physics: Conf. Series **974** (2018) 012039 doi:10.1088/1742-6596/974/1/012039

$$Z(t)=\sum_{s=1}^{p}\left[\Phi_{s0}+\sum_{k=1}^{\lambda_p}\Phi_{sk}W^{(k)}\right]Z(t-s)+e(t) \tag{1}$$

where
$\Phi_{s0}=\text{diag}(\phi_{s0}^{1},\dots,\phi_{s0}^{N})$ and $\Phi_{sk}=\text{diag}(\phi_{sk}^{1},\dots,\ \phi_{sk}^{N})$,
W (weight) are choosen to satisfy $w_{ii}^{(k)}=0$ and $\sum_{i\neq j}w_{ij}^{(k)}=1$.

3. Modified GSTAR Model

The first step to get the GSTAR model is identifying model that include stationary test and spatial and autoregressive order selection. The Box-Cox transformation and Matrix Autocorrelation Function (MACF) plot are employed for testing stationary conditions before GSTAR model is proceeded. The Box-Cox is employed to identify the stationary in varians. If rounded value $\neq 1$, then data is not stationer in varians and transformation must be employed. Whereas MACF plot is employed to identify the stationary in mean. If the data is not stationar, then differencing must be employed [7].

After the data is stationer in mean and varians, then GSTAR order is selected using VAR model. VAR model can be identified from Matrix Partial Autocorrelation Function (MPACF) plot. In this research, 72 in-sample data of the inflation rate is used to modify the GSTAR Model Autoregressive order can be selected by *Schwarz Bayesian Criterion* (SBC) and spatial order is generally restricted on order 1, since the higher order is difficult to be interpreted [7]. Then estimating the parameters of the model that include determination of space weight and autoregressive parameter. In this research, determination of space weight by using inverse distance [8]. Whereas, OLS and Kalman Filter can be used in estimating parameters. The next step after finding significant model is evaluating the residuals to verify normality and white noise of the error. The last step is comparing the RMSE result of the GSTAR and GSTAR-Kalman Filter [9].

MPACF plot from inflation rate in Malang, Probolinggo, and Surabaya is presented in Figure 1. Figure 1 indicates cut off in lag 1,2, 4, and 10. It means that there are several posibility order based on these lag. Then we use SBC as the criterion, we select GSTAR model from the lowest value of SBC.

```
                  Schematic Representation of Partial Autoregression
Variable/
Lag       1   2   3   4   5   6   7   8   9   10  11  12  13  14  15  16  17  18  19  20  21

z1t      ... -.. ... .-. ... ... ... ... ... ..- ... ... ... ... ... ... ... ... ... ... ...
z2t      ... ... ... ... ... ... ... ... ... ... ... ... ... ... ... ... ... ... ... ... ...
z3t      ..- ..- ... ... ... ... ... ... ... ..- ... ... ... ... ... ... ... ... ... ... ...

              + is > 2*std error,  - is < -2*std error,  . is between
```

Figure 1. MPACF plot in three locations

The SBC value is presented in Table 1. Table 1 shows that the smallest SBC value for MA 0 is in the lag 2. It indicates that order AR 2 or $p=2$. Therefore, we obtain GSTAR (2_1) for inflation rate in 3 locations [4].

Table 1. SBC value in any autoregressive order of inflation rate in 3 locations

Lag	MA 0
AR 0	-17,76274
AR 1	-17,85476
AR 2	-18,31683
AR 3	-18,25295
AR 4	-18,25391
AR 5	-17,97148
AR 6	-17,82173
AR 7	-17,44618
AR 8	-17,1605
AR 9	-16,85751
AR 11	-16,65356

4. Parameters Estimation

To estimate the parameters, we must determinate the space weight. In this research, we use the distance inverse space weight (W). Then the parameters are estimated using OLS and Kalman Filter [7].

4.1 Determination of Space Weight

The distance inverse space weight (W) in GSTAR(2_1) are generally taken as follow:

$$W = [w_{ij}] = \begin{bmatrix} 0 & w_{12} & \cdots & w_{1j} \\ w_{21} & 0 & \cdots & w2_j \\ \vdots & \vdots & \ddots & \vdots \\ w_{i1} & w_{i2} & \cdots & w_{ij} \end{bmatrix} \qquad (2)$$

where w_{ij} is distance between $i-th$ location with $j-th$ location. And w_{ij} can be declared as:

$$w_{ij} = \begin{cases} \dfrac{\frac{1}{b_{ij}}}{\sum_{j=1}^{N} \frac{1}{b_{ij}}}, i \neq j \\ 0, i = j \end{cases} \qquad (3)$$

Based on Equation (2) and (3), we can calculate the distance inverse space weight (W) in GSTAR(2_1). The calculation result is expressed as follows:

$$W = \begin{bmatrix} 0 & 0.5148 & 0.4852 \\ 0.5353 & 0 & 0.4647 \\ 0.5202 & 0.4795 & 0 \end{bmatrix} \qquad (4)$$

4.2 Parameter Estimation of GSTAR(2_1) using OLS

Based on Equation (1) with the distance inverse space weight (W) from Equation 4, GSTAR(2_1) model is represented as follow:

$$\begin{bmatrix} z_1(t) \\ z_2(t) \\ z_3(t) \end{bmatrix} = \begin{bmatrix} \Phi_{10}^1 & 0 & 0 \\ 0 & \Phi_{10}^2 & 0 \\ 0 & 0 & \Phi_{10}^3 \end{bmatrix} \begin{bmatrix} z_1(t-1) \\ z_2(t-1) \\ z_3(t-1) \end{bmatrix} + \begin{bmatrix} \Phi_{11}^1 & 0 & 0 \\ 0 & \Phi_{11}^2 & 0 \\ 0 & 0 & \Phi_{11}^3 \end{bmatrix} \begin{bmatrix} 0 & 0,5148 & 0,4852 \\ 0,5353 & 0 & 0,4647 \\ 0,5202 & 0,4795 & 0 \end{bmatrix} \begin{bmatrix} z_1(t-1) \\ z_2(t-1) \\ z_3(t-1) \end{bmatrix}$$
$$+ \begin{bmatrix} \Phi_{20}^1 & 0 & 0 \\ 0 & \Phi_{20}^2 & 0 \\ 0 & 0 & \Phi_{20}^3 \end{bmatrix} \begin{bmatrix} z_1(t-2) \\ z_2(t-2) \\ z_3(t-2) \end{bmatrix}$$
$$+ \begin{bmatrix} \Phi_{21}^1 & 0 & 0 \\ 0 & \Phi_{21}^2 & 0 \\ 0 & 0 & \Phi_{21}^3 \end{bmatrix} \begin{bmatrix} 0 & 0,5148 & 0,4852 \\ 0,5353 & 0 & 0,4647 \\ 0,5202 & 0,4795 & 0 \end{bmatrix} \begin{bmatrix} z_1(t-2) \\ z_2(t-2) \\ z_3(t-2) \end{bmatrix} + \begin{bmatrix} e_1(t) \\ e_2(t) \\ e_3(t) \end{bmatrix} \quad (5)$$

Based on equation (5), the result of parameter value using OLS method can be seen in Table 2. Then these parameters are eliminated using a significance level of 0,05. To eliminate these parameters, we use t-student test with value of t-table is 1.97. The parameter is catagorized as not significant parameter if $|t - value| < t - table$. And the parameter is catagorized as significant parameter if $|t - value| > t - table$.

Table 2. Estimated parameters of GSTAR(2_1) model for inflation rate data

	Coef	SE Coef	T	Conclusion
Φ_{10}^1	-0.1696	0.2431	-0.70	Not significant
Φ_{11}^1	1.2088	0.2718	4.45	Significant
Φ_{20}^1	-0.9206	0.2372	-3.88	Significant
Φ_{21}^1	0.6460	0.2527	2.56	Significant
Φ_{10}^2	1.0855	0.2062	5.26	Significant
Φ_{11}^2	-0.4064	0.2086	-1.95	Not significant
Φ_{20}^2	0.7975	0.2346	3.40	Significant
Φ_{21}^2	-0.6977	0.2183	-3.20	Significant
Φ_{10}^3	-0.5816	0.2504	-2.35	Significant
Φ_{11}^3	1.4074	0.2211	6.36	Significant
Φ_{20}^3	-0.4240	0.2514	-1.69	Not significant
Φ_{21}^3	0.3754	0.2735	1.37	Not significant

Using significant parameters, GSTAR model can be represented as follow

$$\begin{bmatrix} z_1(t) \\ z_2(t) \\ z_3(t) \end{bmatrix} = \begin{bmatrix} 0 & 0 & 0 \\ 0 & \Phi_{10}^2 & 0 \\ 0 & 0 & \Phi_{10}^3 \end{bmatrix} \begin{bmatrix} z_1(t-1) \\ z_2(t-1) \\ z_3(t-1) \end{bmatrix} + \begin{bmatrix} \Phi_{11}^1 & 0 & 0 \\ 0 & 0 & 0 \\ 0 & 0 & \Phi_{11}^3 \end{bmatrix} \begin{bmatrix} 0 & 0,5148 & 0,4852 \\ 0,5353 & 0 & 0,4647 \\ 0,5202 & 0,4795 & 0 \end{bmatrix} \begin{bmatrix} z_1(t-1) \\ z_2(t-1) \\ z_3(t-1) \end{bmatrix} + \begin{bmatrix} \Phi_{20}^1 & 0 & 0 \\ 0 & \Phi_{20}^2 & 0 \\ 0 & 0 & 0 \end{bmatrix} \begin{bmatrix} z_1(t-2) \\ z_2(t-2) \\ z_3(t-2) \end{bmatrix}$$
$$+ \begin{bmatrix} \Phi_{21}^1 & 0 & 0 \\ 0 & \Phi_{21}^2 & 0 \\ 0 & 0 & 0 \end{bmatrix} \begin{bmatrix} 0 & 0,5148 & 0,4852 \\ 0,5353 & 0 & 0,4647 \\ 0,5202 & 0,4795 & 0 \end{bmatrix} \begin{bmatrix} z_1(t-2) \\ z_2(t-2) \\ z_3(t-2) \end{bmatrix} + \begin{bmatrix} e_1(t) \\ e_2(t) \\ e_3(t) \end{bmatrix} \quad (6)$$

Based on equation (6), in the same way, the value of significant parameters are estimated. Estimated significant parameters are represented in Table 3.

Table 3. Estimated significant paremeters of GSTAR(2_1) model for inflation rate data

Parameter	Coef	SE Coef	t-value	Conclusion
Φ_{10}^{1}	1,0400	0,1233	8,43	Significant
Φ_{20}^{1}	-0,9213	0,2363	-3,90	Significant
Φ_{21}^{1}	0,6644	0,2504	2,65	Significant
Φ_{10}^{2}	0,7501	0,1158	6,48	Significant
Φ_{20}^{2}	0,8788	0,2356	3,73	Significant
Φ_{21}^{2}	-0,7875	0,2178	-3,62	Significant
Φ_{10}^{3}	-0,5572	0,2430	-2,29	Significant
Φ_{11}^{3}	1,3849	0,2169	6,38	Significant

4.3 Diagnostic Check

After finding the significant model of GSTAR, The last step is checking the white noise and normality of the error [10]. MACF plot can be used to identify the assumption of white noise. MACF plot of the error is represented in Figure 2.

```
                    Schematic Representation of Cross Correlations
Variable/
Lag       0   1   2   3   4   5   6   7   8   9   10  11  12  13  14  15  16  17  18  19  20

resy1t    +++ --- ... ... ... ... ... ... ... ... ... ... ... ... ... ... ... ... ... ... ...
resy2t    +++ --. ... ... ... ... ... +.. .-. ... ... ... ... ... ... ... ... ... ... ... .-.
resy3t    +++ --- ... ... ... ... ... ... ... ... ... ... ... ... ... ... ... ... ... ... ...

              + is > 2*std error,  - is < -2*std error,  . is between

                              The VARMAX Procedure

                    Schematic Representation of Cross Correlations
Variable/
Lag       21  22  23  24  25  26  27  28  29  30  31  32  33  34  35  36  37  38  39  40  41

resy1t    ... ... ... ... ... ... ... ... ... ... ... ... ... ... ... ... ... ... ... ... ...
resy2t    ... ... ... ... ... ... ... ... ... ... ... ... ... ... ... ... ... ... ... ... ...
resy3t    ... ... ... ... ... ... ... ... ... ... ... ... ... ... ... ... ... ... ... ... ...

              + is > 2*std error,  - is < -2*std error,  . is between

                    Schematic Representation of Cross Correlations
Variable/
Lag       42  43  44  45  46  47  48  49  50  51  52  53  54  55  56  57  58  59  60  61  62

resy1t    ... ... ... ... ... ... ... ... ... ... ... ... ... ... ... ... ... ... ... ... ...
resy2t    ... ... ... ... ... ... ... ... ... ... ... ... ... ... ... ... ... ... ... ... ...
resy3t    ... ... ... ... ... ... ... ... ... ... ... ... ... ... ... ... ... ... ... ... ...

              + is > 2*std error,  - is < -2*std error,  . is between

          Schematic Representation of Cross Correlations
          Variable/
          Lag       63   64   65   66   67   68

          resy1t    ...  ...  ...  ...  ...  ...
          resy2t    ...  ...  ...  ...  ...  ...
          resy3t    ...  ...  ...  ...  ...  ...
```

Figure 2. MACF plot of error inflation rate in 3 locations

Based on Figure 2, error of the GSTAR(2_1) model using distance inverse space weight is suitable for white noise assumption. It is represented from the sign (.) that is more than sign (+) and (-) [10]. Whereas, we can use q-q plot to check the normality of the error. This plot is represented in Figure 3. Figure 3 shows that the plot tends to form into a straight line. We can conclude that error of the GSTAR(2_1) model is normal multivariate distribution. Because of these error, white noise and normal multivariate distribution test are fulfilled, then GSTAR(2_1) with distance inverse space weight is the best model for inflation rate in 3 locations.

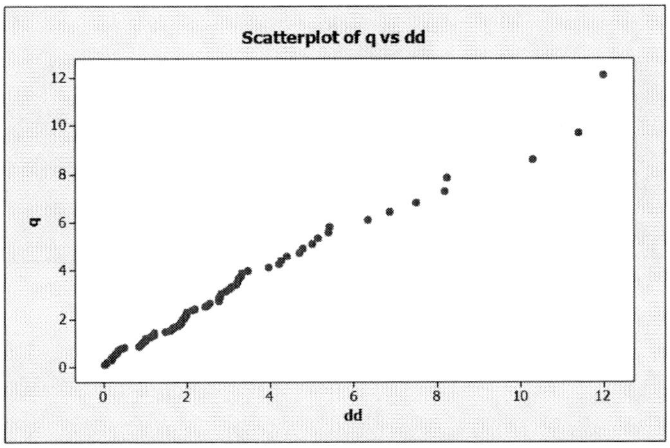

Figure 3. Q-Q plot of the error STAR(2_1)

4.4 Parameter Estimation of GSTAR(2_1) using Kalman Filter

After building the best GSTAR model, Kalman Filter is applied to estimate the GSTAR parameters. In this research, parameters estimated using Kalman Filter are GSTAR parameters that have been estimated using OLS and significant, ie Φ^1_{10}, Φ^1_{20}, Φ^1_{21}, Φ^2_{10}, Φ^2_{20}, Φ^2_{21}, Φ^3_{10}, dan Φ^3_{11}. Equation (6) can be modified in to sistem model as follows

$$\begin{bmatrix} z_1(t) \\ \Phi^1_{11} \\ \Phi^1_{20} \\ \Phi^1_{21} \\ z_2(t) \\ \Phi^2_{10} \\ \Phi^2_{20} \\ \Phi^2_{21} \\ z_3(t) \\ \Phi^3_{10} \\ \Phi^3_{11} \end{bmatrix}_{t+1} = \begin{bmatrix} 0 & a_{11} & a_{12} & a_{13} & 0 & 0 & 0 & 0 & 0 & 0 & 0 \\ 0 & 1 & 0 & 0 & 0 & 0 & 0 & 0 & 0 & 0 & 0 \\ 0 & 0 & 1 & 0 & 0 & 0 & 0 & 0 & 0 & 0 & 0 \\ 0 & 0 & 0 & 1 & 0 & 0 & 0 & 0 & 0 & 0 & 0 \\ 0 & 0 & 0 & 0 & 0 & a_{21} & a_{22} & a_{23} & 0 & 0 & 0 \\ 0 & 0 & 0 & 0 & 0 & 1 & 0 & 0 & 0 & 0 & 0 \\ 0 & 0 & 0 & 0 & 0 & 0 & 1 & 0 & 0 & 0 & 0 \\ 0 & 0 & 0 & 0 & 0 & 0 & 0 & 1 & 0 & 0 & 0 \\ 0 & 0 & 0 & 0 & 0 & 0 & 0 & 0 & 0 & a_{21} & a_{22} \\ 0 & 0 & 0 & 0 & 0 & 0 & 0 & 0 & 0 & 1 & 0 \\ 0 & 0 & 0 & 0 & 0 & 0 & 0 & 0 & 0 & 0 & 1 \end{bmatrix} \begin{bmatrix} z_1(t) \\ \Phi^1_{11} \\ \Phi^1_{20} \\ \Phi^1_{21} \\ z_2(t) \\ \Phi^2_{10} \\ \Phi^2_{20} \\ \Phi^2_{21} \\ z_3(t) \\ \Phi^3_{10} \\ \Phi^3_{11} \end{bmatrix}_t$$

where

$$a_{12} = 0{,}5148 z_2(t-1) + 0{,}4852 z_3(t-1)$$
$$a_{13} = z_1(t-2)$$
$$a_{14} = 0{,}5148 z_2(t-2) + 0{,}4852 z_3(t-2)$$
$$a_{56} = z_1(t-1)$$
$$a_{57} = z_2(t-2)$$
$$a_{58} = 0{,}5353 z_2(t-2) + 0{,}4647 z_3(t-2)$$
$$a_{910} = z_3(t-1)$$
$$a_{911} = 0{,}5205 z_1(t-1) + 0{,}4795 z_2(t-1)$$

Then using Measurement model of Kalman Filter that is declared as: $c_t = Hx_t + v_t$, It can be written as follows.

$$\mathbf{c}_t = \begin{bmatrix} 1 & 0 & 0 & 0 & 0 & 0 & 0 & 0 & 0 & 0 & 0 \\ 0 & 0 & 0 & 0 & 1 & 0 & 0 & 0 & 0 & 0 & 0 \\ 0 & 0 & 0 & 0 & 0 & 0 & 0 & 0 & 1 & 0 & 0 \end{bmatrix} \begin{bmatrix} z_1(t) \\ \Phi^1_{11} \\ \Phi^1_{20} \\ \Phi^1_{21} \\ z_2(t) \\ \Phi^2_{10} \\ \Phi^2_{20} \\ \Phi^2_{21} \\ z_3(t) \\ \Phi^3_{10} \\ \Phi^3_{11} \end{bmatrix} + v_t \quad (7)$$

We can determine matrix H in Equation (7) by considering the data that we have. Because we only have the data of $z_1(t), z_2(t)$, and $z_3(t)$, then matrix H can be written as in Equation (7).

After modifying the model system and measurement model, then do initialization. Initial values for Z_1, Z_2, and Z_3 are taken from the first data of the inflation rate that already stationer. Whereas the initial value of the variance of the noise given $Q = 0{,}3$ dan $R = 0{,}3$. For initial $\hat{x}_0$ and covariant values are given as follows:

$\hat{x}_0 = [0.5787 \quad 1.1234 \quad -0.8787 \quad 0.7667 \quad 0.7394 \quad -0.7012 \quad 0.8099 \quad -0.8097 \quad 0.6925 \quad -0.6767 \quad 1.2348]^T$.

The initial value $\hat{x}_0$ can be determined from the data that we have in the first period in each location. Beside that, we also determine the initial value $\hat{x}_0$ from predicted value of the parameter that we calculate using OLS method.

With $P_0 = I_{11\times 11}$, and $Q_k = I_{11\times 11} \times Q$, Then for Prediction Stage are given from these equation:

$$\hat{x}^-_{t+1} = A_t x + B_t u_t$$

$$P^-_{t+1} = A_t P_t A_t^T + G_t Q_t G_t^T$$

After do Prediction Stage, we do Measurement update. Measurement update is also called Correction Stage. In this stage, given Kalman Gain as this equation:

$$K_{t+1} = P^-_{t+1} M^T_{t+1} (M_{t+1} P^-_{t+1} M^T_{t+1} + R_{t+1})^{-1}$$

Then, $\hat{x}_{t+1}$ value is estimated using $\hat{x}^-_{t+1}$ value from Prediction Stage, then we get:

$$\hat{x}_{t+1} = \hat{x}^-_{t+1} + K_{t+1}(Z_{t+1} - M_{t+1}\hat{x}^-_{t+1})$$

P_{t+1} value can be estimated using P^-_{t+1} value from Prediction Stage, then it can be written:

$$P_{t+1} = (I - K_{t+1} H_{t+1}) P^-_{t+1}$$

Parameters estimation using Kalman Filter of the inflation rate in 3 locations are represented in Table 4. Table 4 shows the value of each paramaters of GSTAR(2_1) that is estimated using Kalman Filter. Then these values are substituted in to Equation (6). The equation using parameters value from Table 4 is used to compare the forecasting result with parameters that is estimated using OLS.

5. Comparison Prediction The Rate Inflation Using GSTAR and GSTAR-Kalman Filter

Comparison between GSTAR and GSTAR-Kalman Filter model is used to see the efficiency of these models. Directly we can compare and see the most efficiency models from the graphic that we can see in Figure 4. But we can fix it mathematically using RMSE value. These models are compared using RMSE value from out-sample data. The most efficiency model is the model with the smallest RMSE. RMSE can be calculated using this equation [9]:

$$RMSE = \sqrt{MSE} = \sqrt{\frac{1}{N}\sum_{t=1}^{N}(Y_t - \hat{Y}_t)^2} \tag{8}$$

Table 4. Parameter estimation of GSTAR model using Kalman Filter

Model	Parameter	Coefficient
GSTAR(2_1)	Φ^1_{11}	0,9229
	Φ^1_{20}	-1,0012
	Φ^1_{21}	0,7623
	Φ^2_{10}	0,9072
	Φ^2_{20}	-0,2844
	Φ^2_{21}	-0,2849
	Φ^3_{10}	-0,5984
	Φ^3_{11}	1,4200

From Figure 4, we can see that prediction data using GSTAR-Kalman Filter is closer with factual data than prediction data using GSTAR. This is indicated by the GSTAR-Kalman Filter graph that is more keep up the factual data graph than GSTAR graph. To determine the most efficiency model in forecasting the inflation rate, we use RMSE value. Based on Equation (8), the result of RMSE calculation of this research can be represented in Table 5.

Table 5. Nilai RMSE *out-sample* di setiap lokasi

Lokasi	GSTAR-*OLS*	GSTAR-Kalman Filter
Malang	0,0661	0,0132
Probolinggo	0,0351	0,0132
Surabaya	0,0597	0,0132

From Table 5, RMSE of GSTAR-Kalman Filter have same value in each location. And also, RMSE of GSTAR-Kalman Filter is smaller than GSTAR in each locations. Then we can conclude that GSTAR-Kalman Filter is more efficiency in forecasting the rate inflation than GSTAR in each locations.

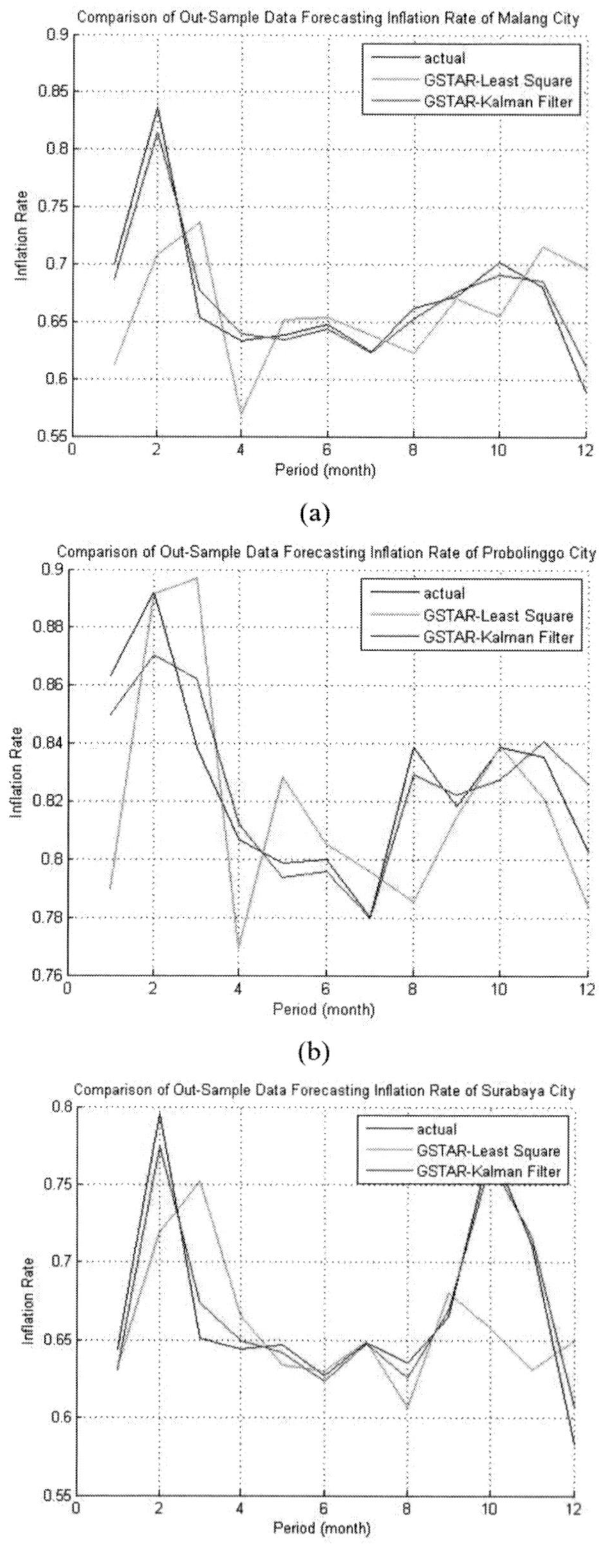

Figure 4. Comparison between data actual and prediction data of inflation rate in 3 locations

6. Conclusion

Based on the results of this research, can be conclude that the best model of inflation rate in Malang, Probolinggo, and Surabaya is GSTAR(2_1). The results show that RMSE value in each location of GSTAR-Kalman Filter is smaller than GSTAR that was estimated using OLS. It means that parameter estimation of GSTAR model using distance inverse space weight can be done optimally by using Kalman Filter.

References

[1] Artikel Ekonomi. (2015). "Artikelsiana, Artikel Belajar dan Bermanfaat". Dalam http://www.artikelsiana.com/2015/02/pengertian-inflasi-jenis-dampak-penyebab.html?m=1.

[2] Kurniawati. (2016). Perbandingan Penerapan Model GSTAR dengan Pembobot Inverse Jarak dan Normalisasi Korelasi Silang pada Laju Inflasi Kota Surakarta, Yogyakarta, dan Surabaya. Tugas Akhir. Universitas Sebelas Maret, Surakarta.

[3] G. Welch, & G.Bishop. (2006). *An Introduction to the Filter Kalman.* Chapel Hill: Department of Computer Science, University of North Carolina.

[4] I.F. Hamsyah. (2015). "Perbandingan GSTAR dan ARIMA Filter Kalman dalam Perbaikan Hasil Prediksi Debit Air Sungai Brantas". Tugas Akhir. Departemen Matematika. Institut Teknologi Sepuluh Nopember, Surabaya.

[5] M. H. Pamungkas. (2015). "Estimasi Parameter Model ARIMA Menggunakan *Kalman Filter* untuk Peramalan Permintaan Darah (Studi Kasus: UTD PMI Surabaya)". Tugas Akhir. Departemen Matematika. Institut Teknologi Sepuluh Nopember, Surabaya.

[6] P. Febritasari. (2016). "Estimasi Inflasi Wilayah Kerja KPwBI Malang Menggunakan ARIMA-Filter Kalman dan VAR-Filter Kalman". Tugas Akhir. Departemen Matematika. Institut Teknologi Sepuluh Nopember, Surabaya.

[7] D. U. Wutsqa, Suhartono, & B. Sutijo. (2012). "Aplikasi Model *Generalized Space Time Autoregressive* Pada Data Pencemaran Udara di Kota Surabaya". Jounal *Pythagoras,* Vol.7, No.2.

[8] Mansoer, A. Shaliha. (2016). "Pemodelan *Seasonal Generalized Space Time Autoregressive* (SGSTAR) (Studi Kasus: produksi Padi di Kabupaten Demak, Kabupaten Boyolali, dan Kabupaten Grobogan)". *Jurnal Gaussian*, 593-6-2.

[9] Makridakis, McGee, dan W. Wheelright. (1999). Metode dan Aplikasi Peramalan. Edisi kedua. Terj. Andriyanto, U.S. Jakarta: Bina Rupa Aksara.

[10] B.J. Cromwell, et al. (1994). *Multivariate Test For Time Series Models.* United State of America: Sage Publication, Inc

IOP Conf. Series: Journal of Physics: Conf. Series **974** (2018) 012040 doi:10.1088/1742-6596/974/1/012040

Demand forecasting of electricity in Indonesia with limited historical data

Mujiati Dwi Kartikasari, Arif Rohmad Prayogi

Department of Statistics, Faculty of Mathematics and Natural Sciences, Universitas Islam Indonesia, Jalan Kaliurang Km 14.5 Sleman, Yogyakarta, Indonesia 55584

E-mail: mujiatikartikasari@uii.ac.id, arifrohmad31@gmail.com

Abstract. Demand forecasting of electricity is an important activity for electrical agents to know the description of electricity demand in future. Prediction of demand electricity can be done using time series models. In this paper, double moving average model, Holt's exponential smoothing model, and grey model GM(1,1) are used to predict electricity demand in Indonesia under the condition of limited historical data. The result shows that grey model GM(1,1) has the smallest value of MAE (mean absolute error), MSE (mean squared error), and MAPE (mean absolute percentage error).

1. Introduction

Every year electricity demand in Indonesia will increase to follow economic growth. Because of this, predicting of electricity demand is very necessary to know the description of electricity demand in today and future. Thus, the Indonesian electrical agents can determine the plan to provide capacities required of power plant.

In order to predict electricity demand in Indonesia, data of electricity demand are collected at several points in time. The kind of this data can be predicted using time series analysis. Many researches have predicted electricity demand using time series analysis. Erdogdu (2007) used cointegration and ARIMA to predict electricity demand in Turkey [1]. Ismail et al. (2009) used ARIMA, SARIMA, and rule based models to predict peak load electricity demand in Malaysia [2]. Kandananond (2011) used artificial neural network (ANN) approach to predict electricity demand in Thailand [3]. Goel et al. (2014) used ARIMA, multiple regression, and trend seasonality model to predict electricity demand in New Delhi [4].

Most of previous researches used large data in construction of time series models to predict electricity demand so there is no problem in selecting the time series models to be used. While the data used in this paper are limited, then only a few time series models can be used. Some suitable time series models under the condition of small data points for model construction are summarized in Table 1 [5].

Table 1. Several time series models with limited historical data.

Time Series Models	Data requirement
Moving average	2 to 30
Exponential smoothing	5 to 10
Grey model	4

To overcome the predicting process of electricity demand with limited historical data, in this paper, we will use three time series models summarized in Table 1. The three models are compared using MAE, MSE, and MAPE.

Following this section, in the next section we present the theoretical basis of the three time series models. The measure of comparison for each model will be provided in Section 3. Then the data are analysed in Section 4. In Section 5, we will draw a conclusion based on analysis result.

2. Time series models

The time series models which will be used in this paper are moving average, exponential smoothing, and grey model. The theories of those models are presented as follows.

2.1. Moving average

Moving average is one of time series models which can be used for smoothing past history data [6]. There are several kinds of moving average models, namely simple moving average, double moving average, and higher-order moving average. This study uses double moving average model. The double moving average model is able to cope with significant trend [6].

The double moving average is a moving average of a moving average. We denote $MA(M \times N)$ as an M-period MA of N-period MA. The general procedure of double moving average is described as follows [6]:

$$s'_t = \frac{(x_t + x_{t-1} + x_{t-2} + \cdots + x_{t-N+1})}{N}, \tag{1}$$

$$s''_t = \frac{(s'_t + s'_{t-1} + s'_{t-2} + \cdots + s'_{t-M+1})}{M}, \tag{2}$$

$$a_t = s'_t + (s'_t - s''_t) = 2s'_t - s''_t, \tag{3}$$

$$b_t = \frac{2}{M-1}(s'_t - s''_t), \tag{4}$$

$$\hat{x}_{t+m} = a_t + b_t(m). \tag{5}$$

The notation of s'_t denotes the simple moving average in the period of t, s''_t denotes the double moving average in the period of t, a_t denotes the adjustment of the single moving average in the period of t, and b_t denotes the estimate of trend from one time period the next. The value of prediction for m periods ahead is denoted by $\hat{x}_{t+m}$.

2.2. Exponential smoothing

Exponential smoothing is a model which use an exponentially weighted smoother for smoothing past history data. There are three kinds of common exponential smoothing models, namely simple exponential smoothing, Holt's exponential smoothing, and Holt-Winters exponential smoothing model. The Holt's exponential smoothing model which is a model frequently used to handle data with linear trend is applied in this study. The Holt's exponential smoothing model requires two smoothing constants that are α as smoothing constant for the data and β as smoothing constant for trend estimate. The formula for Holt's exponential smoothing model is described as follows [6]:

$$s_t = \alpha x_t + (1-\alpha)(s_{t-1} + b_{t-1}), \quad 0 < \alpha < 1, \tag{6}$$

$$b_t = \beta(s_t - s_{t-1}) + (1-\beta) b_{t-1}, \quad 0 < \beta < 1. \tag{7}$$

The prediction value for m periods ahead can be calculated by

$$\hat{x}_{t+m} = s_t + b_t(m). \tag{8}$$

The recursive computation process in Holt's exponential smoothing model uses initial value $s_1 = x_1$ and $b_1 = x_2 - x_1$.

2.3. Grey Model

Grey model which will be described in this part is grey model GM(1,1). The GM(1,1) is one of the most common of grey prediction. This model is a part of grey system theory [7] and is known as a first order model with one variable. The GM(1,1) model is widely applied in predicting of time series data. The procedure of GM(1,1) is divided into three steps which is cited from Dang et al. [5] and is described as follows.

The first step is to arrange an original sequence $x^{(0)}(i)$ as follows:

$$x^{(0)} = \{x^{(0)}(1), x^{(0)}(2), x^{(0)}(3), \cdots, x^{(0)}(i), \cdots, x^{(0)}(n)\}. \tag{9}$$

The notation of $x^{(0)}$ denotes the value for the time period $i (i = 1, 2, \cdots, n)$.

The second step is to generate sequence $x^{(1)}$ by a one time accumulated generating operation (1-AGO) based on the original sequence $x^{(0)}$. The expression of sequence $x^{(1)}$ is

$$x^{(1)} = \{x^{(1)}(1), x^{(1)}(2), x^{(1)}(3), \cdots, x^{(1)}(i), \cdots, x^{(1)}(n)\} \tag{10}$$

where, $x^{(1)}(i) = \sum\limits_{j=1}^{i} x^{(0)}(j)$.

The third step of GM(1,1) procedure is to form an equation of first-order differential with one variable. The equation is defined by

$$\frac{dx^{(1)}}{dt} + ux^{(1)} = v \tag{11}$$

where u and v are the developing coefficient and the grey input coefficient, respectively. By using the least square method, the coefficient of u and v can be determined. The formula for determining the coefficient of u and v is expressed as follows:

$$[u, v]^T = \left(P^T P\right)^{-1} P^T Q \tag{12}$$

where,

$$P = \begin{bmatrix} -\left(x^{(1)}(1) + x^{(1)}(2)\right)/2 & 1 \\ -\left(x^{(1)}(2) + x^{(1)}(3)\right)/2 & 1 \\ \vdots & \vdots \\ -\left(x^{(1)}(n-1) + x^{(1)}(n)\right)/2 & 1 \end{bmatrix}$$

and

$$Q = \left[x^{(0)}(2), x^{(0)}(3), \cdots, x^{(0)}(n)\right]^T.$$

After the values of u and v coefficient are known, the time response function of the GM(1,1) can be determined by

$$\hat{x}^{(1)}(i) = \left[x^{(0)}(1) - \frac{v}{u}\right] e^{-u(i-1)} + \frac{v}{u} \quad (i = 2, 3, \cdots, n). \tag{13}$$

The prediction value which is calculated by GM(1,1) procedure is denoted by $\hat{x}^{(0)}$. It can be calculated by the operation of a one time inverse accumulated generating operation (1-IAGO) based on the time response function of the GM(1,1) $\hat{x}^{(1)}(i)$. The expression of predicted value $\hat{x}^{(0)}$ is

$$\hat{x}^{(0)}(i) = \hat{x}^{(1)}(i) - \hat{x}^{(1)}(i-1) \quad (i = 2, 3, \cdots, n) \tag{14}$$

with $\hat{x}^{(0)}(1) = \hat{x}^{(1)}(1)$.

3. Evaluation of Models

There are many kinds of measure which can be used to evaluate the performance of time series models. In this paper, we will use mean absolute error (MAE), mean squared error (MSE), and mean absolute percentage error (MAPE) to compare the three time series models. The formula of MAE, MSE, and MAPE are defined as follows [6]:

$$MAE = \frac{1}{n}\sum_{t=1}^{n} |x_t - \hat{x}_t|, \tag{15}$$

$$MSE = \frac{1}{n}\sum_{t=1}^{n} (x_t - \hat{x}_t)^2, \tag{16}$$

$$MAPE = \frac{1}{n}\sum_{t=1}^{n} \left|\frac{x_t - \hat{x}_t}{x_t}\right| \times 100\%. \tag{17}$$

4. Results

Data used in this paper are data of electricity consumption in Indonesia obtained from report of Directorate General of Electricity Indonesia from 2007 to 2015 [8, 9, 10, 11]. Statistical data of electricity consumption are presented in Table 2.

Table 2. Electricity Consumption in Indonesia (MWh).

Year	2007	2008	2009	2010	2011	2012	2013	2104	2015
Consumption	129,019	129,019	151,334	165,969	178,279	194,289	208,935	221,296	232,520

Based on the data shown in Table 2, we analyse this data using three time series models that are double moving average, Holt's exponential smoothing, and GM(1,1) model. The predicted results are reported in Figure 1.

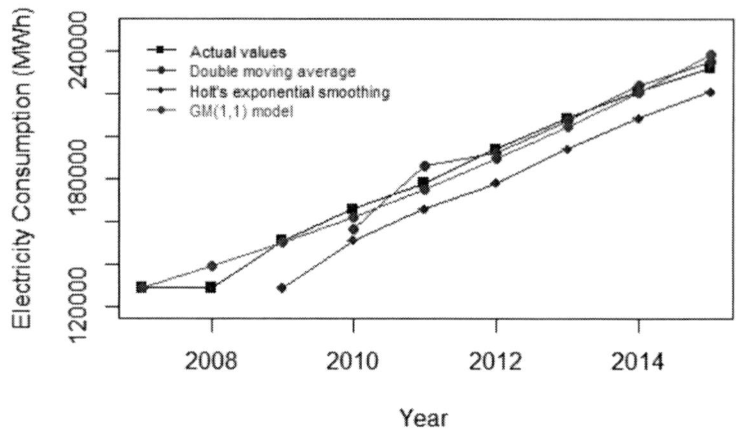

Figure 1. Predicted results of electricity demand in Indonesia.

Figure 1 shows the comparison between the prediction data using double moving average model, Holt's exponential smoothing model, and GM(1,1) model and the actual value. From figure 1 we can see that the prediction data from the double moving average and the GM(1,1) model have closer distance to the actual value compared to Holt's exponential smoothing model. To compare the performance of three models accurately, Table 3 provides the value of the MAE, MSE, and MAPE of each model.

Table 3. Performance of each model for predicting electricity demand in Indonesia.

Model	Evaluation		
	MAE	MSE	MAPE
Double moving average model	4,444.542	28,711,727	2.399113%
Holt's exponential smoothing model	4,664.472	74,449,210	2.840955%
Grey model GM(1,1)	3,642.755	22,894,478	2.156962%

According to the results of the models performance shown in Table 3, GM(1,1) model gives a fewer value of MAE, MSE, and MAPE than double moving average and Holt's exponential smoothing model. Thus, we can say that the GM(1,1) model is more appropriately used than the other models to predict electricity demand in Indonesia

5. Conclusion

We have compared the prediction model of the electricity demand in indonesia using double moving average, Holt's exponential smoothing, and grey model GM(1,1) under the condition of limited historical data. Based on the result, grey model GM(1,1) has the smallest value of MAE, MSE, and MAPE compared to the other models. We conclude that grey model GM(1,1) is the best model in the case studies of electricity demand prediction in Indonesia.

References

[1] Erdogdu E 2007 Electricity demand analysis using cointegration and ARIMA modelling: a case study of Turkey *Munich Personal RePEc Archive.* MPRA Paper No. 19099
[2] Ismail Z, Yahya A and Mahpol KA 2009 Forecasting peak load electricity demand using statistics and rule based approach *American J. Appl. Sciences.* **6(8)** 1618-25
[3] Kandananond K 2011 Forecasting electricity demand in Thailand with an artificial neural network approach *Energies.* **4** 1246-57
[4] Goel A and Goel A 2014 Regression based forecast of electricity demand of New Delhi *Int. J. Scientific and Research Publications.* **4**
[5] Dang HS, Huang YF, Wang CN and Nguyen TMT 2016 An application of the short term forecasting with limited data in the healtcare traveling industry *Sustainability.* **8** 1037
[6] Makridakis S, Wheelwright SC and McGee VE 1983 *Forecasting: methods and applications Second Edition* (Canada: John Wiley & Sons, Inc)
[7] Julong D 1989 Introduction to Grey System Theory *The Journal of Grey System.* **1** 1-24
[8] Directorate General of Electricity 2013 *The Book of Electricity Statistics Number 26 - 2013* (Jakarta: Directorate General of Electricity)
[9] Directorate General of Electricity 2014 *The Book of Electricity Statistics Number 27 - 2014* (Jakarta: Directorate General of Electricity)
[10] Directorate General of Electricity 2015 *The Book of Electricity Statistics Number 28 - 2015* (Jakarta: Directorate General of Electricity)
[11] Directorate General of Electricity 2016 *The Book of Electricity Statistics Number 29 - 2016* (Jakarta: Directorate General of Electricity)

IOP Conf. Series: Journal of Physics: Conf. Series **974** (2018) 012041 doi:10.1088/1742-6596/974/1/012041

Application of Bayesian configural frequency analysis (BCFA) to determine characteristics user and non-user motor X

Muhamad Iqbal Mawardi[1], **Septiadi Padmadisastra**[2], **Bertho Tantular**[3]
[1,2,3] Statistics Department, Faculty of Mathematics and Science, Universitas Padjadjaran, Indonesia
[1] Email: 1992iqbal@gmail.com

Abstract. Configural Frequency Analysis is a method for cell-wise testing in contingency tables for exploratory search type and antitype, that can see the existence of discrepancy on the model by existence of a significant difference between the frequency of observation and frequency of expectation. This analysis focuses on whether or not the interaction among categories from different variables, and not the interaction among variables. One of the extensions of CFA method is Bayesian CFA, this alternative method pursue the same goal as frequentist version of CFA with the advantage that adjustment of the experiment-wise significance level α is not necessary and test whether groups of types and antitypes form composite types or composite antitypes. Hence, this research will present the concept of the Bayesian CFA and how it works for the real data. The data on this paper is based on case studies in a company about decrease Brand Awareness & Image motor X on Top Of Mind Unit indicator in Cirebon City for user 30.8% and non user 9.8%. From the result of B-CFA have four characteristics from deviation, one of the four characteristics above that is the configuration 2212 need more attention by company to determine promotion strategy to maintain and improve Top Of Mind Unit in Cirebon City.

1. Introduction

Configural Frequency Analysis (CFA) is a statistical method that allows one to determine whether unexpected events are significantly different than expected. This analysis focuses on whether or not the interaction among categories from different variables is exist and not the interaction among the variables. The goal of CFA is to detect configuration pattern in the data whether experiencing deviation based on discrepancies from the model, which is marked by the existence of difference beetween the frequency of observation and expectation. In addition, CFA (Von Eye, 2002) is a method for identifying the configuration of categorical variables in multidimensional cross-tables. In CFA, deviation on the configurations will occur when an event more often or more rarely appears from what has been previously expected. An event occurred more often than what has been previously expected is called type, while an event rarely occurred from what has been previously expected is called antitype. The types of data used in CFAs are category pairs. This is based on the notion of configuration (Lienert in Von Eye 2002) which is a category pair that describes a cell of a cross-table.

CFA has been develop intensively since its first presentation in 1968 and is now among the more popular methods of data analysis. One of the reason for the increasing popularity of CFA is that the results of this method of data analysis are deemed easy to interpret. CFA is performed in five steps. The first step involves selecting a base model. The second step involves the selection of a concept of deviation from independence. The third step involves the selection of significance test. The fourth step involves estimation (or determining) the expected cell frequencies, performing the significance test,

IOP Conf. Series: Journal of Physics: Conf. Series **974** (2018) 012041 doi:10.1088/1742-6596/974/1/012041

and identifying those configurations that constitute types or antitypes. The fifth step involve interpreting types or antitypes using substantive background information.

Consider (Gutiérrez-Peña and von Eye, 2000) the cross-classification of $d \geq 2$ categorical variables. Let M be the vector of the model frequencies in this cross-classification, and m_i the observed frequency in cell i where i goes over all cells in the cross-classification. Let $\hat{m}_i$ be the estimated expected cell frequency for cell i and π_i the population probability for cell i. A log-frequency model for this cross-classification can be described as log $M = \mathbf{X\lambda}$, where M is the vector is the vector of model cell frequencies, X is the design matrix and $\boldsymbol{\lambda}$ is the parameter vector. CFA ask whether the model prediction, $\hat{m}_i$, deviates significantly from the observed value m_i. Most popular are the z-test $z = m_i - \hat{m}_i/\sqrt{n\hat{\pi}_i(1-\hat{\pi}_i)}$, where the π_i are estimated by $\hat{m}_i/n$, and Pearson X^2 component test, $X_i^2 = (m_i - \hat{m}_i)^2/\hat{m}_i$, (df = 1). The hypothesis H_0: $E(m_i - \hat{m}_i) = 0$ holds, one applies either of these or one of the other test discussed in the context CFA under the appropriate measure for protection of the experiment-wise α. The most frequently used methods for α protection are Bonferroni method, where an adjusted α is calculated as $\alpha^* = \alpha/t$, where α is the significance threshold, α = 0.05 and t is the number of test. Cases in cell I are said to belong to *type* if $E(m_i - \hat{m}_i) > 0$ and antitype if $E(m_i - \hat{m}_i) < 0$. Bayesian CFA is one of the extensions of the CFA frequentist methods. Bayesian CFA (Gutiérrez-Peña, 2012) defines types and antitypes in terms of the true (unknown) values of the parameters, which can be estimated but will never be observed or fully known. One can (1) search for types and antitypes as before with the advantage that adjustment of the experiment-wise significance level α is not necessary; and (2) test whether groups of types and antitypes form composite types or composite antitypes. Both CFA frequentist and Bayesian CFA are capable of assigning probabilities to patterns of types and antitypes, and thus allow for the comparison of such patterns by means of relative probabilities.

An interesting feature of the Bayesian approach is that it allows us to calculate the posterior probability of any specific pattern of types and antitypes in a cross-classification. Given a particular base model, the posterior distribution of π induces a probability distribution on the set of all possible patterns. In practice pattern in a neighbourhood of the particular pattern suggested by an exploratory analysis which looks at each cell individually. Different to CFA frequentist, from Bayesian viewpoint the posterior distribution of π produces joint probability statements concerning possible patterns of types and antitypes in a given group of cells. There is no need for Bonferroni-type corrections within the Bayesian framework.

2. Bayesian CFA

2.1 Prior and Posterior Distribution

In this section we introduce some notation (Gutiérrez-Peña and von Eye, 2000). Let R denote the total number of cells in the cross classification and let π_i denote the population probability for cell i (i = 1,…,R). Finnaly let $\boldsymbol{\pi}$ be the vector of such probabilities. Under multinomial sampling, the vector of observed frequencies, $m = (m_1,\ldots,m_i)$ can be regarded as an observation from a (R-1) dimentional multinomial distribution with index $N = \sum_i m_i$ and unknown parameter vector $\boldsymbol{\pi}$. From Bayesian point of view, all prior belief concerning the value of $\boldsymbol{\pi}$ must be described in terms of a *prior distribution*. The usual conjugate prior for the multinomial parameter Dirichlet distribution. This distribution is characterized by a parameter vector $\boldsymbol{\beta} = (\beta_1,\ldots,\beta_R)$ such that $E(\pi_i) = \beta_i/\beta_*$, where $\beta_* = \sum \beta_i$. The value of β_* is interpreted as a hypothetical prior sample size and determines the strength of the information contained the prior: a small β_* implies vague prior information whereas a large β_*suggest strong prior beliefs about $\boldsymbol{\pi}$. The posterior distribution is also Dirichlet with parameter $\beta_m = (m_1 + \beta_1,\ldots,m_R + \beta_R)$. This distribution contains all the available information about the population proportion $\boldsymbol{\pi}$ conditional on the observed frequencies.

In the absence of prior information, also called a non informative prior will typically be used. On of the most widely used methods to derive non informative priors is the called Jeffreys' rule, which is in this case yields for the multinomial parameter is precisely the Dirichlet distribution with parameter $\beta = (1/2,\ldots,1/2)$. One of the nice fitures of this prior is that it is conjugate (close under sampling).

Keeping in mind the interpretation of β_* as a prior sample size, the quantity $i = \beta_*/(N + \beta_*)$ can be regarded as the proportion of the total information that is contributed by the prior. Since $E(\pi_i) = \beta_i/\beta_*$, the individual values of β_* should be chosen according to the prior belief concerning $E(\pi_i)$.

The Dirichlet prior provides a simple way to elicit prior information concerning prior population probabilities, but it may not be flexible enough to describe specific prior information about other parameter such as interactions (Knuiman and Speed, 1998).

Any base model imposes constraints on the space of possible values of π, in other words under the base model, the population proportion for cell i is given by $\pi_i^* = f_i(\pi)$ for some function f_i. Consider a 2×12 cross-classification and base model which state that the three variables are independent, then:

$$f_1(\pi) = (\pi_1 + \pi_2)\cdot(\pi_1 + \pi_3 + \pi_5 + \pi_7 + \pi_9 + \pi_{11} + \pi_{13} + \pi_{15} + \pi_{17} + \pi_{19} + \pi_{21} + \pi_{23})$$
$$f_2(\pi) = (\pi_1 + \pi_2)\cdot(\pi_2 + \pi_4 + \pi_6 + \pi_8 + \pi_{10} + \pi_{12} + \pi_{14} + \pi_{16} + \pi_{18} + \pi_{20} + \pi_{22} + \pi_{24})$$
$$f_3(\pi) = (\pi_3 + \pi_4)\cdot(\pi_1 + \pi_3 + \pi_5 + \pi_7 + \pi_9 + \pi_{11} + \pi_{13} + \pi_{15} + \pi_{17} + \pi_{19} + \pi_{21} + \pi_{23})$$
$$\vdots$$
$$f_{24}(\pi) = (\pi_{23} + \pi_{24})\cdot(\pi_2 + \pi_4 + \pi_6 + \pi_8 + \pi_{10} + \pi_{12} + \pi_{14} + \pi_{16} + \pi_{18} + \pi_{20} + \pi_{22} + \pi_{24})$$

We can calculate the posterior probability of any event involving the population proportions, from posterior distribution of π. In particular, that posterior distribution induces a posterior distribution on the vector $\pi^*$of model based proportion. Even if $\pi_i \neq \pi_i^*$, we would be unwilling to classify cell i as type or antitype unless $\pi_i - \pi_i^*$ was significantly greater than zero or less than zero.

Any base model imposes constraints on the possible values of $\boldsymbol{\pi}$. In other words, under the base model the population probability of cell i is given by $\pi_i^* = f_i(\boldsymbol{\pi})$ for some functions f_i. The base model can be tested on the basis of the posterior distribution of $\delta = \sum_i log\,(\pi_i/\pi_i^*)\pi_i$. This quantity is always nonnegative and is zero if and only if the base model is correct. Posterior distributions of δ is not available in close form, it can easily be obtained from that π using Monte Carlo techniques, that posterior distributions concentrated near zero support the base model, where as posterior distributions located away from zero lead to rejection of the base model.

2.2 Types and Antitypes from Bayesian Perspective

If we knew (Gutiérrez-Peña, 2012) the actual value of $\boldsymbol{\pi}$, then cell i could be regarded as a type if $\boldsymbol{\pi_i} > \boldsymbol{\pi_i^*}$, and as an antitype if $\boldsymbol{\pi_i} < \boldsymbol{\pi_i^*}$. However, even if $\boldsymbol{\pi_i} \neq \boldsymbol{\pi_i^*}$, we would be unwilling to classify cell i as a type and antitype unless $\boldsymbol{\pi_i} - \boldsymbol{\pi_i^*}$ was significantly greater and less than zero. This suggests the following definition of types and antitypes: cell i is regarded as a type if and only if $\boldsymbol{u_i} < \boldsymbol{\pi_i} - \boldsymbol{\pi_i^*}$ and antitype in and only if $\boldsymbol{\pi_i} - \boldsymbol{\pi_i^*} < \boldsymbol{l_i}$, where $\boldsymbol{u_i}$ and $\boldsymbol{l_i}$ are suitable threshold values (von Eye and Gutiérrez-Peña, 2004). From the posterior distribution of $\boldsymbol{\pi}$ we can compute the posterior probability of cell i being a type namely, Pr $(\boldsymbol{u_i} < \boldsymbol{\pi_i} - \boldsymbol{\pi_i^*})$.

2.3 Patterns Types and Antitypes

We now turn to the Bayesian analysis (von Eye and Gutiérrez-Peña, 2004) of cross classification under product multinomial sampling. In order to simplify the exposition, we shall only consider discrimination between two group on the basis of two explanatory variables arranged in a $a \times b$ cross classification for each of the groups. More general situations can be handled in a similar fashion. For example a 3 × 2 cross classification, then such possible patterns include (T for type, A for antitype, and N for neither)

$$\begin{pmatrix} T & T \\ T & T \\ T & T \end{pmatrix} \cdots \begin{pmatrix} T & A \\ A & T \\ T & A \end{pmatrix} \cdots \begin{pmatrix} N & T \\ T & N \\ N & T \end{pmatrix} \cdots \begin{pmatrix} N & A \\ A & N \\ T & T \end{pmatrix} \cdots \begin{pmatrix} A & A \\ A & A \\ A & A \end{pmatrix}.$$

A Bayesian solution to the CFA problem would be to report the most probability pattern. For problems of moderate size, the number of all possible patterns may be too large for direct implementation of this approach to be feasible. The calculations of all the posterior probabilities reported based on the simulation of 10000 samples from the Dirichlet distribution of π (von Eye and Gutiérrez-Peña, 2000).

3. Bayesian Analysis of the 'Top of Mind Motor X' Data

In this section, we present the data used are XYZ Company survey on Brand Awareness & Image characteristics in Cirebon City as many as 190 respondents. There are 4 factors to look at:

1. Top Of Mind
   Top of Mind X motors are divided into two categories: X (1) and Competitor (2).
2. Jobs
   The Jobs is divided into three categories: Student (1), Employee (2), Others (3).
3. Spending per month
   Spending per month is divided into two categories: ≤ Rp. 1.250.000 (1) and> Rp. 1.250.000 (2).
4. Motor cycle Users
   Motor cycle users are divided into two categories: Users (motor cycle users) (1) and Non User (potential motor cycle users) (2).

Now consider the base model:

$$log\,\mu_{ijkl} = \lambda_0 + \lambda_i^A + \lambda_j^B + \lambda_k^C + \lambda_l^D + \lambda_{ij}^{AB} + \lambda_{ik}^{AC} + \lambda_{il}^{AD} + \lambda_{jk}^{BC} + \lambda_{jl}^{BD} + \lambda_{kl}^{CD} + \lambda_{ijk}^{ABC} + \lambda_{ijl}^{ABD} + \lambda_{ikl}^{ACD} + \lambda_{jkl}^{BCD} + \lambda_{ijkl}^{ABCD} \quad (1)$$

Where μ_{ijkl} is the model frequency for cell *ijkl*, with λ_0 = intercept, λ_i^A= job, λ_j^B = spending per month, λ_k^C = motor cycle users, λ_l^D= top of mind and the other is interaction between variables. Suppose now we have independent (non informative) Dirichlet priors so that corresponding posterior distributions are also Dirichlet distribution. In order to calculate the posterior probability that cell *ijkl* is a type, we use a Monte Carlo approach. The base model can be tested on the basis of the posterior distribution of

$$\delta = \sum_i \sum_j \sum_k \sum_l \log\left(\frac{\pi_i}{\pi_i^*}\right)\pi_i \quad (2)$$

Table 1 shows the results of the Bayesian CFA of the top of mind data, we use R Software and in Figure 1, we show a histogram of the posterior distribution of δ. This distribution puts most of its mass away from zero, indicating significant discrepancies between the data and the expectancies under the assumption of independence. Notice that both the classical and the Bayesian CFA lead to the same results in this case. The posterior probability of the pattern provided by the two analyses is 0.126. The posterior distribution over the space of all possible patterns is much more concentrated around the pattern obtained from the preliminary analysis.

Table 1. Bayesian CFA for top of mind data

Configuration	Obs	Pr(Type)	Pr(Neither)	Pr(Antitype)	Type/Antitype
1111	14	0.0559	0.9341	0.0100	
1121	2	0.0100	0.9341	0.0559	
1211	4	0.0163	0.9420	0.0417	
1221	3	0.0417	0.9420	0.0163	
2111	12	0.0324	0.9469	0.0207	
2121	2	0.0207	0.9469	0.0324	
2211	18	0.1012	0.8926	0.0062	
2221	7	0.0062	0.8926	0.1012	
3111	4	0.2700	0.7293	0.0007	
3121	13	0.0007	0.7293	0.2700	
3211	16	0.1600	0.8379	0.0021	

3221	8	0.0021	0.8379	0.1600	
1112	10	0.0000	0.2064	0.7936	Antitype
1122	2	0.7936	0.2064	0.0000	Type
1212	3	0.0002	0.7682	0.2316	
1222	1	0.2316	0.7682	0.0002	
2112	5	0.0061	0.9175	0.0764	
2122	0	0.0764	0.9175	0.0061	
2212	31	0.6389	0.3611	0.0000	Type
2222	11	0.0000	0.3611	0.6389	Antitype
3112	5	0.0317	0.9512	0.0171	
3122	3	0.0171	0.9512	0.0317	
3212	13	0.2175	0.7816	0.0009	
3222	3	0.0009	0.7816	0.2175	

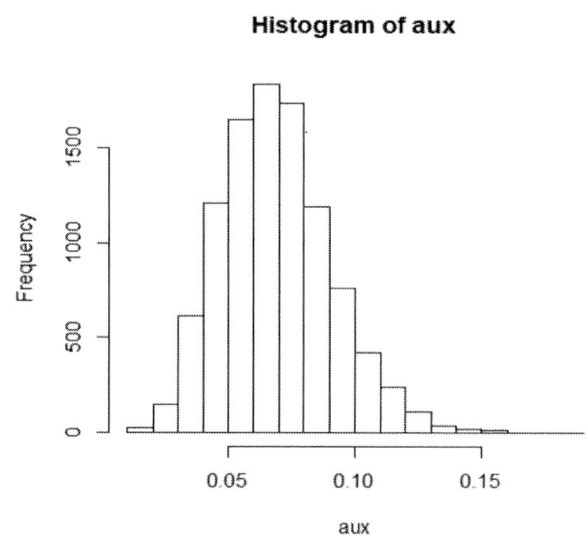

Figure 1. Posterior distribution of δ for the top of mind data

4. Conclusion

Bayesian Configural Frequency Analysis (B-CFA) is capable of assigning probabilities to patterns of types and antitypes, and thus allows for the comparison of such patterns by means of relative probabilities. Recommendations can be derived as to which approach to select for data analysis, when single types and antitypes are of interest and the observed cells frequencies are deemed valid for the estimation of expected cell frequencies. The Bayesian approach presented allows for joint inferential statements about groups of cells and does not require Bonferroni type arguments.

There are eight characteristics that have been determined from deviation, based on the analysis type means observed value is more often than expected value appear on:

1122: respondents with job as student, spending per month $\leq$ Rp. 1.250.000,00, as a non user, and Top of Mind competitor

2212: respondents with job as employee, spending per month > Rp. 1.250.000,00, as a user, and Top of Mind competitor.

Whereas, antitype means observed value is less often than expected value appear on:

1112: respondents with job as student, spending per month ≤ Rp. 1.250.000,00, as a user, and Top of Mind competitor.

2222: respondents with job as employee, spending per month > Rp. 1.250.000,00, as a non user, and Top of Mind competitor.

From that descriptions, recommendations for XYZ Company is need to pay special attention to this characteristics especially 2212, this will be very useful to determine strategy to maintain and improve Top Of Mind Unit in Cirebon City.

5. References

[1] Gutiérrez-Peña, E. and Von Eye, A. (2000). A Bayesian Approach to Configural Frequency Analysis. Journal of Mathematical Sociology, 24, 151-174.

[2] Gutiérrez-Peña, E. (2012). Bayesian predictive Configural Frequency Analysis. Journal Psychological Test and Assessment Modeling, 54, 285-292.

[3] Knuiman, M. W. and Speed, T. P. (1988) Incorporating prior information into the analysis of contingency tables, Biometrics, 44, pp. 1061-1071.

[4] Mawardi, M. I. 2014. Application of *Configural Frequency Analysis (CFA)* to Determine Characteristics *User* dan *Non User* Motor X in Cirebon City (Case Study PT XYZ). Undergraduate Thesis. Statistics Department, Faculty of Mathematics and Science, State University of Padjadjaran, Bandung, Indonesia.

[4] Von Eye, A. 2002. Configural Frequency Analysis: Methods, Models, and Aplications. Lawrence Erlbaum Associates: London.

[5] Von Eye, A. and Brandtstadter, J. (1997). Configural Frequency Analysis as a searching device for possible causal relationships. Methods of Psychological Research - Online, 2, 1-23

[6] Von Eye, A. And Gutiérrez-Peña, E (2004). Configural frequency Analysis – the search for extreme cells. Journal of Applied Statistics, 31, 981–997.

IOP Conf. Series: Journal of Physics: Conf. Series **974** (2018) 012042 doi:10.1088/1742-6596/974/1/012042

Analytical and numerical solution for wave reflection from a porous wave absorber

Ikha Magdalena

Industrial & Financial Mathematics Research Group, Fac. of Mathematics and Natural Sciences, Institut Teknologi Bandung, Jalan Ganesha 10, Bandung, West Java, Indonesia

E-mail: ikha.magdalena@math.itb.ac.id

Marian P. Roque

Institute of Mathematics College of Science, University of the Philippines, Diliman, Quezon City, 1101 Philippines

E-mail: roque@math.upd.edu.ph

Abstract. In this paper, wave reflection from a porous wave absorber is investigated theoretically and numerically. The equations that we used are based on shallow water type model. Modification of motion inside the absorber is by including linearized friction term in momentum equation and introducing a filtered velocity. Here, an analytical solution for wave reflection coefficient from a porous wave absorber over a flat bottom is derived. Numerically, we solve the equations using the finite volume method on a staggered grid. To validate our numerical model, comparison of the numerical reflection coefficient is made against the analytical solution. Further, we implement our numerical scheme to study the evolution of surface waves pass through a porous absorber over varied bottom topography.

1. Introduction

Porous wave absorber offers an excellent solution for protecting harbours against the action of incoming waves or resonance phenomena. Prediction of the reflection and transmission from porous absorber plays a significant role in the assessment of the wave conditions in a harbour. A numerical wave model is valuable when it comes to a quick investigation to account this phenomenon. Several researchers have investigated the advantages of porous structure on reducing wave amplitude, such as in [1], [2], [3], [4], [5] and [6]. They studied submerged type porous structure but here we focus on emerged porous structure. Several researchers have investigated the advantages of a porous breakwater on reducing wave amplitude using an experimental approach such as [7] and [8]. The mathematical model often to use is a model from potential theory. Some researchers who studied this problem using potential theory are [9] and [10]. But the difficulty arises when we work with potential theory due to the complexity of the equation. Using potential theory makes the equations difficult to solve analytically or numerically. In this research, we propose a modification of shallow water equation involving the existent of porous structure to study the transmission and reflection wave with investigating the effect of incoming waves and various wave absorber characteristics. Working with shallow water type model makes us solve the equation relatively easier than using potential theory.

There are six sections in this paper. We start with an introduction. The governing equations are briefly presented in the second section. In the third section, the analytical solution for wave reflection and transmission coefficients are derived. In the following section, the numerical methods were described. A staggered finite volume method is introduced. In the fifth section, the numerical results are presented and compared with analytical solutions. Conclusions are outlined in the last section.

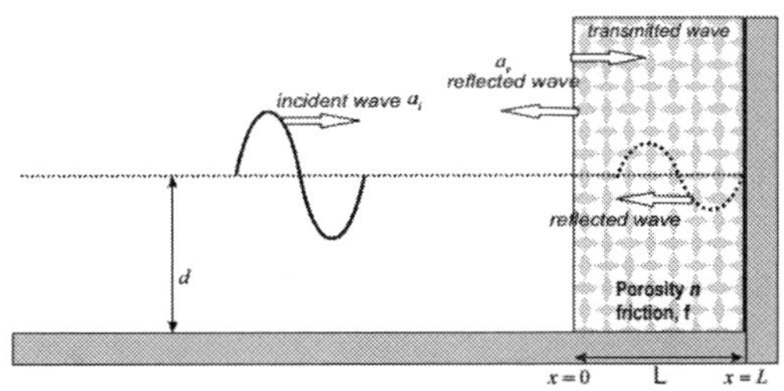

Figure 1. Sketch of the domain

2. Governing equation

Here, we consider wave elevation in a domain as depicted in Figure 1. For that purpose, we denote a domain with porous media as $\Omega_2 = 0 \leq x < L$, whereas the region without porous structure is denoted by Ω_1 for $x < 0$. The governing equation that we used in Ω_1 is shallow water equations over a flat bottom d, read as:

$$\eta_t + du_x = 0, \tag{1}$$

$$u_t + g\eta_x = 0, \tag{2}$$

in which η and u are surface elevation and fluid velocity, respectively. The equations inside the porous wave absorber are modification of a shallow water type model. The adjustment is by adding friction term c_f and filtered velocity $u' = \frac{u}{n}$ written in:

$$\eta_t + \frac{d}{n}u_x = 0, \tag{3}$$

$$\frac{1}{n}u_t + g\eta_x + f\frac{\omega}{n}u = 0,, \tag{4}$$

in which f is friction coefficient and ω is radian frequency.

Further, we recapitulate the governing equations for the whole domain are as follows:

$$\eta_t + (d\frac{u}{N})_x = 0, \tag{5}$$

Analytical and numerical solution for wave reflection from a porous wave absorber

$$\frac{1}{N}u_t + g\eta_x + F_l\frac{\omega}{N}u = 0, \tag{6}$$

with piecewise constant functions of N and F as follows

$$N = \begin{cases} 1, & \text{if } x \in \Omega_1, \\ n, & \text{if } x \in \Omega_2 \end{cases} \tag{7}$$

and

$$F_l = \begin{cases} 0, & \text{if } x \in \Omega_1, \\ f, & \text{if } x \in \Omega_2. \end{cases} \tag{8}$$

3. Wave reflection and transmission coefficient

In this section, we derive wave reflection and transmission coefficient from a porous wave absorber. In this study, we will limit ourselves to the study of small wave amplitude in a shallow water area. Consider a harmonic wave with certain frequency ω in porous domain read as:

$$\eta(x,t) = F(x)\exp(-i\omega t), \tag{9}$$

$$u(x,t) = G(x)\exp(-i\omega t). \tag{10}$$

Substituting equations (9, 10) into (3, 4) will yield:

$$F_{xx} + \frac{\omega^2}{gd}(1 - if)F = 0, \tag{11}$$

$$G(x) = \frac{gn}{\omega}\frac{1}{i+f}F_x. \tag{12}$$

Solution of equation (11) is $F(x) = a_1e^{-i\kappa x} + a_2e^{i\kappa x}$, where the wave number in porous media κ follows the following dispersion relation:

$$\kappa^2 = \frac{\omega^2}{gd}(1 - if). \tag{13}$$

A formulation for surface elevation is then:

$$\eta(x,t) = a_1e^{i(\kappa x - \omega t)} + a_2e^{-i(\kappa x + \omega t)},$$

where a_1 is the right running waves which is our transmitted waves and a_2 is the left running waves that is our reflected wave after hitting a wall. Differentiate $F(x)$ with respect to x and substitute the result into (12) will give us

$$G(x) = \frac{gn}{\omega}\frac{i\kappa}{i+f}\left(a_1e^{-i\kappa x} - a_2e^{i\kappa x}\right). \tag{14}$$

Next, substituting (14) into Eqn. (10) will yield:

$$u(x,t) = \sqrt{\frac{g}{d}}\frac{n}{\sqrt{1-if}}\left(a_1e^{-i(\kappa x - \omega t)} - a_2e^{i(\kappa x + \omega t)}\right). \tag{15}$$

Consider an incoming wave from free water area propagates to the right. After passing through the porous media and hit a seawall, the wave is fully reflected. So, in the upstream free region Ω_1 there will be right running wave $e^{-i(kx-\omega t)}$ and left running wave $e^{i(kx+\omega t)}$, where k denotes wave number in free region. The right running wave has amplitude a_i and left running wave has amplitude a_r. In general, surface elevation in the whole domain can be formulated as:

$$\eta(x,t) = e^{-i\omega t}\begin{cases} a_i e^{-ikx} + a_r e^{ikx}, & \text{in } x < 0 \\ a_1 e^{-i\kappa x} + a_2 e^{i\kappa x}, & \text{in } 0 \leq x < L \end{cases} \tag{16}$$

with k follows dispersion relation $\frac{\omega^2}{gk} = kd$. And the relation between κ and k is $\kappa^2 = k^2(1-if)$.

In free region, traveling waves of the linear SWE has to satisfy the following relation $u(x,t) = \pm\sqrt{\frac{g}{d}}\eta(x,t)$, with positive sign for right running wave and negative sign for left running wave. In porous domain the relation is $u(x,t) = \pm\sqrt{\frac{g}{d}}\frac{n}{\sqrt{1-if}}\eta(x,t)$. Consider $\eta(x,t)$ formulated as (16), suitable anzat for $u(x,t)$ is as follows:

$$u(x,t) = e^{-i\omega t}\begin{cases} \sqrt{\frac{g}{d}}(a_i e^{-ikx} - a_r e^{ikx}), & \text{in } x < 0 \\ \sqrt{\frac{g}{d}}\frac{n}{\sqrt{1-if}}(a_1 e^{-i\kappa x} - a_2 e^{i\kappa x}), & \text{in } 0 \leq x < L. \end{cases} \tag{17}$$

Along the interface $x = 0$ free surface flow admits continuity of free surface η and horizontal momentum du. At the end of our domain $x = L$, hard wall boundary condition is applied $u(L,t) = 0$.

Further, the coefficients a_1, a_2 and a_r can be found by matching η and du at the interface and applying hard wall boundary. Continuity of surface elevation η and horizontal flux du at $x = 0$, will yield

$$a_1 + a_2 = a_i + a_r, \tag{18}$$

and

$$\epsilon(a_1 - a_2) = a_i - a_r, \tag{19}$$

with $\epsilon = \frac{n}{\sqrt{1-if}}$. The hard wall boundary condition at $x = L$ for $n \neq 0$ will yield

$$\epsilon(a_1 e^{-i\kappa L} - a_2 e^{i\kappa L}) = 0, \tag{20}$$

From equations (18) and (19) we can express a_1 and a_2 in terms of a_r:

$$a_1 = \frac{2e^{i\kappa L}}{(1-\epsilon)e^{i\kappa L} + (1+\epsilon)e^{-i\kappa L}}a_r, \tag{21}$$

$$a_2 = \frac{2e^{-i\kappa L}}{(1-\epsilon)e^{i\kappa L} + (1+\epsilon)e^{-i\kappa L}}a_r. \tag{22}$$

Substituting the results above for a_1 and a_2 into equations (20), will give us one equations in two variables a_i and a_r. Eliminating a_i and a_r will give us wave reflection coefficients of the porous structure:

$$K_R = \left|\frac{a_r}{a_i}\right| = \left|\frac{(1-\epsilon)e^{i\kappa L} + (1+\epsilon)e^{-i\kappa L}}{(1+\epsilon)e^{i\kappa L} + (1-\epsilon)e^{-i\kappa L}}\right|. \tag{23}$$

Moreover, we also can obtain the wave transmission coefficient that is achieved from the ratio between a_1 and a_i :

$$K_T = \left|\frac{a_1}{a_i}\right| = \left|\frac{2e^{i\kappa L}}{(1+\epsilon)e^{i\kappa L} + (1-\epsilon)e^{-i\kappa L}}\right|. \tag{24}$$

Analytical and numerical solution for wave reflection from a porous wave absorber

Transmission and reflection coefficient are both depend on the characteristic of porous media such as porosity n, friction f, and length L of the porous media. In this absorber porous structure, the incident wave scattered because it enters a porous media. Formula

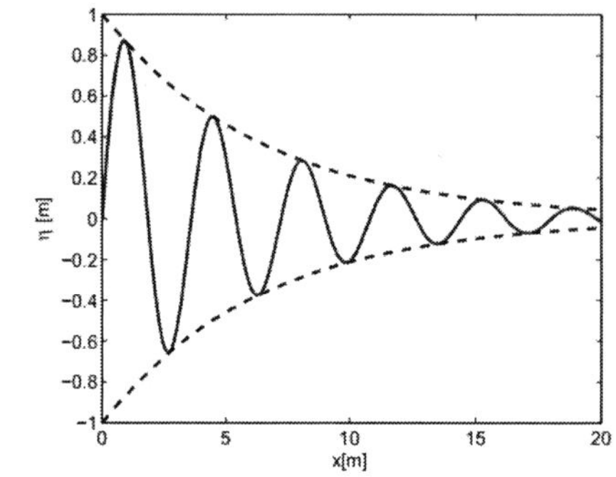

Figure 2. Solid line is the curve of $\eta(x,t) = \exp^{-i(kx-\omega t)}$ at certain time for $\kappa = 1.7443 - 0.1557i$ and its envelope $|\eta(x,t)|$.

(23) is tested for the following limiting case. Case $n \to 0$, the porous media becomes a solid wall and we obtain $K_R \to 1$ which mean perfect reflection and $K_T \to 0$ no transmission. Further, for the case $n \to 1$ and $f \to 0$, in which the porous media becomes a free region, the dispersion relation (13) reduces to the well-known dispersion relation for gravity wave: $\frac{\omega^2}{gk} = kd$, as we expect.

Taking parameter values $\omega = 3\pi$, $d = 3$, $g = 9.81$, $n = 0.8, f = 0.18$, dispersion relation (13) will give us a complex value wave number $\kappa = 1.7443 - 0.1557i$. A monochromatic wave $\exp^{-i(\kappa x - \omega t)}$ with negative imaginary part $\Im(\kappa)$ will undergo amplitude reduction, see Figure 2. Let $K_T = |\eta(x,t)| = \exp \Im(\kappa)x$, the term K_T denotes amplitude reduction of incident wave as a function of x, the horizontal length of an emerged porous breakwater. It also denotes the ratio between wave transmission amplitude and incident wave amplitude or wave transmission coefficient. It is clear that the profile of wave transmission coefficient depends strongly on the complex wave number κ, and hence on the dispersion relation (13). Parameters involved in (13) are wave frequency ω, gravitational acceleration g, porosity n, friction coefficient f, and water depth d.

Next, we study wave reflection and transmission coefficient concerning porous structure parameter such as porosity n, wavelength L, and friction coefficient f. Here, we analyze the dependence of K_R and K_T on those parameters. Figure 4 shows curves of K_R and K_T with respect to non-dimensional variable kL for several values of n. For all computations follow we take $g = 9.8$ and $f = 1$. We observe that the longer porous media does not directly mean larger K_R, after a certain length of the porous structure, the wave reflection stay steady in a certain value follows this formula:

$$\lim_{x \to \infty} K_R = \frac{1-\epsilon}{1+\epsilon}.$$

Analytical and numerical solution for wave reflection from a porous wave absorber

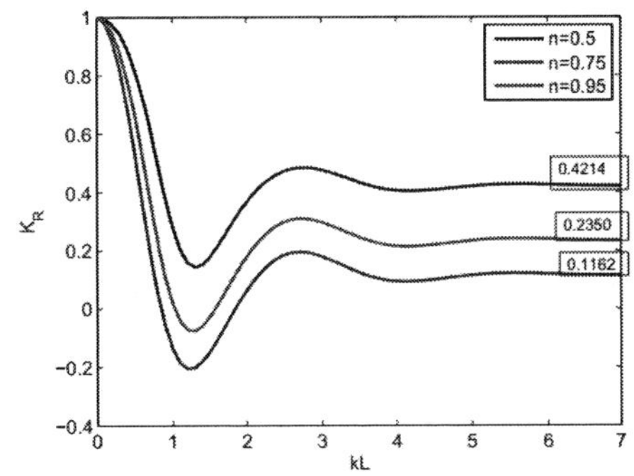

Figure 3. Curves of K_R w.r.t kL for fixed values of the friction coefficient $f = 1$.

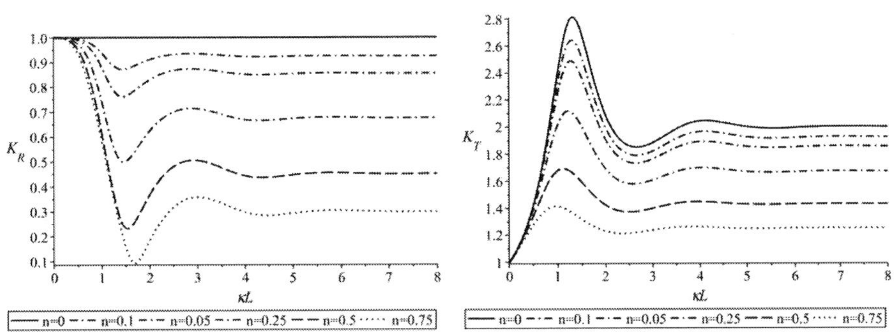

Figure 4. (Left) Curves of K_R w.r.t kL for fixed values of the friction coefficient $f = 1$. (Right) Curves of K_T w.r.t. kL for for fixed values of the friction coefficient $f = 1$.

For relatively small value of kL, we observe an oscillating behavior. The same behavior of K_R curve is also found by W. Sulisz [11] and I. Magdalena et.al [12]. We can also conclude that smaller n will yield smaller K_T.

4. A staggered finite volume method

In this section, a numerical finite volume method on a staggered grid will be implemented for simulating an incident wave passing through an emerged porous media. We will use the numerical computations to confirm the analytical results. Consider equation for waves pass through a porous structure (5, 6) in domain $[0, L]$. We discretize the porous domain in a staggered way $0 = x_{1/2}, x_1, \ldots, x_{Nx+1/2} = L$. Mass conservation (5) is approximated at a cell centred at x_i whereas momentum conservation (6) is approximated at a cell centred at $x_{i+1/2}$, see Figure 6. Approximate equations are then

$$\frac{\eta_i^{n+1} - \eta_i^n}{\Delta t} + \frac{d}{n} \frac{u|_{i+1/2}^n - u|_{i-1/2}^n}{\Delta x} = 0 \tag{25}$$

Analytical and numerical solution for wave reflection from a porous wave absorber

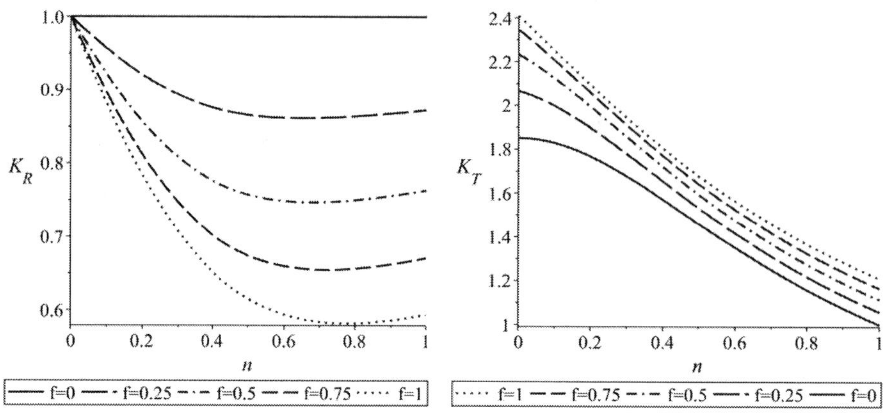

Figure 5. (Left) Curves of K_R w.r.t n for fixed values of $kL = 1$. (Right) Curves of K_T w.r.t. n for fixed values of $kL = 1$.

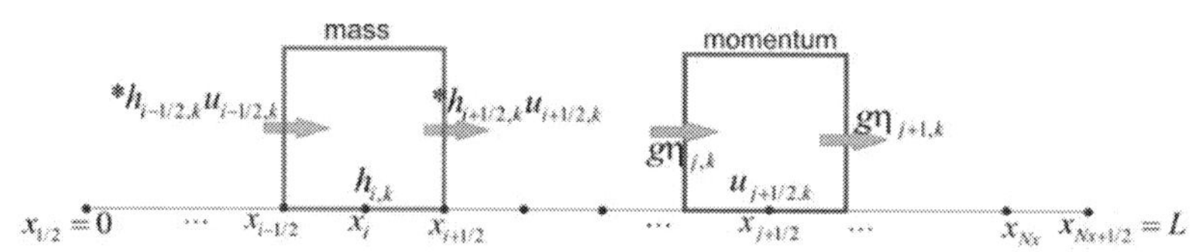

Figure 6. Illustration of staggered grid.

$$\frac{\frac{1}{n}u_{i+1/2}^{n+1} - \frac{1}{n}u_{i+1/2}^{n}}{\Delta t} + g\frac{\eta_{i+1}^{n+1} - \eta_i^{n+1}}{\Delta x} + f\frac{\omega}{n}u_{i+1/2}^{n+1} = 0. \tag{26}$$

The above approach is known as the finite volume method on a staggered grid. This discretization for free region is described extensively in S.R. Pudjaprasetya and I. Magdalena [13] and I. Magdalena et al. [14]. In this setting, values of η will be computed at every full grid points x_i, with $i = 1, 2, ..., Nx$ using mass conservation (25). Velocity u will be computed at every staggered grid points $x_{i+\frac{1}{2}}$, with $i = 1, 2, ..., Nx - 1$ using momentum equation (26). Note that the friction term $f\frac{\omega}{n}u$ is calculated implicitly in order to avoid more restricted stability condition. Implementing Von Neumann stability analysis, we obtain Courant-Friedrichs-Lewy condition for (25,26) which is $\sqrt{gd}\frac{\Delta t}{\Delta x} \leq 1$, where d is the flat bottom depth.

Further, for simulating the gravity waves in free water area, the approximate equations are just (25) and (26) with $n = 1$ and $f = 0$. The resulting scheme is free from numerical damping error, see S.R. Pudjaprasetya and I. Magdalena [13] for details.

5. Numerical simulation

In this section, we will implement the above scheme to simulate wave interaction with a porous structure. For simulation, we take a computational domain $-30 < x < 30$. We take $g = 9.81$ and a constant depth $d = 10$. Along the left and right boundaries, we apply absorbing boundary. The initial condition is the following hump:

$$\eta(x,0) = 1e^{-(0.5x-15)^2} \tag{27}$$

$$u(x,0) = \sqrt{\frac{g}{d}}\eta(x,0). \tag{28}$$

Along the right boundary, we apply a hard wall.

We first test the no porous case $n = 1$ and without friction $f = 0$, for which equations (3) and (4) reduce to the shallow water equations without porous structure. Numerical simulation of (25) and (26) will yield the wave travels to the right undisturbed in shape and hit the wall, it is fully reflected by the wall, as we expect, see Figure 7 (Left). Next, we will simulate the evolution of a wave initially being in a free water

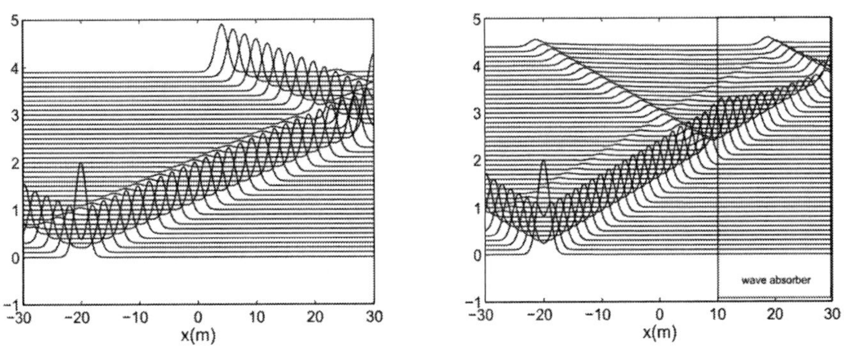

Figure 7. (Left) Wave propagates in a free domain and hit a wall in the right. (Right) Wave interactions with an emerged porous media.

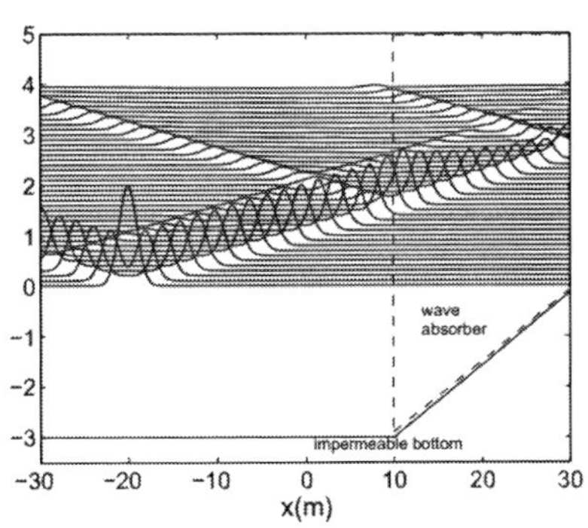

Figure 8. (Left) Wave propagates pass through porous media and hit a wall in the right.

region. It travels to the right and enters a porous region. In the porous region, the wave moves further to the right and hit the wall. The porous structure with parameters $n = 0.8, f = 0.18$ is installed in $10 < x < 30$. The numerical simulation is given in Figure 7 (Right). When the waves hit the emerged porous media, it will split into reflected and transmitted waves. In the porous media, transmitted wave is reduced. When the wave travels further to the right and hit the wall, the waves reflect again. The first and second reflected wave can be seen in in Figure 7 (Right). Moreover, we implement our numerical scheme to general case where the porous absorber is not on the flat bottom but on a sloping bottom, see Figure 8.

5.1. Comparison with Analytical Solution

In this section, we will make a comparison of our numerical result with the analytical wave reflection coefficient formula for various variables such as porosity, length of the porous structure, and friction coefficient. Further, we will show that our numerical reflected wave profile confirms the analytical K_R formula. For numerical computations, we take the parameters used in Section 3, and we use $\Delta x = 0.1$, and $\Delta t = \Delta x/\sqrt{gd}$. The surface profile is plotted in Figure 9. We use Healy's formula to calculate reflection wave: $K_R = \frac{a_{max}-a_{min}}{a_{max}+a_{min}}$, as stated in [15]. Wave reflection and transmission coefficient,

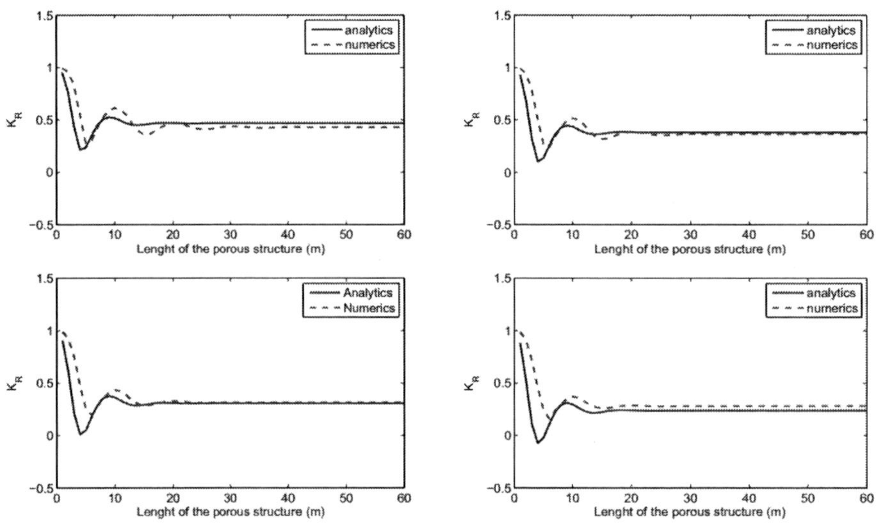

Figure 9. Solid line is the analytical wave reflection coefficient K_R from equation (24). Dash curves are surface elevation in porous media. Left upper: n=0.45, right upper: n=0.55, left lower: n=0.65, right lower: n=0.75

(23) and (24) depend on porosity n, friction coefficient f, length of the porous structure L, and wave number k, and κ. We make another comparison, the dependence of K_T with n and f. For the computation, we take parameter $\omega = 3\pi$ and $L = 10$. We plot the curve of wave transmission coefficient K_T with respect to porosity n for several values of f. From Figure 10, we conclude that for a certain porosity n, larger friction coefficient, will lead to smaller wave transmission coefficient. The numerical results are in good agreement with the analytical result, especially for n greater than 0.8.

REFERENCES

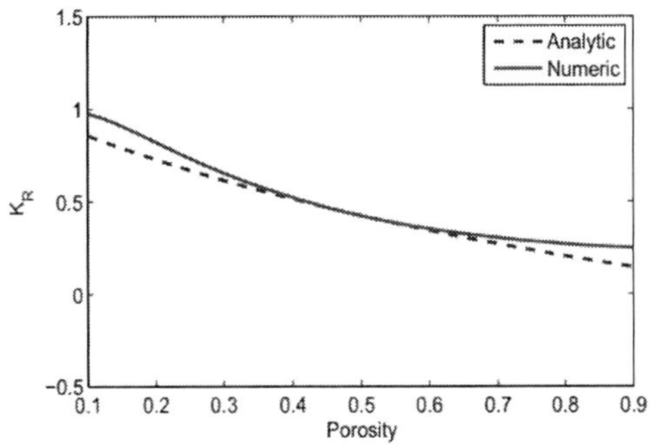

Figure 10. Comparison between wave relection coefficient from numerics and analytic as function of porosity n for a certain friction coefficient f. Solid lines: analytical result, solid lines: numerical result.

6. Conclusions

We have derived dispersion relation that explain diffusive mechanism of waves because of the porous structure. Further, analytical solution for wave reflection coefficients is obtained. To solve our equation numerically, we applied Finite volume method on a staggered grid that is free from damping error. We show that our numerical wave reflection confirms the analytical solution.

References

[1] Calabrese M, Vicinanza D and Buccino M 2008 *Ocean Engineering* **35** 1015 – 1028 ISSN 0029-8018 URL `http://www.sciencedirect.com/science/article/pii/S0029801808000577`

[2] Chwang A T, and Chan A T 1998 *Annual Review of Fluid Mechanics* **30** 53–84 (*Preprint* `https://doi.org/10.1146/annurev.fluid.30.1.53`) URL `https://doi.org/10.1146/annurev.fluid.30.1.53`

[3] Dalrymple R A, Losada M A and Martin P A 1991 *Journal of Fluid Mechanics* **224** 625644

[4] Gu Z and Wang H 1991 *Coastal Engineering* **15** 497–524

[5] Gu Z and Wang H 1992 *Coastal Engineering* **90** 1189–1198

[6] Pudjaprasetya S and Magdalena I 2013 *Applied Mathematical Sciences* **7** 2925–2937

[7] 2003 *Proceeding of The 1st Coastal, Estuary and Offshore Engineering Speciality Conference* **185** 1 – 10 ISSN 1570-6443

[8] Fernando H, Samarawickrama S, Balasubramanian S, Hettiarachchi S and Voropayev S 2008 *Journal of Hydro-environment Research* **1** 187 – 194 ISSN 1570-6443

REFERENCES

[9] Tsai C P, Chen H B and Lee F C 2006 *Ocean Engineering* **33** 1623 – 1643 ISSN 0029-8018 URL `http://www.sciencedirect.com/science/article/pii/S002980180500257X`

[10] Zhao Y P, Bi C W, Liu Y X, Dong G H and Gui F K 2014 *Engineering Applications of Computational Fluid Mechanics* **8** 116–126 (*Preprint* `http://dx.doi.org/10.1080/19942060.2014.11015502`) URL `http://dx.doi.org/10.1080/19942060.2014.11015502`

[11] Sulisz W 1985 *Coastal Engineering* **9** 371–386

[12] Magdalena I, Pudjaprasetya S R and Wiryanto L H 2015 *Advances in Applied Mathematics and Mechanics* **6** 680–692

[13] Pudjaprasetya S and Magdalena I 2014 *East Asia Journal on Applied Mathematics* **4** 152–165

[14] Magdalena I, Erwina N and Pudjaprasetya S 2015 *J. Sci. Comput* **3** 867–874

[15] Dean R and Dalrymple R 1991 *Water Wave Mechanics for Engineers and Scientist* (World Scientific)

Web usability evaluation on BloobIS website by using hallway usability testing method and ISO 9241:11

Tony Dwi Susanto[1], Anisa Ingesti Prasetyo[2], Hanim Maria Astuti[3]

[1,2,3]Department of Information System, Faculty of Industrial Technology, Institut Teknologi Sepuluh Nopember (ITS), Surabaya, 60111, Indonesia

E-mail: tonydwisusanto@is.its.ac.id

Abstract. At this moment, the need for web as an information media is highly important. Not only confined in the infotainment area, government, and education, but health as well uses the web media as a medium for providing information effectively. BloobIS is a web based application which integrates blood supply and distribution information at the Blood Transfusion Unit. Knowing how easy information is on BloobIS is marked by how convenient the website is used. Up until now, the BloobIS website is nearing completion but testing has not yet been performed to users and is on the testing and development phase in the Development Life Cycle software. Thus, an evaluation namely the quality control software which focuses on the perspective of BloobIs web usability is required. Hallway Usability Testing and ISO 9241:11 are the methods chosen to measure the BloobIS application usability. The expected outputs of the quality control software focusing on the usability evaluation are being able to rectify the usability deficiencies on the BloobIs web and provide recommendations to develop the web as a basic BloobIS web quality upgraed which sets a goal to amplify the satisfaction of web users based on usability factors in ISO 9241:11.

1. Introduction

BloobIS is a web based application which integrates blood supply and distribution information at the Blood Transfusion Unit [1]. The objective of creating BloobIS is assisting patients who are in dire of blood to immediately attain them. Moreover, on the doctor/hospital party, BloobIS is constructed to quicken doctors in finding blood supply information at every Blood Transfusion Unit province wise so that requesting blood can be done quickly. BloobIS website is nearing completion but testing has not yet been performed to the society. Testing and measurement of the BloobIS application are two extremely significant activities. Knowing how easy information is on BloobIS is marked by how convenient the website is used. Thus, an evaluation namely the quality control software which focuses on the perspective of BloobIs web usability is required.

Based on the points above, this study is made to implement the Quality Control software which focuses on the perspective of BloobIS web usability before this website is even released. BloobIS website usability testing is performed to assess and ensure usability of the web is effective, efficient, and satisfying. Testing needs to be done because the quality and consumer acceptance of the product depend mostly on the characteristics of ease of use.

In performing the BloobIS web usability measurement, the correct method is a necessity. Various methods can be applied to identify the problem of usability. In this case, the methods used to measure

BloobIS web are Hallway Usability Testing method and a perspective approach towards ISO 9241:11 [2].

Based on the ISO 9241:11 perspective, factors which affect the quality of usability are effectiveness, efficiency, and satisfaction. Therefore, to discover those factors affecting the quality of usability, performance user and usability framework are tested in order to observe the effect of three aforementioned factors.

BloobIS web usability evaluation method is performed by phases and samples based on the Hallway Usability Testing which undertakes evaluation twice with the first evaluation being improving the web before collecting a total of 3 samples to categorizes BloobIS users respectively. The before and after results of web usability evaluation are going to be compared to check whether there is an increase in usability quality along with the improvement of BloobIS.

2. Literature study

Usability is the ability of use especially to know how far the system is easily to be used or in other terms, user friendly. For a system to be categorized as user friendly, the system is composed by a number of factors narrowed down into five components such as learnability, efficiency, memorability, system errors, and satisfaction.

Those five components can be translated into components based on the definition from ISO 9241:11. The effectiveness component in ISO relates with learnability, memorability, and few errors in the Nielsen component. Meanwhile, the efficiency component in ISO is related with efficiency in the Nielsen component. Satisfaction component in ISO involves satisfaction in the Nielsen component.

2.1. Hallway usability testing

Some people think that usability testing requires high expenses and is complex. In performing usability testing, time can also be an issue because of how long it may take. These statements are incorrect. A complex usability testing is an extravagance of resource. The best outcome comes from a testing not more than 5 users and running a number of small tests. A research by Tom Landauer and Jacob Nielsen shows usability problems discovered in the usability testing.

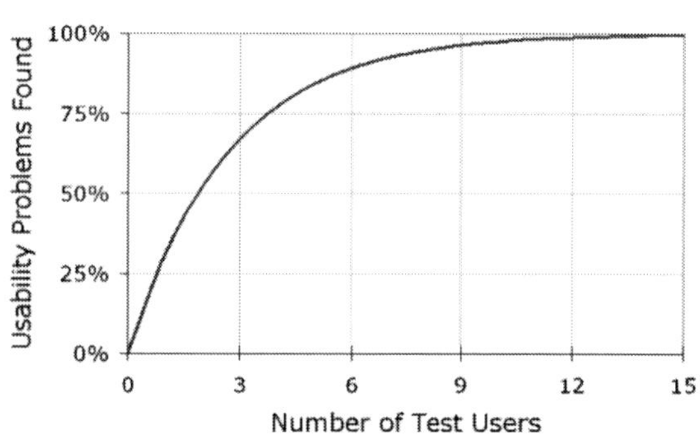

Figure 1. Hallway usability testing curve.

The curve above points out that data (problems) which are founded by testing the first user is one third of all the existing problems. When gathering data from the second user is performed, it will show that the second user performed some of the activities similar to the first user. The third user will perform many activities which are already observed by users one and two. User number three will result in a number of new small data, even not as many as users one and two. When adding the number of users, the usability problem value will consistently drop from the previous values. Until after the fifth user, running another test will waste time with the reason discovering and observing repetitive findings. Testing is performed not only to record usability weaknesses but to improve design issues as well. Thus, fixing the issue is done by a process named re-design. After the new design has been made in accordance with the testing findings from five users, issues surrounding usability are expected to be solved. The conditions in running some various group tests are as follows:

- 3-4 users users from each category if testing two group users
- 3 users from each category if testing three or more group users (3 users at minimum to assure diversed behaviors in a group is covered).

2.2. ISO 9241:11

A standardized ISO which defines the level of usability of a web used by the users to attain a specific objective effectively (effectiveness), efficiently (efficient), and satisfying (satisfaction). As for usability measures which bases on ISO 9241-11: Guidance on usability, is as follows:

Table 1. Usability measures ISO 9241:11.

Factor	Information
Effectiveness	Measures the level of accuracy and perfection performed by a user in completing a task for a specific objective.
Efficiency	Measures the ease of use from a system time punctuality perspective in providing data and information needed by the user.
Satisfaction	Measures how satisfied the users are towards the use of system/application when the user is interacting with system to achieve the objective.

2.3. User performance evaluation

User performance evaluation is a measurement by observing the behavior of respondent or user. Performance measurement is used to derive quantitative data regarding subject performance when users are performing tasks during usability testing.

1. Efficiency measurement
   Time needed in using the system for the first time. Measurements of efficiency are responsive (time efficiency), accurate (target efficiency), affordable (cost efficiency), and convenient (power efficiency).
2. Effectiveness measurement
   Shackel defined effectiveness as performance of the user using the related system. Performance is then measured by the time needed by the user to complete a task and the number of errors performed. In this research, effectiveness measurement is performed by observing the completion time in every phase of the task needed by every user in completing the task as a whole with the following formula:

$$\frac{\text{Respondents able to complete the task}}{\text{Total respondents}} \text{ x } 100\%$$

2.4. Usability questionnaire

In this study, performing measurement is based on the assessment of users to the efficiency, effectiveness, and satisfactions factors and done by filling in the questionnaire consisting of questions (respondents are asked to assess on the provided 5 scale Likert). Afterwards, the result of the questionnaire will be analyzed for usability questionnaire purposes which are able to interpret the effect of the three usability factors (effectiveness, efficiency, and satisfaction) towards the quality of usability.

2.5. Conceptual model and hypothesis
In accordance with ISO 9241:11 used in this study, the following conceptual model consisting of the usability behavior from the Bloobis prototype quality is proposed. The model represents that effectiveness, efficiency, and satisfaction are variables in usability behaviors in the prototype of BloobIS.

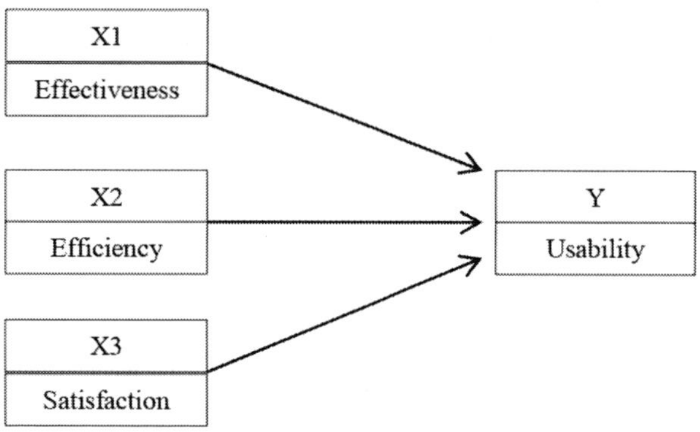

Figure 2. Conceptual model.

Afterwards, the hypotheses derived from this model are as follows:
- H1 → Effectiveness from the usage of BloobIS has a significant effect to the BloobIS prototype usability
- H2→ Effeciency from the usage of BloobIS has a significant effect to the BloobIS prototype usability
- H3→ Satisfaction from the usage of BloobIS has a significant effect to the BloobIS prototype usability.

2.6. Data processing method
In order for the data collected to be beneficial, data must be processed and analyzed foremost until a decision making basis can be made. The purpose of data processing method is interpreting and drawing a conclusion from the collected data.

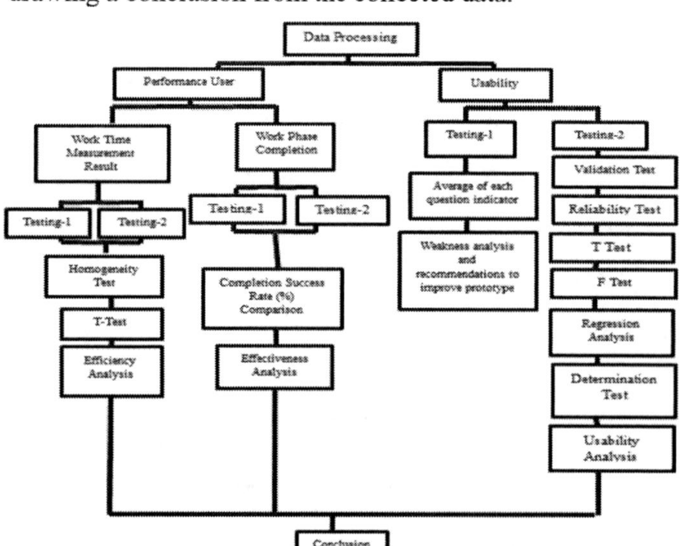

Figure 3. Data processing method.

3. Result and analysis

3.1. Data processing
Testing uses data taken from 9 respondents (testing-1) and 40 respondents (testing-2) which are performed at three locations namely Blood Transfusion Unit, hospital, and Information System department. This testing process employs a questionnaire consisting of 30 questions and tasks which ought to be done by every respondent. Processing data is performed by using SPSS 17.0.

3.2. Testing-1 problems
Some findings surrounding the weaknesses in the usability of the BloobIS prototype is discovered at the testing-1 evaluation. The aforementioned findings of the BloobIS interface user weaknesses are as follows:

a. Lies a number of information which should not have been displayed (related to personal data information),
b. Inconsistency of language used in the BloobIS prototype (currently uses Indonesian and English language),
c. Unnecessary funtions and disrupting the comfortability of the user,
d. "Search" feature does not ease the user, "Search" feature is advised to be made more specific,
e. Some features have not yet existed to ease the user,
f. Inconsistent font and writing,
g. Form detail which should not be editable is the contrary,
h. "Back" feature is non-existent on some website pages,
i. There is no feedback when the user has performed an activity. For example, clicking the link on one of the content title blood donor and loading navigation appears,
j. Server time is different with the actual time,
k. IP and last login time have not been able to update automatically,
l. "Delete" button does not function to delete blood records,
m. Statistics diagrams on the dashboard menu do not pop up,
n. "Help" feature is not yet available, that users are not able to learn every aspect related to the system without the help of BloobIS developers.

3.3. Completion (work phase success)

Table 2. Testing-1 & testing-2 work phase success difference.

User	Work phase	Completion success rate %	
		Testing 1	Testing 2
Admin	Phase 1	67 %	100%
	Phase 2	67 %	100%
	Phase 3	67 %	67%
	Phase 4	100 %	100%
	Phase 5	100 %	100%
	Phase 6	100 %	100 %
Hospital	Phase 1	67%	100%
	Phase 2	100%	67%
	Phase 3	67%	100%
	Phase 4	100%	100%
Donor	Phase 1	100%	100%
	Phase 2	100%	100%
	Phase 3	100%	100%

3.4. Work phase completion time

Work time measurement is done by every respondent measured with a time unit of seconds. Data of work time measurement result can be seen on the following table.

Table 3. Work phase completion time difference.

User	Phase	Testing-1	Testing-2
Admin	1	80.33	44.67
	2	71.67	49
	3	62	45.33
	4	76	47
	5	54.33	30.33
	6	34.33	17
Hospital	1	62.00	38.67
	2	61.33	55.33
	3	52.33	50.67
	4	63.67	60.67
Donor	1	95.7	62
	2	30.7	15
	3	26	20.33

3.5. BloobIS improvement

Improvement results of the Bloobis prototype based on the evaluation performed in testing I and recommendations can be seen on the following table.

Table 4. BloobIS prototype improvement.

Number	Problem	Prototype improvement
1	Lies a number of information which should not have been displayed (related to personal data information)	☑
2	Inconsistency of language used in the BloobIS prototype (currently uses Indonesian and English language)	☑
3	Unnecessary funtions and disrupting the comfortability of the user	☑
4	"Search" feature does not ease the user. "Search" feature is advised to be made more specific.	☐
5	Some features have not yet existed to ease the user.	☐
6	Inconsistent font and writing	☑
7	Form detail which should not be editable is the contrary.	☑
8	"Back" feature is non-existent on some website pages.	☐
9	There is no feedback when the user has performed an activity. For example, clicking the link on one of the content title blood donor and loading navigation appears.	☐
10	Server time is different with the actual time.	☐
11	IP and last login time have not been able to update automatically.	☑
12	"Delete" button does not function to delete blood records.	☑
13	Statistics diagrams on the dashboard menu do not pop up.	☑
14	"Help" feature is not yet available, that users are not able to learn every aspect related to the system without the help of BloobIS developers.	☑

3.6. T-test testing

T-test testing is performed by implementing α=5%. If p value (sig) < 0.05, then the process is concluded that a significant work time difference lies between testing-1 and testing-2 at the respective work phase [3]. Based on the t-test testing, the significant work time difference is only visible at the admin user on phases 1 and 6. There are no significant work time differences on every phase in categorized users hospital and donor.

Table 5. T-test testing result.

User	Phase	P-value
PMI Admin	Phase 1	0.009
	Phase 2	0.215
	Phase 3	0.385
	Phase 4	0.337
	Phase 5	0.314
	Phase 6	0.03
Hospital	Phase 1	0.068
	Phase 2	0.452
	Phase 3	0.878
	Phase 4	0.818
Donor	Phase 1	0.215
	Phase 2	0.342
	Phase 3	0.235

T test is used to partially test each variable. The result can be seen on the Table 6 coefficients and sig column (significance). If the probability value of t or significance < 0.05, then the process can be concluded that there lies an effect between the independent variable towards the dependent variable partially. The t test result indicates that the three usability factors have an effect towards usability quality partially.

Table 6. T-test result.

Partial Effect	Calculated t > t-table (0.2353)	Information
Effectiveness Variability towards Usability	0.976>0.2353	H1 is accepted
Efficiency Variability towards Usability	1.588>0.2353	H1 is accepted
Satisfaction Variability towards Usability	0.802>0.2353	H1 is accepted

3.7. F-test

F test is used to know the effects independent variables (Effectiveness, Efficiency, Satisfaction) have towards the dependent variable (Usability) simultaneously [4]. Significant is defined by a relationship taking place can apply to the population. In this study, the significance level is 0.05 (5%).

The result from the probability value is 0.000 < 0.05, then there lies a conclusion that a significant effect simultaneously affects the independent (Effectiveness, Efficiency, and Satisfaction) and dependent (Usability) variables.

Table 7. F-test result.

ANOVA					
Model	**Sum of squares**	**df**	**Mean square**	**F**	**Sig.**
Regression	2.500	3	.833	24.308	.000*
Residual	1.234	36	.34		
Total	3.734	39			

a. Predictors: (Constant), Mean_Satisfaction, Mean_Efficiency, Mean_Effectiveness
b. Dependent variable: Mean_Usability

3.8. Regression

Multiple linear regression analysis is used to know how significant the effectiveness, efficiency, and satisfaction variables effects partially and or continuously towards usability [5]. Based on the result derived from the regression equation, it can be seen that the entirety of dependent variables (effectiveness, efficiency, satisfaction) have a positive effect on usability using the regression equation as follows:

$$Y = 0.930 + 0.291X1 + 0.361X2 + 0.196X3$$

3.9. Determination

Determination coefficient is applied to find out how substantial the dependent variables have an effect on the independent variable.

Table 8. Determination test result.

Model summary								
				Change statistics				
R	**R square**	**Adjusted R square**	**Std. Error of the estimate**	**R square change**	**F change**	**df1**	**df2**	**Sig. F change**
.818	.669	.642	.1851382	.669	24.308	3	36	.000
Predictors: (Constant), Mean_Satisfaction, Mean_Efficiency, Mean_Effectiveness								

From the calculation result using SPSS program, a conclusion is obtained that the determination coefficient (adjusted R2) is 0.642. This means an amount of 64.2% of BloobIS usability can be explained by the effectiveness, efficiency, and satisfaction variables while 35.8% of BloobIS usability is affected by other variables which are not researched upon in this study.

3.10. Correlation

Based on the result below, it can be discovered that the hypothesis from the framework usability which consists of effectiveness, efficiency, and satisfaction has a positive effect towards BloobIS usability. Therefore, the decision of the initially formulated hypothesis is accepted.

Table 9. Correlation test result.

Variable Correlation Closeness	**Pearson Correlation**	**Correlation Coefficient Interpretation**
Effectiveness Variability towards Usability	0.798	Strong
Efficiency Variability towards Usability	0.792	Strong
Satisfaction Variability towards Usability	0.772	Strong

4. Conclusion

As for some points which can be concluded related to this study is to increase the quality of BloobIS web usability as well as providing recommendations to continuously develop BloobIS by testing the BloobIS prototype based on the ISO 9241:11 perspective and Hallway usability testing. Based on the ISO 9241:11 perspective, factors which affect usability quality are effectiveness, efficiency, and satisfaction. Thus, to know the effects those factors have towards usability quality is done by user performance testing and usability framework testing to monitor the effects which the three factors inflict.

BloobIS web usability evaluation method is performed using phases and samples referring to the hallway usability testing where evaluation is done twice by firstly improving the web before the second evaluation is commensed using 3 respondents as samples for each category of the BloobIS user. The usability web evaluation result before and after improvements are compared to check whether there is an improvement in usability quality with the BloobIS improvement.

Data in the execution of evaluation is obtained by the task completion result in experimental scenarios and questionnaires where the performance user measurement is able to interpret the web usability quality based on the effect of the three usability factors (effectiveness, efficiency, satisfaction) towards usability quality.

Result from BloobIs web usability evaluation is that BloobIS fulfills the three usability factors (effectiveness, efficiency, satisfaction). With the improvements implemented in the evaluation, improving usability quality is proven. Based on the evaluation which has been done then the recommendation to improve BloobIS usability quality is by performing improvements toward some functions and the BloobIS interface based on the problems, complaints, and recommendations for the users. Other than that, the developer ought to monitor the three usability factors in developing BloobIS to improve the quality of usability form the BloobIS prototype.

References

[1] Zulkifli, Nurul Ilmi 2016 Usability Testing Sistem Informasi Pendonor Darah (Studi Kasus di Unit Pelayanan Transfusi Darah RSUP Dr. Sardjito Yogyakarta) *Journal of Information Systems for Public Health* Vol 1(1) pp 10-15

[2] Jokela, Timo 2003 The standard of user-centered design and the standard definition of usability: analyzing ISO 13407 against ISO 9241-11 *In Proc CLIHC* pp 53-60

[3] Kim TK 2015 T test as a parametric statistic *Korean Journal of Anesthesiology* Vol 68(6) pp 540-546

[4] Ning-zhong Shi, Jian Tao 2009 *Statistical Hypothesis Testing: Theory And Methods* (World Scientific Publishing Co Pte Ltd. Singapore)

[5] Gülden Kaya Uyanık, Neşe Güler 2013 A Study on Multiple Linear Regression Analysis *In Procedia - Social and Behavioral Sciences* Vol 106 pp 234-240

Construction of mathematical models the parachute jumper with change position acrobatic

Asmianto Asmianto, Hariyanto Hariyanto and Iis Herisman
Department of mathematics, Faculty of mathematics and natural science,
Institut Teknologi Sepuluh Nopember (ITS)
Jl. Arief Rahman Hakim, Surabaya 60111 Indonesia
e-mail: asmianto57@gmail.com

Abstract. Construction of mathematical models the movement of a parachutist during the air using newton's II law is $\Sigma F = ma$. Position parachutist after exiting the plane immediately unfurled his body so as to create a large air resistance. The presence of air drag resulting movement indirectly parachutist moves down vertically downwards but also shifting toward horizontal and form a parabolic trajectory. Parachutist speed getting down increased until eventually the parachutist reaches terminal velocity it's the position where the air drag is equal to the gravitational force (gravity) jumpers. In this paper is assumed to be parachutist with 91.6 kg mass (including equipment) jumping from a plane at an altitude of 3.000 meters and reach a height of parachutist ± 1000 meters with velocity $\pm 57\ m/s$. So the parachutist have to be clever in taking account of everything, because if just a little too late can dangerous the safety of the parachutist.

1. Introduction

Skydiving is one of the most extreme sports. Beside that, the parachute is also commonly used to save the life of a pilot and distribute soldiers in remote places . Now this is a typical skydiving used is at an altitude of 4000 meters by opening the parachute at an altitude of 800 meters off the ground. Parachuting lasted less than a minute. [1]

In doing parachuting, before making the jump parachutist always pay attention to the air pressure, minimum height early jump and the minimum distance from a point that will be targeted. Because, it was very influential on the safety and speed of the parachutist. to account for this, many of the scientists who developed the numerical model of the parachute problem. The power given by the air resistance is very influential with the speed of a parachute. The altitude of the plane to start a jump should also attention, that the parachutist could land on the desired point. [2]

Mathematical model for the motion of parachute shaped based on Newton's *II* Law $\sum F = ma$. Where m is the mass of the parachutist, a is the acceleration, and F is the sum of F_g and grafitasi style of drag air F_d. Therefore, F $= F_g + F_d$. on air drag there is drag coefficient which are influenced by the magnitude of the Reynolds number. Reynold number depends also on the shape of the canopy used by parachutist. In this paper is the canopy in the shape of a half ball with Reynolds number $Re > 10^3$ and with a coefficient of drag $C_D = 1{,}33$.[4]

Based on the above problems then suggested the paper about, how the movement and position of sitting in the air when the parachutist started earlier jumped from the plane to the position of the parachutist ready to open the parachute. In this mathematical models it's will be easier to know the various influences before making the jump. So the parachutist can jump safely and land at a position already determined.

2. Research and methodology

The establishment of mathematical model of parachutist with acrobat changes the position used newton's II. The second law explains how the velocity of an object changes when it is subjected to an external force. The law defines a force to be equal to change in momentum (mass times velocity) per change in time. Newton also developed the calculus of mathematics, and the "changes" expressed in the second law are most accurately defined in differential forms. (Calculus can also be used to determine the velocity and location variations experienced by an object subjected to an external force.) For an object with a constant mass m, the second law states that the force F is the product of an object's mass and its acceleration a. in the formation of the model there are several steps including assumption of the position of the parachutist after jumping from the plane, the parachutist during motion analysis in air and differensial the formula of speed and position with respect to time and analyze the movement of the parachutist started from motion free fall from the airplane in a position ready to open the parachute. After that,establishment of mathematical model of parachuting approached from the second law of newton, start from parachutist free fall from the airplane, until the parachutist in a position ready to open the parachute. Finally, the result mathematical modelling are simulated and MATLAB will be used as a simulator. After the obtained model equation of parachutist, then will be created for equation of algorithms are simulated.

3. Analysis and discussion

Construction of mathematical model on the parachutist will be contructed is starting to get out of the plane in a position ready to open the parachute. In this paper assumed parachutist doing motion free fall from an airplane and instantly make her spread her down position (Figure 1) so that it created a large air resistance.

Figure 1: Position of parachutist in the air

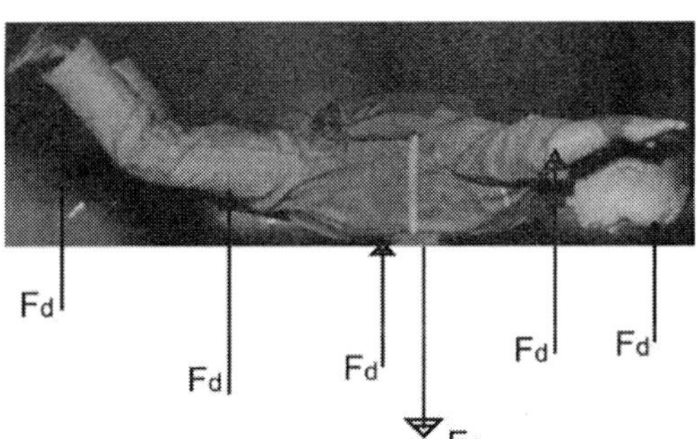

Figure 2: Forces that works on parachutist during in air

By the time the parachutist began jumping from the plane then Newton's law II can be applied and in this case assumed parachutist doing motion free fall. During his time in the air doing translational

IOP Conf. Series: Journal of Physics: Conf. Series **974** (2018) 012044 doi:10.1088/1742-6596/974/1/012044

motion of parachutist and rotation at the position until ready to open the parachute. Mathematically it can be written as follows:

$$\sum F = ma$$
$$\sum F = F_g + F_d$$
$$F_g + F_d = ma$$

F_g is the gravitational force of parachutists that the direction is always toward the center of the earth while the F_d is the air drag (Force drag) the direction is always opposite to the movement of the parachutist.
m is the mass of the whole, it's mass parachutist with the mass of equipment under by parachutists (including parachute).
Acceleration is the first derivative of speed against time, so that it can be written as follows:

$$F_g + F_d = m\frac{dv}{dt}$$

So, by the equation of velocity with respect to time as follows:

$$v(t) = \frac{mg - ce^{-\frac{k}{m}t}}{k}$$

To get the equation with respect to time, with the position of the integration equations of velocity with respect to time.
So obtained

$$x(t) = \frac{mg}{k}\left(t + \frac{m}{k}e^{-\frac{k}{m}t}\right) + c$$

By making the assumption $x(0) = x_0$ equation obtained a position with respect to time, parachutist

$$x(t) = x_0 + \frac{mg}{k}t + \frac{m^2 g}{k^2}\left(e^{-\frac{k}{m}t} - 1\right)$$

4. Simulation and analysis

In this chapter the sub carried out simulations of the equations of motion of parachutist that has already obtained from previous calculations. From the results of the simulation will be done analysis of the movement of the parachutist before the parachute opened in the air.

- Simulation 1

The movement of the parachutist until terminal velocity :
Parameters:

x_0	$= 3000\ m$	ρ	$= 1.2\ kg/m^3$	t	$= 500\ sekon$
m	$= 91.6\ kg$	C_{d1}	$= 0.16$		
g	$= 9.81\ m/s^2$	C_{d2}	$= 39.6$		

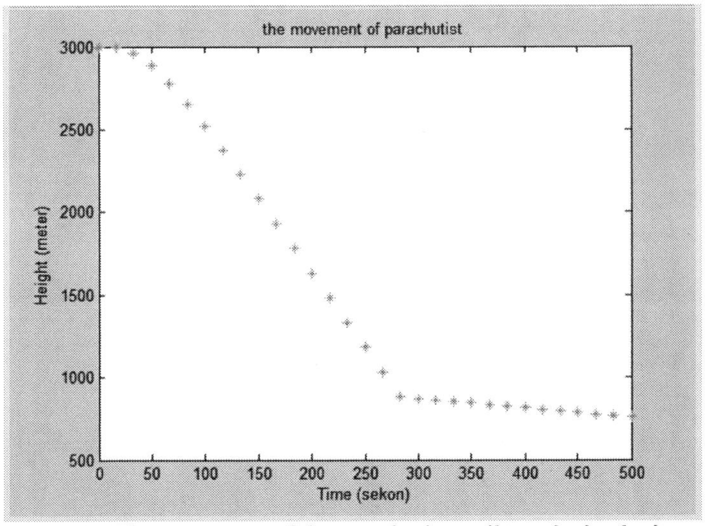

Figure 3 : The movement of the parachutist until terminal velocity

- Simulation 2

  Parameters:

x_0	$= 1000\ m$	ρ	$= 1.2\ kg/m^3$	t	$= 60\ sekon$
m	$= 91.6\ kg$	C_{d1}	$= 0.16$		
g	$= 9.81\ m/s^2$	C_{d2}	$= 39.6$		

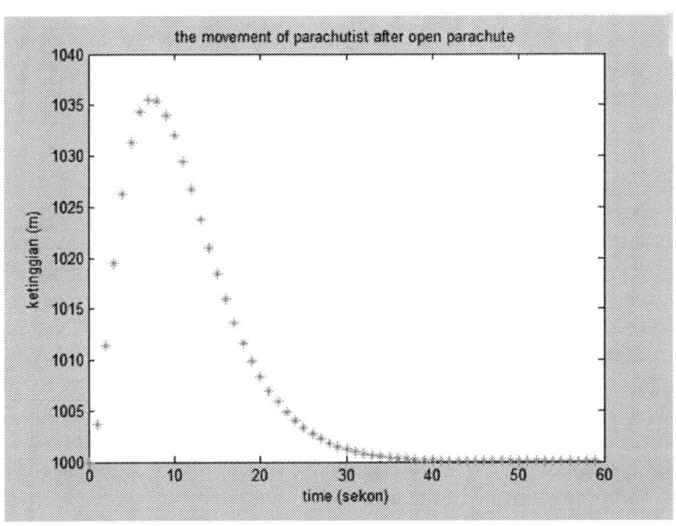

Figure 4: The movement of the parachutist after the parachute opened

IOP Conf. Series: Journal of Physics: Conf. Series **974** (2018) 012044 doi:10.1088/1742-6596/974/1/012044

- Simulation 3

Parameters:

x_0	$= 3000\ m$	ρ	$= 1.2\ kg/m^3$	t	$= 60\ sekon$
m	$= 91.6\ kg$	C_d	$= 0.25$	g	$= 9.81\ m/s^2$

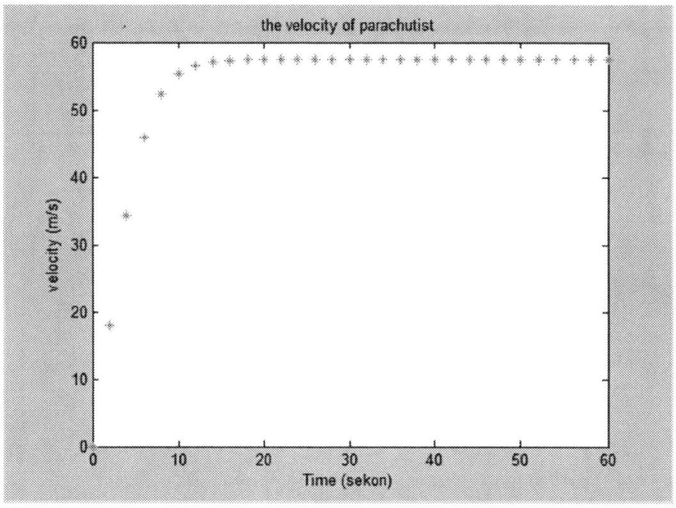

Figure 5 : The velocity of the parachutist before the parachute opened

From the results of the simulation of the movement of the parachutist before and after the parachute opened can be seen that the movement of the parachutist formed a parabolic trajectory. This is caused by the presence of air drag created by the body when the parachutist in the air so that the movement of the parachutist is not likely to fall free.

❖ Analysis of the terminal velocity parachutist

Terminal velocity is speed skydiving where great value force weight objects (W) of parachutist equals great value force of drag or air resistance (F_d). Gravity which causes the parachutist moved vertically down one point valuation will be the same with the air drag. because the gravity constant value where as the air drag increases its value. Gravity has a constant value, because it only affected by mass and gravity. While their drag value varies according to the speed of the parachutist. At the time of parachutist jumping from a plane, the force that causes the parachutist move is gravity. Because of the gravitational acceleration of parachutist have gravity then the object will move with the speed of the faster all the time. But at the moment there are also free fall parachutist drag air is affected by the speed of the parachutist. So on one side of the gravity increased speed of objects but on the other hand with increasing speeds also adds drag. Air drag is a force of anchoring that is caused by fluid flow against the moving object. With the increase of the air drag and gravity then someday there will be equilibrium, meaning that large air drag equals huge gravity In according with the laws of Newton II, then the force acting on an object will be equal to 0. And in according with the laws of Newton, when I force acting on an object is equal to 0 then the objects tend to remain stationary or moving with constant speed. [3]

5. Conclusion

Based on the results of the analysis and discussion in the previous chapter, some things can be summed up as follows:

1. Mathematical Model of movement of a parachutist with the change of the position are as follows:

$$v(t) = \frac{mg}{k}(1 - e^{-\frac{k}{m}t})$$

$$x(t) = x_0 + \frac{mg}{k}t + \frac{m^2g}{k^2}\left(e^{-\frac{k}{m}t} - 1\right)$$

2. Safe of time used parachutist for opening the parachute if he jumped from a plane at an altitude of 3000 meters was before in altitude of about 1000 metres ± and certainly with fixed pay attention to surroundings.
3. Large terminal velocity depends on the weight of the parachutist and the position of the body during the parachutist in the air.

References

[1] Moniuszko, J. (2010). **Modellling dynamics and aerodynamics tests of a sport parachute jumper during flight in sitfly position.** Polnad. Vol. 12 No. 3

[2] Ru-Yi, T. (2013). **Characteristic Analysis of Flying Process of Dispensed Object with Deceleration Parachute.** Vol. 3. No. 1

[3] Long, Lyle N.(1999). **Velocity dependence of Aerodynamic drag.** Univercity park. Washington.

[4] Meade, Douglas B. & Struthers, Allan A. (1999). **Differential Equations in the New Millenium: the Parachute Problem.** Dept.of Mathematic University of South Corolina, Colombia.

Binomial tree method for pricing a regime-switching volatility stock loans

Endah R.M. Putri, Muhammad S Zamani, Daryono B Utomo

Mathematics Departement, Institut Teknologi Sepuluh Nopember Surabaya Indonesia

E-mail: endahrmp@matematika.its.ac.id

Abstract. Binomial model with regime switching may represents the price of stock loan which follows the stochastic process. Stock loan is one of alternative that appeal investors to get the liquidity without selling the stock. The stock loan mechanism resembles that of American call option when someone can exercise any time during the contract period. From the resembles both of mechanism, determination price of stock loan can be interpreted from the model of American call option. The simulation result shows the behavior of the price of stock loan under a regime-switching with respect to various interest rate and maturity.

1. Introduction

Stock loan contracts with two parties involved, the lender and borrower, can be considered as an alternative of investment. Using the stock as the collateral, the stock owners as the borrower can get liquidity without selling their stocks. During the life of the contract, the lender holds the stock and the borrower gets the stock back as he paybacks the loan. As the stock price decreases and reaches the loan amount then the borrower fails to payback. The borrower pledges the stock to the lender and just walks away - this is called non-recourse [1].

The analytic valuation of stock loans as a perpetual contract is proposed by Xia and Zhou [1]. The perpetual scheme is used to approach a long tenor contract of stock loan. However, the tenor of stock loans contract is finite. A finite maturity of stock loan valuation has been studied by Lu and Putri [2] which presented a semi-analytic method. The study evaluated the stock loans in three dividend distributions in Dai and Xu [3]. Further, the optimal exit strategy is discussed in Dai and Xu [3] and is implemented using binomial method under the Black-Scholes model. As the Black-Scholes' assumes a constant volatility, the model is likely to inaccurate for pricing.

The change in economics is an inevitability which is represented by the volatility values in the mathematical model [4]. Therefore some attempts to present better valuations of non-constant volatility stock loans are presented: perpetual stock loans with a regime-switching volatility under the Black-Scholes model [5], a Levy process based stock loans valuation [6], and stock loans with stochastic volatility model [7].

A regime-switching volatility model has advantages for its flexibility to apply and its easier procedure than solving a non-linear PDE based model [8]. Unlike the perpetual stock loans with a regime-switching volatility which has analytic solution, the finite maturity stock loans with regime-switching has not been discussed intensively so far in literature. Therefore, this paper discusses about the finite maturity stock loans with a regime-switching volatility. The method used for valuing such a contract is a binomial method for its simplicity and flexibility.

This paper is organized as follows. Section 2 discusses about the binomial model of stock loans with constant volatility and Section 3 presents the contracts with a regime-switching volatiliy. Numerical simulation in Section 4 and concluding remark in Section 5.

2. Binomial tree method for stock loans with constant volatility

The random dynamic of stock price as the collateral of the stock loans contract is assumed to follow a stochastic differential equation,

$$dS = (r - \delta)\, S\, dt + \sigma S\, dW_t \tag{1}$$

where dW_t is a Wiener process defined for $t \in [0, \infty)$, $r \geq 0$ is the risk-free interest rate, $\delta \geq 0$ is the continuous dividend yield, and σ is the volatility of the stock. Equation 1 as a continuous random walk model is approximated by a discrete random walk model in binomial method [9].

Following the common procedure to obtain the valuation of options in [10], the factor of up and down are determined as a limiting case of binomial distribution and the factors are $u = e^{\sigma\sqrt{\Delta T}}$, dan $d = e^{-\sigma\sqrt{\Delta T}}$ [11].

The risk neutral probability of the stock price's increase or decrease are defined by p and $q = 1 - p$ respectively,

$$p = \frac{e^{rT} - d}{u - d},\ q = \frac{u - e^{rT}}{u - d}.$$

Subsequently, the discrete formula price of stock loan without a regime- switching volatility with binomial model is

$$V_0 = max\left\{\frac{1}{\rho}\left(q^u V_T^u + q^d V_T^d\right), max\left(S_n - qe^{\gamma T}, 0\right)\right\}$$

where q is the loan amount and γ is the loan interest.

3. Stock loans with regime-switching volatility

A regime-switching Markov chain is a stochastic process which represents the state transitions. In this paper, the stock loans assume to have different states of economy those are growing economy and recession economy. Suppose that α_t is a continous Markov chain that describes difference states that occur in a realistic market and can be expressed as follows:

$$\alpha_t = \begin{cases} 1, \text{ if the state of economy is growing / bull state.} \\ 2, \text{ if the state of economy is in recession / bear state.} \end{cases}$$

Based on a continuous Markov chain, the transition matrix probability is formed as $P_{ij}(t) = P\{\alpha(t+s) = j|\alpha(s) = i\}$ that is current state i and next state j. Let $M = \{1, 2\}$ and $\alpha_t \in M$, the transition matrix is written as $P = \begin{bmatrix} p_{11} & p_{12} \\ p_{21} & p_{22} \end{bmatrix}$ and $0 \leq p_{ij} \leq 1$, $\sum_{j=1}^{2} p_{ij} = 1$ with $i, j \in M$. The economy conditions are represented by different volatilities, and the volatility value of each state is $\sigma^k = \sigma(\alpha_t = k)$ with $k = 1, 2$. Briefly, we use σ_1 and σ_2 to represent the economic in growth and in recession respectively.

3.1. Binomial model for stock loans without deviden in a regime- switching

For constructing a multistep binomial model of stock loans, we divide [0,T] into $[t_i, t_{i+1}]$, $i = 0, 1, 2, ..., N - 1$, with $t_i = i\Delta t$ and $\Delta t = \frac{T}{N}$. The dynamic of stock price is constructed in binomial tree under a regime-swicthing scheme with α^θ and $\theta = 1, 2$ is the state of market.

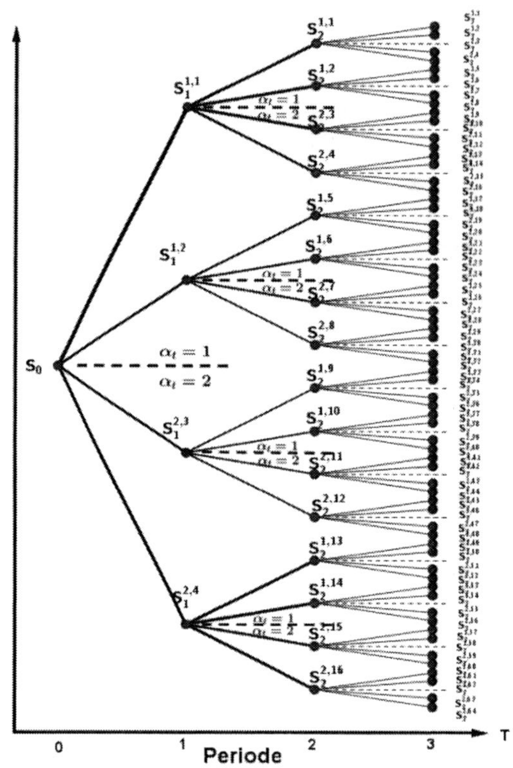

Figure 1: A multi-period binomial tree model for stock price in a two-state regime-switching

Subsequently, the stock price at time n is $S_n^{\alpha^\theta,\theta}$ with $1 \le \theta \le 4^n, 0 \le n \le N-1, \alpha^\theta = 1, 2$. It means that one of four values can be taken at t_{n+1}, that is:

- If $\alpha^\theta = 1$ then obtained $S_{n+1}^{1,4\theta-3} = S_n^\theta u^1$; $S_{n+1}^{1,4\theta-2} = S_n^\theta d^1$
- If $\alpha^\theta = 2$ then obtained $S_{n+1}^{2,4\theta-1} = S_n^\theta u^2$; $S_{n+1}^{2,4\theta} = S_n^\theta d^2$

The multi-period binomial model for stock price movement in a two-state regime switching volatility scheme is described in Figure 1.

The next step is constructing the stock loans value starting with the payoff value at time n is $V_n^{\alpha^\theta,\theta}$. Moving backward from the payoff value, we can calculate each node of the tree using,

$$V_N^{\alpha^\theta,\theta} = \left(S_N^{\alpha^\theta,\theta} - qe^{\gamma T}\right) \text{with } 1 \le \theta \le 4^N, \alpha^\theta = 1, 2,$$

and no-arbitrage assumption implies that

$$V_n^{\alpha^\theta,\theta} \ge max\left(S_N^{\alpha^\theta,\theta} - qe^{\gamma T}, 0\right) \text{ with } 1 \le \theta \le 4^n, 0 \le n \le N, \alpha^\theta = 1, 2$$

The binomial tree of the stock loans value is described in Figure 2.

Given $V_{n+1}^{\alpha^i,i}$ with $1 \le i \le 4^{n+1}, 0 \le n \le N-1, \alpha^i = 1, 2$ and for each θ such that $1 \le \theta \le 4^n$, the recursion formula should fulfill the following requirement:

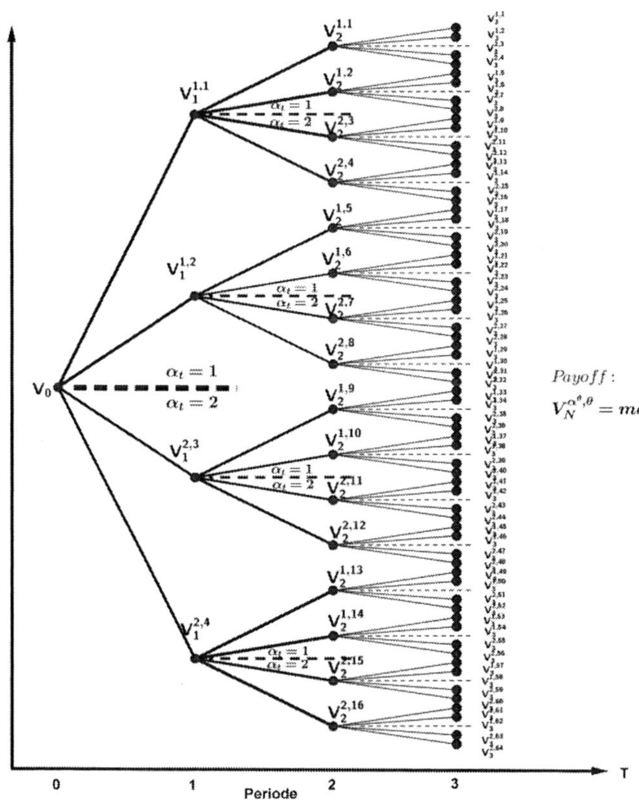

Figure 2: A multi-period binomial tree model for stock loans in a two-state regime-switching

- if $\alpha^\theta = 1$, it is obtained,

$$V_n^{1,\theta} = max\Big\{\frac{1}{\rho}\Big(p_{11}q^{1,u}V_{n+1}^{1,4\theta-3} + p_{11}q^{1,d}V_{n+1}^{1,4\theta-2} + p_{12}q^{2,u}V_{n+1}^{2,4\theta-1} + p_{12}q^{2,d}V_{n+1}^{2,4\theta}\Big), max\left(S_n^{1,\theta} - qe^{\gamma n\Delta t}, 0\right)\Big\} \tag{2}$$

- if $\alpha^\theta = 2$, it is obtained,

$$V_n^{2,\theta} = max\Big\{\frac{1}{\rho}\Big(p_{21}q^{1,u}V_{n+1}^{1,4\theta-3} + p_{21}q^{1,d}V_{n+1}^{1,4\theta-2} + p_{22}q^{2,u}V_{n+1}^{2,4\theta-1} + p_{22}q^{2,d}V_{n+1}^{2,4\theta}\Big), max\left(S_n^{2,\theta} - qe^{\gamma n\Delta t}, 0\right)\Big\} \tag{3}$$

Figure 1 and 2 show a more complicated tree than standar binomial tree which implies a longer computational time. In order to save the computational time, we propose a modification to the tree in the mentioned figures. The same nodes in Figure 1 and 2are packed or recombined into a single node. The recombined tree can be rearranged as shown in Figure 3. The recombined tree then is used in the simulations to show the behavior of the stock loans for various parameters' values.

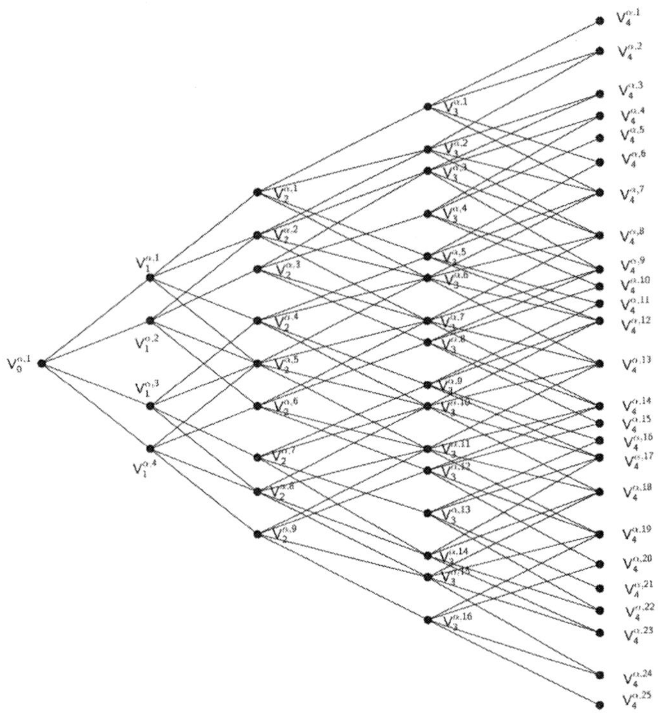

Figure 3: A multi-period binomial tree model for stock loans in a two-state regime-switching

For the recombined tree, the value of stock loans then are calculated based on the following formulas. Knowing that the stock price at time t with $1 \leq \theta \leq (t+1)^2$ is,

$$\begin{aligned} S_{t+1}^{\theta+\lfloor\frac{\theta-1}{t+1}\rfloor} &= S_t^{\theta} u^1 \\ S_{t+1}^{\theta+1+\lfloor\frac{\theta-1}{t+1}\rfloor} &= S_t^{\theta} u^2 \\ S_{t+1}^{\theta+t+2+\lfloor\frac{\theta-1}{t+1}\rfloor} &= S_t^{\theta} d^2 \\ S_{t+1}^{\theta+t+3\lfloor\frac{\theta-1}{t+1}\rfloor} &= S_t^{\theta} d^1 \end{aligned}$$

Subsequently, the stock loan values can be calculated by Equation 4

$$\begin{aligned} V_t^{1,\theta} = max\Big\{ \frac{1}{\rho}\Big(& p_{11}q^{1,u}V_{t+1}^{1,\theta+\lfloor\frac{\theta-1}{t+1}\rfloor} + p_{12}q^{2,u}V_{t+1}^{2,\theta+1+\lfloor\frac{\theta-1}{t+1}\rfloor} \\ & + p_{12}q^{2,d}V_{t+1}^{2,\theta+t+2+\lfloor\frac{\theta-1}{t+1}\rfloor} + p_{11}q^{1,d}V_{t+1}^{1,\theta+t+3+\lfloor\frac{\theta-1}{t+1}\rfloor}\Big), max\left(S_t^{1,\theta} - qe^{\gamma t}, 0\right)\Big\} \end{aligned} \tag{4}$$

4. Numerical results and discussions

We present some numerical simulations for determining the price of stock loans without dividend in a regime-switching volatility. Different from the American call options which has no optimality if there is no dividend paid, the stock loans still has optimal exercise boundary without the dividend paid in the stocks. The parameter values used for the following simulations are stock price $S_0 = 100$, loan amount $q = 100$, stock loans maturity $T = 1$, transition probability matrix

IOP Conf. Series: Journal of Physics: Conf. Series **974** (2018) 012045 doi:10.1088/1742-6596/974/1/012045

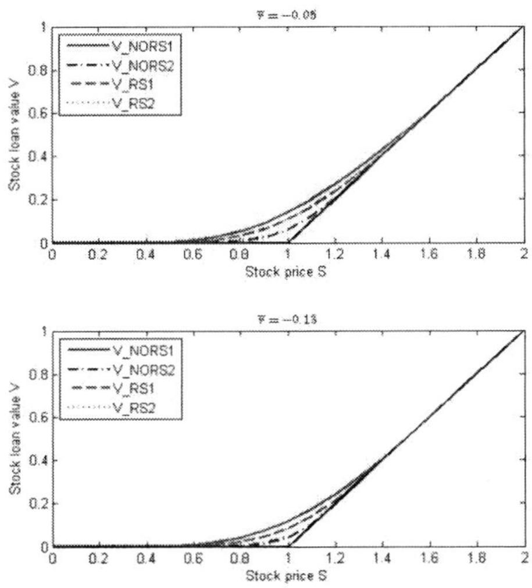

Figure 4: Stock loans value for various accumulated interest rate $\bar{r}$

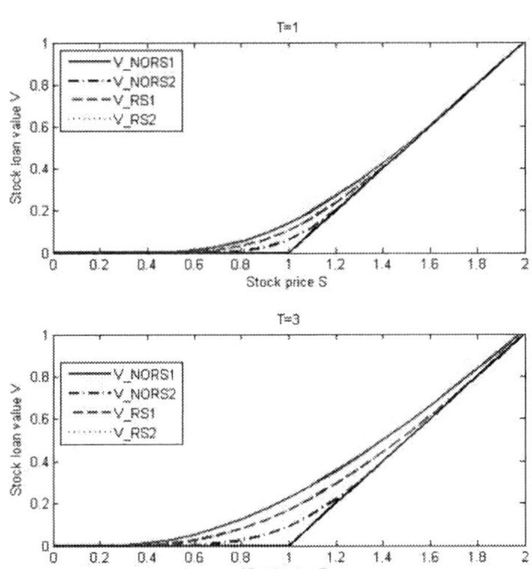

Figure 5: Stock loans value for various maturity T

$$P = \begin{bmatrix} 0.6 & 0.4 \\ 0.3 & 0.7 \end{bmatrix}.$$

For various accumulated interest rate ($\bar{r} = r - \gamma$) values, the stock loans value are shown by Figure 4. The results show that the values increase with respect to the riskless interest rate. The non-switching stock loan values bound the switching ones. In financial, the stock loans contract obviously has lower values as the accumulated interest rate has increased (driven by rising loan interests).

Stock loans contract values with different maturity is presented in Figure 5. A longer stock loan contract seems to have higher values than the shorter ones. The stock loan values with a regime-switching lies between the non-switching ones.

IOP Conf. Series: Journal of Physics: Conf. Series **974** (2018) 012045 doi:10.1088/1742-6596/974/1/012045

5. Conclusion

The behavior of the stock loans values under a regime-switching framework shows that the values decrease with respect to the accumulated interest rate, driven by loan interest rate. For a longer contract, the value of a stock loan is higher.

The stock loans value with switching are in between the non-switching ones. The reason is that the regime-switching framework implies a reduction to the over-valued or under-valued valuation in the non-switching scheme caused by different economic conditions.

6. Acknowledgments

Authors wishing to acknowledge Mathematics Department of Institut Teknologi Sepuluh Nopember for the funding of this research by scheme Department Grant.

References

[1] Xia J and Zhou X Y 2007 *Mathematical Finance* **17** ISSN 1467-9965

[2] Lu X and Putri E R M 2013 *submitted*

[3] Dai M and Xu Z Q 2011 *Mathematical Finance* **21** ISSN 1467-9965

[4] Scruggs J T and Nardari F 2005 Why Does Stock Market Volatility Change Over Time? A Time-Varying Variance Decomposition for Stock Returns EFA 2005 Moscow Meetings

[5] Zhang Q and Zhou X Y 2009 *SIAM Journal on Control and Optimization* **3** 1229–1250

[6] Liang Z, Nie Y and Zhao L 2012 Valuation of Infinite Maturity Stock Loans with Geometric Levy Model in A Risk-Neutral Framework Pre-print

[7] Wong T W and Wong H Y 2012 *Journal of Mathematical Analysis and Applications* **394** 337–346

[8] Liu R H, Zhang Q and Yin G 2006 *Journal of Applied Mathematics and Stochastic Analysis* 1–22

[9] Jiang L 2005 *Mathematical Modelling and Methods of Option Pricing* first edition ed (Singapore:World Scientific)

[10] Wilmott P, Howison S and Dewynne J 1995 *The Mathematics of Financial Derivatives: A Student Introduction* first edition ed (Cambridge University Press,)

[11] Cox J C, Ross S A and Rubinstein M 1979 *Journal of Financial Economics* **3** 229–263

IOP Conf. Series: Journal of Physics: Conf. Series **974** (2018) 012046 doi:10.1088/1742-6596/974/1/012046

Rapid mapping of landslide disaster using UAV-photogrammetry

A B Cahyono and R A Zayd

Department of Geomatics Engineering, Faculty of Civil - Environmental Engineering & Earth Sciences, Institut Teknologi Sepuluh Nopember

* agungbc@geodesy.its.ac.id

Abstract. Unmanned Aerial Vehicle (UAV) systems offered many advantages in several mapping applications such as slope mapping, geohazard studies, etc. This study utilizes UAV system for landslide disaster occurred in Jombang Regency, East Java. This study concentrates on type of rotor-wing UAV, that is because rotor wing units are stable and able to capture images easily. Aerial photograph were acquired in the form of strips which followed the procedure of acquiring aerial photograph where taken 60 photos. Secondary data of ground control points using GPS Geodetic and check points established using Total Station technique was used. The digital camera was calibrated using close range photogrammetric software and the recovered camera calibration parameters were then used in the processing of digital images. All the aerial photographs were processed using digital photogrammetric software and the output in the form of orthophoto was produced. The final result shows a 1: 1500 scale orthophoto map from the data processing with SfM algorithm with GSD accuracy of 3.45 cm. And the calculated volume of contour line delineation of 10527.03 m^3. The result is significantly different from the result of terrestrial methode equal to 964.67 m^3 or 8.4% of the difference of both.

1. Introduction

Photogrammetry is a process to obtain an information of an object by measurements on an aerial photograph result of the object. While photo interpretation is defined as an extraction of qualitative information about aerial photographs of an object by human visual analysis and photographic evaluation. The new terminology using Unmanned Aerial Vehicle or called as UAV is a supportive platform for photogrammetric measurements. This standard UAV makes it possible to track position and orientation of sensors that are implemented in local systems or global coordinates [3]. UAV have a high effectiveness in the term a lower cost and faster time than the conventional vehicles. It is suitable to apply for mapping in a small scope. The UAV-Based system has not been optimally usedfor topography mapping, since it less accurate than terrestrial mapping survey. This because in general, the UAV system is obtained by non-metric camera which has some geometric limitation. However, the use of UAV has some advantages, such as: (a) cost-effective, (b) the accuracy can be improved by camera calibration, (c) fast in mapping production, and (d). The UAV platform has less weight that make it more portable [2]. UAV platforms are nowadays a valuable source of data for inspection, surveillance, mapping and 3D modeling issues. New applications in the short- and close-range domain are introduced, being the UAVs a low-cost alternatives to the classical manned aerial photogrammetry [5].

This research will experiment and analyze of rapid mapping using UAV at landslide disaster location. One of landslide disaster that happened in Indonesia was landslide disaster in Jombang Regency, East Java [5]. The data used from aerial photography of non-metric camera set on the rotary-wing UAV. Data obtained is photo data that can be processed into ortophoto map. The final result of this research is analizing of the Digital Surface Model (DSM) data and the landslide volume calculation using SfM algorithm.

2. UAV-photogrammetry System

Currently drones have a variety of shapes, sizes, configurations, and functions. The presence of a camera device with a lightweight and compact size but has good quality. DJI Phantom Aerial UAV Drone Quadcopter one that weighs not more than 2 kilograms where there is already a camera installed safely and neatly. The drone can be controlled by remote control within 1000 meters. Ready To Fly: Complete fuselage ready to be directly flown consists of Aircraft Kit, servo ESCMotor, BrushlessTx/Transmitter/RemotRx Receiver, Propane Rover, Adapter, Propeler, Battery Charger and AutoPilot. Phantom is a drone that have models waypoint navigation and can now be used for photogrammetry. This quadcopter uses the latest IMU, flight control stabilization technology to fly super still along with a 4k stabilized integrated gimbal and camera. This will create accurate point clouds and perfect 3D maps stitched together using software. Waypoint navigation is very important for creating accurate 3D photogrammetry images. The Dji Phantom 3 Advance used the "Draw Waypoints" for it's autonomous programmed flight.

Figure 1. Diagram and drawing of Dji Phantom 3 Advance [1]

This system consists of Battery & Propellers Included : 1280 g, Diagonal Size (Propellers Excluded) : 350 mm, Max Flight Time : Approx. 23 minutes, Satellite Positioning Systems : GPS/GLONASS, Sensor : 1/2.3" CMOS, Effective pixels: 12.4 M (total pixels: 12.76 M), Camera Sony EXMOR FC300S, Focal length : 3,64 mm. Lens : FOV 94° 20 mm (35 mm format equivalent), f/2.8, Electronic Shutter Speed : 8 - 1/8000 s, Image Size : 4000 × 3000 and Range distances of up to 3.1 miles (5 kilometers). This type of drone is known as Rotary-wing or Vertical Take-Off and Landing type which is currently commonly used by the public because it does not require special skill in controlling it as well as a lot of market with affordable price. But now it has been equipped with various high-tech features that make it possible to do photogrammetric mapping with ease. Among them is the ability to autopilot the waypoint tracking that will make it easier when making the flight path. As a complement in this system also has many types of data processing from the Open Source (VSFM, MeshLab, etc.) to the license software (AgiSoft, Pix4D, Menci, etc.) to get the results of orthophoto.

3. Data Acquisition

Data acquisition begins with terrestrial measurements. Retrieval of coordinate data for photo GCP and terrestrial measurement control points used the topcon GPS Geodetic tool for ± 1 hour of each point. The number of points taken was about 6 points (2 control points and 4 GCP points). For height spot measurement is using tachymetry method, there are divided into two-way framework, i.e. basic measurement and detailed measurement. The basic framework or point control is a new point created by the author. Object control point is measured by GPS Geodetic with detail point measurement using Total Station. Observation of detail points taken are angle data, distance and elevation.

Table 1. Table showing control point between GCP Total Station and GPS Geodetic.

Control Point (name)	Coordinates			
	X-UTM (meter)	Y-UTM (meter)	Elevasi (meter)	Note
Base	9153136.352	648068.402	122.325	Musholla
BM-01	9153191.402	648123.283	125.996	TS-1/GPS
BM-02	9153201.676	648128.231	126.966	TS-2/GPS
BM-03	9153094.941	648138.961	125.433	Home Resident
BM-04	9153121.707	648049.012	122.514	Musholla Corner
BM-05	9153129.140	648018.297	121.787	School Corner

The measurement result shows that the coordinate of the measured area was 0,398 ha. The coordinates result can be made as DSM contours and landslide areas contours. Each color in DSM and each different contour line show the different heights. The next procedure is the acquisition of aerial photograph data. UAV-photogrammetry is divided into two-way frameworks in the field, i.e. take photo data using UAV-Quadcopter and GCP measurement with GPS Geodetic tool. The GCP coordinate data of photo data taken was about 60 photos. Data processed to obtain orthophoto map using image processing software with SfM algorithm with processing time are: align: 1 hour 35 minutes, dense cloud: 3 hours 29 minutes and mesh: 30 minutes

4. Results and Discussion

Coordinates obtained by photographs can be made as contour and DSM to determine the surface shape of the measurement area. The measurement appearance of rough surfaces, this is due to the many results of the cloud point of the UAV varies due to variations of altitude. Also the appearance of contours that seem short and dashed, due to the value of significant altitude variations are different.

Table 2. Table showing comparison implementation of UAV-photogrammetry and terrestrial.

No	Activity	UAV photogrammtery	Terrestrial (Total Sta)	Note
1	Tools used	Drone/UAV multicopter	Total Station	
2	Personels	2	4	
3	Time retrieval in the field*	65 minutes	175 minutes	**Time required without GPS measurement*
4	Data obtained	Aerial Photo	Angle, Distance, Elev.	
5	Numbers of GCP	4	2	

IOP Conf. Series: Journal of Physics: Conf. Series **974** (2018) 012046 doi:10.1088/1742-6596/974/1/012046

The differences of the two methods used in this study between UAV-photogrammetry and terrestrial method showed table above. The UAV-photogrammetry method requires fewer personnel and faster data retrieval time than the terrestrial method.

Table 3. Table showing comparison volume of UAV-photogrammetry and terrestrial.

No	Elements	UAV photogrammtery	Terrestrial (Total Sta)	Note
1	Spot height (point)	8967	260	
2	Volume (m^3)	10527.1	11491.7	
3	Difference (m^3)	964.7		
4	Procentage (%)	8.4		

Volume differences can be caused by differences in data retrieval methods and also by the differences in the number of measurement points.The differences of measurement points affect volume obtained. The volume obtained by UAV-photogrammetry with 8947 points are 10527.032 m^3 and 11491.708 m^3 by terrestrial method with 260 points then the volume difference is about 964.676 m^3. This study done a point reduction of the UAV-photogrammetry to determine the effect of point count measurement to the volume. The point reduction density was divided into two ways, i.e. 1. based on the nearest point and has the same coordinates with terrestrial point, 2. the points taken have an interval of one meter from the highest to the lowest point. Comparison of points can be seen in Table 6. Table 6 shows that the number of points greatly affected to the volume results.

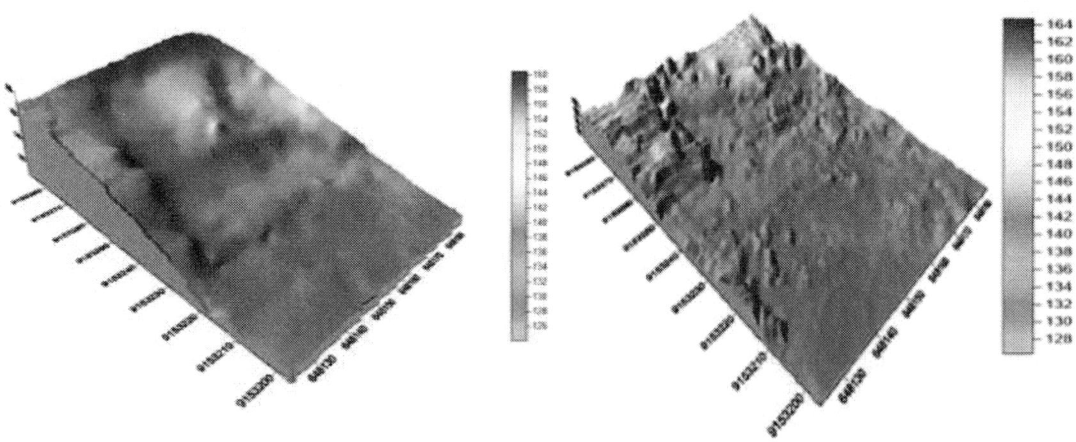

(a) (b)

Figure 2. Comparison of the DSM Form of the terrestrial Method (a) and UAV-photogrammetry (b)

Point differences affected not only on volume quantities but also on the shape and topography. The difference can be seen in DSM picture and contour. Figure 2 shows that the terrestrial method has a smooth surface shape and also regular elevation differences. While the UAV-photogrammetry method has a rough surface shape and much small piles of soil formed so that it resembled the original object. In addition, contour shapes with 0.15 m intervals of both methods have different line densities and also differences in the height or points elevation of both methods. The volume was not only affected by the number of points and different topographic shapes but also the height of the surface measurement. The smooth and regular elevation changes of the surface (contour) from high to low where it describes the landslide form occurred in the area.

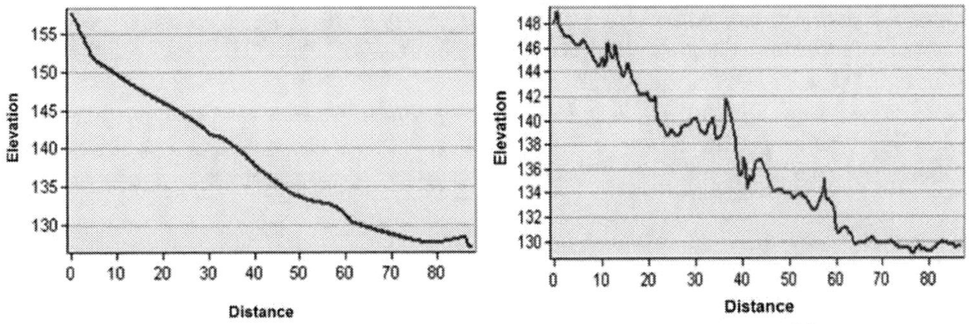

(a) (b)

Figure 3. Comparison of the elongated profile graph of the terrestrial method (a) and UAV-photogrammetry (b)

Figure 3 shows the wavy and irregular elevation changes of the surface (contour) from high to low where it described the landslide form occurred in the area. It was caused UAV-photogrammetry method can obtain very dense points. The UAV-photogrammetry and Total Station measurements have different surface heights. In the same elongated line, the highest point on the UAV-photogrammetry is at an elevation of 151 m while at a total station of 158 m. In addition, it also can be seen the surface obtained by both methods are different. The terrestrial method has a smooth surface while the UAV-photogrammetry surface is wavy.

5. Conclusion

A orthophoto of landslide disaster map has been successfully performed using UAV-photogrammetry from 60 photos at scale of 1: 1500 and GSD accuracy of 3.45 cm. In this study the overall time required to obtain an orthophoto map is +/-10 hours including data acquisition, data processing and visualization.

There is a difference between the results of the two methods to determine the volume of the landslide in Ngrimbi village located in Baren and Jombang Regency. The volume and coordinates point obtained by two calculation methods (UAV-photogrammetry and terrestrial method) are 10527.032 m^3 by UAV-photogrammetry and 11491.708 m^3 by terrestrial method with 964.676 m^3 or 8.4% of the difference of both methods. It was caused by the surface difference and the number of points obtained by both methods. This study shows that UAV technology can present disaster maps quickly and accurately. Although that is used is a rotary-wing UAV type is relatively easy operation (does not require runway/VTOL mode), low-cost and available in ready-to-fly unit.

References

[1] Canon. 2014. http://plazakamera.com/.
[2] Catur Aries Rokhmana and Susilo Utomo. The Low-Cost UAV-Based Remote Sensing System Capabilities for Large Scale Cadaster Mapping. IOP Conference Series: Earth and Environmental Science, Volume 47, Number 1
[3] Eisenbeiss, H. 2009. *UAV Photogrammetry.* Zürich. ETH Zürich.
[4] Mikhail, E. M and Bethel, J. S., *Introduction to Modern Photogrammetry.* John Wiley & Sons Inc., 2001.

[5] Remondino, F. et.all. 2011. Uav Photogrammetry For Mapping And 3d Modeling – Current Status And Future Perspectives. International Archives of the Photogrammetry, Remote Sensing and Spatial Information Sciences, Vol. XXXVIII-1/C22 UAV-g 2011, Conference on Unmanned Aerial Vehicle in Geomatics, Zurich, Switzerland.

[6] Saeno. (2014, 1 28). *Sedang tidur 14 Warga Tertimbun Tanah Longsor*. Retrieved from www.kabar24.com, http://www.kabar24.com/nasional/ read/20140128/65/ 209681/

[7] Udin, W.S and Ahmad, A. 2012. *The Potential Use of Rotor Wing Unmanned Aerial Vehicle for Large Scale Stream Mapping*. International Journal of Scientific & Engineering Research Volume 3, Issue 12, December-2012.

[8] Wolf, P. R. dan Dewitt, B. A., *Elements of Photogrammetry: with Application in GIS*. Columbus: McGraw Hill, 2004

Stock price prediction using geometric Brownian motion

Farida Agustini W, Ika Restu Affianti, Endah RM Putri
Department of Mathematics, Faculty of Mathematics and Science, Institut Teknologi Sepuluh Nopember (ITS), Jl. Arief Rahman Hakim, Surabaya 60111 Indonesia

E-mail: agustini.farida54@gmail.com

Abstract. Geometric Brownian motion is a mathematical model for predicting the future price of stock. The phase that done before stock price prediction is determine stock expected price formulation and determine the confidence level of 95%. On stock price prediction using geometric Brownian Motion model, the algorithm starts from calculating the value of return, followed by estimating value of volatility and drift, obtain the stock price forecast, calculating the forecast MAPE, calculating the stock expected price and calculating the confidence level of 95%. Based on the research, the output analysis shows that geometric Brownian motion model is the prediction technique with high rate of accuracy. It is proven with forecast MAPE value ≤ 20%.

1. Introduction

Capital market are markets for various long term financial instruments that can be traded. Capital markets are vital for fuction of country economy since capital markets generate two functions, first as decision variable for investor or as a facility for companies to get funding from investors, second as capital markets become facility for society to invest on financial instruments and one of them is stock [1].

Stock price index is indicator for stock price movement. The index is one of the references for investor to invest in capital markets, especially stock. One of the stock price indexes in Indonesia Stock Exchange (IDX) is Indeks Harga Saham Gabungan (IHSG) or Jakarta Composite Index. To be able to describe reasonable market condition, IHSG use all companies registered as index calculation components [2].

Stock is one of the popular instruments in financial market. It is because stock able to give high rate of profit return which interest the investors. However, stock trading has high rate of risk. The fluctuation of stock price affect the investor's decision on investing their capital. Mistake in decision making will result loss for investor. Thus, to minimize the high risk investor needs information as a reference for decision making of which stock they should buy, sell, and maintain[3].

There are 2 factors that has significant influence in stock price modelling, which previous state of stock that influence the current stock price and stock response towards latest information of stock [4]. Based on those factors, it can be concluded that the change of stock price follows the Markov chain. The Markov chain process is a stochastic process which make current price has influence to forecast the future stock price.

Forecasting is the best method to predict the future of stock price [5]. However, the rate of loss risk using this method is relatively still high due to the innacurate forecast output. The forecasting method used for forecast the fututre stock closing price for short term investment with low rate of. One of the model that can be used for forecasting stock price is geometric Brownian motion model or as known

IOP Conf. Series: Journal of Physics: Conf. Series **974** (2018) 012047 doi:10.1088/1742-6596/974/1/012047

as Wiener process. Geometric Brownian motion model is stochastic model with continous time, where the random variable follows the Brownian motion [5].

On the previous research the concept of geometric Brownian motion has been descibed by Dmouj [4]. Based on [4] it is described the concept of random walk, Brownian motion and analytical solution of model geometric Brownian motion model. In review [4] stated that the forecast of stock close price is develop using confident level and mean function of lognormal distribution. Other research that implement geometric Brownian motion model is research conducted by Omar dan Jaffar [5]. Their research forecasted the stock close price for several small companies registered in Malaysia stock exchange. The forecast is limited to short term investment. In their research it is proven that geometric Brownian motion model is accurate in forecasting the stock close price for two weeks period. It is proven by the small value of Mean Absolute Percetage Error (MAPE).

Based on that background, this research forecast the stock price of Jakarta Composite Index using the geometric Brownian motion. With utilizing the daily stock close price during January 2014 to December 2014 to forecast the stock price of January 2015. The stages for forecasting the stock price are calculating return value, Estimating the parameter, result collection of stock price forecast, then calculating the MAPE value. In this research 4 forecasts are obtained using geometric Brownian motion. Based on analysis and discussion, the MAPE value $\leq$20%.

2. Research Methodology

The methodology in conducting this research as follows:

2.1. Data Collection

At this stage stock price data collection is conducted by using the source of yahoo finance. The stock data that is used are several stock prices under the Jakarta Composite Index, which are: Charoen Pokphand Indonesia Tbk, Harum Energy Tbk, Media Nusantara Citra Tbk, PP London Sumatra Indonesia Tbk, Vale Indonesia Tbk, Indo Tambangraya Megah Tbk, Indocement Tunggal Prakasa Tbk. The stock price data that is used are daily stock close price during January 2014 to December 2014. Then the normality test is conducted on the stock close price data using SPSS software with Kolmogorov-Smirnov test.

2.2. Literature Study

At this stage, reference theories are collected for supporting fundamental, that is regarding the stock price, random walk, Brownian motion, proses stokastik, lemma Ito and geometric Brownian motion.

2.3. Stock Price Forecasting

At this stage the expected stock price formulation is conducted, the formulation of 95% confidence level and stock price forecasting using geometric Brownian motion. In forecasting the stock price the common formula of volality and log volatility is used.

To obtain expected stock price with 95% confidence level, stages conducted as follow:

a. Determine pdf lognormal.
b. Determine the lognormal distribution mean function.
c. Determine new stock price model.
d. Determine 95% confidence level.

After the new stock price model is obtained then stock close price forecast is conducted on seven stock. The stages in stock price forecasting as follow:

a. Normality test.
b. Calculate stock return.
c. Estimation of drift value (μ) and volatility (σ).
d. Forecast of stock price.
e. Calculate MAPE value.
f. Calculate 95% confidence level.

2.4. Conclusion

At this stage obtaining conclusion from analysis and discussion is done.

2.5. Simulation and Report Writing

At this stage simulation from the research is done and report writing follows after the simulation.

2.6. A subsection

Some text.

3. Research Output and Discussion

3.1. Pdf lognormal

At this section pdf lognormal distribution is obtained from pdf normal. For example, q is random variable which lognormal distributed then $v = ln\, q$ a is random variable which normally distributed with mean μ and variance σ^2 thus defining pdf from variable v as follows:

$$f(v) = \begin{cases} \frac{1}{\sigma\sqrt{2\pi}} e^{-\frac{(v-\mu)^2}{2\sigma^2}}, for -\infty < v < \infty, \\ 0 \qquad , for\ other\ v. \end{cases} \tag{1}$$

Using the pdf from variable v pdf from variable q is obtained. Based on one – one randomize variable theorem, result variable q pdf is obtained from the following equation [4]:

$$h(q) = \frac{f(v)dv}{dq}. \tag{2}$$

With substitution $v = ln(q)$,obtain

$$dv = d(\ln q),$$
$$= \frac{1}{q} dq.$$

The with substitution of $dv = \frac{1}{q} dq$ to equation (2), obtain

$$h(q) = \frac{f(v)\, dv}{dq}, \tag{3}$$

$$= \frac{\frac{1}{\sigma\sqrt{2\pi}} e^{\left(-\frac{1}{2}\left(\frac{\ln q-\mu}{\sigma}\right)^2\right)} \frac{1}{q} dq}{dq},$$

Thius pdf lognormal from variable q as follow [4]:

$$h(q) = \frac{1}{q\sigma\sqrt{2\pi}} e^{\left(-\frac{1}{2}\left(\frac{\ln q-\mu}{\sigma}\right)^2\right)}, \tag{4}$$

where:

μ : mean distribution of variable q lognormal

σ^2 : variance of lognormal sitribution of variable q.

3.2. Lognormal mean distribution

Using the lognormal distribution pdf which obtained from equation (4) then finding the mean of lognormal distribution is conductedsela. Lognormal distribution meand from variable Mean q is defined as [4]:

$$E(q) = \int_{-\infty}^{+\infty} q h(q) dq. \tag{5}$$

$$= \int_{-\infty}^{+\infty} \frac{1}{\sigma\sqrt{2\pi}} e^{\left(-\frac{1}{2}\left(\frac{\ln q-\mu}{\sigma}\right)^2\right)} dq. \tag{6}$$

Suppose if $y = \ln q - \mu$ then $dy = \frac{1}{q} dq$. Integral $\ln q = -\infty$ become $y = -\infty$ and $\ln q = +\infty$ become $y = +\infty$. Thus $E(q)$can be written as:

$$E(e^{y+\mu}) = \int_{-\infty}^{+\infty} \frac{1}{\sigma\sqrt{2\pi}} e^{y+\mu}\, e^{-\frac{y^2}{2\sigma^2}} dy, \tag{7}$$

$$= e^{\mu} e^{\frac{\sigma^2}{2}} \int_{-\infty}^{+\infty} \frac{1}{\sigma\sqrt{2\pi}}\, e^{\frac{-(y-\sigma^2)^2}{2\sigma^2}} dy. \tag{8}$$

IOP Conf. Series: Journal of Physics: Conf. Series **974** (2018) 012047 doi:10.1088/1742-6596/974/1/012047

Next, Suppose if $z = \frac{y-\sigma^2}{\sigma}$ then $dz = \frac{1}{\sigma}\, dy$. Integral $y = -\infty$ become $z = -\infty$ and $y = +\infty$ become $z = +\infty$. Thus $E(e^{y+\mu})$ can be written as:

$$E\left(e^{z\sigma+\sigma^2+\mu}\right) = e^{\mu+\frac{\sigma^2}{2}} \int_{-\infty}^{+\infty} \frac{1}{\sqrt{2\pi}} e^{\frac{-z^2}{2}}\, dz. \tag{9}$$

Integral on equation (9) is pdf from standard normal distribution and has value of 1, Thus mean fuction of variable q lognormal distribution is

$$E(q) = e^{\mu+\frac{\sigma^2}{2}}, \tag{10}$$

where:

μ : mean distribution of variable q lognormal

σ^2 : variance of lognormal sitribution of variable q.

3.3. Expected stock price

At this section expected stock price formulation is conducted. Based on [4], equation

$$\ln S_t = \ln S_0 + \left(\mu - \frac{1}{2}\sigma^2\right)t + \sigma B_t, \tag{11}$$

is Brownian motion with drift, where B_t is Brownian motion in time t with $\mu = 0$ and has value of $\varepsilon\sqrt{t}$ [4]. Whereas Brownian motion definition with drift as follow [4]:

$$B_t = \mu t + \sigma W_t, \tag{12}$$

where t represents time and W_t adalah is random walk process with scale for big value of n and has value of $\varepsilon\sqrt{t}$ [4]. Thus equation (11) and (12) can be written as :

$$\begin{gathered} \ln S_t = \ln S_0 + \left(\mu - \frac{1}{2}\sigma^2\right)t + \sigma B_t. \\ \equiv \\ B_t = \mu t + \sigma W_t. \end{gathered} \tag{13}$$

First step to obtain expected stock price equation is by calculating mean value and variance of Brownian motion with drift. Elaboration of the first step as follow

3.3.1. Calculate mean value of Brownian motion with drift

$$\begin{aligned} E(B_t) &= E(\mu t + \sigma W_t), \\ &= E(\mu t) + E(\sigma W_t), \\ &= \mu t + E(\sigma W_t). \end{aligned} \tag{14}$$

Because $E(W_t) = 0$, then

$$E(B_t) = \mu t. \tag{15}$$

3.3.2. Calculate variance value of Brownian motion with drift

$$\begin{aligned} Var(B_t) &= Var(\mu t + \sigma W_t), \\ &= E[(\mu t + \sigma W_t)^2] - [E(\mu t + \sigma W_t)]^2. \\ E[(\mu t + \sigma W_t)^2] &= E[(\mu t)^2] + E(2\mu t\sigma W_t) + E[(\sigma W_t)^2], \\ &= (\mu t)^2 + 2\mu t\sigma E(W_t) + \sigma^2 E\left(W_t^{\,2}\right), \\ &= (\mu t)^2 + 0 + \sigma^2 E(W_t^{\,2}). \end{aligned} \tag{16}$$

Based on definition of variance:

$$\begin{aligned} Var(W_t) &= E\left(W_t^{\,2}\right) - [E(W_t)]^2, \\ E\left(W_t^{\,2}\right) &= Var(W_t) + [E(W_t)]^2, \\ &= t, \end{aligned} \tag{17}$$

Then:

$$\begin{aligned} E[(\mu t + \sigma W_t)^2] &= (\mu t)^2 + \sigma^2 t. \\ [E(\mu t + \sigma W_t)]^2 &= [E(\mu t) + E(\sigma W_t)]^2, \\ &= [\mu t + \sigma E(W_t)]^2 \\ &= (\mu t)^2, \end{aligned} \tag{18}$$

Thus obtain:

$$Var(B_t) = E[(\mu t + \sigma W_t)^2] - [E(\mu t + \sigma W_t)]^2, \\ = (\mu t)^2 + \sigma^2 t - (\mu t)^2, \\ = \sigma^2 t. \tag{19}$$

Based on the verification above, Brownian motion with drift is normal distributed equation with mean μt and variance $\sigma^2 t$ thus $ln\, S_t$ normally distributed with mean $ln\, S_{t-1} + \left(\mu - \frac{\sigma^2}{2}\right) t$ and variance $\sigma^2 t$ [4]. Thus expected $E(S_t)$ future stock price when t is

$$E(S_t) = e^{\ln S_0 + \left(\mu - \frac{\sigma^2}{2}\right)t + \sigma^2 t}, \\ = e^{\ln S_0} e^{\left(\mu - \frac{\sigma^2}{2}\right)t + \sigma^2 t}, \\ = S_0 e^{\left(\left(\mu + \frac{\sigma^2}{2}\right)t\right)} \tag{20}$$

where:

S_0 : Actual beginning stock price
μ : drift of stock price
σ : volatility of stock price.

3.4. Confidence Level

For testing the forecast accuracy of geometric Brownian motion then 95% confidence level is implemented. Based on the following equation:

$$\ln S_t = \ln S_0 + \left(\mu - \tfrac{1}{2}\sigma^2\right) t + \sigma B_t. \tag{21}$$

With S_t is actial stock price when t, μ and σ are estimated parameter from big sample from population. Thus with 95% confidence level shows the actual stock price in 95% confidence level.

By using the parameter of mean $ln\, S_0 + \left(\mu - \frac{\sigma^2}{2}\right) t$ and variance of $\sigma^2 t$ which obtained form previous discussion of expected stock price sub chapter, so that 95% confidence level and variable $ln\, S_t$ as follow:

$$\ln S_0 + \left(\mu - \frac{\sigma^2}{2}\right) t - 1{,}96\sigma\sqrt{t} \le ln\, S_t \le \ln S_0 + \left(\mu - \frac{\sigma^2}{2}\right) + 1{,}96\sigma\sqrt{t}, \\ e^{\ln S_0 + \left(\mu - \frac{\sigma^2}{2}\right)t - 1{,}96\sigma\sqrt{t}} \le S_t \le e^{\ln S_0 + \left(\mu - \frac{\sigma^2}{2}\right) + 1{,}96\sigma\sqrt{t}}, \tag{22}$$

where:

S_0 : beginning stock price
S_t : stock price when t
μ : stock price drift
σ : Stock price volatility

3.5. Normality test

At this stage, normality test of stock price data during January 1st 2014 to December 31st 2014 period is conducted. Normality test is conducted to find whether the stock data is normally distributed or not.

Based on Kolmogorov – Smirnov normality test, shown $p \ge 0{,}05$, so it can be concluded that the stock price is normally distributed and feasible to do stock price forecast on the data.

3.6. Stock price forecasting

At this stage stock price is forecasted on several registered companies, they are stock of Charoen Pokphand Indonesia Tbk, Harum Energy Tbk, Media Nusantara Citra Tbk, PP London Sumatra Indonesia Tbk, Vale Indonesia Tbk, Indo Tambangraya Megah Tbk and Indocement Tunggal Prakasa Tbk. Data used to estimate the is daily stock close price during January 1st 2014 to December 31st 2014 period [6], whereas forecasting is done for January 1st 2015 period.

IOP Conf. Series: Journal of Physics: Conf. Series **974** (2018) 012047 doi:10.1088/1742-6596/974/1/012047

Hypothesis Test Summary

	Null Hypothesis	Test	Sig.	Decision
1	The distribution of charoen is normal with mean 0,00 and standard deviation 0,02.	One-Sample Kolmogorov-Smirnov Test	,176	Retain the null hypothesis.
2	The distribution of Harum is normal with mean -0,00 and standard deviation 0,02.	One-Sample Kolmogorov-Smirnov Test	,067	Retain the null hypothesis.
3	The distribution of MNC is normal with mean 0,00 and standard deviation 0,02.	One-Sample Kolmogorov-Smirnov Test	,078	Retain the null hypothesis.
4	The distribution of PP is normal with mean 0,00 and standard deviation 0,02.	One-Sample Kolmogorov-Smirnov Test	,260	Retain the null hypothesis.
5	The distribution of Vale is normal with mean 0,00 and standard deviation 0,02.	One-Sample Kolmogorov-Smirnov Test	,115	Retain the null hypothesis.
6	The distribution of Indo is normal with mean -0,00 and standard deviation 0,02.	One-Sample Kolmogorov-Smirnov Test	,169	Retain the null hypothesis.
7	The distribution of Indocement is normal with mean 0,00 and standard deviation 0,02.	One-Sample Kolmogorov-Smirnov Test	,073	Retain the null hypothesis.

Asymptotic significances are displayed. The significance level is ,05.

Figure 1. Normality test

These are the stages in stock price forecasting.

3.6.1. Calculating stock return

At this stage monthly stock return calculation is conducted during January 2014 to December 2014 period using the following stock return equation [7]:

$$R_t = \frac{S_t - S_{t-1}}{S_{t-1}}, \tag{23}$$

where:

R_t : stock return when t

S_t : stock price when t

S_{t-1} : stock price when $t-1$.

3.6.2. Drift value (μ) and volatility (σ) estimation

At this stage drift and volatility are estimated. Drift value and volatility are constant parameter of stock that is used for forecasting January 2015 stock price. Drift formulation as defined [7]:

$$\mu = \frac{1}{M\delta t}\sum_{t=1}^{M} R_t, \tag{24}$$

where:

μ : drift

R_t : stock return

M : amount of stock return.

To obtain drift value, mean calculation of drift value is required monthly from January 2014 to December 2014.

After the drift value is obtained, then volatility value is calculated. The volatility formulation that is used in this research is volatility common formula and log volatility. The equation of volatility and log volatility as defined [7]:

$$\sigma_1 = \sqrt{\frac{1}{(M-1)\delta t}\sum_{t=1}^{M}(R_t - \bar{R})^2}\ , \tag{25}$$

$$\sigma_2 = \sqrt{\frac{1}{(N-1)\delta t}\sum_{t=2}^{N}(\log S_t - \log S_{t-1})^2}, \tag{26}$$

where:
σ_1 : volatility
σ_2 : log volatility
$\bar{R}$: mean of stock return
M : amount of stock return data
N : amount of stock data.

3.6.3. Stock price forecasting
At this stage stock price forecasting is conducted on seven stock close price on January 2015 using geometric Brownian motion model. Forecast 1 and Forecast 2 using geometric Brownian motion model by implementing common volattility and log volatility equation. On forecast 1 and forecast 2 to forecast stock price on *t* time use the beginning actual stock price on $t - 1$ time. Equation implemented for forecast 1 and forecast 2 as [4]:

$$F_t = S_{t-1} + e^{\left(\mu - \frac{1}{2}\sigma^2\right)t + \sigma B_t}, \tag{27}$$

where:
F_t : stock price forecast when t
S_{t-1} : actual stock price when $t - 1$
μ : drift
σ : volatility, where $\sigma = \sigma_1$ or $\sigma = \sigma_2$
B_t : $\varepsilon\sqrt{t}$

Where on forecast 3 and forecast 4 geometric Brownian motion model with common volatility equation and log volatility, however to forecast stock price when t use the beginning value which taken from forecast $t - 1$. Equation used for forecast 3 and forecast 4 as follow:

$$F_t = F_{t-1} + e^{\left(\mu - \frac{1}{2}\sigma^2\right)t + \sigma B_t}, \tag{28}$$

with $F_0$is actual stock price.

3.6.4. MAPE calculation
To determine the forecast accuracy can be determined by calculating MAPE of forecast 1, forecast 2, forecast 3, and forecast 4. MAPE equation as defined [10]:

$$MAPE = \frac{\sum_{t=1}^{N}\frac{|S_t - F(S_t)|}{S_t}}{N}, \tag{29}$$

where:
S_t : actual stock price when t
$F(S_t)$: stock price forecast when t
N : amount of stock price data.

On figure 2 and figure 3 show the graph of stock price forecast of Media Nusantara Citra Tbk using geometric Brownian motion.

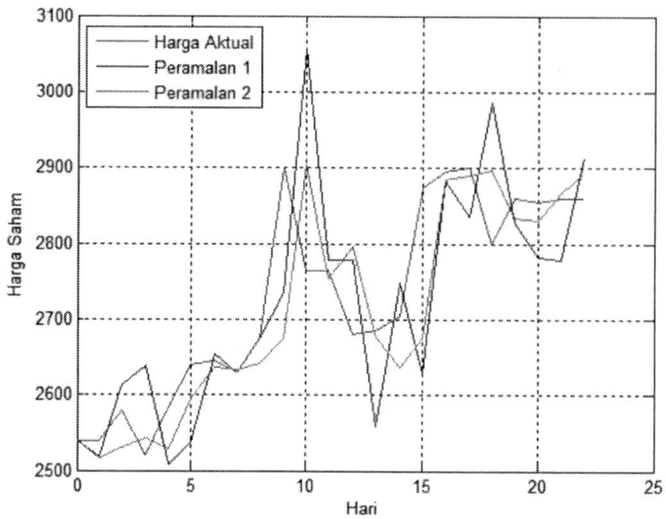

Figure 2. Forecast graph based on beginning value which taken from previous actual stock price

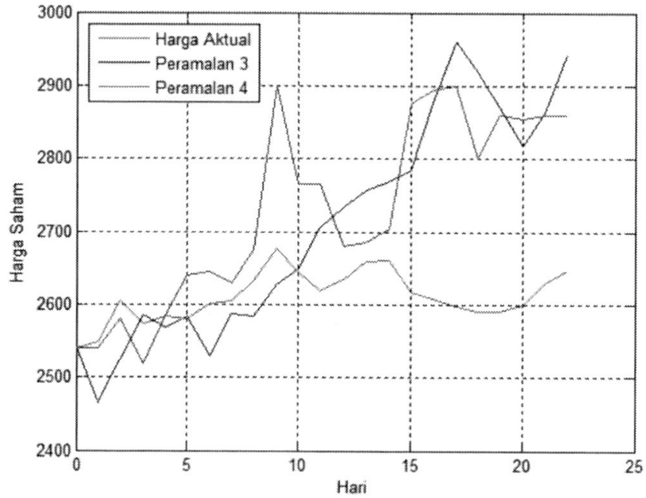

Figure 3. Forecast graph based on beginning value which taken ftom previous forecast

On table 1 shown the MAPE value of forecasts.

Table 1. Forecast MAPE

Stock	MAPE value of each forecast			
	Forecast 1	Forecast 2	Forecast 3	Forecast 4
Charoen Pokphand Indonesia	2,0935%	1,1675%	2,9320%	1,6933%
Harum Energy	2,2033%	1,4982%	3,9128%	3,3974%
Media Nusantara Citra	3,0529%	1,9875%	2,6096%	4,6496%
PP London Sumatra Indonesia	1,8726%	1,6479%	3,0614%	2,9718%
Vale Indonesia	1,7580%	1,5921%	4,7672%	4,3637%
Indo Tambang-raya Megah	2,7305%	2,1767%	4,6661%	3,5049%
Indocement Tunggal Prakasa	2,5349%	1,9187%	4,1217%	4,8249%

Based on [4], with equation

$$F_t = F_{t-1} + e^{\left(\mu-\frac{1}{2}\sigma^2\right)t+\sigma B_t}, \tag{30}$$

where $F_0 = S_0$ is actual stock price and $B_t = \varepsilon\sqrt{t}$ [4]. Simualtion is made for stock price forecasting with 1000 realization of trajectory that may be from geometric Brownian motion, with every realization of trajectory usig the equation of $F_t = F_{t-1} + e^{\left(\mu-\frac{1}{2}\sigma^2\right)t+\sigma B_t}$ with 22 iterations. Suppose if i is every trajectory realization that may be from geometric Brownian motion. Thus for $i = 1$

$$F_1 = S_0 + e^{\left(\mu-\frac{1}{2}\sigma^2\right)t+\sigma\varepsilon\sqrt{t}}, \tag{31}$$

$$F_2 = F_1 + e^{\left(\mu-\frac{1}{2}\sigma^2\right)t+\sigma\varepsilon\sqrt{t}}, \tag{32}$$

$$\vdots$$

$$F_{22} = F_{21} + e^{\left(\mu-\frac{1}{2}\sigma^2\right)t+\sigma\varepsilon\sqrt{t}}. \tag{33}$$

For $i = 2$

$$F_1 = S_0 + e^{\left(\mu-\frac{1}{2}\sigma^2\right)t+\sigma\varepsilon\sqrt{t}}, \tag{34}$$

$$F_2 = F_1 + e^{\left(\mu-\frac{1}{2}\sigma^2\right)t+\sigma\varepsilon\sqrt{t}}, \tag{35}$$

$$\vdots$$

$$F_{22} = F_{21} + e^{\left(\mu-\frac{1}{2}\sigma^2\right)t+\sigma\varepsilon\sqrt{t}}. \tag{36}$$

For $i = 3$

$$F_1 = S_0 + e^{\left(\mu-\frac{1}{2}\sigma^2\right)t+\sigma\varepsilon\sqrt{t}}, \tag{37}$$

$$F_2 = F_1 + e^{\left(\mu-\frac{1}{2}\sigma^2\right)t+\sigma\varepsilon\sqrt{t}}, \tag{38}$$

$$\vdots$$

$$F_{22} = F_{21} + e^{\left(\mu-\frac{1}{2}\sigma^2\right)t+\sigma\varepsilon\sqrt{t}}. \tag{39}$$

Replication above is done until $i = 1000$.

On figure 4 and figure 5 the simulation shown simualtion output of stock price forecast of Media Nusantara Citra Tbk until $i = 1000$ using common volatility equation and log volatility.

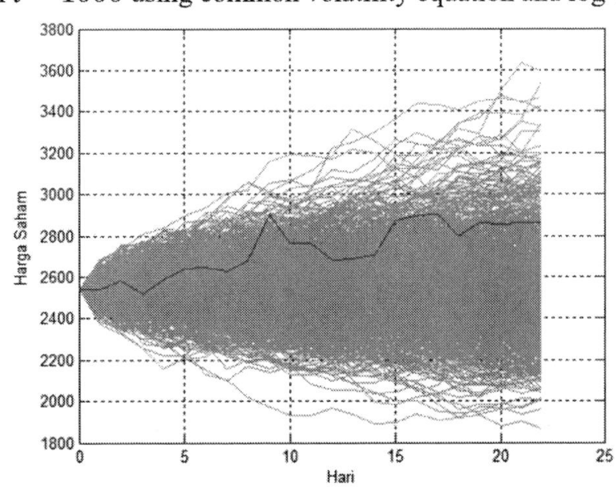

Figure 4. Stock price simulation model using common volatility equation

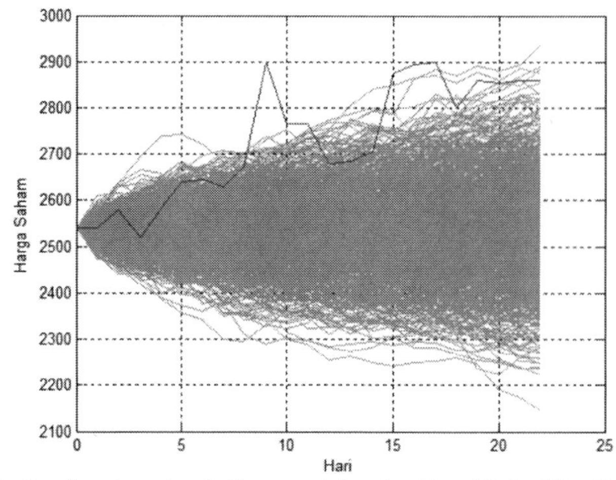

Figure 5. Stock price simulation model using Log Volatility Equation

3.7. Simulation Model Analysis

Based on *geometric* Brownian motion simulation model with 1000 possible trajectory, At this stage calculation of expected stock price and 95% confidence level for each stock price data is conducted. Based on [4], to calculate expected stock price when $t = 22$ use the equation (7), as follow:

$$E(S_t) = S_0 e^{\left(\left(\mu + \frac{\sigma^2}{2}\right)t\right)}. \tag{40}$$

Wherwas to calculate 95% confidence level from stock price required the following equation (8):

$$e^{\ln S_0 + \left(\mu - \frac{\sigma^2}{2}\right)t - 1{,}96\sigma\sqrt{t}} \le S_t \le e^{\ln S_0 + \left(\mu - \frac{\sigma^2}{2}\right) + 1{,}96\sigma\sqrt{t}}. \tag{41}$$

On table 2 and table 3 shown the confidence levelof 95% from seven stock price data.

Table 2. Confidence level of stock price with common volatility equation

Stock	95% Confidence Level			
	$E(S_t)$	L	S_t	U
Charoen Pokphand Indonesia	3853	3231	3955	4527
Harum Energy	1606	1533	1520	1722
Media Nusantara Citra	2555	2097	2860	3055
PP London Sumatra Indonesia	1914	1563	1840	2299
Vale Indonesia	3802	3036	3450	4007
Indo Tambangraya Megah	14820	12026	16750	18155
Indocement Tunggal Prakasa	25648	21442	23000	30214

Table 3. Confidence level of stock price using log volatility equation

Stock	95% Confidence Level			
	$E(S_t)$	L	S_t	U
Charoen Pokphand Indonesia	3841	3560	3955	4132
Harum Energy	1601	1476	1520	1730
Media Nusantara Citra	2545	2336	2860	2763
PP London Sumatra Indonesia	1907	1737	1840	2071
Vale Indonesia	3784	3434	3450	4150
Indo Tambangraya Megah	14759	13479	16750	16094
Indocement Tunggal Prakasa	25569	23673	23000	27535

Based on table 2 and table 3, 95% confidence level using log volatility equation tends to have narrow interval compared to 95% confidence level using common volatility equation. Thus on the simulation there is several actual stock price located outside trajectory realization that may be from geometric Brownian motion.

4. Conclusions

Based on the analysis and discussion, it can be concluded that short term forecasting which are forecast 1 and forecast 2 have MAPE value of $\leq$ **20%**. Whereas long term forecast which are forecast 3 and forecast 4 are dissatisfactory because actual stock price located outside green line or outside the area of forecast value.

References

[1] "*http://www.idx.co.id/id/id-id/beranda/informasi/bagiinvestor/pengan tarpasarmodal.aspx*" accessed on 6 February 2015.
[2] "*http://www.idx.co.id/id/id-id/beranda/informasi/bagiinvestor/indeks. aspx*" accessed on 6 February 2015.
[3] "*http://www.idx.co.id/id/id-id/beranda/informasi/bagiinvestor/saham. aspx*" accessed on 6 February 2015.
[4] Dmouj, A. 2006. "*Stock price modelling:Theory and Practice*". Vrije Universiteit Faculty of sciences Amsterdam, The Netherlands.
[5] Omar, A., Jaffar, M.M., 2014. "*Forecasting Share Price of Small Size Companies in Bursa Malaysia Using Geometric Brownian Motion*". An Journal International, Applied Mathematics & Information Sciences 8, No, **1**:107-112. Faculty of Computer and Mathematical Sciences, Universiti Teknologi MARA, 40450 Selangor, Malaysia.
[6] "*finance.yahoo.com*" accessed on 12 January 2015.
[7] Willmot, P., 2007. "Introduces Quabtitative Finance". 2nd Edition, John Wiley & Son, Ltd, Chichester.
[8] Lawrence, K. D., Klimberg R. K., & Lawrence S. M. 2009. "*Fundamentals of forecasting using excel*". Industrial Press Inc, America.

Optimization of planting pattern plan in Logung irrigation area using linear program

Wasis Wardoyo, Setyono

Civil Engineering Department, Faculty of Civil Engineering and Planning, Institut Teknologi Sepuluh Nopember (ITS), Jl. Arief Rahman Hakim, Surabaya 60111

E-mail: w4515wardoyo@gmail.com

Abstract. Logung irrigation area is located in Kudus Regency, Central Java Province, Indonesia. Irrigation area with 2810 Ha of extent is getting water supply from Logung dam. Yet, the utilization of water at Logung dam is not optimal and the distribution of water is still not evenly distributed. Therefore, this study will discuss about the optimization of irrigation water utilization based on the beginning of plant season. This optimization begins with the analysis of hydrology, climatology and river discharge in order to determine the irrigation water needs. After determining irrigation water needs, six alternatives of planting patterns with the different early planting periods, i.e. 1st November, 2nd November, 3rd November, 1st December, 2nd December, and 3rd December with the planting pattern of rice-secondary crop-sugarcane is introduced. It is continued by the analysis of water distribution conducted using linear program assisted by POM-Quantity method for Windows 3 with the reliable discharge limit and the available land area. Output of this calculation are to determine the land area that can be planted based on the type of plant and growing season, and to obtaine the profits of harvest yields. Based on the optimum area of each plant species with 6 alternatives, the most optimum area was obtained at the early planting periods on 3rd December with the production profit of Rp 113.397.338.854,- with the planting pattern of rice/ beans / sugarcane-rice / beans / sugarcane-beans / sugarcane.

1. Introduction

Indonesia is one of the country in the world that has a large agriculture field. Horticultural crops have important and strategic roles in national and regional development, such as improving food sustainability, gross domestic product (GDP), income sources, and regional and national economies. Agriculture has become an attractive factor for upstream and downstream industry growth that huge contribution to national economic growth. One of the irigation area that should support this task is Logung Irrigation Area. This located in the Logung River, Kudus Regency, Central Java. Logung Irrigation Area, which has an area of 2810 Ha gets water supply from Logung River. Recently, the performance of this irrigation area tend to decrease because of uneven distribution of water. It becomes even worst by unpredictable rainy seasons that cause drought during the dry season and flooded during the rainy season. Therefore, such a method is introduced in order to solve this problem and to increase the crops productions, especially during dry season.

One of the method introduced to improve agricultural products in Logung irrigation area is by regulating a best way of water distribution. This method is done based on linier programming approach. By this methods, the water demand can be adjusted depend upon the water availability in Logung Dam. The aplicability of linier programming in case of water resources optimization has already introduced by Frizzone et.al [1], Jakovskis [2] Wafa Dipalla [3], Anieting [4] Pearson and McRoberts [5] and Suhardono[6]. The water demmand is determined by planting pattern. Optimal planting pattern arrangement simulated based on the type of plants and water availability. The planting pattern will be optimum when the optimum agricultural product is achieved. Thus, this study will discuss the optimization of planting pattern in Logung irrigation area using linier programming.

1.1. Problem analysis
The Logung irrigation area is potential to increase its agriculture products. One of the capital of this area to be improved is the availability of water and the characteristic of catchment area [7, 8]. The water availability in Logung catchment area is slightly fluctuated between dry and rainy season. By introducing some scenario based on starting time of planting, area of planting, type of planting and net field rate of each crops, then the irrigation water needs of each alternative planting pattern in the Logung irrigation area can be determined. Furthermore, the maximum profit production of each the planting pattern is calculated. The better planting pattern is determined based on the value of production's profit where the higher will be the better.

1.2. Aims
There are three main aims of this study i.e. adjusting the water availability, calculating the irrigation water needs and calculating the maximum profit production of each planting pattern. Six starting time of planting pattern is introduced: 1st November, 2nd November, 3rd November, 1st December, 2nd December, and 3rd December respect to planting period of padi. The water availability is adjusted based on hydrological data, topographycal data and some supported data. The maximum profit production is found by multiplying the amount of each product and the unit price of each crops.

1.3. Limitation of analysis
On analysing this case, some limitation is done i.e. Scope of analysis in this study are: the data used are secondary data in the field, his study is only analyzing water for irrigation and not considering sedimentation, all irrigation channels are assumed in a good condition so that there is no water loss due to channel damage or wiretapping and selected planting pattern is rice, beans, and sugarcane.

2. Methods
This part explaines the step on reaching the aim of research. Starting with collecting, reading and understanding some related references, follows by data collection, application of some formula and data analysis of water need. Finally it ended with optimization step, that consist of variation on the beginning of planting pattern and calulation of harvest yields benefit.

2.1. Literature review
On reaching the aim of study, some basic knowledge should be understood well. The basic knowledge has been learned on undergraduate lectures that consist of hydrology, hydraulics, irrigation and optimazion stage aspects [9]. The basic design of channel and its operation sourced to design criterium of irrigation channel [10], while the operation of Logung Dam is stated on Design of Logung Dam [11]. The aplicability of linier programe on optimation of water resources case is already proved by many experts [1,2,3,4,5,6] as mentions above.

2.2. Data collection and analysis
The data used in this study consist of Catchment and Irrigation area data, Hydrology and Hydraulic Data. Firstly the data of the catchment area and Logung river. The most important thing is the data of discharge. The 10 days measurement data of discharge from 2006 up to 2015 are available. Based on this data then the discharge distribution can be determined. Total area of irrigation field as the study area is an important parameter to be known, besides the type of crops, crops limitation and crop's water need . By knowning this, the scenario of optimation can be determined. The climatology data is also needed to calculate water balance not only in the field but also in irrigation intake and dam.

2.3. Optimization of planting patterns
Based on the result of water needs analysis from each alternative of planting pattern and reliable volume, it becomes the input of Linear Program to obtain the optimum planting pattern. The stage of finding optimum planting pattern is shown on Figure 1.

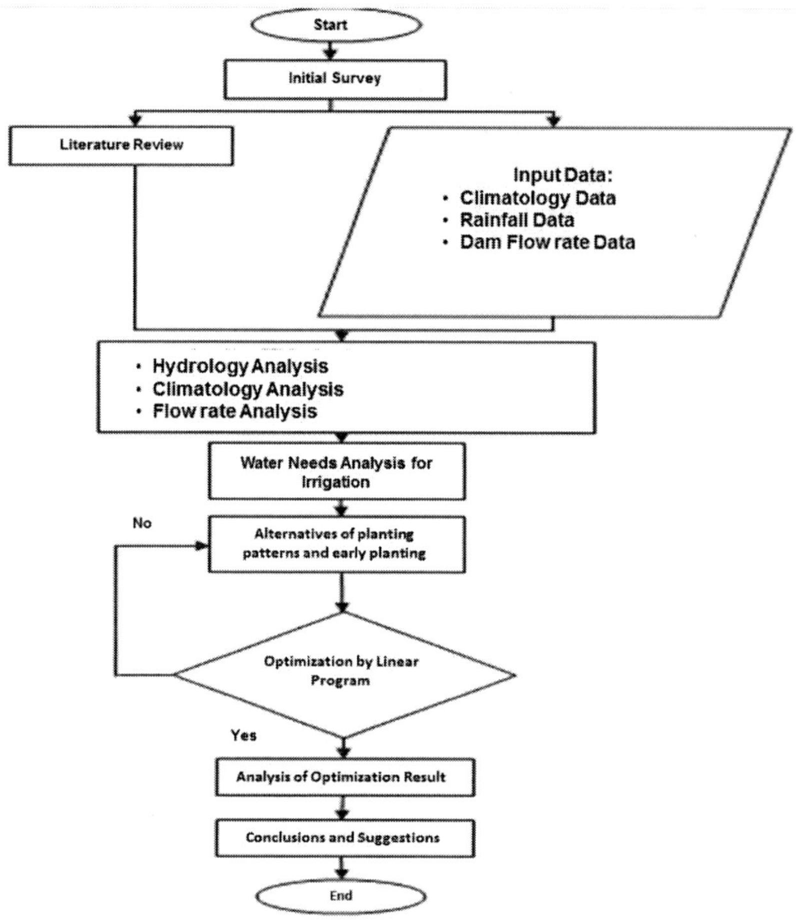

Figure 1. Optimum planting pattern

3. Results and discussion

After calculation, some result should be analysis in order to support the optimization model. The outcme analysis are reliable discharge, water balance, waterneed for irrigation and harvest yields.

3.1. Hydrological analysis

This part explaines the process of hydrologic analysis. Based on discharge data of Logung River (**Table 1**) that collected from measurements, it can be analysis the caracters of discharge. Refers to this characters, then distribution type and design discharge can be determined. By knowing design discharge then prediction of discharge can be estimated

3.1.1. Calculation of the reliable discharge

The discharge data was collected from the discharge measurement in Logung River on 2006 up to 2015.

Table 1. Logung River discharge data for 10 days period (m^3/sec)

Month		Year									
		2006	2007	2008	2009	2010	2011	2012	2013	2014	2015
January	1	1,242	1,237	0,998	2,200	6,930	15,730	13,927	4,657	13,818	11,212
	2	0,995	0,664	0,585	6,100	7,480	17,866	10,236	1,530	16,404	9,296
	3	1,513	0,681	0,540	9,827	8,000	12,521	4,583	5,643	32,768	8,522
February	1	3,278	0,380	0,605	9,400	8,780	14,837	2,477	0,000	13,489	8,774
	2	2,860	0,445	0,591	8,456	6,060	19,147	1,820	0,000	9,661	8,602
	3	2,231	0,613	0,593	7,275	6,450	18,547	1,349	3,611	9,520	9,125
March	1	2,322	0,809	0,611	6,380	2,350	27,767	1,068	0,105	10,161	8,567
	2	1,176	0,519	0,556	4,900	0,680	22,619	0,676	0,014	11,118	5,837
	3	0,907	0,390	0,425	5,064	0,800	17,696	0,383	1,705	9,851	7,134
April	1	0,862	0,366	0,574	6,380	0,740	16,180	0,234	3,973	13,030	5,770
	2	0,838	0,422	0,469	5,350	0,770	15,223	0,091	4,177	6,560	10,126
	3	0,815	0,435	0,483	5,360	0,730	14,033	0,029	0,010	6,025	10,598
May	1	0,714	0,433	0,518	5,510	0,680	12,892	0,000	10,913	7,949	8,256
	2	0,652	0,430	0,628	5,770	0,820	15,444	0,000	8,167	9,012	8,509
	3	0,619	0,388	0,369	4,527	0,864	13,256	0,000	1,857	10,494	10,472
June	1	0,556	0,361	0,389	5,280	0,380	11,504	0,419	1,809	7,400	11,098
	2	0,521	0,378	0,483	5,290	0,070	10,236	0,401	1,140	7,400	11,165
	3	0,500	0,390	0,570	0,700	0,010	8,643	0,364	0,823	14,518	10,829
July	1	0,480	0,377	0,516	0,350	0,000	5,561	0,348	0,585	8,656	12,181
	2	0,472	0,359	0,433	0,000	0,000	2,552	0,345	0,471	9,226	12,181
	3	0,459	0,350	0,445	0,000	0,364	1,383	0,344	0,423	9,520	12,181
August	1	0,430	0,368	0,482	0,000	0,500	0,295	0,366	8,256	9,701	12,181
	2	0,416	0,382	0,518	0,000	0,680	0,166	0,425	8,649	10,301	12,181
	3	0,407	0,350	0,528	3,055	0,900	0,166	0,434	8,613	10,829	12,181
Sept	1	0,409	0,376	0,573	0,170	0,870	0,166	0,463	10,306	8,233	10,829
	2	0,444	0,380	0,648	0,100	0,580	0,166	0,403	10,829	8,958	10,829
	3	0,458	0,380	0,694	0,100	0,580	0,166	0,352	10,829	9,482	10,829
October	1	0,443	0,380	0,711	0,100	0,490	0,000	0,357	8,256	9,520	10,144
	2	0,496	0,380	0,828	0,100	1,230	0,150	0,307	7,405	10,104	9,858
	3	0,533	0,380	0,917	0,300	1,409	1,409	0,306	6,613	10,169	9,520
Nov	1	0,586	0,376	0,901	0,180	1,740	5,641	2,353	5,935	10,367	9,520
	2	0,559	0,343	1,057	0,200	3,340	10,589	2,476	5,878	10,829	9,520
	3	0,779	0,346	1,377	0,650	10,420	9,446	3,049	5,373	9,913	9,520
Dec	1	0,860	0,368	1,017	2,200	0,800	14,366	8,221	7,151	10,828	12,898
	2	1,839	0,530	1,075	2,200	1,970	11,722	7,137	11,420	7,596	12,268
	3	0,000	0,662	2,412	2,200	3,491	13,092	7,377	13,997	8,865	11,134

Source: Calculation Result

The disharge data was sorted from the largest to the smallest to determine the probability of being fulfilled by 80% or unfulfilled by 20%.

Table 2. Calculation records of reliable discharge (m^3/dt)

Rank	Month																	
	January			February			March			April			May			June		
	1	2	3	1	2	3	1	2	3	1	2	3	1	2	3	1	2	3
1	15,730	17,866	32,768	14,837	19,147	18,547	27,767	22,619	17,696	16,180	15,223	14,033	12,892	15,444	13,256	11,504	11,165	14,518
2	13,927	16,404	12,521	13,489	9,661	9,520	10,161	11,118	9,851	13,030	10,126	10,598	10,913	9,012	10,494	11,098	10,236	10,829
3	13,818	10,236	9,827	9,400	8,602	9,125	8,567	5,837	7,134	6,380	6,560	6,025	8,256	8,509	10,472	7,400	7,400	8,643
4	11,212	9,296	8,522	8,780	8,456	7,275	6,380	4,900	5,064	5,770	5,350	5,360	7,949	8,167	4,527	5,280	5,290	0,823
5	6,930	7,480	8,000	8,774	6,060	6,450	2,350	1,176	1,705	3,973	4,177	0,815	5,510	5,770	1,857	1,809	1,140	0,700
6	4,657	6,100	5,643	3,278	2,860	3,611	2,322	0,680	0,907	0,862	0,838	0,730	0,714	0,820	0,864	0,556	0,521	0,570
7	2,200	1,530	4,583	2,477	1,820	2,231	1,068	0,676	0,800	0,740	0,770	0,483	0,680	0,652	0,619	0,419	0,483	0,500
8	1,242	0,995	1,513	0,605	0,591	1,349	0,809	0,556	0,425	0,574	0,469	0,435	0,518	0,628	0,388	0,389	0,401	0,390
9	1,237	0,664	0,681	0,380	0,445	0,613	0,611	0,519	0,390	0,366	0,422	0,029	0,433	0,430	0,369	0,380	0,378	0,364
10	0,998	0,585	0,540	0,000	0,000	0,593	0,105	0,014	0,383	0,234	0,091	0,010	0,000	0,000	0,000	0,361	0,070	0,010

Rank	Month																	
	July			August			September			October			November			December		
	1	2	3	1	2	3	1	2	3	1	2	3	1	2	3	1	2	3
1	12,181	12,181	12,181	12,181	12,181	12,181	10,829	10,829	10,829	10,144	10,104	10,169	10,367	10,829	10,420	14,366	12,268	13,997
2	8,656	9,226	9,520	9,701	10,301	10,829	10,306	10,829	10,829	9,520	9,858	9,520	9,520	10,589	9,913	12,898	11,722	13,092
3	5,561	2,552	1,383	8,256	8,649	8,613	8,233	8,958	9,482	8,256	7,405	6,613	5,935	9,520	9,520	10,828	11,420	11,134
4	0,585	0,472	0,459	0,500	0,680	3,055	0,870	0,648	0,694	0,711	1,230	1,409	5,641	5,878	9,446	8,221	7,596	8,865
5	0,516	0,471	0,445	0,482	0,518	0,900	0,573	0,580	0,580	0,490	0,828	1,409	2,353	3,340	5,373	7,151	7,137	7,377
6	0,480	0,433	0,423	0,430	0,425	0,528	0,463	0,444	0,458	0,443	0,496	0,917	1,740	2,476	3,049	2,200	2,200	3,491
7	0,377	0,359	0,364	0,368	0,416	0,434	0,409	0,403	0,380	0,380	0,380	0,533	0,901	1,057	1,377	1,017	1,970	2,412
8	0,350	0,345	0,350	0,366	0,382	0,407	0,376	0,380	0,352	0,357	0,307	0,380	0,586	0,559	0,779	0,860	1,839	2,200
9	0,348	0,000	0,344	0,295	0,166	0,350	0,170	0,166	0,166	0,100	0,150	0,306	0,376	0,343	0,650	0,800	1,075	0,662
10	0,000	0,000	0,000	0,000	0,000	0,166	0,166	0,100	0,100	0,000	0,100	0,300	0,180	0,200	0,346	0,368	0,530	0,000

Source: Calculation Result

3.1.2. Water balance

Climatological calculation is calculated in order to determine the magnitude of plant evapotranspiration. The parameters should be consider on this calculation are air temperature, wind speed, relative humidity and duration of solar irradiance. The available data of climatology is from 1993 up to 2014. The result of calculation called as the potential evapotranspiration is shown at Table 3. Based on this table, the water balance can be estimated.

3.2. Water needs for irrigation

3.2.1. Effective Rainfall Calculations

Firstly, the rainfall station should be available on study area. There are three rainfall station : Rahwatu Station, Gembong Station, Tanjungrejo Station. To know the amount of surface runoff, the area of rainfall influence can be determined using Thiessen Polygon method. If the land use and rainfall intensity are known, then free surface discharge can be calculated.

Table 3. Climatological data and potential evaporation calculation every month in 1993 - 2014

No	Data	Unit	Month											
			Jan	Feb	Mar	Apr	May	Jun	Jul	Aug	Sep	Oct	Nov	Dec
I	Data													
1	Average Temperature (T)	C	23,09	23,15	23,49	23,78	23,74	23,34	22,94	23,16	23,90	24,56	24,36	23,58
2	Average Solar Irradiance	%	19,05	21,14	33,91	35,55	49,27	47,14	55,14	60,23	61,27	59,14	42,59	24,41
3	Average Relative Humidity (RH)	%	82,41	82,05	82,32	81,86	81,64	20,77	78,68	76,68	72,64	72,41	78,91	83,27
4	Wind Velocity	km/hr	2,62	3,31	2,01	1,04	0,68	0,72	0,86	0,83	0,99	1,04	0,81	1,79
		km/day	62,95	79,53	48,33	24,98	16,36	17,24	20,73	19,96	23,67	24,87	19,53	42,98
II	Calculation													
1	Saturated Vapor Pressure (ea)	mbar	28,25	28,36	28,93	29,42	29,35	28,68	27,99	28,37	29,63	30,86	30,48	29,09
2	Real Vapor Pressure (ed)	mbar	23,28	23,26	23,81	24,08	23,96	23,16	22,02	21,76	21,52	22,35	24,05	24,22
3	Vapor Pressure Difference (ea-ed)	mbar	4,97	5,09	5,12	5,34	5,39	5,51	5,97	6,62	8,11	8,52	6,43	4,87
4	Wind Function (f(u))	km/day	0,44	0,48	0,40	0,34	0,31	0,32	0,33	0,32	0,33	0,34	0,32	0,39
5	W		0,74	0,74	0,74	0,75	0,75	0,74	0,73	0,74	0,75	0,76	0,76	0,74
6	Weighting Factor (1-W)		0,26	0,26	0,26	0,25	0,25	0,26	0,27	0,26	0,25	0,24	0,24	0,26
7	Extraterrestrial Radiation (Ra)	mm/day	15,80	16,00	15,60	14,70	13,40	12,80	13,10	14,00	15,00	15,70	15,80	15,70
8	Shortwave Radiation (Rs)	mm/day	5,45	5,69	6,54	6,29	6,65	6,22	6,89	7,72	8,35	8,57	7,31	5,84
9	Net Shortwave Radiation (Rns)	mm/day	1,36	1,42	1,64	1,57	1,66	1,55	1,72	1,93	2,09	2,14	1,83	1,46
10	Real Vapor Pressure Function (f(ed))		0,13	0,13	0,13	0,12	0,12	0,13	0,13	0,13	0,14	0,13	0,12	0,12
11	Radiation Function (f(n/N))		0,27	0,29	0,41	0,42	0,54	0,52	0,60	0,64	0,65	0,63	0,48	0,32
12	Temperature Function (f(T))		15,27	15,28	15,37	15,45	15,43	15,34	15,23	15,29	15,48	15,64	15,59	15,40
13	Net Longwave Radiation (Rnl)	mm/day	0,53	0,57	0,78	0,80	1,05	1,03	1,21	1,32	1,37	1,31	0,94	0,61
14	Net Radiation (Rn)	mm/day	0,83	0,86	0,86	0,77	0,62	0,52	0,51	0,61	0,72	0,84	0,89	0,85
15	Correction Factor (c)		1,10	1,10	1,00	1,00	0,95	0,95	1,00	1,00	1,10	1,10	1,15	1,15
16	Potential Evapotranspiration (Eto)	mm/day	1,30	1,40	1,17	1,03	0,84	0,80	0,90	1,01	1,34	1,46	1,35	1,29

Source: Calculation Result

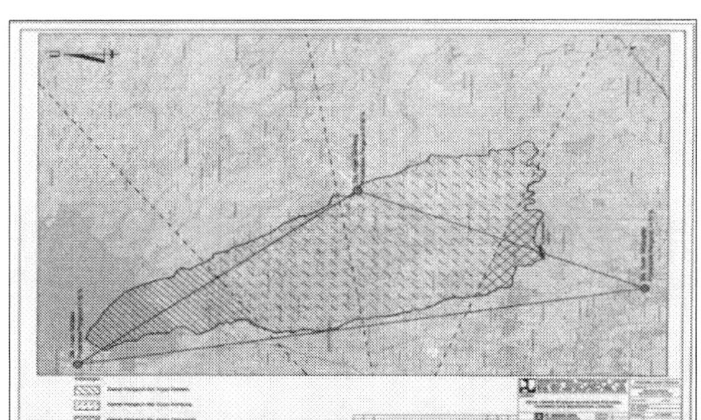

Figure 2. Map of Thiessen polygon
(Source: Brantas river basin organization or balai besar wilayah sungai brantas)

Table 4. Weighting factor calculation

Station Name	Area (Ha)	Percentage (%)
Rahwatu	452,595	16
Gembong	2176,498	78
Tanjungrejo	180,908	6
Total	2810	100

Source: Calculation Result

The average rainfall calculation using Thiessen polygon in 2014 was shown in the following table:

Table 5. Average rainfall on 2014 (mm)

Year	Station Name	Weighting (Factor)	January			February			March			April		
			Jan-01	Jan-02	Jan-03	Feb-01	Feb-02	Feb-03	Mar-01	Mar-02	Mar-03	Apr-01	Apr-02	Apr-03
2014	Rahwatu	16	258,45	95,63	214,02	105,72	78,86	114,88	167,42	167,78	183,98	73,75	143,92	53,11
	Gembong	78	82,42	145,21	133,6	73,84	93,54	126,57	141,24	98,76	166,24	96,66	163,88	116,37
	Tanjungrejo	6	41	36	108	276	92	69	37	74	0	16	4	30
	Average	100	108,9348	131,6439	146,2048	92,72822	92,01181	122,2465	140,1581	109,2703	160,0572	88,7437	152,0109	101,7842

Year	Station Name	Weighting (Factor)	May			June			July			August		
			Mei-01	Mei-02	Mei-03	Jun-01	Jun-02	Jun-03	Jul-01	Jul-02	Jul-03	Aug-1	Aug-2	Aug-3
2014	Rahwatu	16	39,72	93,26	29,55	45,83	5,42	225,93	24,18	26,63	41,31	29,62	41,97	18,24
	Gembong	78	50,33	46,91	7,03	64,96	0,49	16,06	3,95	10,44	24,03	81,16	32,66	11,3
	Tanjungrejo	6	10	38	0	0	2	0	0	0	0	0	0	0
	Average	100	46,52795	54,27068	10,27491	58,34629	1,386168	48,98954	6,993561	12,47993	25,50647	68,44519	32,38347	11,8033

Year	Station Name	Weighting (Factor)	September			October			November			December		
			Sep-01	Sep-02	Sep-03	Okt-01	Okt-02	Okt-03	Nov-01	Nov-02	Nov-03	Des-01	Des-02	Des-03
2014	Rahwatu	16	17,39	1,87	23,6	21,79	71,79	83,57	195,89	158,51	168,72	137,58	218,99	325,58
	Gembong	78	31,25	0	27,22	63,1	40,74	58,32	134,88	49,65	36,12	97,55	125,63	187,05
	Tanjungrejo	6	0	0	10	34	26	8	30	40	66	18	22	309
	Average	100	27,31826	0,301193	25,80052	55,20392	45,22853	59,73061	139,3033	67,06886	59,7622	99,85154	135,2517	219,0841

Source: Calculation Result

Table 6. Average rainfall data records (mm)

Rank	Effective Rainfall (mm)																	
	January			February			March			April			May			June		
	I	II	III	I	II	III	I	II	III	I	II	III	I	II	III	I	II	III
1	164,78	360,59	215,32	300,65	161,86	170,61	204,54	120,82	160,06	138,35	164,71	118,51	83,44	101,66	45,31	66,44	35,52	143,49
2	136,75	131,64	155,79	246,83	104,09	150,39	187,26	112,05	159,90	88,74	1520,10	101,78	75,98	60,88	20,48	65,33	11,03	121,33
3	136,22	126,16	146,79	225,49	97,16	122,25	152,74	109,27	100,40	60,73	140,97	88,22	70,62	54,27	16,79	60,42	10,98	71,16
4	133,28	124,89	146,24	180,19	92,46	119,00	140,16	98,53	96,45	53,51	137,03	81,15	42,53	50,33	14,62	58,35	2,95	56,99
5	114,56	124,69	138,99	179,54	92,01	109,04	131,02	94,97	93,77	53,49	134,44	59,76	42,63	33,31	14,32	46,27	2,53	48,99
6	108,93	123,49	109,61	110,06	61,82	88,85	125,55	63,14	78,45	37,30	117,06	49,91	38,16	31,42	10,57	36,25	2,37	33,15
7	108,63	101,43	107,18	92,73	61,78	88,58	123,94	66,49	66,59	32,24	111,15	39,36	27,98	27,75	10,27	35,00	1,39	28,38
8	84,45	54,72	81,26	79,86	41,78	87,65	104,35	52,02	52,02	30,57	105,68	35,07	24,75	27,52	7,15	31,79	0,58	15,59
9	83,19	46,68	77,07	36,93	26,60	82,28	84,10	15,82	15,82	25,92	101,87	22,90	14,43	9,83	0,00	25,63	0,00	0,00
10	57,75	15,11	17,76	36,02	16,38	47,73	31,61	13,91	13,91	11,43	16,65	9,02	0,16	3,26	0,00	0,00	0,00	0,00

IOP Conf. Series: Journal of Physics: Conf. Series **974** (2018) 012048 doi:10.1088/1742-6596/974/1/012048

Rank	July			August			September			October			November			December		
	Effective Rainfall (mm)																	
	I	II	III	I	II	III	I	II	III	I	II	III	I	II	III	I	II	III
1	95,18	58,10	46,51	68,45	52,58	18,68	27,32	31,36	31,54	128,20	102,86	138,26	189,43	116,07	165,48	164,78	180,26	282,20
2	63,10	35,61	29,94	46,62	49,16	14,31	21,00	20,87	25,80	118,23	74,61	97,14	180,95	96,32	106,05	118,29	146,44	219,08
3	36,86	34,22	25,51	37,00	32,38	11,80	13,55	14,88	25,50	96,60	45,23	63,36	178,53	67,06	105,87	99,85	137,04	212,23
4	25,80	23,03	19,74	28,83	22,18	9,52	8,70	6,11	19,80	57,24	45,21	59,73	139,30	62,90	94,15	82,22	135,25	177,60
5	22,00	12,48	15,50	15,15	11,66	8,23	7,65	0,64	9,23	55,20	31,64	58,82	138,92	56,86	93,43	78,90	127,55	173,87
6	8,59	4,44	15,16	10,73	11,22	7,44	6,61	5,00	3,91	49,03	30,03	57,20	136,28	55,27	60,84	15,82	120,83	157,88
7	7,77	3,14	7,83	9,83	7,89	1,39	0,00	0,30	3,25	14,12	23,21	49,38	118,10	55,03	59,76	64,03	115,24	106,26
8	6,99	0,00	1,79	2,68	0,00	0,00	0,00	0,00	0,00	0,00	11,39	48,17	111,77	54,67	58,30	50,34	108,52	88,17
9	0,00	0,00	0,00	0,00	0,00	0,00	0,00	0,00	0,00	0,00	0,32	0,00	9,33	48,09	41,02	47,15	91,03	53,96
10	0,00	0,00	0,00	0,00	0,00	0,00	0,00	0,00	0,00	0,00	0,00	0,00	0,00	4,65	19,68	31,69	62,09	17,97

Source: Calculation Result

After obtaining the average rainfall, an effective rainfall calculation for rice, secondary crops, and sugarcane are calculated. The result of calculation can be seen at Table 7.

Table 7. Effective rainfall calculations for rice crops (mm/day)

Rank	January			February			March			April			May			June		
	Effective Rainfall (mm)																	
	I	II	III	I	II	III	I	II	III	I	II	III	I	II	III	I	II	III
1	164,78	360,59	215,32	300,65	161,86	170,61	204,54	120,82	160,06	138,35	164,71	118,51	83,44	101,66	45,31	66,44	35,52	143,49
2	136,75	131,64	155,79	246,83	104,09	150,39	187,26	112,05	159,90	88,74	1520,10	101,78	75,98	60,88	20,48	65,33	11,03	121,33
3	136,22	126,16	146,79	225,49	97,16	122,25	152,74	109,27	100,40	60,73	140,97	88,22	70,62	54,27	16,79	60,42	10,98	71,16
4	133,28	124,89	146,24	180,19	92,46	119,00	140,16	98,53	96,45	53,51	137,03	81,15	42,53	50,33	14,62	58,35	2,95	56,99
5	114,56	124,69	138,99	179,54	92,01	109,04	131,02	94,97	93,77	53,49	134,44	59,76	42,63	33,31	14,32	46,27	2,53	48,99
6	108,93	123,49	109,61	110,06	61,82	88,85	125,55	63,14	78,45	37,30	117,06	49,91	38,16	31,42	10,57	36,25	2,37	33,15
7	108,63	101,43	107,18	92,73	61,78	88,58	123,94	66,49	66,59	32,24	111,15	39,36	27,98	27,75	10,27	35,00	1,39	28,38
8	84,45	54,72	81,26	79,86	41,78	87,65	104,35	52,02	52,02	30,57	105,68	35,07	24,75	27,52	7,15	31,79	0,58	15,59
9	83,19	46,68	77,07	36,93	26,60	82,28	84,10	15,82	15,82	25,92	101,87	22,90	14,43	9,83	0,00	25,63	0,00	0,00
10	57,75	15,11	17,76	36,02	16,38	47,73	31,61	13,91	13,91	11,43	16,65	9,02	0,16	3,26	0,00	0,00	0,00	0,00
re padi	5,91	3,83	5,69	5,59	2,92	6,14	7,30	3,64	2,93	2,14	7,40	2,46	1,73	1,93	0,50	2,23	0,04	0,88

Rank	July			August			September			October			November			December		
	Effective Rainfall (mm)																	
	I	II	III	I	II	III	I	II	III	I	II	III	I	II	III	I	II	III
1	95,18	58,10	46,51	68,45	52,58	18,68	27,32	31,36	31,54	128,20	102,86	138,26	189,43	116,07	165,48	164,78	180,26	282,20
2	63,10	35,61	29,94	46,62	49,16	14,31	21,00	20,87	25,80	118,23	74,61	97,14	180,95	96,32	106,05	118,29	146,44	219,08
3	36,86	34,22	25,51	37,00	32,38	11,80	13,55	14,88	25,50	96,60	45,23	63,36	178,53	67,06	105,87	99,85	137,04	212,23
4	25,80	23,03	19,74	28,83	22,18	9,52	8,70	6,11	19,80	57,24	45,21	59,73	139,30	62,90	94,15	82,22	135,25	177,60
5	22,00	12,48	15,50	15,15	11,66	8,23	7,65	0,64	9,23	55,20	31,64	58,82	138,92	56,86	93,43	78,90	127,55	173,87
6	8,59	4,44	15,16	10,73	11,22	7,44	6,61	5,00	3,91	49,03	30,03	57,20	136,28	55,27	60,84	15,82	120,83	157,88
7	7,77	3,14	7,83	9,83	7,89	1,39	0,00	0,30	3,25	14,12	23,21	49,38	118,10	55,03	59,76	64,03	115,24	106,26
8	6,99	0,00	1,79	2,68	0,00	0,00	0,00	0,00	0,00	0,00	11,39	48,17	111,77	54,67	58,30	50,34	108,52	88,17
9	0,00	0,00	0,00	0,00	0,00	0,00	0,00	0,00	0,00	0,00	0,32	0,00	9,33	48,09	41,02	47,15	91,03	53,96
10	0,00	0,00	0,00	0,00	0,00	0,00	0,00	0,00	0,00	0,00	0,00	0,00	0,00	4,65	19,68	31,69	62,09	17,97
re padi	0,49	0,00	0,13	0,19	0,00	0,00	0,00	0,00	0,00	0,00	0,80	3,37	7,82	3,83	4,08	3,52	7,60	6,17

Source: Calculation Result

Table 8. Effective rainfall calculations for secondary crops (mm/day)

Month	Periods	50% re 80	Re	Eto	Re Pol	Re Pol	
		mm/10 days	mm/month	mm/month	mm/month	mm/month	
	Jan-01	42,22					1,93
January	Jan-02	27,36	110,21	39,14	58,04	1,93	1,93
	Jan-03	40,63					1,93
	Feb-01	39,93					1,86
Feb	Feb-02	20,86	104,62	42,08	55,83	1,86	1,86
	Feb-03	43,82					1,86
	Mar-01	52,18					1,75
March	Mar-02	26,01	99,11	34,98	52,45	1,75	1,75
	Mar-03	20,92					1,75
	Apr-01	15,29					1,53
April	Apr-02	52,84	86,66	30,76	45,75	1,53	1,53
	Apr-03	17,54					1,53
	Mei-01	12,38					0,57
May	Mei-02	13,76	29,71	25,27	17,21	0,57	0,57
	Mei-03	3,58					0,57
	Jun-01	15,90					0,44
June	Jun-02	0,29	22,48	23,96	13,05	0,44	0,44
	Jun-02	6,30					0,44
	Jul-01	3,50					0,04
July	Jul-02	0,00	4,39	26,91	1,28	0,04	0,04
	Jul-03	0,89					0,04
	Aug-1	1,34					0,00
Aug	Aug-2	0,00	1,34	30,17	0,00	0,00	0,00
	Aug-3	0,00					0,00
	Sep-01	0,00					0,00
Sep	Sep-02	0,00	0,00	40,04	0,00	0,00	0,00
	Sep-03	0,00					0,00
	Okt-01	0,00					0,60
Oct	Okt-02	5,69	29,78	43,72	17,96	0,60	0,60
	Okt-03	24,08					0,60
	Nov-01	55,88					1,97
Nov	Nov-02	27,33	112,37	40,59	59,21	1,97	1,97
	Nov-03	29,15					1,97
	Des-01	25,17					2,13
Dec	Des-02	54,26	123,52	38,61	63,97	2,13	2,13
	Des-03	44,09					2,13

Source: Calculation Result

Table 9. Effective rainfall calculations for sugarcane crops (mm/day)

Month	Periods	50% re 80	Re	Eto	Re Pol	Re Pol	
		mm/10 days	mm/month	mm/month	mm/month	mm/month	
1	2	3	4	5	6	7	8
	Jan-01	50,67					2,53
January	Jan-02	32,89	132,25	39,14	75,96	2,53	2,53
	Jan-03	48,76					2,53
	Feb-01	47,92					2,44
Feb	Feb-02	25,03	125,54	42,08	73,10	2,44	2,44
	Feb-03	52,59					2,44
	Mar-01	62,61					2,29
March	Mar-02	31,21	118,93	34,98	68,69	2,29	2,29
	Mar-03	25,11					2,29
	Apr-01	18,34					2,00
April	Apr-02	63,41	102,79	30,76	59,99	2,00	2,00
	Apr-03	21,04					2,00
	Mei-01	14,85					0,76
May	Mei-02	16,51	35,65	25,27	22,89	0,76	0,76
	Mei-03	4,29					0,76
	Jun-01	19,07					0,58
June	Jun-02	0,35	26,98	23,96	17,48	0,58	0,58
	Jun-02	7,56					0,58
	Jul-01	4,20					0,07
July	Jul-02	0,00	5,27	26,91	2,19	0,07	0,07
	Jul-03	1,07					0,07
	Aug-1	1,61					0,00
Aug	Aug-2	0,00	1,61	30,17	0,00	0,00	0,00
	Aug-3	0,00					0,00
	Sep-01	0,00					0,00
Sep	Sep-02	0,00	0,00	40,04	0,00	0,00	0,00
	Sep-03	0,00					0,00
	Okt-01	0,00					0,80
Oct	Okt-02	6,83	35,73	43,72	23,89	0,80	0,80
	Okt-03	28,90					0,80
	Nov-01	67,06					2,58
Nov	Nov-02	32,80	134,84	40,59	77,48	2,58	2,58
	Nov-03	34,98					2,58
	Des-01	30,20					2,79
Dec	Des-02	65,11	148,22	38,61	83,67	2,79	2,79
	Des-03	52,90					2,79

Source: Calculation Result

3.2.2. Land preparation

Water needs for land preparations determine the maximum demand of irrigation water in the new irrigated area. Affecting factors of the water demand for land preparation are evapotranspiration and location. It is calculated using methods developed by Van de Goor and Zijlstra (1968). The water need for land preparation is shown in Table 10.

IOP Conf. Series: Journal of Physics: Conf. Series **974** (2018) 012048 doi:10.1088/1742-6596/974/1/012048

Table 10. Calculation of water needs for land preparation (lt /s / ha)

No	Parameter	Unit	Month											
			Jan	Feb	Mar	Apr	May	Jun	Jul	Aug	Sep	Oct	Nov	Dec
1	Eto	mm/day	1,30	1,40	0,17	0,13	0,84	0,80	0,90	1,01	1,34	1,46	1,35	1,29
2	Eo = 1.1 xEto	mm/day	1,44	1,54	1,28	1,13	0,93	0,88	0,99	1,11	1,47	1,60	1,49	1,42
3	Perlokasi	mm/day	2,00	2,00	2,00	2,00	2,00	2,00	2,00	2,00	2,00	2,00	2,00	2,00
4	M = Eo + P	mm/day	3,44	3,54	3,28	3,13	2,93	2,88	2,99	3,11	3,47	3,60	3,49	3,42
5	T	day	31,00	29,00	31,00	30,00	31,00	30,00	31,00	31,00	30,00	31,00	30,00	31,00
6	S	mm/day	300,00	300,00	300,00	300,00	300,00	300,00	300,00	300,00	300,00	300,00	300,00	300,00
7	K = MxT/S		0,35	0,34	0,34	0,31	0,30	0,31	0,32	0,35	0,37	0,37	0,35	0,35
8	IR = (Mxe^k)/(e^k-1)	mm/day	11,50	12,22	11,41	11,65	11,21	11,25	11,31	11,83	11,59	11,59	11,85	11,19
		l/s/ha	1,33	1,41	1,32	1,35	1,30	1,33	1,30	1,31	1,37	1,34	1,37	1,33

Source: Calculation Result

3.2.3. *Water needs for planting*

Table 11. Water needs of rice crops at the beginning of planting on 1st november

Planting Season	Month	Periods	Eto	Re	P	WLR	Rice, 1st November Corps Coefficient				Etc	NFR		DR
			Month	mm/day	mm/day	mm/day	c1	c2	c3	c4	mm/day	mm/day	l/s/ha	l/s/ha
1	2	3	4	5	6	7	8	9	10	11	12	13	14	15
Rainy Season	Nov	1	1,35	0,00	2,00		LP	LP	LP	LP	11,85	13,85	1,60	2,47
		2	1,35	0,80	2,00		1,10	LP	LP	LP	11,85	13,05	1,51	2,32
		3	1,35	3,37	2,00		1,10	1,10	LP	LP	11,85	10,48	1,21	1,87
	Dec	1	1,29	7,82	2,00	1,00	1,10	1,10	1,10	1,10	1,42	-3,40	-0,39	0,00
		2	1,29	3,83	2,00	2,20	1,05	1,10	1,10	1,08	1,40	1,77	0,20	0,32
		3	1,29	4,08	2,00	2,20	1,05	1,05	1,10	1,07	1,38	1,50	0,17	0,27
	Jan	1	1,30	3,52	2,00	2,00	1,05	1,05	1,05	1,05	1,37	2,04	0,24	0,36
		2	1,30	7,60	2,00	2,00	0,95	1,05	1,05	1,02	1,32	-2,08	-0,24	0,00
		3	1,30	6,17	2,00	1,00	0,95	0,95	1,05	0,98	1,28	-1,89	-0,22	0,00
	Feb	1	1,40	5,91	2,00		0,00	0,95	0,95	0,63	0,89	-3,02	-0,35	0,00
		2	1,40	3,93	2,00			0,00	0,95	0,48	0,67	-1,17	-0,13	0,00
		3	1,40	5,69	2,00				0,00	0,00	0,00	0,00	0,00	0,00
Dry Season I	Mar	1	1,17	5,59	2,00		LP	LP	LP	LP	11,41	7,82	0,91	1,39
		2	1,17	2,92	2,00		1,10	LP	LP	LP	11,41	10,49	1,21	1,87
		3	1,17	6,14	2,00	1,00	1,10	1,10	LP	LP	11,41	8,27	0,96	1,47
	Apr	1	1,03	7,30	2,00	2,20	1,10	1,10	1,10	1,10	1,31	-1,97	-0,20	0,00
		2	1,03	3,64	2,00	2,00	1,05	1,10	1,10	1,08	1,12	1,47	0,17	0,26
		3	1,03	2,93	2,00	2,20	1,05	1,05	1,10	1,07	1,10	2,37	0,27	0,42
	May	1	0,84	2,14	2,00	2,20	1,05	1,05	1,05	1,05	0,88	2,94	0,34	0,52
		2	0,84	7,40	2,00	2,20	0,95	1,05	1,05	1,02	0,85	-2,34	-0,27	0,00
		3	0,84	2,46	2,00	1,00	0,00	0,95	1,05	0,67	0,56	1,10	0,13	0,20
	Jun	1	0,80	1,73	2,00		0,00	0,00	0,95	0,32	0,25	0,52	0,06	0,09
		2	0,80	1,93	2,00			0,00	0,00	0,00	0,00	0,07	0,01	0,01
		3	0,80	0,50	2,00				0,00	0,00	0,00	1,50	0,17	0,27
Dry Season II	Jul	1	0,90	2,23	2,00		LP	LP	LP	LP	11,25	11,02	1,28	7,96
		2	0,90	0,04	2,00		1,10	LP	LP	LP	11,25	13,21	1,53	2,35
		3	0,90	0,88	2,00		1,10	1,10	LP	LP	11,25	12,37	1,43	2,20
	Aug	1	1,01	0,49	2,00	1,10	1,10	1,10	1,10	1,10	1,11	3,72	0,43	0,66
		2	1,01	0,00	2,00	2,20	1,05	1,10	1,10	1,08	1,03	5,29	0,61	0,94
		3	1,01	0,13	2,00	2,20	1,05	1,05	1,10	1,07	1,08	5,15	0,60	0,92
	Sep	1	1,34	0,19	2,00	2,20	1,05	1,05	1,05	1,05	1,41	5,42	0,63	0,96
		2	1,34	0,00	2,00	2,20	0,95	1,05	1,05	1,02	0,36	5,56	0,64	0,99
		3	1,34	0,00	2,00	1,10	0,00	0,95	1,05	0,67	0,89	3,99	0,46	0,71
	Oct	1	1,46	0,00	2,00		0,00	0,00	0,95	0,32	0,46	2,46	0,28	0,44
		2	1,46	0,00	2,00			0,00	0,00	0,00	0,00	2,00	0,23	0,36
		3	1,46	0,00	2,00				0,00	0,00	0,00	2,00	0,23	0,36

Source: Calculation Result

The alternative calculations for planting pattern in the optimization study are as follows:

- Alternative 1: the beginning of planting on 1st November
- Alternative 2:the beginning of planting on 2nd November
- Alternative 3: the beginning of planting on 3rd November
- Alternative 4: the beginning of planting on 1st December
- Alternative 5: the beginning of planting on 2nd December
- Alternative 6: the beginning of planting on 3rd December

Table 11, Table 12 and Table 13 show the water need in difference time of the beginning of planting time.

Table 12. Water needs of secondary crops at the beginning of planting on 1st November

Planting Season	Month	Periods	Eto	Re	Secondary Corps, 1st November Corps Coefficient				Etc	NFR		DR
			Month	mm/day	c1	c2	c3	c4	mm/day	mm/day	l/s/ha	l/s/ha
1	2	3	4	5	6	7	8	9	10	11	12	13
		1	1,35	1,97	0,50			0,50	0,68	-1,30	-0,15	0,00
	Nov	2	1,35	1,97	0,63	0,50		0,57	0,76	-1,21	-0,14	0,00
		3	1,35	1,97	0,75	0,63	0,50	0,63	0,85	-1,13	-0,13	0,00
		1	1,29	2,13	1,00	0,75	0,63	0,79	1,02	-1,11	-0,13	0,00
	Dec	2	1,29	2,13	1,00	1,00	0,75	0,92	1,18	-0,95	-0,11	0,00
Rainy Season		3	1,29	2,13	1,00	1,00	1,00	1,00	1,29	-0,84	-0,10	0,00
		1	1,30	1,93	0,82	1,00	1,00	0,94	1,22	-0,71	-0,08	0,00
	Jan	2	1,30	1,93	0,64	0,82	1,00	0,82	1,07	-0,81	-0,10	0,00
		3	1,30	1,93	0,45	0,64	0,82	0,64	0,83	-1,11	-0,13	0,00
		1	1,40	1,86		0,45	0,64	0,55	0,76	-1,10	-0,13	0,00
	Feb	2	1,40	1,86			0,45	0,45	0,63	-1,23	-0,14	0,00
		3	1,40	1,86				0,00	0,00	0,00	0,00	0,00
		1	1,17	1,75	0,50			0,50	0,59	-1,16	-0,13	0,00
	Mar	2	1,17	1,75	0,63	0,50		0,57	0,66	-1,09	-0,13	0,00
		3	1,17	1,75	0,75	0,63	0,50	0,63	0,73	-1,02	-0,12	0,00
		1	1,03	1,53	1,00	0,75	0,63	0,79	0,82	-0,71	-0,08	0,00
	Apr	2	1,03	1,53	1,00	1,00	0,75	0,92	0,94	-0,58	-0,07	0,00
Dry Season I		3	1,03	1,53	1,00	1,00	1,00	1,00	1,03	-0,50	-0,06	0,00
		1	0,84	0,57	0,82	1,00	1,00	0,94	0,79	0,22	0,02	0,04
	May	2	0,84	0,57	0,64	0,82	1,00	0,82	0,69	0,12	0,01	0,02
		3	0,84	0,57	0,45	0,64	0,82	0,64	0,53	-0,04	0,00	0,00
		1	0,80	0,44		0,45	0,64	0,55	0,44	0,00	0,00	0,00
	Jun	2	0,80	0,44			0,45	0,45	0,36	-0,08	-0,01	0,00
		3	0,80	0,44				0,00	0,00	-0,44	-0,05	0,00
		1	0,90	0,04	0,50			0,50	0,45	0,41	0,05	0,07
	Jul	2	0,90	0,04	0,63	0,50		0,57	0,21	0,47	0,05	0,08
		3	0,90	0,04	0,75	0,63	0,50	0,63	0,56	0,52	0,06	0,09
		1	1,01	0,00	1,00	0,75	0,63	0,79	0,80	0,80	0,09	0,14
	Aug	2	1,01	0,00	1,00	1,00	0,75	0,92	0,93	0,93	0,11	0,16
Dry Season II		3	1,01	0,00	1,00	1,00	1,00	1,00	1,01	1,01	0,12	0,18
		1	1,34	0,00	0,82	1,00	1,00	0,94	1,26	1,26	0,15	0,22
	Sep	2	1,34	0,00	0,64	0,82	1,00	0,82	1,10	1,10	0,13	0,20
		3	1,34	0,00	0,45	0,64	0,82	0,64	0,85	0,85	0,10	0,15
		1	1,46	0,60		0,45	0,64	0,55	0,80	0,20	0,02	0,04
	Oct	2	1,46	0,60			0,45	0,45	0,66	0,06	0,01	0,01
		3	1,46	0,60				0,00	0,00	-0,60	-0,07	0,00

Source: Calculation Result

IOP Conf. Series: Journal of Physics: Conf. Series **974** (2018) 012048 doi:10.1088/1742-6596/974/1/012048

Table 13. Water needs of sugarcane crops at the beginning of planting on 1st November

Planting Season	Month	Periods	Eto	Re	Sugarcane, 1st November Corps Coefficient				Etc	NFR		DR
			Month	mm/day	c1	c2	c3	c4	mm/day	mm/day	l/s/ha	l/s/ha
1	2	3	4	5	6	7	8	9	10	11	12	13
Rainy Season	Nov	1	1,35	2,58	0,55	0,60	0,60	0,58	0,79	-1,80	-0,21	0,00
		2	1,35	2,58	0,55	0,55	0,60	0,57	0,77	-1,82	-0,21	0,00
		3	1,35	2,58	0,55	0,55	0,55	0,55	0,74	-1,84	-0,21	0,00
	Dec	1	1,29	2,79	0,80	0,55	0,55	0,63	0,82	0,00	0,00	0,00
		2	1,29	2,79	0,80	0,80	0,55	0,81	0,92	-1,86	-0,22	0,00
		3	1,29	2,79	0,80	0,80	0,80	8,00	1,03	-1,76	-0,20	0,00
	Jan	1	1,30	2,53	0,90	0,80	0,80	0,83	1,08	0,00	0,00	0,00
		2	1,30	2,53	0,95	0,90	0,80	0,88	1,15	-1,38	-0,16	0,00
		3	1,30	2,53	1,00	0,95	0,90	0,95	1,24	-1,30	-0,15	0,00
	Feb	1	1,40	2,44	1,00	1,00	0,95	0,98	1,38	-1,06	-0,12	0,00
		2	1,40	2,44	1,00	1,00	1,00	1,00	1,40	-1,04	-0,12	0,00
		3	1,40	2,44	1,00	1,00	1,00	1,00	1,40	-1,04	-0,12	0,00
Dry Season I	Mar	1	1,17	2,29	1,05	1,00	1,00	1,02	1,19	-1,10	-0,13	0,00
		2	1,17	2,29	1,05	1,05	1,00	1,03	1,21	-1,08	-0,13	0,00
		3	1,17	2,29	1,05	1,05	1,05	1,05	1,23	-1,06	-0,12	0,00
	Apr	1	1,03	2,00	1,05	1,05	1,05	1,05	1,08	-0,92	-0,11	0,00
		2	1,03	2,00	1,05	1,05	1,05	1,05	1,08	-0,92	-0,11	0,00
		3	1,03	2,00	1,05	1,05	1,05	1,05	1,08	-0,92	-0,11	0,00
	May	1	0,84	0,76	1,05	1,05	1,05	1,05	0,88	0,12	0,01	0,02
		2	0,84	0,76	1,05	1,05	1,05	1,05	0,88	0,12	0,01	0,02
		3	0,84	0,76	1,05	1,05	1,05	1,05	0,88	0,12	0,01	0,02
	Jun	1	0,80	0,58	1,05	1,05	1,05	1,05	0,84	0,03	0,03	0,05
		2	0,80	0,58	1,05	1,05	1,05	1,05	0,84	0,26	0,03	0,05
		3	0,80	0,58	1,05	1,05	1,05	1,05	0,84	0,26	0,03	0,05
Dry Season II	Jul	1	0,90	0,07	1,05	1,05	1,05	1,05	0,95	0,87	0,10	0,16
		2	0,90	0,07	1,05	1,05	1,05	1,05	0,95	0,87	0,10	0,16
		3	0,90	0,07	1,05	1,05	1,05	1,05	0,95	0,87	0,10	0,16
	Aug	1	1,01	0,00	1,05	1,05	1,05	1,05	1,06	1,06	0,12	0,19
		2	1,01	0,00	1,05	1,05	1,05	1,05	1,06	1,06	0,12	0,19
		3	1,01	0,00	1,05	1,05	1,05	1,05	1,06	1,06	0,12	0,19
	Sep	1	1,34	0,00	0,80	1,05	1,05	0,97	1,30	1,30	0,15	0,23
		2	1,34	0,00	0,80	0,80	1,05	0,88	1,18	1,18	0,14	0,21
		3	1,34	0,00	0,80	0,80	0,80	0,80	1,07	1,07	0,12	0,19
	Oct	1	1,46	0,00	0,60	0,80	0,80	0,73	1,07	0,27	0,03	0,06
		2	1,46	0,80	0,60	0,60	0,80	0,67	0,97	0,18	0,02	0,03
		3	1,46	0,80	0,60	0,60	0,60	0,60	0,88	0,08	0,01	0,01

Source: Calculation Result

3.3. Optimization of irrigation area

3.3.1. Optimization Model

The optimization modeling equation uses a linear equation or called as *linear programming*. The modeling is to solve problems in Logung irrigation area, Kudus Regency based on steps as follows:

1. Determining the optimization model
2. Determining the optimized variables, i.e. land area for each type of plant at each season
3. Determining the limits value on the modeling (based on calculations)
4. Optimization model Arrangement
5. Optimization process (in this study using POM-QM application program for Windows 3)
6. Analysis of optimization results (based on maximum profit and intensity of planting)

3.3.2. Analysis of harvest yield

Harvest yields are the result of net income from crops harvesting by farmers. It is define as net income, that can be found as the difference between the gross production and production cost. Table 14 shows the basic calculation of harvest yields for Kudus Area.

Table 14. Analysis of harvest yields in Kudus area

No	Detail	Rice	Corn	Sugarcane
1	Product Price (Rp/ton)	Rp4.000.000	Rp4.000.000	Rp380.000
2	Productivity (ton)	7	6,4	110
3	Production (Rp/Ha)	Rp28.000.000	Rp25.600.000	Rp41.800.000
4	Production Cost (Rp/Ha)	Rp13.580.000	Rp12.100.000	Rp30.000.000
5	Profitabilty (Rp/Ha)	Rp14.420.000	Rp13.500.000	Rp11.800.000

Source: semarang.bisnis.com, tabloidsinartani.com, isknews.com

3.3.3. Mathematical Model for Optimization

The linier programe classified the parameter of optimation as functional purpose and function constraints. Two approaches are used as basic assumption : maximize based on land area and water need. The optimization model equations used in this study are as follows:

Functional purpose
Maximize based on land area

$$Z = Xp1 + Xw1 + Xp2 + Xw2 + Xp3 + Xw3 + Xt \quad (1)$$

where:
Xp1 = Land area for rice crops in the rainy season (Ha)
Xw1 = Land area for secondary crops in the rainy season (Ha)
Xp2 = Land area for rice crops in the 1st dry season (Ha)
Xw2 = Land area for secondary crops in the 1st dry season (Ha)
Xp3 = Area of land for rice crops in the 2nd dry season (Ha)
Xw3 = Land area for secondary crops in the 2nd dry season (Ha)
Xt = Land area for sugarcane crops in once growing season (Ha)

Function constraints
• Reliable discharge:

$$Vp1.Xp1 + Vw1.Xw1 + Vt.Xt \leq Q1 \text{ (periods 1 – 12)} \quad (2)$$

$$Vp2.Xp2 + Vw2.Xw2 + Vt.Xt \leq Q2 \text{ (periods 13 – 24)} \quad (3)$$

$$Vp3.Xp3 + Vw3.Xw3 + Vt.Xt \leq Q3 \text{ (periods 25 – 36)} \quad (4)$$

where,
Vpi = water needs of rice crops in each season (lt/s/Ha)
Vwi = water needs of secondary crops in each season (lt/s/Ha)
Vt = water needs of sugarcane crops in each season (lt/s/Ha)

• Maximum area

$$Xp1 + Xw1 + Xt \leq A \text{ total} \quad (5)$$

$$Xp2 + Xw2 + Xt \leq A \text{ total} \quad (6)$$

$$Xp3 + Xw3 + Xt \leq A \text{ total} \quad (7)$$

where, A total = 2810 Ha

• Sugarcane

$$Xt \geq Xte \quad (8)$$

where, Xte = Minimum required sugarcane area (141 Ha)

IOP Conf. Series: Journal of Physics: Conf. Series **974** (2018) 012048 doi:10.1088/1742-6596/974/1/012048

•Non-negativity

$$Xp1\ ,\ Xw1\ ,\ Xp2\ ,\ Xw2\ ,\ Xp3\ ,\ Xw3\ ,\ Xt\ \geq 0 \qquad (9)$$

3.3.4. Analysis of Optimization Results

The optimization analysis is based on two objectives, i.e. the maximum area and maximum profit. Iterative analysis is conducted using POM-QM for Windows 3.

The iteration data result using POM-QM program for Windows 3 obtained the optimum area data on each plant type with each alternative planting pattern. Then, it can be known the planting intensity in every planting season. Moreover, the extent area data will obtain the results of agricultural production annually.

Table 16 shows that the greatest profit of production is obtained in alternative 6 with the early planting periods on the 3rd December.

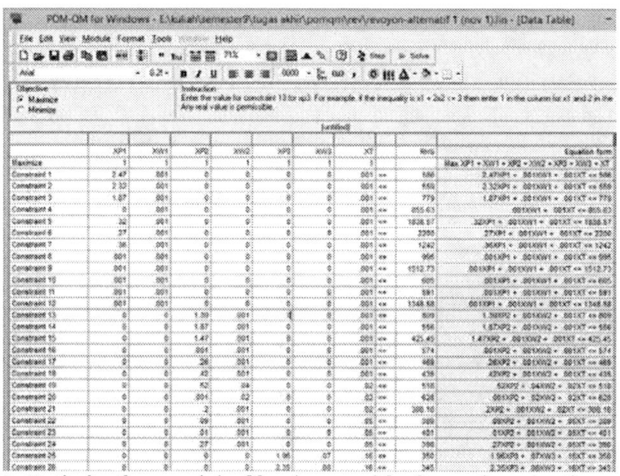

Figure 3. Optimization model of land area with alternative planting pattern 1
Source: Input POM-QM for Windows 3

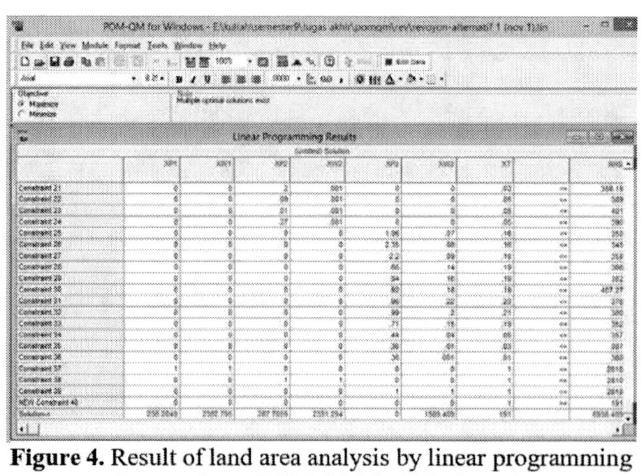

Figure 4. Result of land area analysis by linear programming
Source: Input POM-QM for Windows 3

Table 15. Planting intensity based on land area optimization results

Alternatives	Planting Season	Area						Plant Intensity						
		Rice		Secondary Crops		Sugarcane		Rice		Secondary Crops		Sugarcane		Total
		Ha		Ha		Ha		%		%		%		%
	MH	236		2383		191		8,41		84,8		6,8		
1	MK1	288	524	2331	6223	191	191	10,24	18,64	82,96	221,48	6,8	20,39	260,51
	MK2	0		1509		191		0		53,72		6,8		
	MH	335		2284		191		11,91		81,29		6,8		
2	MK1	226	561	2393	6062	191	191	8,05	19,96	85,16	215,71	6,8	20,39	256,06
	MK2	0		1384		191		0		49,27		6,8		
	MH	722		1897		191		25,69		67,51		6,8		
3	MK1	290	1012	2330	5422	191	191	10,31	36	82,9	192,97	6,8	20,39	249,36
	MK2	0		1196		191		0		42,55		6,8		
	MH	577		2042		191		20,54		72,66		6,8		
4	MK1	202	779	2417	5637	191	191	7,18	27,72	86,02	200,62	6,8	20,39	248,73
	MK2	0		1178		191		0		41,94		6,8		
	MH	331		2288		191		11,77		81,44		6,8		
5	MK1	186	517	2433	6449	191	191	6,63	18,39	86,57	229,52	6,8	20,39	268,3
	MK2	0		1728		191		0		61,51		6,8		
	MH	350		2269		191		12,47		80,74		6,8		
6	MK1	190	540	2429	7254	191	191	6,76	19,22	86,45	258,13	6,8	20,39	297,75
	MK2	0		2556		191		0		90,95		6,8		
	MH	1852		0		26		65,9		0		0,93		
Existing	MK1	1852	3704	0	1852	26	26	65,9	131,8	0	65,9	0,93	2,78	200,47
	MK2	0		1852		26		0		65,9		0,93		

Source: Calculation Result

Table 16. Profit results based on land area optimization results

Alternatives	Planting Season	Area						Plant Intensity			
		Rice		Secondary Crops		Sugarcane		Rice	Secondary Crops	Sugarcane	Total
		Ha		Ha		Ha		Rp 16970300/Ha	Rp 9083000/Ha	Rp 40232800/Ha	Rp
	MH	236		2383		191					
1	MK1	288	524	2331	6223	191	191	Rp7.554.793.736	Rp84.017.223.000	Rp7.684.464.800	Rp99.256.481.536
	MK2	0		1509		191					
	MH	335		2284		191					
2	MK1	226	561	2393	6062	191	191	Rp8.087.320.010	Rp81.830.709.000	Rp7.684.464.800	Rp97.602.493.810
	MK2	0		1384		191					
	MH	722		1897		191					
3	MK1	290	1012	2330	5422	191	191	Rp14.586.944.666	Rp73.202.215.050	Rp7.684.464.800	Rp95.473.624.516
	MK2	0		1196		191					
	MH	577		2042		191					
4	MK1	202	779	2417	5637	191	191	Rp11.234.049.526	Rp76.104.211.500	Rp7.684.464.800	Rp95.022.725.826
	MK2	0		1178		191					
	MH	331		2288		191					
5	MK1	186	517	2433	6449	191	191	Rp7.453.535.054	Rp87.067.494.000	Rp7.684.464.800	Rp102.205.493.854
	MK2	0		1728		191					
	MH	350		7254		191					
6	MK1	190	2269	2429	191	191	191	Rp7.789.881.554	Rp97.922.992.500	Rp7.684.464.800	Rp113.397.338.854
	MK2	0		2556		191					
	MH	1852		0		26					
Existing	MK1	1852	3704	0	1852	26	26	Rp62.849.506.050	Rp16.819.445.250	Rp1.046.052.800	Rp80.715.004.100
	MK2	0		1852		26					

Source: Calculation Result

4. Conclusion and Suggestion

4.1. Conclusions

The conclusions in this study are as follows:

1. Based on the analysis of Logung river discharge data, it can be concluded that the reliable discharge was obtained with 80% probability of reliability. The highest discharge at 80% reliability is 1.8386 m3/s and 0.307 m3/s in the lowest discharge,
2. This study has been analyzed with six alternatives of planting pattern, i.e. 1st November, 2nd November, 3rd November, 1st December, 2nd December, and 3rd December. These alternatives are used as a basic of the water demand calculation for rice, secondary crops, and sugarcane.
3. The extent calculation of the iteration results by POM-QM program for Windows 3 has been obtained from six alternatives of planting patterns. The maximum value is obtained about 8366,77 Ha at the early planting periods in 3rd December with the planting pattern of rice/ sugarcane-beans / sugarcane-rice / sugarcane-beans / sugarcane. The obtained maximum profit value is Rp 113.397.338.854,00.

4.2. Suggestions

The suggestions for the further study are as follows:

1. The application of this optimization study results should be done by approaches and socialization about how to farm well, how to choose a superior seed, how to deal with pest attacks, and other things to farmers in the Logung Irrigation area, Kudus Regency.
2. Based on the data of each alternative existing planting pattern, in the dry season II can't be planted rice. The amount of water at the beginning of the rainy season is very high. It can be used for the irrigation area expansion to obtain the maximum harvest.
3. There should be maintenance and management of irrigation buildings such as weirs and irrigation channels to minimize water loss due to irrigation buildings damage. In addition, it needs a controlling in order to no illegal tapping.

References

[1] Frizzone J.A.; Coelho R.D ; Dourado-Neto D and R. Soliani, *Linear programming model to optimize the water resource use in irrigation projects: an application to the Senator Nilo Coelho Project,* Scientific Agriculture **Vol 54**, June 1997.

[2] Jacovkis P.M., Gradowczyk H., Freisztav A.M. and E. G. Tabak, *A Linear Programming Approach to Water-Resources Optimization, ZOR - Methods and Models of Operations Research,* pp 33:341-362, 1989

[3] Difallah W., Belkacem_Draoui K., and F. Bounaama, *Linear Optimization Model for Efficient Use of Irrigation Water, International Journal of Agronomy*, **Vol. 2017**, Article ID 5353648, 2017,

[4] Anieting A.E, Ezugwu I.O and S. Ologun, *Application of Linear Programming Technique in the Determination of Optimum Production Capacity,* IOSR Journal of Mathematics (IOSR-JM) e-ISSN: 2278-5728.**Volume 5**, Issue 6, pp 62-65, March April 2013

[5] Pearson L. and McRoberts N., *Watershed Management Conference*, ASCE, Madison Wisconsin, United Stated, August 2010

[6] Suhardono, *Optimasi Penggunaan Lahan Pertanian dengan Program Linier*, Thesis, Master Degree, Universitas Brawijaya, 2012.

[7] Balai PSDA Seluna, *Rainfall data of Logung Catchment Area*, Kudus, 2015.

[8] Setiono, *Optimasi Rencana Pola Tanam pada daerah sungai logung, Kudus, dengan menggunakan program linier*, Final Assigment,Undergraduated, Department of Civil Enginering, ITS, 2016.

[9] Anwar N., *Analisa Sistem Untuk Teknik Sipil*, Handout, Department of Civil Enginering ITS,

Surabaya, 2001
[10] Direktorat Jenderal Sumber Daya Air, Departemen Pekerjaan Umum, *Standard of Irigation Design Criterium – Irigation Networking*, 2010.
[11] Hidayat A.W and Indriyono Sukawi, *Perencanaan Embung Logung, Jekulo, Kudus*, Undergraduated Final Assignment, Universitas Diponegoro, 2007.

IOP Conf. Series: Journal of Physics: Conf. Series **974** (2018) 012049 doi:10.1088/1742-6596/974/1/012049

Dynamic game balancing implementation using adaptive algorithm in mobile-based Safari Indonesia game

Anny Yuniarti, Novita Nata Wardanie and Imam Kuswardayan

Department of Informatics, Faculty of Information Technology, Institut Teknologi Sepuluh Nopember (ITS), Jl. Arief Rahman Hakim, Surabaya 60111 Indonesia

anny@if.its.ac.id

Abstract. In developing a game there is one method that should be applied to maintain the interest of players, namely dynamic game balancing. Dynamic game balancing is a process to match a player's playing style with the behaviour, attributes, and game environment. This study applies dynamic game balancing using adaptive algorithm in scrolling shooter game type called Safari Indonesia which developed using Unity. The game of this type is portrayed by a fighter aircraft character trying to defend itself from insistent enemy attacks. This classic game is chosen to implement adaptive algorithms because it has quite complex attributes to be developed using dynamic game balancing. Tests conducted by distributing questionnaires to a number of players indicate that this method managed to reduce frustration and increase the pleasure factor in playing.

1. Introduction

The easier it is for someone to create a game is one of the supporting factors that becomes the reason for the increasing number of mobile device games that continue to grow every day. Game developers must have innovations so that the game they create can be in demand by consumers and not quickly abandoned. Games generally have a level or difficulty level (easy, medium, and difficult) that can be played or selected directly by the user. On the average, the level of difficulty in traditional games runs linearly, in the sense that the higher the level, then, the higher the level of difficulty that will be faced by players. This level of difficulty already has its own set of attributes that the developer has set statically. Not infrequently the players considered a level in the game is too easy, thus making the people who play bored. Meanwhile, there is also the type of player who feels the game is too difficult to play, thus making the player feel frustrated. In the development of games that uses the concept of game balancing, players will be challenged based on their ability to play. Players are not forced to follow the flow of games that tend to increase the level of difficulty and make it frustrating. So it can be assumed in general, that the entertainment value of a balanced game is higher than games that tend to be too easy or too difficult [1].

The increase in game development industry, especially in Indonesia, is accompanied by the level of enthusiasm of the public on the ownership of smartphones, tablets, or other mobile devices. Many developers are increasingly motivated to create an exciting and value-added game for mobile device users. Therefore, this paper proposed a two-dimensional mobile game scrolling shooter type that introduces the Indonesian culture called Safari Indonesia, using dynamic game balancing as a method to balance the game pattern of players with the complexity of the game so as to maintain the interest of players in playing.

2. Literature review

2.1. Dynamic game balancing

Dynamic game balancing or also called dynamic difficulty adaptation is one of the mechanisms in the game that aims to provide a good level design to challenge the players [1]. The level design is made as good as possible by the developer to keep the player interested in the game being played. Various characteristics of players in play, requires a developer to create a way for players to not feel bored or frustrated with the game played. Therefore, balancing the game is required. Balancing the game consists of changing the parameters, scenarios, and behavior of the game environment in order to prevent players from being bored in playing [2].

2.2. Adaptive algorithm

Adaptive algorithm is a collection of instructions to show that a function can adapt to changing circumstances or surrounding environments. Many examples can be given in terms of adaptive algorithm utilization, among others, on radar systems for false alarm rate detection that may occur, the utilization of adaptive algorithms is also done on machine learning process and optimization, data compression, and signal processing.

Adaptive ability has two interrelated meanings. First, it is easy to change. Second, adaptive means learning and transforming [3]. The first meaning means to adapt from one form to another, from one predetermined state to another predetermined state. Whereas in the second meaning, adaptation occurs when there is learning from the tests that have been done before, and there is a transformation to other circumstances that have not been known before. The adaptation can be applied to the game in several ways, among others are:

1. Player character
2. Non-player character on the game
3. Game environment

The easiest method to understand the three ways to adapt the game above is to change the difficulty level of non-player characters in the game. Non-player characters can be used to provide clues or as a support for players in realizing the needs and desires of play [4].

2.3. Player modelling

Player modeling becomes one of the simplest, flexible and lightweight approaches to be a solution in developing adaptive games [5]. The basic idea of game modeling is quite simple, that is, a game saves every player profile that contains abilities, weaknesses, skills, and other characters. As the game progresses, updates are made to the model according to the player's interaction with the game.

2.4. Scrolling shooter game

Scrolling shooter game, can also be called *shoot 'em up*, is a type of game in which a player controls a character, generally in the form of a fighter aircraft, to defend against enemy attacks in large numbers by shooting them. In this type of game scrolling shooter, players can control the character by moving it to a minimum of two directions (top down or right left) or to the four directions in general. In the game, not infrequently players also perform activities to collect power-ups used to develop character and defense.

3. Analysis and modelling

3.1. General description of the system

IOP Conf. Series: Journal of Physics: Conf. Series **974** (2018) 012049 doi:10.1088/1742-6596/974/1/012049

Safari Indonesia game to be built is a scrolling shooter manifest game in which there is a single player who try to defend themselves from enemy attacks that continue to come periodically. The game will be developed in two modes, adventure and endless. The adventure mode is a game developed with static level configuration, while the endless mode is a mode developed by implementing dynamic game balancing using adaptive algorithm.

3.2. Functional requirements

Functional requirements are a description of the information that occurs on the system to be built. Safari Indonesia game apps have four functional requirements as outlined in Table 1.

Table 1. Functional requirement list.

Functional ID	Functional requirements
F-01	Play the game with endless mode
F-02	Play the game with adventure mode
F-03	Perform character updates
F-04	View the status of the game

3.3. Player modelling design

The main part of the player modeling system is the model itself. The model is actually a collection of numerical attributes or traits that describe the playing style of each player [5]. The player modeling discussed in this section focuses on the design of endless modes that implement dynamic game balancing using adaptive algorithms.

In modeling and matching the difficulty level of the game with player game patterns, enemies in endless fashion games have behaviors that are compiled into a variable hereinafter called v. Behavior referred to in variable v, consists of:

$$speed = speed_0 \tag{1a}$$

$$shotIntensity = shotIntensity_0 \tag{1b}$$

$$enemyNumbers = enemyNumbers_0 \tag{1c}$$

This game has three levels of difficulty which hereinafter called player type.

$$playerType = \{easy, medium, difficult\} \tag{2}$$

Furthermore, each type of player that has a constant multiplier p is useful for matching characters to be done later. Multipliers for each type are set as follows:

$$p(easy) = 0.85 \tag{3a}$$

$$p(medium) = 1.05 \tag{3b}$$

$$p(difficult) = 1.3 \tag{3c}$$

The next design is modeling for player characters that will be observed against it. Player character modeling is implemented in the form of numeric numbers ranging from 0 to 1 as shown in Table 2.

When a player starts the game in an endless fashion, the player will be initialized with a moderate player type and has a player modeling attribute value that is in the range of medium player type minimal value up to the maximum value of the medium player type. The initialization values are performed with equation (4).

After getting the middle value of each character, the values are summed to the player's initial performance when starting the game. The equation for summing the middle values of each character is shown in equation (5).

Table 2. Player modeling.

Player character	Easy		Medium		Hard	
	Min	Max	Min	Max	Min	Max
Accuracy	0.0	0.2	0.2	0.6	0.6	1.0
Health variation	0.6	1.0	0.3	0.6	0.0	0.3
Number of defeated enemies	0.0	0.3	0.3	0.7	0.7	1.0
Total	0.6	1.5	0.8	1.9	1.3	2.3

$$Character_i = \frac{CharacterMinimal_i + CharacterMaximum}{2} \quad (4)$$

$$Player's\ initial\ performance = \sum_{i=1}^{n} Character_i \quad (5)$$

3.4. Adaptive algorithm of Safari Indonesia game

The design of the algorithm formulated in this paper is shown in pseudocode Figure 1. The learning rate constant used in this algorithm has a range of values 0 - 1. The smaller the value, it takes more attributes to change the player model. The range of learning rate constants recommended by reference [5] is 0.1 - 0.3.

```
α ← learning rate
playerType0 ← player type initialization
ci ← (CharacterMinimal_i + CharacterMaximum_i) / 2 (The
number of i characters becomes the player performance
initialization)
while (true)
  count the observed character (ci_obs)
  ci ← ci + α x (ci_obs – ci)
  performance = Σ_{i=1}^{n} ci
  if performance ∈ [MIN^tipe , MAX^tipe] then
    playerType[level] ← playerType()
  end if
  if playerType[level] ≠ playerType[level – 1] then
    v ← match(playerType[level])
  else
  maintain player behavior
  end if
  save the performance of each level
end while
```

Figure 1. The adaptive algorithm pseudocode.

4. Testing

4.1. Game control test

Game control is an important element in the game that can support the user's comfort in playing the game. The results of the game control tests in Figures 2 and 3 show a player who moves the game characters of an aircraft in all directions to fire a bullet to the enemy. This shows that the implementation of game controls in Safari Indonesia Games runs well.

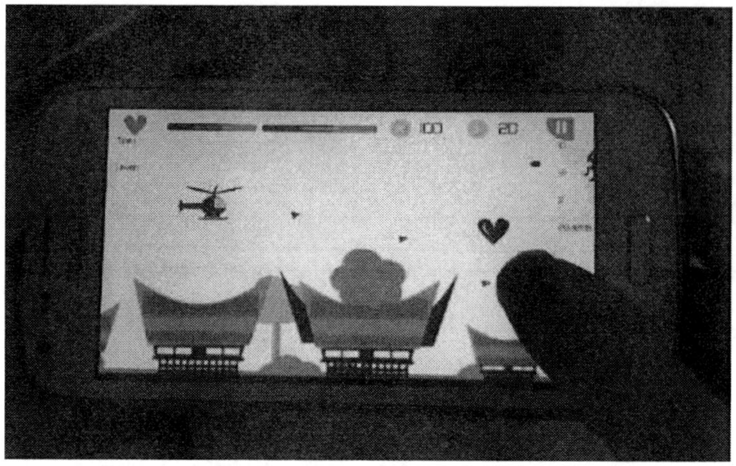

Figure 2. Game control testing in endless mode.

Figure 3. Game control testing in adventure mode.

Table 3. Player performance test results.

Level	Performance	Type	Accuracy	Live	Defeated enemy
1	1.3500	2	0.29	0.33	0.29
2	1.2626	1	0.51	0.67	0.50
3	1.3452	1	1.00	1.00	1.00
4	1.6762	2	1.00	0.71	1.00
5	1.8838	2	0.98	1.00	0.96
6	2.1070	3	0.96	0.63	0.91
7	2.1849	3	0.92	0.67	0.87
8	2.2394	3	0.72	0.50	0.77
9	2.1188	3	0.68	0.36	0.69
10	2.0983	3	0.50	0.68	0.56

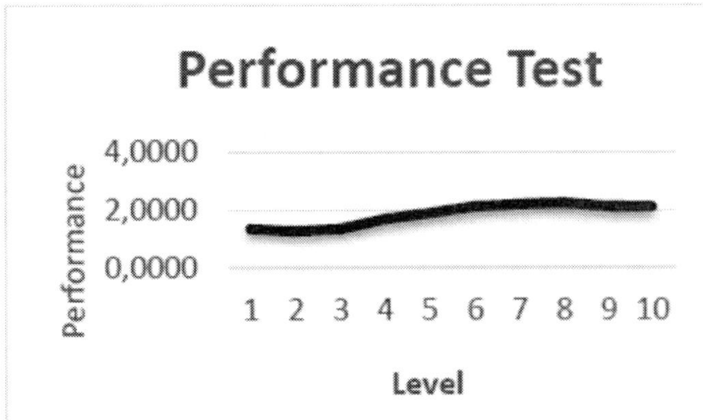

Figure 4. Player performance test graph.

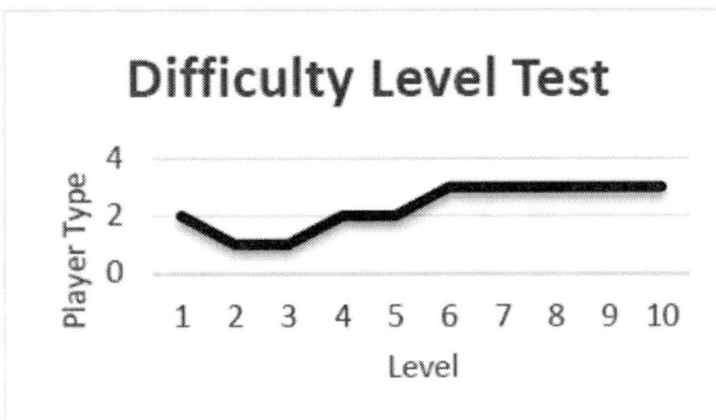

Figure 5. Player difficulty level test graph.

4.2. Performance Test

Performance test results in the calculation of attributes that characterize the player model and the resulting performance of the sum of the attributes. Performance is what will determine the type of player.

The experiments shown in Table 3 indicate that the player has decreased performance with a value of 1.2626 at level 2 which causes the player type to change in type 1. The player successfully improves the performance on the next level, and changes the type of player in type 2 at level 4. Next, players experience increased performance that causes players to change to type 3 at level 6.

Experiments to test the performance of players in Safari Indonesia Game prove that the implementation of adaptive algorithms applied to balance the ability of players with difficulty level of the game is declared successful (Figure 4 and 5).

4.3. Functional requirements test

The functional requirements test of the game is performed by the black box method in which testing is performed by simply observing the results of running programs and checking each use case that has been implemented in the game. Table 4 is a breakdown of the results of functional requirements testing.

4.4. User Testing

User testing is done to test important factors in the game. Questionnaires and factors tested in the questionnaire analysis refer to reference [6]. Table 5 shows the relationship between the factors tested in the game with the statement on the questionnaire.

Table 4. Functional requirement test result.

Functional ID	Functional requirements	Test result
F-01	Play the game with endless mode	Successful
F-02	Play the game with adventure mode	Successful
F-03	Perform character updates	Successful
F-04	View the status of the game	Successful

Table 5. Relationship between important factors in the game with statements on the questionnaire.

Questionnaire number	Factors
1, 4, 5	Fun
2, 3	Frustration
6 – 38	Important elements in the game
6 – 12, 25, 28	Game controls
13 – 18	Game facilities
26 – 31	Game environment
32 – 37, 39 [a]	Rules

[a] Statement of questionnaire number 39 is only intended for endless fashion games.

Table 6. Safari Indonesia game correspondent details.

	Correspondent			
Total	Men	Women	Beginner	Skilled
18	8	10	8	10
Percentage	44%	56%	44%	56%

Table 7. Results of questionnaire analysis of beginner players.

Factor	**Endless Mode**		**Adventure Mode**		**Deviation (%)**
	Total	**Average**	**Total**	**Average**	
Fun	130	5.4166	120	5	8
Frustration	51	3.1875	51	3.1875	0
Important elements	1249	4.7310	1216	4.6060	2.67
Game controls	337	4.6805	334	4.6388	0.89
Game facilities	208	4.4583	196	4.0833	8.78
Game environment	240	5	241	4.4629	11.35
Rules	284	5.1428	229	4.0892	22.82

Table 8. Comparison of playing duration.

Player ID	Player Type	Playing Duration (Minute)	
		Adventure Mode	Endless Mode
1	Advanced	5	10
2	Advanced	4	19
3	Advanced	5	6
4	Beginner	4	19
5	Beginner	5	13
6	Advanced	6	11
7	Beginner	4	11
8	Beginner	5	16
9	Beginner	4	5
10	Advanced	6	10
11	Advanced	5	13
12	Beginner	3	10
13	Advanced	7	13
14	Advanced	7	6
15	Advanced	3	8
16	Advanced	5	16
17	Beginner	4	13
18	Beginner	5	8
Average		5	12

User testing of the Safari Indonesia game was conducted by 18 correspondents. Those 18 people consists of 8 men and 10 women with the age range between 20-22 years. 18 correspondents are divided into two groups of players, advanced and beginners based on the experience of playing on a similar game (Table 6). Testing performed by users with a beginner type player for both endless and adventure modes can be seen in Table 7.

The results of the analysis of the questionnaires filled by beginner players show that the level of fun in endless mode game that applied the dynamic game balancing method using adaptive algorithm has a higher value than the adventure mode game, while the frustration rate of both modes has the same value. The average value of other tested factors such as important elements in the game, game controls, game facilities, game environments, and rules show higher points for endless modes.

Next, the results of the questionnaire analysis of the advanced players show the level of fun in the endless mode is still higher than the adventure mode. However, the frustration rate in endless mode games has significantly lower value than the adventure mode. This indicates that the player can feel frustrated if the level of difficulty at a level does not match with his ability, causing the players to stop to play. Other factors analyzed in the questionnaire showed higher values in endless mode, except for one factor that is an endless mode game facility that has a lower value than the adventure mode.

Table 8 shows the result of comparison of level completion by each correspondent in endless and adventure mode. The results of the final recapitulation of all correspondents show a higher value in the endless mode. This proves that players can last longer to play a game that implements the dynamic game balancing method in it.

5. Conclusions

From the process of working on this research which through the stage of design, implementation, and testing, the following conclusions are obtained:

1. Safari Indonesia game successfully implemented the rules of the game of scrolling shooter type games well.
2. Game controls for Safari Indonesia games have been tailored to the requirements of mobile devices to make it easier for players to play on the Android platform.
3. Indonesia Safari game that has implemented the concept of dynamic game balancing using adaptive algorithm, successfully balancing between player game pattern as measured by accuracy, number of defeated enemies, and amount of health left with game environment especially enemy characters implemented on speed behavior, shot intensity, and number of enemies at each level.
4. Based on the user testing process, endless mode has a higher level of fun than adventure mode. According to an advanced player endless mode also has a low frustration value. This shows the achievement of the goal of the implementation of dynamic game balancing which is to reduce the frustration and boredom felt by players when playing a game.

References

[1] Andrade G, Ramalho G, Gomes A Sandro and Corruble V 2006 *AAAI conference on artificial intelligence and interactive digital entertainment (California),* p 3-8
[2] Koster R 2004 *Theory of Fun for Game Design* (Phoenix: Paraglyph Press)
[3] Charles D and Black M 2004 *Proc. of the Int. Conf. on Computer Games: Artificial Intelligence, Design and Education*, p 29-35
[4] Charles D, Kerr A, McNeill M, McAlister M, Black M, Kcklich J, Moore A and Stringer K 2005 *Proc. of the digital games research conference*, p 100
[5] Houlette R 2004 Player modeling for adaptive games *AI Game Programming Wisdom 2*
[6] Calvillo-Gámez E H, Cairns P and Cox AL 2010 Assessing the core elements of the gaming experience *Evaluating User Experience in Games: Human-Computer Interaction Series* ed Bernhaupt R (London: Springer)

IOP Conf. Series: Journal of Physics: Conf. Series **974** (2018) 012050 doi:10.1088/1742-6596/974/1/012050

Optimal control of predator-prey mathematical model with infection and harvesting on prey

Diva Amalia R. U.[1], **Fatmawati**[1*], **Windarto**[1], **Didik Khusnul Arif**[2]

[1]Department of Mathematics, Faculty of Science and Technology, Universitas Airlangga, Surabaya 60115, Indonesia
[2]Department of Mathematics, Faculty of Mathematics, Computing and Data Science, Institut Teknologi Sepuluh Nopember, Surabaya 60111, Indonesia.

E-mail: fatmawati@fst.unair.ac.id

Abstract. This paper presents a predator-prey mathematical model with infection and harvesting on prey. The infection and harvesting only occur on the prey population and it assumed that the prey infection would not infect predator population. We analysed the mathematical model of predator-prey with infection and harvesting in prey. Optimal control, which is a prevention of the prey infection, also applied in the model and denoted as U. The purpose of the control is to increase the susceptible prey. The analytical result showed that the model has five equilibriums, namely the extinction equilibrium (E_0), the infection free and predator extinction equilibrium (E_1), the infection free equilibrium (E_2), the predator extinction equilibrium (E_3), and the coexistence equilibrium (E_4). The extinction equilibrium (E_0) is not stable. The infection free and predator extinction equilibrium (E_1), the infection free equilibrium (E_2), also the predator extinction equilibrium (E_3), are locally asymptotically stable with some certain conditions. The coexistence equilibrium (E_4) tends to be locally asymptotically stable. Afterwards, by using the Maximum Pontryagin Principle, we obtained the existence of optimal control U. From numerical simulation, we can conclude that the control could increase the population of susceptible prey and decrease the infected prey.

1. Introduction

Ecosystem consists of two components, namely the components of biotic (alive) and abiotic (not alive). The ecosystem itself is divided into two, namely natural ecosystems and artificial ecosystems. Natural ecosystems are ecosystems that form naturally without human intervention. In the other hand, artificial ecosystems occur with human intervention. Within the ecosystem, there is a reciprocal relationship or interaction between the living and the environment. Interactions that occur can be mutualism, commensalism, parasitism, and predation. This will all be studied in ecology. Ecology is a branch of biological science that studies about living things and their habitats [1].

In ecology there is the term food chain. The food chain is the transfer of energy from the organism at a tropical level to the next tropical level in the eating and eaten event in a specific order. Food chains are arranged in tropical levels, where one tropical level includes all organisms or species that share the same position in the food chain. In a food chain there are at least two species: predators and prey [2]. The occurrence of this food chain event will affect the population of a species, because there is interaction in the form of predation in a food chain.

Another problem that might occur in a population is an epidemic. Epidemic is when an illness or an outbreak occurs at a higher-than-expected frequency [3]. Epidemic events can occur in any population of living creatures, including populations in the food chain. One of the examples of epidemics that occur in real life is the outbreak of White Spot Disease (WSD) which attacks the penaid shrimp (*Penaeidae sp*) in a pond. Pond is an example of artificial ecosystem. The prevalence of this disease is caused by White Spot Syndrome Virus (WSSV). This disease is caused by a lack of oxygen levels in water, bad water quality, and poor aquatic environments [4].

The predation interactions in the ecosystem and the presence of epidemics that occur in the population, affect the balance of the ecosystem. The balance can be achieved if the average number of populations of predator and prey populations that are interacting in proportion. But the epidemic in the population will affect the balance of predators and prey. Therefore, we should find a way to cope with the occurrence of epidemics, one of which is to provide control in the form of prevention of the occurrence of epidemics in the prey. As mentioned above, WSD disease greatly affects the development of penaid shrimp populations, while penaid shrimps (*Penaeidae sp*) they have birds as their natural predators such as the cangak birds (*Ardea cinera rectirostris*) and blekok birds (*Ardeola ralloides*) [5].

Several previous studies have developed predator-prey mathematical model with infected prey [6] and the application of optimal controls in predator-prey system [7]. Furthermore, there is also the addition of harvesting factor in the predator-prey system with infected prey in attempt to balance the predator-prey populations [8]. But in real life, harvesting somehow is not an easy thing to do because it will be less efficient, especially if the prey population is a lot of small creatures such as shrimp. Therefore, another effort besides harvesting is needed to handle the infection. One of the efforts is by giving prevention to the infection. For the example, providing prevention to WSD infection by giving the prey population a liquid extract of mangrove tree to attempt to boost the shrimp's immune against the WSD. Also, by giving a natural based liquid extract to a pond will be safer to environment compared to other chemical fluid drugs [9]. This effort is able to do in artificial ecosystem because the area is still accessible to human. In this paper, the mathematical model that developed by authors in [8] is modified by inserting an optimal control that is the prevention of infection in prey population.

2. The Predator-prey mathematical model with infection and harvesting on prey

Below are the assumptions that are used in formation of predator-prey mathematical model with infection and harvesting on prey:

a. The predator-prey mathematical model with infection and harvesting in prey involves three sub populations:
   i. $x_1(t)$ is the number of susceptible prey population at t.
   ii. $x_2(t)$ is the number of infected prey population at t.
   iii. $y(t)$ is the number of predator population at t.
b. The infection occurs because of a kind of virus and it is spread among prey population according to the *S-I-S (Susceptible-Infected-Susceptible)* model.
c. The susceptible prey population growth is following logistic model.
d. The infected prey population is harvested.
e. The predator population eats both type of prey.
f. The Holling type II response function is applied in the predation of susceptible prey and Holling type I response function is applied on infected prey.
g. Predator could not be infected.
h. When the prey population extinct, predator will experience natural death.

The parameter description of the model is given in Table 1.

Table 1. The parameter of predator-prey mathematical model

Parameter	Explanation
a	Logistic growth rate of susceptible prey
k	Environmental carrying capacity
α	Rate of contact between susceptible prey and infected prey
β	Rate of transformation from infected prey to susceptible prey
p_1	Predation rate on susceptible prey
p_2	Predation rate on infected prey
h	Rate of harvesting of infected prey
c_1	Conversion efficiency on susceptible prey
c_2	Conversion efficiency on infected prey
s	Half saturation constant
m	Natural death rate of predator

The predator-prey model with infection and harvesting on prey is presented as follows:

$$\frac{dx_1}{dt} = ax_1\left(1-\frac{x_1}{k}\right) - \alpha x_1 x_2 + \beta x_2 - \frac{p_1 x_1 y}{1+sx_1} \quad (1)$$

$$\frac{dx_2}{dt} = \alpha x_1 x_2 - \beta x_2 - p_2 x_2 y - hx_2 \quad (2)$$

$$\frac{dy}{dt} = \frac{c_1 p_1 x_1 y}{1+sx_1} + c_2 p_2 x_2 y - my \quad (3)$$

with $x_1, x_2, y \geq 0$ and all of the parameters are non-negative.

3. Analysis of the model

The Predator-prey mathematical model with infection and harvesting in prey has five equilibrium points, namely the extinction equilibrium $(\boldsymbol{E_0})$, the infection free and predator extinction equilibrium $(\boldsymbol{E_1})$, the infection free equilibrium $(\boldsymbol{E_2})$, the predator extinction equilibrium $(\boldsymbol{E_3})$, and the coexistence equilibrium $(\boldsymbol{E_4})$ which will be mentioned as follows.

a. The extinction equilibrium (E_0)

$$E_0 = \left(x_{1_0}, x_{2_0}, y_0\right) = (0,0,0).$$

b. The infection free and predator extinction equilibrium (E_1)

$$E_1 = \left(x_{1_1}, 0, 0\right) = (k, 0, 0).$$

c. The infection free equilibrium (E_2)

$$E_2 = \left(x_{1_2}, 0, y_2\right) = \left(\frac{m}{c_1 p_1 - ms}, 0, \frac{ac_1(kc_1p_1 - kms - m)}{k(c_1p_1 - ms)^2}\right),$$

which exists if

i. $c_1 p_1 > ms$

ii. $kc_1 p_1 > m(ks+1)$.

d. The predator extinction equilibrium (E_3)

$$E_3 = \left(x_{1_3}, x_{2_3}, 0\right) = \left(\frac{\beta+h}{\alpha}, \frac{a(k\alpha\beta + k\alpha h - \beta^2 - 2\beta h - h^2)}{kh\alpha^2}, 0\right),$$

which exists if

$$k(\alpha\beta + \alpha h) > \beta^2 + 2\beta h + h^2.$$

e. The coexistence equilibrium (E_4)

$$E_4 = ({x_1}^*, {x_2}^*, y^*) = \left(\frac{\beta + p_2 y^* + h}{\alpha}, \left(m - \frac{c_1 p_1 \beta + c_2 p_2 y^* + c_1 p_1 y^*}{\alpha + s(\beta + p_2 + y^* + h}\right)\frac{1}{c_2 p_2}, \frac{a{x_1}^*\left(\frac{{x_1}^*}{k} - 1\right) - {x_2}^* h}{p_2 {x_2}^* + \frac{p_1 {x_1}^*}{1+s{x_1}^*}}\right),$$

which exists if

i. $m > \frac{c_1p_1\beta+c_2p_2y^*+c_1p_1y^*}{\alpha+s(\beta+p_2+y^*+h}$

ii. $\frac{ax_1^{*2}}{k} > x_2^*h + ax_1^*.$

To determine the local stability of the equilibrium points, it is necessary to linearize the predator-prey mathematical model in the presence of infection and harvesting in the prey using the Jacobian matrix.

a. Stability of the extinction equilibrium (E_0)
Linearizing the model (1)-(3) near the equilibrium E_0 gives eigenvalues: a, $-\beta - h$, and $-m$. Since there is a positive eigenvalue, the equilibrium is unstable.

b. Stability of the infection free and predator extinction equilibrium (E_1)
Linearizing the model (1)-(3) near the equilibrium E_1 gives eigenvalues: $-a$, $\alpha k - \beta - h$, and $\frac{c_1p_1k}{1+sk} - m$. The E_1 equilibrium will be locally asymptotically stable if all of eigenvalues are negatives. Therefore, we have these following conditions:

i. $\frac{\alpha k}{\beta+h} < 1$

ii. $\frac{c_1p_1k}{1+sk} < m.$

c. Stability of the disease free equilibrium (E_2)
Linearizing the model (1)-(3) near the equilibrium E_2 gives eigenvalues: K_2, and the roots of this following quadratic equation:

$$\lambda^2 + (-K_1)\lambda + \frac{K_3m}{c_1} = 0,$$

where

$$K_1 = a\left(1 - \frac{2m}{k(c_1p_1-ms)}\right) - \frac{a(kc_1p_1-kms-m)}{kc_1p_1}$$

$$K_2 = \frac{\alpha m}{(c_1p_1 - ms)} - \beta - \frac{ac_1p_1(kc_1p_1 - kms - m)}{k(c_1p_1 - ms)^2} - h$$

$$K_3 = \frac{\alpha(kc_1p_1-kms-m)}{kp_1}.$$

Based on the Routh-Hurwitz criteria, the roots of equation will be negatives, or the real parts will be negatives if only if $-K_1, \frac{K_3m}{c_1} > 0$.

It is observed that the equilibrium E_2 is locally asymptotically stable if

i. $\frac{2am}{k(c_1p_1-ms)} + \frac{a(kc_1p_1-kms-m)}{kc_1p_1} > a$

ii. $kc_1p_1 > m(ks + 1)$.

iii. $\frac{\alpha m}{(c_1p_1-ms)} < \beta + \frac{ac_1p_1(kc_1p_1-kms-m)}{k(c_1p_1-ms)^2} + h.$

d. Stability of the predator extinction equilibrium (E_3)
Linearizing the model (1)-(3) near the equilibrium E_3 gives eigenvalues: $F_3$and the roots of this following quadratic equation:

$$\lambda^2 - F_1\lambda + hF_3 = 0$$

where

$$F_1 = a\left(1 - \frac{2(\beta + h)}{\alpha k}\right) - \left(\frac{a(k\alpha\beta + k\alpha h - \beta^2 - 2\beta h - h^2)}{kh\alpha}\right)$$

$$F_2 = \left(\frac{a(k\alpha\beta + k\alpha h - \beta^2 - 2\beta h - h^2)}{kh\alpha}\right)$$
$$F_3 = \frac{c_1 p_1(\beta + h)}{\alpha + s(\beta + h)} + c_2 p_2 \left(\frac{a(k\alpha\beta + k\alpha h - \beta^2 - 2\beta h - h^2)}{kh\alpha^2}\right) - m.$$

Based on the Routh-Hurwitz criteria, the roots of equation will be negative or the real parts will be negatives if only if $-F_1, F_3 > 0$.
So the equilibrium E_3 is locally asymptotically stable if

i. $\frac{2a(\beta+h)}{\alpha k} + \left(\frac{a(k\alpha\beta+k\alpha h-\beta^2-2\beta h-h^2)}{kh\alpha}\right) > a,$
ii. $k\alpha\beta + k\alpha h > \beta^2 + 2\beta h + h^2$
iii. $\frac{c_1 p_1(\beta+h)}{\alpha+s(\beta+h)} + c_2 p_2 \left(\frac{a(k\alpha\beta+k\alpha h-\beta^2-2\beta h-h^2)}{kh\alpha^2}\right) < m.$

e. Stability of the coexistence equilibrium (E_4)

The stability of coexistence equilibrium (E_4) is difficult to be determined analytically because the equilibrium point does not appear explicitly and it depends on the many variables. Therefore, a numerical approach is needed to determine the stability of the coexistence equilibrium point (E_4) by using phase portrait. The parameter values of the model are given in Table 2 and 3, respectively

Table 2. The initial values for phase portrait.

Initial values	x_1	x_2	y	Colors
1	4	2	2	Blue
2	60	50	40	Red
3	100	60	50	Green

Table 3. Parameter values.

Parameter	Value	Reference
a	16	Assumed
k	100	[8]
α	0.8	[8]
β	0.7	[8]
p_1	0.33	[8]
p_2	0.44	[8]
h	0.7	[8]
c_1	0.04	[8]
c_2	0.04	[8]
s	0.5	[8]
m	0.2	[8]

Figure 1 is the numerical simulation result of phase portrait of the model (1)-(3). Those orbits tend to a same point as time evolves. Thus, based on the numerical simulation we can conclude that the coexistence equilibrium E_4 tends to be asymptotically stable.

IOP Conf. Series: Journal of Physics: Conf. Series **974** (2018) 012050 doi:10.1088/1742-6596/974/1/012050

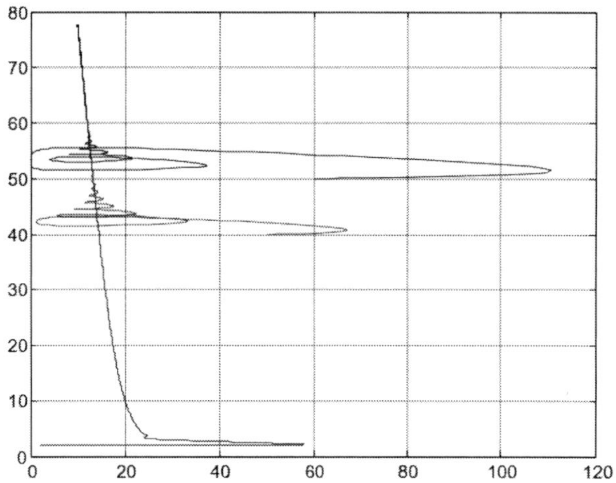

Figure 1. Phase portrait of the model in x_2 and y plane. The horizontal axis represents x_2 variable, whereas the vertical axis represents the y variable.

4. Optimal control problem

The control of predator-prey system is possible to do if there is a certain limit that still can be reached by human. Control that applied in this model is in the form of prevention of infection. Based on that, we can form the predator-prey mathematical model with infection and harvesting on prey that has been modified by control variable as follows:

$$\frac{dx_1}{dt} = ax_1\left(1-\frac{x_1}{k}\right) - (1-U)\alpha x_1 x_2 + \beta x_2 - \frac{p_1 x_1 y}{1+sx_1}, \tag{4}$$

$$\frac{dx_2}{dt} = \alpha(1-U)x_1x_2 - \beta x_2 - p_2 x_2 y - hx_2, \tag{5}$$

$$\frac{dy}{dt} = \frac{c_1 p_1 x_1 y}{1+sx_1} + c_2 p_2 x_2 y - my. \tag{6}$$

The purpose of the optimal control is to maximize the number of susceptible prey, also to minimize the cost of the control. The Maximum Pontryagin Principle [10] is used in this problem.

We consider an optimal control problem with the objective function given by

$$J(U) = \int_0^{t_f}\left(-\omega_1 x_1(t) + \omega_2 U^2(t)\right) dt \tag{7}$$

where ω_1, ω_2 are weighted constants for the state x_1 and the control variable. Our goal is to find an optimal control U such that

$$J(U^*) = \min_\Gamma J(U) \tag{8}$$

where $\Gamma = \{U | 0 \le U \le 1\}$.

The Maximum Pontryagn Principle converts the equations (4)-(6), (7) and (8) into a problem of minimizing pointwise a Hamiltonian H, with respect to U [11], that is

$$H = -\omega_1 x_1 + \omega_2 U^2 + \begin{pmatrix}\delta_1\\ \delta_2\\ \delta_1\end{pmatrix}^T \begin{pmatrix} ax_1\left(1-\frac{x_1}{k}\right) - (1-U)\alpha x_1 x_2 + \beta x_2 - \dfrac{p_1 x_1 y}{1+sx_1} \\ \alpha(1-U)x_1x_2 - \beta x_2 - p_2x_2y - hx_2 \\ \dfrac{c_1 p_1 x_1 y}{1+sx_1} + c_2p_2x_2y - my \end{pmatrix}$$

The variable $\delta_i, i = 1,2,3$ are called adjoint variables satisfying the following co-state equations

$$\dot{\delta}_1 = \frac{\partial H}{\partial x_1} = -\omega_1 + \delta_1\left(a - \frac{2x_1}{k} - (1-U)\alpha x_2 + \left(\frac{p_1 y(1+sx_1) - c_1 x_1 y s}{(1+sx_1)^2}\right)\right)$$
$$+ \delta_2\big((1-U)\alpha x_2\big) + \delta_3\left(\frac{c_1 p_1 y(1+sx_1) - c_1 p_1 x_1 y s}{(1+sx_1)^2}\right)$$

$$\dot{\delta}_2 = \frac{\partial H}{\partial x_2} = \delta_1(-(1-U)\alpha x_1 + \beta) + \delta_2\big((1-U)\alpha x_1 - \beta - p_2 y - h\big) + \delta_3 c_2 p_2 y$$

$$\dot{\delta}_3 = \frac{\partial H}{\partial y} = \frac{\delta_1 p_1 x_1}{1+sx_1} - \delta_2 p_2 x_2 + \delta_3\left(\frac{c_1 p_1 x_1}{1+sx_1} - m\right)$$

where the transversality conditions
$\delta_1(t_f) = \delta_3(t_f) = \delta_3(t_f) = 0.$

The optimal control U can be solve from the optimally condition,

$$\frac{\partial H}{\partial U} = 0.$$

Hence, we obtain

$$U = \frac{(\delta_2 - \delta_1)\alpha x_1 x_2}{2\omega_2}$$

The value of U is in interval between 0 and 1, so that some possibilities are obtained below:

$$U^* = \begin{cases} 0 & \text{, if } \dfrac{(\delta_2 - \delta_1)\alpha x_1 x_2}{2\omega_2} \le 0 \\ \dfrac{(\delta_2 - \delta_1)\alpha x_1 x_2}{2\omega_2} & \text{, if } 0 < \dfrac{(\delta_2 - \delta_1)\alpha x_1 x_2}{2\omega_2} < 1. \\ 1 & \text{, if } \dfrac{(\delta_2 - \delta_1)\alpha x_1 x_2}{2\omega_2} > 1 \end{cases}$$

Hence, we obtain the optimal control variable value as follows

$$U^* = min\left(max\left(0, \frac{(\delta_2 - \delta_1)\alpha x_1 x_2}{2\omega_2}\right), 1\right) \quad (9)$$

Next we discuss the numerical approach of the optimality system. The optimality system is the state and adjoint systems coupled with the optimal control characterization.

5. Numerical simulation

In this section, we present the numerical simulations of model (4)-(6) with and without optimal control. An iterative scheme is used for solving the optimality system. The state equations are solved by the forward Runge-Kutta method of order 4. Then the co-state equations with the transversality conditions are solved by the backward Runge Kutta method of order 4. Finally, the controls are updated by using a convex combination of the previous controls and the value from the characterizations for U^*. This iterative process is stopped when current state, co-state and control values converge sufficiently [12].

The result of numerical simulation will be compared in healthy prey population (x_1) and also in infected prey population (x_2). The simulation will be done based on the initial values and parameter values that shown in Table 2 and Table 3. The numerical simulation results of the model are given as follows.

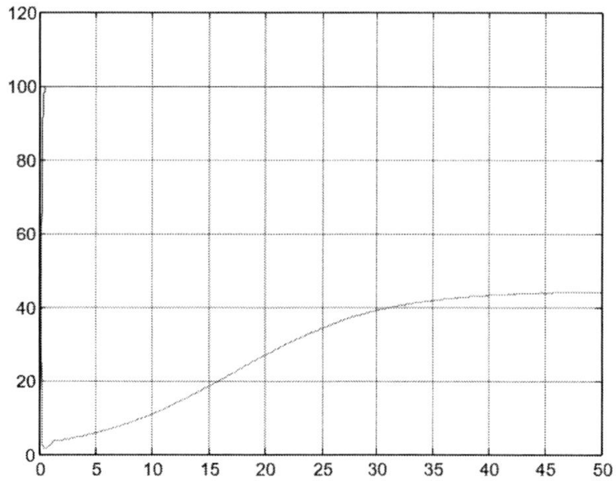

Figure 2. The dynamics of susceptible prey population (x_1). The horizontal axis represents the time index in days, whereas the vertical axis represents the number of population. Solid line (—) represents solution with control, whereas dashed line (--) represents solution without control.

Figure 2 showed that there is a difference in susceptible prey population number between before and after being given control variable. On the 50 days of observation based on the result, it can be seen that there is an increase on susceptible prey population and it will constant after it reaches number of 100 until the end of observation. It indicates that giving control in the form of prevention of infection in prey is quite influential and can be used as an effort to increase the number of susceptible prey population.

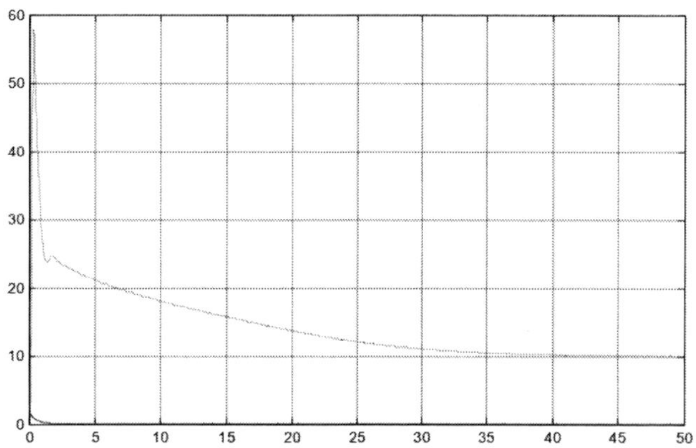

Figure 3. The dynamics of infected prey population (x_2). The horizontal axis represents the time index in days, whereas the vertical axis represents the number of population. Solid line (—) represents solution with control, whereas dashed line (--) represents solution without control.

Figure 3 shows that there is a difference in the number of prey populations infected before and after the addition of control variables. On observations made over 50 days, it showed that there was a decrease in the number of infected prey populations and would then be constant after approaching 0 until the end of the observation. This indicates that giving control in the form of prevention of infection in prey is quite influential and can be used as an effort to increase the number of infected prey population.

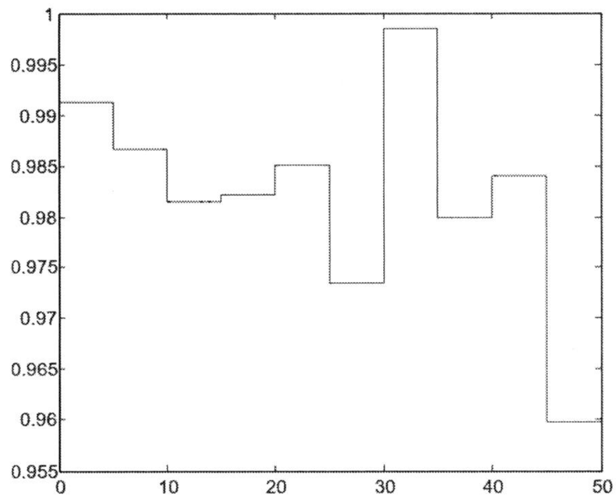

Figure 4. The profile of the optimal control. The horizontal axis represents the time index in days, whereas the vertical axis represents the control variable.

Figure 4 shows the profile of the optimal control U^*. From Figure 4, it can be seen that the control variable U^* on prevention is in the range of 0.9 to 1. The business performed on the first day is 0.99 and continues to decrease until 0.96, until the 30th day. Then on the 31st day the business will rise steadily until it reaches 1 and then decrease again to 0.96 as the last day of observation.

6. Conclusion

Based on the analytical result of predator-prey mathematical model with infection and harvesting on prey, we obtained five equilibriums, namely the extinction equilibrium (E_0), the infection free and predator extinction equilibrium (E_1), the infection free equilibrium (E_2), the predator extinction equilibrium (E_3), and the coexistence equilibrium (E_4). The extinction equilibrium is unstable, whereas the other equilibriums have their condition to be locally asymptotically stable. By using the Pontryagin Maximum Principle, the optimal control theory is then derived analytically. From the numerical simulation result, it is shown that the prevention effort can increase the number of susceptible prey population and decrease the population of infected prey.

Acknowledgments
Part of this research is financially support by the Hibah Riset Mandat, Universitas Airlangga 2017 according to SK Rektor Nomor: 569/UN3/2017.
One of the author, DKA had additional support from Directorate for Higher Education, Ministry of Research, Technology, and Higher Education of Indonesia for the Research Grant "Penelitian Terapan Unggulan Perguruan Tinggi" (PTUPT) Institut teknologi Sepuluh Nopember, 2017 according to SK Rektor Nomor: 010/SP2H/LT/DRPM/IV/2017.

References

[1] Logan D R 2015 *A First Course in Differential Equations* Third Edition (New York: Springer Science+Business Media Inc.)
[2] Jorgensen S E 2009 *Ecosystem Ecology* First Edition (Italy: Elsevier)
[3] Guerrant R L, *Walker* D H, and Weller P F 2011 *Tropical Infectious Diseases: Principles, Pathogens and Practice* Third Edition (China: Saunders Elseviers)
[4] Javier M J 2010 Dynamics of intensive production of shrimp *litopenaeus vannamei* affected by white spot disease *Aquaculture* **300** 113
[5] Chattopadhyay J and Bairagi N 2001 Pelicans at risk in salton sea-an eco-epidemiological model *Ecol. Model.* **136** 103
[6] Chattopadhyay J, Bairagi N and Chaudhuri S 2007 Harvesting as disease control measure in an

eco-epidemiological system-atheoritical study *Math. Biosci.* **217** 134
[7] Kar T K and Gosh B 2012 Sustainibility and optimal control of an exploited prey predator system through provision of alternative food to predator *Biosystems* **109** 220
[8] Sujatha K and Gunasekaran M 2016 Dynamic in a harvested prey-predator model with Susceptible-Infected-Susceptible (SIS) epidemic disease in the prey *Adv. Appl. Math. Biosci.* **7** 23
[9] Wahjuningrum D, Sholeh S H and Nuryati S 2006 Pencegahan infeksi virus White Spot Syndrome Virus (WSSV) pada udang windu (*Penaeus monodon*) dengan Cairan Ekstrak Pohon Mangrove (CEPM) *Avicennia sp.* dan *Sonneratia sp. Jurnal Akuakultur Indonesia* **5** 65
[10] Pontryagin L S, Boltyanskii V G, Gamkrelidze R V and Mishchenko E F 1962 *The Mathematical Theory of Optimal Processes* (New York: John Wiley & Sons)
[11] Naidu D S 2002 *Optimal Control Systems* (New York: CRC PRESS)
[12] Lenhart S and *Workman* J T 2007 *Optimal Control Applied to Biological Models* (New York: Chapman & Hall)

IOP Conf. Series: Journal of Physics: Conf. Series **974** (2018) 012051 doi:10.1088/1742-6596/974/1/012051

Enhancement of Efficiency and Reduction of Grid Thickness Variation on Casting Process with Lean Six Sigma Method

Witantyo and David Setyawan

Engineerig Industrial Systems Laboratory, Department of Mechanical Engineering, faculty of Industrial Technology, Institut Teknologi Sepuluh Nopember (ITS)

Corresponding author: witantyo@gmail.com

Abstract. In a lead acid battery industry, grid casting is a process that has high defect and thickness variation level. DMAIC (Define-Measure-Analyse-Improve-Control) method and its tools will be used to improve the casting process. In the Define stage, it is used project charter and SIPOC (Supplier Input Process Output Customer) method to map the existent problem. In the Measure stage, it is conducted a data retrieval related to the types of defect and the amount of it, also the grid thickness variation that happened. And then the retrieved data is processed and analyzed by using 5 Why's and FMEA method. In the Analyze stage, it is conducted a grid observation that experience fragile and crack type of defect by using microscope showing the amount of oxide Pb inclusion in the grid. Analysis that is used in grid casting process shows the difference of temperature that is too high between the metal fluid and mold temperature, also the corking process that doesn't have standard. The Improve stage is conducted a fixing process which generates the reduction of grid variation thickness level and defect/unit level from 9,184% to 0,492%. In Control stage, it is conducted a new working standard determination and already fixed control process.
Keywords: battery, grid casting, lean six-sigma, DMAIC.

1. Introduction

Casting stage in the process of battery manufacture is a stage that determines the quality of grid plates which will be processed in the next steps. Grid that has been made is transferred to pasting area to attach paste on the plate. However, there are so many defects on the grid production processes. Types of defect that happens on grid are hole, fragile, oblique, miss-trim and crack. Fragile, crack and hole are types of defect that have the highest percentage among other defects that dominantly lowered production level. The grid produced also has a high thickness variation which is not in accordance with the determined specification and creates rejects or disturb the further production processes in the battery making processes. Grid defect problems makes the company could not reach production target for fulfilling the existing demand.

2. Research Methods

This experiment was conducted by using lean six sigma method in the form of DMAIC (Define-Measure-Analyze-Improve-Control)[2]. In every stage on DMAIC method, appropriates supporting tools were used.

2.1. *Define*

Figure 1 show the casting process efficiency. The efficiency level in casting process is measured by calculating the amount of grid production a month divided by ideal amount of grid production at the whole total production time available in that month. The higher defect/reject rate and idle in production would lower the amount of grid production. Because grid was a main component consumed in the following processes, low grid production would affect to the production level of the batteries.

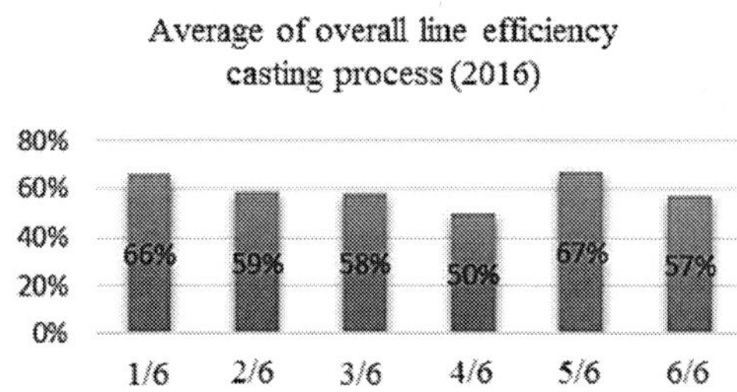

FIGURE 1. Casting Line Efficiency

2.2. Measure

Figure 2 shows the Pareto chart of casting process for June 1st – June 6th 2016 period. It is shown that the highest donator on waste is caused by defect/reject. Rejected grid needs to be reprocessed, so this matter causes wasting. Figure 3 is the amount of defect of production result for each type of defect that was obtained through records by company when casting production process was running for June 13th – September 1st 2016 period. It was shown that only 3 main defect dominates, which are the problem of fragile, holed and cracked grid.

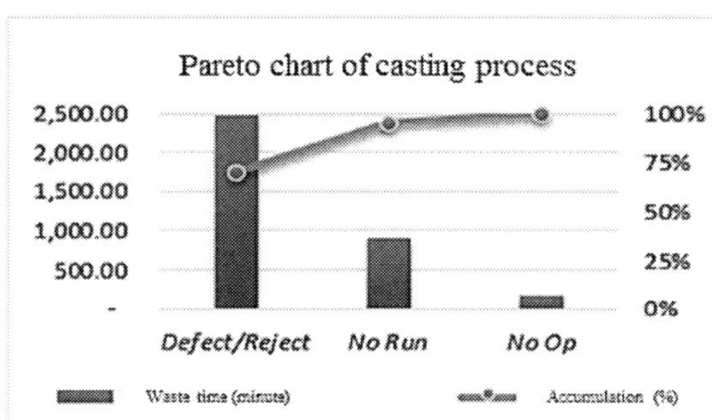

FIGURE 2. Pareto Chart of Casting Process Waste

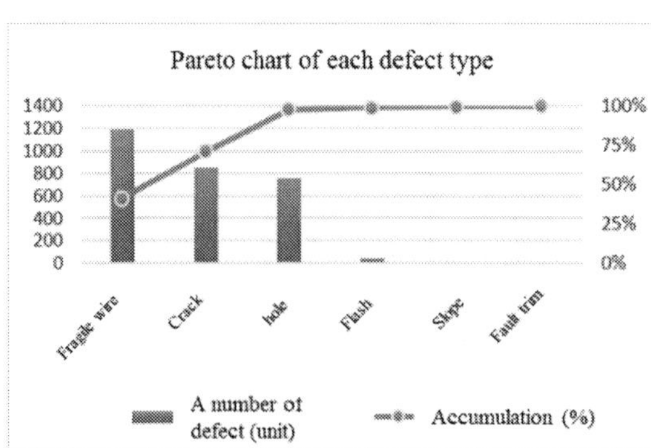

FIGURE 3. Pareto Chart of Each Type of Defect on Casting Process

Table 1 is a retrieved data from company related to the amount of defect that happens for every type of defect in casting process for August 25th, 26th and 31st 2016 period. Grid production level for those 3 days was 30150 grid, with total defect that happened was 2769 grid. From this data, it could be known that the defect/unit level that happened before the improvement process is conducted is 9,184%.

TABLE 1. Defect Level Before Improvement

	Aug 25th	*Aug 26th*	*Aug 31st*
Fragile wire	104	1056	
Crack	851		
Hole			758

Figure 4 shows grid thickness graph for CAL 1.5 type which was measured on November 2016 by taking 30 grids randomly sampled which was measured on four upper sides and four lower sides of the grid. Grid CAL 1.5 type has a thickness specification standard of 1.5 mm with permitted tolerance by the company for production of every grid type is 5% from standard thickness. So, grid CAL 1.5 type has lower tolerance limit of 1.425 mm and upper tolerance limit of 1.575 mm. But, it is shown on the graph that this thickness standard was not achieved and it's even far below the expected specification standard.

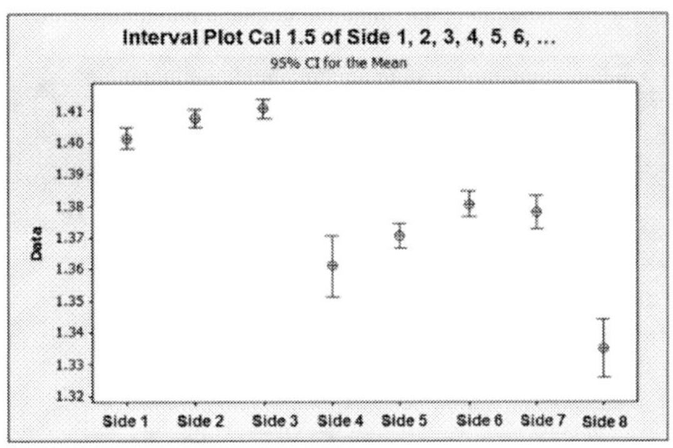

FIGURE 4. Graph of Grid Thickness Data on Every Side for Grid Cal 1.5 Type

2.3. Analyse

The analyze stage is where root cause of the problem inspected. Figure 5 shown defect called flash that retrieved at December 3rd 2016 during casting production. Flash is not a fatal defect. Small flash can still pass the inspection stage and can be used for the next step, but when there are too many flash on the grid, this grid will get into the reject category because it will potentially disturb the pasting process. It makes paste material not attached properly causing hole defect on the grid in pasting process. Grid with many flashes on the main frame or ear/lugs grid is also included into the reject product because too many flashes on the main frame or ear/lugs grid could disturb further processes.

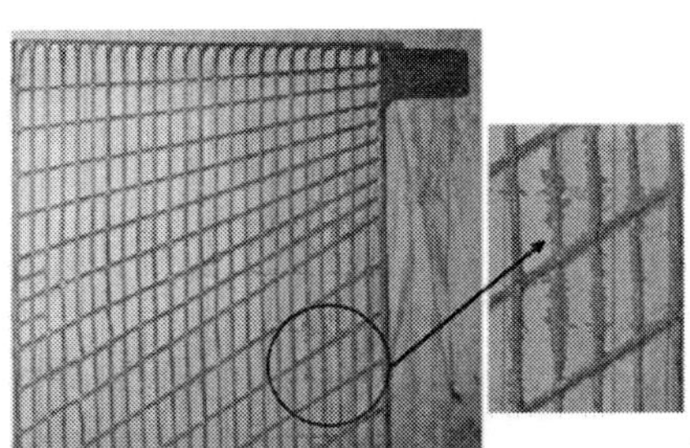

FIGURE 5. Grid of Casting Production Result on December 3rd 2016 Which Had Flash Type of Defect and Also a Fragile Grid

In figure 6, it is shown that the wire grid is not connected, it is called run out. Hole/unfilled type of defect is also included into reject product because it's potentially causing the grid to have different voltage from standard specification because electricity can't flow. It can affect the battery performance. Run out also happens in ear/lugs area that will disturb the welding process. Unfilled/run out type of defect can happen caused by the temperature is overly low on the mold, causing molten metal on the riser on the upper part can't fulfill the mold chamber which starts to freeze and the volume shrinks beforehand.

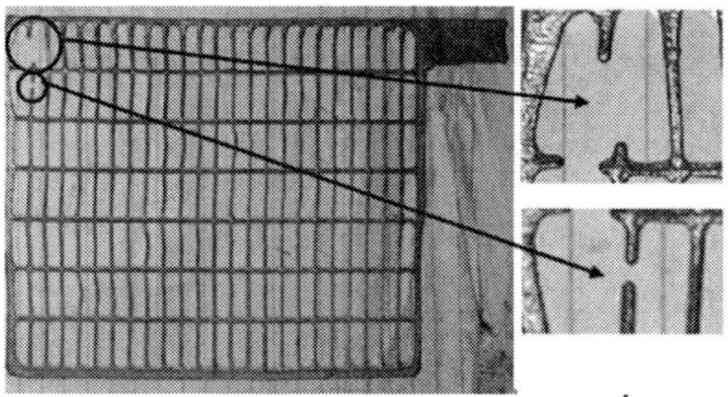

FIGURE 6. Grid As A Result Of Casting Production On December 3rd 2016 Which Has Hole/Unfilled Type Of Defect And Also Fragile Grid

Figure 7 shown the other defect that the wire and the main frame are not merged to each other. It is caused by fragile grid so it was easily break on the wire part or main frame. Grid that had fragile type of defect was taken to be used as a further observation sample and all of the observed samples were taken from the same grid. The observed samples were on the wire part that was broken, on the upper frame and lower frame that weren't broken by using microscope. The observed broken sample wire was not rubbed beforehand to see the broken pattern. The upper and lower parts of the main frame sample were rubbed beforehand because it was aimed to see the inner condition of the grid.

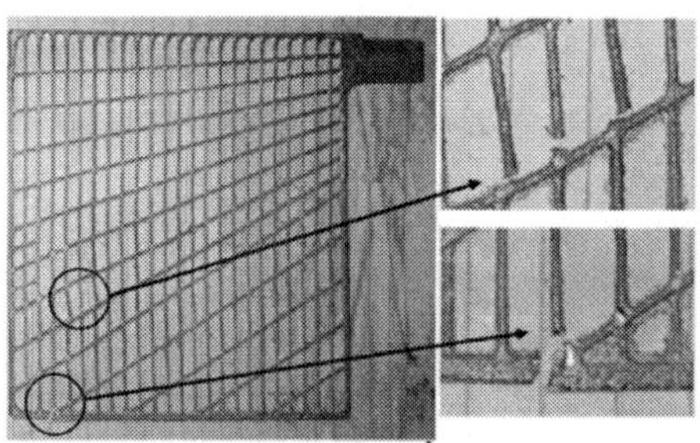

FIGURE 7. Grid Production Result on December 3rd 2016 which had a crack and fragile type of defect

In figure 8, it is shown that the broken pattern on the wire grid a brittle fracture. This brittle fracture could happen due to several things, such as inclusion/impurities in that part so it has different properties with the original properties from the strong and resilient lead-antimony alloy properties [1].

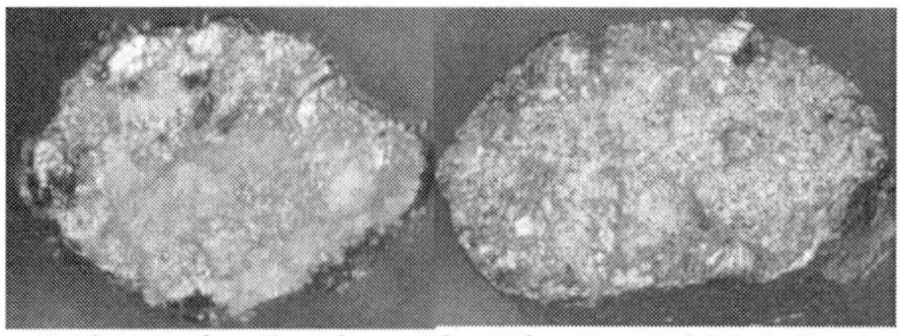

FIGURE 8. Microscope Observation Result on The Wire Part Which Is Broken (Top) And on The Wire Connection Which Is Broken (Bottom), With 100x Magnification

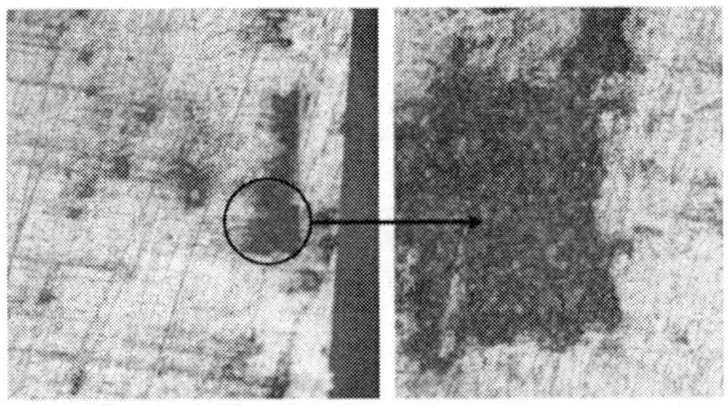

FIGURE 9. Microscope Observation Result on Upper Part of The Main Frame For 200x Magnification (Left) And For 1000x Magnification (Right)

Figure 9 is an observation sample from the upper part of the main frame grid, it is shown that this part has dark/black area. The dark/black area that is evenly spread shows that the inside of the grid has many inclusions. Inclusions on the grid as a result of casting process could be the inclusion that arises due to molten metal which has many inclusions since the beginning of the process, then it comes in to the mold when the grid molding is being conducted. From the observation, this dark/black area has a huge possibility as an oxide from lead-antimony, because of the addition of antimony on lead could increase the oxide level on the molten lead to the air [4].

Table 2 is 5 Why's table that is used to help finding root of the problem that happens in casting process by finding the problem first and then finding the causes gradually until getting root of the problem which is the main cause. This matter could be used to find root of the problem and potential failure that could happen as well. Root of the problem and potential failure will be used to determine the formulation of action plan of casting process improvement.

Table 2. Analysis of Root of the Problem with 5 why's Method

	Why 1	Why 2	Why 3	Why 4
Grid thickness varies	Wrong corking process	No standard/working instruction for corking process		
Grid has crack/fragile type of defect	Inclusion on the grid	High dross level on casting process that is exposed to air	High oxidization level on area that is exposed to air	No cover on melting pot and ladle
Grid has crack/fragile type of defect	High shrinkage level	Overly high temperature gradient	Overly high ladle temperature and overly low mold temperature	
Grid has hole type of defect	Overly fast solidification	Overly high temperature gradient	Overly high ladle temperature and overly low mold temperature	
Grid has hole type of defect	Surface layer of the mold is too thick	Error on corking process	No standard/working instruction for corking process	

2.4. *Improve*

2.4.1. *Changing on Temperature Standard for Casting Process*

Overly high temperature standard usage on melting pot area (460°C), feedline pipe (482°C) and ladle (570°C) needs to be changed considering an overly high temperature usage only aims to make sure that the lead-antimony is in liquid condition, whereas the lead-antimony has already reached the melting point at ± 327°C [3]. The temperature that is used on melting pot, feedline pipe and especially ladle needs to be changed to lower temperature to avoid overly high temperature gradient with mold temperature. Moreover, temperature of the upper mold is supposed to be higher than the lower mold, so the solidification process could be started from the lower mold.

TABLE 3. Recommendation for Temperature Standard Changing on Casting Process

	Old Standard	**Experiment Temperature**	**New Temperature Recommendation**
Melting Pot	460 °C	420 °C	460 °C
Feedline Pipe	482 °C	450 °C	460 °C
Ladle	570 °C	470 °C	460 °C
Cooling Water	40 - 60 °C	40 – 60 °C	40 - 60 °C
Upper Mold	120 – 150 °C	170 - 190 °C	170 -190 °C
Lower Mold	140 – 170 °C	120 - 150 °C	120 - 150 °C

Table 3 shows old temperature standard, experiment temperature and new temperature standard recommendation for each casting process. Experiment temperature is temperature that is used as an experiment when the production process is on the run, the given value is based on analysis and literature study that has been done before. The given new temperature recommendation is based on analysis and literature study that was conducted. The difference of experiment temperature usage between one process to the other is not too high, it is meant to improve the temperature gradient not too high so the shrinkage level which is occurred when the grid is in solidification process won't be too high because the shrinkage level will get lower as well as the temperature gradient that is used. The decrease of temperature gradient is obtained by decreasing the temperature of ladle, increasing the temperature on the upper mold and decreasing the temperature on the lower mold. This temperature gradient is possible to make the solidification process occurs sequentially, that is from the lower part to the upper part. New temperature recommendation on melting pot, feedline pipe and ladle is 460°C, this is meant so the temperature gradient between ladle and mold is not too low and not too high. An overly low temperature gradient between ladle and mold could cause an overly fast solidification level on the mold, so it could complicate the movement of molten metal to fulfill the room on casting process. An overly high temperature gradient on ladle and mold could cause waste of energy and overly high shrinkage level that occurs on casting process. Melting pot uses burner with gas fuel and feedline pipe as well as ladle use electric heater, so the higher the temperature gradient that is used, the more waste of energy as well.

2.4.2. Dross Cleaning on Casting Process

FIGURE 10. Ladle Full of Dross (Left), Already Cleaned Ladle from Dross (Right)

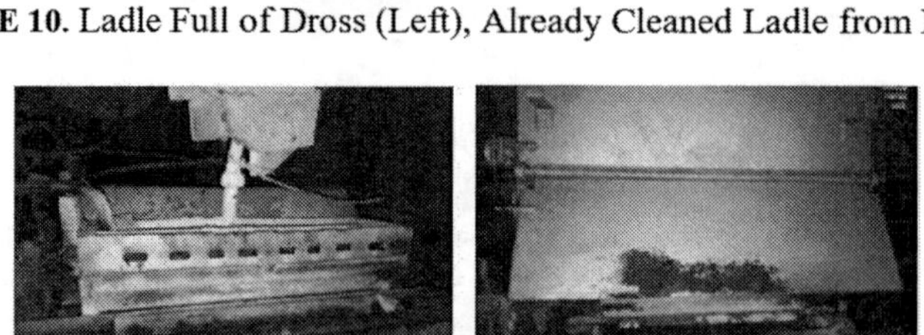

FIGURE 11. Ladle That Hasn't Covered (Left), Ladle that has Already Covered (Right)

Figure 10 is ladle sighting before and after cleaned from dross. Previously, dross that was shaped from the previous casting process, especially on ladle, was not cleaned beforehand before the

IOP Conf. Series: Journal of Physics: Conf. Series **974** (2018) 012051 doi:10.1088/1742-6596/974/1/012051

casting process was started, dross was left to flow and reduced along with the running of casting process. But, from this experiment, it is known that there should be no dross in grid casting process because dross was known to be the cause of inclusion on grid of casting process result, so this matter causes the changing of all properties of grid lead-antimony and also potentially causes the defect problem on grid. By knowing these conditions, dross cleaning on ladle should be done before casting process is started to minimize dross that gets in the grid casting process.

Figure 11 shows ladle that hasn't been covered and ladle that has already covered. Cover giving aims to not to expose the molten metal directly to open air continuously. Cover should be given especially on the ladle part considering ladle is a process right before the molten metal gets in to the mold. It's intended for molten metal in the ladle not to react to open air continuously, because this is potentially causing dross formed, so it is expected that ladle always has low dross level, even there is no dross at all.

2.4.3. Corking Process Improvement

Recommendation on the corking process are (i) Clean properly both part of the mould (movable mold and stationary mould) with brass brush, (ii) Spray the corking solution on each part of the mould with distance for about 50 cm in front of the mold at moderate and constant speed, following the flow just like on the given illustration on Figure 12 perpendicularly with mold (not at an angle), (iii) Pour talk powder evenly on both surface of the mold.

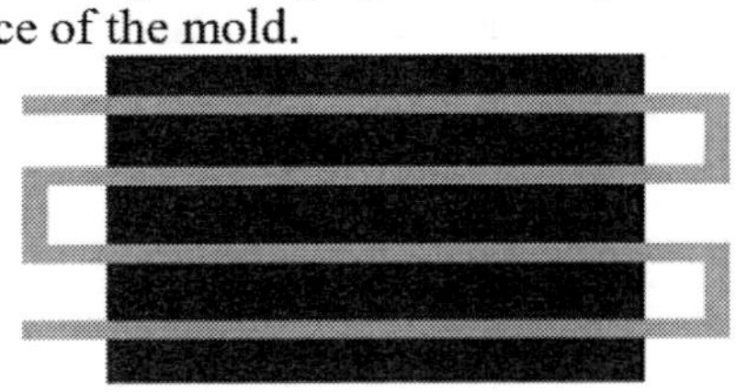

FIGURE 12. Illustration of Recommended Corking Process Movement

2.4.4. After Improvement Data

TABLE 4. Defect Level After Improvement

	Dec 6th	*Dec 10th*	*Dec 11th*	*Dec 17th*	*Dec 18th*
Hole	23	19	21	24	15
Fragile	18	17	17	17	12
Crack	13	16	12	13	16

Table 4 is data of the amount of casting process production result defect which was obtained after action plan has been conducted. It is shown that there is a significant reduction of defect after action plan was conducted. After observation on casting process, it is known that the highest defect level happens when casting process is just started and after stable, the process will generate grid that is relatively has small defect. The reduction of defect level is a result from action plan implementation which is ladle cleaning from dross before casting process is started, temperature changing on casting process according to the new temperature recommendation, changing of corking process and cover addition on ladle. Grid production level for those 5 days is 51000 grids, with the occurrence of total defect is 251 grids. From this data, it is known that defect/units level after improvement process has decreased to 0,492%.

Figure 13 shows graph of grid thickness data on each side which was measured by using the same method after action plan was conducted. From the data above, it is shown that grid thickness variation decreases and grid that was generated has thickness close to the standard and gets in to the specification limit given by company, which is 5% from thickness standard. The reducing of grid thickness variation was obtained by doing the new recommended corking process. The new corking process is possible for the operator to do a uniform layering process on all side of the mold, so it is obtained a relatively uniform grid from casting process production result.

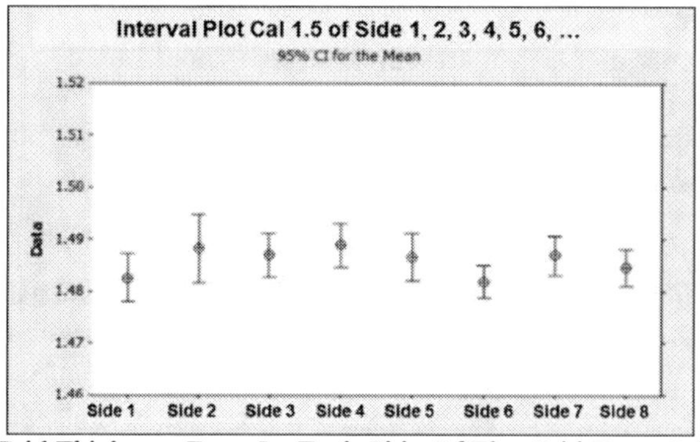

FIGURE 13. Graph Of Grid Thickness Data On Each Side Of The Grid Type Cal 1.5 After Action Plan Is Conducted

Conclusion

The conclusions that was obtained from the experiment about Enhancement of Efficiency and Reduction of Grid Thickness Variation on Casting Process with Lean Six Sigma Method are:

1. Fragile and crack type of defects are caused by the amount of inclusion in casting process so the overall material doesn't have uniform properties, and also caused by an overly high ladle temperature usage and overly low mold temperature so the shrinkage level that happens on solidification process becomes high. Grid that doesn't have uniform structure overall then it can't maintain its structure very well on shrinkage process.
2. Inclusion on the grid is an oxide from lead-antimony which is on the ladle then it gets into the casting process, this causes the molding result doesn't have a uniform property overall.
3. Hole/unfilled type of defect is caused by an overly high temperature gradient between ladle and mold. The temperature of upper mold is too low, causing the solidification process of molten metal will happen too fast, so the molten metal will have obstacles to fulfill the empty space on the mold.
4. There is no standard for corking process, which causes the conducted corking process, is not uniform between one operator to the other. It leads to different thickness on mold layering that could cause grid thickness variation and overly thick corking layer could cause hole type of defect on grid/production result.
5. Overly thin corking layer on the flat side of the mold could cause flash type of defect because it can make that part not covered perfectly when casting process is conducted. Overly thick or overly thin on cavity part of the mold could cause grid thickness variation on casting process result.
6. The given improvement suggestion consists of giving new temperature recommendation, cleaning ladle from dross, giving cover on ladle to avoid dross and recommendation of corking process standard.
7. After action plan has been done, thickness variation on grid of casting process result decreased and the defect level on casting process is decreasing from 9,184% to 0,492%.

References

[1] Casas, J. S. and Sordo, J, "Lead: Chemistry, Analytical Aspects, Environmental Impact, and Health Effects," (Elsevier B.V, Amsterdam, 2006).
[2] V. S. Terra, "Lean Six Sigma International Standards and Global Guidelines," (Taylor and Francis Group, 2015).
[3] W. Hofmann, "Lead and Lead Alloys: Properties and Technology". (Springer-Verlag Berlin Heidelberg, New York).
[4] S. Guruswamy, "Engineering Properties and Applications of Lead Alloys". (Marcel Dekker Inc, New York, 2000).

Decreasing inventory of a cement factory roller mill parts using reliability centered maintenance method

Witantyo and Anita Rindiyah

Engineering Industrial Systems Laboratory, Department of Mechanical Engineering, faculty of Industrial Technology, Institut Teknologi Sepuluh Nopember (ITS)

Corresponding author: witantyo@gmail.com

Abstract. According to data from maintenance planning and control, it was obtained that highest inventory value is non-routine components. Maintenance components are components which procured based on maintenance activities. The problem happens because there is no synchronization between maintenance activities and the components required. Reliability Centered Maintenance method is used to overcome the problem by reevaluating maintenance activities required components. The case chosen is roller mill system because it has the highest unscheduled downtime record. Components required for each maintenance activities will be determined by its failure distribution, so the number of components needed could be predicted. Moreover, those components will be reclassified from routine component to be non-routine component, so the procurement could be carried out regularly. Based on the conducted analysis, failure happens in almost every maintenance task are classified to become scheduled on condition task, scheduled discard task, schedule restoration task and no schedule maintenance. From 87 used components for maintenance activities are evaluated and there 19 components that experience reclassification from non-routine components to routine components. Then the reliability and need of those components were calculated for one-year operation period. Based on this invention, it is suggested to change all of the components in overhaul activity to increase the reliability of roller mill system. Besides, the inventory system should follow maintenance schedule and the number of required components in maintenance activity so the value of procurement will be decreased and the reliability system will increase.

1. Introduction

Inventory is needed for the sake of the production process smoothness in a company. But too much inventories would be a problem. According to a cement factory data, it was obtained that value of 3 types of inventories are so high. They are raw materials for 51 billion, routine components for 227 billion and non-routine components for 647 billion. Raw materials and routine components are fast moving inventories that has been already determined by ERP the minimum-maximum quantity of inventory and time of order on the ERP system while non-routine components are a slow-moving inventory and hard to control because it was purchased based on user request.

The components availability in maintenance activities becomes big issues. In one side, the user assigned to do maintenance activities want their components always on stock when machine break down. On the other hand, Department of Planning Control Supplies want to keep inventories low but always available when is required. Considering the maintenance activities, components availability and inventory are intertwined, so the determination of required components is a must so then the component availability is always maintained but as low as possible.

2. Research Methodology

The object of this research is a system on material processing unit that has the highest downtime frequency which is roller mill system. The related problem about reliability, downtime frequency and the availability of components for maintenance activity should be solved by determining the components needed on the right maintenance task. Reliability Centered Maintenance method will be used to evaluate the whole system to determine the failure distribution each component. Then, the required components will be reclassified from non-routine component to become routine component, so the inventory system can be determined. The expected outcome is to reduce the value of inventory while still keeping the components availability so the maintenance activity could be running smoothly.

The input data in this experiment is a time to failure data from each of the components. By using a software, failures data was processes to the appropriate distribution including known parameters and can be used to calculate the components needed. But before calculating the components need, a component reclassification will be conducted first, following the process in figure 1.

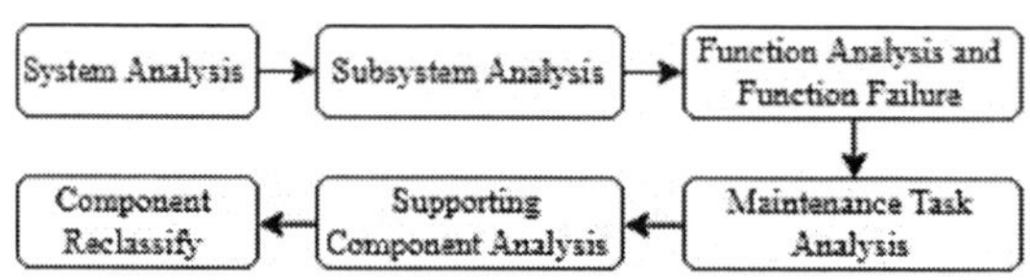

FIGURE 1. Component Reclassification Stage

2.1. System Analysis

The first step is conducting the system analysis. The RCM process analysis is conducted on system level then on component level. It's because of the function failure of a system that usually can be seen from system level first and then it could be determined the supporting function of the system on component level. The system analysis is using a pareto diagram to obtain the system that has the highest downtime.

2.2. Subsystem Analysis

Next is subsystem analysis which is conducted on chosen system. Each of the subsystems were explained their functions, so it can be formed a unity of subsystem that compiles system in a Functional Block Diagram [1].

2.3. Function Analysis and Function Failure

Function and failure function forms from each subsystem were determined and arranged in RCM Information Worksheet [1] just like in table 1.

TABLE 1. RCM Information Worksheet

RCM INFORMATION WORKSHEET			
Function	*Functional Failure*	*Failure Mode*	*Failure Effect*

2.4. Maintenance Task Analysis

After the failure function forms of each subsystem is already known, next the right maintenance task could be determined, using RCM Decision Worksheet[1] just like in table 2.

TABLE 2. RCM Decision Worksheet

Information Reference			Consequence Evaluation				H1	H2	H3	Default Action			Proposed Task
							S1	S2	S3				
							O1	O2	O3				
F	FF	FM	H	S	E	O	N1	N2	N3	H4	H5	S4	

2.5. Supporting Component Analysis

Determine the supporting components during maintenance activity which is obtained from work order system data, then it is arranged in a System Work Breakdown Structure (SWBS)[2] component in each subsystem just like in table 3.

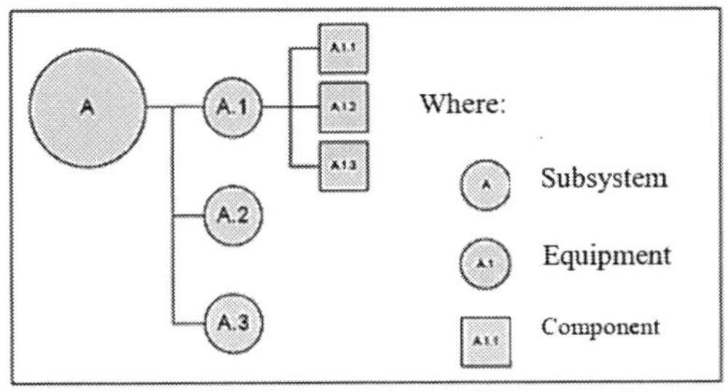

FIGURE 2. Work Breakdown Structure System

2.6. Determine the Component Needs

The number of needed components (N) for a certain period is obtained by using this equation below [3].

$$N = \frac{t}{T} + \frac{K^2-1}{2} + K\sqrt{\frac{t}{To}}\Phi^{-1}(p)$$

Where:

N = the number of needed components
t = the period of planned operation
T = MTTF
K = coefficient variation of MTTF
$\Phi^{-1}(p)$ = service level

3. Analysis

3.1. System Analysis

Based on cement production process in A cement factory., the material processing unit is chosen as an experiment object, so those systems in this unit will be shown in figure 3. Those systems on material processing unit works consecutively from raw material reclaimer process to rolled material. In the running of the process, there have been a downtime for sure, either it's scheduled downtime or unscheduled downtime. Unscheduled downtime is the main source that causing the loss of productivity in most part of the company. Because of that, this experiment will conduct an analysis in the system that has the highest unscheduled downtime which is the roller mill, it is shown in figure 4.

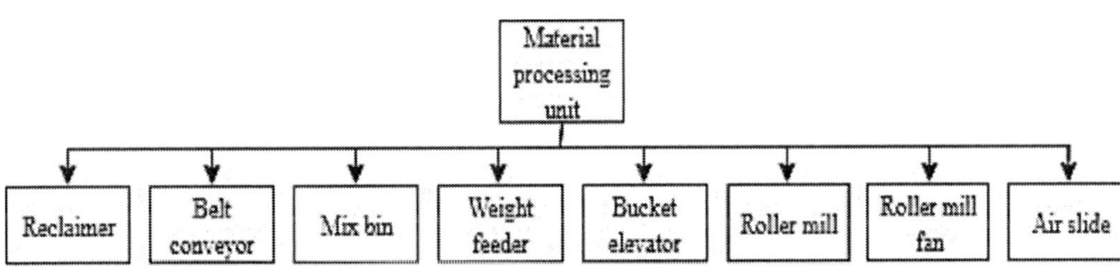

FIGURE 3. Systems on Material Processing Unit

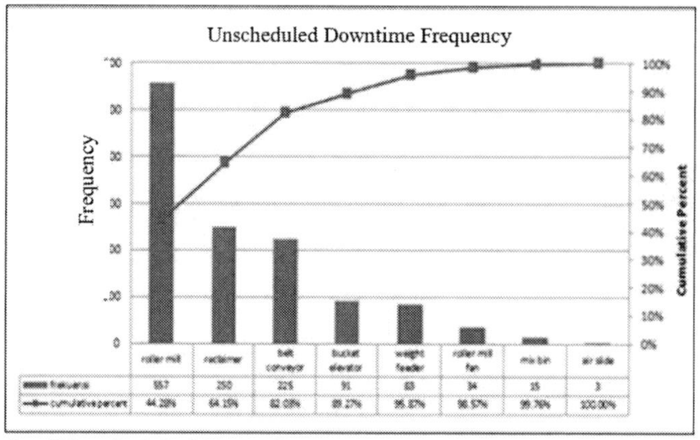

FIGURE 4. Graph of Unscheduled Downtime System Frequency in Plan 1 Material Process Unit

3.2. Subsystem Analysis

Roller mill system in PT. Semen Indonesia is a Vertical Roller Mill which is divided in 4 subsystems as shown in figure 5. The subsystem are :

- Grinding Subsystem
  In this subsystem, the material grinding process occurs which is conducted by roller and mill table.
- Drying and Transporting Subsystem
  In this subsystem, the material transporting or feeding process occurs consecutively using rotary feeder and also the material drying process using louvre ring insert and armour ring.
- Separating Subsystem
  In this subsystem, soft material and coarse material are separated by classifier and reducer classifier.
- Hydraulic Subsystem
  This subsystem functions to control pressure and vibration in roller mill by absorber, rocker arm and bladder accumulator.

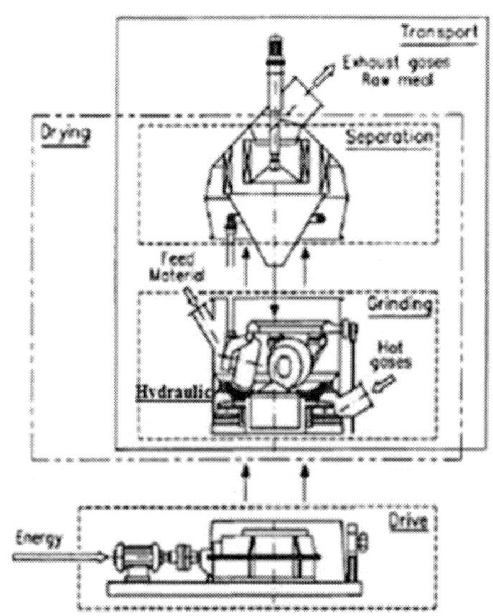

FIGURE 5. Subsystem of Roller Mill

Based on the description in the previous subsystem, Functional Block Diagram (FBD) was developed according to figure 6.

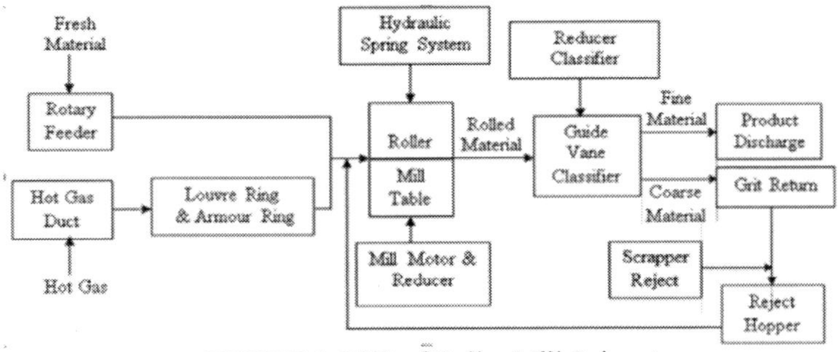

FIGURE 6. FBD of Roller Mill Subsystem

This process is started by the material entering through rotary feeder and hot gas entering through hot gas duct which is directed by louvre ring and armor ring to the roller mill. Inside the roller mill occurs the milling process in mill table and 4 rollers so it produces rolled material. In this milling process, the vibration and pressure will be controlled by Hydraulic Spring System. The rolled material will be separated based on its size by guide vane classifier. If the size of the rolled material is already fit to the standard (fine material), then it will be released to product discharge while the rolled material that hasn't fulfilled the standard (coarse material) will be put in to the reject hopper by scrapper reject through grit return classifier. This coarse material then will be put back to the roller mill to be milled again until the size is fitted to the standard.

3.3. *Function Analysis and Function Failure*

Analysis on Separating Subsystem which contains function failure, function failure modus and effect of function failure on each of the subsystems that will be shown in table 3.

TABLE 3. RCM Information Worksheet Separating Subsystem

RCM Information Worksheet			
Function	Functional Failure	Failure Mode	Failure Effect
1. Controlling the material size which is released from roller mill after milling process by separating the material (fine material and coarse material)	A. Blade rotation is not maximum	1. Wear on the rotor blade	The produced materials are too coarse Too many rejected material
		2. Wear on guide vane	
		3. Wear on outer seal and inner seal	Bad lubrication on bearing buffer
		4. Misalignment on coupling	High vibration
		5. Unbalanced blade classifier	High vibration

3.4. *Maintenance Task Analysis*

Maintenance Task analysis on separating subsystem is shown in table 4 below.

TABLE 4. RCM Decision Worksheet Separating Subsystem

Information Reference			Consequence Evaluation				H1 S1 O1	H2 S2 O2	H3 S3 O3	Default Action			Proposed Task
F	FF	FM	H	S	E	O	N1	N2	N3	H4	H5	S4	
1	A	1	Y	N	N	Y	N	N	Y				Blade replacement when overhaul, scheduled discard task
1	A	2	Y	N	N	Y	N	N	Y				Guide vane replacement when overhaul, scheduled discard task
1	A	3	Y	N	N	Y	N	Y					Replacement or service when overhaul, scheduled restoration task
1	A	4	Y	N	N	N	Y						Checking or service when scheduled

IOP Conf. Series: Journal of Physics: Conf. Series **974** (2018) 012052 doi:10.1088/1742-6596/974/1/012052

													on condition task
1	A	5	Y	N	N	N	Y						Checking or service when scheduled on condition task

3.5. Supporting Component Analysis
Supporting component analysis in Separating Subsystem is shown in table 5.

TABLE 5. SWBS Separating Subsystem

SEPARATING SUBSYSTEM			
KODE	***EQUIPMENT***	**KODE**	**KOMPONEN**
C.3	*CLASSIFIER*	C.3.1	*BLADE SEGMENT LINER*
		C.3.2	*BLADE ROTOR BLADE*
		C.3.3	*SEAL UPPER-OUTER*
		C.3.4	*COUPLING*
		C.3.5	*COUPLING FLAK*
		C.3.6	*SQUARE BAR*
C.4	*REDUCER CLASSIFIER*	C.4.1	*BEARING*
		C.4.2	*BEARING*
		C.4.3	*SLEEVE ADAPTER*
		C.4.4	*BEARNG*
		C.4.5	*BEARING*
		C.4.6	*BEARING*
		C.4.7	*BEARING*
		C.4.8	*BEARING*
		C.4.9	*BEARING*
		C.4.10	*SEAL OIL LOW SPEED*
		C.4.11	*SEAL OIL HIGH SPEED*

3.6. Component Reclassification
According to table 5, components that experiencing reclassification are seal upper-outer and rotor blade. Rotor blade is used as material separator after milling process so this component often experiencing wear and tear which is cause by rotor blade is operated on high rotation speed and often experiencing impact with coarse rolled material. If there is a failure on this component, it will cause so many product reject because of the blade's rotation is not in its maximum condition. Seal upper router is used as a gap closer between the rotor blade and guide vane. The damage of this component usually caused by the time wearing and also it often experiences an impact process with material. Component reclassification on Separating Subsystem is shown in the following table 6.

TABLE 6. Separating Subsystem Reclassification

SEPARATING SUBSYSTEM			
EQUIPMENT	COMPONENT	CLASSIFICATION	RECLASSIFICATION
CLASSIFIER	BLADE SEGMENT LINER	NON-ROUTINE	NON-ROUTINE
	BLADE ROTOR BLADE	NON-ROUTINE	ROUTINE
	SEAL UPPER-ROUTER	NON-ROUTINE	ROUTINE
	COUPLING	NON-ROUTINE	NON-ROUTINE
	COUPLING FLAK	NON-ROUTINE	NON-ROUTINE
	SQUARE BAR	NON-ROUTINE	NON-ROUTINE
REDUCER CLASSIFIER	BEARING	NON-ROUTINE	NON-ROUTINE
	BEARING	NON-ROUTINE	NON-ROUTINE
	SLEEVE ADAPTER	NON-ROUTINE	NON-ROUTINE
	BEARING	NON-ROUTINE	NON-ROUTINE
	BEARING	NON-ROUTINE	NON-ROUTINE
	BEARING	NON-ROUTINE	NON-ROUTINE
	BEARING	NON-ROUTINE	NON-ROUTINE
	BEARING	NON-ROUTINE	NON-ROUTINE
	BEARING	NON-ROUTINE	NON-ROUTINE
	SEAL OIL LOW SPEED	NON-ROUTINE	NON-ROUTINE
	SEAL OIL HIGH SPEED	NON-ROUTINE	NON-ROUTINE

3.7. Failure Distribution and Components Needs
In this stage, the rotor blade failure time data in table 7 was inputted in reliasoft Weibull software so it was obtained parameter which is used for component needs calculation.

TABLE 7. Rotor Blade Failure Data

Waktu Kegagalan	Interval Kegagalan (hari)
25/6/2011	230
11/12/2011	401
15/1/2013	505
4/6/2014	

Based on the data test table above, the adjusted distribution is normal distribution with parameter μ= 378,7 and σ = 167,858, so the Cumulative Density Function graph and failure rate against time is shown in figure 7. The failure probability value = 1 for t>700 days, so this component ascertained will be fail if the operation is exceeding the mentioned time. While in figure 7, the failure rate value is larger for longer operation time so this component could be categorized in wear out phase. The condition where it occurs function reduction and failure rate enhancement which are in line with the increasing wear age.

Besides, the inputted parameter to calculate the number of component needs is MTTF (T) = 378,7, σ(T) = 167,8 and K = 0,44, so the number of need that should be available for 365 days of operation period with 95% of confidence level for 1,3 components plus the average of need per year for 13,5 components, so the total need of rotor blade that should be available is 14,8 $\approx$ 15 components.

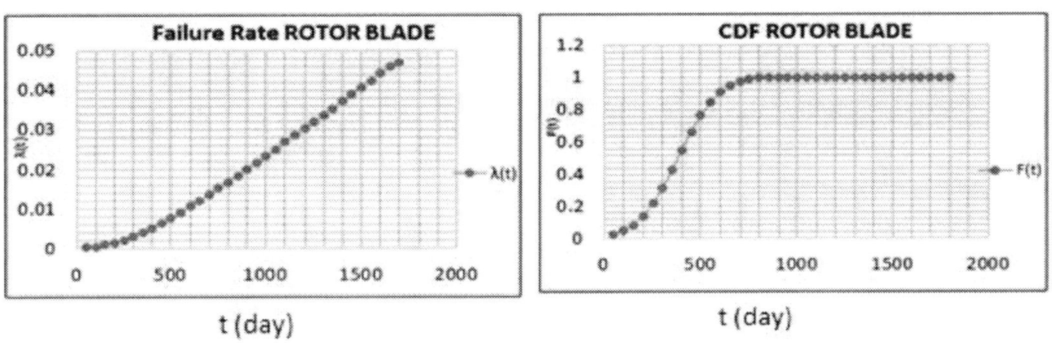

FIGURE 7. CDF Rotor Blade Graph (left) and Rotor Blade Failure Rate Graph (right)

Discussion and Conclusion

In roller mill, the component replacement is conducted on the damaged component only or it could be said breakdown maintenance system. This makes the reliability of roller mill becomes lower and cause high cost of component inventory. If a higher reliability system is expected, it's better to do predictive/preventive maintenance so the component procurement could follow the overhaul schedule and it's much easier to predict. Therefore the conclusions are:

1. The system that has the highest total of unscheduled downtime is roller mill system. This system is divided into 4 subsystems, which are grinding subsystem, drying and transporting subsystem, separating subsystem and hydraulic subsystem.
2. In roller mill system, it was obtained that 87 components are used for maintenance activity. From these components, there are 19 components that experiencing reclassification from non-routine component to routine component.
3. There are 9 components which are in wear out phase marked with components condition starts to deteriorate and 10 components which are in burn-in phase caused by design or production error, installation error or operator error. The components that are in wear out phase need components replacement to reduce the occurrence of fatal damage probability while the components that are in burn-in phase need the correct installation and maintenance so its reliability could be reobtained.
4. Procurement for 19 components that are experiencing reclassification shouldn't be done regularly in a certain time interval, but those components will be held regularly on conditional which means

the procurement is only conducted when it will be conducted service activity or overhaul to avoid overstock and stock out components by still considering the procurement lead time.

References

[1] Barai, R M., Barve, P S., Harde, A V., and Kadam, A D.2012. Reliability Centered Maintenance Methodology for Goliath Crane of Transmission Tower.IOSR Journal of Mechanical and Civil Engineering.

[2] Moubray, John. 1997. *Reliability Centered Maintenance.* 2nd Edition. New York: Industrial Press Inc.

[3] Callom, Frank L., Mustafa, Salim G., and Sheikh, Anwar K. 1990. Strategies in Spare Part Management Using a Reliability Engineering Approach. Saudi Arabia.

[4] Sutikno, Andik. 2013. *Vertikal Raw Mill di Pabrik Tuban.* Tuban, Indonesia : Pemeliharaan Roller Mill 1-2, PT. Semen Indonesia.

Stress analysis on passenger deck due to modification from passenger ship to vehicle-carrying ship

A Zubaydi[1], S H Sujiatanti[2*], T R Hariyanto[3]

[1,2,3]Department of Naval Architecture, Faculty of Marine Technology, Institut Teknologi Sepuluh Nopember (ITS), Indonesia

E-mail: zubaydi@na.its.ac.id, *septia.rahmadianto@gmail.com (corresponding author)

Abstract. Stress is a basic concept in learning about material mechanism. The main focus that needs to be brought to attention in analyzing stress is strength, which is the structural capacity to carry or distribute loads. The structural capacity not only measured by comparing the maximum stress with the material's yield strength but also with the permissible stress required by the Indonesian Classification Bureau (BKI), which certainly makes it much safer. This final project analyzes stress in passenger deck that experiences modification due to load changes, from passenger load to vehicle one, carrying: 6-wheels truck with maximum weight of 14 tons, a passenger car with maximum weight of 3.5 tons, and a motorcycle with maximum weight of 0.4 tons. The deck structure is modelled using finite element software. The boundary conditions given to the structural model are fix and simple constraint. The load that works on this deck is the deck load which comes from the vehicles on deck with three vehicles' arrangement plans. After that, software modelling is conducted for analysis purpose. Analysis result shows a variation of maximum stress that occurs i.e. 135 N/mm2, 133 N/mm2, and 152 N/mm2. Those maximum stresses will not affect the structure of passenger deck's because the maximum stress that occurs indicates smaller value compared to the Indonesian Classification Bureau's permissible stress (175 N/mm2) as well as the material's yield strength (235 N/mm2). Thus, the structural strength of passenger deck is shown to be capable of carrying the weight of vehicles in accordance with the three vehicles' arrangement plans.

1. Introduction

Engineering may be freely defined as knowledge application for general purpose in life. In order to fulfill such mission, engineers design quite an amount of objects to serve the society's needs. Factors that need to be reviewed in design mentioned above covers several aspects such as usage, strength, appearance, economy, and environment protection. In learning the material mechanism, the main design that needs to be brought to attention is strength, which is an object's capacity to carry or distribute loads. The example of objects which need to be able to carry weight are building, machine, containers, truck, airplane, ship, etc. For simpler explanation, we will refer to all those objects as structure. A structure is each and every object that has to be able to withstand, carry, or distribute loads. In order to review such principle, a basic concept of material mechanism (stress and strain) is necessary.

Stress is defined as a comparison between the working force on a structure with the structure's cross-sectional area. Stress also shows the force strength that can cause shape changes in a structure. Changes in the structure's shape, when compared to the initial size of that object, can be concluded as strain. So, the relationship between stress and strain is quite closely related, illustrated in a stress-strain curve. That curve indicates the nature of a structure's: elastic or plastic. Then, the difference between elastic and plastic is characterized by yield strength with the value of 235 N/mm² for structure with normal steel material.

Ship owner who wants to modified their ship from the passenger ship become the vehicle-carrying ship should considering the structural strength of the deck againts the changing of the load. The modification of the passenger ship become the vehicle-carrying ship would affect on the deck structural strength. Therefore, in this research would be analyzed the deck strength againts passenger loads and vehicle loads [1].

2. Ship Structural Strength

Ship, as a thin-walled building, is a three-dimensional object which is bounded by plane or curved fields. Ship, other than being a building, is also a transportation that carries passengers and stuffs across the sea or river (like boat or other smaller ships). Big ship usually provides smaller ship such as lifeboat. Meanwhile, in English terms, the definition is separated between larger ships and smaller boats [2].

Construction system is a combination of several constructions which are connected to one another, forming one unity to with hold the received force in accordance to the main purpose of building such system. Construction system in this ship deck which becomes the object in this research covers the longitudinal framing system where the majority of construction system consists of profile and plat.

Meanwhile, in ship's structure known as ship construction that has a construction system. Framing system is divided into 2 (two) main types: transverse framing system and longitudinal framing system. From these 2 (two) main framing systems, another type of framing system is also known, called mixed framing system [2].

2.1. Normal stress and strain

The most basic concept in material mechanism is stress and strain. The effects of force and moment that occurs on block is providing deflection that is perpendicular to the longitudinal axis of the stem, resulting in normal stress as well as friction in each cross-section of stem perpendicular to the stem axis [2].

The amount of average stress in a plane can be defined as the force intensity which occurs in that plane. Thus, systematically-speaking, average normal stress can be declared in the following equation:

$$\sigma = \frac{P}{A} \tag{1}$$

where:
σ = Average Normal Stress (N/m2)
P = Occurring Force (N)
A = Area of Object (m2

A straight rod will experience length change if it's axially burdened, becoming longer if it experiences pull/drag and shorter if it experiences pressure. In accordance with this concept, extension per length unit is called strain [2].

The relation above if written into equation (2) is as following:

$$\varepsilon = \frac{\delta}{L} \tag{2}$$

where:
ε = Strain
δ = Extension of object's length (m)
L = Length of object (m)

2.2. Stress and strain on elastic object

Nearly all engineering materials have certain nature, which is elasticity. If a material is subjected to an external force, then that material will experience a shape change, known as deformation. As long as the applied force does not exceed certain boundary, when the external force is lifted then the material will back to its initial shape.

In this particular research, objects subjected to external force are always considered as perfectly elastic, meaning the object will always go back to its initial shape when the external force is removed [3]. In finite method formulation, analyzing stress and the relation between strain and displacement is very essential. Displacement, which is denoted by "u" and "v" is function of "u" coordinate = u (x,y,z), v = v (x,y,z), w = w (x,y,z) [3].

The relation between strain – displacement can be generally defined in the following equation (3):

$$\varepsilon_x = \frac{\partial u}{\partial x} \; ; \; \varepsilon_y = \frac{\partial v}{\partial y} \; ; \; \varepsilon_z = \frac{\partial w}{\partial z} \tag{3}$$

Where u, v, and w are translation in the direction of x, y, and z. Shear strain of γxy , γxy , etc is declared in the following equation (4), (5), and (6), respectively:

$$\gamma_{xy} = \frac{\partial u}{\partial y} + \frac{\partial v}{\partial x} = \gamma_{yx} \tag{4}$$

$$\gamma_{yz} = \frac{\partial v}{\partial z} + \frac{\partial w}{\partial y} = \gamma_{zy} \tag{5}$$

$$\gamma_{zx} = \frac{\partial w}{\partial x} + \frac{\partial u}{\partial z} = \gamma_{xz} \tag{6}$$

The relation between stress – strain for isotropic material, which is derived from theory of elasticity, is as written in the following equation:

$$\varepsilon_x = \frac{\sigma_x - v\sigma_y - v\sigma_z}{E} , \; \gamma_{xy} = \frac{\tau_{xy}}{G} \tag{7}$$

$$\varepsilon_y = \frac{-v\sigma_x + \sigma_y - v\sigma_z}{E} , \; \gamma_{yz} = \frac{\tau_{yz}}{G} \tag{8}$$

$$\varepsilon_z = \frac{-v\sigma_x - v\sigma_y + \sigma_z}{E} , \; \gamma_{zx} = \frac{\tau_{zx}}{G} \tag{9}$$

with

$$G = \frac{E}{2\,(1+v)} \tag{10}$$

Where:
E = young modulus,
G = shear modulus,
υ = poison ratios.

3. Finite Element Analysis

The ship used as a reference in this research is a passenger ship that was particularly modified due to changes in the passenger load into a vehicle. One particular dock that got to be the material for this research is the freeboard deck. This freeboard deck is located under the upper deck. The vessel data required for modeling are the main measurements of the said vessel consist of a multitude of terms and definitions specifically related to ships, which includes the length between perpendiculars (LBP/LPP), width (B), draft (T), height (H), and speed. The principal dimension of ship is the following:

Length Overall	134,86	m
Length Construction	125,00	m
Breadth (moulded)	20,00	m
Depth (moulded)	14,09	m
Draft (moulded)	6,95	m
Service speed	17,00	knots

3.1. Modeling of passenger deck construction

The ship's passenger deck construction modeling use finite element software. The modeling of a construction must be done in such a way that the model can actually represent the real condition. In general, the modeling of a construction consists of several stages [4].

a. Creating Ship Geometry
   Platform and profile modeling can be done using two approaches: Bottom-Up (modeling a progression from the start point to volume) and Top-Up (a combination of various small and large entities into a single model).
b. Giving Material Properties
   Once the stage for geometry is done then the next step is defining the material properties. The aim for this is to create a model that looks like it is made from material that is close to the material in reality. This modeling is done with the premise that the ship is made with grade A steel 36 with a modulus of 20 Gpa elasticity and that the position ratio is 0.3.
c. Meshing
   Generally, meshing is a stage to discredit the geometry model into elements up to and nodal points that are to be analyzed, shown as in **Error! Reference source not found.**.

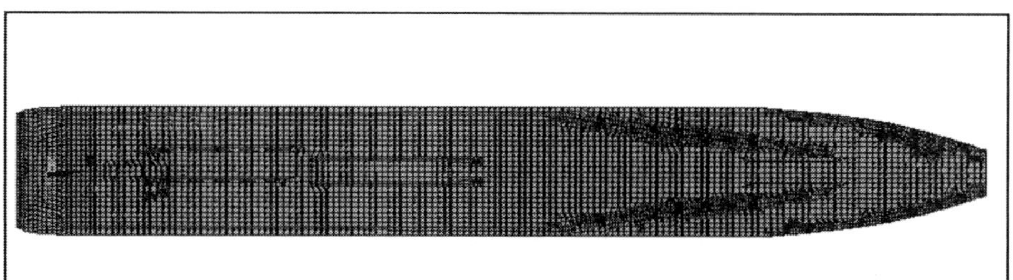

Figure 1. Finite Element (FE) model of deck construction.

3.2. Model constraint

The assumption of constraints in finite element calculations should be arranged in such a way that can be as closely as possible with the real conditions. This way the structure model can really represent the real conditions that occur.

The constraints used in this calculation are simple and fix constraint. The constraint at the end of longitudinals and plates are assumed to use fix constraint, while the area of the bulkhead is assumed to use a simple constraint. The course of the binding is according to Table 1 as follows [5]:

Table 1. Constraint on FE model.

Constraint	Translation			Rotation		
	Dx	Dy	Dz	Rx	Ry	Rz
All DOF (fix)	fix	fix	fix	fix	fix	fix
Dy (Simple)		fix				

3.3. Loading

In this part of deck construction analysis, the applied load type is pressure load. The load occuring on the deck is the vehicle load resulting from the weight of the vehicle that is distributed in accordance to the design of vehicle's location.

The types of vehicles in this study are consisting of freight cars in the form of six-wheeled trucks, four-wheeled passenger cars, and motorcycles. In accordance to the data described in the previous sub-section, the weight of deck load can be obtained using equation (11) as explained below:

$$W = m \cdot g \qquad (kN) \tag{11}$$

where:

W = weight of vehicle
m = mass of vehicles
g = gravitys acceleration

From the equation (11) the value of W from each vehicles can be obtained. The next part is to find out the unity load of each vehicle using the following equation (12):

$$P = W/A \qquad (kN/m^2) \tag{12}$$

where:

P = pressure on area
W = weight of vehicle
A = cross-sectional area

There will be 3 (three) variations of the vehicle's arrangement plan as the loading scenario, all being done on the deck. The vehicle's arrangement plan is to be explained in Table 2.

Table 2. Vehicle Arrangement plan on the deck.

Variations of Vehicle Arrangement Plan	Vehicle Arrangement		
	Rear	**Middle**	**Front**
1	Truck	Car	Motorcycle
2	Motorcycle	Truck	Car
3	Car	Motorcycle	Truck

The load on the deck of this ship consists of 6-wheeled trucks load, in which the weight varies from 10.605 tons, 9.99 tons and 9.147 tons, passenger cars with maximum weight of 3.5 tons and motorcycles with a maximum weight of 0.4 tons. In accordance with the calculations performed, then the loading value is obtained as shown in Table 3 and Table 4, respectively.

Table 3. Loading.

Vehicle	Maximum mass (ton)	Gravitation (m/s^2)	Load (kN)	P x l (m^2)
Truck	14	9,81	137,34	6,1 x 2
Car	3,5	9,81	34.335	5 x 2
Motorcycle	0,4	9,81	3.924	2 x 0,8

IOP Conf. Series: Journal of Physics: Conf. Series **974** (2018) 012053 doi:10.1088/1742-6596/974/1/012053

Table 4. Load per area.

Vehicle	Load (kN)	Area (m^2)	Load per area (kN/m^2)
Truck	137,34	12.2	11,257377
Car	34.335	10	3,4335
Motorcycle	3.924	1.6	2,4525

4. Result and Discussion

The analysis of deck structure uses static analysis. The analysis stage is applied to the model in the form of elements in accordance with the loading and constraining that is given to the model. At this stage finite element software runs the process of analyzing the model that has been generated in the form of finite element and has been given certain limit and load conditions.

Once the analysis process is done, the output obtained is the maximum stress value that occurs in each and every vehicle arrangement plan. Figure 2 shows example of one of the results of the deck construction analysis.

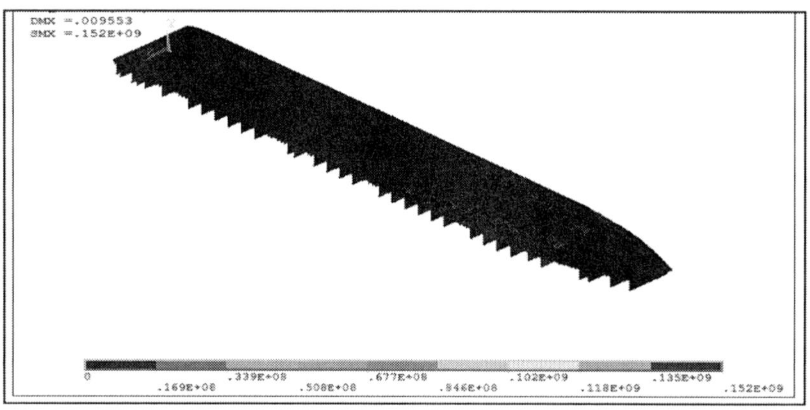

Figure 2. Result of deck strength analysis.

Based on the analysis results, the maximum stress that occurs in each vehicle arrangement plan will be reviewed and examined. The following results of the maximum stress that occurs are shown in Table 5 and Table 6.

Table 5. The comparison between maximum stress and permissible stress.

Vehicle Arrangement Plan	Maximum Stress σ_{maks} (N/mm^2)	Permissible Stress σ_{perm} (N/mm^2)	$\sigma_{maks} < \sigma_{perm}$
1	135	175	Accepted
2	133	175	Accepted
3	152	175	Accepted

Table 6 The Comparison Between Maximum Stress and Yield Strength

Vehicle Arrangement Plan	Maximum Stress σ_{maks} (N/mm^2)	Yield Strength (N/mm^2)	σ_{maks} < *Yield Strength*
1	135	235	Accepted
2	133	235	Accepted
3	152	235	Accepted

IOP Conf. Series: Journal of Physics: Conf. Series **974** (2018) 012053 doi:10.1088/1742-6596/974/1/012053

5. Conclusion

Based on the analysis results shown in Table 5 and Table 6, it can be concluded that the maximum stresses occur in deck structure are less than both permissible stress and yield strength of material. Therefore, it can be assumed that the strength of the passenger deck structure is capable for vehicle loads in accordance to vehicle arrangement plan.

References

[1] Hasanudin, J H Chen, Modification of the Intact Stability Criteria to Assess the Ship Survivability from Capsizing, Procedia Earth and Planetary Science vol.14 pp 64-75, (2015)

[2] R Taggart, Ship Design and Construction, the Society of Naval Architects and Marine Engineers (SNAME), (1980)

[3] Y Bai, Marine Structural Design, Elsevier, Oxford, (2003)

[4] T Abramowski, K Żelazny, T Szelangiewicz, Numerical Analysis of Influence of Ship Hull Form Modification on Ship Resistance and Propulsion Characteristics, POLISH MARITIME RESEARCH 1(64) Vol 17; pp. 10-13, (2010)

[5] DNV, Rules For Classification of Passenger and Dry Cargo Ships, DNV Rules Part 5 Chapter 2, (2016)

Construction strength analysis of landing craft tank conversion to passenger ship using finite element method

Mohammad Nurul Misbah[1] **, Dony Setyawan**[2] **, Wisnu Murti Dananjaya**[3]

[1,2,3]Department of Naval Architecture, Faculty of Marine Technology, Institut Teknologi Sepuluh Nopember (ITS), Surabaya 60111 Indonesia

E-mail: mnmisbah@na.its.ac.id

Abstract. This research aims to determine the longitudinal strength of passenger ship which was converted from Landing Craft Tank with 54 m of length as stated by BKI (Biro Klasifikasi Indonesia/ Indonesian Classification Bureau). Verification of strength value is done to 4 (four) loading conditions which are (1) empty load condition during sagging wave, (2) empty load condition during hogging wave, (3) full load condition during sagging wave and (4) full load condition during hogging wave. Analysis is done using Finite Element Analysis (FEA) software by modeling the entire part of passenger ship and its loading condition. The back and upfront part of ship centerline were used as the boundary condition. From that analysis it can be concluded that the maximum stress for load condition (1) is 72,393 MPa, 74,792 MPa for load condition (2), 129,92 MPa for load condition (3), and 132,4 MPa for load condition (4). Longitudinal strength of passenger ship fulfilled the criteria of empty load condition having smaller stress value than allowable stress which is 90 MPa, and during full load condition with smaller stress value than allowable stress which is 150 MPa. Analysis on longitudinal strength comparison with entire ship plate thickness variation of ± 2 mm from initial plate was also done during this research. From this research it can be concluded that plate thickness reduction causes the value of longitudinal strength to decrease, while plate thickness addition causes the value of longitudinal strength to increase.

1. Introduction

Refering to Directorate General of Sea Transportation Decree Letter SK.885/AP.005/DRJD/2015 issued on March 19th 2015 about Prohibition of Landing Craft Tank (LCT) Usage as Ferry Boat causes many LCT in Indonesia unable to operate as it should be. Company who operates LCT as ferry boat would be sanctioned according to Law Number 17 Year 2008 Article 289 on Shipping, and Government Regulation Number 20 Year 2010 on Sea Transportation. According to Ministry of Transportation Decree Number 39 Year 2015 on Standard Procedure of Ferry Boat Passenger, and Directorate General of Land Transportation Decree Letter SK. 4608/AP.005/DRJD/2012 Year 2012 on Standard Procedure of Ferry Boat, the ship that carries passenger and also motor vehicle should at least have 2 (two) ramp doors which would be used as entrance & exit path of vehicle, also the ship should has double bottom.

Many occurrences of sinking LCT ship in Indonesia waters were the reason behind law regulations as mentioned above. It is because LCT ship was not designed and produced to become passenger ship. Meanwhile, numerous LCT ships are still operating as ferry boat for passenger and also freight carrier inter-island in Indonesia. Automatically it would affect all LCT ships and causes disadvantage to its owner. In order to make the ship operates as it used to, it needs to be converted to become passenger ship. However, to ensure that the ship could fulfill its function well and safely, the conversion should be done with precise analysis, most importantly on the calculation of longitudinal strength.

This research will be focused on finding out the importance of internal load in the form of cargo and superstructure to the entire load condition. An analysis of longitudinal strength is also in order as the ship was the conversion result of Landing Craft Tank. Generally the ship production process has been planned and calculated in accordance to regulations. However in this research since the passenger ship is a conversion result from LCT, longitudinal strength re-calculation is needed as the load condition is different from the previous ship type. Load condition differences will be the consideration on how often inspection should be done in order to prevent structural failure. Stress analysis of the passenger ship will be done with finite element method.

2. Literature study

2.1. Definition of landing craft tank

Landing Craft Tank is an assault craft to land tank on the seashore. This kind of ship existed during the time of World War II and was used by the United States Navy and the Royal Navy of United Kingdom. The US Navy later used it for other purposes during the time of Korean and Vietnam Wars. The ship was known as LCT at the time of World War II.

Figure 1. Trisna Dwitya LCT (source: MarineTraffic.com).

Nowadays, while still using the LCT name, there are many ships operating commercially in the waters and river way to carry various big freights (e.g. as dump truck, dozer, excavator, construction equipment, steel structure, boiler, turbine engine, rig equipment, transformer, project material, etc.) all over Indonesia, specifically to mining site or project location which happened to be located in islands or near to river way. LCT is also used as ferry boat for inter-islands transportation. Another function of LCT is to carry fresh water and distribute fossil fuel to fulfil demands in remote areas around Indonesia. (Wikipedia, 2016).

2.2. Definition of passenger ship

Passenger ship is the kind of ship which uses permanently built motor as its propulsion force. To improve efficiency and serve broader purpose, the ship could be in the form of Ro-Ro ship, or in the form of ferry ship for short course (Wikipedia, 2016).

IOP Conf. Series: Journal of Physics: Conf. Series **974** (2018) 012054 doi:10.1088/1742-6596/974/1/012054

Figure 2. Conversion result of LCT Trisna Dwitya.

2.3. Definition of longitudinal strength
Longitudinal strength is a ship strength measured lengthwise to sustain the freight load and the vessel load itself. The calculation is done accordingly to the vessel dimension and scantling (profile and plate size) used in the vessel. The inertia force of scantling will be calculated to obtain stress and moment value as the result of freight and wave loads. Under the article it is assumed that moment distribution of gravitation load and the buoyancy lengthwise of vessel fulfilled the second stability requirement which is the center of gravity and buoyancy is located on the same vertical line (Santosa, 2013).

2.4. Finite element method
Finite element method is a numerical method used to complete problems in engineering such as geometric, loads, and complex properties of material. These kinds of things are not easy to be solved with mathematical analysis solution. An approach using finite element method utilizes information on the node. The process of determining node is called discretization, a diversification of system into smaller parts, in which problem solving is done and later will be collected as a whole (DNV-GL, 2015).

2.5. Hogging conditions
In this research, the ship shape structure deformation caused by wave load could be explained by assuming the vessel moves on regular wave, where wave length is equal to the vessel length. This will caused vertical moment to happen. Assuming the hull as a beam, then the possibilities of condition are hogging condition.

Hogging condition is a kind of ship deformation forming a convex shape. Hull girder is supported on midship with the wave crest. In this condition, even though ship total weight is equal to buoyancy, there is buoyancy overbalance found on the midship and overweight found on bow and stern parts. As a result, the ship tends to move downward on the bow and move upward on the midship.

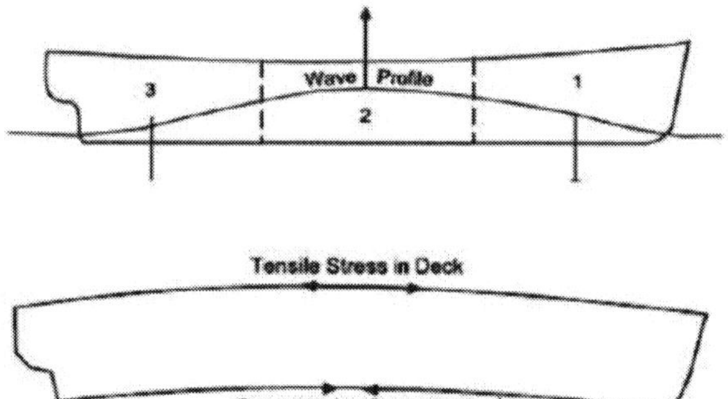

Figure 3. Hogging condition (source: Barrass, 1999).

2.6. Sagging condition

Sagging condition is a kind of ship deformation forming a concave shape. Hull girder is supported on stern and bow with 2 wave crests. There are overweight on midship and buoyancy overbalance on bow and stern. As a result, the ship tends to move upward on the bow and downward on the midship.

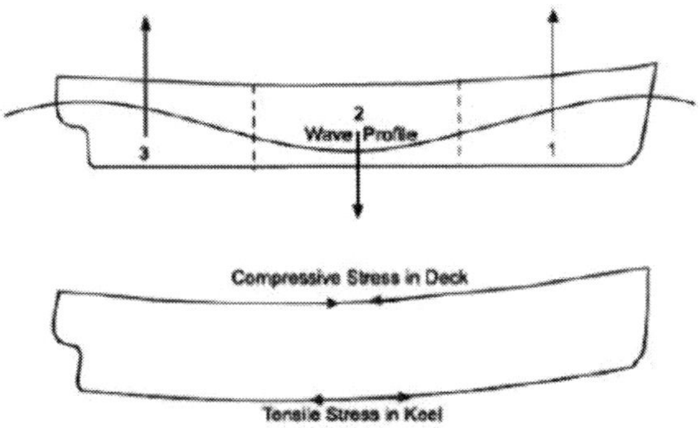

Figure 4. Sagging condition (source: Barrass, 1999).

3. Methodology

In this step a modeling of ship structure will be done in order to analysis longitudinal strength of the ship using finite element method. Modeling of entire ship section will be done using finite element analysis software (FEA software). The objective of structure modeling is to obtain the value of ship stress with 4 variation of load conditions.

3.1. Geometric modelling

In this research, several modeling methods were used to make geometric image of passenger ship according to existing data. It should be noted that during structure modeling for global analysis, surface model (area) is more recommended as opposed to solid model (volume). According to literature about ship global modeling based on class and other references, modeling method is determined as follow:

- Girder, modeled with 2D surface for the web section and 1D line for the face/ flange
- Stiffener, modeled with 1D line
- Plate (for hull and separator) modeled with 2D surface (Figure 6)

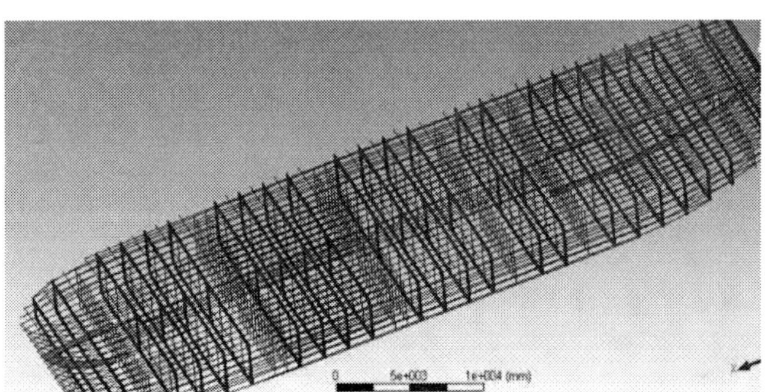

Figure 5. Girders and stiffeners modeling.

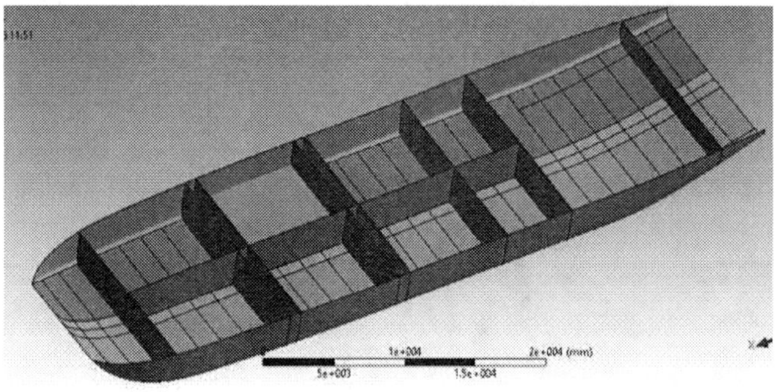

Figure 6. Hull and separator modeling.

3.2. Meshing

The type of meshing used in this research is coarse mesh, where element dimension is relatively big (not smooth). The reason behind choosing this kind of meshing is because for analysis which results in global meshing calculation, coarse mesh is the most effective, as also recommended by class authority. Element dimension is limited to the size of scaffold which is 600mm due to researcher's computer limited capability. Too much fine mesh will overload the computer and causes error.

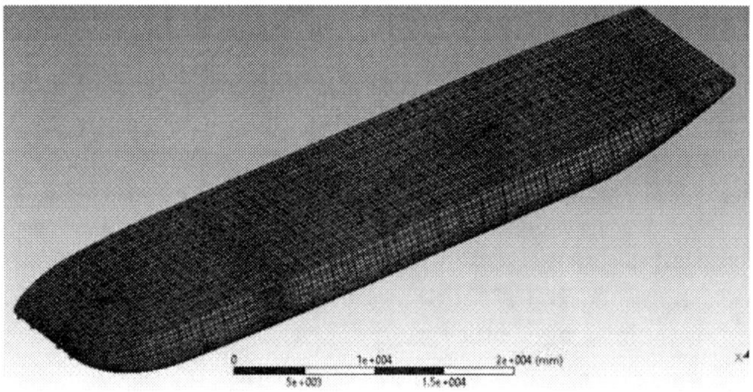

Figure 7. Full model mesh.

3.3. Boundary condition

Boundary condition in the form of support structure is needed in FEA. If there is no boundary condition then the FEA software would not run properly and resulting in an invalid value. The boundary condition in this research is in the form of 2 support structures. The first point is located behind ship centerline, and the second point is located in front of ship centerline. Degree of Freedom (fixed) on both point heads to the Z-axis (vertical).

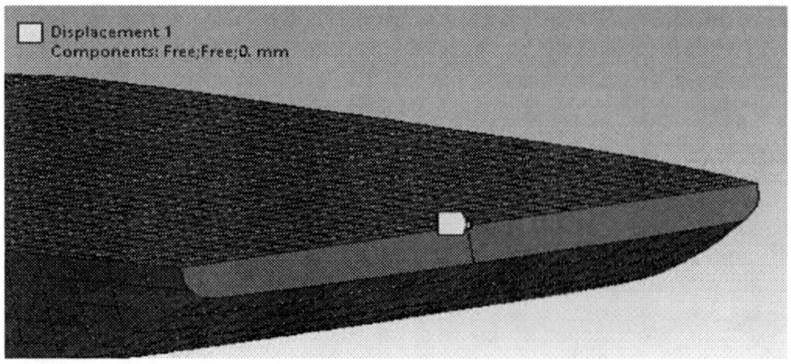

Figure 8. Boundary condition on the rear side.

IOP Conf. Series: Journal of Physics: Conf. Series **974** (2018) 012054 doi:10.1088/1742-6596/974/1/012054

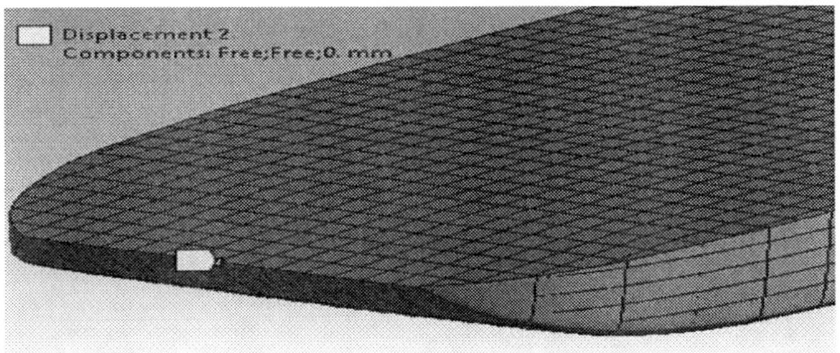

Figure 9. Boundary condition on the front side.

3.4. Loading

As mentioned before, all loading conditions should be in balance position. In order to achieve balance condition, downward working load should have equal value with the upward working loads. The downward working load consists of the weight of steel and freight. The upward working load is buoyancy or the hydrostatic pressure. This research uses both downward working load and upward working load.

The downward working load consists of freight load, tanks load, machine load, and superstructure load 1, 2, and 3. The upward working load consists of hydrostatic load, which refers to the actual wave condition. Loadings are given in the form of pressure on the surface.

3.5. Loading condition variation

In this research there are 4 (four) loading conditions analyzed with finite element method to obtain primary respond in the form of longitudinal stress. The following are the 4 (four) loading conditions variation:

- Case 1: the condition in which passenger ship is in sagging wave condition with no freight load inside.
- Case 2: the condition in which passenger ship is in hogging wave condition with no freight load inside.
- Case 3: the condition in which passenger ship is in sagging wave condition with freight load inside.
- Case 4: the condition in which passenger ship is in hogging wave condition with freight load inside.

4. Result and analysis

4.1. Reaction forces verification

Upon going through result analysis, verification to the result is needed by overviewing the result with boundary condition as mentioned before. Ideally, all boundary condition should have zero reaction to ensure the load working downward is equal to the load working upward, as in reality there is no such thing happening on floating structure like passenger ship. For research purpose, the value of reaction which will be focused on is the force reacting vertically, or in other word, the force in Z-axis as the main forces working on the structure is the forces working upward and downward. The following are the value of reaction for each boundary condition:

Table 1. Calculation of reaction value for boundary condition case 1 zero sagging load.

Item	Reaction force (N)	Weight (Ton)	Percentage (%)
Reaction force 1	141950	14.470	1.440%
Reaction force 2	-142200	-14.495	-1.442%

Table 2. Calculation of reaction value for boundary condition case 2 zero hogging load.

Item	Reaction force (N)	Weight (Ton)	Percentage (%)
Reaction force 1	127250	12.971	1.291%
Reaction force 2	-127370	-12.984	-1.292%

Table 3. Calculation of reaction value for boundary condition case 3 full sagging load.

Item	Reaction force (N)	Weight (Ton)	Percentage (%)
Reaction force 1	221840	22.614	1.485%
Reaction force 2	-221850	-22.615	-1.485%

Table 4. Calculation of reaction value for boundary condition case 4 full hogging load.

Item	Reaction force (N)	Weight (Ton)	Percentage (%)
Reaction force 1	210380	21.445	1.408%
Reaction force 2	-209990	-21.406	-1.405%

4.2. Stress

After verification of boundary condition reaction is done, the next thing to do is analysis of stress result. The following are the result of stress presented in 4 (four) cases:

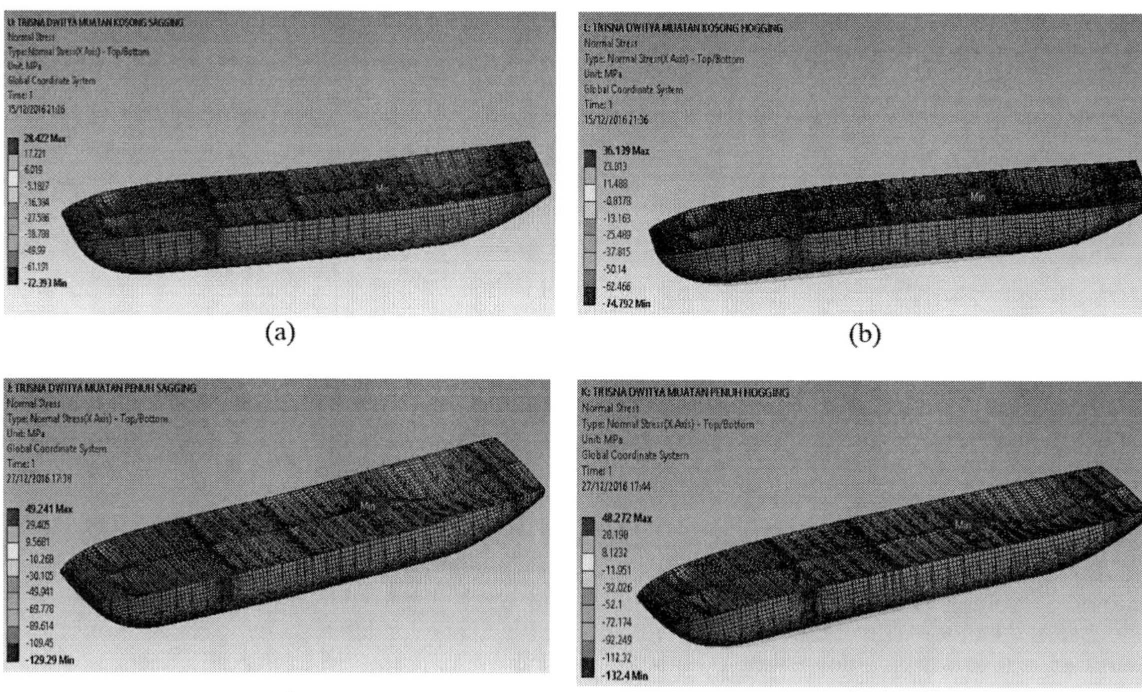

Figure 10. Stress results; (a) case 1, (b) case 2, (c) case 3, (d) case 4.

4.3. Finite element with plate thickness variation analysis result

After software running and analyzing stress result of 4 (four) load conditions variation, the next step is to do stress analysis to the passenger ship if the ship is under ± 2 mm variation of plate thickness from the initial thickness. Plate thickness addition analysis is done to measure the escalation of longitudinal strength after plat thickness alteration, and also to be used as reference for ship plate reparation. Plat thickness reduction analysis is done to measure the reduction of longitudinal strength after plat thickness alteration, and also to be used as reference when the entire part of the ship has corroded as a result of aging and ship condition during sailing.

In the research, explanation about longitudinal strength analysis with plate thickness variation is also shown in graph. The objective is to facilitate research analysis understanding. Analysis result of this sub-chapter is divided into 2 (two) sections of analysis, which is analysis based on ship plate thickness and analysis based on ship freight condition. The following is the longitudinal strength analysis result with ± 1 mm plate thickness variation as shown in graph:

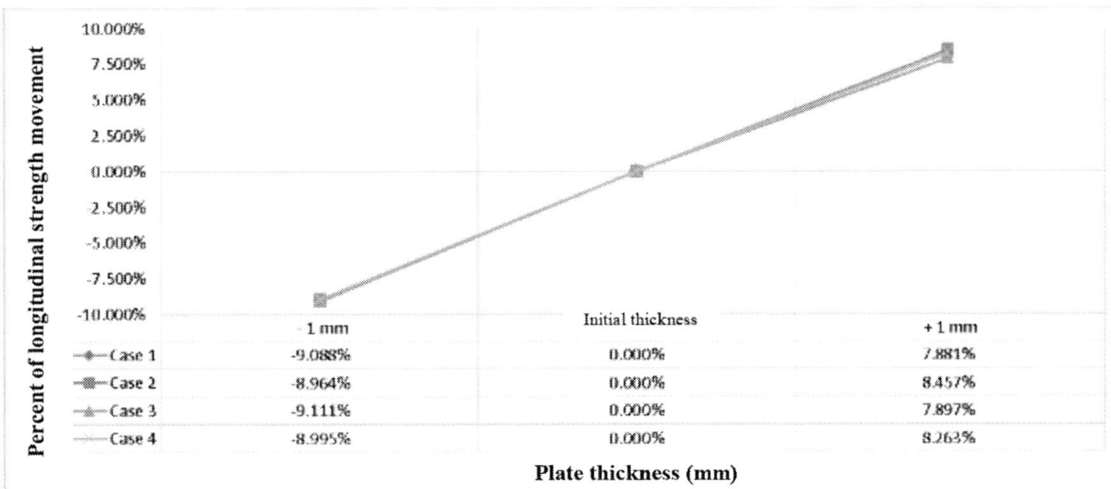

Figure 11. Graph of ± 1mm plate thickness variation.

Below is the longitudinal strength analysis result with ± 2 mm plate thickness variation as shown in graph:

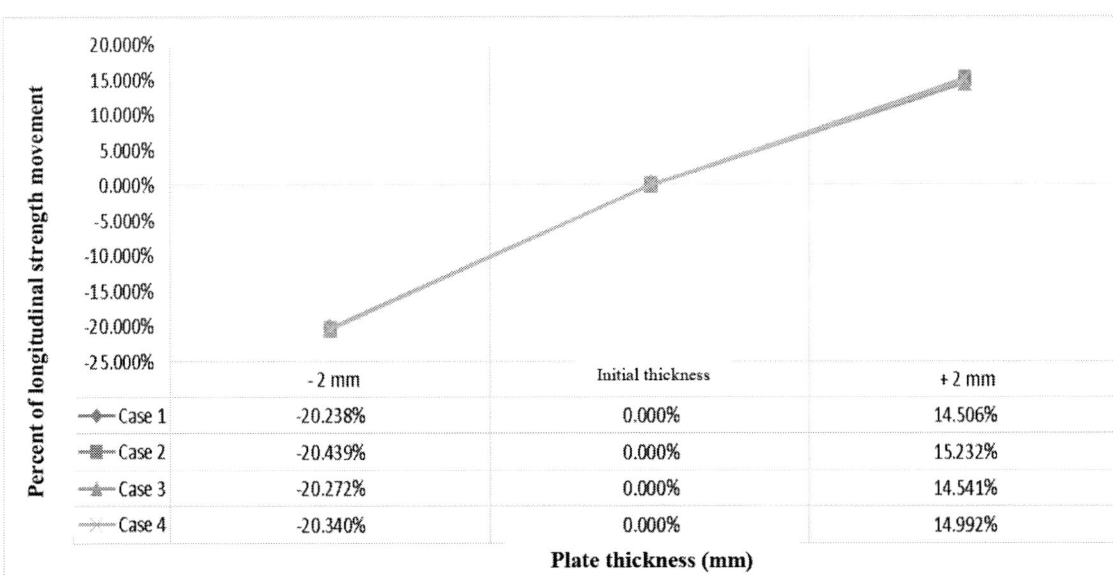

Figure 12. Graph of ± 2 mm plate thickness variation.

Analysis based on freight condition will be divided into 2 topics, which are empty load condition and full load condition.

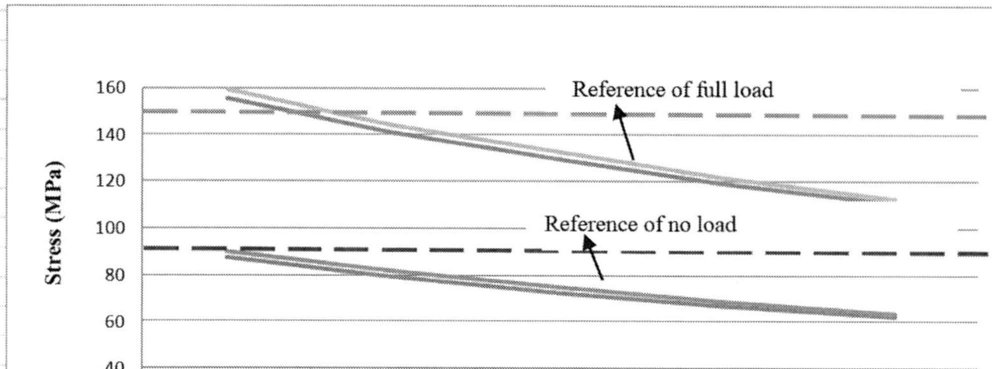

	- 2 mm	- 1 mm	Initial thickness	+ 1 mm	+ 2 mm
Case 1	87.044	78.972	72.393	66.688	61.892
Case 2	90.079	81.496	74.792	68.467	63.4
Case 3	155.5	141.07	129.29	119.08	110.49
Case 4	159.33	144.31	132.4	121.46	112.55

Plate thickness (mm)

Figure 13. Graph of longitudinal strength stress.

5. Conclusion

Referring to ship load modelling, the results can be summarized into points as follow:

1. Maximum stress on all 4(four) load conditions happened on parallel middle body each with the value of 72,393 MPa, 74,792 MPa, 129,29 MPa, and 132,40 MPa.
2. Reduction of plate thickness about -1 mm to -2 mm from the initial thickness also caused the value of longitudinal strength to decrease. The analysis could be used as reference when the entire part of the ship has corroded as a result of aging and ship condition during sailing.
3. Plate thickness addition about +1 mm to + 2 mm from the initial plate also caused the value of longitudinal strength to increase. When the entire part of the ship has corroded to the point of critical conditionas a result of aging, the analysis could be used as reference for plate maintenance and reparation schedule. and ship condition during sailing.
4. Stress value on plate thickness variation of -2 mm from initial plate during empty load condition is still accepted even though the value is higher than allowable stress which is 90,079 MPa for hogging wave, because the stress value is still under the material yield strength which is 235 MPa.
5. Stress value on plate thickness variation of -2 mm from initial plate during full load condition is still accepted even though the value is higher than allowable stress which is 155,50 MPa for sagging wave, because the stress value is still under the material yield strength which is 235 MPa.

References

[1] Barras 1999 *Ship Stability for Master and Mates* (Oxford: Elseiver)
[2] DNV-GL 2015 *Class Guideline – Finite Element Analysis* (DNV-GL)
[3] Lee, H.-H 2012 *Finite Element Simulations with ANSYS Workbench 14* (United States of America: SDC Publications)
[4] Santosa, B 2013 *Diktat Kekuatan Kapal* (Surabaya: Institut Teknologi Sepuluh Nopember)

IOP Conf. Series: Journal of Physics: Conf. Series **974** (2018) 012055 doi:10.1088/1742-6596/974/1/012055

Designing neuro-fuzzy controller for electromagnetic anti-lock braking system (ABS) on electric vehicle

Josaphat Pramudijanto[1], **AndriAshfahani**[2], and **Rian Lukito**
Department of Electrical Engineering Otomation, Faculty of Vocational, Sepuluh Nopember Institute of Technology (ITS).
[1] E-mail: jos@ee.its.ac.id.
[2] E-mail: andriashfahani@ee.its.ac.id

Abstract. Anti-lock braking system (ABS) is used on vehicles to keep the wheels unlocked in sudden break (inside braking) and minimalize the stop distance of the vehicle. The problem of it when sudden break is the wheels locked so the vehicle steering couldn't be controlled. The designed ABS system will be applied on ABS simulator using the electromagnetic braking. In normal condition or in condition without braking, longitudinal velocity of the vehicle will be equal with the velocity of wheel rotation, so the slip ratio will be 0 (0%) and if the velocity of wheel rotation is 0 (in locked condition) then the wheels will be slip 1 (100%). ABS system will keep the value of slip ratio so it will be 0.2 (20%). In this final assignment, the method that is used is Neuro-Fuzzy method to control the slip value on the wheels. The input is the expectable slip and the output is slip from plant. The learning algorithm which is used is Backpropagation that will work by feedforward to get actual output and work by feedback to get error value with target output. The network that was made based on fuzzy mechanism which are fuzzification, inference and defuzzification, Neuro-fuzzy controller can reduce overshoot plant respond to 43.2% compared to plant respond without controller by open loop.

1. Introduction

Braking is one of the most important part of the vehicle system. The correct way of braking is not doing an inside braking (sudden) because it can put the vehicle and other vehicles on the back in danger. But, if in emergency situation can block the speed of vehicle and put the driver in danger, then the driver should do the inside braking (sudden). In inside braking, wheels could lock so vehicle steering can not be controlled and this could cause the vehicle to slip. One of the system that can keep or control that brake is Anti-Lock Braking System (ABS). ABS system can control slip on vehicle's wheels so the wheels couldn't get locked when on brake. Slip on the wheels depends on the comparison between the difference of longitudinal velocity of the vehicle and the wheel speed with longitudinal velocity of the vehicle [1]. Electromagnetic braking is a braking system that is using an electromagnetic force to decelerate shaft motion of the vehicle. The electromagnetic braking that is used is microelectromagnetic clutch brake type, utilizing the electromagnetic field which is generated by coil because of the current going through the coil. The amount of magnetic field that is generated is directly proportional with the amount of current that is generated by coil. The force of magnetic field

IOP Conf. Series: Journal of Physics: Conf. Series **974** (2018) 012055 doi:10.1088/1742-6596/974/1/012055

that is generated is opposite with the direction of the wheel shaft rotation, so it can decelerate the velocity of vehicle.

A conventional controller, such PID, can not fulfill the need of ABS system because it couldn't find the parameter of controller dynamically and couldn't reduce the effect of disturbance of the system. Neuro-fuzzy controller is a smart control which is a combination of neural network algorithm and fuzzy control. Neural network algorithm is a supervised learning algorithm that is using neural network as a learning network. This network will be trained continuously by updating the value weighting on each neuron in hidden layer in feedforward and feedback.

2. System Description

2.1. Vehicle Dynamic [2]

The dynamic motion of a vehicle has 3 motion axes, they are longitudinal axis, vertical axis and transverse axis. Each of these axes has a rotation axis, they are roll, pitch and yaw. One of the force that works on the vehicle from starting to move until comes to a stop is a braking force that is in opposite direction with the direction of vehicle. When the braking force works, then the vehicle dynamic will change on 3 axes. In this discussion, the vehicle dynamic that will be discussed is on the longitudinal axis. On longitudinal axis, the braking effect will cause changes in 2 parameters, which are the velocity of vehicle's longitudinal and the wheel rotation velocity. The rotation velocity and longitudinal velocity will be slip or brake slip. The slip value needs to b controlled in 15-30% so the vehicle can be driven in an inside brake (sudden) [6]. Slip happens in each wheel in a vehicle. The dynamic force that happens when there's a braking on a wheel is the forces that work in the wheel. When braking, the working force is braking force, a normal vehicle force and road friction force. When the vehicle wheel locks (slip value 1) then the vehicle can't be moved on lateral axis (axis transverse), so the vehicle can't be moved either right or left to avoid an obstacle or an object that stands in front of the vehicle.

2.2. ABS Simulator [1]

ABS Simulator is a device that is used to test slip on a wheel without having to run the wheel on the road. On ABS simulator, the wheel will be installed on another wheel (tangent to each other) rotated with certain velocity. The velocity of the wheel will be represented by the upper wheel and the velocity of the bottom wheel will be represented by longitudinal velocity of the vehicle. ABS simulator only calculates longitudinal dynamic of the vehicle and not considering vertical dynamic and lateral (sides) of vehicle. The free wheel speed represents the car wheel speed (ω) and the bottom wheel speed that is rotated by DC motor represents longitudinal velocity of the car (v). Figure 1 is block diagram of the ABS simulator system design that was made:

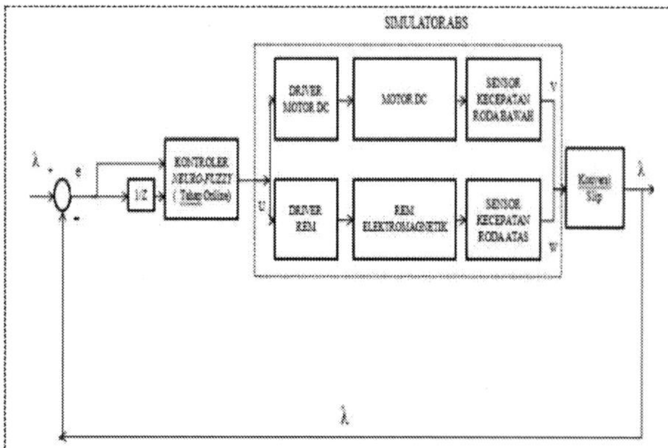

Figure 1 ABS Block Diagram.

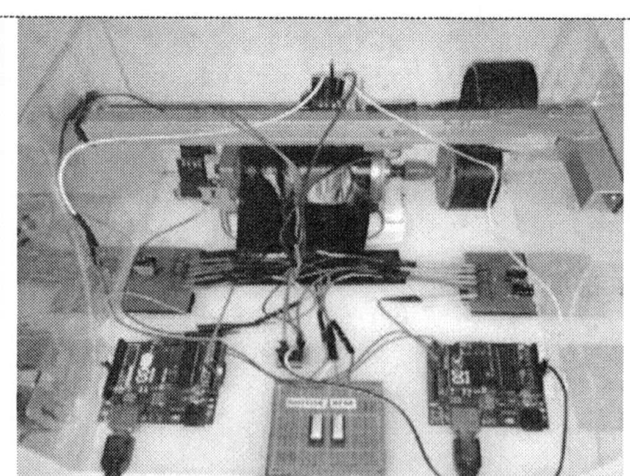

Figure 2 Physical form of Plant.

Based on figure 1, the ABS system itself consists of an expected slip as an input, computer as a controller block, Arduino Uno interface, ABS simulator and output in the form of a result calculation slip in Arduino Uno where the processed data was obtained from upper and bottom wheel speed sensor. ABS simulator plant is an upper wheel that is connected with an electromagnetic brake, and a bottom wheel that is connected with DC motor. Actuator on ABS simulator is an electromagnetic brake that managed the amount of brake presentation. The workflow of ABS system block diagram is the Arduino Uno will send the slip value through serial communication to computer, then the computer will process error signal using Neuro-fuzzy algorithm on Matlab program. The error signal process is ran by offline through general learning. After offline training has been conducted then the adjusted weight that has been obtained is inserted to the program by online stage. In online stage the error signal goes to the controller and generates control signal to ABS simulator plant through Arduino Uno as interface. ABS simulator design consist of hardware designing and software designing. Figure 2 shows the physical form of ABS simulator plant

3. System Designing

3.1. Plant Identification

Identification process aims to looking for parameter of a plant [3]. Parameter that has been obtained is used to make mathematical model from a plant by evaluating input output measuring data. Plant parameter identification is conducted dynamically by changing the bottom wheel velocity and pwm brake value of the upper wheel simultaneously. ABS plant parameter is a slip as output plant and the amount of pwm brake as input plant. The mathematical model of plant is calculated with stochastic approach structural model. Stochastic approach structural model that has been used is Auto Regressive Exogeneus structure (ARX).

Identification system process is conducted by getting slip data on several points of bottom wheel speed and several points of pwm brake value of upper wheel. The data acquisition was obtained by increasing the bottom wheel speed slowly and on each certain velocity was given a brake with several pwm value. The bottom wheel speed was increased every 10% pwm, so there are 10 points of velocity as a data sample. The bottom wheel speed was increased starts from 10% to 100%, so it was obtained the connection between the bottom wheel speed and the changing of duty cycle (PWM) of DC motor that was connected one shaft with bottom wheel as shown in Table 1 below:

Table 1 The Connection of Bottom Wheel Speed and DC Motor PWM

DC Motor PWM	Bottom Wheel Speed (RPM)
10%	573
20%	1224
30%	1621
40%	1854
50%	1994
60%	2097
70%	2155
80%	2173
90%	2220
100%	2313

Every point of bottom wheel speed is given pwm brake from 0% to 100% (PWM brake is increased every 10%). When braking is conducted, then the bottom wheel has come to a stop, after that it can be ran again. The obtained data is a slip value on several velocity range of the bottom wheel and range of pwm brake value of the upper wheel. The work point of slip is the slip average that was

IOP Conf. Series: Journal of Physics: Conf. Series **974** (2018) 012055 doi:10.1088/1742-6596/974/1/012055

obtained on one point of working brake to one point of bottom wheel speed. The process data is slip data of bottom wheel speed and upper wheel speed when it starts to decelerate. When the pwm brake value is high, the upper part of the wheel had a chance to lock (not running) until the lower part of the wheel stops rotating. Here is the connection between the pwm brake changing of the upper wheel. Every point of bottom wheel speed is given 0% to 100% of pwm brake (PWM brake is increased every 10%). When braking is conducted, then the bottom wheel has come to a stop, after that it could be ran again. The obtained data is a slip value of several bottom wheel speed range and pwm brake value range of the upper wheel. Slip work point is the average of the obtained slip at one-point brake work to one-point bottom wheel speed. The processed data is slip data of the bottom wheel speed and upper wheel speed when it stars to decelerate. On the high pwm brake of the upper part of the wheel had a chance to lock (not rotating) until the lower part of the wheel stops rotating. Table 2 below is the connection between the changing of pwm brake on the upper wheel:

Table 2 The connection Pwm Brake and Slip Working Point

Upper Wheel Pwm Brake	Bottom Wheel Speed (RPM)		
	2313 (100%)	2220 (90%)	2193 (80%)
10%	0.12934	0.08476	0.16076
20%	0.14616	0.12543	0.17403
30%	0.17498	0.13771	0.21248
40%	0.23304	0.24226	0.28326
50%	0.48866	0.50346	0.44012
60%	0.66523	0.68928	0.66226
70%	0.76070	0.78347	0.74383
80%	0.82027	0.80893	0.79062
90%	0.81499	0.84319	0.87207
100%	0.83794	0.86017	0.86869

From the obtained data, it could be determined brake working point against slip for each bottom wheel speed as shown in Table 2, from the data in Table 2, its shown that working slip point has value more than 0.2 when pwm brake percentage is above 40% for 3 points of bottom wheel speed. In this final assignment, the author chose 100% (maximum) velocity for bottom wheel speed. Here is the graphic plot of the connection brake working point against slip for 100% pwm bottom wheel speed:

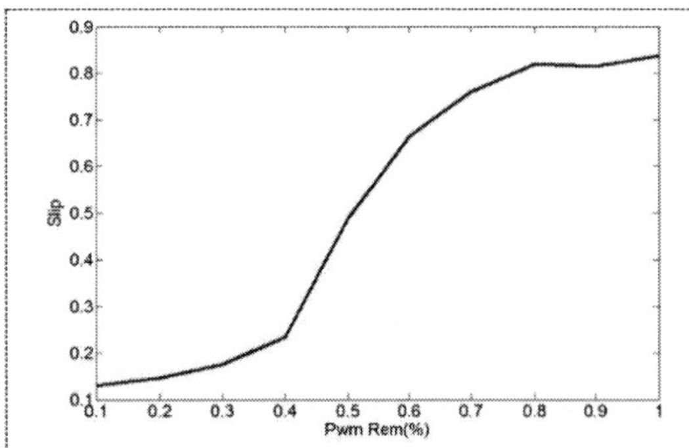

Figure 3 Brake Plant vs Slip Model.

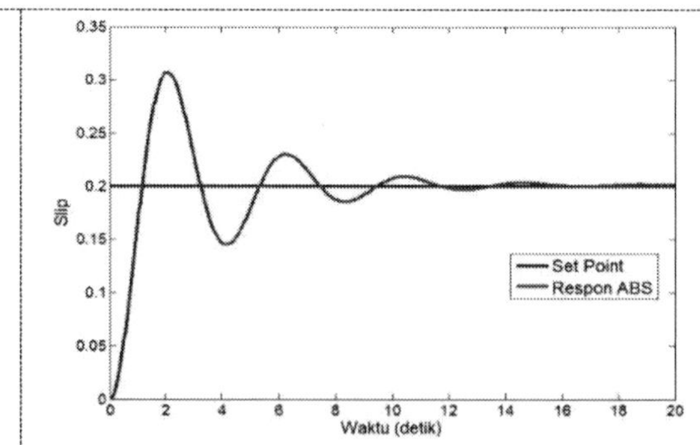

Figure 4 ABS System Response Open Loop without Controller

Based on figure3, the data of brake working point against slip, will be modeled plant model approach by stochastic approach structure which is Modeling with auto regressive exougeneus (ARX)

using toolbox system identification Simulink. For choosing the model, na, nb and nk (delay) should be determined beforehand to determine the model order. Choosing model order based on the estimated error that was generated after putting in several model orders. There are 3 model orders in this process, they are model order 2, model order 3 and model order 4. Model order 3 which was obtained already represents the ABS plant mode. The value of RMSE from estimated error plant model is as big as or about 6.64% and the deviation standard value estimated error plant model is 0.0393. The obtained plant model is already in a district form with z domain:

$$G(z) = \frac{\mathbf{0.0021184z^2}}{\mathbf{z^3 - 1{,}9771z^2 + 0{,}97671z + 0{,}0024989}} \qquad (1)$$

After the plant model was obtained, next thing to do is to test plant model by open loop without controller to see the transient characteristic of system response. The given test signal is step unit signal of 0.2. This 0.2 value is the expectable slip value when the friction coefficient is on maximum on several road condition. In figure 4, there will be system response from test result by open loop:

3.2. Designing Neuro-Fuzzy Controller [4]

Neuro-fuzzy algorithm that is used is fuzzy modelling network algorithm (FMN). FMN has 3 different types, they are type I, type II and type III. FMN will identify the fuzzy rules and the membership function automatically by nerve network weights modification through backpropagation algorithm. In FMN type I usingcivil inference system, where the consequent of fuzzy inference system is a membership function. In FMN type II using civil inference system, where the consequent of fuzzy inference system is a single value (constant). In FMN type III using sugeno fuzzy inference system, where the consequent of fuzzy inference system using linear equation. In controller design using FMN type II to manage the slip value on ABS plant. The neuro-fuzzy algorithm calculation is divided in to 2 stages, they are forward propagation stage and backpropagation stage. Figure 5 is an architecture of neuro-fuzzy network of FMN type II:

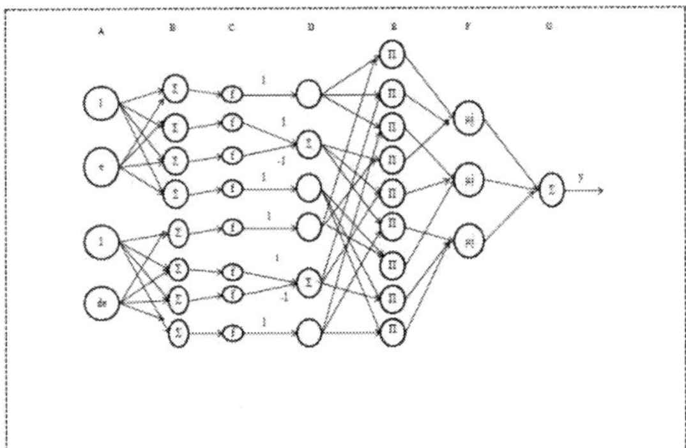

Figure 5 Neuro-Fuzzy Network [4].

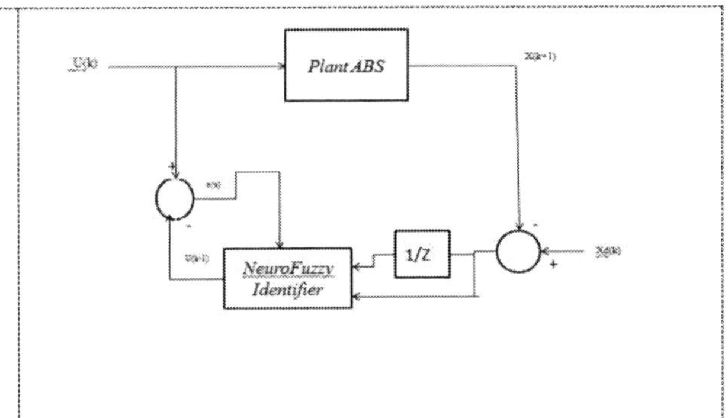

Figure 6 Block Diagram of General Learning

Offline data training stages uses general learning and application stage or online stage. From the designed neuro-fuzzy model, general learning represents the invers dynamic of a plant. In offline stage, it is conducted data training to obtained minimum weight that is corresponding to the arc of neuro-fuzzy network by determine minimum error target and the amount of maximum iteration, as long as the error value is larger than the minimum error target, then iteration can still be proceeded. Figure 6 is block diagram of offline training using general learning.

Based on Figure 6, the target signal is the difference between control signal (u(k)) with estimated control signal (U(k)). For control signal value learning function could be a scalar value so the control signal value (u(k)) is a control signal value when the slip value is expected (set point). Based on the plant model in Figure 6, it is obtained the control signal value on this equation:

$$X_d(k) = 0.2 \quad (2)$$

$$u(k) = 0.345 \quad (3)$$

The signal training neuro-fuzzy offline data is conducted with maximum amount of iteration (maximum epoh) 100 and minimum error target 10^{-6}. In training data stage, the calculation of neuro-fuzzy consists of forward propagation and backpropagation. The error value in each iteration consists of forward propagation and backpropagation process will be squared then rooted. The error value in each iteration will always be compared with minimum error target until the error value is smaller than minimum error target.

The FMN type II network structure as shown in figure 7 consists of 7 layers. 7 layers of the network are one layer of input, 5 hidden layers and one layer of output. In this part, it will be explained that one of the stages from neuro-fuzzy algorithm calculation is forward propagation stage. Forward propagation stage works from layer A (input layer) to layer G (output layer). Table 3 below is the basic of fuzzy rules in the 6th layer:

Table 3 Basic Rule of Neuro-Fuzzy Controller

U		Delta *Error*		
		N	Z	P
Error	N	N	N	Z
	Z	N	Z	P
	P	Z	P	P

Backpropagation stage will be conducted after forward propagation stage by revising the weight on neuro-fuzzy network. After the data training has been conducted by offline with forward propagation and back propagation calculation stage with error target parameter 10^{-6}; learning rate 0.01; and maximum iteration (maximum epoh) is 100, then it is obtained the graph connection between the error value of each iteration (epoh) such as in figure 7 below:

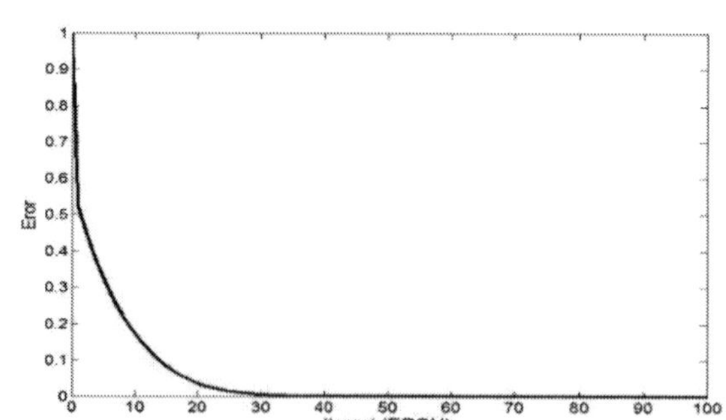

Figure 7 Training Result

The stage after offline data training is online simulation stage. The corresponding weight will be obtained after the training process is finish with error minimum parameter 10^{-6} and the amount of maximum iteration is 100. Training is finish when the error value has reached 0.000005094, although this value is still above 10^{-6}. But in this value, the error value is relatively not changing and not passing through 10^{-6}. So the iteration stops on the error value 0.000005094 on 100th iteration (epoh). So on

100thepoh will be obtained the corresponding weights with the neuro-fuzzy network arc that will be the constant weight in online simulation stage. Table 4 below is the weight of training results:

Table 4 Weight of Training Results.

Weight	Weight Value
Wdcde1	0.0957
Wcde2	0.4736
Wcek1	0.5462
Wcek2	0.7070
Wgde	0.0468
Wgek	0.4276
Wout	[0.4895; 0.8396 ; 0.5820]

After weight from neuro-fuzzy network arc has obtained then the next step is putting in those weight values to the neuro-fuzzy algorithm without backpropagation stage. Calculation stage that is used is only forward propagation stage.

4. Testing and Analysis

4.1. System Simulation

Simulation is conducted by testing the plant model by close loop against the set point in the form of signal step. Designing the neuro-fuzzy controller consists of 2 stages: offline data training and online data training. Plant model testing that will be analyzed is when the online stage which is the weights from offline training result has already put in to the neuro-fuzzy algorithm without backpropagation calculation. The given signal test is a step unit signal for 0.2 which represents the expectable slip value. In figure 8 will be shown a system response using neuro-fuzzy controller.

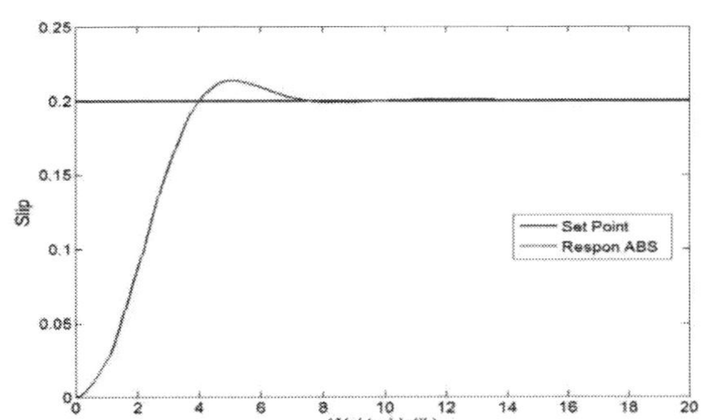

Figure 8 ABS System Response using Neuro-Fuzzy Controller

From the system response above in figure 8, it was obtained the transient response data and steady state system response will be shown in table 5 below:

Table 5 Transient Response Data using Neuro-Fuzzy Controller

Rise Time	*Settling time*$\pm$5 % (sec)	*Maximum Overshoot* (%)	*Ess*(%)
4	8	6.8	0

System response with neuro-fuzzy controller that has been obtained will be compared with system response with PID controller as a comparison. The designed PID controller based on trial-error mechanism. The trial-error mechanism was conducted by tuning on Kp, Ki, and Kd parameters. In the first step, tuning was conducted by changing Kp parameter to be larger, then changing Ki parameter and Kd parameter. In figure 9 is system response comparison with PID controller and Neuro-Fuzzy controller using one of the tunings, which are Kp value tuned for 5, Ki value tuned for 4 and Kd value tuned for 4.

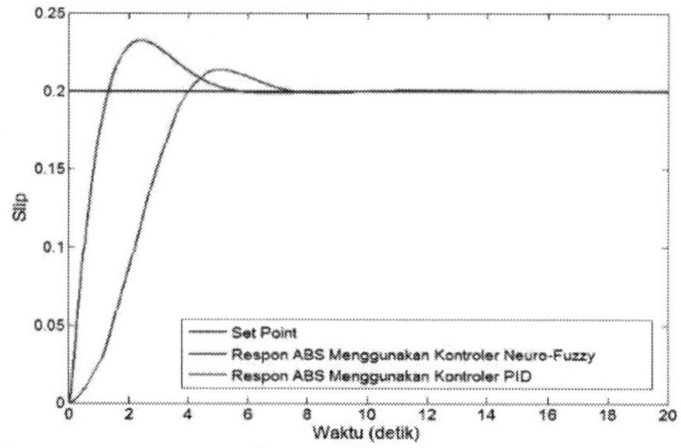

Figure 9 ABS System Response Comparison with PID Controller and Neuro-Fuzzy Controller

Response specification in table 6 will describe the tuning results for Kp, Ki and Kd which has been conducted against the transient system response:

Table 6 Tuning Results for Kp, Ki and Kd Parameters

Kp	Ki	Kd	*tr*(second)	*ts* ±5%	*Mp*(%)	*Ess*(%)
5	3	4	1.56	8	11.25	0
5	4	4	1.35	8.5	16.1	0
5	4	3	1.11	5.76	13.15	0
5	3	4	1.29	8.25	8.25	0
5	3	1	0.6	6.2	23.5	0

Based on table 6, the average rise time of PID controller is much better compares to the rise time of neuro-fuzzy controller which means the neuro-fuzzy controller is not capable to give a better rise time compared to PID controller. Neuro-fuzzy controller is capable to give settling time value twice larger for system response without controller. Neuro-fuzzy controller has settling time value that doesn't have much difference with several PID controller parameter tuning results, which means the neuro-fuzzy controller generally is more capable to give much better settling time value compared to PID controller. Neuro-fuzzy controller is capable on reducing the overshoot in system response without controller for 43.2%, which means the neuro-fuzzy controller is capable to reduce the overshoot in system response significantly. Neuro-fuzzy controller also has a smaller overshoot value compared to PID controller. Neuro-fuzzy and PID controller can fix the system because they have steady state error approaching 0%.

4.2. Implementation Result

Implementation is conducted on ABS real plant simulator with Matlab software. The implementation result for 2300 rpm velocity (100%) is shown in figure 10.

IOP Conf. Series: Journal of Physics: Conf. Series **974** (2018) 012055 doi:10.1088/1742-6596/974/1/012055

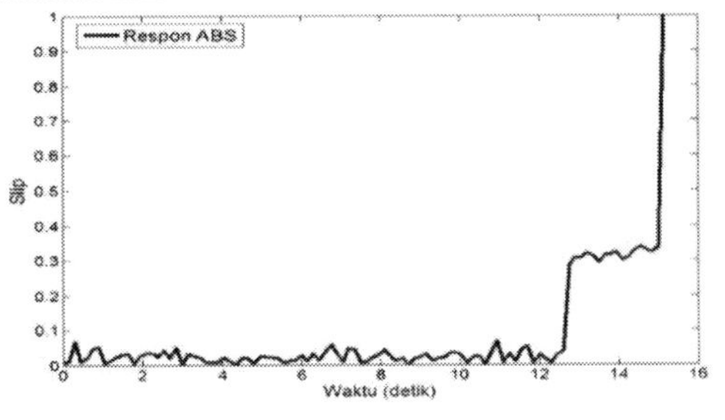

Figure 10 ABS Response Implementation Result.

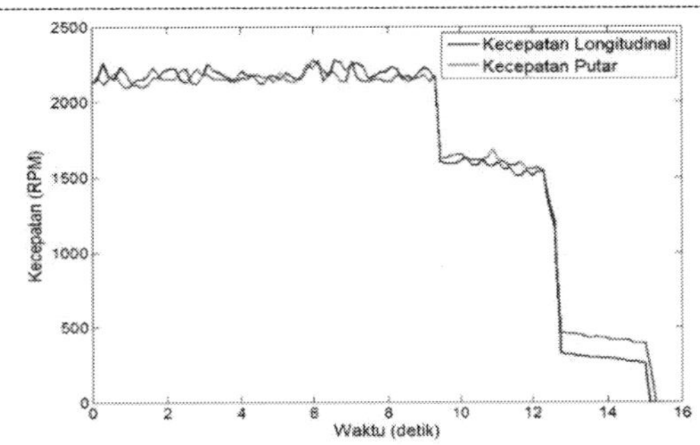

Figure 11 Longitudinal Velocity Response and Rotation Speed

Based on ABS system response on figure 10 shows that neuro-fuzzy controller can work in a braking condition. When on braking the slip value can't follow the reference signal value for 0.2 but it can keep the upper wheel not to lock or keeping the slip value not 1. From system response on figure 10 has rise time value for 0.2 seconds and error steady state for 50%. When the braking is conducted, there is a longitudinal velocity and rotation speed change as an effect from the controllable slip value. Here is the velocity response when the neuro-fuzzy controller works, which will be shown in figure 11.

The reduction of longitudinal velocity and rotation speed also have flatter slope compared to the reduction of velocity without ABS system. The implementation also conducted in 2200 rpm velocity (90%). In 2200 rpm velocity, slip value that is generated can follow the reference signal for 0.2. Figure 12 below is ABS system response in 2200 rpm velocity:

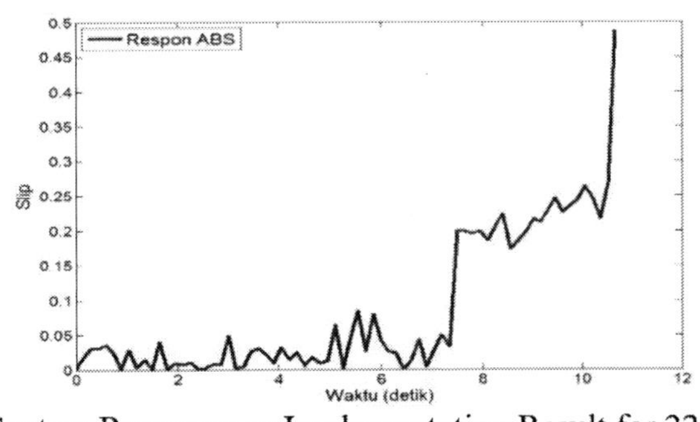

Figure 12 ABS System Response as Implementation Result for 2200 RPM Velocity

Based on figure 12, the ABS system response can follow reference signal for 0.2 when braking is conducted. Bu it has a steady state error value (ess) which tends to increase until the upper wheel and the bottom wheel stop. Neuro-fuzzy controller can work for 2200 rpm velocity because it can keep the upper wheel stays unlock when the braking is conducted. The system response is 1 when the upper and bottom wheel has already stopped, it's because of the effect of slip calculation which has 0/0 value when the upper and bottom wheel has stopped.

5. Conclusion

Based on plant model test result using neuro-fuzzy controller by simulation and implementation result on ABS real plant using neuro-fuzzy controller, it can be concluded to several points, which are:

- Neuro-fuzzy controller can reduce the overshoot system response for 43,2% and accelerate the time response twice faster to reach stable condition
- The designed neuro-fuzzy controller can work when the implementation for bottom wheel speed is 2200 and 2300 rpm.
- For 2300 rpm wheel speed, the neuro-fuzzy controller has an error steady state value (ess) for 50% and rise time for 0.2 seconds. Slip value that is generated is still in an acceptable range in some road conditions, which is 0.15-0.3 [6]. For 2200 rpm velocity can follow signal reference approximately 0.2 for 0.5 second in initial braking.

6. References

[1] M. Smeja, 2010. *Modelling and Identificaion of Antilock Braking System (ABS)*, Slovakia: AGH University.
[2] K.Reif, 2014. *Brakes, Brake Control and Driver Assistance Systems Function*, Springer, vol.VIII, pp. 275-306.
[3] Wibowo, B.P, 2014. "Traction Control Pada Parallel Hybrid Electric Vehicle (HEV) Menggunakan Metode Kontrol Neuro-Fuzzy Prediktif", *Tugas Akhir*, Jurusan Teknik Elektro, Institut Teknologi Sepuluh Nopember, Surabaya.
[4] JSR Jang dan Sun CT, 1997. *Neuro-Fuzzy and Soft Computing*, London: Prentice Hall.
[5] Kusumadewi, Sri, Neuro-Fuzzy, 2006. *Integrasi Sistem Fuzzy dan Jaringan Syaraf*, Yogyakarta: Graha Ilmu.
[6] Sobottka, Christian, 1996. "Optimal Fuzzy Logic Control For an Anti-Lock Braking System", *IEEE International Conference On Control Applications*, New York.

Analysis of time series for postal shipments in Regional VII East Java Indonesia

DE Kusrini[1], B S S Ulama[1] and L Aridinanti[1]

[1]Statistika Bisnis Department, Institut Teknologi Sepuluh Nopember, Surabaya, Indonesia.

E-mail: dwi_endah@statistika.its.ac.id

Abstract. The change of number delivery goods through PT. Pos Regional VII East Java Indonesia indicates that the trend of increasing and decreasing the delivery of documents and non-documents in PT. Pos Regional VII East Java Indonesia is strongly influenced by conditions outside of PT. Pos Regional VII East Java Indonesia so that the prediction the number of document and non-documents requires a model that can accommodate it. Based on the time series plot monthly data fluctuations occur from 2013-2016 then the model is done using ARIMA or seasonal ARIMA and selected the best model based on the smallest AIC value. The results of data analysis about the number of shipments on each product sent through the Sub-Regional Postal Office VII East Java indicates that there are 5 post offices of 26 post offices entering the territory. The largest number of shipments is available on the PPB (Paket Pos Biasa is regular package shipment/non-document) and SKH (Surat Kilat Khusus is Special Express Mail/document) products. The time series model generated is largely a Random walk model meaning that the number of shipment in the future is influenced by random effects that are difficult to predict. Some are AR and MA models, except for Express shipment products with Malang post office destination which has seasonal ARIMA model on lag 6 and 12. This means that the number of items in the following month is affected by the number of items in the previous 6 months.

Keywords: ARIMA, documents, non-documents, shipments, post office.

1. Introduction

The average increase of internet user households in Indonesia is 3.5% per year as cellular phone users in Indonesia grow by 2% per year (Statistics Indonesia, 2016) encourages shifting the way Indonesians shop from the direct transaction system (in the market/shop) to indirect transaction online, it encourages PT. Pos Indonesia to cooperate with some online shopping service providers such as Matahari mall, Ali Express, etc. (PT. Pos Indonesia Annual Report, 2015). These conditions affect the magnitude of the types of documents and non-documents through the Post, which based on the annual report of PT. Pos Indonesia (2015) delivery of goods through packages has increased since 2010 from 14,944,000 units to 25,241,000 packages in 2014, while standard letter delivery actually decreased from 64.267.000 units in 2010 to 60.472.000 units in the Year 2014. This change indicates that the

trend of increasing and decreasing the delivery of documents and documents in PT. Pos Indonesia is strongly influenced by conditions outside of PT. Pos Indonesia so that the prediction the number of document submissions and non-documents requires a model that can accommodate it.
ARIMA models are, in theory, the most general class of models for forecasting a time series which can be made to be "stationary" by differencing (if necessary), perhaps in conjunction with nonlinear transformations such as logging or deflating (if necessary). A random variable that is a time series is stationary if its statistical properties are all constant over time. A stationary series has no trend, its variations around its mean have a constant amplitude, and it wiggles in a consistent fashion, i.e., its short-term random time patterns always look the same in a statistical sense. The latter condition means that its *autocorrelations* (correlations with its own prior deviations from the mean) remain constant over time, or equivalently, that its *power spectrum* remains constant over time. A random variable of this form can be viewed (as usual) as a combination of signal and noise, and the signal (if one is apparent) could be a pattern of fast or slow mean reversion, or sinusoidal oscillation, or rapid alternation in sign, and it could also have a seasonal component. An ARIMA model can be viewed as a "filter" that tries to separate the signal from the noise, and the signal is then extrapolated into the future to obtain forecasts.

Based on the above background, the purpose of this reaearch are to obtain and analyze the best time series model of document delivery and non-document and review the forecasting of documents and non-documents based on the ARIMA model obtained in the next period

2. Materials and Methods

2.1. ARIMA Model

The acronym ARIMA stands for Auto-Regressive Integrated Moving Average. Lags of the stationeries series in the forecasting equation are called "autoregressive" terms, lags of the forecast errors are called "moving average" terms, and a time series which needs to be differenced to be made stationary is said to be an "integrated" version of a stationary series. Random-walk and random-trend models, autoregressive models, and exponential smoothing models are all special cases of ARIMA models.

A non seasonal ARIMA model is classified as an "ARIMA(p,d,q)" model, where:

- **p** is the number of autoregressive terms,
- **d** is the number of no seasonal differences needed for stationery, and
- **q** is the number of lagged forecast errors in the prediction equation.

The forecasting equation is constructed as follows. First, let y denote the dth difference of Y, which means:
If d=0: $y_t = Y_t$
If d=1: $y_t = Y_t - Y_{t-1}$
If d=2: $y_t = (Y_t - Y_{t-1}) - (Y_{t-1} - Y_{t-2}) = Y_t - 2Y_{t-1} + Y_{t-2}$
Note that the second difference of Y (the d=2 case) is not the difference from 2 periods ago. Rather, it is the *first-difference-of-the-first difference*, which is the discrete analog of a second derivative, i.e., the local acceleration of the series rather than its local trend.
In terms of y, the general forecasting equation is:

$$\hat{y}_t = \mu + \phi_1 y_{t-1} + \ldots + \phi_p y_{t-p} - \theta_1 e_{t-1} - \ldots - \theta_q e_{t-q} \quad (1)$$

Here the moving average parameters (θ's) are defined so that their signs are negative in the equation, following the convention introduced by Box and Jenkins. Some authors define them so that they have plus signs instead. When actual numbers are plugged into the equation, there is no ambiguity, but it's important to know which convention your software uses when you are reading the output. Often the parameters are denoted there by AR(1), AR(2), …, and MA(1), MA(2), … etc..

To identify the appropriate ARIMA model for y_t, we begin by determining the order of differencing (d) needing to stationary the series and remove the gross features of seasonality, perhaps in conjunction with a variance-stabilizing transformation such as logging or deflating. If we stop at this point and predict that the differenced series is constant, we have merely fitted a random walk or random trend model. However, the stationeries series may still have auto correlated errors, suggesting that some number of AR terms ($p \geq 1$) and/or some number MA terms ($q \geq 1$) are also needed in the forecasting equation.

2.2. PT.Pos Products

PT. Pos Indonesia (Persero) has an extensive network of 4,800 online post offices (http://www.posindonesia.co.id). The number of service points reached 58,700 points in the form of post office, postal agency, Mobile Postal Service, and others. Innovation continues to be carried out by Pos Indonesia, among others, Post shop development which is a retail business development implemented to change the image of conventional office into a modern post office with one stop shopping service pattern, namely Postal services in the form of mail delivery service, package, financial services, stamps, philately products and others. Online Shopping services Pos Indonesia also provides e-commerce services and other services through myPos and m-pospay applications.

In general, PT.Pos Indonesia's products are divided into three categories:

1. Letters and Packages
   In this category there are some products such as: Admailpos, Express Mail Service (EMS), Philatelic, Paketpos, Posexpress, Special Poskilat, Ordinary Letter (Standard).
2. Financial Services
   In this category there are several products such as: Bank Channelling, Fund Distribution, Giropos, Weselpos, Post Pay.
3. Retail in cooperation with ANTAM gold
4. Logistic Integration, Customized according to customer's request which include services: Transporting, Warehousing, Freight forwarding, Distribution of goods and Supply chain management.
5. International post include: EMS, international fast Package, Export Package, International registered mail, International Pos Post, International air mail.

In this research the products used in the analysis are Admailpos product, Express Mail Service (EMS), Paketpos, Posexpress, Special Poskilat, Ordinary Letter (Standard). Preparation of products based on document category and non document can not be done directly because there are several categories of products whose data can not be differentiated whether included in the document or non document category.

2.3. Methodology for Data Modelling

Sources of datas in this research were taken from the data delivery notes (documents and non-documents) or product services letter and package services received by PT.Pos regional VIII East Java for 5 years starting from 2013 to 2017. The types of products used for research data are:

1. PPB is Ordinary Package Shipment (non document)
2. SKH is a Special Express Mail (document)
3. PPKH is a Special Express Package Delivery (non-document)
4. XP is Express (non documents and documents)
5. EMS is Express Mail Service (document)

The datas used in this research are in the monthly data series (with bag for unit of data) per each delivery destination (26 post offices throughout East Java), so that modelling will be done at each delivery destination. After we were doing the modelling then we were forecasting for the next few months.

3. Results

The results of the research divided into three parts:

1. Descriptive analysis of the number of shipments based on the post office of destination of the products
2. Time series data modelling based on products type and post office of products destination
3. Forecasting based on the model obtained for each type of products and post office of destination of the mail.

The results can be seen in the section 3.1, 3.2 and 3.3.

3.1. Descriptive analysis of the number of shipments

According to Figure 1, there are five post offices that have a total larger product shipment than other post offices (more than 1,200,000 bags). While the other 19 post offices only get total shipments of products under 200,000 bags)

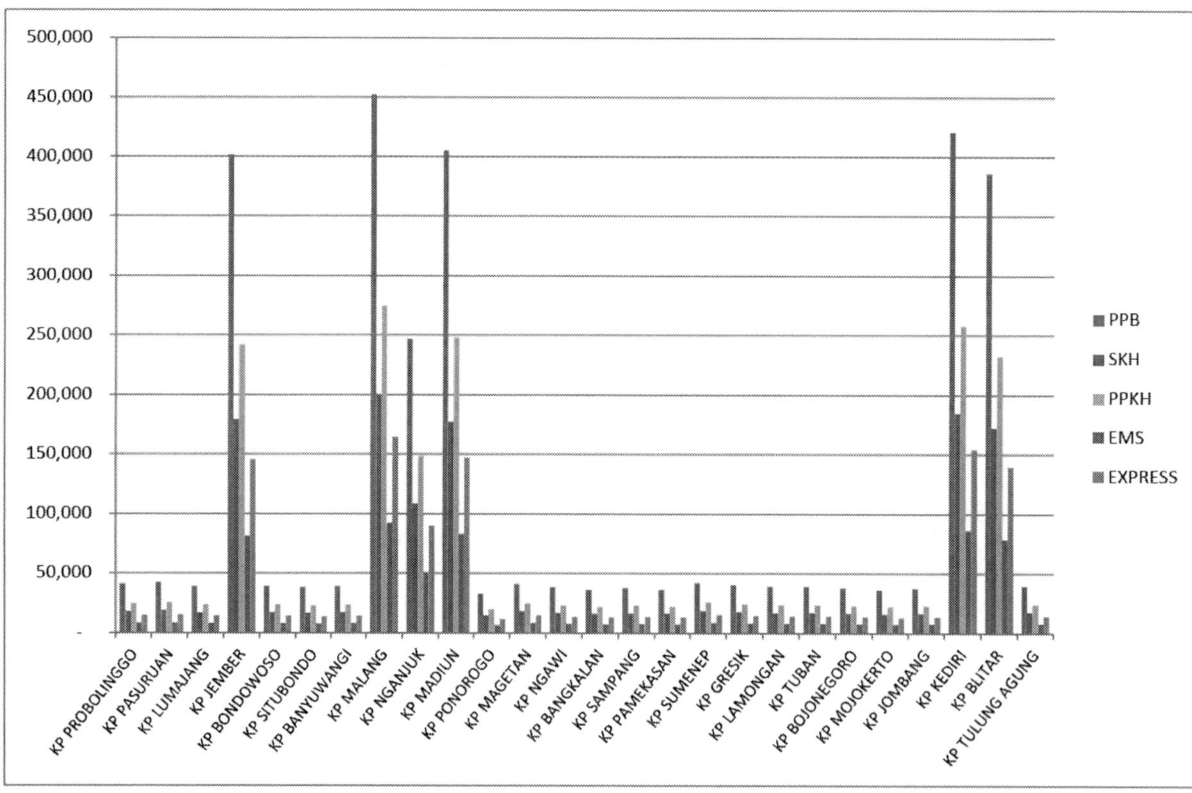

Figure 1. Bar charts the number of bags of shipment by products and post offices destination

Five post offices that get the most posts are Post Office KP Malang, Post Office KP Kediri, Post Office KP Madiun, Post Office KP Jember and Post Office KP Blitar. The largest number of mails for the five Post Offices is not only on the total products but also on all types of products, so for Post

Office Sub-Regional VII East Java can make this information to provide more space for the products delivered there than the office another destination post.

3.2. Time series data modeling

Based on the time series plot monthly data fluctuations occur from 2013-2016 and due to the cause of the fluctuation of unknown data, then the modeling is done using ARIMA or seasonal ARIMA model and selected the best model based on the smallest AIC value. Based on the ACF and PACF plots and testing the significance of the parameters and the coefficient reduction to obtain the appropriate model, the corresponding model is shown in full in Table 1.

Table 1. The Time series model for the shipment of products based on the post office of destination of the mail

Post Office	Products				
	PPB	**SKH**	**PPKH**	**EMS**	**EXPRESS**
Probolinggo	ARIMA (0,1,0)	ARIMA (0,1,0)	ARIMA (0,1,0)	ARIMA (0,1,0)	ARIMA (0,1,0)
Pasuruan	ARIMA (0,1,0)	ARIMA (0,1,0)	ARIMA (0,1,0)	ARIMA (0,1,0)	ARIMA (0,1,0)
Lumajang	ARIMA (0,1,0)	ARIMA (0,1,0)	ARIMA (0,1,0)	ARIMA (0,1,0)	ARIMA (0,1,0)
Jember	ARIMA (0,1,0)	ARIMA (0,1,0)	ARIMA (0,1,0)	ARIMA (0,1,0)	ARIMA (0,1,0)
Bondowoso	ARIMA (0,1,0)	ARIMA (0,1,0)	ARIMA (0,1,0)	ARIMA (0,1,0)	ARIMA (0,1,0)
Situbondo	ARIMA (0,0,1)	ARIMA (2,1,1)	ARIMA (0,0,1)	ARIMA (0,0,1)	ARIMA (0,0,1)
Banyuwangi	ARIMA (0,1,0)	ARIMA (0,0,1)	ARIMA (0,1,0)	ARIMA (0,1,0)	ARIMA (0,1,0)
Malang	ARIMA (0,0,1)	ARIMA (0,0,1)	ARIMA (0,1,0)	ARIMA (0,1,0)	ARIMA ([6,10],0,[6,12])
Nganjuk	ARIMA (0,1,0)	ARIMA (0,1,0)	ARIMA (0,1,0)	ARIMA (0,1,0)	ARIMA (0,1,0)
Madiun	ARIMA (0,1,0)	ARIMA (0,0,2)	ARIMA (0,1,0)	ARIMA (0,1,0)	ARIMA (0,1,0)
Ponorogo	ARIMA (0,1,0)	ARIMA (0,1,0)	ARIMA (0,1,0)	ARIMA (0,1,0)	ARIMA (0,1,0)
Magetan	ARIMA (0,0,1)	ARIMA (0,0,1)	ARIMA (0,0,1)	ARIMA (2,0,2)	ARIMA (0,0,1)
Ngawi	ARIMA (0,0,1)	ARIMA (0,0,1)	ARIMA (0,1,0)	ARIMA (0,0,1)	ARIMA (0,0,1)
Bangkalan	ARIMA (0,1,0)	ARIMA (0,1,0)	ARIMA (0,1,0)	ARIMA (0,1,0)	ARIMA (0,1,0)
Sampang	ARIMA (0,1,0)	ARIMA (0,1,0)	ARIMA (0,0,1)	ARIMA (0,1,0)	ARIMA (0,1,0)
Pamekasan	ARIMA (0,1,0)	ARIMA (0,1,0)	ARIMA (0,1,0)	ARIMA (0,0,1)	ARIMA (0,1,0)
Sumenep	ARIMA (0,0,1)	ARIMA (0,0,1)	ARIMA (0,1,0)	ARIMA (0,1,0)	ARIMA (0,1,0)
Gresik	ARIMA (0,1,0)	ARIMA (0,1,0)	ARIMA (0,1,0)	ARIMA (0,1,0)	ARIMA (0,1,0)
Lamongan	ARIMA (0,1,0)	ARIMA (0,1,0)	ARIMA (0,1,0)	ARIMA (0,1,0)	ARIMA (0,1,0)
Tuban	ARIMA (0,1,0)	ARIMA (0,1,0)	ARIMA (0,1,0)	ARIMA (0,1,0)	ARIMA (0,1,0)
Bojonegoro	ARIMA (0,1,0)	ARIMA (0,1,0)	ARIMA (0,1,0)	ARIMA (0,1,0)	ARIMA (0,1,0)
Mojokerto	ARIMA (0,1,0)	ARIMA (0,1,0)	ARIMA (0,1,0)	ARIMA (0,1,0)	ARIMA (0,1,0)
Jombang	ARIMA (0,0,1)	ARIMA (0,0,1)	ARIMA (0,1,0)	ARIMA (0,1,0)	ARIMA (0,1,0)
Kediri	ARIMA (0,1,0)	ARIMA (0,0,1)	ARIMA (0,1,0)	ARIMA (0,1,0)	ARIMA (0,1,0)
Blitar	ARIMA (0,0,1)	ARIMA (1,0,0)	ARIMA (0,0,1)	ARIMA (0,0,1)	ARIMA (0,0,1)
Tulungagung	ARIMA (0,1,0)	ARIMA (0,1,0)	ARIMA (1,0,0)	ARIMA (0,1,0)	ARIMA (1,0,0)

Most time series models are produced in the form of a Random walk where a random walk is defined as a process where the current value of a variable is composed of the past value plus an error term defined as a white noise (a normal variable with zero mean and variance one). The implication of a process of this type is that the best prediction of y for next period is the current value, or in other

IOP Conf. Series: Journal of Physics: Conf. Series **974** (2018) 012056 doi:10.1088/1742-6596/974/1/012056

words the process does not allow to predict the change $(y_t - y_{t-1})$. That is, the change of y is absolutely random. It can be shown that the mean of a random walk process is constant but its variance is not. Therefore a random walk process is non stationary, and its variance increases with t. In practice, the presence of a random walk process makes the forecast process very simple since all the future values of y_{t+s} for $s > 0$, is simply y_t.

Table 2. Time series model without random walk

Post Office (KP)	Products				
	PPB	SKH	PPKH	EMS	EXPRESS
Situbondo	$Z_t = 792,75 + a_t + 0,33a_{t-1}$	$Z_t = 0,49Z_{t-1} + 0,11Z_{t-2} + 0,40Z_{t-3} + a_t + 0,76a_{t-1}$	$Z_t = 476,74 + a_t + 0,32a_{t-1}$	$Z_t = 160,48 + a_t + 0,36a_{t-1}$	$Z_t = 287,6 + a_t + 0,29a_{t-1}$
Banyuwangi		$Z_t = 358,89 + a_t + 0,36a_{t-1}$			
Malang	$Z_t = 9399,84 + a_t + 0,39a_{t-1}$	$Z_t = 4142,8 + a_t + 0,49a_{t-1}$			$Z_t = 6837,348 + Z_{t-6} + a_t - 0,58129a_{t-6}.$
Madiun		$Z_t = 3750,65 + a_t + 0,31a_{t-1} + 0,36a_{t-2}$			
Magetan	$Z_t = 863,18 + a_t + 0,49a_{t-1}$	$Z_t = 384,99 + a_t + 0,46a_{t-1}$	$Z_t = 520,18 + a_t + 0,5a_{t-1}$	$Z_t = 175,46 + 0,15Z_{t-1} - 0,83Z_{t-2} + a_t + 0,57a_{t-1} - 0,84a_{t-2}$	$Z_t = 311,69 + a_t + 0,51a_{t-1}$
Ngawi	$Z_t = 793,42 + a_t + 0,36a_{t-1}$	$Z_t = 353,92 + a_t + 0,4a_{t-1}$	$Z_t = 160,98 + a_t + 0,35a_{t-1}$	$Z_t = 160,98 + a_t + 0,35a_{t-1}$	$Z_t = 287,91 + a_t + 0,33a_{t-1}$
Sampang			$Z_t = 486,05 + a_t + 0,31a_{t-1}$	$Z_t = 153,37 + a_t + 0,33a_{t-1}$	
Sumenep	$Z_t = 883,99 + a_t + 0,33a_{t-1}$				
Jombang	$Z_t = 772,58 + a_t + 0,65a_{t-1}$	$Z_t = 339,35 + a_t + 0,72a_{t-1}$			
Kediri		$Z_t = 3854,67 + a_t + 0,29a_{t-1}$			
Blitar	$Z_t = 7948,82 + a_t + 0,47a_{t-1}$	$Z_t = 3558 - 0,38Z_{t-1} + a_t$	$Z_t = 4797,11 + a_t + 0,45a_{t-1}$	$Z_t = 1618,78 + a_t + 0,5a_{t-1}$	$Z_t = 2879,1 + a_t + 0,49a_{t-1}$
Tulung-agung			$Z_t = 504,73 + 0,31Z_{t-1} + a_t$		$Z_t = 299,81 + 0,25Z_{t-1} + a_t$

Based on Table 2 it can be concluded that only express shipment products with the destination of Malang post office have a seasonal ARIMA model in which the number of deliveries in the future is affected by shipment in the previous 6 months. The rest are AR and MA models in addition to the previously described random walk model.

3.3. Forecasting

The model already obtained in the previous section is used to predict the number of items by type of postal product and post office. Forecasting is done by using outsampel data to forecast data up to 3 months ahead. The forecasting results are shown in Table 3.

Forecasting results obtained are expected to provide information to the Regional Postal Coordinator VII East Java to make the preparation of a more adequate place and workforce.

Table 3. Forecasting number of bags for every products and post office destination for three months in 2017

Post Office (KP)	Month	PPB	SKH	PPKH	EMS	EXPRESS
Probolinggo	October	853	374	515	173	310
	November	853	374	515	173	310
	December	853	374	515	173	310
Pasuruan	October	881	395	527	177	316
	November	881	395	527	177	316
	December	881	395	527	177	316
Lumajang	October	814	354	496	166	296
	November	814	354	496	166	296
	December	814	354	496	166	296
Jember	October	8,363	3,729	5,041	1,699	3,032
	November	8,363	3,729	5,041	1,699	3,032
	December	8,363	3,729	5,041	1,699	3,032
Bondowoso	October	820	359	499	167	297
	November	820	359	499	167	297
	December	820	359	499	167	297
Situbondo	October	793	412	477	160	288
	November	793	409	477	160	288
	December	793	411	477	160	288
Banyuwangi	October	814	359	491	165	295
	November	814	359	491	165	295
	December	814	359	491	165	295
Malang	October	9,400	4,143	5,718	1,922	2,832
	November	9,400	4,143	5,718	1,922	1,964
	December	9,400	4,143	5,718	1,922	5,540
Nganjuk	October	729	325	439	148	264
	November	729	325	439	148	264
	December	729	325	439	148	264
Madiun	October	8,441	3,751	5,162	1,732	3,068
	November	8,441	3,751	5,162	1,732	3,068
	December	8,441	3,751	5,162	1,732	3,068
Ponorogo	October	683	309	413	138.54	245
	November	683	309	413	138.54	245
	December	683	309	413	138.54	245
Magetan	October	863	385	520	197	312
	November	863	385	520	163	312
	December	863	385	520	156	312
Ngawi	October	793	354	481	161	289
	November	793	354	481	161	289
	December	793	354	481	161	289

IOP Conf. Series: Journal of Physics: Conf. Series **974** (2018) 012056 doi:10.1088/1742-6596/974/1/012056

Post Office (KP)	Month	PPB	SKH	PPKH	EMS	EXPRESS
Bangkalan	October	757	338	456	153	274
	November	757	338	456	153	274
	December	757	338	456	153	274
Sampang	October	787	344	486	161	287
	November	787	344	486	161	287
	December	787	344	486	161	287
Pamekasan	October	761	343	458	153	275
	November	761	343	458	153	275
	December	761	343	458	153	275
Sumenep	October	884	392	537	180	320
	November	884	392	537	180	320
	December	884	392	537	180	320
Gresik	October	840	369	510	171	305
	November	840	369	510	171	305
	December	840	369	510	171	305
Lamongan	October	812	357	495	166	295
	November	812	357	495	166	295
	December	812	357	495	166	295
Tuban	October	815	359	492	165.35	298
	November	815	359	492	165.35	298
	December	815	359	492	165.35	298
Bojonegoro	October	786	346	476	159.90	286
	November	786	346	476	159.90	286
	December	786	346	476	159.90	286
Mojokerto	October	754	334	458	153.63	273
	November	754	334	458	153.63	273
	December	754	334	458	153.63	273
Jombang	October	773	339	475	159	282
	November	773	339	475	159	282
	December	773	339	475	159	282
Kediri	October	8,773	3,855	5,372	1,800	3,210
	November	8,773	3,855	5,372	1,800	3,210
	December	8,773	3,855	5,372	1,800	3,210
Blitar	October	7,949	3,558	4,797	1,619	2,879
	November	7,949	3,558	4,797	1,619	2,879
	December	7,949	3,558	4,797	1,619	2,879
Tulung agung	October	821	364	505	168	300
	November	821	364	505	168	300
	December	821	364	505	168	300
TOTAL		179,280	79,635	108,890	36,565	65,223

4. Discussion

The results of data analysis of the number of shipments on each product sent through the Sub-Regional Postal Office VII indicates that there are 5 post offices of 26 post offices entering the territory. The largest number of shipments is available on the PPB (non-document) and SKH (document) products.

The time series model generated most are random walk models. The random walk model explain that the number of shipment is influenced by random effects in the future, that are difficult to predict, Some models are AR and MA, except for Express shipment products with Malang post office destination which has seasonal ARIMA model on lag 6 and 12. This means that the number of items in the following month is affected by the number of items in the previous 6 months.

There are several things that can be suggested in this research are:

a. The data unit used is still in the form of pockets rather than the number of letters for documents or the number of units of goods for non-documents
b. Time units of data used also in monthly rather than weekly or daily units so the models are less sensitive to capture changes in data behaviour.

5. Acknowlegment

This research was supported by the Ministry of Research and High Education, Indonesia (Kemenristek Dikti) under Research Local ITS (Department Grant 2017) and the Statistika Bisnis Department, Faculty of Vocation Institut Teknologi Sepuluh Nopember Surabaya Indonesia.

6. References

[1]. Al-Zeaud H A 2011 Modelling and Forecasting Volatility using ARIMA model European *Journal of Economics Finance & Administrative Science* Issue 35 p 109-125 2
[2]. Azad A K and Mahsin M 2011 Forecasting Exchange Rates of Bangladesh using ANN and ARIMA models: A comparative research*International Journal of Advanced Engineering Science & Technologies* vol 10 Issue 1 p 031-036. 3
[3]. Contreras J, Espinola R, Nogales F J and Conejo A J 2003 ARIMA models to predict Next Day Electricity Prices *IFEE Transactions on power system* vol 18 no 3 p 1014-1020 4
[4]. Datta K 2011 ARIMA Forecasting of Inflation in the Bangladesh Economy *The IUP Journal of Bank Managemen* vol 10 no 4 p 7-15 5
[5]. Kumar K, Yadav A K, Singh M P, Hassan H and Jain V K 2004 Forecasting Daily Maximum Surface Ozone 6 Concentrations in Brunei Darussalam An ARIMA Modelling Approach *Journal of Air and Waste Management and Association* vol 54 p 809- 814. 7
[6]. Liv Q, Liu X, Jiang B, and Yang W 2011 Forecasting incidence of hemorrhagic fever with renal syndrome in China using ARIMA model Biomed Central p 1-7 8
[7]. Merh N, Saxena VP and Pardasani KR 2011 Next Day Stock market Forecasting: An Application of ANN & ARIMA *The IUP Journal of Applied Science* vol 17 no 1 p 70-84 9
[8]. Uko A K, Nkoro E 2012 Inflation Forecasts with ARIMA Vector Autoregressive & Error Correction Models in Nigeria *European Journal of Economic Finance and Administrative Science* Issue 50 p 71- 87

IOP Conf. Series: Journal of Physics: Conf. Series **974** (2018) 012057 doi:10.1088/1742-6596/974/1/012057

Application of optimal control strategies to HIV-malaria co-infection dynamics

Fatmawati*, **Windarto**, **Lathifah Hanif**

Department of Mathematics, Faculty of Science and Technology, Universitas Airlangga, Surabaya 60115, Indonesia

E-mail: `fatmawati@fst.unair.ac.id`, `fatma47unair@gmail.com`

Abstract. This paper presents a mathematical model of HIV and malaria co-infection transmission dynamics. Optimal control strategies such as malaria preventive, anti-malaria and antiretroviral (ARV) treatments are considered into the model to reduce the co-infection. First, we studied the existence and stability of equilibria of the presented model without control variables. The model has four equilibria, namely the disease-free equilibrium, the HIV endemic equilibrium, the malaria endemic equilibrium, and the co-infection equilibrium. We also obtain two basic reproduction ratios corresponding to the diseases. It was found that the disease-free equilibrium is locally asymptotically stable whenever their respective basic reproduction numbers are less than one. We also conducted a sensitivity analysis to determine the dominant factor controlling the transmission. sic reproduction numbers are less than one. We also conducted a sensitivity analysis to determine the dominant factor controlling the transmission. Then, the optimal control theory for the model was derived analytically by using Pontryagin Maximum Principle. Numerical simulations of the optimal control strategies are also performed to illustrate the results. From the numerical results, we conclude that the best strategy is to combine the malaria prevention and ARV treatments in order to reduce malaria and HIV co-infection populations.

1. Introduction

HIV (Human Immunodeficiency Virus) is the infectious diseases that can develop into AIDS (Acquired Immune Deficiency Syndrome). HIV infects the immune system and weakens human defence system againsts infections and some types of cancer. A weakened immune in patients with HIV/AIDS will lead to a variety of bacteria/viruses are easy to infect the body (opportunistic infections). Until now, HIV/AIDS is untreatable disease, but the effective antiretroviral (ARV) drugs could control the disease. Moreover, a HIV patient with antiretroviral therapy could can enjoy healthy, long and productive lives [1].

Malaria is a contagious disease that is still a major health problem in the world. The disease is caused by parasites that are transmitted to people through the bites of infected female Anopheles mosquitoes. The female mosquito Anopheles require human blood for their reproductive process. In 2015, WHO reported that there are 91 countries and areas had ongoing malaria transmission. Malaria is preventable and curable, and increased efforts are dramatically reducing the malaria burden in many places [2].

Until now, malaria has not been regarded as an opportunistic infection of HIV disease. However, the weakening of the immune system due to HIV infection affects an increase in malaria infections infected people. Malaria and HIV co-infection has caused more than 2 million deaths every year [3]. In areas highly endemic malaria, HIV-infected individuals more at risk of developing severe malaria. Some reports also recommend that antimalarial therapy failure might be more common in HIV-infected adults with low CD4-cell counts compared to those not infected with HIV. People who are stricken with acute malaria increases viral replication so that the increase in HIV viral load in patients with HIV [4].

Mathematical models are useful to analyze the transmission dynamics of HIV and malaria co-infection. The dynamics of HIV and malaria co-infection spread have been conducted in the literature [5, 6, 7]. For instance, the authors in [5] constructed and analyzed a simple mathematical model to study the co-infection of HIV and malaria. Mukandavire et al. [6] formulated and analyzed a realistic mathematical model for HIV-malaria co-infection, which combines the key epidemiological and biological features of each of the two diseases. Nyabadza et al. [7] developed a mathematical model to describe population dynamics of HIV/AIDS and malaria co-infections.

The optimal control strategies have been employed in the study of epidemiological models such as HIV, TB, Malaria, HCV, co-infection TB-HIV and Malaria-Cholera diseases dynamics [8, 9, 10, 11, 12, 13, 14]. Very little studies have been applied the optimal control to HIV and malaria co-infection model that conducted researchers. Recently, the authors in [15] have used optimal control strategies for a nonlinear dynamical system to describe the dynamics and effects of HIV-malaria co-infection in a workplace. In this paper, we developed a mathematical model of the HIV and malaria co-infection transmission dynamics of incorporate malaria preventive, anti-malaria treatment and ARV treatment for HIV as control optimal strategies.

The remainder of this paper is organized as follows. In Section 2, we propose a mathematical model of HIV and malaria co-infection transmission with controls on malaria prevention, anti-malaria and ARV treatments. In Section 3, we analyze the model without controls and perform sensitivity analysis of the basic reproduction numbers. In the Section 4, we present the optimal control analysis. In Section 5, we perform some numerical simulations to illustrate the purpose of the treatments. The conclusion of is presented in the last section.

2. Model formulation

In general, the population is classified into two classes namely human (host) population and the mosquito (vector) population. The human population was assumed to be homogeneous and closed. The total human population at time t, denoted by $N(t)$, is classified into five classes, namely, the class of susceptible population $(S(t))$, the class of the infected with malaria only $(I_m(t))$ and susceptible to HIV, the class of the infected with HIV only and susceptible to malaria $(I_h(t))$, the class of the infected with HIV and malaria both $(I_{hm}(t))$, and the class of the infected with AIDS population $(A(t))$. Hence, the total human population $N(t) = S(t) + I_m(t) + I_h(t) + I_{hm}(t) + A(t)$.

We also assume that the susceptible cannot get HIV and malaria infection simultaneously at the same time. The class of individuals with AIDS $(A(t))$ and dually infected individuals $(I_{hm}(t))$ are isolated, so that they cannot infect anyone.

The total vector population at time t, denoted by $N_v(t)$, is classified into two classes, namely, the class of susceptible vector population $(S_v(t))$ and the class of infected vector $(I_v(t))$. Thus, $N_v(t) = S_v(t) + I_v(t)$.

We suppose the malaria prevention $(u_1(t))$, anti-malaria treatment $(u_2(t))$ and the ARV treatment $(u_3(t))$ as the control efforts to reduce the HIV and malaria infection respectively. The control functions u_1, u_2 and u_3 are defined on interval $[0, t_f]$, where $0 \leq u_i(t) \leq 1$, $t \in [0, t_f], i = 1, 2, 3$ and t_f denotes the final time of the controls.

the transmission diagram for deriving our model was shown in the Figure 1.

IOP Conf. Series: Journal of Physics: Conf. Series **974** (2018) 012057 doi:10.1088/1742-6596/974/1/012057

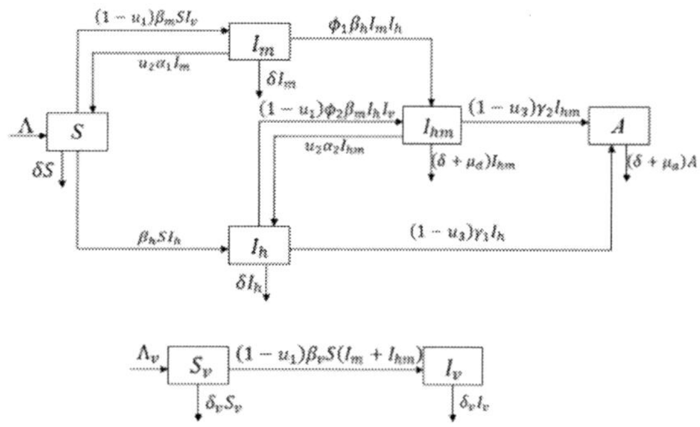

Figure 1. HIV and malaria co-infection transmission diagram.

The model is as follows.

$$
\begin{aligned}
\frac{dS}{dt} &= \Lambda - (1-u_1)\beta_m S I_v - \beta_h S I_h + u_2\alpha_1 I_m - \delta S,\\
\frac{dI_m}{dt} &= (1-u_1)\beta_m S I_v - u_2\alpha_1 I_m - \phi_1\beta_h I_m I_h - \delta I_m,\\
\frac{dI_h}{dt} &= \beta_h S I_h + u_2\alpha_2 I_{hm} - (1-u_1)\phi_2\beta_m I_h I_v - (1-u_3)\gamma_1 I_h - \delta I_h,\\
\frac{dI_{hm}}{dt} &= \phi_1\beta_h I_m I_h + (1-u_1)\phi_2\beta_m I_h I_v - u_2\alpha_2 I_{hm} - (\delta+\mu_d)I_{hm} - (1-u_3)\gamma_2 I_{hm},\\
\frac{dA}{dt} &= (1-u_3)\gamma_1 I_h + (1-u_3)\gamma_2 I_{hm} - (\delta+\mu_a)A,\\
\frac{dS_v}{dt} &= \Lambda_v - (1-u_1)\beta_v S_v(I_m + I_{hm}) - \delta_v S_v,\\
\frac{dI_v}{dt} &= (1-u_1)\beta_v S_v(I_m + I_{hm}) - \delta_v I_v.
\end{aligned} \tag{1}
$$

Table 1. Parameters of model

Description	Parameter
Infection rate for HIV	β_h
Progression rate from malaria to HIV-malaria co-infection	ϕ_1
Progression rate from HIV to HIV-malaria co-infection	ϕ_2
Recovery rate from malaria	α_1
Recovery rate from malaria among HIV-malaria	α_2
Progression rate from HIV only to AIDS	γ_1
Progression rate from HIV-malaria co-infection to AIDS	γ_2
Disease AIDS induced death rate	μ_a
Disease HIV-malaria induced death rate	μ_d
Recruitment rate into the host population	Λ
Recruitment rate into the vector population	Λ_v
Infection rate for host	β_m
Infection rate for vector	β_v
Natural death rate for host	δ
Natural death rate for vector	δ_v

IOP Conf. Series: Journal of Physics: Conf. Series **974** (2018) 012057 doi:10.1088/1742-6596/974/1/012057

The region of biological interest of the model (1) is

$$\hbar = \hbar_h \times \hbar_v \subset \Re^5_+ \times \Re^2_+,$$

with

$$\hbar_h = \left\{(S(t), I_m(t), I_h(t), I_{hm}(t), A(t)) \in \Re^5_+ : 0 \leq N_h(t) \leq \frac{\Lambda}{\delta}\right\},$$

and

$$\hbar_v = \left\{(S_v(t), I_v(t)) \in \Re^2_+ : 0 \leq N_v(t) \leq \frac{\Lambda_v}{\delta_v}\right\}.$$

It is assumed that all parameters used in the model (1) are non-negative. The description of the parameters is given in Table 1. The region $\hbar$ is positively invariant. Hence, model (1) is well-posed in this region since any vector fields on the boundary will not move to the exterior region. So, if it was given any non-negative initial condition in $\Re^7_+$, then the solution was defined for any time $t \geq 0$ and the solution remain in the region. We desire to minimize the number of a HIV and malaria co-infection while keeping the costs of applying malaria preventive, anti-malaria and ARV treatment controls as low as possible. The cost function is defined as

$$J(u_1, u_2, u_3) = \int_0^{t_f} \left(I_m + I_{hm} + A + I_v + \frac{c_1}{2} u_1^2 + \frac{c_2}{2} u_2^2 + \frac{c_3}{2} u_3^2\right) dt, \tag{2}$$

where c_1, c_2 and c_3 are the weighting constants for malaria preventive, anti-malaria and ARV treatment efforts, respectively. We take a quadratic form for measuring the control cost [8, 10]. The term $c_1u_1^2, c_2u_2^2$ and $c_3u_3^2$ describe the cost associated with the malaria preventive, anti-malaria and ARV treatment controls respectively. The cost of malaria preventive is associated with the costs of vaccination, spraying of insecticide, personal protection and insecticide treated bed nets. Larger values of c_1, c_2 and c_3 will imply more expensive implementation cost for malaria preventive, anti-malaria and ARV treatment efforts. Our goal is to find an optimal control pair u_1^*, u_2^* and u_3^* such that

$$J(u_1^*, u_2^*, u_3^*) = \min_{\Gamma} J(u_1, u_2, u_3), \tag{3}$$

where $\Gamma = \{(u_1, u_2, u_3) | 0 \leq u_i \leq 1, i = 1, 2, 3\}$.

3. Model and sensitivity analysis

Consider the mathematical model in eq. (1) without the control functions u_1, u_2 and u_3. Let

$$\begin{aligned} R_m &= \frac{\sqrt{\beta_m \beta_v \Lambda \Lambda_v}}{\delta \delta_v} \\ R_h &= \frac{\beta_h \Lambda}{\delta(\delta + \gamma_1)}. \end{aligned}$$

Parameters R_m and R_h are basic reproduction ratio corresponding to the malaria infection, and HIV infection, respectively. These ratios define number of secondary cases of primary case throughout the infective period because of the type of infection [16, 17].

By setting $u_1 = u_2 = u_3 = 0$, the mathematical model in eq. (1) has four equilibria (to the coordinate $(S, I_m, I_h, I_{hm}, A, S_v, I_v)$, namely,

1. The disease-free equilibrium $E_0 = (\frac{\Lambda}{\delta}, 0, 0, 0, 0, \frac{\Lambda_v}{\delta_v}, 0)$.
2. The malaria endemic equilibrium $E_m = \left(\frac{\Lambda}{\beta_m I_v+\delta}, \frac{\beta_m \Lambda I_v}{\delta(\beta_m I_v+\delta)}, 0, 0, 0, \frac{\Lambda_v}{\beta_v I_m+\delta_v}, \frac{\beta_v \Lambda_v I_m}{\delta_v(\beta_v I_m+\delta_v)}\right)$.
3. The HIV endemic equilibrium $E_h = \left(\frac{\gamma_1+\delta}{\beta_h}, 0, \delta(\gamma_1+\delta)(R_h-1), 0, \frac{\gamma_1 I_h}{\delta+\mu_a}, \frac{\Lambda_v}{\delta_v}, 0\right)$. The equilibrium E_h exists if $R_h > 1$.
4. The HIV-malaria endemic equilibrium $E_{hm} = (S^*, I_m^*, I_h^*, I_{hm}^*, A^*, S_v^*, I_v^*)$, with

$$\begin{aligned}
S^* &= \frac{\phi_2\beta_m I_v + \gamma_1 + \delta}{\beta_h}, \\
I_m^* &= \frac{\beta_m S I_v}{\phi_1\beta_h I_h + \delta}, \\
I_h^* &= \frac{\Lambda - \beta_m S I_v - \delta S}{\beta_h S}, \\
I_{hm}^* &= \frac{\phi_1\beta_h I_m I_h + \phi_2\beta_m I_h I_v}{\mu_d + \delta + \gamma_2}, \\
A^* &= \frac{\gamma_1 I_h + \gamma_2 I_{hm}}{\delta + \mu_a}, \\
S_v^* &= \frac{\Lambda_v}{\beta_v(I_m + I_{hm}) + \delta_v}, \\
I_v^* &= \frac{\beta_v S_v(I_m + I_{hm})}{\delta_v}.
\end{aligned}$$

The equilibrium E_{hm} exists if $R_h > 1$ and $\delta(\delta+\gamma_1)(R_h-1) > \phi_2\beta_m^2 I_v^2 + \beta_m(\delta+\gamma_1)I_v + \delta\phi_2\beta_m I_v$.

The following theorem give the stability criteria of the disease-free equilibrium.

Theorem 1 *The disease-free equilibrium E_0 is locally asymptotically stable whenever $R_m, R_h < 1$. Furthermore, the equilibrium E_0 is unstable if $R_m, R_h > 1$.*

Proof. Linearizing the mathematical model in eq. (1) near the equilibrium E_0 gives eigenvalues $-\delta, -\delta_v, -(\delta+\mu_a), -(\delta+\mu_d+\gamma_2), -(\delta+\gamma_1)(R_h-1)$ and the roots of quadratic equation $x^2 + (\delta+\delta_v)x + \delta\delta_v(1-T_m) = 0$, with $T_m = \frac{\beta_m\beta_v\Lambda\Lambda_v}{\delta^2\delta_v^2}$. The quadratic equation have negative roots if $T_m < 1$ or equivalently $R_m < 1$. It is clear that all of the eigenvalues are negative if $R_h < 1$ and $R_m < 1$. So, if $R_m, R_h < 1$, then the equilibrium E_0 is locally asymptotic stable. Otherwise, it is unstable.

Next, we analyze the sensitivity of the basic reproduction numbers R_m and R_h to the parameters in the presented model. The purpose of this analysis was to find the parameters that have the greatest effects on the reproduction numbers. By using the method in [18], we derived the analytical expression for sensitivity index of the R_m and R_h to each parameters.

The sensitivity indices of a variable, R_0, that depends differentially on a parameter, l, is defined as

$$\Upsilon_l^{R_0} := \frac{\partial R_0}{\partial l} \times \frac{l}{R_0}. \tag{4}$$

Now, by using the parameter values in the Table 2, we have the following results in Table 3. For example, the sensitivity index of R_m with respect to β_m is

$$\Upsilon_{\beta_m}^{R_m} := \frac{\partial R_m}{\partial \beta_m} \times \frac{\beta_m}{R_m} = 0.5. \tag{5}$$

The sensitivity indices of R_m and R_h with respect to other parameters such as Λ, Λ_v, β_v, δ, δ_v, β_h, γ_1, can be derived in the same way as (5). The interpretation of the sensitivity index is as follow. Since $\Upsilon_{\beta_m}^{R_m} = 0.5$, that is to say, increasing (or decreasing) infection rate for malaria, β_m, by 10%, increases (or decreases) the reproduction number R_m by 5%. Thus, increasing (or decreasing) natural death rate δ by 10% decreases (increases) R_m by 10%.

Table 2. Parameter values.

Parameter	Value	Ref.
Λ	500/year	Assumed
δ	0.02/year	[13]
β_h	0.00031/year	Assumed
β_m	0.00045/year	Assumed
β_v	0.00035/year	Assumed
γ_1	0.01/year	Assumed
γ_2	0.05/year	Assumed
Λ_v	5000/year	Assumed
δ_v	0.1429/year	[18]
ϕ_1	1.1/year	Assumed
ϕ_2	1.15/year	Assumed
μ_a	0.02/year	Assumed
μ_d	0.03/year	Assumed
α_1	0.2/year	Assumed
α_2	0.2/year	Assumed

Table 3. Sensitivity indices to parameter for model (1).

Parameter	Sensitivity index (R_m)	Parameter	Sensitivity index (R_h)
Λ	0.5	Λ	1
Λ_v	0.5	β_h	1
β_m	0.5	δ	-1.67
β_v	0.5	γ_1	-0.33
δ	-1		
δ_v	-1		

4. Mathematical analysis of optimal control

In this section, we consider the mathematical model in eq. (1) with its control functions u_1, u_2 and u_3. We describe the cost function in eq. (2) for the model in eq. (1). The necessary conditions to find the optimal controls u_1^*, u_2^* and u_3^* such that condition (3) with constraint model (1) must satisfy come from the Pontriyagin Maximum Principle [19]. The principle converts the equations (1) - (3) into minimizing Hamiltonian function H problem with respect (u_1, u_2, u_3), that is

$$H(S, I_m, I_h, I_{hm}, A, S_v, I_v, u_1, u_2, u_3, \lambda_1, \lambda_2, \cdots, \lambda_7) =$$
$$I_m + I_{hm} + A + I_v + \frac{c_1}{2}u_1^2 + \frac{c_2}{2}u_2^2 + \frac{c_3}{2}u_3^2 + \sum_{i=1}^{7}\lambda_i g_i,$$

where g_i is the right hand side of the mathematical model in eq. (1) which is the i-th state variable equation. The variables λ_i, $i = 1, 2, \ldots, 7$, are called adjoint variables satisfying the following co-state equations

$$\begin{aligned}
\frac{d\lambda_1}{dt} &= (\lambda_1 - \lambda_2)(1-u_1)\beta_m I_v + (\lambda_1 - \lambda_3)\beta_h I_h + \lambda_1\delta, \\
\frac{d\lambda_2}{dt} &= -1 + (\lambda_2 - \lambda_1)\alpha_1 u_2 + (\lambda_2 - \lambda_4)\phi_1\beta_h I_h + (\lambda_6 - \lambda_7)(1-u_1)\beta_v S_v + \lambda_2\delta, \\
\frac{d\lambda_3}{dt} &= (\lambda_1 - \lambda_3)\beta_h S + (\lambda_2 - \lambda_4)\phi_1\beta_h I_h + (\lambda_3 - \lambda_4)(1-u_1)\phi_2\beta_m I_v \\
&\quad + (\lambda_3 - \lambda_5)(1-u_3)\gamma_1 + \lambda_3\delta, \\
\frac{d\lambda_4}{dt} &= -1 + (\lambda_4 - \lambda_3)u_2\alpha_2 + (\lambda_4 - \lambda_5)(1-u_3)\gamma_2 + (\lambda_6 - \lambda_7)(1-u_1)\beta_v S_v + \\
&\quad \lambda_4(\delta + \mu_d), \qquad (6) \\
\frac{d\lambda_5}{dt} &= -1 + \lambda_5(\delta + \mu_d), \\
\frac{d\lambda_6}{dt} &= (\lambda_6 - \lambda_7)(1-u_1)\beta_v(I_h + I_{hm}) + \lambda_6\delta_v, \\
\frac{d\lambda_7}{dt} &= -1 + (\lambda_1 - \lambda_2)(1-u_1)\beta_m S + (\lambda_3 - \lambda_4)(1-u_1)\phi_2\beta_m I_h + \lambda_7\delta_v,
\end{aligned}$$

where the transversality conditions $\lambda_i(t_f) = 0$, $i = 1, \ldots, 7$.

By using Pontryagin's Maximum Principle and the existence result for the optimal control pairs, the steps to obtain the optimal controls $u = (u_1^*, u_2^*, u_3^*)$ are as following [20, 21].

(i) Minimize the Hamiltonian function H with respect to u, that is $\frac{\partial H}{\partial u} = 0$ which is the stationary condition. Hence, we find

$$u_1^* = \begin{cases} 0 & \text{for } u_1 \le 0 \\ \frac{(\lambda_4 - \lambda_3)\phi_2\beta_m I_v I_h + (\lambda_2 - \lambda_1)\beta_m S I_v + (\lambda_7 - \lambda_6)\beta_v S_v (I_h + I_{hm})}{c_1} & \text{for } 0 < u_1 < 1 \\ 1 & \text{for } u_1 \ge 1 \end{cases}$$

$$u_2^* = \begin{cases} 0 & \text{for } u_2 \le 0 \\ \frac{(\lambda_4 - \lambda_3)I_{hm}\alpha_2 + (\lambda_2 - \lambda_1)\alpha_1 I_h}{c_2} & \text{for } 0 < u_2 < 1 \\ 1 & \text{for } u_2 \ge 1 \end{cases}$$

$$u_3^* = \begin{cases} 0 & \text{for } u_3 \le 0 \\ \frac{(\lambda_5 - \lambda_3)\gamma_1 I_h + (\lambda_5 - \lambda_4)\gamma_2 I_{hm}}{c_3} & \text{for } 0 < u_3 < 1 \\ 1 & \text{for } u_3 \ge 1 \end{cases}$$

(ii) Solve the state system $\dot{x}(t) = \frac{\partial H}{\partial \lambda}$ which is the mathematical model in eq. (1), where $x = (S, I_m, I_h, I_{hm}, A, S_v, I_v)$, $\lambda = (\lambda_1, \lambda_2, ..., \lambda_7)$ and the initial condition $x(0)$.

(iii) Solve the co-state system $\dot{\lambda}(t) = -\frac{\partial H}{\partial x}$ which is the system in eq. (6) with the final condition $\lambda_i(t_f) = 0$, $i = 1, \ldots, 7$.

Henceforth, we find the following theorem.

Theorem 2 *The optimal controls* (u_1^*, u_2^*, u_3^*) *that minimizes the objective function* $J(u_1, u_2, u_3)$ *on* Γ *is given by*

$$\begin{aligned} u_1^* &= \max\left\{0, \min\left(1, \frac{(\lambda_4 - \lambda_3)\phi_2\beta_m I_v I_h + (\lambda_2 - \lambda_1)\beta_m S I_v + (\lambda_7 - \lambda_6)\beta_v S_v(I_h + I_{hm})}{c_1}\right)\right\} \\ u_2^* &= \max\left\{0, \min\left(1, \frac{(\lambda_4 - \lambda_3)I_{hm}\alpha_2 + (\lambda_2 - \lambda_1)\alpha_1 I_h}{c_2}\right)\right\}, \\ u_3^* &= \max\left\{0, \min\left(1, \frac{(\lambda_5 - \lambda_3)\gamma_1 I_h + (\lambda_5 - \lambda_4)\gamma_2 I_{hm}}{c_3}\right)\right\} \end{aligned}$$

where λ_i, $i = 1, ..., 7$ *is the solution of the co-state equations (6) with the transversality conditions* $\lambda_i(t_f) = 0$, $i = 1, \ldots, 7$.

Substituting the optimal control (u_1^*, u_2^*, u_3^*) which is achieved from the state system in eq. (1) and the co-state system in eq. (6), we find an optimal system. The solutions of the optimality system will be solved numerically for some choices of the parameter. Because of lack of data, most of parameter values are assumed within realistic ranges for a typical scenario.

5. Numerical simulation

In this section, we perform some numerical simulations of the presented model in eq. (1) with and without optimal control. The optimal control strategy is found by the iterative method of the fourth order Runge-Kutta method. The state equations are initially solved by the forward Runge-Kutta method of the fourth order. Then, by using the backward Runge-Kutta method of the fourth order, we solved the co-state equations with the transversality conditions. The controls are updated by using a convex combination of the previous controls and the value from the characterizations of u_1^*, u_2^* and u_3^*. This process is reiterated and the iteration is ended if the current state, the adjoint, and the control values converge sufficiently [22].

We present four scenarios. In the first situations, we study combination of the malaria preventive and the anti-malaria treatment as control strategy. In the second scenario, we consider combination of the malaria preventive and ARV treatment controls. In the third scenario, we use the combination of the anti-malaria and ARV treatment controls. In the last one, we use the combination of the malaria preventive, anti-malaria and ARV treatment as control strategy. Parameters used in these simulations given in Table 2. In these simulations, we use initial condition $(S(0), I_m(0), I_h(0), I_{hm}(0), A(0), S_v(0), I_(0)) = (500, 100, 50, 30, 50, 5000, 100)$, weighting constants $c_1 = 20$, $c_2 = 50$, $c_3 = 50$.

5.1. First scenario

In this scenario, we apply the malaria prevention control u_1 and the anti-malaria treatment control u_2 to optimize the objective function J, while setting the ARV control u_3 to zero. For this scenario, we found in the Figure 2-3 that the population of malaria infected (I_m) and HIV-malaria co-infection (I_{hm}) decreases, while the population increases when there is no any controls. Similarly, the result in the Figure 4-5 show that the number of AIDS infected (A) and mosquito infected (I_v) a reduction in the number infected compared with the case without control. Hence, the malaria prevention and the anti-malaria treatment control gives a significant influence in controlling infected malaria, HIV-malaria co-infection and AIDS infected. The profile of the optimal treatment controls u_1^* and u_2^* for this scenario is presented in the Figure 6.

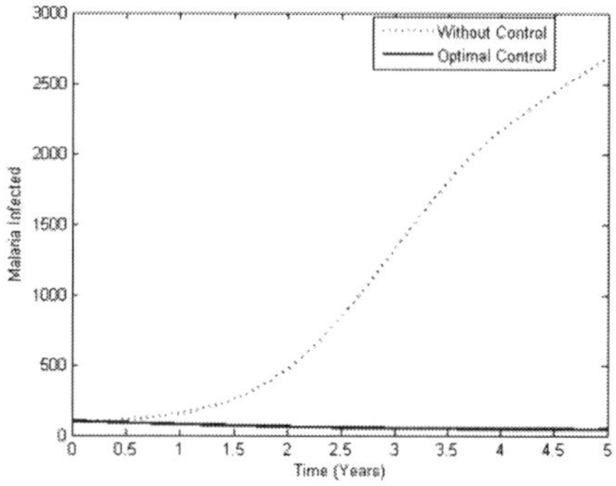

Figure 2. The dynamics of I_m using the controls u_1^* and u_2^*.

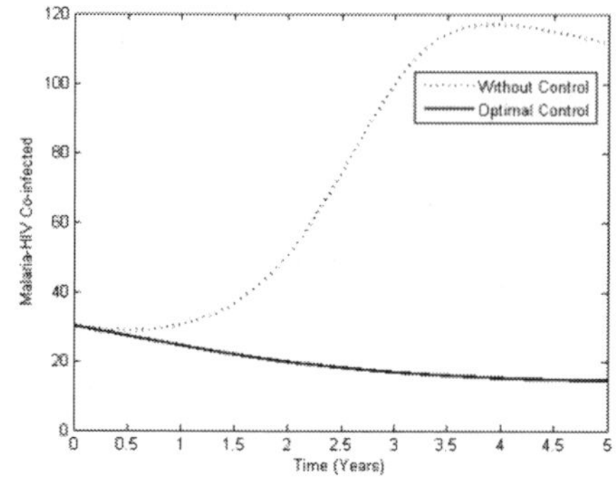

Figure 3. The dynamics of I_{hm} using the controls u_1^* and u_2^*.

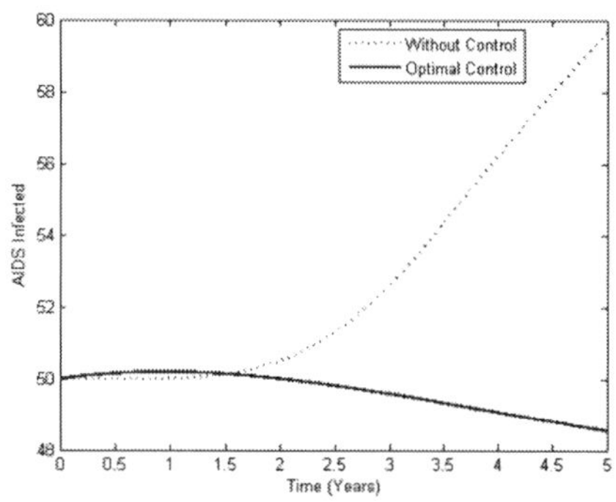

Figure 4. The dynamics of A using the controls u_1^* and u_2^*.

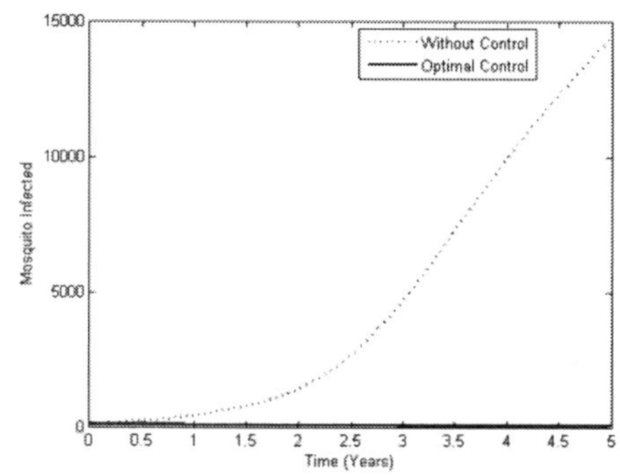

Figure 5. The dynamics of I_v using the controls u_1^* and u_2^*.

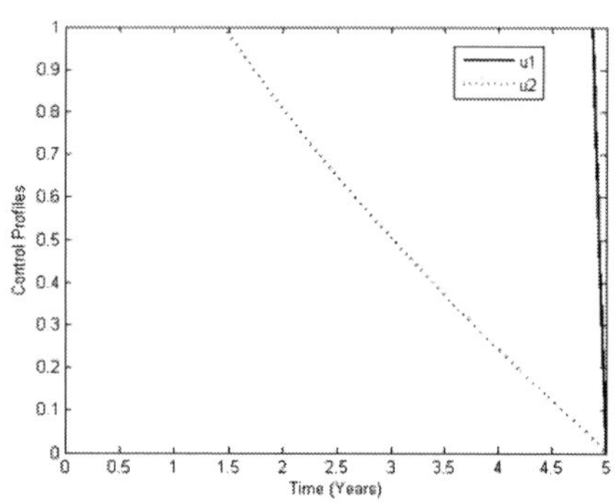

Figure 6. The profile of the optimal controls u_1^* and u_2^*.

5.2. Second scenario

In the second scenario, we utilize the malaria prevention control u_1 and the ARV treatment control u_3 to optimize the objective function J, while set the anti-malaria treatment control u_2 to zero. The infected populations dynamics of this scenario are presented in Figure 7-10. We observe in Figure 7-8 that this control strategy results in a significant decrease in the number of malaria infected (I_m) and HIV-malaria co-infection (I_{hm}) compared with the case without control. Also in Figure 9-10, this control strategy results in a significant decrease in the number of AIDS infected (A) and mosquito infected (I_v) as against an increase in the uncontrolled case. The result shown in the Figure 7-10 clearly suggests that this scenario is very effective in the control of the number of the infected. The control profiles of the malaria prevention and ARV treatment is shown in the Figure 11.

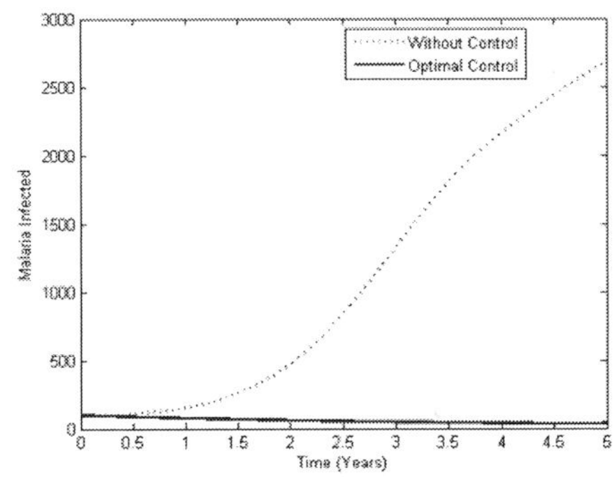

Figure 7. The dynamics of I_m using the controls u_1^* and u_3^*.

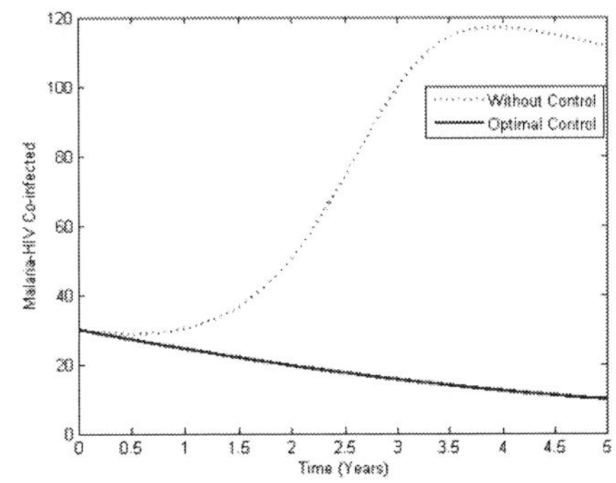

Figure 8. The dynamics of I_{hm} using the controls u_1^* and u_3^*.

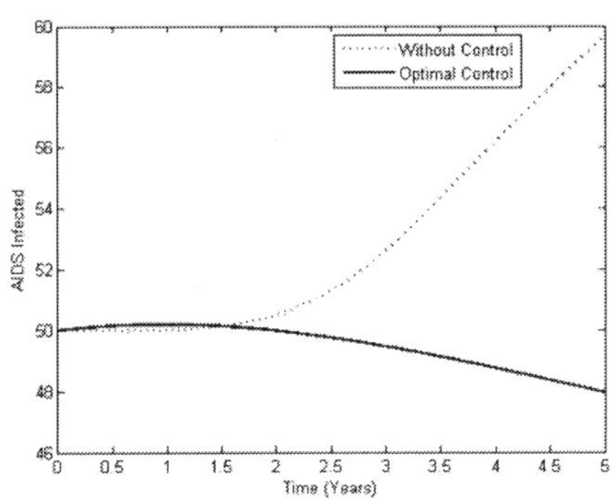

Figure 9. The dynamics of A using the controls u_1^* and u_3^*.

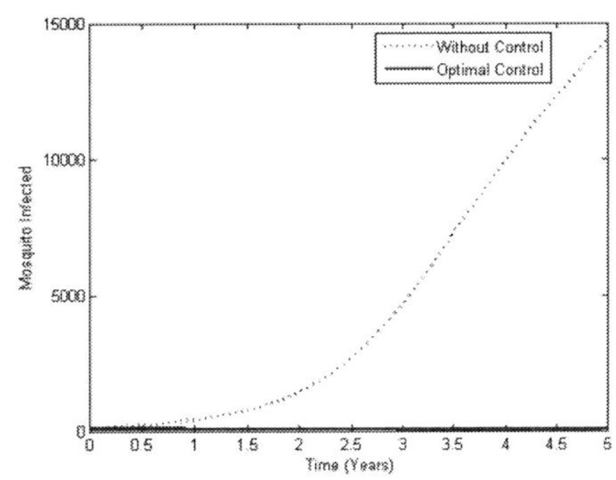

Figure 10. The dynamics of I_v using the controls u_1^* and u_3^*.

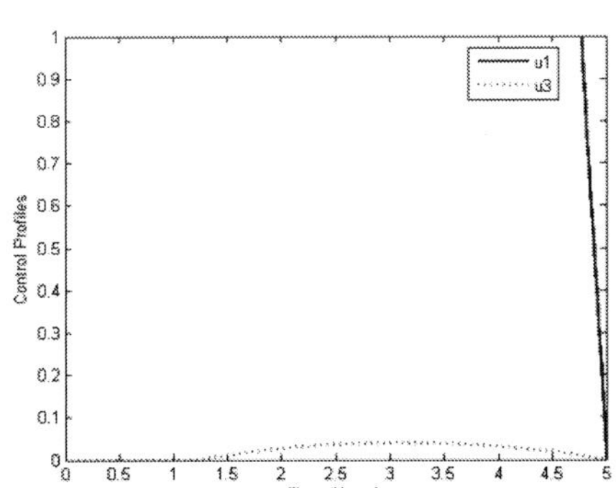

Figure 11. The profile of the optimal controls u_1^* and u_3^*.

IOP Conf. Series: Journal of Physics: Conf. Series **974** (2018) 012057 doi:10.1088/1742-6596/974/1/012057

5.3. Third scenario

In the third scenario, we active controls u_2 and u_3 on the anti-malaria and ARV treatment to optimize the objective function J whereas the malaria prevention control u_1 is set to zero. The dynamics of the infected populations of this scenario are given in Figure 12-15. We observe in Figure 12 that the malaria infected populations are lesser when the control strategy is used than when the control strategy is not implemented. It was also shown in Figure 13 that the number of HIV-malaria co-infection populations increases with this control strategy compared to the number without control. Figure 14-15 depict the number of AIDS infected and mosquito infected that the populations is higher with this control scenario than the cases without control. Figure 16 shows the control profiles of the anti-malaria (u_2) and ARV treatment (u_3) in which the control u_2 is given maximum in almost 5 years, while the the optimal control u_3 starts and remain at the lower bound in 5 years. The results show that the anti-malaria treatment give the positive impact to reduce the malaria infected, while the mosquito infected tend to increases using this strategy. Then, the HIV-malaria co-infection and AIDS infection populations also increase with this strategy due to there is no significant ARV treatment intervention.

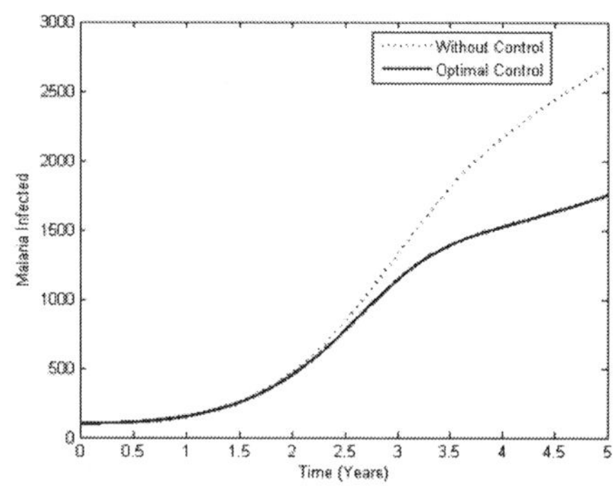

Figure 12. The dynamics of I_m using the controls u_2^* and u_3^*.

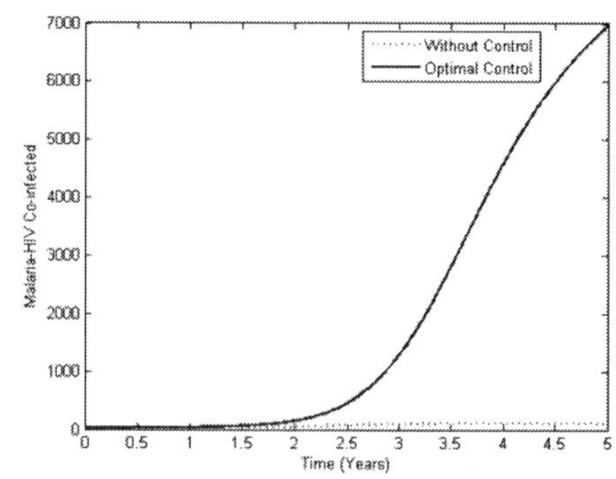

Figure 13. The dynamics of I_{hm} using the controls u_2^* and u_3^*.

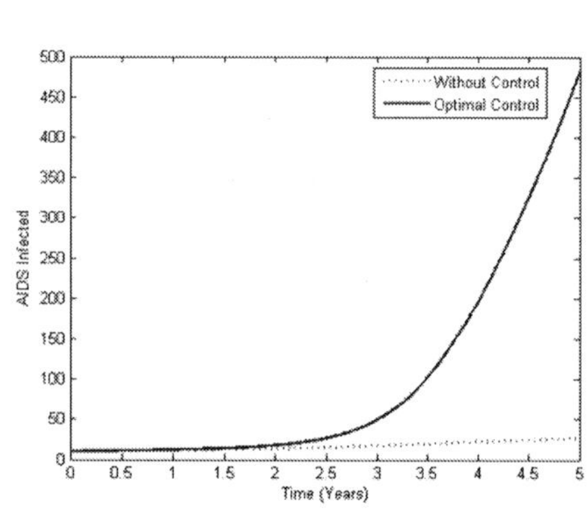

Figure 14. The dynamics of A using the controls u_2^* and u_3^*.

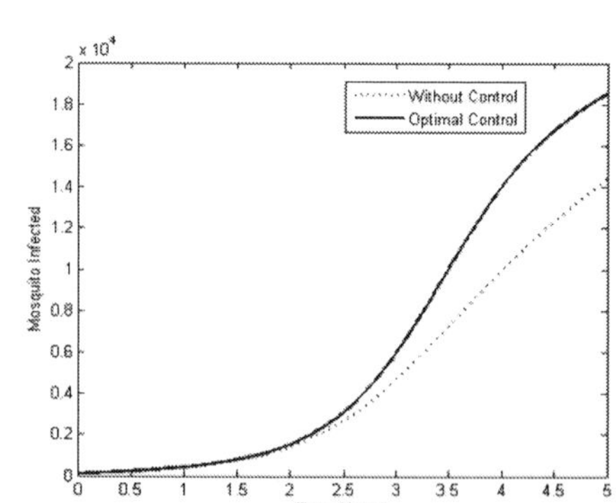

Figure 15. The dynamics of I_v using the controls u_2^* and u_3^*.

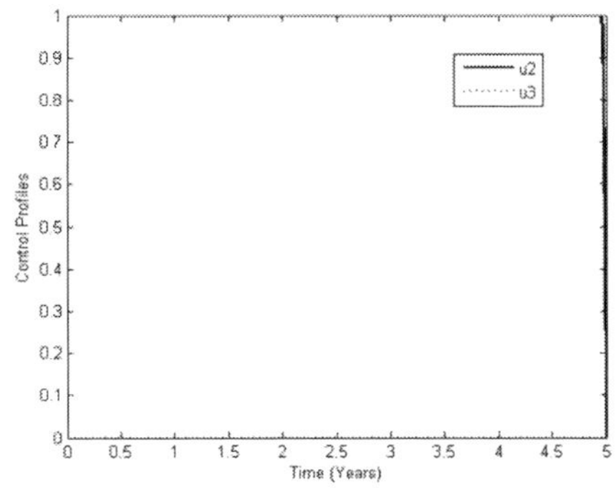

Figure 16. The profile of the optimal controls u_2^* and u_3^*.

5.4. Fourth scenario

In this scenario, all the controls (u_1, u_2 and u_3) are used to optimize the objective function J. The dynamics of the malaria infected, malaria-HIV co-infection, AIDS infection and mosquito infected are given in Figure 17-20. For this strategy, we observed in Figure 17-18 that the control strategies resulted in a decrease in the number of the malaria infected and HIV-malaria co-infection compared to the number without control. A similar reduction is observed in Figure 19-20 for AIDS infected and mosquito infected in the control strategy, while an increased number for the uncontrolled case resulted.

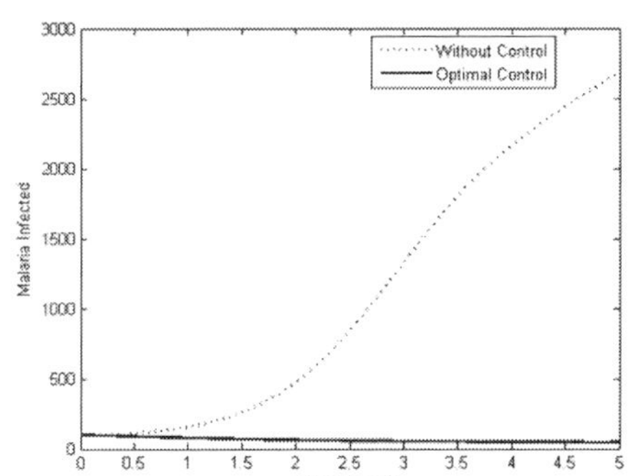

Figure 17. The dynamics of I_m using the controls u_1^*, u_2^* and u_3^*.

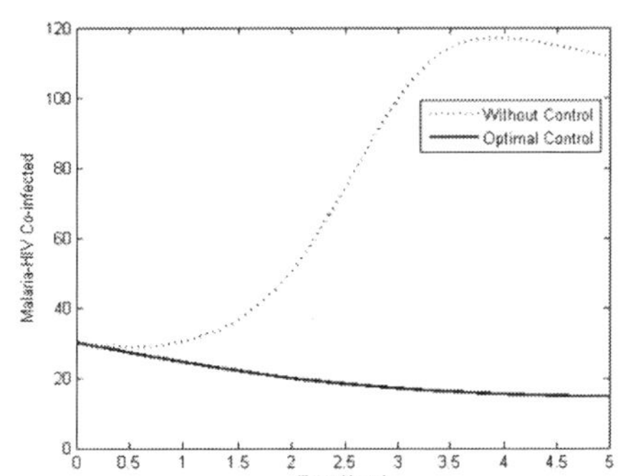

Figure 18. The dynamics of I_{hm} using the controls u_1^*, u_2^* and u_3^*.

The profiles of the optimal malaria prevention u_1^*, anti-malaria treatment control u_2^* and ARV control u_3^* of this scenario is given in Figure 21. To reduce HIV and malaria co-infection in 5 years, the malaria prevention should be given intensively almost 5 years before dropping gradually until reaching the lower bound in the end 5th year. Meanwhile, anti-malaria treatment should be needed intensive effort during 1.5 years before finally dropping to its lower bound. While ARV treatment control decreases from 0.1 (10%) at the beginning time before finally dropping to its lower bound at the end of the intervention.

Based on the numerical results, the optimal control scenarios are arranged from the least to the costly based on the total cost function J. Table 4 shows the optimal values of the objective function J in the four scenarios. From Table 4, we conclude that the combination of the malaria

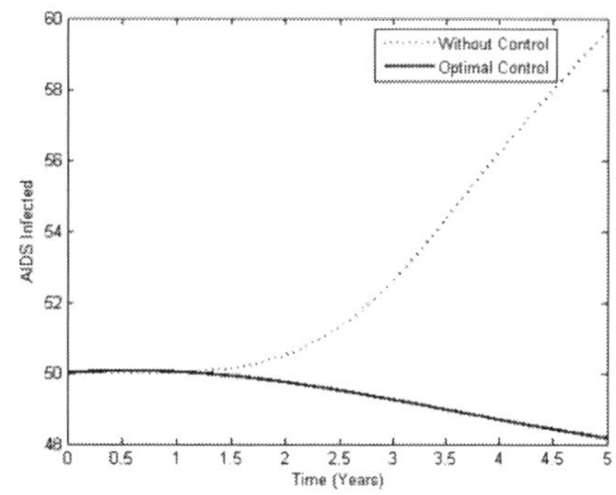

Figure 19. The dynamics of A using the controls u_1^*, u_2^* and u_3^*

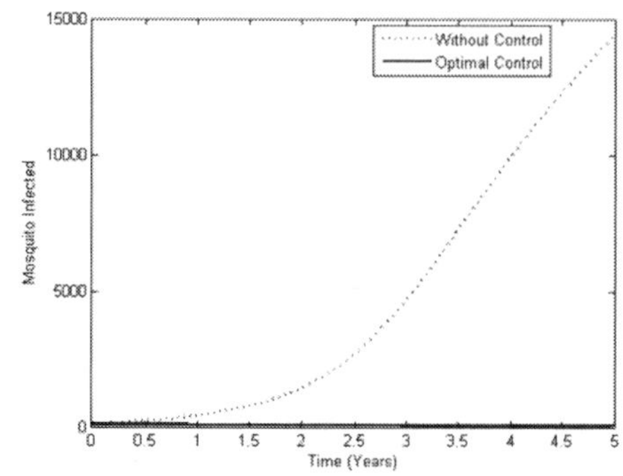

Figure 20. The dynamics of I_v using the controls u_1^*, u_2^* and u_3^*

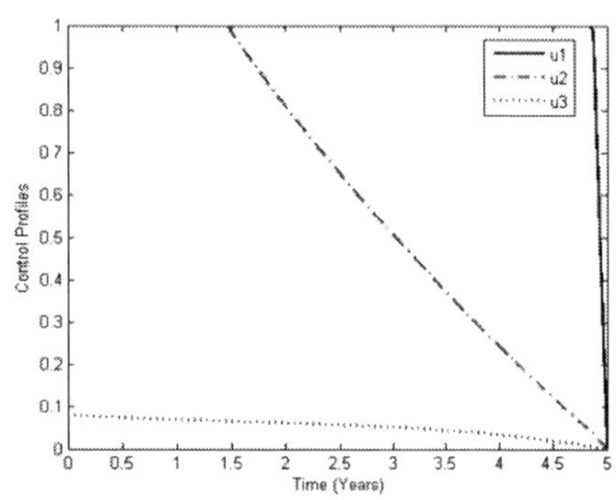

Figure 21. The profile of the optimal controls u_1^*, u_2^* and u_3^*

Table 4. Optimal values of the cost functional.

Scenario of the optimal control	Total cost functional J
u_1^*, and u_3^*	1.0359×10^3
u_1^*, u_2^*, and u_3^*	1.1234×10^3
u_1^*, and u_2^*	1.1237×10^3
u_2^*, and u_3^*	4.5127×10^4

preventive (u_1^*) and ARV treatment (u_3^*) have the least cost function followed by combination all controls (u_1^*, u_2^*, and u_3^*) to reduce the number of malaria infected, HIV-malaria co-infection, AIDS infected and mosquito infected.

6. Conclusion

In this paper, we have developed a deterministic mathematical model for the spread of malaria and HIV co-infection that incorporates malaria prevention, anti-malaria and ARV treatment as optimal control strategies. For the model without controls, we obtain two thresholds R_m and R_h which are basic reproduction ratios for the malaria and HIV diseases respectively. These ratios determine the existence and stability of the equilibria of the model. If the thresholds are less than unity then the diseases free equilibrium is locally asymptotically stable. Finally, the conditions for existence of optimal control is studied analytically using the Pontryagin Maximum

Principle. From our numerical analysis of the optimal control indicates that the best strategy is to combine the malaria prevention and ARV treatments followed by combination all controls in order to reduce malaria and HIV co-infection.

Acknowledgements
Parts of this research is funded by Directorate for the Higher Education, Ministry of Research, Technology, and Higher Education of Indonesia for the Research Grant "Penelitian Unggulan Perguruan Tinggi" (PUPT) Universitas Airlangga 2016 according to SK Rektor Nomor: 304/SP2H/LT/DRPM/II/2016.

References

[1] World Health Organization 2016 HIV/AIDS *Geneva: World Health Organization* [Online] Available from: http://www.who.int/mediacentre/factsheets/fs360/en/ [Accesed on 22nd Desember, 2016]
[2] World Health Organization 2016 Malaria *Geneva: World Health Organization* [Online] Available from: http://www.who.int/mediacentre/factsheets/fs094/en/ [Accesed on 22nd Desember, 2016]
[3] World Health Organization 2016 Malaria in HIV AIDS patients *Geneva: World Health Organization* [Online] Available from: http://www.who.int/malaria/areas/high_risk_groups/hiv_aids_patients/en/ [Accesed on 22nd Desember, 2016]
[4] van Geertruyden J P 2014 Interaction between malaria and human immunodeciency virus onno *J. Vac. Sci. Technol.Clin Microbiol Infect* **20** 278
[5] Barley K, Murillo D, Roudenko S, Tameru A M, and Tatum S 2012 A mathematical model of HIV and malaria co-infection in Sub-Saharan Africa *J. AIDS Clinic Res.* **3** 7
[6] Mukandavire Z, Gumel A B, Garira W, Tchuenche J M 2009 Mathematical analysis of a model for HIV-malaria co-infection *Math. Biosci. Eng.* **6** 333
[7] Nyabadza F, Bekele B T, Rua M A, Malonza D M, Chiduku N, and Kgosimore M 2015 The implication of HIV treatment on the HIV-malaria coinfection dynamic: a modeling perspective, *BioMed Res. Int.* **2015** article ID 659651.
[8] Okosun K O, Makinde O D, Takaidza I 2013 Impact of optimal control on the treatment of HIV/AIDS and screening of unaware infectives *Appl. Math. Modelling* **37** 3802.
[9] Ahmadin, Fatmawati 2014 Mathematical modeling of drug resistance in tuberculosis transmission and optimal control treatment *Appl. Math. Sci.* **8** 4547
[10] Makinde O D, Okosun K O 2011 Impact of chemo-therapy on optimal control of malaria disease with infected immigrants *Biosystems* **104** 32
[11] Fatmawati, Tasman H, 2015 An optimal control strategy to reduce the spread of malaria resistance *Math. Biosci.* **262** 73
[12] Mushayabasa S, 2017 Dynamics of HCV in the presence of optimal bleaching *Differ. Equ. Dyn. Syst.* **25** 101.
[13] Fatmawati, Tasman H, 2016 An optimal treatment control of TB-HIV coinfection *Int. J. Math. Math. Sci.* **2016** Article ID 8261208.
[14] Okosun K O, Makinde O D, 2014 A co-infection model of malaria and cholera diseases with optimal control *Math. Biosci.* **258** 19.
[15] Seidu B, Makinde O D, Seini I Y 2015 Mathematical analysis of the effects of HIV-malaria co-infection on workplace productivity, *Acta Biotheor.* **63** 151.
[16] Diekmann O, Heesterbeek J A P, Metz J A J 1990 On the definition and the computation of the basic reproduction ratio R_0 in models for infectious diseases in heterogenous populations *J. Math. Biol.* **28** 362
[17] Diekmann O, Heesterbeek J A P 2000 *Mathematical Epidemiology of Infectious Diseases, Model Building, Analysis and Interpretation* (New York: John Wiley & Son)
[18] Chitnis N, Hyman J M, Cushing J M 2008 Determining important parameters in the spread of malaria through the sensitivity analysis of a mathematical model *Bull. Math. Biol.* **70** 1272
[19] Pontryagin L S, Boltyanskii V G, Gamkrelidze R V, Mishchenko E F 1962 *The Mathematical Theory of Optimal Processes* (New York: Wiley)
[20] Lewis F L, Syrmos V L 1995 *Optimal Control* (New York: John Wiley & Sons)
[21] Naidu D S 2002 *Optimal Control Systems* (New York: CRC PRESS)
[22] Lenhart S, Workman J T 2007 *Optimal Control Applied to Biological Models* (London: John Chapman & Hall)

IOP Conf. Series: Journal of Physics: Conf. Series **974** (2018) 012058 doi:10.1088/1742-6596/974/1/012058

Convergence of sequences in $\ell_2(P)$ with respect to a partial metric

Annisa Rahmita Soemarsono and Mahmud Yunus

Department of Mathematics, Institut Teknologi Sepuluh Nopember, Surabaya, Indonesia

E-mail: annisarahmitas@gmail.com, yunusm@matematika.its.ac.id

Abstract. Partial metric spaces are generalization of metric space. The distance from a point to itself need not be zero in partial metric space. By the properties of metric and partial metric space, we have the analogue of the two spaces. Using the analogue, we construct sequences in $\ell_2(P)$ with respect to a partial metric. We then investigate the convergence of sequences in $\ell_2(P)$. In this work, we obtain that the convergence of sequences in $\ell_2(\mathbb{N})$ can be established in $\ell_2(P)$ with respect to a partial metric.

1. Introduction

Partial metric spaces are generalization of metric spaces. The concept of partial metric space is introduced by Matthews [2]. He defines that the distance from a point to itself need not be zero. From the properties of partial metric, Matthews has an analogue of metric and partial metric space by defining a function. The function is induced by a partial metric such that it is a metric. The results of Matthews's research are extended by Heckmann [3] to construct sequences in partial metric space. He need partially ordered set (P) for constructing sequences in a partial metric space. Then, Waszkiewicz [4] continue the study of partially ordered set in partial metric space.

From the last researches presented in [2], [3] and [4], Kadak, et al [1] construct sequences in partial metric space with respect to partial ordering. In their research, we know that convergence of sequences in metric space is useful for investigating the convergence of sequences in partial metric space. If sequences converge to a point in a metric space with usual metric d^p induced by a partial metric p, then we can identify the convergence of sequences in a partial metric space.

Space $\ell_2(\mathbb{N})$ is one of well-known metric spaces containing convergent sequences [6]. In this paper, using the analogue of metric and partial metric space, we ensure that sequences in the metric space $\ell_2(\mathbb{N})$ also converge in $\ell_2(P)$ with respect to a partial metric.

2. Preliminaries

In this section, we give the concepts of partial metric and partially ordered set required to identify the convergence of sequences in $\ell_2(P)$ with respect to a partial metric space.

2.1. Partial metric space

Matthews [2] introduces a concept which the distance from a point to itself need not be zero. The concept is a generalization of metric space, and is called as partial metric space. There is a slight difference between metric and partial metric space. One of metric axioms state that the distance $d(x, y)$ of two points x and y must be zero if $x = y$. This condition is not always true in partial metric space. It makes an axiom of partial metric for which if there are two same points $x = y$, then $p(x, y)$ need not be zero. By extending the metric axioms, Matthews [2] defines partial metric space.

Definition 2.1 *A function $p : X \times X \to \mathbb{R}^+ \cup \{0\}$ is a partial metric on a nonempty set X if for $x, y, z \in X$ satisfies the following axioms:*
(P1) $x = y \Leftrightarrow p(x, x) = p(x, y) = p(y, y)$;
(P2) $p(x, x) \leq p(x, y)$;
(P3) $p(x, y) = p(y, x)$;
(P4) $p(x, z) \leq p(x, y) + p(y, z) - p(y, y)$.
A pair (X, p) is called a partial metric space such that X is a nonempty set and p is a partial metric on X.

One of the examples is defined by a function $p : \mathbb{R}^+ \cup \{0\} \times \mathbb{R}^+ \cup \{0\} \to \mathbb{R}^+ \cup \{0\}$ with $p(x, y) = \max\{x, y\}, \forall x, y \in \mathbb{R}^+ \cup \{0\}$. The partial metric is required to show that a partial metric can induce a metric.

2.2. Analogue of metric and partial metric

Partial metric space is a generalization of metric space, therefore some properties of metric space remain exist in partial metric space. This condition makes an analogue of metric and partial metric. The analogue is showed by a function d^p induced by a partial metric p with $p(x, y) = \max\{x, y\}$. As consequence, the function d^p is a metric on X [2].

Theorem 2.2 *If p is a partial metric on a nonempty set X, then the function $d^p : X \times X \to \mathbb{R}$ with*

$$d^p(x, y) = 2p(x, y) - p(x, x) - p(y, y), \forall x, y \in X \tag{1}$$

is a metric on X.

The function represents the distance of two points in the usual case as $d^p(x, y) = |x - y|$. By the analogue of the two spaces, we can ensure the convergence of sequences in a partial metric space.

2.3. Partially ordered set

Matthews [2] defines a partially ordered set in a partial metric space.

Definition 2.3 *A partially ordered set (poset) is a pair $(X, \sqsubseteq)$ such that $\sqsubseteq$ is a partial ordering on a nonempty set X. For each partial metric $p : X \times X \to \mathbb{R}^+ \cup \{0\}$ where $\sqsubseteq_p$ is a binary relation over X such that $x \sqsubseteq_p y$ if and only if $p(x, x) = p(x, y)$ for all $x, y \in X$.*

By Definition 2.3, it can be proved that $\sqsubseteq_p$ is a partial ordering for each partial metric p. Accordingly, the binary relation $\sqsubseteq_p$ satisfies the following properties:
(i) $x \sqsubseteq_p y$ (reflexivity);
(ii) If $x \sqsubseteq_p y$ and $y \sqsubseteq_p x$, then $x = y$ (antisymmetry) ;
(iii) If $x \sqsubseteq_p y$ and $y \sqsubseteq_p z$, then $x \sqsubseteq_p z$ (transitivity).

Matthews [2] also define the partial metric $\max\{a, b\}$ (or $\min\{a, b\}$) over the nonnegative reals with the partial ordering $\sqsubseteq_{\max}$ (or $\sqsubseteq_{\min}$). For intervals $[a, b] \sqsubseteq_p [c, d]$ if and only if $[c, d]$ is a subset of $[a, b]$.

2.4. Space $\ell_2(\mathbb{N})$

In the space $\ell_2(\mathbb{N})$, the distance function $d : \ell_2(\mathbb{N}) \times \ell_2(\mathbb{N}) \to \mathbb{R}$ defined as follow [6]:

$$d(x, y) = \left(\sum_{i=1}^{\infty} |x_i - y_i|^2 \right)^{1/2}, \forall x, y \in \ell_2(\mathbb{N}). \tag{2}$$

The space $\ell_2(\mathbb{N})$ is one of metric spaces containing convergent sequences.

2.5. Space $\ell_2(P)$

Kadak, et al [1] define a distance function p_q induced by the metric d^p defined in Equation 1.

Proposition 2.4 *A distance function $p_q : l_q(P) \times l_q(P) \to \mathbb{R}^+ \cup \{0\}$ for $1 \leq q < \infty$ defined by*

$$p_q(x, y) = \left(\sum_{i=0}^{\infty} d^p(x_i, y_i)^q \right)^{1/q} \tag{3}$$

where $x, y \in l_q(P)$, is a partial metric. A pair $(l_q(P), p_q)$ is a partial metric space.

As consequence of Proposition 2.4, we get a pair $(\ell_2(P), p_2)$ is a partial metric space with the partial metric $p_2 : \ell_2(P) \times \ell_2(P) \to \mathbb{R}^+ \cup \{0\}$ defined by

$$p_2(x, y) = \left(\sum_{i=0}^{\infty} d^p(x_i, y_i)^2 \right)^{1/2}, \forall x, y \in \ell_2(P). \tag{4}$$

3. Results and Discussions

In this section, we present how to investigate the convergence of sequences in $\ell_2(P)$ with respect to a partial metric space. By the analogue of metric and partial metric, we can identify the convergence of sequences in a partial metric space with respect to metric such that we also can ensure the convergence of sequences in $\ell_2(P)$ with respect to a partial metric.

3.1. Convergence of sequences in a partial metric space with respect to a metric

A sequence $\{x_n\}$ in a metric space (X, d^p) converges to a point $x \in X$ if for every $\epsilon > 0$, there exists an $n_0 \in \mathbb{N}$ such that $d^p(x_n, x) < \epsilon$ for all $n > n_0$. By the analogue of metric and partial metric, we have $\{x_n\}$ converges to a point $x \in X$ in a partial metric space (X, p) with respect to d^p.

Lemma 3.1 *If a sequence $\{x_n\}$ converges to a point $x \in X$ in a metric space (X, d^p), then $\{x_n\}$ converges to a point $x \in X$ in a partial metric space (X, p).*

Proof. On the previous section, we know that d^p is a metric induced by a partial metric p. If $\{x_n\}$ is a sequence in a metric space (X, d^p) and converges to a point $x \in X$, then for every $\epsilon > 0$, there exists an $n_0 \in \mathbb{N}$ such that for all $n > n_0$,

$$d^p(x_n, x) = 2p(x_n, x) - p(x_n, x_n) - p(x, x) < \epsilon. \tag{5}$$

We know that p is a partial metric, therefore it satisfies the axioms (P2) and (P3). Consequently, $0 \leq p(x_n, x_n) \leq p(x_n, x)$ and $0 \leq p(x, x) \leq p(x, x_n) = p(x_n, x)$. Furthermore, Equation 5 is only satisfied if $p(x_n, x) < \epsilon$. Hence, for every $\epsilon > 0$, there exists an $n_0 \in \mathbb{N}$ such that for all $n > n_0$,

$$p(x_n, x) < \epsilon. \tag{6}$$

IOP Conf. Series: Journal of Physics: Conf. Series **974** (2018) 012058 doi:10.1088/1742-6596/974/1/012058

From Equation 6, we get $\{x_n\}$ also converges in a partial metric space (X, p) if $\{x_n\}$ converges in a metric space (X, d^p). In the other words, $\{x_n\}$ converges in a partial metric space (X, p) with respect to d^p. □

We identify that Equation 1 can be generalized. From Equation 1, we obtain that the coefficients of the right-hand side are fixed i.e the coefficients of $p(x, y)$, $p(x, x)$ and $p(y, y)$ are 2, 1 and 1. Then, we generalize Equation 1 by choosing the constants $a, b, c \in \mathbb{R}$ such that $a \geq b + c$, where a is the coefficient of $p(x, y)$, b is the coefficient of $p(x, x)$ and c is the coefficient of $p(y, y)$. Moreover, we can induce the metric d^p using the constants a, b and c.

Theorem 3.2 *If p is a partial metric on a nonempty set X, the function $d^p : X \times X \to \mathbb{R}$ defined by*

$$d^p(x, y) = ap(x, y) - bp(x, x) - cp(y, y), \forall x, y \in X \tag{7}$$

is a metric on X with $a, b, c \in \mathbb{R}$ such that $a \geq b + c$.

Proof. We use the contradiction to prove the theorem. Choose $a, b, c \in \mathbb{R}$ such that $a < b + c$. On Theorem 3.2, we know that p is a partial metric on X, therefore it satisfies (P2) and (P3). Moreover, $0 \leq p(x, x) \leq p(x, y)$ and $0 \leq p(y, y) \leq p(x, y)$. If $p(x, x) = p(x, y) = p(y, y)$, then

$$d^p(x, y) = ap(x, y) - bp(x, y) - cp(x, y) = (a - (b + c))p(x, y). \tag{8}$$

In Equation 8, we obtain that $d^p(x, y) < 0$. This condition contradicts with the fact that p is a partial metric. It proves that the metric d^p can be induced by a partial metric p if a is the coefficient of $p(x, y)$, b is the coefficient of $p(x, x)$ and c is the coefficient of $p(y, y)$ such that $a \geq b + c$ for $a, b, c \in \mathbb{R}$. In the other words, we show that the coefficients of the right-hand side of Equation 1 are not fixed. □

From Theorem 3.2, we obtain that a sequence $\{x_n\}$ converges to a point $x \in X$ in a partial metric space (X, p) with respect to d^p defined in Therorem 3.2.

3.2. Convergence of sequences in $\ell_2(P)$ with respect to a partial metric

In Theorem 2.2, we know that the function d^p represents the distance of two points in the usual case where $d^p(x, y) = |x - y|$. As consequence, we denote the usual metric d defined in Equation 2 as the metric d^p . Moreover, by Equation 2, we obtain that $(\ell_2(\mathbb{N}), d^p)$ is a metric space containing convergent sequences.

Let $\{x_n\}$ is a sequence in a metric space $(\ell_2(\mathbb{N}), d^p)$. We say that $\{x_n\}$ converges to a point $a \in \ell_2(\mathbb{N})$ if for every $\epsilon > 0$, there exists an $n_0 \in \mathbb{N}$ such that $d^p(x_n, a) < \epsilon$ for all $n > n_0$. From Lemma 3.1, we can ensure that $\{x_n\}$ also converges in a partial metric space $(\ell_2(P), p_2)$ with respect to a partial metric p_2 defined in Equation 4.

The principle of space $\ell_2(P)$ is almost same with $\ell_2(\mathbb{N})$. If the sequences in $\ell_2(\mathbb{N})$ are formed by elements of $\mathbb{N}$, then the sequences in $\ell_2(P)$ are formed by elements of partially ordered set P. We need to choose a partially ordered set in a partial metric space.

Proposition 3.3 *A closed interval $[a, b]$ with the partial ordering $\sqsubseteq_p$ denoted as a pair $([a, b], \sqsubseteq_p)$, is a partially ordered set for a partial metric $p : [a, b] \times [a, b] \to \mathbb{R}^+ \cup \{0\}$ defined by*

$$p([s, t], [u, v]) = \max\{t, v\} - \min\{s, u\}, \tag{9}$$

for all $[s, t], [u, v] \in \{[a, b] : a \leq b\}$.

Proof. By Definition 2.3, we show that P defined in Proposition 3.3 is a partially ordered set. First, we show that the partial ordering $\sqsubseteq_p$ satisfies reflexivity. Let $x = [s, t]$, therefore $p(x, x) = t - s = p(x, x)$. It represents that $x \sqsubseteq_p x$ for all $x \in \{[a, b] : a \leq b\}$.

IOP Conf. Series: Journal of Physics: Conf. Series **974** (2018) 012058 doi:10.1088/1742-6596/974/1/012058

Second, we present that the partial ordering $\sqsubseteq_p$ satisfies antisymmetry. For all $x, y \in \{[a,b] : a \leq b\}$ where $x = [s,t]$ and $y = [u,v]$, if $x \sqsubseteq_p y$, then $p(x,x) = t - s = p(x,y)$. It is satisfied only if $s \leq u$ and $t \geq v$. If $y \sqsubseteq_p x$, then $p(y,y) = v - u = p(y,x)$. We know that p defined in Equation 9 is a partial metric such that it satisfies (P3). As consequence, we get $p(y,y) = v - u = p(x,y)$. It is only satisfied if $s \geq u$ and $t \leq v$. Furthermore, we obtain that $s = u$ and $t = v$. It represents that $x = [s,t] = [u,v] = y$.

Third, we indicate that the partial ordering $\sqsubseteq_p$ satisfies transitivity. For all $x, y, z \in \{[a,b] : a \leq b\}$ where $x = [s,t]$, $y = [u,v]$ and $z = [m,n]$, if $x \sqsubseteq_p y$, then $p(x,x) = t - s = p(x,y)$. It is satisfied only if $s \leq u$ and $t \geq v$. If $y \sqsubseteq_p z$, then $p(y,y) = v - u = p(y,z)$. It is only satisfied if $u \leq m$ and $v \geq n$. Therefore, we get $s \leq m$ and $t \geq n$ such that $p(x,z) = t - s = p(x,x)$. Consequently, we get $x \sqsubseteq_p z$. □

Let $\{x_n\}$ is a sequence in $\ell_2(P)$ with respect to a partial metric defined in Equation 9. First, we investigate that a metric can be induced by a partial metric. By the analogue of metric and partial metric in Equation 1, we obtain that the function d^p is a metric induced by partial metric p defined in Equation 9.

Theorem 3.4 *A function $d^p : \mathbb{R} \times \mathbb{R} \to \mathbb{R}$ for all $x, y \in \mathbb{R}$ defined by*

$$d^p(x,y) = |x - y|, \tag{10}$$

is a metric induced by a partial metric $p : [a,b] \times [a,b] \to \mathbb{R}^+ \cup \{0\}$ for all $[s,t], [u,v] \in \{[a,b] \,|a \leq b\}$, where

$$p([s,t],[u,v]) = \max\{t,v\} - \min\{s,u\}. \tag{11}$$

Proof. It is obvious that the function d^p defined in Equation 10 is a metric on $\mathbb{R}$. Now, we show that the metric d^p can be induced by a partial metric p defined in Equation 11. From Equation 1, we obtain that

$$d^p(x,y) = 2\Big(\max\{t,v\} - \min\{s,u\}\Big) - (t-s) - (v-u) \tag{12}$$

There is four cases of Equation 12.

- Case 1:
  For $[s,t] \cap [u,v] = \emptyset$, we have $d^p(x,y) = (u-s) - (t-v)$ where $s < u$ and $t < v$ and $d^p(x,y) = (s-u) - (v-t)$ where $s > u$ and $t > v$.
- Case 2:
  For $[s,t] \cap [u,v] \neq \emptyset$, we have $d^p(x,y) = (u-s) - (t-v)$ where $s < u$ and $t < v$ and $d^p(x,y) = (s-u) - (v-t)$ where $s > u$ and $t > v$.
- Case 3:
  For $[s,t] \subseteq [u,v]$, we have $d^p(x,y) = (s-u) - (t-v)$ where $s > u$ and $t < v$.
- Case 4:
  For $[u,v] \subseteq [s,t]$, we have $d^p(x,y) = (u-s) - (v-t)$ where $s < u$ and $t > v$.

From the four cases above, we have $d^p(x,y) = |s-u| - |t-v|$ for all $s,t,u,v \in \mathbb{R}$. We consider that $|s-u| - |t-v| = |t-v| - |s-u|$, therefore $d^p(x,y) = |s-u| - |t-v| = |t-v| - |s-u|$.

Let $x = |s-u| \in \mathbb{R}$ and $y = |t-v| \in \mathbb{R}$, then we obtain that $d^p(x,y) = x - y = y - x = |x-y|$. It represents that the metric d^p defined in Equation 10 can be induced by a partial metric p defined in Equation 11. □

Let $\{x_n\}$ is a sequence in $\ell_2(P)$ where P is partially ordered set defined in Proposition 3.3. We know that $\{x_n\}$ converges to a point $a \in \ell_2(\mathbb{N})$ such that for every $\epsilon > 0$, there exists an $n_0 \in \mathbb{N}$ satisfying $d^p(x_n, a) < \epsilon$ for all $n > n_0$. Consequently, for every $\epsilon > 0$, there exists an $n_0 \in P$ such that $p_2(x_n, a) < \epsilon$ for all $n > n_0$. It concludes that $\{x_n\}$ also converges to a point $a \in \ell_2(P)$ with respect to a partial metric p_2 induced by partial metric p defined in Equation 11.

IOP Conf. Series: Journal of Physics: Conf. Series **974** (2018) 012058 doi:10.1088/1742-6596/974/1/012058

Corollary 3.5 *A sequence $\{x_n\}$ is called to converge in the partial metric space $(\ell_2(P), p_2)$ if for every $\epsilon > 0$, there exists an $n_0 \in P$ such that $p_2(x_n, x) < \epsilon$ for all $n > n_0$.*

4. Concluding Remarks

We have presented that the convergence of sequences in a metric space can be constructed to a partial metric space. We notice the analogue of metric and partial metric to ensure the convergence of sequences in a partial metric space using a partial metric with respect to partial ordering. For a particular case, we show that the convergence of sequences in the metric space $(\ell_2(\mathbb{N}), d^p)$ can be established in the partial metric space $(\ell_2(P), p_2)$.

Acknowledgments

Acknowledgements This paper is part of our work for Master Program at Department of Mathematics, Institut Teknologi Sepuluh Nopember, Surabaya, Indonesia to be finished on December 2017.

References

[1] Kadak U, Basar F and Efe H 2013 *Some Partial Metric Spaces of Sequences and Functions* vol 19 (Turkey: ICSRS Publication) p 10
[2] Matthews S 1994 *Partial Metric Topology* vol 728 (New York: Annals of the new York Academy of Sciences) p 183
[3] Heckmann R 1997 *Approximation of Metric Spaces by Partial Metric Spaces* (FB 14-Informatik, Universitä des Saarlandes)
[4] Kopperman R, Matthews S and Pajoohesh H 2004 *Partial Metrizability in Value Quantales* vol 5 (New York: Universidad Politécnica de Valencia) p 115
[5] Waszkiewicz P 2004 *Partial Metrizability of Continuous Posets* (Institute of Computer Science, Jagiellonian University, ul. Nawojki 11)
[6] Teschl G 2010 *Topics in Real and Functional Analysis* (Austria: Fakultät für Mathematik, Oskar-Mogenstern-Platz 1, Universität Wien)
[7] Matthews S, Bukatin M 2012 *An Intelligent Theory of Cost for Partial Metric Spaces* (New York: Artificial General Intelligence: 5th International Conference) p 168

IOP Conf. Series: Journal of Physics: Conf. Series **974** (2018) 012059 doi:10.1088/1742-6596/974/1/012059

Tracking and people counting using Particle Filter Method

D R Sulistyaningrum[1] , B Setiyono[2] and M S Rizky[3]
[1,2,3]Department of Mathematics, Institut Teknologi Sepuluh November
Campus ITS, Sukolilo, Surabaya, 60111, Indonesia

E-mail: dwiratna@matematika.its.ac.id

Abstract. In recent years, technology has developed quite rapidly, especially in the field of object tracking. Moreover, if the object under study is a person and the number of people a lot. The purpose of this research is to apply Particle Filter method for tracking and counting people in certain area. Tracking people will be rather difficult if there are some obstacles, one of which is occlusion. The stages of tracking and people counting scheme in this study include pre-processing, segmentation using Gaussian Mixture Model (GMM), tracking using particle filter, and counting based on centroid. The Particle Filter method uses the estimated motion included in the model used. The test results show that the tracking and people counting can be done well with an average accuracy of 89.33% and 77.33% respectively from six videos test data. In the process of tracking people, the results are good if there is partial occlusion and no occlusion

1. Introduction

The development of technology has increased rapidly in recent years. One is the use of surveillance cameras to oversee the activities of people in certain areas, such as highways, parking lots and offices. The surveillance camera is usually always equipped with an object tracking module to know the movement of the object. Visual object tracking has attracted considerable attention because there are so many events that require object tracking, whether it is object detection, analyzing people's activities and interactions between people and computers. For example, supervision within the school environment, traffic, office and others [1].

Tracking objects is one way to simplify people's work. Usually tracking and counting objects, especially people used by monitoring activities that exist in those environments [2]. Based on the object under study, object tracking can be grouped into one object tracking and tracking multiple objects [3].

Tracking one object or multiple objects is relatively simple from tracking multiple objects with poor background conditions [4]. There has been a lot of research related to tracking, whether it's an object or a lot of objects. The study aims to obtain effective tracking, ie through the detection of objects, techniques, or different algorithms such as Kalman filter, mean shift and particle filter [5]. In person tracking, Jin et al. [6] divides body parts into 3 representations of histogram, color and HOG, and are tracked separately. But the method only tracks objects singly and cannot overcome occlusion. Habibi et al [7] used adaptive particle filter for tracking of small object. Although the formulated particle filter kernel shows better results than a typical particle filter, it cannot track many objects and cannot overcome occlusion. Cabido et al. [8] formulates a tracking algorithm that combines a particle filter with a memetic algorithm. Although it can track many objects, but not yet able to handle occlusion.

In this study, tracking and people counting using Particle Filter Method has been performed. Experimental results show that the model used successfully tracking many objects and the occlusion condition.

2. Methods

2.1. Particle filter for tracking and people counting

Figure 1 shows the block diagram of the process of tracking and people counting using particle filter method applied in this study. The process of tracking object consists of four process: pre-processing, segmentation, tracking object, and counting.

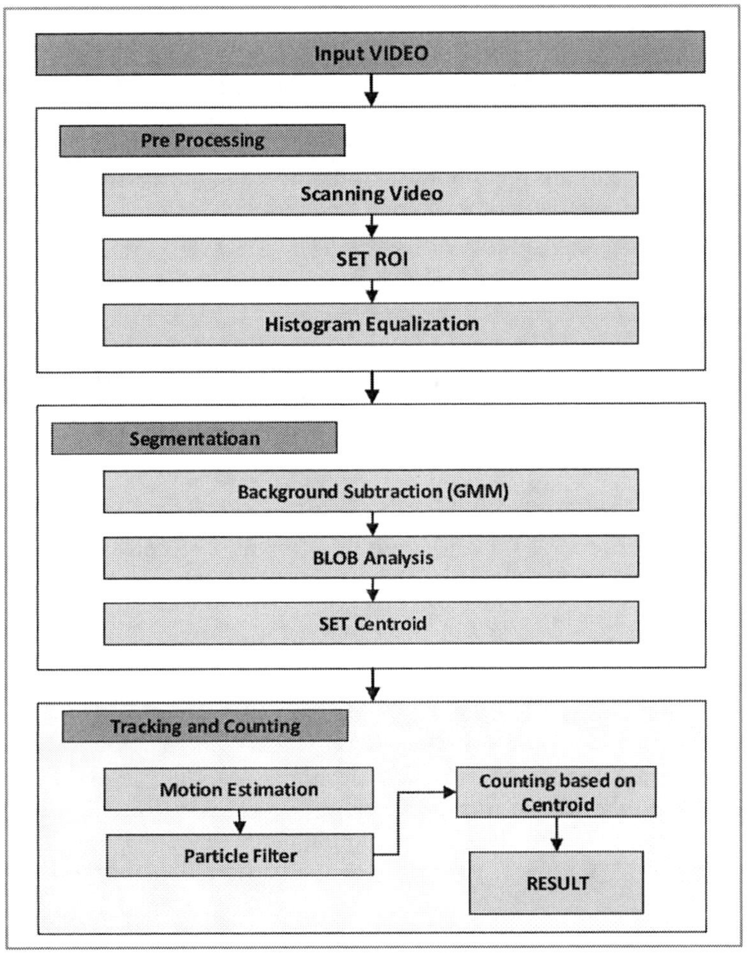

Figure 1. Scheme of the tracking and counting system

The input video is a color video of scene. At the stage of pre-processing the input video is processed by video processing operations that is scanning video, set ROI, and Histogram Equalization. Results of pre-processing is input to the segmentation stage with *Gaussian Mixture Model* (GMM) and BLOB analysis. The stage of tracking consist of motion estimation and particle filter. Counting of people based on based on the amount of the centroid on ROI.

2.2. Gaussian mixture model

Gaussian mixture model (GMM) is popular method that has been employed to segmentation background and non-background. That problem is call background subtraction. The GMM method based on a parametric model in which the probability density function of the gray levels in the image.

GMM is an appropriate method for various conditions contained in the image, such as the image background is always static, multimodal, or that contain noise (disturbance or unwanted objects contained in the image). In the GMM process clustering algorithm required to classify any pixels that include foreground or background. GMM models are formed from time-based pixel color data. The result of the model will be 2 parts, a model that reflects the background and non-background model. The number of GMM models used affects the number of background models. The more GMM models are used, the more the background model is owned by a pixel.

Step by step of GMM method:

1. Select the ROI area of the frame
2. Build GMM models. GMM models are formed from time-based pixel color data. The result of the model will be 2 parts, a model that reflects the background and non-background model.
3. Match each frame with pre-established GMM model frame.
4. Check whether the model is appropriate or not, if not then have to create a new model and return to step three, if it is appropriate it will proceed to the next step.
5. Update the GMM parameters then normalize the weights.
6. Sort the mode based on the weight, then select the GMM that will be the background and foreground.

2.3. Particle filter

Particle filter is a state space method for applying Bayesian filters. The main idea is to approximate the posterior probability distribution by the particles. Each particle represents a state of the object hypothesis, with a discrete sampling corresponding to probability (weight). The particles are usually done resampling to relieve particle degeneration. The efficiency and accuracy of particle filters for tracking depends on the distribution and effective model of observation for particle weighting [9].

Define an object state vector as $X = (x, y)$ with (x, y) being the center of the object. The state space model of the object being tracked is:

$$X_{t+1} = f(X_t, \mu_t) \qquad (1)$$

$$Z_t = g(X_t, \xi_t) \qquad (2)$$

where f is a dynamic model, g is a observation model, μ_t and ξ_t are the process and observation noise, respectively.

The dynamic models and observation models used in this study are similar to those used in [9]. The tracking process is done on the second frame because the object initialization has been done on the first frame. Then to get the motion vector added to the motion model, motion estimation is performed on two consecutive frames (N-1 and N frames). Furthermore the particle filter method is applied until the last frame.

2.4. Block matching algorithm

Block matching is one of the motion estimation algorithms. The purpose of the motion estimation technique is to obtain a motion vector (displacement or velocity) for each pixel in the image. This algorithm estimates movement on block to block. The area of movement is assumed to be constant with a square block area and represented by a single motion vector on each block [10]. Each block in the frame is now matched to the block in the previous frame based on a specific cost function. Some of the most popular and inexpensive computational cost functions are Mean Absolute Difference (MAD) as in Equation 3.

$$MAD = \frac{1}{N^2}\sum_{i=0}^{N-1}\sum_{j=0}^{N-1}\left|C_{ij} - R_{ij}\right| \tag{3}$$

where N being the macro size of the block, C_{ij} is the pixel on the current block macro and R_{ij} is the pixel on the block macro being referenced. Figure 1 shows the steps in the block matching algorithm.

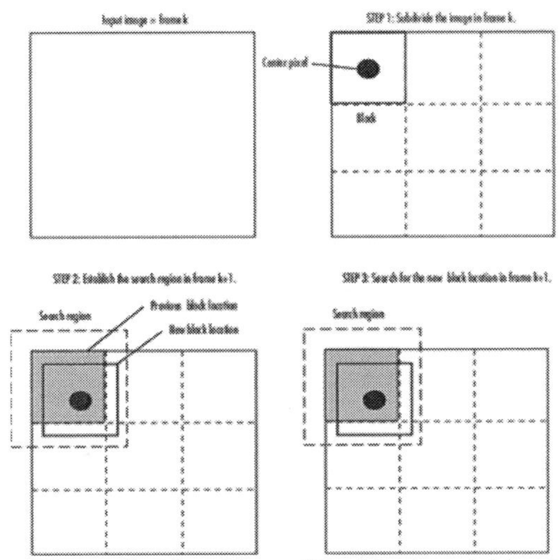

Figure 2. Algorithm of the block matching [10]

3. Experimental result

In the experiments, we used recorded videos of an outdoor scene. There are 6 videos used as test materials, Video has different number of frame. Table 1 shows the detailed information about the six videos.

Table 1. Test Data

No	Name	Number of Frame	Description
1	Video1.avi	78	Two people with occlusion, walk
2	Video2.avi	58	Two people without occlusion, walk
3	Video3.avi	97	Seven people with partial occlusion, walk
4	Video4.avi	26	One people, run
5	Video5.avi	84	One person with full occlusion, walk
6	Video6.avi	100	Two people walk concurrently

Figures 3, 4, 5, 6 show some examples of the results of tracking experiments and people counting.

IOP Conf. Series: Journal of Physics: Conf. Series **974** (2018) 012059 doi:10.1088/1742-6596/974/1/012059

Figure 3. Result of video1.avi

Figure 4. Result of video2.avi

Figure 5. Result of video3.avi

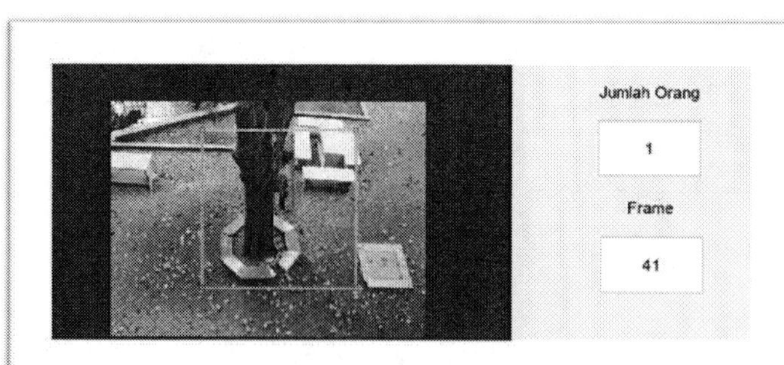

Figure 6. Result of video5.avi

Testing on video1.avi and video2.avi gives result of tracking and people's counting are good accurate with few errors. In the test on video3.avi the results obtained are not good because of partial occlusion so that people cannot be counted. Experiments on video4.avi obtained the result that the particle filter method can perform the object tracking very well. Tests on video6.avi obtained results that the system can do tracking but when two people walk together then the person is considered an object. Result of tracking and people counting for each of the videos are presented in Table 2 and Table 3.

Table 2. The results of tracking

Video	Number of Frame	Number of frames traced	Number of frames not traced	Percentage
Video1.avi	78	71	7	91 %
Video2.avi	58	58	0	100 %
Video3.avi	97	67	30	69 %
Video4.avi	26	25	1	96 %
Video5.avi	84	74	10	88 %
Video6.avi	100	92	8	92 %

Table 3. The results of people counting

Video	Number of Frame	Number of frame counting True	Number of frame counting False	Percentage
Video1.avi	78	71	7	91 %
Video2.avi	58	58	0	100 %
Video3.avi	97	67	30	69 %
Video4.avi	26	21	5	81%
Video5.avi	84	74	10	88 %
Video6.avi	100	35	65	35 %

4. Conclusion

In this study, particle filter method has been successfully applied to tracking and people counting. The tracking and counting done by several stages: video reading, pre-processing, segmentation using Gaussian Mixture Model(GMM), tracking using particle filter, and counting based on centroid.

The results show that the system can tracking and people counting with an accuracy rate of 89.33% and 77.33% respectively from six videos test data. This research has successfully tracked and counted people in the ROI area with good performance when there is no occlusion, it shows the results of experiment video1.avi, video2.avi, and video 4.avi. The Particle Filter has not been able to track properly if the observed people are running side by side based on the video5.avi test. In the process of tracking people, the results are good if there is partial occlusion and no occlusion.

References

[1] Naushad Ali M.M, Abdullah-Al-Wadud M and Lee, S.L et al. 2014 Multiple Object Tracking With Partial Occlusion Handling Using Salient Feature Points *Information Sciences* **278** 448–465.

[2] Lefloch D, et al. 2008 Real-time people counting system using a single video camera *Proc. SPIE 6811, Real-Time Image Processing* **681109**

[3] Amer E, Dubois E and Mitiche A. 2003 Real-time system for high-level video representations: application to video surveillance *SPIE International Symposium on Electronic Imaging* **5022** 530-541

[4] Hou Y.Land Pang G.K.H. 2011 People counting and human detection in a challenging situation, *IEEE. Trans.sys.man* and *cybernetics* **41(1)** 24-33

[5] Yang H, Shao L, Zheng F, Wang L and Song Z. 2011 Recent advances and trends in visual tracking: a review *Neurocomputing* **74** 3823–3831

[6] Jin L., Cheng J. and Huang H 2010 Human tracking in the complicated background by particle filter using color-histogram and HoG *IEEE International Symposium on Intelligent Signal Processing and Communication* **2010 1-4**

[7] Habibi Y,Sulistyaningrum D.R and Setiyono B 2017 A new algorithm for small object tracking based on super-resolution technique AIP Conference Proceedings **1867(020024)**

[8] Cabido R., Montemayor A.S. and PantrigoJ.J. 2012 High performance memetic algorithm particle filter for multiple object tracking on modern GPUs *Soft. Comput.* **16(2)** 217–230.

[9] Huang y and Llach J. 2008 Tracking the Small Object through Clutter with Adaptive Particle Filter *Proc. Int'I. Conf. Audio Language Image Process* **2008** 357-362

[10] Wang Y., Ostermann J., and Zhang Y-Q. 2002. *Video Processing and Communication* (Upper Saddle River, N.J: Prentice Hall)

GSTAR-SUR Modeling With Calendar Variations And Intervention To Forecast Outflow Of Currencies In Java Indonesia

M S Akbar[1,2]Setiawan[1], Suhartono[1], B N Ruchjana[3], and M A A Riyadi[4]
[1]Department of Statistics, Institut Teknologi Sepuluh Nopember (ITS)
[2]PhD Student,Department of Statistics, Institut Teknologi Sepuluh Nopember (ITS)
[4]Master Student,Department of Statistics, Institut Teknologi Sepuluh Nopember (ITS)
Arief Rahman Hakim Street, Surabaya 60111, Indonesia
[3]Department of Mathematics, Faculty of Mathematics and Natural Sciences Universitas Padjadjaran (UNPAD), Bandung, Indonesia. Raya Bandung Street-Sumedang Km. 21 Jatinangor Sumedang 45363, Indonesia

E-mail: [1)] m_syahid_a@statistika.its.ac.id; [2)]setiawan@statistika.its.ac.id; [3)] suhartono@statistika.its.ac.id; [4)] budinr@unpad.ac.id;
[5)] m.alfan.alfian.riyadi@gmail.com

Abstract. Ordinary Least Squares (OLS) is general method to estimates Generalized Space Time Autoregressive (GSTAR) parameters. But in some cases, the residuals of GSTAR are correlated between location. If OLS is applied to this case, then the estimators are inefficient. Generalized Least Squares (GLS) is a method used in Seemingly Unrelated Regression (SUR) model. This method estimated parameters of some models with residuals between equations are correlated. Simulation study shows that GSTAR with GLS method for estimating parameters (GSTAR-SUR) is more efficient than GSTAR-OLS method. The purpose of this research is to apply GSTAR-SUR with calendar variation and intervention as exogenous variable (GSTARX-SUR) for forecast outflow of currency in Java, Indonesia. As a result, GSTARX-SUR provides better performance than GSTARX-OLS.

1. Introduction

There are many applications of time series analysis with location effects. For example tea productions, daily oil productions, air pollution measurements per hour at several cities [1] [2] [3]. These examples constitute to spatial time series, which is indexed data by time and location. In spatial applications, the locations can be correlated each other. The Space Time Autoregressive (STAR) method is one of the time series methods involving location.

STAR model was first developed by Cliff and Ord [4]. The STAR model applies to a homogeneous location, because all locations have the same time for series time and space time values. Ruchjana developed STAR model for parameter values of each location can be different, both time series and space time parameter values. The development of this model was known as Generalized Space Time Autoregression (GSTAR). Research about GSTAR has been applied in various fields, such as Application of Generalized Space-Time Autoregressive Model on GDP Data in West European Countries [5], Forecasting oil production data at volcanic layer Jatibarang, West Java Indonesia [6]. In forecasting research, there are many forecasting cases involving exogenous variables matric and non-metric. Forecasting models that involve exogenous variables are usually coupled with the suffix - X, such that ARIMAX or VARIMAX. The GSTAR method also has been studied method GSTAR-X, that is GSTARX-SUR Model for Spatio-Temporal Data Forecasting [7]. This research involves intervention as exogenous variable in cases of inflation in four cities of East Java.

This article will discuss forecasting with the GSTAR method that involves the effects of calendar variations and interventions simultaneously on exogenous variables. The application of this method is to forecast rupiah outflow in five provinces of Java island, there are Jakarta, West Java, Central Java, Yogyakarta, and East Java. As an exogenous variables, the case of rupiah outflow on Java island influenced by Eid Fitr (calendar variation) and government policy in January 2007 and January 2010. So that forecasting of rupiah outflow data consists of two exogenous variables, there are calendar variation and intervention. The aim of this research is to apply GSTAR-SUR with calendar variation and intervention as exogenous variable (GSTARX-SUR) for forecast outflow of currency in five provinces of Java, Indonesia.

2. Method

In this section, we describe the statistical methods used to estimate GSTAR model parameters and the concept of calendar variations and interventions.

2.1. Generalized Space Time Autoregression (GSTAR)

The Space Time Autoregressive (STAR) model was first developed by Cliff and Ord [4]. Ruchjana developed STAR model into GSTAR. GSTAR is more flexible spatial weights when compared with STAR [2]. GSTAR model with λp spatial order and p-time order (GSTAR (λ_p, p)) is [7]:

$$\boldsymbol{Z}(t)=\sum_{s=1}^{p}\left(\boldsymbol{\Phi}_{s0}+\sum_{k=1}^{\lambda_s}\boldsymbol{\Phi}_{sk}\mathbf{W}^{(k)}\right)\boldsymbol{Z}(t-s)+\boldsymbol{a}(t), \tag{1}$$

with $\boldsymbol{\Phi}_{s0}=\mathrm{diag}(\phi_{s0}^{(1)},\cdots,\phi_{s0}^{(N)})$, $\boldsymbol{\Phi}_{sk}=\mathrm{diag}(\phi_{sk}^{(1)},\cdots,\phi_{sk}^{(N)})$, weight $\mathbf{W}^{(k)}$, a(t) error model that are identic, independent and normal distribution with zero mean and covariance Σ. If spatial order 1, time order 1 and three locations ($\lambda_p = 1, p = 1, N = 3$) then equation (1) becomes [6]:

$$\boldsymbol{Z}(t)=\boldsymbol{\Phi}_{10}\boldsymbol{Z}(t-1)+\boldsymbol{\Phi}_{11}\mathbf{W}^{(1)}\ \boldsymbol{Z}(t-1)+\boldsymbol{a}(t). \tag{2}$$

Equation (2) can be written in matrix form:

$$\begin{pmatrix}\boldsymbol{Z}_1(t)\\ \boldsymbol{Z}_2(t)\\ \boldsymbol{Z}_3(t)\end{pmatrix}=\begin{pmatrix}\phi_{10}^{(1)}&0&0\\0&\phi_{20}^{(2)}&0\\0&0&\phi_{30}^{(3)}\end{pmatrix}\begin{pmatrix}\boldsymbol{Z}_1(t-1)\\ \boldsymbol{Z}_2(t-1)\\ \boldsymbol{Z}_3(t-1)\end{pmatrix}+\begin{pmatrix}\phi_{11}^{(1)}&0&0\\0&\phi_{21}^{(2)}&0\\0&0&\phi_{31}^{(3)}\end{pmatrix}\begin{pmatrix}0&w_{12}&w_{13}\\w_{21}&0&w_{23}\\w_{31}&w_{32}&0\end{pmatrix}\begin{pmatrix}\boldsymbol{Z}_1(t-1)\\ \boldsymbol{Z}_2(t-1)\\ \boldsymbol{Z}_3(t-1)\end{pmatrix}+\begin{pmatrix}\boldsymbol{a}_1(t)\\ \boldsymbol{a}_2(t)\\ \boldsymbol{a}_3(t)\end{pmatrix}. \tag{3}$$

Equation (3) can be written like linear model:

$$\boldsymbol{Y}=\mathbf{X}\boldsymbol{\beta}+\boldsymbol{a}\,. \tag{4}$$

So $\boldsymbol{\beta}$ can be obtained by OLS method given by:

$$\hat{\boldsymbol{\beta}}_{\text{GSTAR-OLS}}=(\mathbf{X}'\mathbf{X})^{-1}\,\mathbf{X}'\boldsymbol{Y}. \tag{5}$$

If each locations are correlated, then estimations GSTAR parameters can be done using GLS method. Because its produce efficient estimators. This will be proven in simulation result. Estimated GSTAR parameters using GLS is [8]:

$$\hat{\boldsymbol{\beta}}_{\text{GSTAR-SUR}}=(\mathbf{X}'\boldsymbol{\Omega}^{-1}\mathbf{X})^{-1}\,\mathbf{X}'\boldsymbol{\Omega}^{-1}\boldsymbol{Y}. \tag{6}$$

Impact of weight on GSTAR is significant. Therefore weight that is used are uniform weight and normalization cross correlation.

2.2. Calender variation and intervention

Calendar effect in time series was first researched by Liu (1980) and Cleveland and Devlin (1980). Liu discussed effects of calendar variations on ARIMA model identification and applied it to Taiwan highway traffic volumes [9]. Cleveland and Devlin identifies the effects of calendar variations with two methods, namely spectrum analyzes and time domain graphical displays. This method applies to international airline passengers series [10]. In 2010 Lee and Suhartono forecasted the sale of Muslim clothing from Muslim garment companies in Indonesia. This study used Eid Fitr as effect of calendar variation [11]. This is indicated by the high level of Bank Indonesia expenditure during Eid Fitr as shown in Figure 2. Date of Eid Fitr from 2003 to 2011 as shown in Table 1.

Table 1. Date of Eid Fitr since 2003 - 2011

Year	Date	Year	Date
2003	25-26 November	2008	01-02 October
2004	14-15 November	2009	21-22 September
2005	03-04 November	2010	10-11 September
2006	23-24 October	2011	30-31August
2007	12-13 October		

Calendar variation model can be written as linear regression equation:

$$y_t = \beta_0 + \beta_1 V_{1,t} + \beta_2 V_{2,t} + \cdots + \beta_p V_{p,t} + e_t \,, \tag{7}$$

with $V_{p,t}$ as dummy variable for p calendar variation effects. Identification of calendar variation effects can be seen from time series plot data. e_t is error model that are identic, independent and normal distribution with zero mean and covariance σ_e^2.

The intervention model is a model used for external events beyond estimates as well as internal events that are expected to affect predicted variables. The general form of the intervention model is [12]:

$$Z_t = \frac{\omega_s(B)B^b}{\delta_r(B)} I_t + N_t \,, \tag{8}$$

with $\omega_s(B) = (\omega_0 - \omega_1 B - \omega_2 B^2 - \ldots - \omega_s B^s)$, $\delta_r(B) = (1 - \delta_1 B - \delta_2 B^2 - \ldots - \delta_r B^r)$, I_t = intervention variables, N_t = error model that are identic, independent and normal distribution with mean **0** and covariance σ_N^2. In general, there are two types of intervention variables, which are *step* function and *pulse* function. The step function is an intervention that occurs over a long period of time. The notation of step function is:

$$I_t = S_t = \begin{cases} 0, & t < T. \\ 1, & t \geq T. \end{cases} \tag{9}$$

Pulse function is an intervention that affects the time of intervention and does not continue at a later time. Pulse function notation is given by:

$$I_t = P_t = \begin{cases} 0, & t \neq T. \\ 1, & t = T. \end{cases} \tag{10}$$

3. Result

There are two results in this research, which are simulation study and real case. Simulation studies are used to show that parameter estimation using GLS method is more efficient than OLS. Application of GSTAR-X SUR model for the accuracy of forecast of outflow data of rupiah in each propince in Java of island.

3.1. Simulation

The simulation study was designed by generating two types of error, ie same variance, no correlation (1) and same variance, correlation (2) for three locations. Each variance covariance matrix and simulation parameter coefficients are:

$$\Sigma_{(1)} = \begin{pmatrix} 1 & 0 & 0 \\ 0 & 1 & 0 \\ 0 & 0 & 1 \end{pmatrix} \quad \Sigma_{(2)} = \begin{pmatrix} 1 & 0.5 & 0.45 \\ 0.5 & 1 & 0.55 \\ 0.45 & 0.55 & 1 \end{pmatrix} \quad \phi = \begin{pmatrix} 0.7 & 0 & 0 \\ 0 & 0.6 & 0 \\ 0 & 0 & 0.8 \end{pmatrix}$$

The simulation process was iterated 1000 times. So that obtained 1000 parameters for each location. Based on Figure 1, if between each location are not correlated, the estimated value of GSTAR-OLS (Blue line) and GSTAR-SUR (Red line) parameters are same (as Figure 1 (a)). If between location are correlated then GSTAR-SUR more efficient. It is shown that standard deviation of GSTAR-SUR is smaller than GSTAR-OLS (as Figure 1 (b)).

3.2. Case study

Case study uses Rupiah outflow data in Java with five regions since January 2003 until December 2012, which are Jakarta, West Java, Central Java, Yogyakarta, and East Java. Data are devided into in-sample and out-sample. January 2003 until December 2011 as in-sample data and January 2012 until December 2012 as out-sample data. Figure 2 shows two kind of events, which are calendar variation effect on Eid Fitr and intervention in January 2007 and January 2010. Therefore forecasting Rupiah outflow are devided into two stages. Stage one is modeling calendar variation effect on Eid Fitr and intervention as predictors with regression GLS. R-square for five models are shown at Table 1.
Residuals from stage one are input for stage two. Stage two builds GSTAR model for forecasting outflow data in 2012. Table 3 shows that there are high correlation between each locations. Identification of GSTAR model use cross correlation function (MCCF) matrix as Figure 3. It shows that autoregressive order is GSTAR(1,[1,12]). The first order of autoregressive is chosen with the consideration that the outflow data is affected by the conditions of the previous month. In this study, we use two kinds of weight, which are uniform and normalization cross correlation. Estimated GSTAR model parameters use two methods, which are OLS (GSTAR-OLS) and GLS (GSTAR-SUR). Criteria of model evaluation use Mean Absolute Percentage Error (MAPE).

Table 2. R-Square Model for Five Location

Jakarta	West Java	Central Java	Yogyakarta	East Java
0.533	0.699	0.742	0.610	0.680

Table 3. Correlation Between Location Outflow Rupiah

Location	Jakarta	West Java	Central Java	Yogyakarta
West Java	0.786			
Central Java	0.747	0.890		
Yogyakarta	0.736	0.758	0.806	
East Java	0.815	0.931	0.883	0.778

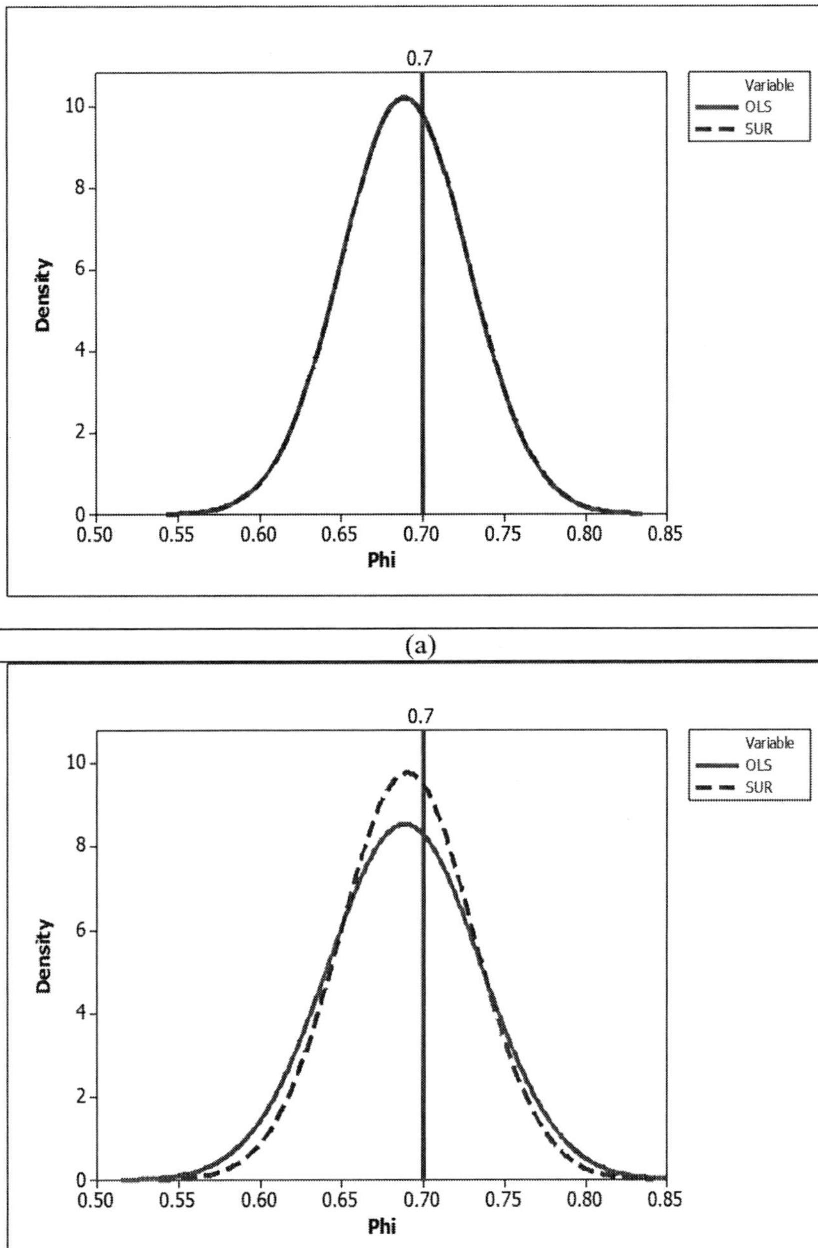

(a)

(b)

Figure 1. Histogram Parameters ϕ_{10} GSTAR(1,1) with 1000 iterations,
(a) Type Variance Covariance Matrix 1 and (b) Type Variance Covariance Matrix 2

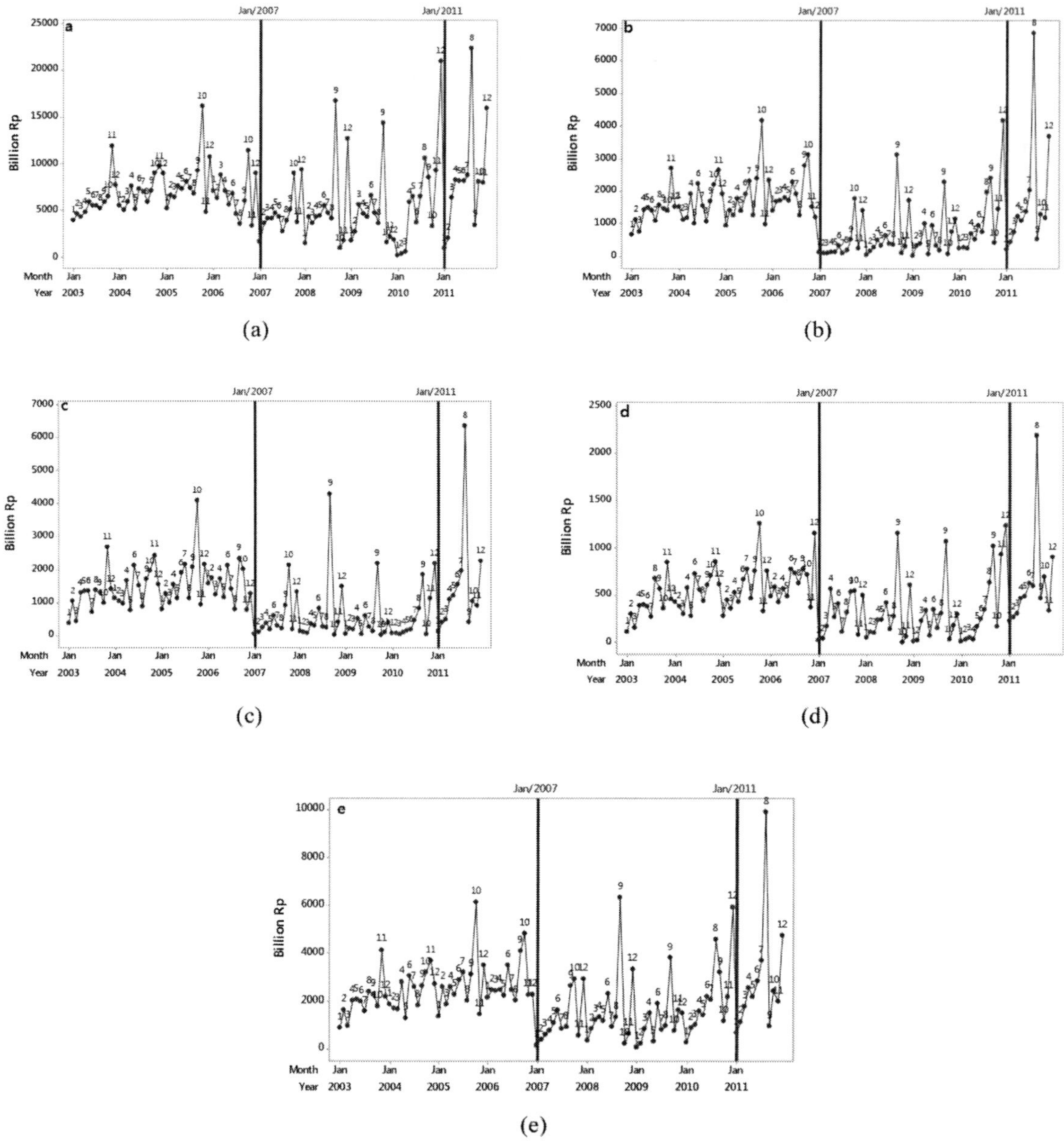

Figure 2. Time Series Plot Outflow Rupiah in Five Locations,
(a) Jakarta, (b) West Java, (c) Central Java, (d) Yogyakarta, (e) East Java

Table 4 shows in general that the best model is GSTARX-SUR (1, [1,12]) with normalization cross correlation weight. This is indicated by the three smallest MAPE value.

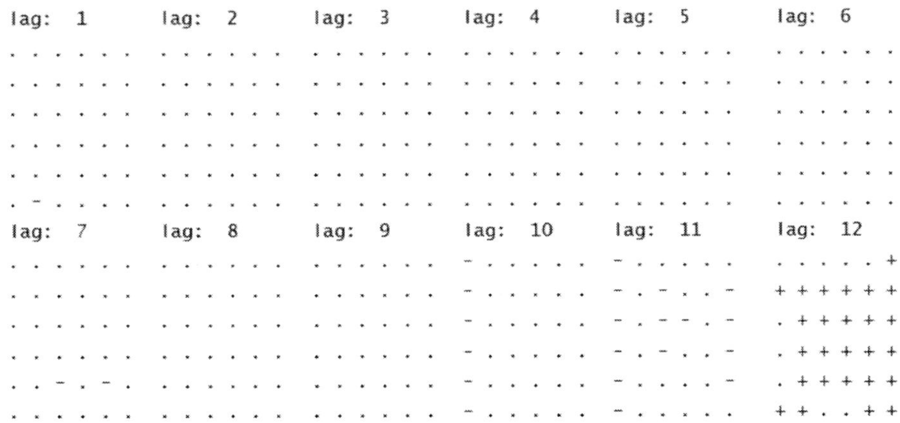

Figure 3.MCCF Residuals Stage-One Rupiah Outflow in Five Locations

Table 4. MAPE Out Sample Data Between GSTARX-OLS and GSTARX-SUR

Method	Weight	Jakarta	West Java	Central Java	Yogyakarta	East Java
GSTARX-OLS	Normalization	2.880	2.342	3.439	2.609	1.839*
	Uniform	2.528*	2.302	2.838	2.521	2.008
GSTARX-SUR	Normalization	2.568	2.255*	2.591*	2.288*	1.909
	Uniform	2.690	2.415	2.854	2.511	2.032

4. Conclusion

Based on the result of research, it is found that simulation of GSTARX-SUR estimator is more efficient than GSTAR-OLS. This is indicated by the standard error value of GSTARX-SUR estimators are smaller than GSTAR-OLS in cases residuals correlated between locations. In the case study of rupiah outflow forecast in five provinces in Java with effect of calendar variation (Eid Fitr) and intervention (government policy january 2007 and january 2010), GSTAR-X SUR with normalization weights are more accurate in West Java, Central Java and Yogyakarta provinces. GSTAR-X OLS with normalization weight produces more accurate in East Java province. GSTAR-X OLS with uniform weight produces more accuracy in the province of Jakarta. This is indicated by a lower MAPE value. Further research is development of GSTAR-X model by considering exogenous variable that are metric. In addition, other comparative studies in other areas of forecasting need to validate the proposed model.

5. Acknowledgments

I would like to thanks to The Directorate General of Higher Education, The Ministry of Education and Culture of the Republic of Indonesia for BPPS grant and Intitut Teknologi Sepuluh Nopember for financial support.

References

[1] Mukhaiyar U and Pasaribu U S 2012 A New Procedure for Generalized STAR Modeling using IAcM Approach, *ITB Journal of Science*, **vol. 44 A**, **no. 2**, pp. 179-192.

[2] Ruchjana B N 2002 *A Generalized Space-Time Autoregressive Model and Its Application to Oil Production*, Ph.D. Thesis, ITB, Bandung.

[3] Ruchjana B N, Abdullah A S, Toharudin T, and Jaya I G N M 2013 Clustering Spatial on the GSTAR Model for Replacement New Oil Well, in *Padjadjaran International Physics Symposium 2013 (PIPS-2013)*, Bandung, pp. 205-210.

[4] Cliff A D and Ord J K 1975 Model Building and the Analysis of Spatial Pattern in Human Geography, *Journal of the Royal Statistical Society, Series B (Methodological)*, **vol. 37**, **no. 3**, pp. 297-348.

[5] Nurhayati N, Pasaribu U S and Neswan O 2012 Application of Generalized Space-Time Autoregressive Model on GDP Data in West European Countries, *Journal of Probability and Statistics*, **vol. 2012**, pp. 1-16.

[6] Ruchjana B N, Borovkova S A, and Lopuhaa H P 2012, Least Squares Estimation of Generalized Space Time Autoregressive (GSTAR) Model and Its Properties, *The 5th International Conference on Research and Education in Mathematics*, pp. American Institute of Physics pp. 61-64.

[7] Suhartono, Wahyuningrum S R, Setiawan, and Akbar M S 2016 GSTARX-GLS Model fo Spatio-Temporal Data forecasting, *Malaysian Journal Of Mathematical Sciences*, **vol. 10(S)** , pp. 91-103.

[8] Setiawan, Suhartono, and Prastuti M. 2016 S-GSTAR-SUR Model for Seasonal Spatio Temporal Data Forecasting, *Malaysian Journal of Mathematical Sciences*, **vol. 10(S)**, pp. 53-65,.

[9] Liu L. M 1980 Analysis of Time Series with Calender Effects," *Management Science*, **Vol.26, No.1**.

[10] Cleveland W S and Devlin S J 1980 Calendar Effects in Monthly Time Series: Detection by Spectrum Analysis and Graphical Methods, *Journal of the American Statistical Association*, **vol. 75, no. 371**, pp. 487-496.

[11] Lee M H, Suhartono, and Hamzah N A 2010 Calender Variation Model Based on ARIMAX for Forecasting Sales Data with Ramadhan Effect, *Proceedings of The Regional Conference on Statistical Sciences 2010 (RCSS'10), Malaysia Institute of Statistics, Faculty of Computer and Mathematical Sciences, Universiti Teknologi MARA (UiTM), Malaysia*, pp. hal349-361.

[12] Wei W W 2006 *Time Series Analysis Univariate and Multivariate (Second Edition)*. USA: Pearson Education, Inc.,

Study of parameters of the nearest neighbour shared algorithm on clustering documents

Alvida Mustika Rukmi [1], Daryono Budi Utomo[2], Neni Imro'atus Sholikhah[3]
Mathematics Department
Institut Teknologi Sepuluh Nopember (ITS) Indonesia

E-mail: alvidamr@gmail.com[1], daryono@matematika.its.ac.id[2], nenisholikhah@gmail.com[3]

Abstract. Abstract. Document clustering is one way of automatically managing documents, extracting of document topics and fastly filtering information. Preprocess of clustering documents processed by textmining consists of: keyword extraction using Rapid Automatic Keyphrase Extraction (RAKE) and making the document as concept vector using Latent Semantic Analysis (LSA). Furthermore, the clustering process is done so that the documents with the similarity of the topic are in the same cluster, based on the preprocesing by textmining performed. Shared Nearest Neighbour (SNN) algorithm is a clustering method based on the number of "nearest neighbors" shared. The parameters in the SNN Algorithm consist of: k nearest neighbor documents, ε shared nearest neighbor documents and *MinT* minimum number of similar documents, which can form a cluster. Characteristics The SNN algorithm is based on shared 'neighbor' properties. Each cluster is formed by keywords that are shared by the documents. SNN algorithm allows a cluster can be built more than one keyword, if the value of the frequency of appearing keywords in document is also high. Determination of parameter values on SNN algorithm affects document clustering results. The higher parameter value k, will increase the number of neighbor documents from each document, cause similarity of neighboring documents are lower. The accuracy of each cluster is also low. The higher parameter value ε, caused each document catch only neighbor documents that have a high similarity to build a cluster. It also causes more unclassified documents (noise). The higher the *MinT* parameter value cause the number of clusters will decrease, since the number of similar documents can not form clusters if less than *MinT*. Parameter in the SNN Algorithm determine performance of clustering result and the amount of noise (unclustered documents). The Silhouette coeffisient shows almost the same result in many experiments , above 0.9, which means that SNN algorithm works well with different parameter values.

1. Introduction

Document clustering is to recognize clusters or groups of documents that have the same attribute value [1]. One of the goals of document clustering is to help search information quickly and accurately to document data. The development of information technology, requires a tool to enables representation of the information in accordance with user request.

Text Mining is part of the document clustering, and is the indexing process for database in documents searching. The Text Mining process includes: document keyword extraction and concept extraction. In this paper, the extraction of keyword documents using the Rapid Automatic Keyphrase Extraction (RAKE) method [2]. The RAKE method considers the word association by computing the matrix of blurring with one word against another. The matrix is used to measure the candidate's score and then ranking it [3]. Taking a keyword candidate using an induvidu document-based approach without relying on document collections, thus speeding up computational time. The process of keyword extraction using the RAKE method has 5 main stages: keyword candidate extraction, calculating the co-ocurrence matrix, calculating the value of the ratio, calculating the basic feature value, and selecting the keyword with the highest feature value. Extraction concept using Latent Semantic Analysis method. The process of concept vector formation with LSA method has two stages: construct a weighted matrix of tf-idf and apply Singular Value Decomposition (SVD) method to the matrix formed. The LSA is a method for finding linkages and similarities between documents, fragments and words that appear in documents [4]. Weighted tf-idf performs calculations that describe how important the term (term) in a document. The SVD method is used to minimize the matrix dimension [5]. The result of the SVD decomposition process will give birth to a vector of document concepts.

The clustering algorithm do how to construct clusters of equal size and density, how to cope with noise and how to determine the number of clusters so that the clustering results presenting the members of a cluster bear a high resemblance compared to the different cluster members.

The clustering constraint is generally if the data sets are in large dimensions, the accuracy of the clustering results will decrease, because the similarity between the data objects is higher, making it more difficult for the clustering process [6]. Clustering with the SNN algorithm overcomes it, by determining the nearest residence of all data objects, then the new equality values between the data objects are determined by the number of shared states. The disadvantage of the SNN algorithm is having to specify the tagar parameter value resulting in the desired clustering [7].

In a study conducted by Adriano Moreira, Maribel Y. Santos and Sofia Carneiro stated that the SNN clustering results gave better results for certain datasets than the k-Means algorithm [8]. SNN algorithm requires 3 input parameters as minimum threshold value to build cluster, ie k number of nearest neighbor; ε the number of shared neighbors, and MinT the number of cluster members. After specifying the input parameters, the dokume cluster finds the nearest neighbor of each document. Then the similarity between the two documents is calculated based on the nearest neighbor ε together. By using a similarity measure, the density of each document is calculated as the number of shared neighbors of two documents. Next, the documents are classified as core points, if the density of the document is greater than the core point threshold [9]. Giving optimal parameter values in clustering, capable of handling data in various sizes, controlling cluster quality, and able to handle noise.

Document clustering is to recognize clusters or groups of documents that have the same attribute value [1]. One of the benefits of clustering documents is to help search information quickly and accurately to document data. The development of the world of information, requires a tool capable of presenting the search results information in accordance with user demand.

Text Mining is part of the document clustering, and is the indexing process for database searches in the form of documents. The Text Mining process includes: document keyword extraction and concept extraction.

In this paper the extraction of keyword documents using the Rapid Automatic Keyphrase Extraxction (RAKE) method [2]. The RAKE method considers the word association by computing the matrix of blurring with one word against another. The matrix is used to measure the candidate's score and then ranking it [3]. Taking a keyword candidate using an induvidu document based approach is independent of document collections, thereby speeding up computational time. The process of keyword extraction using the RAKE method has 5 main stages: keyword candidate extraction, calculating the co-ocurrence matrix, calculating the value of the ratio, calculating the basic feature value, and selecting the keyword with the highest feature value.

While the extraction concept using Latent Semantic Analysis method. The process of concept vector formation with LSA method has two stages: construct a weighted matrix of tf-idf and apply Singular Value Decomposition (SVD) method to the matrix formed. The LSA is a method for finding linkages and similarities between documents, fragments and words that appear in documents

[4]. Weighted tf-idf performs calculations that describe how important the term (term) in a document. The SVD method is used to minimize the matrix dimension [5].
The clustering algorithm is often confronted with how to construct clusters of equal size and density, how to cope with noise and how to determine the number of clusters so that the clustering results presenting the members of a cluster bear a high resemblance compared to the different cluster members.

The clustering constraint is generally if the data sets are in large dimensions, the accuracy of the clustering results will decrease, because the similarity between the data objects is higher, making it more difficult for the clustering process [6]. Clustering with the SNN algorithm overcomes it, by determining the nearest residence of all data objects, then the new equality values between the data objects are determined by the number of shared states. The disadvantage of the SNN algorithm is having to specify the tagar parameter value resulting in the desired clustering [7].
In a study conducted by Adriano Moreira, Maribel Y. Santos and Sofia Carneiro stated that the SNN clustering results gave better results for certain datasets than the K-Means algorithm [8]. SNN algorithm requires 3 input parameters to construct clusters, ie k number of nearest neighbors, ε density threshold (number of shared neighbors), and $MinT$ number of core points (cluster members). After specifying the input parameters, the dokume cluster finds the nearest neighbor of each document K. Then the similarity between two documents is calculated based on the number of nearest neighbors with it. By using a similarity measure, the density of each document is calculated as the number of shared neighbors of two documents. Next, the documents are classified as core points, if the density of the document is greater than the core point threshold [9]. This study examines the optimal parameter values in clustering that is capable of handling data in various sizes, controlling cluster quality, and able to handle noise.

2. Discussion

2.1 Document keyword extraction using rapid automatic keyphrase extraxction (RAKE) method

The experiment uses 525 documents. Clustering the document uses the research title as input on RAKE. The RAKE method has the following stages:

1. Candidate keyword extraction.
   The extraction of a keyword candidate begins by splitting the text using stopword and punctuation. Suppose that is the i th document of the data. After the document is extracted, we get the keyword candidate $Td_i = \{t_1, t_2, t_3, ..., t_n\}$, where $t_1, t_2, t_3, ..., t_n$ be words, or phrases of keyword candidates
2. Build matrix
   After the keyword candidate is obtained, the next step is to calculate the co-ocurrence matrix. The co-ocurrence matrix represent the frequency of occurrence of a word and keyword phrase
3. Calculating ratios.
   Ratio value is the ratio between word degree and word frequency. The degree of a word is the number of word occurrences in the document plus the number of phrases that contain the word. The degree of the word on the co-ocurrence matrix is obtained from the sum of one column or one line. The word frequency is the number of words in the text. The frequency value can be obtained at the diagonal value of the co-ocurrence matrix. The ratio can be calculated using the following Equation 2.1:

$$rasio = \frac{deg(w)}{freq(w)} \quad (2.1)$$

   where : $deg(w)$ = degree of word
   $freq(w)$ = frequency of word
4. Calculating the value of basic features.
   The basic feature value is the sum of the existing keyword ratios in the keyword candidate. The value is then sorted by the highest value to the lowest. The basic feature values are calculated using the following Equation 2.2

$$Nfd = ratio\,(t) + ratio\,(ft) \quad (2.2)$$

where : *Nfd* = value of basic features
ratio (t) = keyword ratio of *t*
ratio (ft) = phrase ratio containing the keyword of *t*.

2.2 Concept extraction using latent semantic analysis (LSA) method
The steps of concept vector formation are as follows:

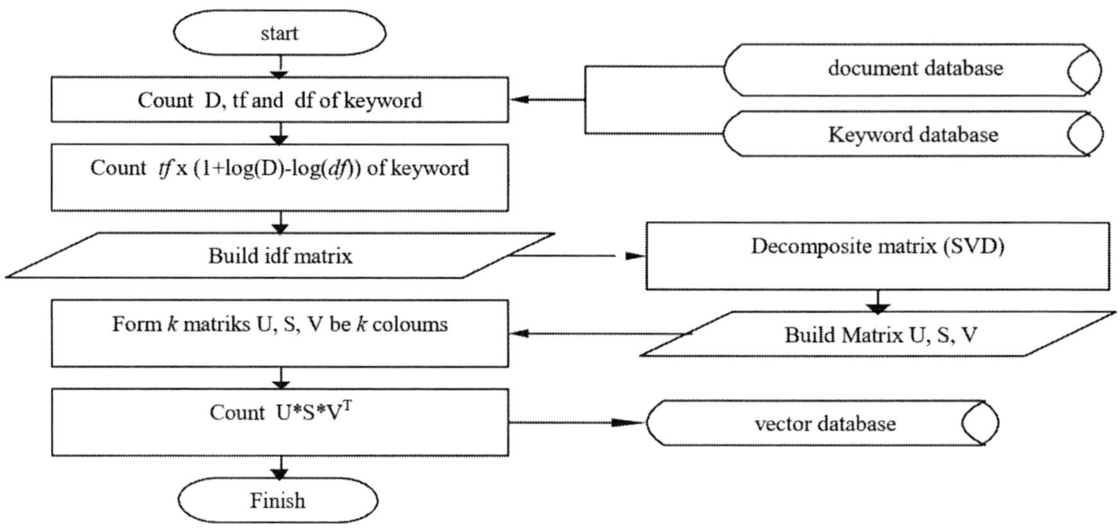

Figure 1. Building vector process.

Term Weighting aims to determine the weight of each term. The calculation of term weight requires two things: Term Frequency (*tf*) and Inverse Document Frequency (*idf*). Term Frequency (*tf*) is the frequency of occurrence of a term (term) in a document. The value of *tf* varies across each document depending on the appearance of a word in a document. The value of *tf* is proportional to the rate of occurrence of the term in the document. The more often the term appears on a document, the greater the value of *tf* in the document and the less the term appears the smaller the value of *tf*. In addition to Term Frequency is also required Inverse Document Frequency (*idf*) on the weighting term. Inverse Document Frequency (*idf*) is the frequency of occurrence of term in the whole document. The Equation 2.3 is calculate the value of *tf-idf* :

$$W_{i,j} = tf_{i,j} \times idf_j = tf_{i,j} \times \log\left(\frac{N}{df_j}\right) \quad (2.3)$$

$$idf = \log\frac{N}{df} \quad (2.4)$$

where :
$D_1,.., D_n$ = document
$W_{i,j}$ = weight i^{th}-*term* to j^{th}-document
$tf_{i,j}$ = number of occurrences of ith-term into j^{th}-document
N = total number of documents
df_j = number of documents containing term-i
idf_j = *Inverse document frequency* contain term-i
SVD will decomposite a matrix into three new matrices of left singular vector matrix, singular value matrix and right singular vector. The SVD of an mxn dimension, then matrix A is as follows

$$:A_{m\times d} = U_{m\times n} S_{n\times n} V^T_{n\times d} \quad (2.5)$$

where :
A = Matrix by mxd dimension

U = left singular vector matrix by mxn dimension
S = singular value matrix by nxn dimension
V = right singular vector dxn dimension

2.3 Clustering by shared nearest neighbor (SNN) algorithm

In this process the extracted concepts of the document are grouped into clusters representing a topic. This clustering uses the SNN method. At this stage insert the input parameter of *k* list of neighbor, ε of radius of neighbor region Nε (p) from a data point p and *MinT* minimum number of interior points in region Nε (p). The following is a drawing of the document cluster flow diagram with the SNN algorithm:

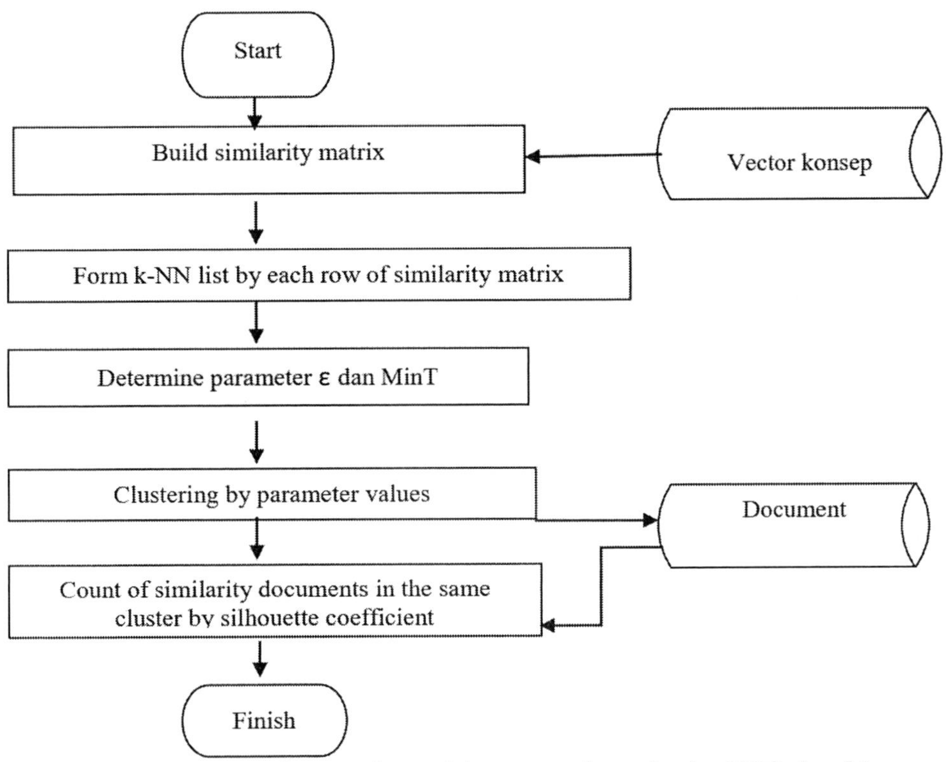

Figure 2. Diagram flow of document clustering by SNN algorithm.

Then, carry out of the result praprocess text mining using by RAKE and LSA and then simulate using parameter in SNN Algorithm to cluster documents. So, characteristic of parameter in SNN algorithm will be study. The experiment were descripted below :

2.3.1 Determine K = 3

Table 1. Document clustering by k=3.

ε	MinT	Total Cluster	Doc Clustered	Doc Noise	Procentage Noise/Total doc	Coefficient Silhoute
2	2	116	413	112	21,3%	0.987840123
2	3	63	307	218	41,5%	0.98666718495585
2	4	41	241	284	45,9%	0.986599355602242
3	2	35	79	446	84,9%	0.9820522612131688
3	3	9	27	498	94,8%	0.9808678419862584

2.3.2 Determine K = 5

Table 2. Document clustering by k=5.

ε	MinT	Total Cluster	Doc Clustered	Doc Noise	Procentage Noise/Total doc	Coefficient Silhoute
2	2	10	514	11	2,09%	0.98926368
2	3	3	500	25	4,76%	0.988637763
2	4	3	500	25	4,76%	0.988637763
3	2	82	342	183	34,8%	0.987840123
3	3	45	268	257	48,9%	0.986534609269849
3	4	29	220	305	58,1%	0.986074886

2.3.3 Determine K = 10

Table 3. Document clustering by k=10.

ε	MinT	Total Cluster	Doc Clustered	Doc Noise	Procentage Noise/Total doc	Coeficient Silhoute
2	2	1	524	1	0,19%	0.991300821
2	4	1	524	1	0,19%	0.991300821
2	6	1	524	1	0,19%	0.991300821
7	2	47	174	351	66,8%	0.985487083
7	4	14	99	426	81,1%	0.985088194
7	6	7	70	456	86,8%	0.9810475153652917

2.3.4 Determine K = 15

Table 4. Document clustering by k=15

ε	MinT	Total Cluster	Doc Clustered	Doc Noise	Procentage Noise/Total doc	Coeficient Silhoute
3	2	1	523	2	0,38%	0.990178031
3	4	1	523	2	0,38%	0.990178031
3	6	1	523	2	0,38%	0.990178031
10	2	46	186	339	64,5%	0.986044764
10	4	14	116	409	77,9%	0.985424659
10	7	8	88	437	83,2%	0.984802256
12	2	21	82	443	84,3%	0.984374682802898
12	6	3	34	491	93,5%	0.98130261110419
12	12	1	15	510	97,1%	0.9804164843922515

The value of k must be less than or equal to ε value. Based on the experimental results, the parameter value of k on documents clustering does not always result in a large number of cluster, depending on the ε and *MinT* values.

Determine k value above 10, if the value of ε is small, tends to produce a single cluster, even any *MinT* value has no effect. While the determination of the small value of k (less than 1% of the amount of data) tends to represent more large number of clustered documents than value of k >1% of the amount of data. When ε approach to k value, it will increase number of clusters according by decreasing of *MinT* value. Giving a large value of ε, will form clusters that contain documents with high topic similarities. This results in each cluster having a high accuracy value because the relevance of the document to the keyword is also high, so that only a few irrelevant documents appear on each cluster indicate a minimum threshold of the number of documents in a cluster. Based on the experiment show that greater *MinT* value will form t less the number of clusters.

The noise document is an unclassified document. The percentage noise shows the percentage ratio of unclassified documents against the total number of documents.This percentage increases when the difference in k value with ε value reduce, moreover *MinT* increases. In the experiments show that smaller ε will give the result smaller percentage of noise. Small noise percentage value means to show more documents that can be clustered .

Silhouette coefficient refers to a method of interpretation and validation of consistency within clusters of data.The silhouette value is a measure of how similar an object is to its own cluster (cohesion) compared to other clusters (separation). The silhouette ranges from −1 to +1, where a high value indicates that the object is well matched to its own cluster and poorly matched to neighboring clusters. If most objects have a high value, then the clustering configuration is appropriate. If many points have a low or negative value, then the clustering configuration may have too many or too few clusters.The Silhouette coefficient shows almost the same result in many experiments , above 0.9, which means that SNN algorithm works well with different parameter values.

3. Conclusion

Here are the conclusions obtained during the clustering process with the SNN algorithm:

1. Each cluster is formed by keywords that are shared by documents in a cluster. SNN algorithm allows a cluster can be built more than one keyword, if the value of the frequency of along with keywords is also high. Thus, a cluster is formed based on the keywords used together by the documents in there.
2. Determination of parameter values on the SNN algorithm will affect the results of document clustering, in the form of:
   a. The value of k must be less than or equal to ε value so it form clustering. The higher k value will increase the number of nearest neighbor documents for each document. It can decrease similarity of neighboring documents. The clustering results show that the low similarity values between documents will be in one cluster.
   b. High ε value cause more noise, but among documents in one cluster show a high similarity. The high ε value results in the document only 'pulling' neighboring documents have a high similarity when forming a cluster.
   c. The *MinT* value limits the number of clusters based on the minimal number of documents that can form a cluster. If the *MinT* value is high, the number of clusters will decrease, since the number of documents less than MinT can not form a cluster.

Determination value of this parameter depends on requirement of a problem. In the document clustering requires a low k value, a high ε value (k> ε)and low *MinT* value. It can represent a good clustering.

References

[1] Peng, Y., Kou, G., Chen, Z., Shi, Y., 2006, Recent Trends in Data Mining (DM) :Document Clustering of DM Publications, *Int'l Conference on Service Systems and Service Management*, vol. 2, pp.
[2] Berry, M.W., dan Kogan, J., 2010, Text Mining : Applications and Theory, *John Willey & Sons.*
[3] Lee, D.L., 1997, Document Ranking and the Vector-Space Model. *Hong Kong University of Science and Technology.*
[4] Luthfiarta, A., Zeniarja, J., Salam, A., 2013, Integrasi Peringkas Dokumen Otomatis Dengan

Latent Semantic Analysi(LSA) pada Peringkas Dokumen Otomatis untuk Proses Clustering Dokumen, *Seminar nasional teknologi informasi & komunikasi terapan.*

[5] Salton G., dan Buckley C., 1998, Term Weighting Approaches in Automatic Text Retrieval,*Information Processing and Management*, vol.24, no. 5, hal :513-523.

[6] Heidelberg, 2005, High Dimensional Shared Narest Neighbor Clustering Algorithm", Lecture Notes on Compuer Science, Publisher Springer Berlin vol.3614, Hal. 494-50

[7] Ertoz L., Steinbach M., dan Kumar V.2002, Finding topics in document, a Shared Nearest Neighbor Approach Clustering and Information Retrieval, *Kluwer Academic Publisher.*

[8] Moreira Adriano, Santos Maribel Y and Carneiro Sofia, 2005, Density-based clustering algorithms – DBSCAN and SNN, *University of Minho – Portugal.*

[9] Jarvis R.A. dan Patrick E. 1973, Clustering Using a Similarity Measure based on Shared Nearest Neighbor, *Proceeding IEEE Transaction on Computer*, vol C-22, hal. 1025-1034.

[10] Rukmi, A.M., Irawan M.I, Hidayat, N., 2009, Pengklasteran Data Kategoris dengan Algoritma Shared Nearest Neighbor, *Institut Teknologi Sepuluh Nopember*,Vol. 6, No. 1

IOP Conf. Series: Journal of Physics: Conf. Series **974** (2018) 012062 doi:10.1088/1742-6596/974/1/012062

Elementary properties of triangle in normed spaces

Deri Triana[1,a] **and Mahmud Yunus**[1,b]
[1]Mathematics Department, Sepuluh Nopember Institute of Technology Surabaya, Indonesia
E-mail: [a]deritriana85@gmail.com and [b]yunusm@matematika.its.ac.id

Abstract. Based on concepts of trigonometric on plane, In this paper we generalized those concept in normed spaces. About the orthogonality concept between two vectors already well known, we are interested to develop elementary properties of triangle, especially the properties of its angle. We propose a non-linear (Wilson) functional to define an angle and explore its properties.

1. Introduction

Before we discuss about triangle in normed spaces, we introduce definition Wilson functional, then we will use it to define an angle as a tool to explore elementary properties of triangle in normed spaces.

Suppose V is a real linear space, and in arbitrary normed spaces $V := (V, \|\cdot\|)$, following [8, 3], we define (Wilson) non-linear functional $\langle \cdot, \cdot \rangle_\bullet : V \times V \to \mathbf{R}$, by

$$\langle x, y \rangle_\bullet := \frac{\|x\|^2 + \|y\|^2 - \|x-y\|^2}{2},$$

for every $x, y \in V$.

For some simple properties of our functional above, see [3, p. 26], and our first result is Cauchy-Schwartz inequality as follow.

Theorem 1.1 *For every $x, y \in V$, we have the following*

$$|\langle x, y \rangle_\bullet| \leq \|x\| \cdot \|y\|.$$

Proof. Suppose $x, y \in V$. Since $\big|\|x\| - \|y\|\big| \leq \|x - y\|$, then

$$\big|\|x\| - \|y\|\big|^2 \leq \|x-y\|^2 \Leftrightarrow \|x\|^2 - 2\|x\| \cdot \|y\| + \|y\|^2 \leq \|x-y\|^2 \Leftrightarrow \langle x, y \rangle_\bullet \leq \|x\| \cdot \|y\|.$$

On the other hand, since $\|x - y\| \leq \|x\| + \|y\|$, then we will obtain

$$\|x-y\|^2 \leq \big(\|x\| + \|y\|\big)^2 \Leftrightarrow \|x-y\|^2 - \|x\|^2 - \|y\|^2 \leq 2\|x\| \cdot \|y\| \Leftrightarrow -\langle x, y \rangle_\bullet \leq \|x\| \cdot \|y\|,$$

and this complete the proof. ■

Note that, our functional $\langle \cdot, \cdot \rangle_\bullet$ is continuous. That is, for every $x, y, z \in V$, we have

$$|\langle x, y \rangle_\bullet - \langle x, z \rangle_\bullet| \leq K_1 \|y - z\|,$$

where $K_1 = K_1(x, y, z) := \|x\| + \|y\| + \|z\|$. Moreover, we always have

IOP Conf. Series: Journal of Physics: Conf. Series **974** (2018) 012062 doi:10.1088/1742-6596/974/1/012062

Theorem 1.2 *For every $x, y, w, z \in V$, we have the following*

$$|\langle x, y\rangle_\bullet - \langle w, z\rangle_\bullet| \le K_2[\|x - w\| + \|y - z\|],$$

with $K_2 = K_2(x, y, w, z) := \|x\| + \|y\| + \|w\| + \|z\|$.

2. Elementary properties of $V_\bullet$

From the above observations, we define an angle between $x, y \in V$, denote by $\angle_\bullet(x, y)$ (see [3, 4, 8], for more informations about this notions, and their elementary properties), by

$$\cos\angle_\bullet(x, y) := \frac{\langle x, y\rangle_\bullet}{\|x\|\|y\|}.$$

In this case we only consider $0 \le \angle_\bullet(x, y) \le \pi$.

For any normed spaces $V := (V, \|\cdot\|)$, we will introduce the notations

$$V_\bullet := (V, \|\cdot\|, \angle_\bullet).$$

In $V_\bullet$, we define the orthogonality between two nonzero vectors as follows. Two nonzero vectors $x, y \in V_\bullet$ are orthogonal, written $x \perp_\bullet y$, if $\langle x, y\rangle_\bullet = 0$. The following is easy, and we omit the detailed proof.

Theorem 2.1 *For every $x, y \in V_\bullet$, we have the following*

(i). $x \perp_\bullet y \;\Leftrightarrow\; x \perp_\bullet (-y) \;\Leftrightarrow\; y \perp_\bullet x \;\Leftrightarrow\; (-x) \perp_\bullet y \;\Leftrightarrow\; (-x) \perp_\bullet (-y)$,

(ii). *If $\{x, y\}$ are linearly dependent, then $\angle_\bullet(x, y) = 0$, or $\angle_\bullet(x, y) = \pi$.*

Proof.

(i). The first equivalences is easy, since for every $x, y \in V_\bullet$, we have

$$x \perp_\bullet y \;\Leftrightarrow\; \|x - y\|^2 = \|x\|^2 + \|y\|^2.$$

(ii). Suppose $y = kx$, for some $k \in \mathbf{R} \setminus \{0\}$. Then

$$\cos\angle_\bullet(x, y) = \frac{\langle x, y\rangle_\bullet}{\|x\|\|y\|} = \frac{\langle x, kx\rangle_\bullet}{\|x\|\|kx\|} = \frac{k}{|k|}. \quad \blacksquare$$

The converse of statement (ii) in Theorem 2.1 is not true, when we consider a sequence spaces $\ell^1(\mathbf{R})$, with its members $x := \langle 1, 1, 2, 0, 0, \ldots\rangle$, and $y := \langle 1, 1, 0, 0, \ldots\rangle$, for $x, y \in \ell^1(\mathbf{R})$.

The following result is a geometrical interpretation of "*parallelogram law*" in any normed spaces $V_\bullet$.

Theorem 2.2 *For every $a, b \in V_\bullet$, we have the following*

$$\angle_\bullet(a, b) + \angle_\bullet(a, -b) = \pi \;\Leftrightarrow\; \|a + b\|^2 + \|a - b\|^2 = 2(\|a\|^2 + \|b\|^2).$$

Proof. Its proof based on the following two keys observations. That is,

$$\angle_\bullet(a, b) + \angle_\bullet(a, -b) = \pi \;\Leftrightarrow\; \cos\angle_\bullet(a, b) + \cos\angle_\bullet(a, -b) = 0, \quad \text{and}$$

$$\cos\angle_\bullet(a, b) + \cos\angle_\bullet(a, -b) = \frac{2\|a\|^2 + 2\|b\|^2 - \|a - b\|^2 - \|a + b\|^2}{2\|a\|\|b\|},$$

and this complete the proof. ■

3. Triangle in $V_\bullet$

For every $a, b, c \in V_\bullet \setminus \{0\}$, we introduce notation $\Delta[a,b,c]$ as a set $\{a,b,c\}$, such that $a+c=b$, and for $\Delta[a,b,c]$ we also consider the following $\angle_\bullet(a,b)$, $\angle_\bullet(c,-a)$, and $\angle_\bullet(b,c)$.

From the above informations, we will have the following

$$\begin{cases} \|a\|^2 = \|b\|^2 + \|c\|^2 - 2\|b\| \cdot \|c\| \cos \angle_\bullet(b,c), \\ \|b\|^2 = \|a\|^2 + \|c\|^2 - 2\|a\| \cdot \|c\| \cos \angle_\bullet(c,-a), \\ \|c\|^2 = \|a\|^2 + \|b\|^2 - 2\|a\| \cdot \|b\| \cos \angle_\bullet(a,b), \end{cases}$$

which we called *Cosine Rule* for $\Delta[a,b,c]$.

With $K := 2\sqrt{s(s-\|a\|)(s-\|b\|)(s-\|c\|)}$, and $2s := \|a\| + \|b\| + \|c\|$, then we will have *Sine Rule* as folows.

$$\|b\|\|c\| \cdot \sin \angle_\bullet(b,c) = \|a\|\|c\| \cdot \sin \angle_\bullet(c,-a) = \|a\|\|b\| \cdot \sin \angle_\bullet(a,b) = K.$$

In $\Delta[a,b,c]$, it is easy to see that our cosine rule is equivalent with the following *Triangle Sides Formula.*

$$\begin{cases} \|a\| = \|b\| \cos \angle_\bullet(a,b) + \|c\| \cos \angle_\bullet(c,-a), \\ \|b\| = \|a\| \cos \angle_\bullet(a,b) + \|c\| \cos \angle_\bullet(b,c), \\ \|c\| = \|a\| \cos \angle_\bullet(c,-a) + \|b\| \cos \angle_\bullet(b,c). \end{cases}$$

Theorem 3.1 *In every $\Delta[a,b,c]$, cosine rule is equivalent with triangle sides formula, and we can derive sine rule from cosine rule.*

4. Elementary properties of angle in $V_\bullet$

In this section we will discuss about the sum of angles in triangle. We start with some special triangle, that is, equilateral triangle.

Suppose in $\Delta[a,b,c]$ we have $\|a\| = \|b\| = \|c\|$, its mean that

$$\cos \angle_\bullet(b,c) := \frac{\|b\|^2 + \|c\|^2 - \|a\|^2}{2\|b\|\|c\|} = \frac{1}{2}.$$

So we will have $\angle_\bullet(b,c) = \angle_\bullet(c,-a) = \angle_\bullet(a,b) = \frac{\pi}{3}$, and for equilateral $\Delta[a,b,c]$, we have

$$\angle_\bullet(a,b) + \angle_\bullet(b,c) + \angle_\bullet(c,-a) = \pi.$$

Now we move to right triangle.

Suppose $\cos \angle_\bullet(b,c) = 0$ or $\angle_\bullet(b,c) = \frac{\pi}{2}$, then $\|a\|^2 = \|b\|^2 + \|c\|^2$. It is easy to see that

$$\cos \angle_\bullet(c,-a) := \frac{\|a\|^2 + \|c\|^2 - \|b\|^2}{2\|a\|\|c\|} = \frac{2\|c\|^2}{2\|a\|\|c\|} = \frac{\|c\|}{\|a\|},$$

and

$$\cos \angle_\bullet(a,b) := \frac{\|a\|^2 + \|b\|^2 - \|c\|^2}{2\|a\|\|b\|} = \frac{2\|b\|^2}{2\|a\|\|b\|} = \frac{\|b\|}{\|a\|}.$$

Since $\|a\|^2 = \|b\|^2 + \|c\|^2$, then $\cos^2 \angle_\bullet(c,-a) + \cos^2 \angle_\bullet(a,b) = 1$ or $\cos^2 \angle_\bullet(c,-a) = 1 - \cos^2 \angle_\bullet(a,b) = \sin^2 \angle_\bullet(a,b)$.

Since $\cos\angle_\bullet(c,-a) = -\sin\angle_\bullet(a,b)$ will arrive to a contradiction, then we will conclude that

$$\cos\angle_\bullet(c,-a) = \sin\angle_\bullet(a,b), \quad \text{and} \quad \angle_\bullet(c,-a) = \pi/2 - \angle_\bullet(a,b).$$

For right $\Delta[a,b,c]$, we will have

$$\angle_\bullet(b,c) + (\angle_\bullet(a,b) + \angle_\bullet(c,-a)) = \frac{\pi}{2} + \frac{\pi}{2} = \pi.$$

Our result below is important for our $\Delta[a,b,c]$, who serve as a generalization of the above results.

Theorem 4.1 *In every* $\Delta[a,b,c]$, *we will have* $\angle_\bullet(a,b) + \angle_\bullet(b,c) + \angle_\bullet(c,-a) = \pi$.

Proof. We see that our conclusion is equivalent with the following

$$\begin{aligned} &\angle_\bullet(a,b) + \angle_\bullet(b,c) + \angle_\bullet(c,-a) = \pi \;\Leftrightarrow\; \cos(\angle_\bullet(a,b) + \angle_\bullet(b,c)) = \cos(\pi - \angle_\bullet(c,-a)) \\ &\Leftrightarrow\; \cos\angle_\bullet(a,b)\cos\angle_\bullet(b,c) - \sin\angle_\bullet(a,b)\sin\angle_\bullet(b,c) = -\cos\angle_\bullet(c,-a). \end{aligned}$$

Next, we choose $x, y > 0$, such that $\|a\| = x\|c\|$, and $\|b\| = y\|c\|$.
It means that,

$$\begin{cases} x^2 = y^2 + 1 - 2y\cos\angle_\bullet(b,c), \\ y^2 = x^2 + 1 - 2x\cos\angle_\bullet(c,-a), \\ 1 = x^2 + y^2 - 2xy\cos\angle_\bullet(a,b), \end{cases}$$

that serve as a cosine rule in its simpler form.
After some several simplifications,

$$\begin{aligned} &\cos\angle_\bullet(a,b)\cos\angle_\bullet(b,c) - \sin\angle_\bullet(a,b)\sin\angle_\bullet(b,c) + \cos\angle_\bullet(c,-a) \\ &= \frac{x^2+y^2-1}{2xy}\cdot\frac{y^2+1-x^2}{2y} + \frac{x^2+1-y^2}{2x} - \frac{K^2}{\|a\|\|b\|^2\|c\|} \\ &= \frac{y^4-(1-x^2)^2}{4xy^2} + \frac{x^2+1-y^2}{2x} + \frac{[1-(x+y)^2][1-(x-y)^2]}{4xy^2} \\ &= \frac{[1-(x+y)^2][1-(x-y)^2] + y^4 - (1-x^2)^2 + 2y^2(x^2+1-y^2)}{4xy^2} \\ &= 0, \end{aligned}$$

and this completes the proof. ■
From the above elementary result, we have the following. That is, we prove that cosine rule is equivalent with sine rule.

Theorem 4.2 *We can derive cosine rule from sine rule.*

Proof. From $\angle_\bullet(a,b) + \angle_\bullet(b,c) + \angle_\bullet(c,-a) = \pi$, we have

$$\sin\angle_\bullet(b,c) = \sin(\angle_\bullet(a,b) + \angle_\bullet(c,-a)), \quad \cos\angle_\bullet(b,c) = -\cos(\angle_\bullet(a,b) + \angle_\bullet(c,-a)).$$

And so,

$$\begin{aligned} &K^2(\|b\|^2 + \|c\|^2 - \|a\|^2) = \|a\|^2\|b\|^2\|c\|^2[\sin^2\angle_\bullet(c,-a) + \sin^2\angle_\bullet(a,b) - \sin^2\angle_\bullet(b,c)] \\ &= -2\|a\|^2\|b\|^2\|c\|^2\sin\angle_\bullet(a,b)\cdot\sin\angle_\bullet(c,-a)\cdot\cos\angle_\bullet(b,c) \\ &= -2K^2\|b\|\|c\|\cos\angle_\bullet(b,c), \end{aligned}$$

and this completes the proof. ■
From Theorem 3.1 and Theorem 4.1, we have the result as follow.

IOP Conf. Series: Journal of Physics: Conf. Series **974** (2018) 012062 doi:10.1088/1742-6596/974/1/012062

Theorem 4.3 *In every $\Delta[a,b,c]$, the following statements are equivalent*

(a). *Cosine rule*

(b). *Sine Rule*

(c). *Triangle sides formula*

5. Further results

Following [7, 8], we define, for any $a, b \in V_\bullet$,

$$\mathcal{S}_\bullet(a,b) := \{\lambda a + (1-\lambda)b : 0 \le \lambda \le 1\},$$

the line segment joining a and b. We also define its length by $|\mathcal{S}_\bullet(a,b)| := \|b-a\|$.

We denote $[b;a;c]_\bullet$, the angle between $\mathcal{S}_\bullet(a,b)$ and $\mathcal{S}_\bullet(a,c)$, which defined by

$$\cos[b;a;c]_\bullet := \frac{\|b-a\|^2 + \|c-a\|^2 - \|b-c\|^2}{2\|b-a\|\|c-a\|}.$$

Of course, we only consider that $0 \le [b;a;c] \le \pi$.

If we consider the notion of betweenness coined by Pasch in 1882 (see [2] for details), then for every $a,b,c \in V_\bullet$, the following result say that a is between b, and c, if and only if $[b;a;c]_\bullet = \pi$.

In particular, 0 is always between b, and $-b$, for every $b \in V_\bullet \setminus \{0\}$.

Theorem 5.1 *If $b \in \mathcal{S}_\bullet(a,c)$, or $c \in \mathcal{S}_\bullet(a,b)$, then $[b;a;c]_\bullet = 0$. Moreover, we also have $[b;a;c]_\bullet = \pi$, if and only if $\|b-a\| + \|c-a\| = \|b-c\|$. In particular, we always have $[b;0;-b]_\bullet = \pi$.*

Proof. Suppose $b := \lambda a + (1-\lambda)c$, for some $0 < \lambda < 1$. Then

$$\frac{\langle b-a, c-a\rangle_\bullet}{\|b-a\|\|c-a\|} = (1-\lambda)\cdot\frac{\|c-a\|}{\|b-a\|} = 1.$$

The second part of the theorem follow from the equivalence below,

$$\begin{aligned}\|a-c\| + \|a-b\| = \|c-b\| &\Leftrightarrow -1 = \frac{\|b-a\|^2 + \|c-a\|^2 - \|b-c\|^2}{2\|b-a\|\|c-a\|}\\ &\Leftrightarrow [b;a;c]_\bullet = \pi,\end{aligned}$$

and this complete the proof. ■

6. Concluding remark

From the result of this study, we still have another opportunity to extend the above results, but now with the different definition of angle.

Even in any normed spaces $V := (V, \|\cdot\|)$, we still can define another form of angle between $a, b \in V$, by the definition

$$\cos\angle_I(a,b) := \frac{\|a+b\|^2 - \|a-b\|^2}{4\|a\|\|b\|}.$$

From the above definition of angle, we will have another form of cosine rule for $\angle_I$,

$$\begin{cases}\|a\|^2 = \|b+c\|^2 - 4\|b\|\|c\|\cos\angle_I(b,c),\\ \|b\|^2 = \|c-a\|^2 - 4\|c\|\|a\|\cos\angle_I(c,-a),\\ \|c\|^2 = \|a+b\|^2 - 4\|a\|\|b\|\cos\angle_I(a,b).\end{cases}$$

7. Acknowledgement

There are many people that have assisted and supported me during the preparation of this paper, and I would like to express my gratitude to them. First of all, my supervisor Dr. Mahmud Yunus, M.Si who supervise me that make this paper as a part of my Thesis.

Special thanks goes to my family and my friends in the Mathematics Department of ITS, for their supported.

References

[1] W. Benz, *Classical Geometries in Modern Contexts, Geometry of Real Inner Product Spaces*, Birkhauser Springer, Heidelberg, 2012.

[2] L.M. Blumenthal, *Theory and Applications of Distance Geometry*, Chelsea Publishing Company, Bronx New York, 1970.

[3] H. Gunawan, J. Lindiarni, and O. Neswan, "P-,I-,g-, and D-Angles in Normed Spaces", *ITB J. Sci.* **40A** (1) (2008), 24–32.

[4] L.M. Kelly (ed.), *The Geometry of Metric and Linear Spaces*, Lecture Notes in Mathematics, Springer Verlag, New York, 1975.

[5] S.L. Loney, *Plane Trigonometry*, Cambridge University Press, Cambridge, 1893.

[6] J.R. Partington, "Orthogonality in Normed Spaces", *Bull. Austral. Math. Soc.* **33** (1986), 449–455.

[7] M. Searcoid, *Metric Spaces*, Springer Undergraduate Mathematics Series, Springer Verlag London Limited, 2007.

[8] J.E. Valentine, S.G. Wayment, "Wilson Angles in Linear Normed Spaces", *Pacific J. Math.* **36** (1) (1971), 239–243.

IOP Conf. Series: Journal of Physics: Conf. Series **974** (2018) 012063 doi:10.1088/1742-6596/974/1/012063

Application of image processing to calculate the number of fish seeds using raspberry-pi

A Rahmadiansah[1,5], A Kusumawardhani[2], F N Duanto[3] and F Qoonita[4]

[1,2,3,4] Engineering Physics Departments, Sepuluh Nopember Institute of Technology, Jalan Raya ITS, Sukolilo, Surabaya, Indonesia
[5] Center of Excellence Science and Technology Automotive Systems and Controls, Sepuluh Nopember Institute of Technology, Jalan Raya ITS, Sukolilo, Surabaya, Indonesia

E-mail: andi.rahmadiansah@gmail.com

Abstract. Many fish cultivator in Indonesia who suffered losses due to the sale and purchase of fish seeds did not match the agreed amount. The loss is due to the calculation of fish seed still using manual method. To overcome these problems, then in this study designed fish counting system automatically and real-time fish using the image processing based on Raspberry Pi. Used image processing because it can calculate moving objects and eliminate noise. Image processing method used to calculate moving object is virtual loop detector or virtual detector method and the approach used is "double difference image". The "double difference" approach uses information from the previous frame and the next frame to estimate the shape and position of the object. Using these methods and approaches, the results obtained were quite good with an average error of 1.0% for 300 individuals in a test with a virtual detector width of 96 pixels and a slope of 1 degree test plane.

1. Introduction

Many fish farmers in Indonesia who suffered losses as the result from the sale and purchase of fish seeds which does not correspond to the amount agreed. Losses occur due to calculate fish seed, count manually. It is by counting the fish seeds one by one, the possibility of errors is very large such as human error. Maybe the amount of the loss is not great if the amount of the purchased fish seed is not too much. But if the number of fish seed purchased in great numbers, the losses will greatly impact. In addition, the manual counting methods takes quite a long time, It takes approximately 1 hour to the sorting and counting of 500 fish seeds with a size of 1-3 cm,While once a sale range up to 1000 to 3000 fish seeds [1]. In fact there have been many studies that have been conducted on fish seeds count automatically. As an example of research done by [2] and [3]. The sensors used in research [2] using phototransistor. When there is a bubble of air which passes through the pipe, the air bubbles also detected by the sensor [2].While the sensor is used in research [3] i.e. using LED (Light Emitting Diode) and photodiode problems on the research of [3] similar to that performed [2] i.e not able to calculate fish seeds with small size accurately and also not able to distinguish between the seed of the fish with the other objects which enter into tools such as air bubbles.

To get the accurate data, then image processing sensor is developed. the sensor that is used in this method is the sensor of the camera. The camera will capture images of fish seeds that pass through the camera sensor. Then the information in the form of an image of the camera is processed to be able to be counted. Pre-launch stage processing function to improve the quality of the image prior to further processing. Improvement of the quality of the image can be through the process of smoothing and sharpening to eliminate noise in the image [4]. Smoothing technique used to smooth the image, while sharpening technique was used to sharpen the image. After a pre-launch stage processing, the next step is the image segmentation.. the purpose of image segmentation is to distinguish objects to its background. Once separated from the background object, the object is detected will be calculated automatically. The system is expected to be able to calculate the fish seeds of various sizes and also more accurate compared to systems that use sensors and photodiode-LED phototransistor.

2. Methodology

In order for research purposes can be achieved, then there are several stages of research to be done, as can be seen in Figure 1. The main stages that must be done in this research is the image segmentation, filtering, image morphology, and fish seeds calculation.

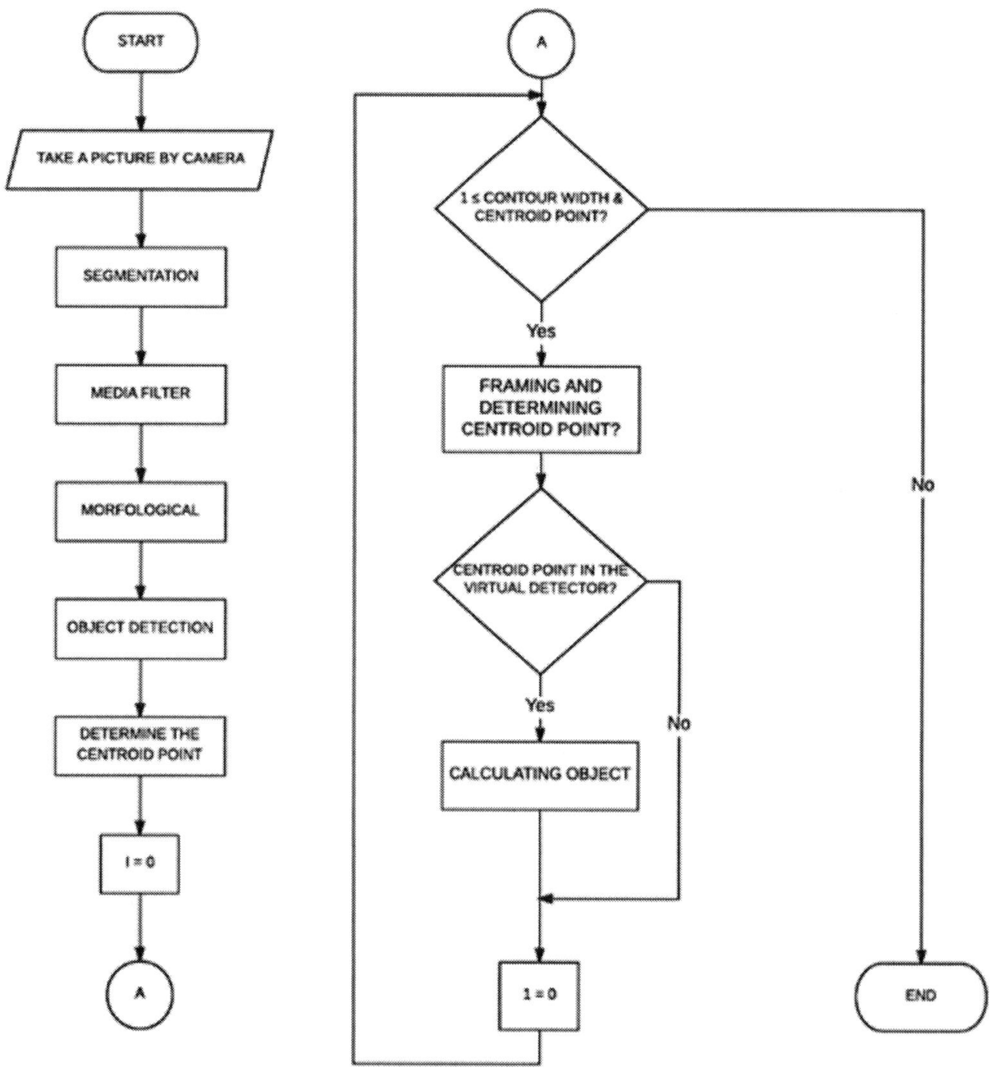

Figure 1. Flow diagram of fish seeds calculation.

Based on Figure 1, the experimental setting of the fish seed counter begins with the fish seed inserted through the inlet mixed with water. the purpose of water use is to prevent the fish from returning or turning around. After that the fish seed passes through the area highlighted by the camera.

In the area highlighted by the camera is used then as image processing to calculate the number of fish seeds entered. At the end of the fish seed has been calculated out through the outlet.

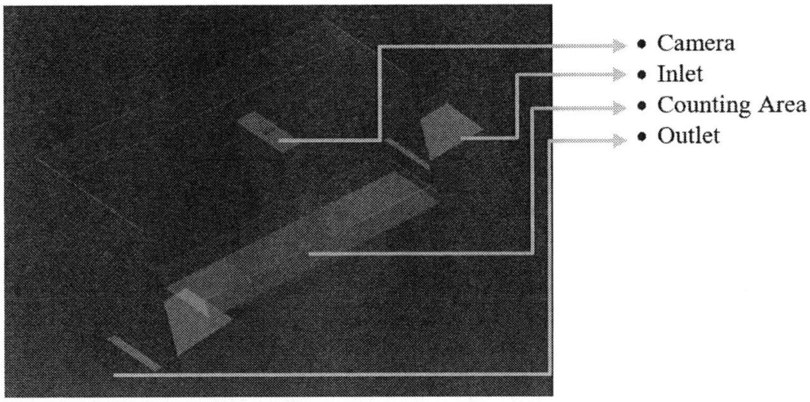

Figure 2. Experimental setting or fish seeds counter.

2.1. Segmentation
Segment object is one of the processes, in the system of image processing which is used to separate objects with the background. Background subtraction were the first stage or also we known as pre-processing, the main thing in the vision based-Application. Technically, a moving foreground extracted from a static background. The background substraction done in absolute terms where the difference every pixel of both images will be acquired and is always positive Because this technique will notice the difference for every pixel in the image, so that the two images must have the same size of data type and it sizes. [8]. As for the background of the process of subtraction algorithm is to use the method model in every pixel background with a mixture of Gaussian distribution of K (K = 3 to 5). Mixed weights represent the proportion of time with these colors remain on the screen. By using this algorithm, the background color of the possibility last longer and are more static.

Background Subtraction adaptive is almost the same as background subtraction methods, It's just a background image constantly updated due to altered conditions, for example in the morning with the daytime then it will causes different intensity. The Intensity differences will affect the threshold value of each pixels . So in need the Background subtraction adaptive methods to do as follows. Soon after you have the original background, then filter the image frame to be more subtle using Gaussian filter, and convert RGB image color type, into grayscale [9].

2.2. Filtering
After image segmentation stage is done, then the next step is to reduce noise. Because the system uses research on image processing then the possibility of noise is very likely to happen. Noises may caused by due to lack of lighting or too dark. It also can be caused by the shadows of the water because of the water undergo the turbulence. Therefore do a filter process, due the noise reduction of the image before the next step of the process. Thus the filter that is used in this research is Median Filter.

2.3. Morphologhy
To refine the shape of the object detected by the morphological operations in this research, the morphological operations that is needed to is as it follows: opening morphology operation, closing morphological operation, dilation operation. Opening morphology operation to smoothen the contour object and reduce pixels beside the smallest pixel outside the object identified, thus it will not affect detection process of moving object. In addition the operation is used as the size of the object is relatively the same after the surgery compared to using morphological erosion. Closing morphology operation which is used in this research able to smoothen the object contour and vanish small hole in

IOP Conf. Series: Journal of Physics: Conf. Series **974** (2018) 012063 doi:10.1088/1742-6596/974/1/012063

the object, in case could fill up the contour wholely, therefore the operation system able to define the centroid point from the moving object. Dilation operation used to combine an object which is separated that caused by imperfect segmentation process.

2.4. Fish Seeds Calculation

The object detection by finding the centroid points objects. For the centroid points could see in the Figure 3 as follows:

(a) (b) (c)

Figure 3. (a) Centroid point, (b) Frame, (c) Blob tracking.

After centroid points is obtained, the next stage is make a frame object. It is intended to display the objects that are being processed and the large dimensions of the frame are the same as the process for seeking centroid points. To track a moving object with detect the object motions, the quantity of the object and the trajectory of the objects. The methods used in this study adopts the *blob tracking* technology, which is used to count every vehicles in a roadway in a camera or video which is recorded perspectives.[9].

After the process of finding the point of the centroid, create frames and blob tracking is done, then the next step is to create virtual detector. Virtual detector that will be used represented as a normal line, where is the line will be the part of the region of interest (ROI) itself. Which will calculate the amount of fish that pass through the normal line, in the absence of error calculation.

The weakness of virtual loop detector traditional methods is the wrong calculation from the object which is not moving, or the object that move more slowly or an object which runs adjacent. [5]. For it is required an additional algorithm which is consists of frame differencing. Traditional Frame differencing contour of the moving object acquired by using the information differences between two frames in a row, and is widely used in segmentation movement because The characteristic of the object detection process its easy to handle. Besides the traditional frame differencing is less sensitive to changes in lighting or dim lighting condition. So it suited to systems with unstable lighting. But this approachment is also have a shortage of such sensitive to the object speed and frame rate of the video. [5].

To overcome those problems [6] we suggest to variation between these two methods, which is named as "the double difference" methods. the difference between the threshold applying frae at time t, t-1, and t+1 thus combined it with AND logic proposing a techniques which is utilize between two frames, which is frame difference and background subtraction [7]. The differences between actual frame and the prior frame, merged with the differences between actual frame and the actual mdel of the background to detect moving objects. The first image of the sequence of frames used as an initial background model and dynamically renewed with the new frame which is appropriate with the segmentation moving objects.This approachment promised an object motion, in case the background model gave the approriate and up-to-date information.

2.5. Double Difference Methods

Double-difference methods applied to get motion regions from video frames and a pose partially on the region where the motion regions is detected. The motion a region where a pixel value changes. A double-difference image is obtained by and operation between successive two difference images. It is assumed that the video frames do not include any moving object except for a fish seed. The double differences images made by three successive frames in an video stream. First, two differences images enerates from corresponding two successive images ('t-1' and 't', 't' and 't+1'). Then binarize the difference images and execute. Operation on these two images, called a resultant binary image a double-difference image.

As a double-difference image is a product of two differences images. It tends to include isolated noise pixels. Therefore, each 4 by 4 pixels in the double-differences image is grouped into one square block. A block is marked true if more than half of the pixels in the block is true. This process not only prevents noise but also reduces the computation cost in the image processing. Then the system removes isolated blocks to get rid of slight changes in the video frames. The motion regions consists of the pixels whose value is true in the remained blocks [10].

3. Analysis and Discussion

The position and the shape of the object in a correct position is a key factor when calculating an accurate object using the methods that is given (virtual loop methods), since it is depends on pixel value. The contour difference in a normal image does not represent the object shape clearly, because it was a coherency of the shape object in different two times. The "double difference image" approach, apply the information from the previous information, in a prior image sequence, and the next frame merely predict the shape and the position of the objects,which gives a more accurate position information.

To get the image of the object is " double the difference"by distinguishing two images resulting from two consecutive images that match. (frame in the sequence time as beside,'t-1', 't', and 't+1'), thus from the differential object binerization is conducted and logical operation AND execute in these two graphic images. The object from the graphic generated, retaining shape and its position at time 'T'. At the same time , to reduce noise in pixels, using opening,closing,dilate morphology operation in binary graphic which have been executed. Virtual wide variation detector used in this research can be seen in Figure 4 below.

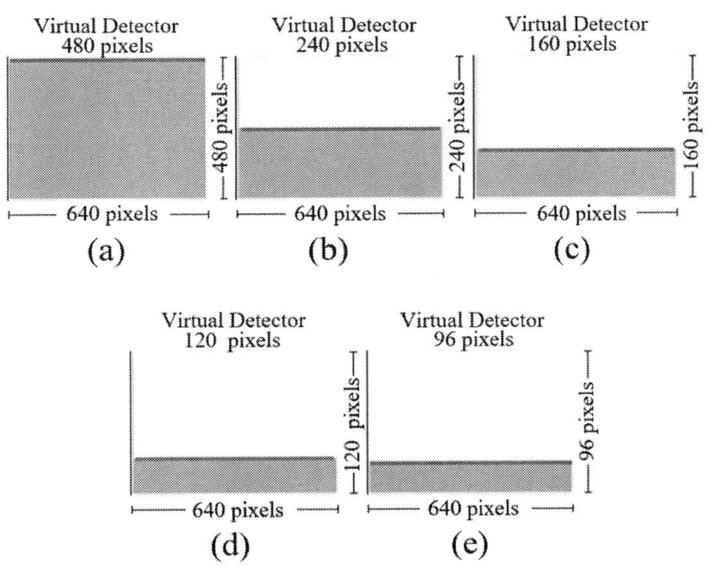

Figure 4. Virtual detector width (a) 480 pixel, (b) 240 pixel, (c)160 pixel, (d)120 pixel, and (e) 96 pixel.

In this chapter will be explained about the data analysis at each test specification. Testing has been done include: double difference testing, slope field testing and accuration testing.

IOP Conf. Series: Journal of Physics: Conf. Series **974** (2018) 012063 doi:10.1088/1742-6596/974/1/012063

3.1. Double Difference Testing

The program used for software testing by using marbles turned out to be unsuitable to calculate fish seeds, because the program uses full virtual detection. So when used to calculate the fish seeds, theres causes an error such as the amount of the fish were counted twice.is therefore necessary the program corresponding to the character of the object in the form of fish seeds using double difference methods. on this test aims to determine the width of the virtual detector, and the number if the fish seed used in this test is 50. The testing process conducted by as much as 5 times a wide virtual detector tested. Thus the virtual detector width that is used in this tests is about 480, 240, 160, 120, and 96 pixels. Following are the results of the testing of the virtual width detector, as shown as in Figure 5.

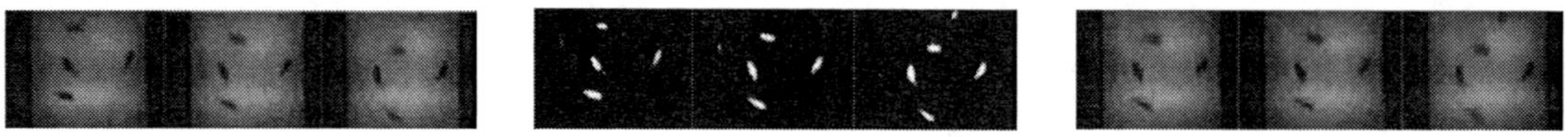

(a) (b) (c)

Figure 5. (a) Original frame, (b) Results of the morphological operations, (c) Results of the virtual detector width 96 pixels.

Table 1. *Virtual detector* width testing

Width of Virtual Detector (pixel)	% Average Error
480	61,2
240	15,2
160	8,0
120	4,4
96	2,8

Based on Table 1, the average of the smallest error in virtual detector width 96 pixels obtained around 2,8 %.

3.2. Slope Field Testing

This test aims to find out the slope of the field trials that yield the smallest error so on the accuracy testing using only one type of slpe field test course. The amount of the fish examined is 50. Testing process conducted by as much as 5 times each tilt of the field test.the slope of the field test to be tested is about 1°, 2°, 3°, 4°, and 5°.

Table 2. Slope field testing process

Tilt (^{o})	% Average Error
1	5,6
2	8,0
3	12,8
4	18,0
5	16,4

Based on Table 2, the average error is obtained the smallest o test fields 1-degree slope, with an average *error* is 5,6 %. The slope field is 1-degree which is used in accuration testing.

IOP Conf. Series: Journal of Physics: Conf. Series **974** (2018) 012063 doi:10.1088/1742-6596/974/1/012063

3.3. Accuration Testing

The accuracy testing comes to find out the accuracy of tool in calculating fish seeds, where based on the experiment that have been done the average calculation time takes 5 minutes for 500 fish seeds. This test done to know the tools accuracy in calculate the fish seeds. Based on the previous testing the tilt of the field and the width of *virtual* detector, used to perform the accuracy testing namely 1° and 96 pixel. Thus the amount of the fish which is used is as the statement follows : 100, 150, 200, 250, 300, 350, 400, 450, and 500 fish seeds. Table 3 perform the average value of *error* in accuracy testing.
The use of the virtual detection is not suitable if applied to the moving objects such as fish, because of the fish seeds is not stable, will make the calculation of fish seeds for twice or more. So if necessary, there is an additional methods. I.e the *double difference* methods, to prevent the calculation of the amount of the fish will be counted twice or more. Based on the experiment that has been done, the use of *virtual* detector with a width of 96 pixels is suitable on this research, because the average errors obtained its small, which is about 2,8 %.

After the *virtual detector* width is obtained, then the next step is to determine the rate of fish seeds that is suitable for use with the *frame rate* 25 *frame*/second. Because if the rate of the fish seeds pace quickly, then the camera is not able to calculate fish seeds that pass through the camera. So to determine the rate of fish seeds with how to set-up the slope of the field test. Based on an experiment that is conducted before, the uses of the slope in 1° does very well on this research, because the average error obtained small i.e 5,6%.

Table 3. *Accuracy testing with various fish amount*

Amount of Fish Seeds	% Average Error
100	3,2
150	2,4
200	1,1
250	1,2
300	1,0
350	4,7
400	3,5
450	4,4
500	4,5

In order the accuration of the tools were acknowledged, then the testing accuracy is performed. Testing is done using the amount of fish seeds which is 100, 150, 200,and 250. With the slope field 1^{0} and the use of *virtual detector* with *96* pixels width. The results obtained are quite good on testing, with a number of 300 fish seeds, and an average error of 1.0 %

There is a lot of factors affect the value of the error in this research, include the presence of fish seeds pace that return to the field test, which causes *bounding* will break and binding again, so the calculation of the fish seeds will be counted twice. Besides due the two fish seeds attached one to another, it will account for only one fish seeds.

4. Conclusions

Based on the results that have been obtained as well as revisiting problems with constraints and objectives, then it can be formulates some conclusions obtained, namely:

- The automatically fish seeds counters and *real time* based image processing, hve successfully designed.
- Based on width testing of *virtual detector,* the best results are obtained on the use of virtual width detector 96 pixels with the value of the average errors of 2.8 %

- Based on the slope testing of the field tests, the average error is obtained the least on the slope of the field test in 1^0 with the average value of error of 5.6%
- Based on the accuracy testing, a tool that has been designed able to produce the smallest average value of error 1.0 % in 300 fish seeds.

References

[1] Yondri S, Efrizon and Suar Z 2010 *Rancang Bangun Alat Penyortir dan Penghitung Bibit Ikan secara Otomatis untuk Meningkatkan Proses Transaksi Jual Beli Bibit Ikan pada Petani Tambak Ikan* (Padang, Indonesia: Unand Repository)

[2] Purbowaskito W 2014 *Rancang Bangun Alat Penghitung Benih Ikan Menggunakan Sensor Optik Fototransistor Berbasis Mikrokontroler Atmel ATMega 8* (Yogyakarta, Indonesia: UGM Repository)

[3] Padiyono 2015 *Penghitung Benih Ikan Lele Otomatis Berbasis Mikrokontroler ATMega 8* (Surakarta, Indonesia: UMS Repository)

[4] Rahmah F, Kusumawardhani A, Setijono H, Hatta A.M and Irwansyah 2015 *Radial Line Method for Rear-View Mirror Distortion Detection* (SPIE Press)

[5] Tursun M and Amrulla G 2013 *A Video Based Real-Time Vehicle Counting System Using Optimized Virtual Loop Method* IEEE Signal Processing and their Applications, pp. 75-78

[6] Mujib F, Rahmadiansah 2012 A *Desain Pencahayaan Lapangan Bulu Tangkis Indoor ITS*, Jurnal TEKNIK POMITS, pp. 1-8

[7] Yoshinari K and Michihiko M 1996 *A Human Motion Estimation Method Using 3-Successive Video Frames* IEEE Virtual Systems and Multimedia, pp.135-140

[8] Migliore D. A, Matteucci M and Naccari M 2006 *A Revaluation of Frame Difference in Fast and Robust Motion Detection* ACM International Wokshop on Video Surveillance and Sensor Networks, pp. 215-218

[9] Solichin A and Harjoko A 2013 *Metode Background Subtraction untuk Deteksi Obyek Pejalan Kaki pada Lingkungan Statis* Seminar Nasional Aplikasi Teknologi Informasi (SNATI), pp. B1-B6

[10] Li D, Liang B and Zhang W 2014 *Real-Time Moving Vehicle Detection, Tracking, and Counting System Implemented with OpenCV*" IEEE, pp. 631-634

[11] Yoshinari K and Michihiko M 1996 *A Human Motion Estimation Method Using 3-Successive Video Frames* IEEE Virtual Systems and Multimedia, pp.135-140

Simulations of linear and Hamming codes using SageMath

Tahta D. Timur[1], Dieky Adzkiya[2] and Soleha[3]

[1,2,3]Department of Mathematics, Institut Teknologi Sepuluh Nopember, Kampus ITS Sukolilo-Surabaya 60111, Indonesia

E-mail: timur.17071@mhs.its.ac.id[1], dieky@matematika.its.ac.id[2], seha_07@matematika.its.ac.id[3]

Abstract. Digital data transmission over a noisy channel could distort the message being transmitted. The goal of coding theory is to ensure data integrity, that is, to find out if and where this noise has distorted the message and what the original message was. Data transmission consists of three stages: encoding, transmission, and decoding. Linear and Hamming codes are codes that we discussed in this work, where encoding algorithms are parity check and generator matrix, and decoding algorithms are nearest neighbor and syndrome. We aim to show that we can simulate these processes using SageMath software, which has built-in class of coding theory in general and linear codes in particular. First we consider the message as a binary vector of size k. This message then will be encoded to a vector with size n using given algorithms. And then a noisy channel with particular value of error probability will be created where the transmission will took place. The last task would be decoding, which will correct and revert the received message back to the original message whenever possible, that is, if the number of error occurred is smaller or equal to the correcting radius of the code. In this paper we will use two types of data for simulations, namely vector and text data.

1. Introduction

Digital communication technologies such as the internet are now considered as primary telecommunication tools by most people around the world. The internet consists of large network of networks, which are interconnected and constantly sending and receiving huge amount of data [1]. Operations of these networking devices are governed by a set of rules i.e. protocols on how to send, receive and process digital signals [1]. These signals are actually binary digits (bits) transmitted through dial-up modems, wires, radio signals, and optic cables. During transmission, these bits could easily be changed due to hardware or software errors. How to detect these errors and to obtain the original message, are the main goals of coding theory [2, 3].

There are two kinds of codes, namely error-detecting and error-correcting codes. The ISBN-10 (International Standard Book Number) code is an error-detecting code. The tenth digit d_{10} can be obtained by $d_{10} = 11 - \{10d_1 + 9d_2 + \ldots + 2d_9 \pmod{11}\}$, such that any changes made to the other digits could be detected, although we could not know exactly which digit has changed [2]. The tenth digit on ISBN-10 code is called *check digit*, that is, an additional digit to help us detect whether a code has changed due to errors. This idea leads to error-correcting codes, which has the abilities of detecting and correcting errors, provided they are less or equal to correcting radius of the code [4]. Some applications of coding theory was shown by [5] on secure

IOP Conf. Series: Journal of Physics: Conf. Series **974** (2018) 012064 doi:10.1088/1742-6596/974/1/012064

video steganography using Hamming code, [6] on lossless data compression, [7] on development of Hamming codes as error-reducing codes, and [8] on biomedical engineering's medical image authentication using Hamming code.

In this paper we will use two types of data for simulations, namely vector and text data. These data then will be represented as binary vectors which will be denoted as bold letters such as $\mathbf{x}, \mathbf{y}, \mathbf{u}$ or $\mathbf{v}$ and written as transposed row matrices such as $\begin{bmatrix}1 & 1 & 0 & 1\end{bmatrix}^T$ or simply as 1101. Matrices will be denoted as italic capital letters such as A or H. Simulations will be performed using SageMath mathematical software [9]. In this work we use `sagetex` package [10] to incorporate SageMath's computations and literate programming onto Latex. Inline SageMath's commands and source codes will be written as typewriter font style such as `for i in range(3): print i`, while its display commands will be written as

```
sage: vector(GF(2), [1,0,0,0])
(1, 0, 0, 0)
```

aiming for clarity of writing. Sections will be organised as follows. Section 2 will shortly discuss some basic theories on linear and Hamming codes, including error-correcting codes, parity check and generator matrix encoders, and nearest neighbor and syndrome decoders. In Section 3, we will perform some simulations on data encoding, transmission, and decoding, with given data and methods using SageMath. We aim to show how message could be changed by errors, how often errors change messages, how to construct codes, how to encode, transmit and decode messages, and investigate properties of codes using SageMath. Finally in Section 4 we will discuss some conclusions based on works we did in preceding sections.

2. Preliminaries

In this section we will discuss some definistions and theories regarding encoding and decoding algorithms used in linear and Hamming codes.

2.1. Parity check encoder

In Section 1 we briefly discussed error-detecting codes such as the ISBN-10 code. In this section we will introduce error-correcting codes and parity check encoder. Digital data transmission consists of three stages: encoding, transmission, and decoding. These processes can be modeled as Figure 1 [4].

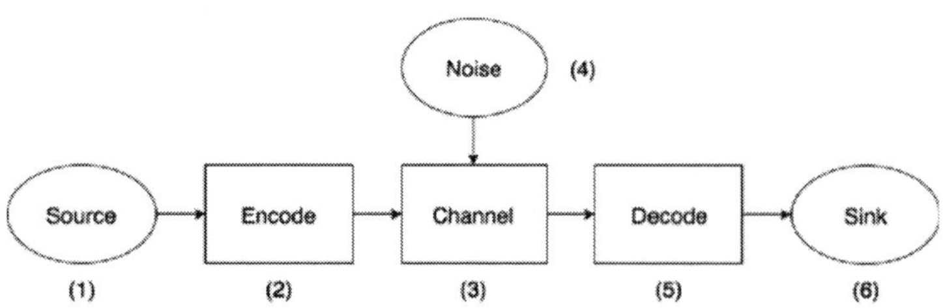

Figure 1. Digital Data Transmission Model.

Part (1) in Figure 1. is the information source. This informations then encoded in (2), and the encoded message is transmitted through channel (3). If noise (4) exists, then (3) is called *noisy channel* and there is a probability of errors which could change the encoded message. After encoded message is received, it will be decoded and corrected if possible in (5), and finally the decoded message will arrive at sink (6). Like on error-detecting codes, we need to add some additional check digits to the message such that any changes made to the other digits will be detected and corrected properly.

IOP Conf. Series: Journal of Physics: Conf. Series **974** (2018) 012064 doi:10.1088/1742-6596/974/1/012064

Let $\mathbb{F}_q$ be a finite field with characteristic q. The set $M \subseteq \mathbb{F}_q^k$ is called *message space*, which its elements are all possible vector messages of length k. The set $C \subseteq \mathbb{F}_q^n$ is called *code* of length n over $\mathbb{F}_q$ which its elements then referred to as *codewords*. The error vector $\mathbf{e}$ is the difference $\mathbf{x}-\mathbf{y}$ between the transmitted codeword $\mathbf{x}$ and the received codeword $\mathbf{y}$, where $\mathbf{x}, \mathbf{y} \in C$. Another concept we need to discuss is the Hamming distance. Hamming distance $d(\mathbf{x}, \mathbf{y})$ between two vectors is the number of different coordinates in $\mathbf{x}$ and $\mathbf{y}$ [2, 4]. We need Hamming distance to define the minimum distance of a code C with respect to Hamming distance. In [2], the minimum distance of code C is defined by

$$d_{min}(C) = \min d(\mathbf{x}, \mathbf{y}) \tag{1}$$

where $\mathbf{x} \neq \mathbf{y}$ for all $\mathbf{x}, \mathbf{y} \in C$.

Definition 2.1 (Linear Codes). Let H be a full rank $(n-k) \times n$ check matrix over $\mathbb{F}_q$. The set

$$C = \{\mathbf{x} : H\mathbf{x}^T = \mathbf{0}, \mathbf{x} \in \mathbb{F}_q^n\}$$

is called a linear code C over $\mathbb{F}_q$ of length n, where k and $n-k$ is the length of message and check digits, respectively [2]. Transmission rate is defined as k/n, and if $q = 2$, then C is called *binary codes.*

From Definition 2.1 we see that linear code C actually is the set of all solutions of matrix equation $H\mathbf{x}^T = \mathbf{0}$. For instance let vector $\begin{bmatrix} x_1 & x_2 & x_3 \end{bmatrix}$ be a message $\mathbf{a}$ and let x_4, x_5, x_6 be check digits. We can obtain these check digits by

$$\begin{aligned} x_1 + x_2 + x_4 &= 0 \\ x_1 + x_3 + x_5 &= 0 \\ x_2 + x_3 + x_6 &= 0 \end{aligned} \tag{2}$$

or equally written as a check matrix

$$H = \begin{bmatrix} 1 & 1 & 0 & 1 & 0 & 0 \\ 1 & 0 & 1 & 0 & 1 & 0 \\ 0 & 1 & 1 & 1 & 0 & 1 \end{bmatrix}.$$

By using Equation (2) or H, message $\mathbf{a} = \begin{bmatrix} 0 & 1 & 0 \end{bmatrix}$ can be encoded to $\begin{bmatrix} 0 & 1 & 0 & 1 & 0 & 1 \end{bmatrix}$. Such method is called parity check encoder.

2.2. Generator matrix encoder

Suppose we have C an (n, k) linear code over $\mathbb{F}_q$ with check matrix H. If $H = (A \mid I_{n-k})$, then we can define Generator Matrix of C as follows [2].

Definition 2.2. Let C be a linear (n, k) code over $\mathbb{F}_q$ with check matrix $H = (A \mid I_{n-k})$. Matrix $G = \left(I_k \mid -A^T\right)$ is called Generator Matrix of C, and codewords $\mathbf{x} \in C$ of length n can be obtained by $\mathbf{x} = \mathbf{a}G$, for all vector messages $\mathbf{a}$ in message space.

Continuing the discussion in previous subsection, with given H we can obtain generator matrix

$$G = \begin{bmatrix} 1 & 0 & 0 & 1 & 1 & 0 \\ 0 & 1 & 0 & 1 & 0 & 1 \\ 0 & 0 & 1 & 0 & 1 & 1 \end{bmatrix}$$

and we can obtain codeword $\mathbf{x} = \mathbf{a}G = \begin{bmatrix} 0 & 1 & 0 \end{bmatrix} G = \begin{bmatrix} 0 & 1 & 0 & 1 & 0 & 1 \end{bmatrix}$.

2.3. Nearest neighbor decoder
We have discussed two encoding algorithms in previous subsections. Now we will continue on one of decoding algorithms called nearest neighbor decoder. Suppose we have a linear code $C \subseteq \mathbb{F}_q^n$. The set $S_r(\mathbf{x})$ is called sphere of radius r with respect to Hamming distance, centered at $\mathbf{x}$. Or we can write

$$S_r(\mathbf{x}) = \{\mathbf{y} \in \mathbb{F}_q^n : d(\mathbf{x}, \mathbf{y}) \leq r, \mathbf{x} \in C\}. \quad (3)$$

This process is also referred to as sphere packing [2, 4], that is, constructing spheres with codewords as its center and contain all vectors within its radius. If vector $\mathbf{y}$ is received and it belongs to precisely one of the spheres, then it will be decoded to its center, namely vector $\mathbf{x}$.

We have to make the radius of these spheres as large as possible to be able to decode codewords as much as possible. On the other hand, the radius have to be small enough such that there are *no* two spheres intersect. If such event occurred, then the codeword that belongs to the intersection cannot be decoded.

Theorem 2.1. *A linear code C with $d_{min}(C) = d$ can detect up to $d - 1$ errors and correct up to $\lfloor \frac{1}{2} d \rfloor$ errors.*

Suppose we transmit a codeword $\mathbf{x}$ and vector $\mathbf{y}$ received. If $\mathbf{y}$ belongs to one of the spheres, say $S_r(\mathbf{x}_1)$, then $\mathbf{y}$ will be decoded to $\mathbf{x}_1$, which is the center of the sphere. This method is called nearest neighbor decoding.

2.4. Syndrome decoder
Let C be a linear (n, k) code over $\mathbb{F}_q$ and consider C as subspace of $\mathbb{F}_q^n$. The set $\mathbb{F}_q^n/C$ contains all cosets $\mathbf{a} + C = \{\mathbf{a} + \mathbf{x} \,|\, \mathbf{x} \in C\}$ where $\mathbf{a} \in \mathbb{F}_q^n$. If a vector $\mathbf{y}$ is received, then it belongs to an exactly one of these cosets [2].

Definition 2.3. Let C be a linear (n, k) code over $\mathbb{F}_q$ with check matrix H. Vector $S(\mathbf{y}) = H\mathbf{y}^T$ of length $n - k$ is the syndrome of $\mathbf{y}$, for any $\mathbf{y} \in \mathbb{F}_q^n$.

Let us consider a linear code C and a codeword $\mathbf{x} \in C$ is transmitted. Suppose a vector $\mathbf{y}$ belongs to a particular coset $\mathbf{e} + C$, where $\mathbf{e}$ is a vector of minimum weight in the coset. If $\mathbf{y}$ is received, then its syndrome $S(\mathbf{y})$ must be equal to $S(\mathbf{e})$ because $H\mathbf{y}^T = H(\mathbf{x} + \mathbf{e})^T = H\mathbf{e}^T$. Thus, $\mathbf{y}$ is decoded to $\mathbf{x} = \mathbf{y} - \mathbf{e}$.

2.5. Hamming code
Hamming code is a special class of linear codes. It is an error-correcting codes with 2 detection radius and 1 correcting radius with high transmission rate [2, 7].

Definition 2.4. Let C_m be a binary code of length $n = 2^m - 1$ where $m \geq 2$. If H is the $m \times (2^m - 1)$ parity check matrix which columns consist of all nonzero binary vectors of length m, then C_m is called the Hamming code.

All encoder and decoder algorithms we discussed earlier also hold for Hamming code because it is inherently a linear code.

3. Simulations of vector and text data
SageMath provides a broad range of class (or object) in coding theory [11], including linear and Hamming codes. These classes can be accessed in SageMath under `codes` class. In this work we will use `codes.LinearCode()` and `codes.HammingCode()` for simulations. Simulations implemented in SageMath using Jupyter Notebook [12].

3.1. Data definition and representation
Data exact definition and representation is needed for simulation programs to work properly. For text data, we will consider a list $T = \{$, a, b, c, d, e, f, g, h, i, j, k, l, m, n, o, p, q, r, s, t, u, v, w, x, y, z, A, B, C, D, E, F, G, H, I, J, K, L, M, N, O, P, Q, R, S, T, U, V, W, X, Y, Z, 1, 2, 3, 4, 5, 6, 7, 8, 9, 0, .$\}$ of both lower and upper case letters, numbers, full stop mark (point) and whitespace with total of 64 elements. Each element of the list will be represented by a binary vector of its index digit, ranging from 0 to 63. For instance, 'c' which has index of 3, has binary vector representation of 000011.

3.2. Simulation workflow
First we will briefly define and discuss a general workflow of simulation processes we about to perform.

```
Require: C, input, channel, encoder, decoder
data ← make binary vector representations of input
procedure TRANSMISSION(C, channel, encoder, decoder)
    for all j ∈ data do
        encoded ← ENCODE(C, j, encoder)
        transmitted ← TRANSMIT(channel, encoded)
        decoded ← DECODE(C, encoded, decoder)
    end for
end procedure
output ← revert binary vectors back to original values
```

First we need to represent the input to its binary vectors form. Then, encoding, transmission and decoding will be performed element-wise. If the output is equal to the input, then it is successfully transmitted. Otherwise, it has errors greater than correcting radius of the code, hence incorrectly decoded.

3.3. Codes and channels construction
First we construct linear code C by using parity check and generator matrices which will be defined as follows:

$$H = \begin{bmatrix} 1 & 1 & 1 & 0 & 0 & 0 & 1 & 0 & 0 & 0 & 0 & 0 \\ 0 & 1 & 1 & 1 & 0 & 0 & 0 & 1 & 0 & 0 & 0 & 0 \\ 0 & 0 & 1 & 1 & 1 & 0 & 0 & 0 & 1 & 0 & 0 & 0 \\ 0 & 0 & 0 & 1 & 1 & 1 & 0 & 0 & 0 & 1 & 0 & 0 \\ 1 & 0 & 1 & 0 & 1 & 0 & 0 & 0 & 0 & 0 & 1 & 0 \\ 0 & 1 & 0 & 1 & 0 & 1 & 0 & 0 & 0 & 0 & 0 & 1 \end{bmatrix}, \quad G = \begin{bmatrix} 1 & 0 & 0 & 0 & 0 & 0 & 1 & 0 & 0 & 0 & 1 & 0 \\ 0 & 1 & 0 & 0 & 0 & 0 & 1 & 1 & 0 & 0 & 0 & 1 \\ 0 & 0 & 1 & 0 & 0 & 0 & 1 & 1 & 1 & 0 & 1 & 0 \\ 0 & 0 & 0 & 1 & 0 & 0 & 0 & 1 & 1 & 1 & 0 & 1 \\ 0 & 0 & 0 & 0 & 1 & 0 & 0 & 0 & 1 & 1 & 1 & 0 \\ 0 & 0 & 0 & 0 & 0 & 1 & 0 & 0 & 0 & 1 & 0 & 1 \end{bmatrix},$$

where H is the parity check matrix and G is the generator matrix. After defining these two matrices, we can then construct linear code C by running

```
sage: C = codes.LinearCode(generator=G)
sage: C
[12, 6] linear code over GF(2)
sage: C.random_element()
(0, 1, 0, 0, 1, 1, 1, 1, 1, 0, 1, 0)
```

in SageMath. In the third command above, we can take a random element of C i.e. a codeword, by running `C.random_element()` which should give an output of random vector of length 12.

Channel construction is provided by `channels.StaticErrorRateChannel` class which can construct channel that creates static errors or randomly choose errors between non-negative integers a and b to our codewords. If otherwise we want a randomly chosen errors instead of static ones, then we need only to replace `E` with a tuple `(a,b)`, where `a` and `b` is the lower and upper bound, respectively. In this section, we use 0, 1 and 3 as values for `E`.

3.4. Encoding and transmission processes

Suppose we have a message vector $msg1 = \begin{bmatrix}1 & 0 & 0 & 1 & 1 & 0\end{bmatrix}$ and $msg2 = \begin{bmatrix}1 & 0 & 1 & 0\end{bmatrix}$. We can encode the first message using parity check encoder (which is named 'Systematic' in SageMath) using `encode()` method. We can use generator matrix encoder on both codes by changing the string value of `encoder_name` to `'GeneratorMatrix'` which will yield the same results.

As we have encoded messages or codewords, now we can transmit them. Transmission in SageMath is conveniently provided by `transmit` method. We can see that given two errors, `encoded1` has changed after transmission.

3.5. Correcting and decoding processes

Suppose the transmitted codewords arrive. Then we can decode these transmitted codewords using `C.decode()` or `C.decode_to_message()` methods. We will consider three cases, which in each case the number of errors is 0, 1 and 3, respectively.

a) For 0-error, it is obvious that `encoded1` will be exactly the same as `transmitted1`, for any given codewords. Hence the value of `decoded1` and `decoded2` will be equal to `msg1` and `msg2`, respectively. Suppose that we use nearest neighbor decoder. There is a sphere $S_1(1010101)$ which contains all vectors $\mathbf{y}$ such that $d(1010101, \mathbf{y}) \leq 1$. Because 1010101 is the center of $S_1(1010101)$, then it will be decoded back correctly to 1010. If we use syndrome decoder, then we must compute the syndrome $S(1010101) = H \begin{bmatrix}1 & 0 & 1 & 0 & 1 & 0 & 1\end{bmatrix}^T = \mathbf{0} = S(0000000)$. Hence, 1010101 will be decoded back correctly to 1010. In other words, if an encoded codeword is transmitted without being changed, then it will be decoded back correctly to its original message.

b) As discussed earlier that C and D has minimum distance of 3. Thus both of them can detect up to 2 errors and correct up to 1 error. vectors `transmitted1` and `transmitted2` will be decoded back correctly as expected. Suppose that we use syndrome decoder. Then we must compute the syndrome $S(0010101) = H \begin{bmatrix}0 & 0 & 1 & 0 & 1 & 0 & 1\end{bmatrix}^T = 100 = S(1000000)$. Hence 0010101 will be corrected as $\mathbf{y} - \mathbf{e} = 0010101 - 1000000 = 1010101$ and decoded back correctly to 1010.

c) Consider transmitted vectors which has 3 errors. As discussed earlier that both codes are unable to detect nor to correct them. Hence it will be incorrectly decoded. Hamming code D has a minimum distance of 3. This means that if both of $\mathbf{x}_1$ and $\mathbf{x}_2$ are codewords in D, then the minimum distance $d(\mathbf{x}_1, \mathbf{x}_2)$ between them would be 3. Because $d(1010101, 1000011) = 3$, then 1000011 actually is another codeword in D. If we compute its syndrome $S(1000011) = \mathbf{0}$, then it will be decoded back *incorrectly* to 1000.

Simulation results using vector data we performed up to this point is presented in Table 1.

3.6. Simulations using text data

In this subsection we will perform two simulations, namely linear code C and Hamming code D using text data. Unlike in previous subsection, we will not provide any literate programming nor inline computations using `sagetex`. Instead, we will present only the simulation results as

Table 1. Vector simulation result using Hamming code D.

n-errors	MESSAGE	ENCODED	TRANSMITTED	DECODED
0	1010	1010101	1010101	1010
1	1010	1010101	0010101	1010
3	1010	1010101	1000011	1000

Table 2. Text Data simulation result using linear code C.

n-errors	DECODED MESSAGE
0	This is a dummy text
1	This is a dummy text
2	Uhis is a dwKFy text
3	NJCKbhKxefZe9g5zoh4l
4	0yIqq7Mntt3YnK7msLV5
5	IHQETccHfBep86XuUU .
6	sRYf6DhAA7GnQfkisYJ5

Table 3. Text Data simulation result using Hamming code D.

n-errors	DECODED MESSAGE
0	This is a dummy text
1	This is a dummy text
2	i T0hadi.0 s.0tu9xdT
3	ehetteyiT0hsTm9uumy
4	.mmu0e.tymxy.dex xea
5	s0eTyt9y TieaTsadhyi
6	eyxt.xt.0.9muuh.sTis

tables with text message "This is a dummy text". Full implementation of the simulations can be obtained in our GitHub repository in `https://github.com/tdtimur/sage3rdicompac2017`.

From Table 2 and Table 3 we can see that once the error exceeds the correcting radius of the code, the resulting decoded message changed considerably. This shows the importance of coding theory in the development of reliability and integrity of digital communication.

4. Conclusion

In this work we show that we can use SageMath for linear and Hamming code simulations. First we define the codes by defining either parity check or generator matrices, then create a channel which randomly choose errors between given interval. We can encode our messages using `encode()` methods to obtain our codewords. These codewords then can be transmitted through our previously created channel by using `transmit()` method. Depending on the number of errors given to codewords, we can decode these transmitted codewords by using either `decode()` or `decode_to_message()` methods. If the number of errors is less than or equal to the code's correcting radius, then the transmitted codeword will be corrected and decoded back correctly to its original message. Otherwise, it will be incorrectly decoded to another codeword or message. Simulation result on text data shows that once the error exceeds the correcting radius of the code, the resulting decoded message could be changed siginificantly. This work can be extended by using different codes or by using different types of data such as image and audio.

References

[1] Kurose J F and Ross K W 2013 *Computer Networking: A Top-Down Approach* (Addison-Wesley)
[2] Lidl R and Pilz G 2012 *Applied abstract algebra* (Springer Science & Business Media)
[3] Berlekamp E R 2015 *Algebraic coding theory* (World Scientific Publishing Co)
[4] Hamming R W 1980 *Coding and Theory* (Prentice-Hall)
[5] Mstafa R J and Elleithy K M 2014 *Systems, Applications and Technology Conference (LISAT), 2014 IEEE Long Island* (IEEE) pp 1–6
[6] Al-Bahadili H 2008 *Computers & Mathematics with Applications* **56** 143–150
[7] Rurik W and Mazumdar A 2016 *Information Theory Workshop (ITW), 2016 IEEE* (IEEE) pp 404–408
[8] Abadi M A M, Danyali H and Dehnavi M N 2010 *Biomedical Engineering (ICBME), 2010 17th Iranian Conference of* (IEEE) pp 1–4
[9] Developers T 2017 Sagemath, the sage mathematics software system (version 8.0)
[10] Drake D *et al.* 2009 *Available from CTAN,/macros/latex/contrib/sagetex*
[11] Joyner D and Miller R 2008 *ACM Communications in Computer Algebra* **42** 74–78
[12] Perez F and Granger B E 2015 Project jupyter: Computational narratives as the engine of collaborative data science Tech. rep. Technical Report. Technical report, Project Jupyter

IOP Conf. Series: Journal of Physics: Conf. Series **974** (2018) 012065 doi:10.1088/1742-6596/974/1/012065

Modelling lecturer performance index of private university in Tulungagung by using survival analysis with multivariate adaptive regression spline

M Hasyim[1] and **D D Prastyo**[2]

[1]Study Program of Mathematics Education, STKIP Tulungagung, Indonesia
[2]Department of Statistics, Institut Teknologi Sepuluh Nopember, Surabaya, Indonesia

E-mail: [1] maylita@stkippgritulungagung.ac.id, [2] dedy-dp@statistika.its.ac.id

Abstract. Survival analysis performs relationship between independent variables and survival time as dependent variable. In fact, not all survival data can be recorded completely by any reasons. In such situation, the data is called censored data. Moreover, several model for survival analysis requires assumptions. One of the approaches in survival analysis is nonparametric that gives more relax assumption. In this research, the nonparametric approach that is employed is Multivariate Regression Adaptive Spline (MARS). This study is aimed to measure the performance of private university's lecturer. The survival time in this study is duration needed by lecturer to obtain their professional certificate. The results show that research activities is a significant factor along with developing courses material, good publication in international or national journal, and activities in research collaboration.

1. Introduction

Survival analysis is one of the statistical methods that are specially used to process the data or cases related to the time duration until an event occurs. The typical data in survival analysis is the existence of censored data [1]. At the beginning of its development, the study about survival is concerned on the probability prediction of average life expectancy. Recently, survival analysis has been developed for identification of risk factors and prognostic factors associated with the maturity of the disease [2] and many other areas.

The survival analysis describes a process that is associated with time, from the time origin until the occurrence of a special occurrence or end point [3]. There several approaches in survival analysis, for example parametric and nonparametric approaches. One of the nonparametric approaches that can be employed is Multivariate Adaptive Regression Splines (MARS) as a nonparametric and semi parametric regression that takes into accounts the covariates with a multivariate approach [4].

Nonparametric regression using MARS does not depend on the assumption of a certain curve shape, has flexibility in high dimensional data, and modelling involves a lot of interaction with a few variables [5]. In survival analysis using MARS approach, the residuals of Cox Proportional Hazard Model (Cox PHM) is treated as dependent variable in MARS modelling [6]. Reference [7] is an example of research that employ survival analysis using MARS approach in Dengue fever cases, where the dependent variable of MARS models is martingale residual for uncensored data. Another study that employs Cox PHM with MARS approach also had been done for the case of survival of heart attack patients in Germany. It shows that the MARS approach gives better results than Cox PHM [6]. Reference [8] employed such approach in digital economic case.

Another nonparametric approach for survival analysis is survival support vector machine, i.e. the modification of support vector machines (SVM) for survival analysis modelling. References [9, 10] developed additive survival SVM and applied it in medical cases. A simulation studies on survival SVM had been done by [11, 12] and applied to cervical cancer study. The survival SVM and survival MARS have different approaches.

In this study, the lecturer performance index or *Indeks Kinerja Dosen* (IKD) is modelled by using survival MARS. Lecturers are professional educators and scientists with the main task of transforming, developing and disseminating science, technology and the arts through education, research and dedication to society. Lecturer performance is defined as the ability to carry out the work or tasks that lecturers have in completing a job. Performance can be interpreted as work presentation, job performance, work achievement, or work result. It also refers to the result or output of a process. Education is aimed to: (1) improve the performance, capability and output of education; (2) facilitate communication and exchange of information about best practice of education with various types of educational institutions; and (3) as tools to improve institutional performance education and guidelines in strategic planning. By using Survival MARS, it will be known significant factors that influence the time needed by lectures to get their professional certificate.

2. Literature review

2.1. *Survival analysis*

Survival analysis is a statistical method that can be used to analyse data that related to start time (time origin) or start point until the specific event happened (end point) or failure event [13]. To determining the survival time, there are three term required: (i) *Time origin (starting point),* the starting time to record, (ii) *Ending event of interest (recent events*), the time when a lecture got the professional certificate, (iii) *Measurement scale for the passage of time* as a limit of the time of incident from the beginning to the end. The scale is measured in days, weeks, months, or years. In this study measuring scale used the time duration when a lecture was in years.

In survival analysis, there is difficulty in observing data is the possibility of some individual which cannot be observed from the start point to the end point. This situation is called the censored data [1] caused by three reasons: (i) *Loss to follow up*, it occurs when a lecture decides to move another university, (ii) *Drop Out*, it occurs at a lecture chooses to pension, (iii) *Termination of study*, it occurs when the research period was ended while the lecture has not get professional certificate yet.

2.2. *Hazard function and survival function*

There are two main functions in survival analysis. They are the survival function and hazard function. Survival function is the basis of survival analysis that refers to an individual opportunity to survive over time t [2]. Suppose T is survival time and has density function $f(t)$, the cumulative distribution is denoted as:

$$F(t) = P(T < t) = \int_0^t f(u)\,du \tag{1}$$

The survival function $S(t)$ is probability of individual life longer than time t as follows:

$$S(t) = Pr(T > t) = 1 - Pr(T \le t) = 1 - F(t) \tag{2}$$

The hazard function for estimating individual probability get an event at time t is described as [1].

$$h(t)dt = \Pr(t \le T < t + \Delta t \mid T \ge t) \tag{3}$$

Relationship between density function, hazard function, and survival function is:

$$h(t) = \frac{f(t)}{S(t)} \tag{4}$$

The relationship of cumulative hazard and survival function is:

$$S(t) = \exp(-\int h(t)\,dt) = \exp(-H(t)) \text{ and } H(t) = -\ln S(t) \tag{5}$$

2.3. *Cox proportion hazard model*

The Cox PHM is proposed by Cox in 1972 [14] and now is widely developed. This modelling is a log-linear relationship between independent variables and hazard function as follows:

$$h(t \mid X = x) = h_0(t)\exp(\beta' x) \tag{6}$$

where $h_0(t)$ baseline hazard which is the form is not specified. The baseline hazard has specific form corresponding to the distribution of survival time.

Distribution approximation used in this study is time needed (started from first time to be a lecturer) by lecturer to get professional certificate. The distribution of survival time can be tested by using minimum Anderson Darling score as follows.

$$A^2 = -n - \frac{1}{n}\sum_{i=1}^{n}(2i-1)\left[\ln(F(t_i)) + \ln(1 - F(t_{n+1-i}))\right] \tag{7}$$

2.4. *Weibull distribution with three parameters*

Given the survival time follows Weibull distribution with three parameters; the baseline hazard has specific form accordingly. The cumulative distribution for Weibull distribution with three parameters is:

$$\boldsymbol{F(t) = 1 - \exp\left\{-\left(\frac{t-\gamma}{\eta}\right)^{\beta}\right\}, t \geq \gamma} \tag{8}$$

with $\boldsymbol{\gamma}$ is location parameter, $\boldsymbol{\eta}$ is scale parameter, $\boldsymbol{\beta}$ is shape parameter, where $\boldsymbol{\eta, \beta > 0}$ and $\boldsymbol{\gamma \leq t <} \infty$. The baseline hazard for survival time that Weibull distribution with three parameters is:

$$\boldsymbol{h_0(\, t|\gamma, \eta, \beta\,) = \frac{\beta(t-\gamma)^{\beta-1}}{\eta^{\beta}}} \tag{9}$$

2.5. *Multivariate adaptive regression splines (MARS)*

Multivariate Adaptive Regression Spline (MARS) is one of the flexible methods for modelling high-dimensional regression data. MARS is a form of extension of the Basis Spline Functions where the number of basis function is the parameters of the model. Some terms that need to be considered in MARS is as follows,

a. *Knots*, the point of a regression line to form a region of a regression function.
b. *Basis Function* (BF), collection of some of the functions that are used to describe the relationship between the response variable and the predictor.
c. *Interaction*, a correlation between variables and the maximum number of interaction (MI) 1, 2, and 3.

MARS model is formulated as follows:

$$f(x) = \alpha_0 + \sum_{m=1}^{M} \alpha_0 \prod_{k=1}^{Km} S_{km}(x_{v(k,m)} - \tau_{km}) + \varepsilon \tag{10}$$

The best model produces smallest GCV which is denoted as:

$$GCV(m) = \frac{ASR}{\left[1 - \frac{C(M)}{n}\right]^2} = \frac{\frac{1}{n}\sum_{i=1}^{n}[y_i - f_M(x_i)]^2}{\left[1 - \frac{C(M)}{n}\right]^2} \tag{11}$$

where $C(M) = trace(B(B^T B)^{-1} B^T) + 1$.

3. Research methodology

Dependent variable of this study is lecturer performance index (IKD). The IKD is measured for both lecturer as civil servant (DPK) and lecturer with permanent status (DTY). Categorical dependent variable has goal to be suitable with MARS binary response. The data for survival function in this study is the time needed by lecturer as DPK or DTY in private university. The origin point is the starting time becomes lecture. The predictors involve education activities (6 indicators), research activities (4 indicators) and public services (3 indicators). Steps of analyses in this study are as follows:

a. Determining the survival data that will be used and eliminate the censored data.
b. Describing the characteristics of lecturer as research object
c. Fitting the distribution of survival data based on smallest score of Anderson-Darling test
d. Determining the form of baseline hazard function
e. Estimating the survival function and cumulative hazard function
f. Modelling survival function with MARS approach

4. Analysis and discussion

The first step of analysis is to fit the distribution of survival data. According to minimum Anderson Darling as reported in Table, the survival time follows Weibull distribution with three parameters such that the baseline hazard is specified accordingly.

Table 1. Anderson Darling Score

No.	Distribution	AD Score
1	2-Parameter Weibull	1.094
2	2-Parameter Lognormal	1.760
3	Exponential	9.235
4	Log logistic	2.043
5	3-Parameters Weibull	1.020
6	3-Parameters Lognormal	2.703
7	3-Parameters Exponential	11.492
8	3-Parameters Log logistic	1.185

The baseline hazard function is estimated as follows:

$$h_0(t\,|\,\hat{\gamma},\hat{\eta},\hat{\beta}) = \frac{\hat{\beta}(t-\hat{\gamma})^{\hat{\beta}-1}}{\hat{\eta}^{\hat{\beta}}} = \frac{3.1503(t-(-1.708))^{6.1503-1}}{3.345^{6.1503}}$$

$$= 0.00365(t+1.708)^{5.1503}$$

According to the baseline hazard function, the calculation of survival function and cumulative hazard function can be made. Estimation of survival function and cumulative hazard are shown on Table 2.

Table 2. Survival Function and Cumulative Hazard

t	$S(t)$	$H(t)$
1	0.998105	0.001897
2	0.925461	0.077463
3	0.462685	0.770708
4	0.016751	1
5	2.57E-07	1

Table 2 indicates the probability a lecture get their professional certificate through time *t*. In general, rate of getting certificate is increase in sequence time. A lecturer will get his/her existence of profession between three and four years, starting from first time as a lecture.
Based on the results of trial and error combination BF, MI, and MO, the combination of which produces minimum GCV value is a combination of 22, 3, 1 with a value of GCV = 0.373 with R^2 = 0.829. Based on the results of this combination, it is known MARS models produced are as follows:

$$\hat{y} = \alpha_0 + \sum_{m=1}^{M} \alpha_m \prod_{k=1}^{K_m} \left[s_{km} \cdot \left(x_{v(k,m)} - t_{km}\right)\right]_+^2 + \varepsilon_i$$

$$= 0.617 + 0.872BF_6 + 0.539BF_7 + 0.463BF_9 + 0.312BF_{16}$$

and

$$h(t) = h_0(t)\exp(\hat{y})$$

$$h(t) = 0.00365\,(1 + 1.708)^{5.1503}\exp(0.617 + 0.872BF_6 + 0.539BF_7 + 0.463BF_9 + 0.312BF_{16})$$

where BF_6 = research activities, BF_7 = submission of article to International/national journal, BF_9 = attending conference activities, and BF_{16}= participating in workshop and training.

The results of survival MARS modelling show that in general, the variables that affect the survival to be professional lecturer are research activities, submission International/national journal, Attend conference activities, workshop and training.

5. Conclusion

Based on the empirical results using Cox PHM MARS approach, which used a combination Basis Functions, Maximum interaction are 22, 3, and 1 with a minimum GCV value was 0.022. Significant predictors influencing the survival time until the professional certificate for lecturer is obtained are research activities, submission to International/national journal, attending conference activities, and participation in workshop and training. Research activities have the largest contribution in the model, followed by submission to international/national journal, attending the conference, and participating in workshop and training.

Acknowledgments
Authors wishing to acknowledge assistance or encouragement from research partner for any advice and idea. Special thanks to Indonesian ministry for research and higher education for the financial support under PEKERTI scheme.

References
[1] Kleinbaum, D.G. 2012. *Survival Analysis*. London: Springer
[2] Lee, E.T. 2003 *Statistical Method for survival Data Analysis*. London: John Willey
[3] Collet, D. 2003. *Modelling Survival Data in Medical Research*. London: Chapman & Hall.
[4] Friedman, J.H. 1991. Estimating Function of Mixed Ordinal and Categorical Variables Using Multivariate Adaptive Regression Splines. Technical Report LCS 107, Statistics Department, Stanford University
[5] Friedman, J.H. 1991. Multivariate Adaptive Regression Spline. *The Annals of Statistics*, 19: 1-141.
[6] Kriner, M. 2007. *Survival Analysis with Multivariate Adaptive Regression Splines*. Dissertation. Munchen University.

[7] Nisa', F.S. and Budiantara, I.N. 2012. Analisis Survival dengan Pendekatan Multivariate Adaptive Regression Spline pada Kasus Demam Berdarah Dengue (DBD). *Jurnal Sains dan Seni ITS*, 1 (1): 318-323.

[8] Irwansyah, E., Nyoman. D.A. and Bekti R.D. 2015. Cox Proportional Hazard with Multivariate Adaptive Regression Spline to Analyse the Product Sales Time in E-Commerce. *International Journal of Applied Mathematics and Statistics*, 53(5).

[9] Van Belle, V., Pelckmans, K., Suykens, J.A. and Van Huffel, S. 2010. Additive Survival Least-Squares Support Vector Machines, *Statistics in Medicine*, 29(2): 296-308.

[10] Van Belle, V., Pelckmans, K., Suykens, J.A. and Van Huffel, S. 2011. Support Vector Methods for Survival Analysis: A Comparison between Ranking and Regression Approaches, *Artificial Intelligence in Medicine*, 53(2): 107-118.

[11] Khoiri, H.A., Prastyo, D.D. and Purnami, S.W. 2017. A Simulation Study and Application of Feature Selection on Survival Least Square Support Vector Machines. *Proceeding of the 13th IMT-GT International Conference on Mathematics, Statistics, and their Applications (ICMSA2017), AIP Conference Proceedings* 1905 (050024).

[12] Khotimah, C., Purnami, S.W., Prastyo, D.D., Chosuvivatwong, V. and Sriplung, H. 2017. Additive Survival Least Square Support Vector Machines: A Simulation Study and Its Application to Cervical Cancer Prediction. *Proceeding of the 13th IMT-GT International Conference on Mathematics, Statistics, and their Applications (ICMSA2017), AIP Conference Proceedings* 1905 (050024).

[13] Le, C.T. 1997. *Applied Survival Analysis*. New York: John Wiley & Sons, Inc.

[14] Cox, D. R. 1972. Regression Model and Live Tables (with discussion), *Journal of The Royal Statistical Society*, 34: 187-220.

IOP Conf. Series: Journal of Physics: Conf. Series **974** (2018) 012066 doi:10.1088/1742-6596/974/1/012066

Steering Angle Control of Car for Dubins Path-tracking Using Model Predictive Control

Dian Kusuma Rahma Putri[1], **Subchan**[2], **Tahiyatul Asfihani**[3]

Department of Mathematics, Faculty of Mathematics and Natural Science, Sepuluh Nopember Institute of Technology(ITS)
Jl. Arief Rahman Hakim, Surabaya 60111 Indonesia

E-mail: diankrp23@gmail.com[1], s.subchan@gmail.com[2], tahiyatul.asfihani@gmail.com[3]

Abstract. Car as one of transportation is inseparable from technological developments. About ten years, there are a lot of research and development on lane keeping system(LKS) which is a system that automaticaly controls the steering to keep the vehicle especially car always on track. This system can be developed for unmanned cars. Unmanned system car requires navigation, guidance and control which is able to direct the vehicle to move toward the desired path. The guidance system is represented by using Dubins-Path that will be controlled by using Model Predictive Control. The control objective is to keep the car's movement that represented by dinamic lateral motion model so car can move according to the path appropriately. The simulation control on the four types of trajectories that generate the value for steering angle and steering angle changes are at the specified interval.

1. Introduction

Rapid technological developments affect many aspect of human life, on of them is transportation field. Car as one of transportation vehicle didn't escape from this technological developments. About ten years, there was many research and development about Lane Keeping System (LKS) which is a system that automatically controls the steering to keep the vehicle, especially car always on track[1]. That system can be further developed into unmanned car.

Unmanned car requires navigation, guidance and control system which are able to direct the vehicle to move toward the desired point[2]. Navigation system usually using Global Positioning System (GPS) which gives the position on earth by using satellite signals. Guidance system is a path planning that result desired trajectory based on initial and final position from navigation system. Path planning can be obtained from several methods, one of them is Dubins Path. Dubins path is an optimal path that satisfy maximum curvature bound between two points with a particular orientation in a field of both CLC(*Circle-Line-Circle*) or CCC(*Circle-Circle-Circle*) path, or combination of both where C is circle arc and L is a line that intersect with C([3],[4]). Dubins path have several advantages including path planning using this method can use for angle in all quadrant both at initial or final point, and this path can optimize the time because it can generate the path curvature[2]. Besides that, Dubins path not require a lot of time in its planning so it can be more optimal and efficient[5], it also requires little itme to achieve the target[6].

The next system that requires is control system which aims to makes the car can move according the desired path. This control system also can be obtained from several methods, one

IOP Conf. Series: Journal of Physics: Conf. Series **974** (2018) 012066 doi:10.1088/1742-6596/974/1/012066

of them is Model Predictive Control (MPC). Basically, MPC use the model explicitly to predict the output process in a horizon time and the objective of control's calculation is to minimize a certain objective function. MPC has several advantages compare to other methods, including its uses to control a multivariable case easily and it can solve the system with the disturbance using simple concept directly during the design process[7].

Research on trajectory-tracking have done before where the track obtained from the GPS that sometimes given waypoint irregularly, so a curve fitting done based on that irregular waypoints. The result of that research is the oscillations value of steering angle decreased significantly using MPC method compared with linear quadratic, especially in the curved trajectory[8].

Based on some of the things that has been said, this research aims to control the car's steering angle in order to move in accordance with the desired trajectory especially in not straight trajectory.

2. Research method

In this chapter are discribed the steps to control the car's steering angle on Dubins-path tracking using Model Predictive Control including determination of Dubins-path, dynamic lateral motion model, discretization of model, formulation of objective function and its constraints.

2.1. Determination of Dubins-path

Determination of Dubins-path required an initial and final position of the car. Both of these positions consists of positions in Cartesian coordinates (x, y), car's heading angle (ψ) and minimum radius of curvature in this case symbolized with R. R obtained from a car's minimum turning radius.

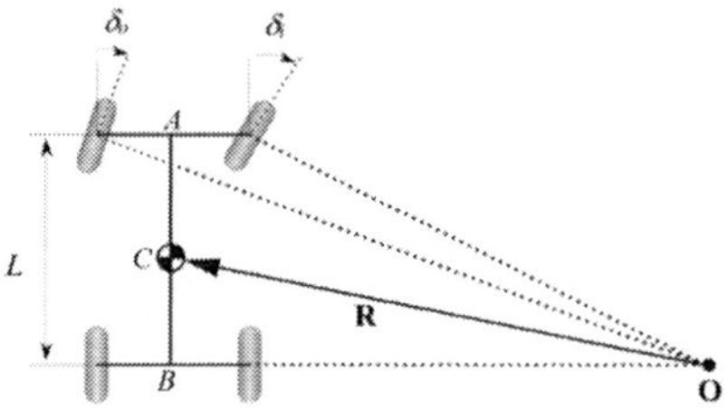

Figure 1. Car's turning geometry[9]

Before obtained the value of R, first thing to do was identifying things that related such as front and rear wheels steering angle that symbolized with δ_f and δ_r, and the center of gravity of the car that given by point C. The outside steering angle (δ_o) and inside steering angle (δ_i) of the front wheels in Fig. 1 is considerered to have same value so it can be symbolized by δ_f. The distance between each point A and B to point C is given by l_f and l_r. So the length of L that defined as the distance between the front and rear wheels is $L = l_f + l_r$. The car's velocity at center of gravity is symbolized by V with angle of slip β. In this paper, assumed that there was not a angle slip or $\beta = 0$ so the car's velocity has same direction with car's heading angle. Suppose that L is very small compared with the R, so the angular velocity can be approximated by[1]

$$\frac{\dot{\psi}}{V} \approx \frac{1}{R} = \frac{\delta}{L}$$

For the steering angle maximum input $\delta = 0.5386 rad$, $l_f = 1.1m$ and $l_r = 1.58m$ result $R = 4.9759 \approx 5m$.

Initial and final position defined after R have been obtained as follows:

$$p_s = \begin{bmatrix} 1100 & 1150 & 180 & 5 \end{bmatrix}$$
$$p_f = \begin{bmatrix} 3200 & 2675 & 180 & 5 \end{bmatrix}$$

The length of each path type given in Table 1 can be obtained using Dubins-algortihm.

Table 1. Dubins-path length table

Dubins-path type	Path length (m)
RSR	2626.7
LSL	2626.7
RSL	2614.6
LSR	2664

Table 1 provides RSL as optimal trajectory path with its length is 2664 m. The path form given on Fig. 2. Minimum value of car's turning radius is too small compared to the trajectory length so it's not visible in the initial form of trajectory. Therefore, magnification of image at the initial and final positions done and shown in Fig. 3.

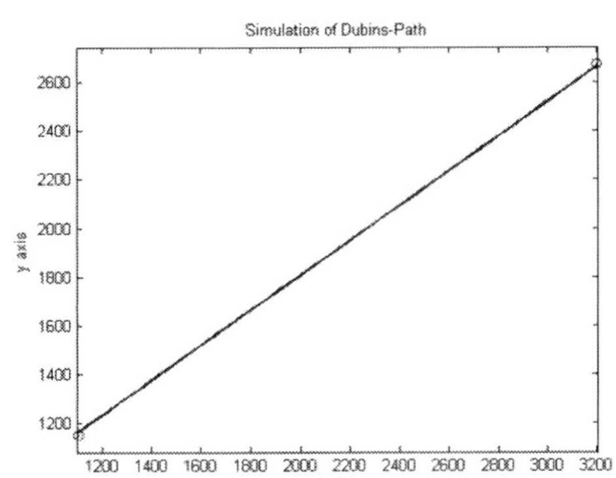

Figure 2. RSL type of Dubins-path

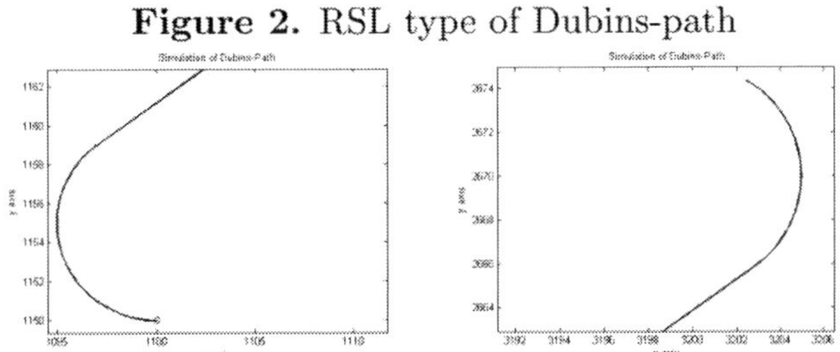

Figure 3. Image magnification of Dubins-Path at initial position (left) and final position (right)

2.2. Dynamic lateral motion model

Dynamic lateral motion model ([1],[9]) is given as follows:

$$\dot{x} = \begin{bmatrix} -\frac{2C_{\alpha f}+2C_{\alpha r}}{mV_x} & -\frac{2C_{\alpha f}-2C_{\alpha r}}{mV_x} - V_x \\ -\frac{2l_f C_{\alpha f}-2l_r C_{\alpha r}}{I_z V_x} & -\frac{2l_f{}^2 C_{\alpha f}+2l_r{}^2 C_{\alpha r}}{I_z V_x} \end{bmatrix} x + \begin{bmatrix} \frac{2C_{\alpha f}}{m} \\ \frac{2l_f C_{\alpha f}}{I_z} \end{bmatrix} u \tag{1}$$

IOP Conf. Series: Journal of Physics: Conf. Series **974** (2018) 012066 doi:10.1088/1742-6596/974/1/012066

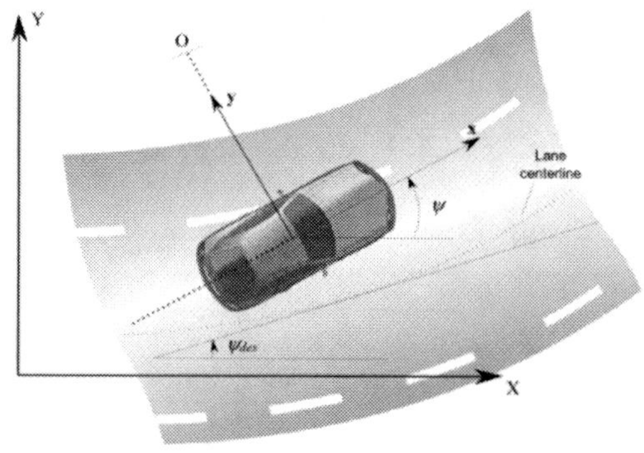

Figure 4. Dynamic lateral motion[9]

with $x = \begin{bmatrix} \dot{y} \\ \dot{\psi} \end{bmatrix}$, $u = \delta$ and output system is $\dot{\psi}$ which can be expressed in matrix form as follows:

$$\bar{y} = \begin{bmatrix} 0 & 1 \end{bmatrix} x$$

and also parameters given as follow:

$\{XYZ\}$: global coordinate plane
$\{xyz\}$: local (vehicle) coordinate plane
m : vehicle total mass
I_z : yaw moment of inertia of vehicle
l : (l_f, l_r) longitudinal distance from c.g. to (front,rear) tires
α : (α_f, α_r) slip angle at (front,rear) wheel tires
C : $(C_{\alpha f}, C_{\alpha r})$ cornering stiffness of (front, rear) tires
δ : steering angle

2.3. Discretization of model
Model in the discrete time is given in equation as follow:

$$x(k+1) = A_d(T)x(k) + B_d(T)u(k)$$

with

$$\begin{aligned} A_d(T) &= e^{At} \\ B_d(T) &= \left(\int_0^T e^{A\lambda} d\lambda \right) B \end{aligned}$$

and

$$e^{AT} = I + At + \frac{1}{2!}A^2t^2 + \frac{1}{3!}A^3t^3 + \cdots + \frac{1}{n!}A^nt^n + \cdots$$

Analytic calculation produces value cutting for $n > 6$ so discretization done using a software to obtain more accurate results. Based on state space equation in (1), discrete state space obtained with a sampling time 0.1 seconds as follow:

$$x(k+1) = \begin{bmatrix} 0.4450 & -1.3734 \\ 0.0431 & 0.4402 \end{bmatrix} x(k) + \begin{bmatrix} 1.6503 \\ 4.5607 \end{bmatrix} u(k)$$

with system output is $\bar{y} = \begin{bmatrix} 0 & 1 \end{bmatrix} x$

IOP Conf. Series: Journal of Physics: Conf. Series **974** (2018) 012066 doi:10.1088/1742-6596/974/1/012066

2.4. Control design using MPC

Control design using MPC was done by determining the objective function and its constraint in quadratic programming form.

A discrete system is given as follow:

$$x(k+1) = Ax(k) + Bu(k) \tag{2}$$
$$y(k) = Cx(k) \tag{3}$$

Controls carried out on the steering angle so that the steering angle changes need to know from time to time during the journey of the vehicle (car). Therefore, the input increment is defined as follows:

$$\Delta u(k) = u(k) - u(k-1)$$

so equation (2) becomes

$$x(k+1) = Ax(k) + Bu(k-1) + B\Delta u(k) \tag{4}$$

Consecutive given prediction horizon and control horizon, N_p and N_c with $N_c = N_p$ so obtainable

$$\begin{aligned} x(k+1|k) &= Ax(k|k) + Bu(k|k) \\ &= Ax(k|k) + Bu(k-1|k) + B\Delta u(k|k) \\ x(k+2|k) &= Ax(k+1|k) + Bu(k|k) + \\ & B\Delta u(k+1|k) \\ &\vdots \\ x(k+N_p|k) &= Ax(k+N_p-1|k) + Bu(k-1|k) + \\ & B\Delta u(k|k) + \cdots + B\Delta u(k+N_p-1|k) \end{aligned}$$

with prediction output as follows:

$$\begin{aligned} \hat{y}(k|k) &= Cx(k|k) \\ \hat{y}(k+1|k) &= Cx(k+1|k) \\ &\vdots \\ \hat{y}(k+N_p|k) &= Cx(k+N_p|k) \end{aligned}$$

Objective function that uses in this research is given as follows:

$$J_{min} = \sum_{i=1}^{N_p} \|\hat{r}(k+i|k) - \hat{y}(k+i|k)\|^2_{Q_{(i)}} + \|\Delta\hat{u}(k+i-1|k)\|^2_{R_{(i-1)}}$$

with the first part of objective function is representaion of error between prediction output $(\hat{y}(k+i|k))$ and reference from Dubins-path $(\hat{r}(k+i|k))$ while the seconde part of objective function is representation of required energy when steering angle changes $(\Delta\hat{u}(k+i|k))$ at $k+i$ time that measure at k time. Objective function changes in quadratic programming form and result this equation bellows:

$$J_{min} = \frac{1}{2} j^T E' j + j^T F \tag{5}$$

with

$$j = \begin{bmatrix} \Delta u(k|k) \\ x(k+1|k) \\ \vdots \\ \Delta u(k+N_p-1|k) \\ x(k+N_p|k) \end{bmatrix}, F = \begin{bmatrix} 0 \\ -2C^T Q_1 \hat{r}(k+1|k) \\ \vdots \\ 0 \\ -2C^T Q_{N_p} \hat{r}(k+N_p|k) \end{bmatrix}, E = \begin{bmatrix} R_0 & 0 & \cdots & 0 & 0 \\ 0 & C^T Q_1 C & \cdots & 0 & 0 \\ \vdots & \vdots & \ddots & \vdots & \vdots \\ 0 & 0 & \cdots & R_{N_p-1} & 0 \\ 0 & 0 & \cdots & 0 & C^T Q_{N_p} C \end{bmatrix}$$

Objective function is given as equation (5) with constraint given as follows:

$$\begin{aligned} \Delta u^{min} \leq & \quad \Delta u & \leq \Delta u^{max} \\ u^{min} \leq & \quad u & \leq u^{max} \end{aligned}$$

That constraint are formulated in quadratic programming form

$$Pj \leq h$$

and matrices that corresponding to the boundary constraints was obtained as follow:

$$P = \begin{bmatrix} M_1 & 0 & 0 & \cdots & 0 & 0 \\ 0 & 0 & M_1 & \cdots & 0 & 0 \\ 0 & 0 & 0 & \cdots & 0 & 0 \\ \vdots & \vdots & \vdots & \ddots & \vdots & \vdots \\ 0 & 0 & 0 & \cdots & M_1 & 0 \\ M_2 & 0 & 0 & \cdots & 0 & 0 \end{bmatrix}, h = \begin{bmatrix} N_1 \\ N_1 \\ N_1 \\ \vdots \\ N_1 \\ N_2 \end{bmatrix}$$

The constraints will be formulated from equation (4) with prediction horizon N_p

$$Yj = b$$

then obtained the corresponding matrices as follows:

$$Y = \begin{bmatrix} -B & I & 0 & 0 & 0 & 0 & \cdots & 0 \\ -B & -A & -B & I & 0 & 0 & \cdots & 0 \\ -B & 0 & -B & -A & -B & I & \cdots & 0 \\ \vdots & \vdots & \vdots & \vdots & \vdots & \ddots & \vdots & \\ -B & 0 & -B & 0 & -B & 0 & \cdots & I \end{bmatrix}, b = \begin{bmatrix} Ax(k|k) + Bu(k-1|k) \\ Bu(k-1|k) \\ Bu(k-1|k) \\ \vdots \\ Bu(k-1|k) \end{bmatrix}$$

In this paper, given the constraints values as follows:

$$\begin{aligned} -0.4987(rad) \leq & \quad \Delta u & \leq 0.4987(rad) \\ -0.5386(rad) \leq & \quad u & \leq 0.5386(rad) \end{aligned}$$

3. Analysis of the experimental results

In this section, there will be simulation using weighting coeffisients $Q = 100$ and $R = 1$. It's mean that objective function more proritize on prediction output that have been controlled accordance with reference which given and ignoring driver comfort despite the large steering angle changes in a short time. Next will be simulated some cases the steering angle control problems on the car.In the simulation, the initial value is defined $x_1(1) = 0.5$ m/s and $x_2(1) = 0$ rad. It also defined the initial value of the steering angle is $u(1) = 0$ rad.

3.1. The influence of trajectory type with same value of prediction horizon

In order To determine the effect of trajectories on the results of MPC, simulation test conducted with a different trajectory and the other parameter in the same value. Simulations were performed for 60 seconds with a sampling time 0.1 seconds.

Table 2. Dubins-path length table on simulation 1

Dubins-path type	Path length (m)
RSR	1796.3
LSL	1844.4
RSL	1826.2
LSR	1814.5

Simulation 1 The first trajectory using initial and final position that defined as follows:

$$p_s = \begin{bmatrix} 10 & 10 & 180 & 5 \end{bmatrix}$$
$$p_f = \begin{bmatrix} 1000 & 1500 & 0 & 5 \end{bmatrix}$$

The length of each path type given in Table 2 can be obtained using Dubins-algortihm.

Table 2 provides RSR (Right-Straight-Right) as optimal trajectory path with its length is 1796.3 m.The simulation results for prediction horizon $N_p = 10$ indicates that the steering angle and steering angle changes do not exceed the constraint limits which shown in Fig. 5 and Fig. 6. Fig. 7 shows that difference value between reference of car's heading angle changes and car's heading angle which was controlled is small or can be said almost same but at some point in the beginning, there is a considerable difference between reference of car's heading angle changes and car's heading angle which was controlled because of trajectory tracking influence.

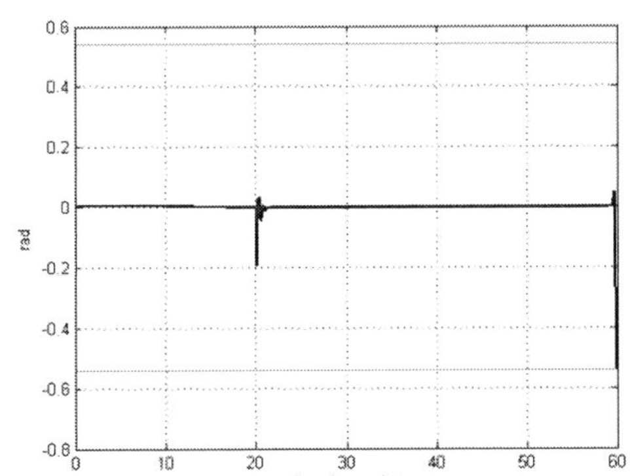

Figure 5. Steering angle which was controlled on RSR type of Dubins-path

Simulation 2 The second trajectory using initial and final position that defined as follows::

$$p_s = \begin{bmatrix} 1100 & 1150 & 180 & 5 \end{bmatrix}$$
$$p_f = \begin{bmatrix} 2600 & 2065 & 180 & 5 \end{bmatrix}$$

The length of each path type given in Table 3 can be obtained using Dubins-algortihm.

Table 3 provides RSL (Right-Straight-Left) as optimal trajectory path with its length is 1777.9 m.The simulation results for prediction horizon $N_p = 10$ indicates that the steering angle and steering angle changes do not exceed the constraint limits which shown in Fig. 8 and Fig. 9. Fig. 10 shows that difference value between reference of car's heading angle changes and car's heading angle which was controlled is small or can be said almost same but at some point in the

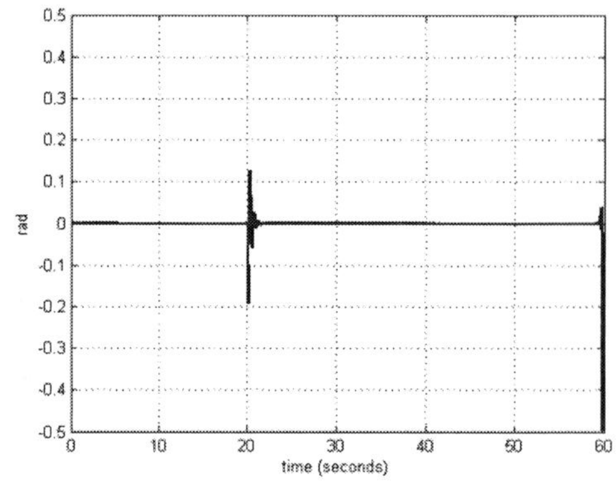

Figure 6. The changes of steering angle which was controlled on RSR type of Dubins-path

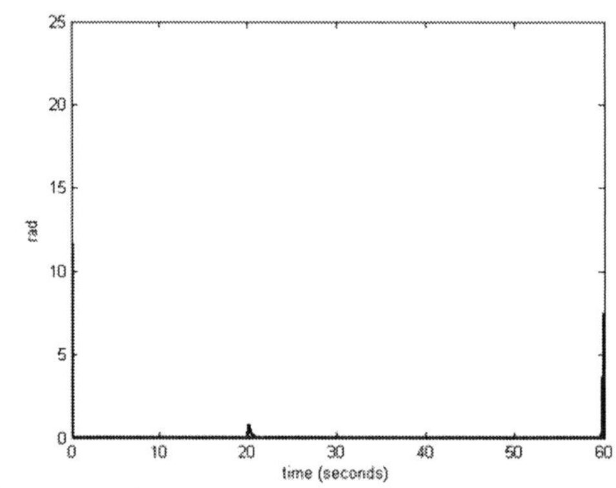

Figure 7. The difference between heading angle of car's which was controlled and its reference on RSR Type of Dubins-path

Table 3. Dubins-path length table on simulation 2

Dubins-path type	Path length (m)
RSR	1788.5
LSL	1788.5
RSL	1777.9
LSR	1825.1

beginning, there is a considerable difference between reference of car's heading angle changes and car's heading angle which was controlled because of trajectory tracking influence.

Simulation 3 The third trajectory using initial and final position that defined as follows:

$$p_s = \begin{bmatrix}10 & 1200 & 120 & 5\end{bmatrix}$$
$$p_f = \begin{bmatrix}200 & 10 & 45 & 5\end{bmatrix}$$

The length of each path type given in Table 4 can be obtained using Dubins-algortihm.

Table 4 provides LSL (Left-Straight-Left) as optimal trajectory path with its length is 1224.1 m.The simulation results for prediction horizon $N_p = 10$ indicates that the steering angle and

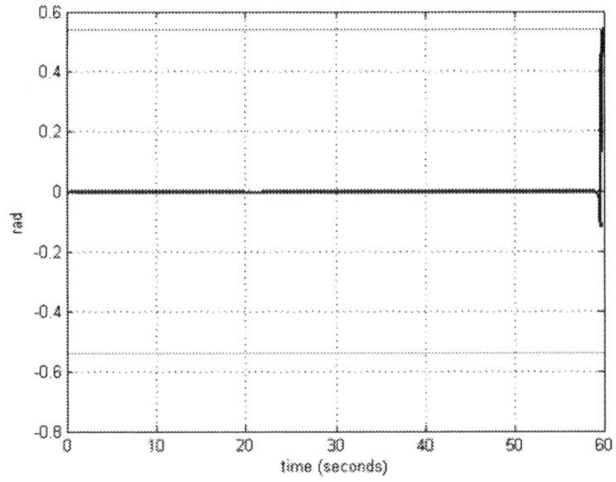

Figure 8. Steering angle which was controlled on RSL type of Dubins-path

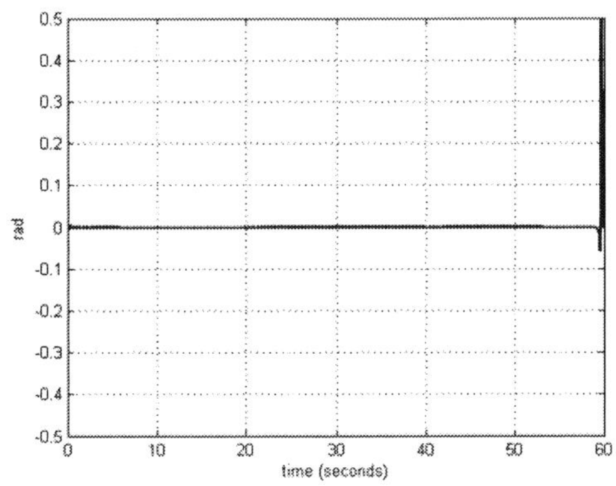

Figure 9. The changes of steering angle which was controlled on RSL type of Dubins-path

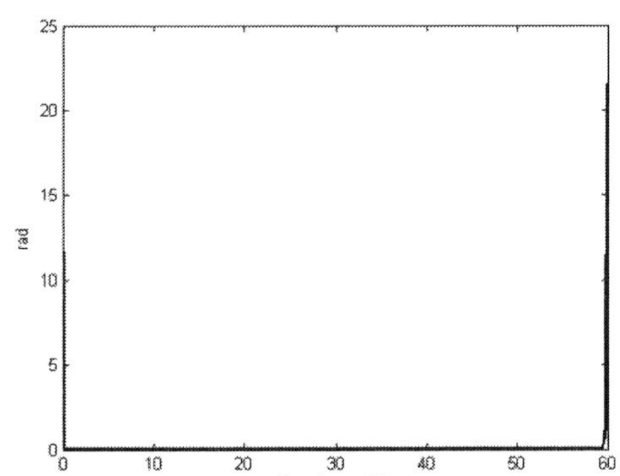

Figure 10. The difference between heading angle of car's which was controlled and its reference on RSL Type of Dubins-path

steering angle changes do not exceed the constraint limits which shown in Fig. 11 and Fig. 12. Fig. 13 shows that difference value between reference of car's heading angle changes and car's heading angle which was controlled is small or can be said almost same but at some point in the beginning, there is a considerable difference between reference of car's heading angle changes and car's heading angle which was controlled because of trajectory tracking influence.

Table 4. Dubins-path length table on simulation 3

Dubins-path type	Path length (m)
RSR	1248.9
LSL	1224.1
RSL	1231.5
LSR	1263.6

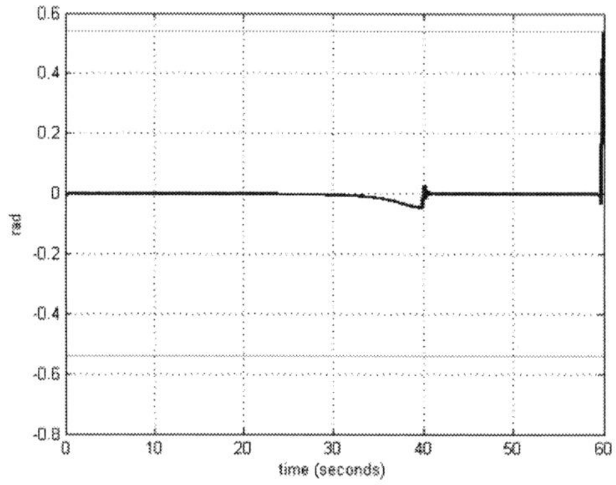

Figure 11. Steering angle which was controlled on LSL type of Dubins-path

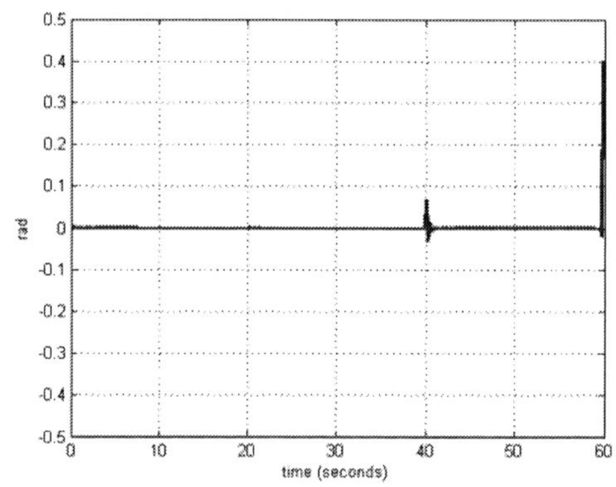

Figure 12. The changes of steering angle which was controlled on LSL type of Dubins-path

Simulation 4 The fourth trajectory using initial and final position that defined as follows::

$$p_s = \begin{bmatrix}1500 & 0 & 90 & 5\end{bmatrix}$$
$$p_f = \begin{bmatrix}0 & 0 & 30 & 5\end{bmatrix}$$

The length of each path type given in Table 5 can be obtained using Dubins-algortihm.

Table 5 provides LSR (Left-Straight-Right) as optimal trajectory path with its length is 1513.5 m.The simulation results for prediction horizon $N_p = 10$ indicates that the steering angle and steering angle changes do not exceed the constraint limits which shown in Fig. 14 and Fig. 15. Fig. 16 shows that difference value between reference of car's heading angle changes and car's heading angle which was controlled is small or can be said almost same but at point $t = 40$, there is a large difference because of trajectory tracking influence.

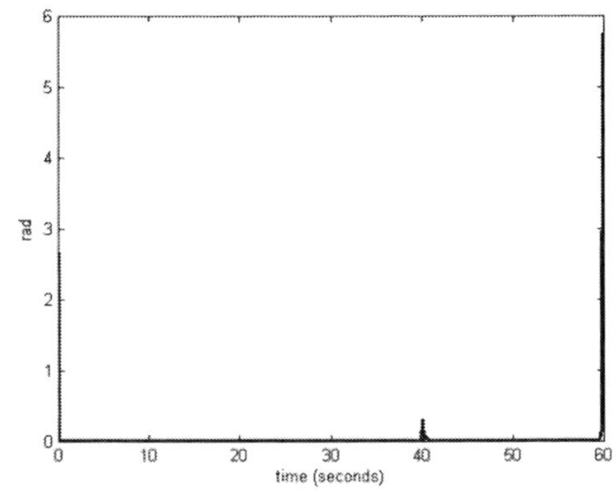

Figure 13. The difference between heading angle of car's which was controlled and its reference on LSL Type of Dubins-path

Table 5. Dubins-path length table on simulation 4

Dubins-path type	Path length (m)
RSR	1539.2
LSL	1523.7
RSL	1549.5
LSR	1513.5

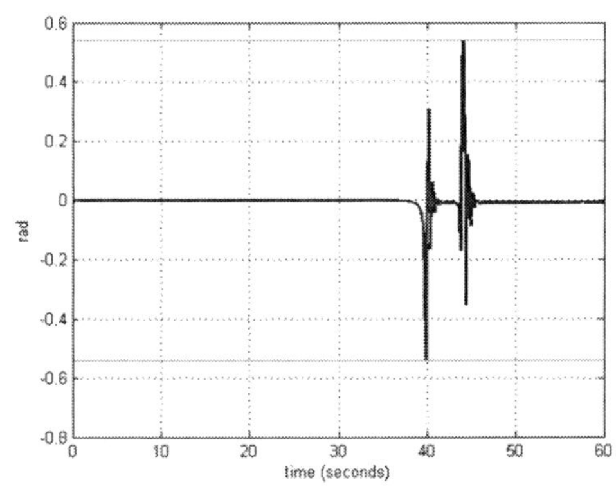

Figure 14. Steering angle which was controlled on LSR type of Dubins-path

3.2. The influence of prediction horizon with the same trajectory type

In order to determine the effect of prediction horizon on the results of MPC, simulation test conducted with a different value of prediction horizon that is 10, 50 and 100. Each value of prediction horizon will be simulated with the same trajectory type which intial and final position defined as follows:

$$\begin{aligned} p_s &= \begin{bmatrix} 1100 & 1150 & 180 & 5 \end{bmatrix} \\ p_f &= \begin{bmatrix} 2600 & 2065 & 180 & 5 \end{bmatrix} \end{aligned}$$

which produces the type and length of the path as in the previous simulations 2. The other parameter at the same value. Simulations were performed for 60 seconds with a sampling time 0.1 seconds. The value of Root-Mean-Square-Error(RMSE) and its computation time for each

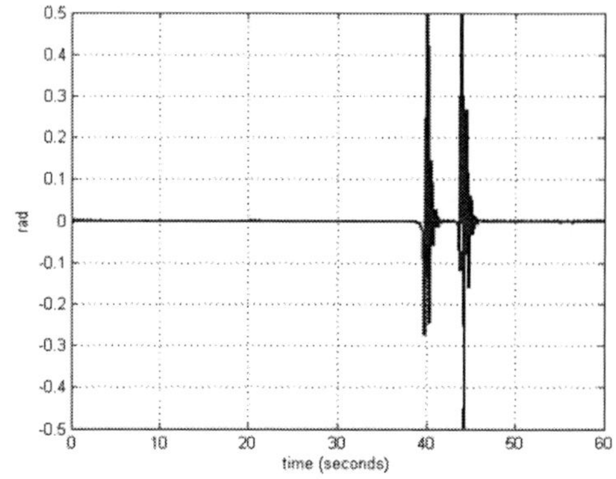

Figure 15. The changes of steering angle which was controlled on LSR type of Dubins-path

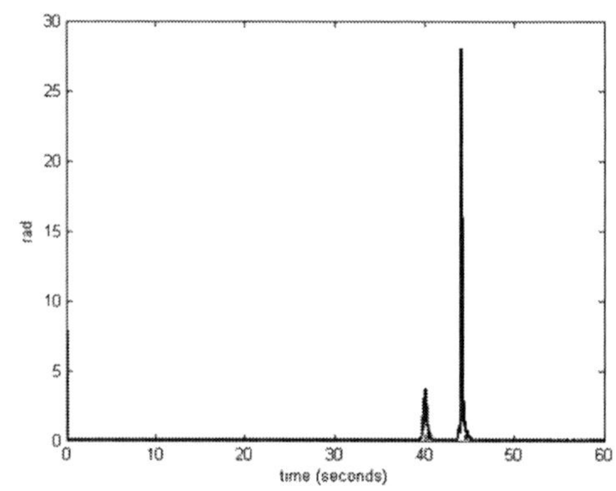

Figure 16. The difference between heading angle of car's which was controlled and its reference on LSR Type of Dubins-path

value of N_p are calculated to determine the effect of N_p value changes.

The result is given on Table 6 which is shown the differences between value of RSME and

Table 6. Error value between reference of car's heading angle and car's heading angle which was controlled and each computation time

Np	RMSE	Computation time (seconds)
10	0.00704210296467999	17.6487602346906
50	0.00706729399940004	244.029521069138
100	0.00706729399939593	357.740033058028

computation time on $N_p = 10$ and $N_p = 50$,while on $N_p = 50$ and $N_p = 100$ the differences between value of RSME is small but there was a large difference on camputation time. $N_p = 10$ is better than $N_p = 50$ and $N_p = 100$ when viewed from the value of RSME and computation time for each N_p.

4. Conclusion

Based on the analysis and simulation of control using MPC, obtained the following conclusions:

(i) Model Predictive Control (MPC) can be applied well in steering angle control of car for Dubins path-tracking. It can be seen from the simulation control on the four types of trajectories that generate the value for steering angle and steering angle changes are at the specified interval.

(ii) The simulation results showed that in this paper which is using coefficient weights $Q = 100$ and $R = 1$, $N_p = 10$ is better than $N_p = 50$ and $N_p = 100$ when viewed from the value of RSME and computation time for each N_p.

References

[1] Rajamani, R 2012. *Vehicle Dynamic and Control* (Springer)
[2] Dewi, N. K 2010 *Perencanaan Lintasan Menggunakan Geometri Dubins Pada Pesawat Udara Nir Awak (PUNA).* (Tugas Akhir Jurusan Matematika. Surabaya, Jawa Timur, Indonesia: ITS)
[3] Tsourdos, A., White, B., and Shanmugavel, M 2011 *Cooperative Path Planning of Unmanned Aerial Vehicles* (WILEY)
[4] Subchan, S., White, B., Tsourdos, A., Shanmugavel, M., and Zbikowski, R 2008 *Dubins Path Planning of Multiple UAVs for Tracking Contaminant Cloud* (Proceedings of the 17th World Congress The International Federation of Automatic Control. Seoul, Korea, pages: 5718-5723)
[5] Mu'alifah, N., Herisman, I., dan Subchan 2013 *Perencanaan Lintasan Dubins-Geometri pada Kapal Tanpa Awak untuk Menghindari Halangan Statis* (JURNAL SAINS DAN SENI Vol. 1, No. 1, halaman:1-6)
[6] Sorbo, H. E 2013 *Vehicle Collision Avoidance System* (Thesis Department of Engineering Cybernetics. Norwegian: NTNU)
[7] Camacho, E. F. and Bordous, C 1995 *Model Predictive Control in the Process Industry* (Springer: Verlag London)
[8] S.J. Jeon, C.M. Kang, S. -H. Lee and C. C. Chung 2015 *GPS Waypoint Fitting and Tracking Using Model Predictive Control* (IEEE Intelligent Vehicles Symposium (IV), pages: 298-303)
[9] Jazar, R., N 2009 *Vehicle Dynamic Theory and Applications* (Springer: Australia)

IOP Conf. Series: Journal of Physics: Conf. Series **974** (2018) 012067 doi:10.1088/1742-6596/974/1/012067

Modelling fourier regression for time series data- a case study: modelling inflation in foods sector in Indonesia

[1]Alan Prahutama, [2]Suparti and [3]Tiani Wahyu Utami

[1,2]Statistics Department, Diponegoro University, Semarang, Indonesia
[3]Statistics Department, Muhammadiyah Semarang University, Semarang, Indonesia

E-mail: alanprahutama@undip.ac.id

Abstract
Regression analysis is an analysis to model the relationship between response variables and predictor variables. The parametric approach to the regression model is very strict with the assumption, but nonparametric regression model isn't need assumption of model. Time series data is the data of a variable that is observed based on a certain time, so if the time series data wanted to be modeled by regression, then we should determined the response and predictor variables first. Determination of the response variable in time series is variable in *t*-th (y_t), while the predictor variable is a significant lag. In nonparametric regression modeling, one developing approach is to use the Fourier series approach. One of the advantages of nonparametric regression approach using Fourier series is able to overcome data having trigonometric distribution. In modeling using Fourier series needs parameter of K. To determine the number of K can be used Generalized Cross Validation method. In inflation modeling for the transportation sector, communication and financial services using Fourier series yields an optimal K of 120 parameters with R-square 99%. Whereas if it was modeled by multiple linear regression yield R-square 90%.

1. Introduction

Regression method is a statistical method to know the relationship between dependent variable with independent variable. Regression method can be done with three approaches such as parametric, semiparametric and nonparametric regression. A nonparametric regression approach is performed if the function of the regression is not known to form the function curve. It has been widely developed, among others, using Spline, Kernel, Local Polynomial, Wavelet, and Fourier. One of the advantages in nonparametric regression approach using Fourier series was able to overcome data having trigonometric spreading, in this case sine and cosine. The data pattern corresponding to the Fourier approach is a repeatable data pattern, which is a repetition of the value of the dependent variable for different independent variables. Research using Fourier series for nonparametric and semiparametric have been developed. Research on the Fourier series includes [1] Bilodeau (1992), then [2] Tripena (2007) reviewed Fourier series estimators on nonparmetric regression, [3] Semiati (2010) developed an estimated nonparametric regression model of the Fourier birespon series. As for semiparametric regression using Fourier series has been developed by [4] Asrini (2011). In the determination of K optimal in Fourier regression using Generalized Cross Validation (GCV).

Time series data is the data of a variable that is observed based on a certain time. Analysis of time series data is usually done by ARIMA method (Autoregressive Integrated Moving Average).

ARIMA method requires assumptions such as stationary of data, residual white noise and normal distribution [5] . The ARIMA modeling concept basically uses the concept of regression. For modeling regressionmodel for time series data can be done by determining response variable and predictor variables. The response variable used is Yt while the predictor variable is variable at Yt-p, with p being significant lag length. So in the regression time series is meant an autoregressive model. In the determination of significant lag lengths, it can be seen using the PACF (Partial Autocorrelation Function) plot [6].

Inflation is an increase in the price of goods continuously where the increase in the price of this goods causes the increase in the price of other goods. The most commonly used indicator to measure the inflation rate is the Consumer Price Index (CPI). Changes in CPI from time to time indicate price movements of packages of goods and services consumed by the community [7]. The Consumer Price Index here means an index that calculates the average of price changes within a period, of a set of prices of goods and services consumed by the population or household for a certain period of time [8]. Type of goods and services are grouped into seven groups, namely foodstuffs; finished food, beverages, and tobacco; housing; clothing; health; education, recreation and sports; transportation, communications and financial services. The following descriptive statistics are presented from seven sector sectors in Indonesia.

Table 1. Descriptive statistics for inflation in 7 sectors from January 2011- August 2017.

Sector	Mean	SE Mean	Stdev	Min	Max
Food stuff	8.81	0.382	4.317	1.452	20.019
Instant Foods, Cigarretes and tobacco	7.071	0.165	1.862	4.25	12.931
Housing, water, Electricity, an Gas	4.933	0.219	2.474	1.183	12.398
Clothing	5.18	0.253	2.858	-0.413	12.267
Health	4.59	0.151	1.707	2.187	9.691
Education, Recreation, and Sport	4.87	0.159	1.799	2.747	10.408
Transportation, Comunnication and Financial services	3.715	0.461	5.215	-6.852	16.333

Based on Table 1, it is seen that the foodstuff group has a high average inflation compared by with other sectors. The standard deviation of inflation in the foodstuff sector is quite high compared to other sectors. This shows that the value is fluctuating. Based on these problems, we want to model the inflation rate in the food sector using Fourier series. The inflation rate in the foodstuff sector is the time series data with series from January 2007 to August 2017.

2. Nonparametric regression of Fourier series

Regression analysis is an analysis of the relationship between the dependent variable with the independent variable. The regression equation model is as follows [9]:

$$y_i = \beta_0 + \beta_1 x_{1i} + \beta_2 x_{2i} + \ldots + \beta_p x_{pi} + \varepsilon_i \quad (1)$$

With y_i is dependent variable in i-th while x is independent variables and ε is residual of model. In parametric regression, residual assumsed that identically, independent and Normal distribution $\varepsilon \sim IIDN(0,\sigma^2)$. Estimation parameters of regression model using OLS (Ordinary Least Square) method or using MLE (Maximum Likelihood Estimation) method.

Nonparametric regression method is a regression method approach where the curve shape of the regression function is unknown. The function curve is assumed to be contained within a certain functional space [10]. Nonparametric regression model is as follows:

$$y_i = \eta(x_i) + \varepsilon_i \tag{2}$$

$\eta(x_i)$ is regression curve that unknow of shape and x_i is independent variable. The curve of estimation of $\eta(x_i)$ was assumed to be smooth on a particular function space.

Fourier series is a trigonometric polynomial function that has a degree of flexibility. This is because the Fourier series is a curve that shows the sine and cosine functions. The functions of the Fourier series are as follows [1]:

$$f(x) = \frac{1}{2}\alpha_0 + \gamma x + \sum_{k=1}^{K} \alpha_k \cos kx \tag{3}$$

Given pairs of data as follows (y_j, x_{ij}) with $j = 1,2,\ldots,m$ is the number of observation and $i = 1,2,\ldots,n$ the number of independent variables. Regression model as follow as:

$$y_j = \beta_0 + \beta_1 x_{1j} + \beta_2 x_{2j} + \cdots + \beta_n x_{nj} + \varepsilon_j . \tag{4}$$

If Eq. (4) could be written as matrix as: $\mathbf{y} = \mathbf{X\beta} + \boldsymbol{\varepsilon}$ then:

$$\mathbf{y} = \begin{bmatrix} y_1 \\ y_2 \\ \vdots \\ y_m \end{bmatrix}_{m\times 1} ; \mathbf{X} = \begin{bmatrix} 1 & x_{11} & x_{21} & \cdots & x_{n1} \\ 1 & x_{12} & x_{22} & \cdots & x_{n2} \\ \vdots & \vdots & \vdots & \ddots & \vdots \\ 1 & x_{1m} & x_{2m} & \cdots & x_{nm} \end{bmatrix}_{m\times(n+1)} ; \boldsymbol{\beta} = \begin{bmatrix} \beta_0 \\ \beta_1 \\ \vdots \\ \beta_n \end{bmatrix}_{(n+1)\times 1} ; \boldsymbol{\varepsilon} = \begin{bmatrix} e_1 \\ e_2 \\ \vdots \\ e_m \end{bmatrix}_{m\times 1}$$

If $\mathbf{y} = \mathbf{f(x)} + \boldsymbol{\varepsilon}$ then

$$\mathbf{f(x)} = \left[f(x_{11}) \;\; f(x_{12}) \;\; \cdots \;\; f(x_{1m}) \;\; f(x_{21}) \;\; f(x_{22}) \;\; \cdots \;\; f(x_{2m}) \;\; \cdots \;\; f(x_{n1}) \;\; f(x_{n2}) \;\; \cdots \;\; f(x_{nm}) \right]^T \tag{5}$$

with $f(x_{ij})$ is an unknown curve of its shape. $f(x_{ij})$ was approached by Fourier series as follows:

$$f(x) = \frac{1}{2}\alpha_0 + \gamma x + \sum_{k=1}^{K} \alpha_k \cos kx \tag{6}$$

so If Eq. (6) subtituted in Eq. (5) can be follows:

$$f(x_{ij}) = \frac{1}{2}\alpha_0 + \gamma_j x_{ij} + (\alpha_{i1}\cos x_{ij} + \alpha_{i2}\cos 2x_{ij} + \ldots + \alpha_{iK}\cos Kx_{ij}) . \tag{7}$$

If $\mathbf{f(x)} = \mathbf{A\theta}$ then

$$\mathbf{A} = \begin{bmatrix} 1 & x_{11} & \cos x_{11} & \cos 2x_{11} & \cdots & \cos Kx_{11} & x_{21} & \cos x_{21} & \cos 2x_{21} & \cdots & \cos Kx_{21} & \cdots & x_{n1} & \cos x_{n1} & \cdots & \cos Kx_{n1} \\ 1 & x_{12} & \cos x_{12} & \cos 2x_{12} & \cdots & \cos Kx_{12} & x_{22} & \cos x_{22} & \cos 2x_{22} & \cdots & \cos Kx_{21} & \cdots & x_{n2} & \cos x_{n2} & \cdots & \cos Kx_{n2} \\ \vdots & \vdots & \vdots & \vdots & \ddots & \vdots & \vdots & \vdots & \vdots & \ddots & \vdots & \ddots & \vdots & \vdots & \ddots & \vdots \\ 1 & x_{1m} & \cos x_{1m} & \cos 2x_{1m} & \cdots & \cos Kx_{1m} & x_{2m} & \cos x_{2m} & \cos 2x_{2m} & \cdots & \cos Kx_{21} & \cdots & x_{nm} & \cos x_{nm} & \cdots & \cos Kx_{nm} \end{bmatrix}_{(n(k+1)\times m)}$$

$$\boldsymbol{\theta} = \left[\phi \;\; \gamma_1 \;\; \alpha_{11} \;\; \alpha_{12} \;\; \cdots \;\; \alpha_{1K} \;\; \gamma_2 \;\; \alpha_{21} \;\; \alpha_{22} \;\; \cdots \;\; \alpha_{2K} \;\; \cdots \;\; \gamma_n \;\; \alpha_{n1} \;\; \alpha_{n2} \;\; \cdots \;\; \alpha_{nK} \right]^T_{1\times(n(K+1)+1)}$$

with $\phi = \frac{n}{2}\alpha_0$, whereas α_0 is constanta parameter.

Estimation of nonparametric regression model is to use OLS (Ordinary Least Square) as follows: $\psi(\boldsymbol{\theta}) = \boldsymbol{\varepsilon}^T\boldsymbol{\varepsilon}$ and $\hat{\boldsymbol{\theta}} = (\mathbf{A}^T\mathbf{A})^{-1}\mathbf{A}^T\mathbf{y}$ is a nonparametric regression estimator using the Fourier series approach.

3. Generalized Cross Validation (GCV)

In nonparametric regression modeling using Fourier series, the thing to note is to determine the value of K. To determine the optimal of K, can be used the Generalized Cross Validation (GCV) method. Optimal K determination will result in a high value. According to Wu and Zhang (1988) the GCV scores were obtained as follows [11]:

$$GCV(K) = \frac{MSE(K)}{\left(n^{-1} trace(\mathbf{I} - \mathbf{A})\right)^2}$$

with $MSE(K) = n^{-1} \sum_{j=1}^{n} (y_j - \hat{y}_j)^2$ .The smallest value of GCV will results optimal K value.

4. Results and discussion

Figure 1 presented time series plot for inflation data of foodstuff in Indonesia period January 2007 to August 2017

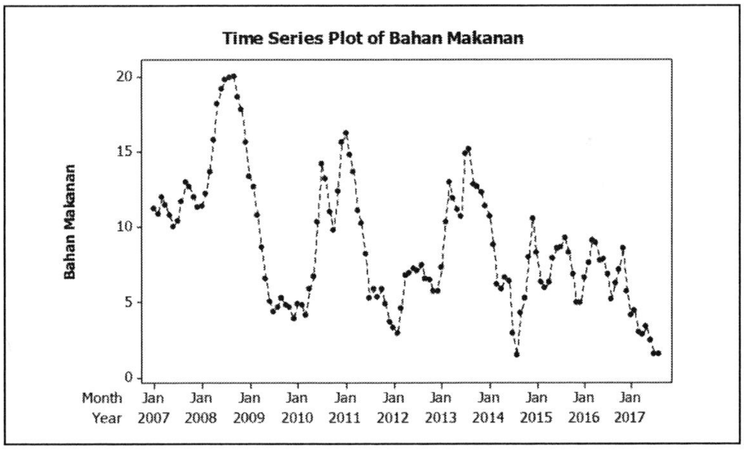

Figure 1. Time Series inflation plots in the foodstuff sector from January 2007 to August 2017

Based on Figure 1 it is seen that the value tends to fluctuate, and in the range of 2008 reached the highest value up to 20. In 2009 It has been decreased, while in 2011 and 201, it has been increased. For the identification of the regression model, the PACF plot is determined first.

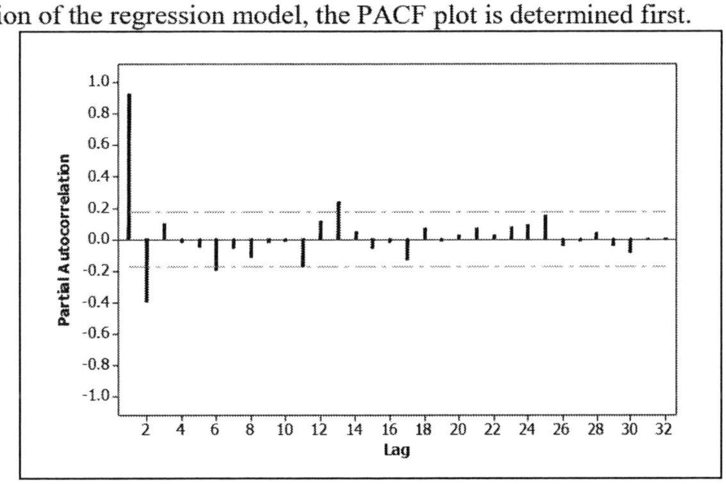

Figure 2. PACF plot of inflation data on the foodstuff sector

According to Fig. 2, the lags which out of bonds were the 1st lag and the 2nd lag. Its show significant lag, so that the predictor variables are Y_{t-1} and Y_{t-2}. However, in the plot of PACF is seen there are some of lag that come out besides lag-1 and 2 are the 6th lag, the 11th and 13th. Therefore it should be modeled by subset of Autoregressive Integrated Moving Average (ARIMA). However, we limit it by ignoring the lag that makes up a subset significantly. So if a plot is made between Y_t with Y_{t-1} and Y_t with Y_{t-2} then the plot is generated as follows:

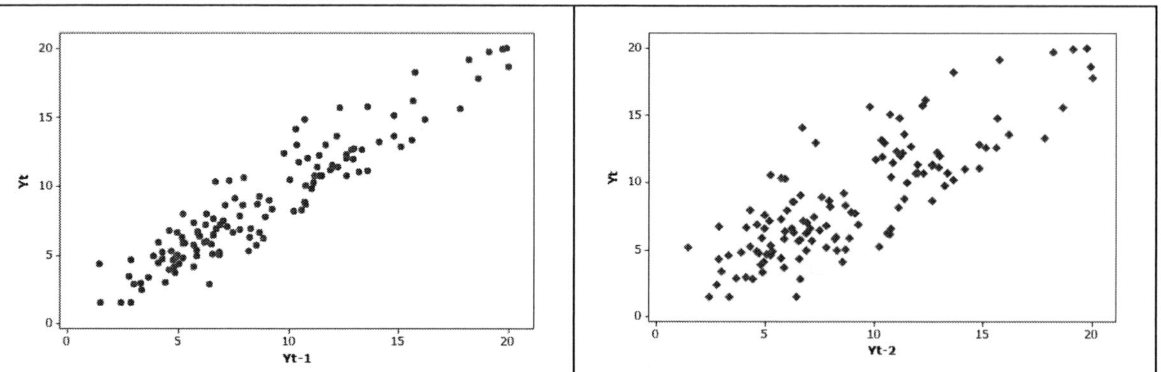

(a) Plot between Y_t and Y_{t-1} (b) Plot between Y_t and Y_{t-2}

Figure 3. Sactter plot between Y_t with Y_{t-1} and Y_{t-2}

It can be seen from Figure 3 that the relationship between Y_t and Y_{t-1} and Y_t with Y_{t-2} is linear, so if we wolud like to model Y_t with Y_{t-1} and Y_{t-2} with parametric approach (multiple regression model), the model can be written as:

$$\hat{Y}_t = 0.718 + 1.38Y_{t-1} - 0.48Y_{t-2}$$

The model yields an R-square value of 90.6%. Based on partial testing and simultaneously showed that the model and parameters are significant. While for testing model assumptions obtained value of Variance in Factor 8.19 (<10) so it shows non-multicollinearity. Obtained Kolmogorov Smirnof residual value of 0.05 indicates that the residual is normally distributed. It will then be seen if the data is modeled using ARIMA with the model identification is AR (2) the model can be follows

$$y_t = \theta_0 + \theta_1 y_{t-1} + \theta_2 y_{t-2} + a_t$$

Where θ is parameter of AR model and a_t is residual, the following values are obtained the data:

Table 2. The parameter model of AR(2)

Parameter	**Coefecient**	**t-value**	**p-value**
Constanta	1.3805	17.36	0.000
θ_1	1.3805	-5.78	0.000
θ_2	-0.4651	6.17	0.000

Based on Table 2 shows multiple regression modeling for time series data in principle is the same as the Autoregressive model in ARIMA. The resulting coefficient parameter values are the same. In the AR (2) model, residual white noise and normal distribution. Then the next analysis we would liked to compare with nonparametric approch, as Fourier series approach. In this model the response variable is Y_t whereas the predictor variable is t, so the model will be searched according to figure 1. First In modeling using Fourier series must be determined K-optimal to be used. To determine the optimal of K using GCV method. Here we get the value of GCV for each K.

Table 3 shows the GCV value for each K, in the table it is seen that the more K values indicate that the GCV value is smaller. So for inflation modeling in the transportation sector, communication and financial services using Fourier series obtained the optimal K is 120. The larger value of K fed model has a high R-square value. This becomes a problem, because there is no tolerance limit for the value of K used. Therefore return again to the researchers how much K is used in the model. In modeling Fourier derives, it is determined that the number of K is used. Determination of K can be trial and error or based on GCV. The number of parameters used in the Fourier model is also dependent on the magnitude of K used. The larger the K used the more complex (no parsimony) the

Fourier model gets. Fourier series uses cosine functions, but the use of sine functions is also allowed in Fourier series modeling.

Table 3. The GCV Value for K

K	GCV	K	GCV	K	GCV	K	GCV	K	GCV	K	GCV
1	55392.52	21	422.9885	41	110.8041	61	44.13997	81	22.15301	101	8.853676
2	24617.16	22	386.9981	42	105.6992	62	42.15303	82	21.51775	102	8.648823
3	13846.88	23	355.4123	43	100.9001	63	40.32212	83	20.97063	103	8.280728
4	8861.702	24	327.2724	44	94.76517	64	38.39719	84	20.48007	104	8.078747
5	6152.765	25	300.1265	45	90.68891	65	37.23815	85	19.99762	105	7.801152
6	4363.353	26	278.1719	46	86.85723	66	36.12314	86	19.54007	106	7.634961
7	3328.926	27	258.655	47	83.25121	67	35.05095	87	19.02408	107	0.392063
8	2627.724	28	241.123	48	79.87412	68	34.04203	88	17.84956	108	0.384897
9	2128.456	29	225.3008	49	76.25085	69	32.56083	89	17.44671	109	0.377837
10	1759.028	30	210.8314	50	66.86628	70	31.62559	90	17.04659	110	0.370086
11	1477.942	31	196.6431	51	63.74293	71	30.67885	91	16.67789	111	0.351825
12	1256.089	32	184.3921	52	61.27269	72	29.84409	92	16.29643	112	0.344592
13	1067.082	33	173.686	53	59.02376	73	29.03067	93	15.73403	113	0.133138
14	927.5844	34	163.8777	54	56.89678	74	28.26108	94	12.38048	114	0.100932
15	815.2073	35	154.8623	55	54.81395	75	26.58002	95	11.76036	115	0.099115
16	722.0605	36	146.4442	56	52.9012	76	25.89412	96	11.47563	116	0.097192
17	643.8038	37	136.8378	57	50.50366	77	25.16988	97	11.24058	117	0.085506
18	571.6414	38	128.5258	58	48.7845	78	24.53604	98	10.98337	118	0.069527
19	512.0645	39	122.1769	59	47.16521	79	23.85823	99	10.73105	119	0.068131
20	464.4549	40	116.2841	60	45.62322	80	23.21715	100	10.20467	**120**	**0.058915**

The estimated Fourier series model for K = 120 is as follows:

Table 4. The Estimation parameter of Fourier model in Inflation Foodstuffs sector with K=120

ϕ	6.146492	α_{19}	-11.4266	α_{39}	0.877294	α_{59}	0.056442	α_{79}	0.091165	α_{99}	-0.406
γ	0.063669	α_{20}	-0.80506	α_{40}	-0.2816	α_{60}	0.034651	α_{80}	-0.09497	α_{100}	-0.523
α_1	-1.15998	α_{21}	0.489422	α_{41}	-0.08272	α_{61}	-0.01181	α_{81}	0.513633	α_{101}	-0.371
α_2	-0.27338	α_{22}	-0.04609	α_{42}	0.183759	α_{62}	-0.97708	α_{82}	-0.83423	α_{102}	-0.233
α_3	0.001185	α_{23}	-0.41283	α_{43}	1.19462	α_{63}	8.91401	α_{83}	-0.21264	α_{103}	0.040
α_4	0.341201	α_{24}	0.36074	α_{44}	1.025097	α_{64}	1.062949	α_{84}	0.081338	α_{104}	-0.073
α_5	-1.31388	α_{25}	10.4323	α_{45}	0.823046	α_{65}	-0.33199	α_{85}	0.025803	α_{105}	-0.015
α_6	-2.70349	α_{26}	-1.2719	α_{46}	0.177653	α_{66}	0.018978	α_{86}	-0.02472	α_{106}	-0.394
α_7	0.83987	α_{27}	-0.19244	α_{47}	0.001068	α_{67}	0.298753	α_{87}	-1.06595	α_{107}	-7.917
α_8	-0.36867	α_{28}	0.025488	α_{48}	-0.42938	α_{68}	-0.00348	α_{88}	4.960176	α_{108}	-0.124
α_9	-0.00943	α_{29}	0.314326	α_{49}	1.463847	α_{69}	-8.57809	α_{89}	-0.36675	α_{109}	0.181
α_{10}	-0.03687	α_{30}	0.217291	α_{50}	3.868109	α_{70}	0.887639	α_{90}	-0.07721	α_{110}	-0.063
α_{11}	-0.41881	α_{31}	-1.19076	α_{51}	-0.12984	α_{71}	0.240574	α_{91}	-0.00424	α_{111}	-0.145
α_{12}	-0.9905	α_{32}	1.248954	α_{52}	0.174751	α_{72}	-0.06586	α_{92}	0.196842	α_{112}	-0.302
α_{13}	-0.222	α_{33}	0.33728	α_{53}	-0.06116	α_{73}	-0.50562	α_{93}	-0.95926	α_{113}	4.391
α_{14}	-0.11921	α_{34}	0.008233	α_{54}	-0.02662	α_{74}	0.099919	α_{94}	-0.11138	α_{114}	-0.394
α_{15}	-0.23047	α_{35}	0.040783	α_{55}	0.521595	α_{75}	1.091575	α_{95}	0.527988	α_{115}	0.008
α_{16}	-0.0234	α_{36}	0.280047	α_{56}	0.860921	α_{76}	-0.59737	α_{96}	0.094404	α_{116}	-0.014

α_{17}	0.167964	α_{37}	-0.92176	α_{57}	-0.22129	α_{77}	-0.06827	α_{97}	-0.05691	α_{117}	0.334
α_{18}	1.508606	α_{38}	3.350155	α_{58}	0.322191	α_{78}	-0.02226	α_{98}	-0.040	α_{118}	-0.335
								α_{119}	0.01	α_{120}	0.038

Obtained average estimator of Fourier series in K=120 is 0.086 with standard deviation 2.23. The value of R-square generated is 99.715% while the MSE value obtained is 0.0526. Due to the optimal K value = 120, of course the Fourier model is very complex because it requires parameters of 122 parameters. The following chart presents the estimation with actual data for some K values.

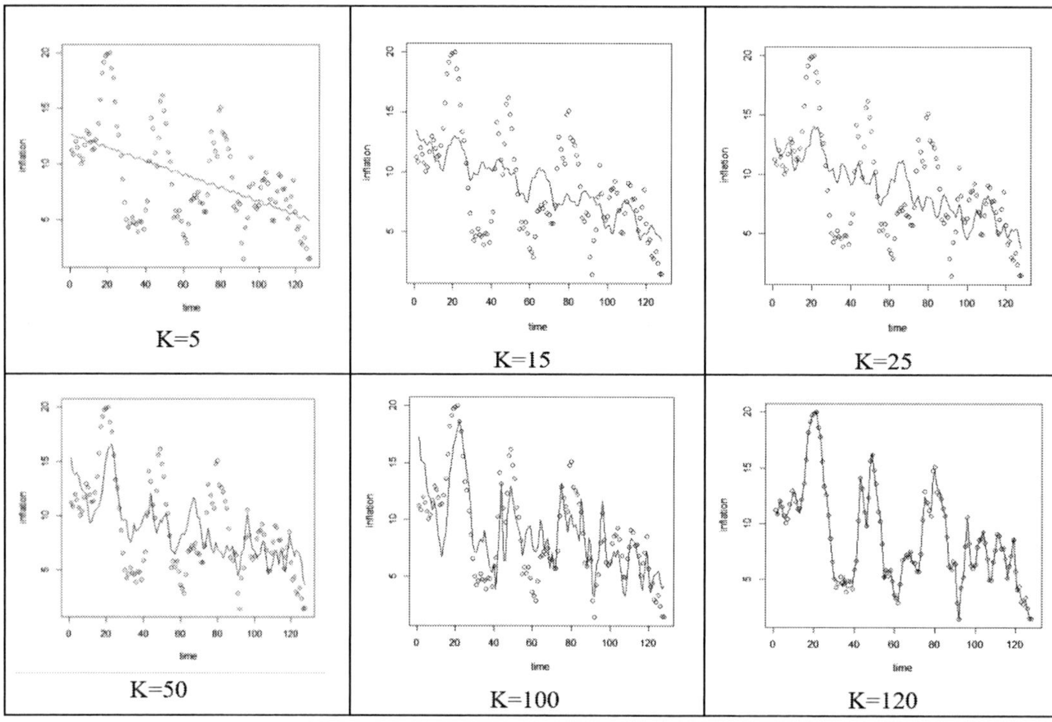

Figure 4. Scatterplot between Y_t and $\hat{Y}_t$ for some of K values with Fourier model

Figure 4 shows that the greater the value of the estimated value of the Fourier curve near the actual data. In the figure it is seen that the estimation curve will search by itself the form of the actual data. It means that the estimation of Fourier model in Foodstuf inflation, will following plot of the data. Estimates using Fourier series result in very complex model estimators instead of parametric approaches. However, the resulting estimation results tend to be better. Next will be modeled Fourier series to Y_t with Y_{t-1} and Y_t with Y_{t-2}. The shape of the intermediate curve Y_t with Y_{t-1} is linear, however it will be approximated by Fourier series. Based on Figure 5 shows that the bigger the value of K then the value of R^2 will be greater. However, for modeling with K = 5 has reached R^2 of 88.19% so that if it is enough then the model for K = 5 is the best model. Although the value of R^2 for K = 30 is greater than K = 5 (not too far), the Fourier model with K = 5 is more parsimonious than K = 30. Based on Figure 4 it is seen that when using a low K, the curve is very rough to model. However if using K is too high (in this case> 120) then the model is too oversmooth. Therefore the K limit taken by the researcher is up to R^2 value above 90%.

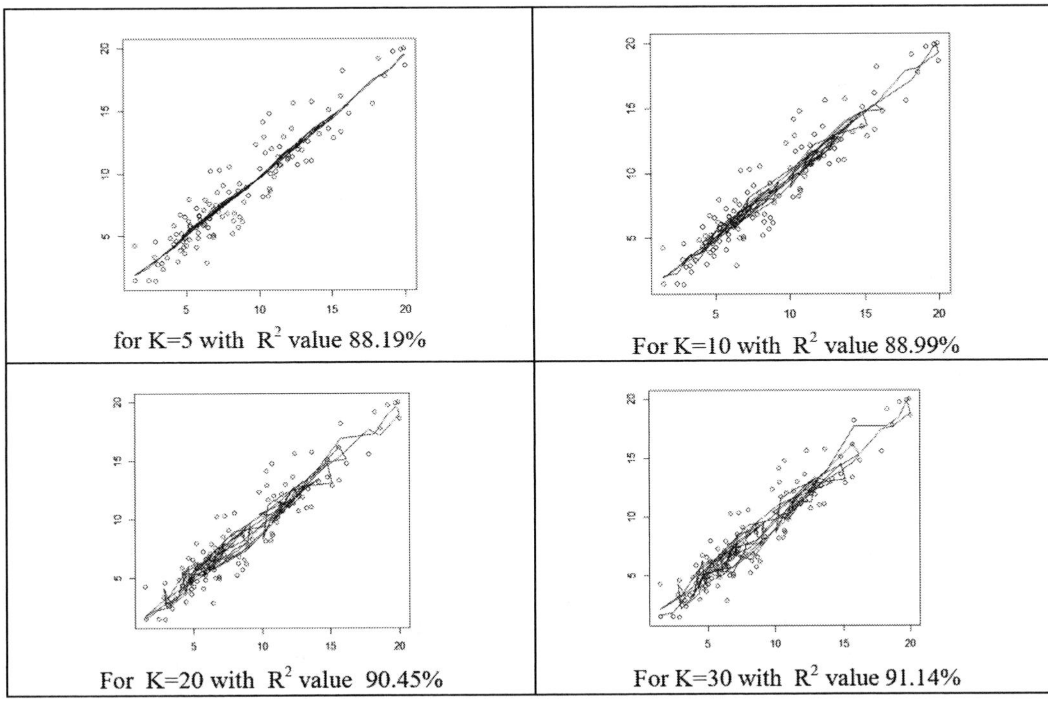

Figure 5. Scatter plot between $\hat{Y}_t$ and Y_t with predictor variable Y_{t-1}

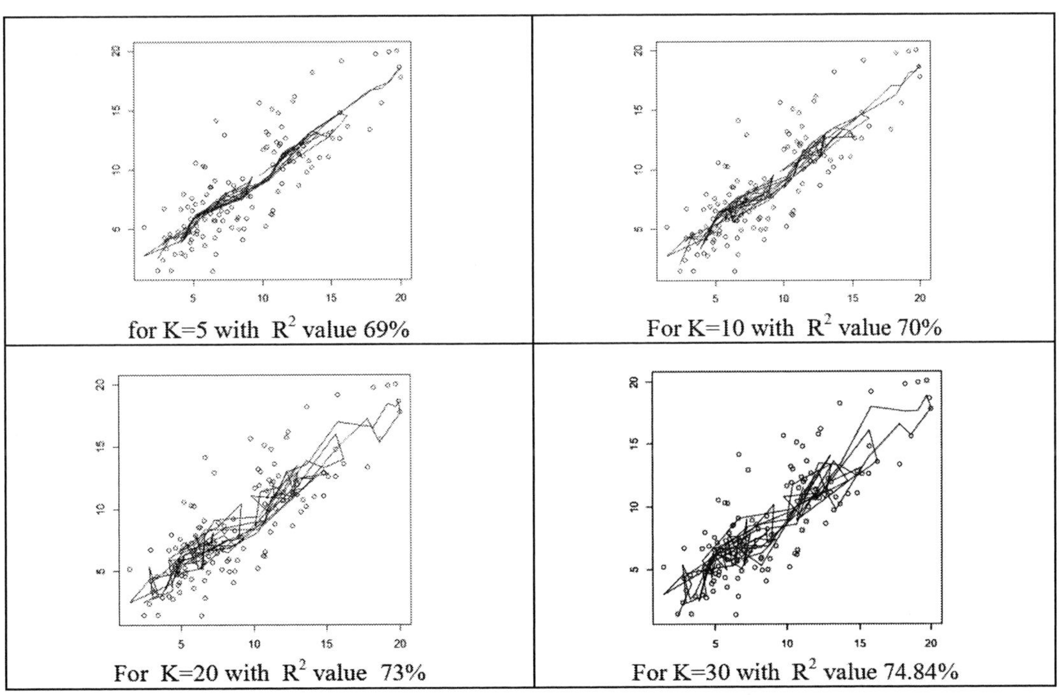

Figure 6. Scatter plot between $\hat{Y}_t$ and Y_t with predictor variable Y_{t-2}

Figure 5 shows the scatterplot between actual and estimated data with predictor variables Y_{t-2}. Based on the figure it is seen that the estimation result is better for the model Y_t with predictor variable Y_{t-1} compared with the predictor variable Y_{t-2}.

5. Conclussion

In nonparametric regression modeling of Fourier series only needs a value of K to estimate the parameters. The determination of K value can be done by using GCV method. In inflation modeling for transportation and communication sector, multiple regression model is able to produce R^2 more than 90%, while for Fourier model able to produce K value almost 99%. However, the model obtained is very complex involving 122 parameters for the Fourier model. However, multiple linear models require only 3 parameters. For regression modeling im Foodstuff inflation based on lag obtained significant lag Y_{t-1} and Y_{t-2}.

Acknowlegments
This paper supported by Diponegoro university as a founder in Research. Thanks to laboratory of Statistics Department Diponegoro University.

Refferences

[1] Bilodeau, M. (1992), "Fourier Smoother and Additive Models", *The Canadian Journal of Statistics*, 3, hal. 257-259.
[2] Tripena, A., and Budiantara, I.N., (2007), "Fourier Estimator in Nonparametric Regression, *International Conference on Natural Sciences and Applied Natural Sciences*, Ahmad Dahlan University, Yogyakarta
[3] Semiati, R., (2010), *Regresi Nonparametrik Deret Fourier Birespon*, Tesis, ITS, Surabaya.
[4] Asrini, L. J., (2012), "Regresi Parametrik Deret Fourier", *Prosiding Seminar Nasional FMIPA Universitas Negeri Surabaya*, hal.77-80, 24 November 2012.
[5] Box, G.E.P., Jenkins, G.M., and Reissel. G.C.,(1994), *Time Series Analysis Forecasting and Control*. 3rd edition. Englewood Cliffs : Prentice Hall.
[6] Wei, W.W.S., (1990), *Time Series Analysis*, Addison-Wisley Publishing Company Inc., United States
[7] Reza Y.S, and Siwei, G., (2010), Effectiveness and Commitment to inflation targeting policy: Evidance from Thailand and Indonesia, *Journal of Asian Economics* 21.113-128
[8] Rizki E. W, Paul M.T, Maximilian J.B.H, (2010), Expectations and the inertia of inflation: The case of Indonesia, *Journal of Policy Modelling,* 33.426-438
[9] Drapper, N.R dan Smith, H. (1992), *Applied Regression Analysis 2nd Edition*, Marcel Dekker, New York.
[10] Eubank, R.L. (1988), *Spline Smoothing and Nonparametric Regression*, Marcel Dekker, New York
[11] Hardle, W., 1990. *Applied Nonparametric Regression.* Cambridge University.

IOP Conf. Series: Journal of Physics: Conf. Series **974** (2018) 012068 doi:10.1088/1742-6596/974/1/012068

Toward the analytical model of SEI formation and development in lithium-ion battery: The case of solution of the nonlinear transport equation

Gunawan Nugroho[1], **Nur Laila Hamidah**[1], **Fu-Ming Wang**[2]

[1] Department of Engineering Physics, Institut Teknologi Sepuluh Nopember, Kampus-ITS-Sukolilo, Surabaya 60111, Indonesia
[2] National Taiwan University of Science and Technology, No. 43, Section 4, Jilong Rd, Da'an District, Taipei City 10607, Taiwan

E-mail: gunawan@ep.its.ac.id, gunawanf317@gmail.com

Abstract. The nonlinear lithium-ion transport equation with variable diffusion coefficient is considered in this research. The solution is produced by transforming the second order spatial variable into the first order differential in, ξ. The method of characteristics is then implemented for covering the initial-boundary conditions. Several interesting results are obtaining after the calculating the generated solution. It is found that dendrite formation is detected after 100 s which is considered as steady operation. In this case, the concentration ratio is smaller than boundary value at the top surface and the ratio of tip and surface current is high. Also the development of dendrite length is investigated to support the insight from the analytical model.

1. Introduction

Lithium ion battery (LIB) becomes the most popular types of rechargeable battery for many electronic applications and it gained increasing attention as an important device for electric vehicles and power storage systems. The LIB performance is measured by several important parameters such as high energy density, no memory effect, high stability, rapid charge performance, low cost and life cycle, respectively [1].

The problem of advancing the lithium battery is significantly affected by the formation of dendritic growth during electro disposition – electro dissolution cycles [2].The SEI (Solid Electrolyte Interphase) formation causes several drawbacks such as density loss due to release of the lithium crystal and even explosion. Hence, an understanding a dendrites evolution is important from the practical point of view. Some experimental and theoretical works were conducted to gain the understanding of the dendrites growth [3,4]. The works conclude that the formation is mainly caused by the current distribution on the lithium surface. Meanwhile, theoretical studies through the solutions of the governing equations often play a special role in the description and prediction of the dendrites growth. They are found able to describe the detailed behavior of the concerning systems and facilitate a proper understanding, paving the way to correct solutions by combining with the experimental data [5].

The main difficulty of the governing equations is the contribution of the variable diffusion which then forms nonlinear terms representing an actual laboratory data and render difficulty to the conventional analysis in general cases. However, it is not surprising that they attract many efforts until recently, since the simplified solution of this nonlinear problems still supplements the considerable results [6]. Concerning to the simplified solutions, there are some works have already been conducted in the literatures. One of them is the simulation of the three discharge/charge rates until the rated capacity is irreversibly lost in side reactions. It is concluded that the degradation is faster due to SEI accumulation in negative electrode [7]. The potential drop across the SEI layer is also investigated by assuming that the capacity fading is driven by solvent reaction at film interphase [8]. It is found that moderate potential drop will substantially increase capacity fade.

The analysis based on the variable porosity model is also implemented for a lithium battery. The work proposes a novel mass balance model to predict the volume fraction variation due to lithium deposition [9]. The model quantifies the loss of unprotected anode and helps the required pre-lithiation. The procedure for obtaining the solutions is derived in a quite complicated way which is based on the implementation of kinetics equations. The resulting expression is then integrated the current density which then compared with the applied current. However, the results are not considerably differs with the simplified analysis.

Accordingly, this work investigates the transient one-dimensional diffusion equation with variable diffusion coefficient. The method of characteristics is utilized by transforming the coordinate of the second order term. The solution is then supplemented by algebraic reaction models to determine the current density where dendrite tip grow at and the dendrite length.

2. The transport equation and its solution

The mass conservation of the transient one-dimensional lithium transport is,

$$\frac{\partial C}{\partial t}=\frac{\partial}{\partial x}\left(D\frac{\partial C}{\partial x}\right)-\frac{i.\Delta t_+}{z_+F} \tag{1a}$$

where, C is a time dependent concentration, D is concentration-dependent diffusion coefficient, t_+ is the Li^+ transport number, z_+ is charge number and i is the current density. In this research the diffusion coefficient is defined by the following equation,

$$D=ae^{-bC} \tag{1b}$$

with $a=2.582\times10^{-5}$ cm^2s^{-1} and $b=2.856$ M^{-1} as determined experimentally [2]. Hence, the transport equation becomes,

$$\frac{\partial C}{\partial t}=\frac{\partial}{\partial x}\left(ae^{-bC}\frac{\partial C}{\partial x}\right)-\frac{i.\Delta t_+}{z_+F} \tag{1c}$$

Equation (1c) is supplemented by the following initial and boundary values,

$$t=0:\quad C=C_i$$

$$\text{At } x=0:\ D\frac{\partial C}{\partial x}=\frac{i_f\left(1-t_+\right)}{nF} \tag{1d}$$

$$x=\delta:\quad C=C_0$$

where i_f is the time-dependent current density on the flat Li surface and C_0 is the bulk concentration.

The solution of equations (1c) & (1d) is then substituted into,

$$\frac{i_t}{i_f}=\left(\frac{C_e}{C_i}\right)^{-\frac{\alpha_C}{n}} \tag{2}$$

as the ratio of the dendrite tip current density to the current density of flat surface. In this case the parameter value is $\alpha_C=0.2-0.4$. Since i_t is defined, the dendrite length can be calculated as,

$$h=\frac{M}{n\rho F}\int_0^t i_t dt \tag{3}$$

The step now is to consider equation (1c), suppose that $B = e^{-bC}$, equation (1c) is then transformed into,

$$\frac{\partial B}{\partial t} = aB\frac{\partial^2 B}{\partial x^2} + b\frac{i.\Delta t_+}{z_+ F}B \tag{4a}$$

Let, $\frac{\partial^2 B}{\partial x^2} = \frac{\partial B}{\partial \xi}$, the solution is,

$$B = e^{a_1 \xi}\left(a_2 \cosh a_4 x + a_3 \sinh a_4 x\right) \tag{4b}$$

Equation (4a) is then becomes,

$$\frac{\partial B}{\partial t} = aB\frac{\partial B}{\partial \xi} + b\frac{i.\Delta t_+}{z_+ F}B \tag{4c}$$

Suppose there is a function $r = r(t,\xi)$ such that, $\frac{dB}{dr} = \frac{\partial t}{\partial r}\frac{\partial B}{\partial t} + \frac{\partial \xi}{\partial r}\frac{\partial B}{\partial \xi}$, the characteristics are expressed as,

$$r = t \text{ and } \xi = -aBr + \xi_0 \text{ or } \xi = -aBt + \xi_0 \tag{5a}$$

The solution of equation (4c) is then,

$$B = B_0 e^{b\frac{i.\Delta t_+}{z_+ F}t} \text{ or } B = F(\xi_0)e^{b\frac{i.\Delta t_+}{z_+ F}t} \tag{5b}$$

Unfortunately the function, ξ_0 cannot be extracted explicitly without choosing $F(\xi_0)$, the simplest case is, $F(\xi_0) = \xi_0$ and equation (5b) becomes,

$$\xi = -a\xi_0 e^{b\frac{i.\Delta t_+}{z_+ F}t}t + \xi_0 \tag{5c}$$

The expression for ξ_0 is then,

$$\xi_0 = \xi\left(1 - ae^{b\frac{i.\Delta t_+}{z_+ F}t}t\right)^{-1} \tag{5d}$$

The step now is to determine ξ by substituting (4b) and (5d) into equation (5b) as,

$$B = \xi\left(1 - ae^{b\frac{i.\Delta t_+}{z_+ F}t}t\right)^{-1} e^{b\frac{i.\Delta t_+}{z_+ F}t} = e^{a_1\xi}\left(a_2 \cosh a_4 x + a_3 \sinh a_4 x\right) \text{ or}$$

$$-a_1\xi e^{-a_1\xi} = \left(at - e^{-b\frac{i.\Delta t_+}{z_+ F}t}\right)a_1\left(a_2 \cosh a_4 x + a_3 \sinh a_4 x\right) \tag{6a}$$

which solution is a Lambert W-function as,

$$\xi = \frac{1}{a_1}W\left[\left(at - e^{-b\frac{i.\Delta t_+}{z_+ F}t}\right)a_1\left(a_2 \cosh a_4 x + a_3 \sinh a_4 x\right)\right] \tag{6b}$$

Therefore, the solution of the diffusion equation is,

$$C = -\frac{1}{b}\ln\left\{\frac{1}{a_1}\left(1 - ae^{b\frac{i.\Delta t_+}{z_+ F}t}t\right)^{-1} e^{b\frac{i.\Delta t_+}{z_+ F}t} W\left[\left(at - e^{-b\frac{i.\Delta t_+}{z_+ F}t}\right)a_1\left(a_2 \cosh a_4 x + a_3 \sinh a_4 x\right)\right]\right\} \tag{6c}$$

Implementing initial condition will produce the following expression,

$$C_i = -\frac{1}{b}\ln\left\{\frac{1}{a_1}W\left[a_1\left(a_2 \cosh a_4 x + a_3 \sinh a_4 x\right)\right]\right\} \text{ or}$$

$$\left(a_2 \cosh a_4 x + a_3 \sinh a_4 x\right) = e^{-bC_i} \exp\left(a_1 e^{-bC_i}\right) \tag{7a}$$

Thus, by assuming that $a_4 = \frac{\pi}{\delta}$, the case of boundary conditions is formulated as in the following,

$$\frac{i_f\left(1-t_+\right)}{nF} = -\frac{1}{b} D\left[\frac{1}{a_1}\frac{a_3}{a_2}\left(1-ae^{b\frac{i.\Delta t_+}{z_+F}t}t\right)^{-1} e^{b\frac{i.\Delta t_+}{z_+F}t} \frac{W\left[\left(at-e^{-b\frac{i.\Delta t_+}{z_+F}t}\right)a_1 a_2\right]}{1+W\left[\left(at-e^{-b\frac{i.\Delta t_+}{z_+F}t}\right)a_1 a_2\right]}\right] \text{ or}$$

$$\left\{\frac{1}{a_1}\left(1-ae^{b\frac{i.\Delta t_+}{z_+F}t}t\right)^{-1} e^{b\frac{i.\Delta t_+}{z_+F}t} W\left[\left(at-e^{-b\frac{i.\Delta t_+}{z_+F}t}\right)a_1 a_2\right]\right\}^{-1}$$

$$\frac{i_f\left(1-t_+\right)}{nF} = -\frac{D}{b}\frac{a_3}{a_2}\left\{1+W\left[\left(at-e^{-b\frac{i.\Delta t_+}{z_+F}t}\right)a_1 a_2\right]\right\}^{-1} \tag{7b}$$

Substituting D into (7b) we have,

$$\frac{i_f\left(1-t_+\right)}{nF} = -\frac{1}{b}\frac{1}{a_1}\frac{a_3}{a_2}\left(1-ae^{b\frac{i.\Delta t_+}{z_+F}t}t\right)^{-1} e^{b\frac{i.\Delta t_+}{z_+F}t} W\left[\left(at-e^{-b\frac{i.\Delta t_+}{z_+F}t}\right)a_1 a_2\right]\left\{1+W\left[\left(at-e^{-b\frac{i.\Delta t_+}{z_+F}t}\right)a_1 a_2\right]\right\}^{-1} \tag{7c}$$

The third boundary condition will produce,

$$C_0 = -\frac{1}{b}\ln\left\{\frac{1}{a_1}\left(1-ae^{b\frac{i.\Delta t_+}{z_+F}t}t\right)^{-1} e^{b\frac{i.\Delta t_+}{z_+F}t} W\left[\left(at-e^{-b\frac{i.\Delta t_+}{z_+F}t}\right)a_1\left(a_2\cosh\pi + a_3\sinh\pi\right)\right]\right\} \tag{7d}$$

The step now is to consider equations (7c) and (7d), perform a_2 from (7d) as,

$$a_2 = -0.0866e^{-bC_0}\exp\left[a_1 e^{-bC_0}\left(e^{-b\frac{i.\Delta t_+}{z_+F}t} - at\right)\right] - a_3 \tag{8a}$$

Substituting the result into (7c), the expression for a_3 is defined in implicit form as,

$$a_3 = \frac{a_2 i_f\left(t_+-1\right)a_1 b}{nF}\left(e^{-b\frac{i.\Delta t_+}{z_+F}t} - at\right)\left[\left\{W\left[\left(at-e^{-b\frac{i.\Delta t_+}{z_+F}t}\right)a_1 a_2\right]\right\}^{-1}+1\right] \tag{8b}$$

The complete solution of the diffusion equation is then,

$$C\left(x,t\right) = -\frac{1}{b}\ln\left\{\left(\frac{\delta}{\pi}\right)^2\left(1-ae^{b\frac{i.\Delta t_+}{z_+F}t}t\right)^{-1} e^{b\frac{i.\Delta t_+}{z_+F}t} W\left[\left(at-e^{-b\frac{i.\Delta t_+}{z_+F}t}\right)a_1\left(a_2\cosh\frac{\pi}{\delta}x + a_3\sinh\frac{\pi}{\delta}x\right)\right]\right\} \tag{8c}$$

Therefore, the ration of the current tip density to the current density on the surface is,

$$\frac{i_t}{i_f} = \left[-\frac{1}{bC_i}\ln\left\{\left(\frac{\delta}{\pi}\right)^2\left(1-ae^{b\frac{i.\Delta t_+}{z_+F}t}t\right)^{-1} e^{b\frac{i.\Delta t_+}{z_+F}t} W\left[\left(at-e^{-b\frac{i.\Delta t_+}{z_+F}t}\right)a_1\left(a_2\cosh\frac{\pi}{\delta}x + a_3\sinh\frac{\pi}{\delta}x\right)\right]\right\}\right]^{-\frac{\alpha_C}{n}} \tag{9a}$$

And the dendrite length is,

IOP Conf. Series: Journal of Physics: Conf. Series **974** (2018) 012068 doi:10.1088/1742-6596/974/1/012068

$$h=-\frac{M}{n\rho F}\int_0^t i_f\left[-\frac{1}{bC_i}\ln\left\{\left(\frac{\delta}{\pi}\right)^2\left(1-ae^{b\frac{i.\Delta t_+}{z_+F}t}t\right)^{-1}e^{b\frac{i.\Delta t_+}{z_+F}t}W\left[\left(at-e^{-b\frac{i.\Delta t_+}{z_+F}t}\right)a_1\left(a_2\cosh\frac{\pi}{\delta}x+a_3\sinh\frac{\pi}{\delta}x\right)\right]\right\}\right]^{-\frac{\alpha_C}{n}}dt \tag{9b}$$

3. The example of dendrite formation and development

SEI is formed on the surface during the first charge and discharge cycle which is the product of solvent molecules decomposition and intercalation into the planes. The presence of SEI will decrease density stability of the LIB. In this study, the SEI formation is influenced by the presence of electrolyte additive, which is parametrized as, α_c. In this section the SEI formation will be evaluated according to the solution of the nonlinear diffusion equation and the parameter values are taken from Alokar [2], except the value of parameter b is adjusted into 18560 M^{-1}

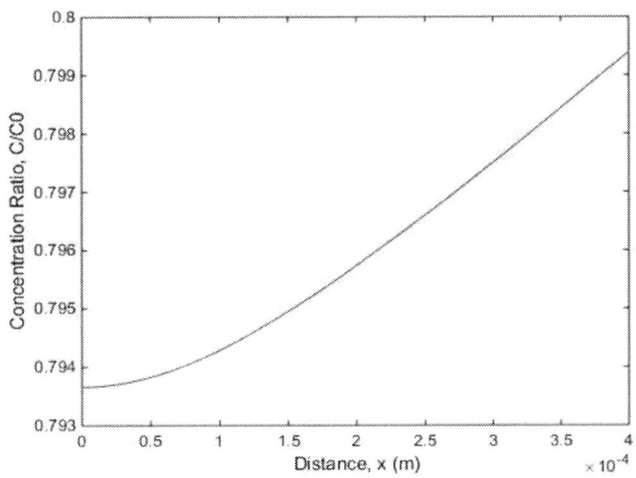

Figure 1. The behavior of the lithium concentration within the layer, δ. The result is at the steady battery operation, 100 s.

Figure 1 depicts the evolution of the lithium concentration within the thin layer, revealing the condition that the concentration is approaching the initial value at the steady operation. The figure also indicates that the SEI formation already affects battery performance since the concentration at the top layer is smaller than the initial value.

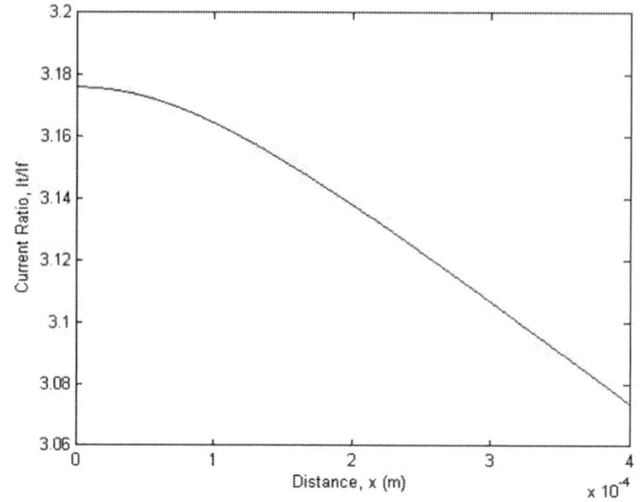

Figure 2. The behavior of tip current within the layer, δ, at the steady battery operation, 100 s.

On the other hand, the result of current ratio shows that the tip current increase with the presence of dendrite (SEI) formation as described in figure 2. The current result is consistent with the concentration since the current on the surface layer is higher than surface current.

Moreover, in order to realize the thickness of SEI formed within the layer, the thickness is also calculated according to the solution of the transport equation. In this case, the value $\alpha_c = 0.2$ is implemented.

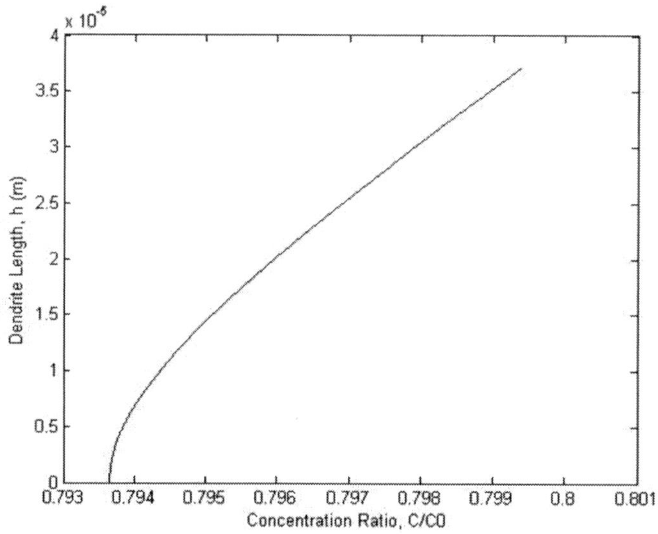

Figure 3. The development of dendrite length for a current density of 0.2 mA cm^{-2} until steady time operation, 100 s.

Figure 3 shows that the SEI formation is higher for higher concentration ratio which indicates the rate of dendrite development is affected by the operating conditions and lower current ratio correspond to faster dendrite formation. However, the experimental data of dendrite length is commonly spread in several values for a single concentration ratio, and the model based on the solution of the nonlinear transport equation can be a valuable insight about the kinetic mechanism in understanding the dendrite growth.

4. Conclusions

The solution of the one-dimensional time dependent transport equation is produced in this research. The variable diffusion coefficient is considered which the parameters are provided by the experimental data from the literatures. The characteristic method is utilized for generating solution in term of Lambert function. The calculation examples in steady state condition show that the concentration ratio is suppressed at the top layer, δ, which indicate the SEI formation. The result is also supported from the tip to surface current ratio which decreasing toward the top layer.

Acknowledgment

The authors are grateful for financial support from the Institut Teknologi Sepuluh Nopember under contract no: 850/PKS/ITS/2017.

References

[1] Hamidah NL and Wang F-M 2013 *National Taiwan University of Science and Technology* 1–155
[2] Alokar R 2013 Mathematical model of the dentritic growth during lithium electrodeposition *Journal of Power Sources* **232** p 23 – 28
[3] Fan L-Z, Wang X-L and Long F 2009 All-solid-state polymer electrolyte with plastic crystal materials for rechargeable lithium-ion battery *Journal of Power Sources* **189** p 775–778
[4] Zhang SS 2006 A review on electrolyte additives for lithium-ion batteries *Journal of Power Sources 169* p 1379–1394
[5] Kasavajjula U, Wang C and Appleby AJ 2007 Nano- and bulk-silicon-based insertion anodes for lithium-ion secondary cells *Journal of Power Sources* **163** p 1003–1039
[6] Chada TS, Suthar B, Rife D, Subramanian VR and Biswas P 2017 Model based of one-dimensional oriented lithium-ion battery electrodes *Journal of Electrochemical Society* **164 (11)** p 3114 – 3121
[7] Rashid M and Gupta A 2014 Mathematical model of combined effect of SEI formation and gas evolution in li-ion batteries *ECS Electrochemistry Letters* **3 (10)** p 95 – 98
[8] Phul S, Deshpande A and Krishnamurthy B 2015 A mathematical model to study the effect of potential drop across the sei layer on the capacity fading of a lithium ion battery *Electrochimica Acta* **164** p 281 – 287
[9] Ashwin TR, McGordon A and Jennings PA 2017 A mass transfer based variable porosity model with particle radius change for a lithium-ion battery *Electrochimica Acta* **232** p 203 – 214

Local adjacency metric dimension of sun graph and stacked book graph

Alifiah Yulisda Badri[1], Darmaji [2]

[1,2]Mathematics, Computing and Data Science, Institut Teknologi Sepuluh Nopember, Surabaya, Indonesia

E-mail: [1]alifiah.yulisda@yahoo.com [2]darmaji@matematika.its.ac.id

Abstract. A graph is a mathematical system consisting of a non-empty set of nodes and a set of empty sides. One of the topics to be studied in graph theory is the metric dimension. Application in the metric dimension is the navigation robot system on a path. Robot moves from one vertex to another vertex in the field by minimizing the errors that occur in translating the instructions (code) obtained from the vertices of that location. To move the robot must give different instructions (code). In order for the robot to move efficiently, the robot must be fast to translate the code of the nodes of the location it passes. so that the location vertex has a minimum distance. However, if the robot must move with the vertex location on a very large field, so the robot can not detect because the distance is too far.[6] In this case, the robot can determine its position by utilizing location vertices based on adjacency. The problem is to find the minimum cardinality of the required location vertex, and where to put, so that the robot can determine its location. The solution to this problem is the dimension of adjacency metric and adjacency metric bases. Rodrguez-Velzquez and Fernau combine the adjacency metric dimensions with local metric dimensions, thus becoming the local adjacency metric dimension. In the local adjacency metric dimension each vertex in the graph may have the same adjacency representation as the terms of the vertices. To obtain the local metric dimension of values in the graph of the Sun and the stacked book graph is used the construction method by considering the representation of each adjacent vertex of the graph.

1. Introduction

The concept of metric dimension was first introduced by F. Harary, et all [3] in On the metric dimension of a graph. Application in the metric dimension is the navigation robot. Research about local metric dimension of graph have done by Okamoto et al.[7], Then, Rodriguez and Fernau [9], continuoued their reseach that is about local adjacency metric dimension of corona graphs. Their research is developing of the concept about adjacency metric dimension of graphs that has introduced by Jannesari and Omoomi [4]. Farthes before. The local adjacency metric dimension is that each vertex on a graph may have the same adjacency representation as the terms neither of the adjacency vertex. Motivated by results in [3], [7], [4], and [9], this paper Local Adjacency Metric dimension of Sun Graph and Stacked Book Graph

2. Preliminaries

This section presents about some definitions and notions that are using in this research. These concepts are taken from [1]. Let $G = (V(G), E(G))$ is a simple, finite and connected graph with a set of vertices $V(G)$ and set of edges $E(G)$. Distance between two vertices u and v on graph

G is the shortest path length from u to v on graph G, denoted by $d(u,v) = d(v,u)$. If the u and v vertex are not connected, then $d(u,v)$ is undefined and listed as $d(u,v) = \infty$. [8]

Two adjacent vertices v and w_k will be write $v \sim w_k v$ and two vertices v and w_k that is not adjacent with $v \nsim w_k$. The distance between two vertices v and w_k in G, $d(v, w_k)$ is the lenght of shortest path joining v and w_k. The adjacency distance between v and w_k denoted by $d_2(v, w_k)$ and defines by [9]

$$d_2(u, w_k) = \begin{cases} 0, & \text{if } v = w_k \\ 1, & \text{if } v \sim w_k \\ 2, & \text{if } v \nsim w_k \end{cases}$$

Let $W = \{w_1, w_2, ..., w_k\} \subseteq V(G)$ be in order set of vertices and v is a vertex in G The adjacency representation of v with respect to W is the ordered k-tuple :

$$r_2(v|W) = (d_2(v, w_1), d_2(v, w_2), d_2(v, w_3), ..., d_2(w, k))$$

W is called a local adjacency resolving set of G if a pair of adjacent distinct vertex in G have different adjacency representations. A minimum local adjacency resolving set for G is local adjacency metric basis of G. Adjacency metric dimension for G, denoted by $dim_{A,l}(G)$ is the cardinality of vertices in local adjacency metric basis for G. Sun graph is defined by a graph cycle C_n corona Complement graph Complete $\bar{K}_m$. In this research for writing the notation of the Sun graph is denoted M $_{n,m}$, where M $_{n,m}$= $C_n \odot$ $\bar{K}_m$ with $n \geq 3$ and $m \geq 1$.[5] Stacked Book Graph is defined by a cartesian $K_{1,n} \times P_m$ with $n \geq 2$, $m \geq 2$. In this research for writing the notation of the stacked book graph is denoted B $_{n,m}$.[2]

3. Main Results

In the following, we present some useful results on the local adjacency metric dimension of sun graph and stacked book graph.

3.1. Local adjacency metric dimension of sun graph

Let $W \subseteq V(M_{n,m})$ be lokal adjacency metric basis. Let sun graph with order $n \geq 2$ and $m \geq 2$.

Theorem 3.1 *Let $G \cong M_{n,m}$ with $n \geq 3$ and $m \geq 1$, then*

$$dim_{A,l}(G) = \begin{cases} 2, & \text{if } n = 3 \\ \left\lfloor 2 + \dfrac{n-4}{3} \right\rfloor, & \text{if } n > 3 \end{cases}$$

Proof 3.2 *For $n = 3$, Let $W = \{v_1, v_2\} \subseteq V(G)$. We will show that W is a local adjacency resolving set of G. The local adjacency of vertices from $V(G) - W$ are as follows :*

$r(v_1|W) = (0,1)$, $r(v_{1m}|W) = (1,2)$
$r(v_2|W) = (1,0)$, $r(v_{2m}|W) = (2,1)$
$r(v_3|W) = (1,1)$, $r(v_{3m}|W) = (2,2)$

As we see that all of the adjacency representation of adjacent vertices are distinct. So $W = \{v_1, v_2\} \subseteq V(G)$ is a local adjacency resolving set for G.

The cardinality of W, $|W| = 2$ is minimum because if $|W| < 2$ have $r(v_2|W) = r(v_3|W)$ and $r(v_2|W) = r(v_{2m}|W)$. Without loss of generality,so we get $dim_{A,l}(G) = 2$ for $n = 3$.

For $n > 3$. Let $W = \left\{ v_1, v_2, v_3, ..., v_{\left\lfloor 2+\frac{n-4}{3} \right\rfloor} \right\} \subseteq V(G)$. *All of the adjacency representation of adjacent vertices are distinct. So $W \subseteq V(G)$ is a local adjacency resolving*

IOP Conf. Series: Journal of Physics: Conf. Series **974** (2018) 012069 doi:10.1088/1742-6596/974/1/012069

set for G. The cardinality of W, $|W| = \left\lfloor 2 + \frac{n-4}{3} \right\rfloor$ *is minimum because if* $|W| < \left\lfloor 2 + \frac{n-4}{3} \right\rfloor$ *have* $r(v_{n-1}|W) = r(v_{(n-1)m}|W)$.

Without loss of generality,so we get $dim_{A,l}(G) = \left\lfloor 2 + \frac{n-4}{3} \right\rfloor$ *for* $n > 3$.

3.2. Local adjacency metric dimension of stacked book graph

Let $W \subseteq V(B_{n,m})$ be lokal adjacency metric basis. Let stacked book graph with order $n \geq 2$ and $m \geq 2$.

Theorem 3.3 *Let* $G \cong B_{n,m}$ *with* $n \geq 2$ *and* $m \geq 2$, *then*

$$dim_{A,l}(G) = \begin{cases} \frac{m}{2}, & \text{if } m \text{ even} \\ \frac{m-1}{2}, & \text{if } m \text{ odd} \end{cases}$$

Proof 3.4 *For* $n \geq 2$, $m \geq 2$, *and m even we will be shown* $dim_{A,l}(G) = \frac{m}{2}$. *To determine local adjacency metric dimension of stacked book graph, then we will look for upper bound and lower bound of stacked book graph.*

a *To determine of upper bound*
Let $W_1 = \{v_1, v_3, v_5, v_7, ..., v_{2p-1}\}$, $p = 1, 2, ..., \frac{m}{2}$ *so that* W_1 *has a different local adjacency distance representation for each vertex in the stacked book graph.* W_1 *is local adjacency resolving set,* $|W_1| = \frac{m}{2}$. W_1 *is local adjacency resolving set but not a local adjacency metric basis because is still a* W_1 *that has a minimum of cardinality. so* $dim_{A,l}(G) \leq \frac{m}{2}$

b *To determine of lower bound*
Let $W_2 = \{v_1, v_3, v_5, v_7, ..., v_{2p-1}\}$, $p = 1, 2, ..., (\frac{m}{2} - 1)$. $|W_2| = \frac{m}{2} - 1$. *Because* $r(v_m|W_2) = r(v_{m-1}|W_2)$ *so* W_2 *is not local adjacency resolving set. Two adjacency vertices has some representation. So determine of lower bound is* $\frac{m}{2} \leq dim_{A,l}(G)$. *Without loss of generality, we have* $\frac{m}{2} \leq dim_{A,l}(G) \leq \frac{m}{2}$. *So it proved that so it proved that* $dim_{A,l}(G) = \frac{m}{2}$

For $n \geq 2$, $m \geq 2$, *and m odd we will be shown* $dim_{A,l}(G) = \frac{m-1}{2}$. *To determine local adjacency metric dimension of stacked book graph, then we will look for upper bound and lower bound of stacked book graph.*

a *To determine of upper bound*
Let $W_1 = \{v_2, v_4, v_6, v_8, ..., v_{2p}\}$, $p = 1, 2, ..., \frac{m-1}{2}$ *so that* W_1 *has a different local adjacency distance representation for each vertex in the stacked book graph.* W_1 *is local adjacency resolving set,* $|W_1| = \frac{m-1}{2}$. W_1 *is local adjacency resolving set but not a local adjacency metric basis because is still a* W_1 *that has a minimum of cardinality. so* $dim_{A,l}(G) \leq \frac{m-1}{2}$

b *To determine of lower bound*
Let $W_2 = \{v_2, v_4, v_6, v_8, ..., v_{2p}\}$, $p = 1, 2, ..., (\frac{m-1}{2} - 1)$. $|W_2| = \frac{m-1}{2} - 1$. *Because* $r(v_m|W_2) = r(v_{m-1}|W_2)$ *and* $r(v_{m-1}|W_2) = r(v_{m-2}|W_2)$ *so* W_2 *is not local adjacency resolving set. Two adjacency vertices has some representation. So determine of lower bound is* $\frac{m-1}{2} \leq dim_{A,l}(G)$. *Without loss of generality, we have* $\frac{m-1}{2} \leq dim_{A,l}(G) \leq \frac{m-1}{2}$. *So it proved that so it proved that* $dim_{A,l}(G) = \frac{m-1}{2}$

4. Conclusion

In this paper we have given result the local adjacencey metric dimension of sun graph and stacked book graph. The result shows that the local adjacencey metric dimension of sun graph as follows:
we have we have $dim_{A,l}(G) = 2$ if $n = 3$ and $dim_{A,l}(G) = \left\lfloor 2 + \frac{n-4}{3} \right\rfloor$ if $n > 3$ and $dim_{A,l}(G) = \frac{m}{2}$ if m even and $dim_{A,l}(G) = \frac{m-1}{2}$ if m odd. Based on our research, we have maximum adjacency distance to determine local adjacency metric basis is 3 vertices.

5. References

[1] Chartrand G, and Lesniak L,"Graphs and Digraph Third Edition",*Chapman and Hall Florida*,2000.
[2] Edi Hidayat, Yayuk W and Inna K,"Dimensi Metrik Ketetanggaan dan Dimensi Metrik Ketetanggaan Lokal Graf Hasil Kali Kartesian antara Graf Tangga dan Graf Lintasan",*Fakultas Sains dan Teknologi*,2014,Universitas Airlangga-Surabaya
[3] F.Harary and R.A Melter, "On the metric dimension of a graph" *Ars Combinatoria* **2**,1976,pp.191195.
[4] Jannesari, M and Omoomi B,"The Metric Dimension of the Lexicographic Product of Graphs",*Discrete Mathematics*,2012,Vol. **312**, Number 22, pp.3349-3356.
[5] Kavitha B and Indra K,"Split and Equitable Domination In Book Graph and Stacked Book Graph",*http://www.ijarcs.info/index.php/Ijarcs/article/viewFile/4475/3953*,14 October 2017.
[6] Khuller S, Raghavachari B and Rosenfeld A,"Landmarks in Graphs",*Discrete Applied Mathematics*,1996, Volume 70, Number **3**, pp.217-229.
[7] Okamoto F, Crosse L, et all,"The Local Metric Dimension of A Graph",*Mathematica Bohemica*,2010, Number **3**, pp. 239-255.
[8] Rinurwati, Herry and Slamin,"On Local Adjacency Metric Dimension of Some Wheel Related Graphs With Pendant Points", *American Institute of Physics,http://aip.scitation.org/doi/pdf/10.1063/1.4994468*, 12 October 2017.
[9] Rodrguez-Velzquez J and Fernau H,"On the (adjacency) Metric Dimension of Corona and Strong Product Graphs and Their Local Variants: Combinatorial and Computational Results",*http://arxiv.org/pdf/1309.2275.pdf*, 13 May 2014.

The partition dimension of cycle books graph

Jaya Santoso[1] **and Darmaji**[2]

[1,2]Department of Mathematics, Institut Teknologi Sepuluh Nopember, Surabaya, Indonesia.

E-mail: [1]pandiangan529@gmail.com [2]darmaji@matematika.its.ac.id

Abstract. Let G be a nontrivial and connected graph with vertex set $V(G)$, edge set $E(G)$ and $S \subseteq V(G)$ with $v \in V(G)$, the distance between v and S is $d(v, S) = min\{d(v, x)|x \in S\}$. For an ordered partition $\Pi = \{S_1, S_2, S_3, ..., S_k\}$ of $V(G)$, the representation of v with respect to Π is defined by $r(v|\Pi) = (d(v, S_1), d(v, S_2), ..., d(v, S_k))$. The partition Π is called a resolving partition of G if all representations of vertices are distinct. The partition dimension $pd(G)$ is the smallest integer k such that G has a resolving partition set with k members. In this research, we will determine the partition dimension of Cycle Books $B_{C_r,m}$. Cycle books graph $B_{C_r,m}$ is a graph consisting of m copies cycle C_r with the common path P_2. It is shown that the partition dimension of cycle books graph, $pd(B_{C_3,m})$ is 3 for $m = 2, 3$, and m for $m \geq 4$. $pd(B_{C_4,m})$ is $3 + 2k$ for $m = 3k + 2$, $4 + 2(k - 1)$ for $m = 3k + 1$, and $3 + 2(k - 1)$ for $m = 3k$. $pd(B_{C_5,m})$ is $m + 1$.

1. Introduction

One of interesting topic in discrete mathematics is graph theory. One of the topics in graph theory is the dimension metric and partition dimension. Chartrand et.al [1] introduced the concept of partition dimension of G. They introduced the same concept of resolving partition with term partition dimension of graph denoted by $pd(G)$. For $S \subseteq V(G)$ with vertex $v \in V(G)$, the distance between v and S is $d(v, S) = min\{d(v, x)|x \in S\}$.

Given $k-$partition $\Pi = \{S_1, S_2, S_3, ..., S_k\}$ of $V(G)$ and vertex $v \in V(G)$. Representation of v with respect to Π defined by ordered $r(v|\Pi) = (d(v, S_1), d(v, S_2), ..., d(v, S_k))$ for every $v \in V(G)$. The minimum of k such that resolving partition Π is partition dimension of G, denoted by $pd(G)$. In this research, we will determine the partition dimension of cycle books $B_{C_r,m}$.

2. Preliminaries

Cycle books graph $B_{C_r,m}$ is a graph consisting of m copies cycle C_r with the common path P_2. In his paper, Ayhan, et al.[2] has discussed the dominating numbers of the book graph B_m. Prasetyo [3] has discussed the metric dimension and partition dimension of stacked book graph $B_{3,3}$. Grigorious, et al.[4] has discussed the partition dimension of a class of circulant graphs. Amrullah et al.[5] has shown that partition dimension of a subdivision of a complete graph. Haryeni et al.[6] has shown that partition dimension of some classes of homogenous disconnected graph.

Useful properties in determining $pd(G)$ are given in the following Lemma 2.1 and Teorema 2.2.

Lemma 2.1 *[7] Let Π be a resolving partition of G and $u, v \in V(G)$. If $d(u, w) = d(v, w)$ for all $w \in V(G) - \{u, v\}$, then u and v belong to distinct elements of Π.*

Theorem 2.2 *[7] Let G be a connected graph of order* $n \geq 2$.
(i.) $pd(G) = 2$ *if and only if* $G = P_n$
(ii.) $pd(G)=n$ *if and only if* $G = K_n$
(iii.) For $n \geq 4$, $pd(G)=n-1$ *if and only if* $G = K_{r,s}, (r, s \geq 1), G = K_r + \overline{K_s}, (r \geq 1, s \geq 2), or G = K_r + (K_1 \bigcup K_s), (r, s \geq 1)$.

3. Result and discussions

In this section, we will show the partition dimension of cycle books graph .

3.1. The partition dimension of cycle books graph $B_{C_3,m}$

Cycle books graph $B_{C_3,m}$ is a graph consisting of m copies cycle C_3 with the common path P_2.

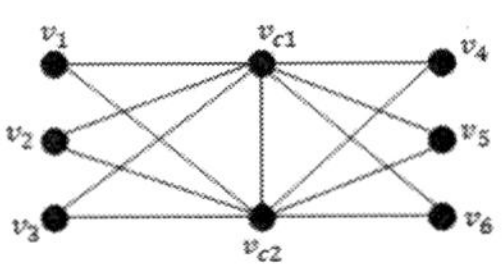

Figure 1. cycle books graph $B_{C_3,m}$

Theorem 3.1 *Let G is cycle books graph* $B_{C_3,m}$ *for* $m = 2, 3, 4, 5, ...,$ *then*

$$pd(G) = \begin{cases} 3 & if\, m = 2, 3 \\ m & if\, m \geq 4 \end{cases}$$

Proof: Consider three cases as follows:

(i) For $m = 2$
Let $\Pi = \{S_1, S_2, S_3\}$ be a partition of $V(G)$ with $S_1 = \{v_{c1}, v_1\}$, $S_2 = \{v_{c2}\}$, and $S_3 = \{v_2\}$. For $V(G)$ will be shown that all vertices G have distinct representation with respect to Π. The representation of all vertices of G are as follows:

$$r(v_{c1}|\Pi) = (0, 1, 1) \qquad r(v_1|\Pi) = (0, 1, 2)$$

$$r(v_{c2}|\Pi) = (1, 0, 1) \qquad r(v_2|\Pi) = (1, 1, 0)$$

Since all vertices of G have distinct representation. So, $\Pi = \{S_1, S_2, S_3\}$ is resolving partition of $V(G)$ then $pd(G) \leq 3$. By Theorem 2.2.(i), we have lower bound $pd(G) \geq 3$. So we get $pd(G) = 3$.

(ii) For $m = 3$
Let $\Pi = \{S_1, S_2, S_3\}$ be a partition of $V(G)$ with $S_1 = \{v_{c1}, v_1\}$, $S_2 = \{v_{c2}, v_2\}$, and $S_3 = \{v_3\}$. For $V(G)$ will be shown that all vertices G have distinct representation with respect to Π. The representation of all vertices of G are as follows:

$$r(v_{c1}|\Pi) = (0, 1, 1) \qquad r(v_1|\Pi) = (0, 1, 2)$$

$$r(v_{c2}|\Pi) = (1, 0, 1) \qquad r(v_2|\Pi) = (1, 1, 0)$$

$$r(v_3|\Pi) = (1, 0, 2)$$

Since all vertices of G have distinct representation. So, $\Pi = \{S_1, S_2, S_3\}$ is resolving partition of $V(G)$ then $pd(G) \leq 3$. By Theorem 2.2.(i), we have lower bound $pd(G) \geq 3$. So we get $pd(G) = 3$.

IOP Conf. Series: Journal of Physics: Conf. Series **974** (2018) 012070 doi:10.1088/1742-6596/974/1/012070

(iii) For $m \geq 4$.
Let $\Pi = \{S_1, S_2, S_3, ..., S_m\}$ be a partition of $V(G)$ with $S_1 = \{v_{c1}, v_1\}$, $S_2 = \{v_{c2}, v_2\}$, $S_j = \{v_j \mid 3 \leq j \leq m\}$ will be shown that all vertices G have distinct representation with respect to Π. The representation of all vertices of G are as follows:

$$r(v_{c1}|\Pi) = (0, \underbrace{1, ..., 1}_{m-1})$$
$$r(v_{c2}|\Pi) = (1, 0, \underbrace{1, ..., 1}_{m-2})$$
$$r(v_1|\Pi) = (0, 1, \underbrace{2, ..., 2}_{m-2})$$
$$r(v_2|\Pi) = (1, 0, \underbrace{2, ..., 2}_{m-2})$$
$$r(v_3|\Pi) = (1, 1, 0, \underbrace{2, ..., 2}_{m-3})$$
$$r(v_j|\Pi) = (1, 1, \underbrace{2, ..., 2}_{j-3}, 2, \underbrace{2, ..., 2}_{m-j}); 4 \leq j \leq m$$

Since all vertices of G have distinct representation. So, $\Pi = \{S_1, S_2, S_3, ..., S_m\}$ is resolving partition of $V(G)$ then $pd(G) \leq m$.
The lower bound of the partition dimension in cycle books graph $B_{C_3,m}$ can be obtained by using Lemma 2.1. According to Lemma 2.1 vertex in common path v_{c1} and v_{c2} must be in a different partition class, and other vertices are $v_j; 1 \leq j \leq m$ must be in a different partition class. Therefore, by placing the vertex v_{c1} and one other vertex (which is not v_{c2}) in a partition class (say S_1), and placing the vertex v_{c2} and one vertex others are in a partition class (say S_2), then the other vertex are each placed in a different partition class (say S_3,S_4, S_5, ..., S_m), then obtained the construction of the minimum resolving partition with cardinality m or $pd(B_{C_3,m}) \geq m$. So we get $pd(G) = m$.

3.2. The partition dimension of cycle books graph $B_{C_4,m}$
Cycle books graph $B_{C_4,m}$ is a graph consisting of m copies cycle C_4 with the common path P_2.

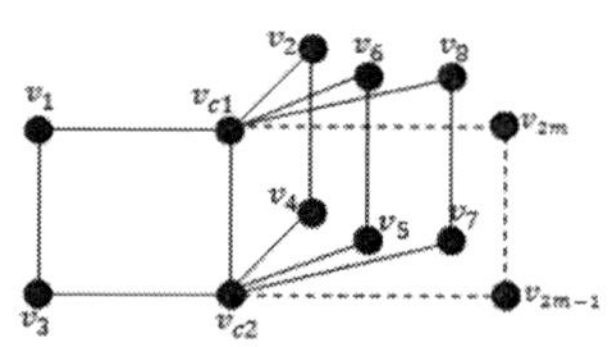

Figure 2. cycle books graph $B_{C_4,m}$

Theorem 3.2 *Let G is cycle books graph $B_{C_4,m}$ for $m = 2, 3, 4, 5, ...,$ then*

$$pd(G) = \begin{cases} 3 + 2k & \text{if } m = 3k + 2 \\ 4 + 2(k-1) & \text{if } m = 3k + 1 \\ 3 + 2(k-1) & \text{if } m = 3k \end{cases}$$

with $k = 0, 1, 2, 3, ...$

Proof: Consider three cases as follows:

(i) For $m = 3k$.
Let $\Pi = \{S_1, S_2, S_3, ..., S_{3+2(k-1)}\}$ be a partition of $V(G)$ with $S_1 = \{v_{c1}, v_1, v_2\}$, $S_2 = \{v_{c2}, v_3\}$, $S_3 = \{v_4, v_5, v_6\}$, $S_4 = \{v_7, v_8, v_9\}$ ..., $S_{3+2(k-1)} = \{v_{2m-2}, v_{2m-1}, v_{2m}\}$. For $V(G)$ will be shown that all vertices G have distinct representation with respect to Π. The representation of all vertices of G are as follows:
$r(v_{c1}|\Pi) = (0, 1, 1, 1, 1, ..., 1, 1)$
$r(v_1|\Pi) = (0, 1, 2, 2, 2, ..., 2, 2)$
$r(v_2|\Pi) = (0, 2, 1, 2, 2, ..., 2, 2)$
$r(v_{c2}|\Pi) = (1, 0, 1, 1, 1, ..., 1, 1)$
$r(v_3|\Pi) = (1, 0, 2, 2, 2, ..., 2, 2)$
$r(v_4|\Pi) = (1, 1, 0, 2, 2, ..., 2, 2)$
$r(v_5|\Pi) = (2, 1, 0, 2, 2, ..., 2, 2)$
$r(v_6|\Pi) = (1, 2, 0, 2, 2, ..., 2, 2)$
$r(v_7|\Pi) = (2, 1, 2, 0, 2, ..., 2, 2)$
$r(v_8|\Pi) = (1, 2, 2, 0, 2, ..., 2, 2)$
$r(v_9|\Pi) = (1, 2, 2, 0, 1, .., 2, 2)$
$r(v_{10}|\Pi) = (2, 1, 2, 1, 0, .., 2, 2)$
$\vdots$
$r(v_{2m-2}|\Pi) = (2, 1, 2, 2, 2, ..., 1, 0)$
$r(v_{2m-1}|\Pi) = (2, 1, 2, 2, 2, ..., 2, 0)$
$r(v_{2m}|\Pi) = (1, 2, 2, 2, 2, ..., 2, 0)$
Since all vertices of G have distinct representation. So, $\Pi = \{S_1, S_2, S_3, ..., S_{3+2(k-1)}\}$ is resolving partition of $V(G)$ then $pd(G) \leq 3 + 2(k-1)$.
The lower bound of the partition dimension, will be shown that the resolving partition of graph $B_{C_4,m}$ has cardinality less than $3 + 2(k-1)$, say $2 + 2(k-1)$. Without loss of generality, let $\Pi = \{S_1, S_2, S_3, ..., S_{2+2(k-1)}\}$ be a partition of $V(G)$ with $S_1 = \{v_{c1}, v_1, v_2\}$, $S_2 = \{v_{c2}, v_3\}$, $S_3 = \{v_4, v_5, v_6\}$, $S_4 = \{v_7, v_8, v_9\}$ ..., $S_{2+2(k-1)} = \{v_{2m-5}, v_{2m-4}, v_{2m-3}, v_{2m-2}, v_{2m-1}, v_{2m}\}$. Then, we obtain $r(v_{2m-5}|\Pi) = r(v_{2m-2}|\Pi) = r(v_{2m-1}|\Pi)$, $r(v_{2m-4}|\Pi) = r(v_{2m-3}|\Pi) = r(v_{2m}|\Pi)$. So $\Pi = \{S_1, S_2, S_3, ..., S_{2+2(k-1)}$ is not a resolving partition of $V(G)$ then $pd(G) \geq 3 + 2(k-1)$. So we get $pd(G) = 3 + 2(k-1)$.

(ii) For $m = 3k + 1$.
Let $\Pi = \{S_1, S_2, S_3, ..., S_{4+2(k-1)}\}$ be a partition of $V(G)$ with $S_1 = \{v_{c1}, v_1, v_2\}$, $S_2 = \{v_{c2}, v_3\}$, $S_3 = \{v_4, v_5, v_6\}$, $S_4 = \{v_7, v_8, v_9\}$ ..., $S_{4+2(k-1)} = \{v_{2m-1}, v_{2m}\}$. For $V(G)$ will be shown that all vertices G have distinct representation with respect to Π. The representation of all vertices of G are as follows:
$r(v_{c1}|\Pi) = (0, 1, 1, 1, 1, ..., 1, 1)$
$r(v_1|\Pi) = (0, 1, 2, 2, 2, ..., 2, 2)$
$r(v_2|\Pi) = (0, 2, 1, 2, 2, ..., 2, 2)$
$r(v_{c2}|\Pi) = (1, 0, 1, 1, 1, ..., 1, 1)$
$r(v_3|\Pi) = (1, 0, 2, 2, 2, ..., 2, 2)$
$r(v_4|\Pi) = (1, 1, 0, 2, 2, ..., 2, 2)$
$r(v_5|\Pi) = (2, 1, 0, 2, 2, ..., 2, 2)$
$r(v_6|\Pi) = (1, 2, 0, 2, 2, ..., 2, 2)$
$r(v_7|\Pi) = (2, 1, 2, 0, 2, ..., 2, 2)$
$r(v_8|\Pi) = (1, 2, 2, 0, 2, ..., 2, 2)$
$r(v_9|\Pi) = (1, 2, 2, 0, 1, .., 2, 2)$
$r(v_{10}|\Pi) = (2, 1, 2, 1, 0, .., 2, 2)$
$\vdots$
$r(v_{2m-1}|\Pi) = (2, 1, 2, 2, 2, ..., 2, 0)$

$r(v_{2m}|\Pi) = (1, 2, 2, 2, 2, ..., 2, 0)$

Since all vertices of G have distinct representation. So, $\Pi = \{S_1, S_2, S_3, ..., S_{4+2(k-1)}\}$ is resolving partition of $V(G)$ then $pd(G) \leq 4 + 2(k-1)$.

The lower bound of the partition dimension, will be shown that the resolving partition of graph $B_{C_4,m}$ has cardinality less than $4 + 2(k-1)$, say $3 + 2(k-1)$. Without loss of generality, let $\Pi = \{S_1, S_2, S_3, ..., S_{3+2(k-1)}\}$ be a partition of $V(G)$ with $S_1 = \{v_{c1}, v_1, v_2\}$, $S_2 = \{v_{c2}, v_3\}$, $S_3 = \{v_4, v_5, v_6\}$, $S_4 = \{v_7, v_8, v_9\}$ ..., $S_{3+2(k-1)} = \{v_{2m-4}, v_{2m-3}, v_{2m-2}, v_{2m-1}, v_{2m}\}$. Then, we obtain $r(v_{2m-4}|\Pi) = r(v_{2m-3}|\Pi) = r(v_{2m-1}|\Pi)$, $r(v_{2m-2}|\Pi) = r(v_{2m}|\Pi)$. So $\Pi = \{S_1, S_2, S_3, ..., S_{3+2(k-1)}$ is not a resolving partition of $V(G)$ then $pd(G) \geq 4 + 2(k-1)$. So we get $pd(G) = 4 + 2(k-1)$.

(iii) For $m = 3k + 2$

Cycle books graph $B_{C_4,m}$ has $2m + 2$ vertices (say $v_{c1}, v_{c2}, v_1, v_2, ..., v_{2m}$). Let $\Pi = \{S_1, S_2, S_3, ..., S_{3+2k}\}$ be a partition of $V(G)$ with $S_1 = \{v_{c1}, v_1, v_2\}$, $S_2 = \{v_{c2}, v_3\}$, $S_3 = \{v_4, v_5, v_6\}$, ..., $S_{3+2k} = \{v_{2m}\}$. For $V(G)$ will be shown that all vertices G have distinct representation with respect to Π. The representation of all vertices of G are as follows:

$r(v_{c1}|\Pi) = (0, 1, 1, 1, 1, ..., 1, 1, 2)$
$r(v_1|\Pi) = (0, 1, 2, 2, 2, ..., 2, 2, 3)$
$r(v_2|\Pi) = (0, 2, 1, 2, 2, ..., 2, 2, 3)$
$r(v_{c2}|\Pi) = (1, 0, 1, 1, 1, ..., 1, 1, 1)$
$r(v_3|\Pi) = (1, 0, 2, 2, 2, ..., 2, 2, 2)$
$r(v_4|\Pi) = (1, 1, 0, 2, 2, ..., 2, 2, 2)$
$r(v_5|\Pi) = (2, 1, 0, 2, 2, ..., 2, 2, 2)$
$r(v_6|\Pi) = (1, 2, 0, 2, 2, ..., 2, 2, 3)$
$r(v_7|\Pi) = (2, 1, 2, 0, 2, ..., 2, 2, 2)$
$r(v_8|\Pi) = (1, 2, 2, 0, 2, ..., 2, 2, 3)$
$r(v_9|\Pi) = (1, 2, 2, 0, 1, .., 2, 2, 3)$
$\vdots$
$r(v_{2m}|\Pi) = (2, 1, 2, 2, 2, ..., 1, 0)$

Since all vertices of G have distinct representation. So, $\Pi = \{S_1, S_2, S_3, ..., S_{3+2k}\}$ is resolving partition of $V(G)$ then $pd(G) \leq 3 + 2k$.

The lower bound of the partition dimension of cycle books graph $B_{C_4,m}$ will be shown that the resolving partition of graph $B_{C_4,m}$ has cardinality less than $3 + 2k$, say $2 + 2k$. Without loss of generality, let $\Pi = \{S_1, S_2, S_3, ..., S_{2+2k}\}$ be a partition of $V(G)$ with $S_1 = \{v_{c1}, v_1, v_2\}$, $S_2 = \{v_{c2}, v_3\}$, $S_3 = \{v_4, v_5, v_6\}$, ..., $S_{2+2k} = \{v_{2m-3}, v_{2m-2}, v_{2m-1}, v_{2m}\}$. Then, we obtain $r(v_{2m-3}|\Pi) = r(v_{2m}|\Pi)$, $r(v_{2m-2}|\Pi) = r(v_{2m-1}|\Pi)$. So $\Pi = \{S_1, S_2, S_3, ..., S_{2+2k}\}$ is not a resolving partition of $V(G)$ then $pd(G) \geq 3 + 2k$. So we get $pd(G) = 3 + 2k$.

3.3. The partition dimension of cycle books graph $B_{C_5,m}$

Cycle books graph $B_{C_5,m}$ is a graph consisting of m copies Cycle C_5 with the common path P_2. Cycle books graph $B_{C_5,n}$ has $3m + 2$ vertices (say $v_{c1}, v_{c2}, v_1, v_2, ..., v_{3m}$).

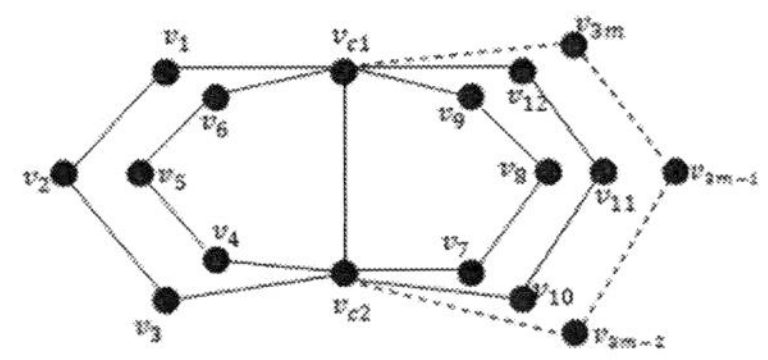

Figure 3. cycle books graph $B_{C_5,m}$

Theorem 3.3 *Let G is cycle books graph $B_{C_5,m}$ for $m = 2, 3, 4, 5, ...,$ then $pd(G) = m + 1$*

Proof: The upper bound of the partition dimension of cycle books graph $B_{C_5,m}$ can be obtained by constructing a resolving partition Π. Let $\Pi = \{S_1, S_2, S_3, ..., S_m, S_{m+1}\}$ be a partition of $V(G)$ with $S_1 = \{v_{c1}, v_1, v_2\}$, $S_2 = \{v_{c2}, v_3\}$, $S_3 = \{v_4, v_5, v_6\}$, ..., $S_m = \{v_{3m-5}, v_{3m-4}, v_{3m-3}\}$, $S_{m+1} = \{v_{3m-2}, v_{3m-1}, v_{3m}\}$. For $V(G)$ will be shown that all vertices G have distinct representation with respect to Π. The representation of all vertices of G are as follows:

$r(v_{c1}|\Pi) = (0, 1, 1, 1, 1, ..., 1, 1)$
$r(v_1|\Pi) = (0, 2, 2, 2, 2, ..., 2, 2)$
$r(v_2|\Pi) = (0, 1, 3, 3, 3, ..., 3, 3)$
$r(v_{c2}|\Pi) = (1, 0, 1, 1, 1, ..., 1, 1)$
$r(v_3|\Pi) = (1, 0, 2, 2, 2, ..., 2, 2)$
$r(v_4|\Pi) = (2, 1, 0, 2, 2, ..., 2, 2)$
$r(v_5|\Pi) = (2, 2, 0, 3, 3, ..., 3, 3)$
$r(v_6|\Pi) = (1, 2, 0, 2, 2, ..., 2, 2)$
$r(v_7|\Pi) = (2, 1, 2, 0, 2, ..., 2, 2)$
$r(v_8|\Pi) = (2, 2, 3, 0, 3, ..., 3, 3)$
$r(v_9|\Pi) = (1, 2, 2, 0, 2, .., 2, 2)$
$r(v_{10}|\Pi) = (2, 1, 2, 2, 0, .., 2, 2)$
$\vdots$
$r(v_{3m-2}|\Pi) = (2, 1, 2, 2, 2, ..., 2, 0)$
$r(v_{3m-1}|\Pi) = (2, 2, 3, 3, 3, ..., 3, 0)$
$r(v_{3m}|\Pi) = (1, 2, 2, 2, 2, ..., 2, 0)$

Since all vertices of G have distinct representation. So, $\Pi = \{S_1, S_2, S_3, ..., S_m, S_{m+1}\}$ is resolving partition of $V(G)$ then $pd(G) \leq m + 1$.

The lower bound of the partition dimension of cycle books graph $B_{C_5,m}$ will be shown that the resolving partition of graph $B_{C_5,m}$ has cardinality less than $m + 1$, say m . Without loss of generality, let $\Pi = \{S_1, S_2, S_3, ..., S_m\}$ be a partition of $V(G)$ with $S_1 = \{v_{c1}, v_1, v_2\}$, $S_2 = \{v_{c2}, v_3\}$, $S_3 = \{v_4, v_5, v_6\}$, ..., $S_m = \{v_{3m-5}, v_{3m-4}, v_{3m-3}, v_{3m-2}, v_{3m-1}, v_{3m}\}$. Then, we obtain $r(v_{3m-5}|\Pi) = r(v_{3m-2}|\Pi)$, $r(v_{3m-4}|\Pi) = r(v_{3m-1}|\Pi)$ and $r(v_{3m-3}|\Pi) = r(v_{3m}|\Pi)$. So $\Pi = \{S_1, S_2, S_3, ..., S_m\}$ is not a resolving partition of $V(G)$ then $pd(G) \geq m + 1$. So we get $pd(G) = m + 1$.

4. Conclusion

In this paper, we have the partition dimension of cycle books graf as follows: The partition dimension of cycle books graf, $pd(B_{C_3,m})$ is 3 for $m = 2$, and m for $m \geq 3$. $pd(B_{C_4,m})$ is $3 + 2k$ for $m = 3k + 2$, $4 + 2(k - 1)$ for $m = 3k + 1$, and $3 + 2(k - 1)$ for $m = 3k$. $pd(B_{C_5,m})$ is $m + 1$.

References

[1] Chartrand G, Salehi E and Zhang P 2000 The partition dimension of a graphs *Aequationes Math* No.59 45-54

[2] Ayhan A and Omar A 2010 Determination and testing the domination numbers of tadpole graph, book graph and staced book graph using matlab, *College of Basic Education Researchers Journal* **10(1)** 491-501.
[3] Prasetyo E 2013 *Dimensi metrik dan dimensi partisi graf buku bertumpuk* $B_{3,n}$ (Thesis) Departemen Matematika, Fakultas Sains dan Teknologi, Universitas Airlangga-Surabaya.
[4] Grigorious C, Stephen S, Rajan B, Miller M and William A 2014 On the partition dimension of a class of circulant graphs *Information Processing Letters*, **114(7)** 353-356
[5] Amrullah, Baskoro E, Simanjuntak R and Uttunggadewa S 2015 The partition dimension of a subdivision of a complete graph *Procedia Computer Science* **74** 53-59
[6] Haryeni D and Baskoro E 2015 Partition dimension of some classes of homogenous disconnected graphs *Procedia Computer Science* **74** 73-78
[7] Chartrand G, Eroh L, Johnson M and Oellermann O 2000 Resolvability in graphs and the metric dimension of a graph *Discrete Appl. Math* No.105 99-113

IOP Conf. Series: Journal of Physics: Conf. Series **974** (2018) 012071 doi:10.1088/1742-6596/974/1/012071

Estimation of three-dimensional radar tracking using modified extended kalman filter

Prima Aditya, Erna Apriliani, Didik Khusnul Arif, and Komar Baihaqi

Mathematics Department, Institut Teknologi Sepuluh Nopember Surabaya, Indonesia

E-mail: `paulusprima32@gmail.com` ; `{april,didik,komar}@matematika.its.ac.id`

Abstract. Kalman filter is an estimation method by combining data and mathematical models then developed be extended Kalman filter to handle nonlinear systems. Three-dimensional radar tracking is one of example of nonlinear system. In this paper developed a modification method of extended Kalman filter from the direct decline of the three-dimensional radar tracking case. The development of this filter algorithm can solve the three-dimensional radar measurements in the case proposed in this case the target measured by radar with distance r, azimuth angle θ, and the elevation angle ϕ. Artificial covariance and mean adjusted directly on the three-dimensional radar system. Simulations result show that the proposed formulation is effective in the calculation of nonlinear measurement compared with extended Kalman filter with the value error at 0.77% until 1.15%.

1. Introduction

Kalman filter known as linear quadratic estimation, is an algorithm that uses a series of measurements observed over time, containing statistical noise and other inaccuracies, and produces estimates of unknown variables that tend to be more accurate than those based on a single measurement alone. Extensions to the method have also been developed, the extended Kalman filter is nonlinear version of Kalman filter which used for nonlinear systems. Algorithm is almost same with Kalman filter but in the extended Kalman filter, the state transition and observation models don't need to be linear functions of the state but may instead be differentiable functions[1]. Extended Kalman filter can be used for three-dimensional radar tracking system (one of example of nonlinear systems). Radar tracking control system that has been designed according to the needs with the level of accurate measurement there will be a noise. By reducing the noise in the measurement system and needed an approach that is more accurate than ever before. The approach taken is estimated to determine the level of noise [2]. Based on the results of research conducted by Jang Song and in 2001 on "Improved Kalman Filter Design for Three-Dimensional Radar Tracking" [3], is obtained new filter to overcome the problem of radar tracking. In that paper, the level of measurement accuracy compared to the method extended Kalman filter with new methods improved of design Kalman filter, and the results of the simulation indicated that the improved of Kalman filter is a better result. Extended Kalman Filter are also used in research Nousheen in 2015 [4] that discussed abaout it can not be used on the system parameters and / or measurement that is not linear. In the topic of this research, studied the relevant journal in which the estimate is not a linear measurement of three-dimensional radar tracking system. There are many research on Kalman filter such as control position of underwater vehicle [5], also mobile robot position [6], heat conduction and its development [7], about air pollution [8], also stirred tank reactor [9], extended preferred radar tracking [10] and the last is estimation of reduced model [11].

2. Dynamical model

Case of three-dimensional radar tracking is sensor measures the target in three dimension space including the range (distance) r, azimuth angle (turning angle) θ, and elevation angle (elevation angle) ϕ (cf. Fig.1)

Range r is the measured distance of target from the center of the tracker, then the azimuth angle θ shows direction of the tracker rotation motion in the direction of the x-axis and y-axis, and the elevation angle ϕ indicates the direction of tracker's motion elevation in the direction of x, y to z axis (the development of 2D radar) [10]. Then the equation can be described as follows:

$$r = \sqrt{x^2 + y^2 + z^2} \tag{1}$$

$$\theta = \tan^{-1} \frac{y}{x} \tag{2}$$

$$\phi = \tan^{-1} \frac{z}{(x^2 + y^2)^{\frac{1}{2}}} \tag{3}$$

Equation (1)-(3) describe the measurement model in this research. Notice that the three equations (1)-(3) are obtained from the spherical shape. Then we transform (1)-(3) into the shape of cartesian coordinates by using the following transformation:

$$x = r \cos\theta \cos\phi \tag{4}$$

$$y = r \sin\theta \cos\phi \tag{5}$$

$$z = r \sin\phi \tag{6}$$

where x, y and z is the scalar component of the cartesian coordinate measurement that is described as a position of a moving target in three-dimensional space [1]. Using the chain rule,

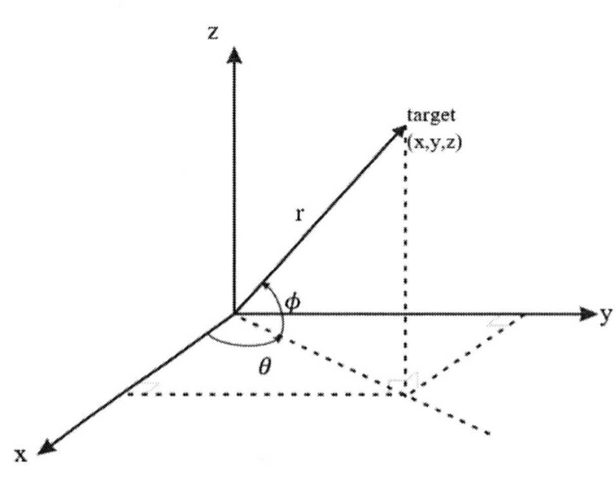

Figure 1. Radar measures r, θ, ϕ.

IOP Conf. Series: Journal of Physics: Conf. Series **974** (2018) 012071 doi:10.1088/1742-6596/974/1/012071

the derivation of (4)-(6) w.r.t. time is as follows:

$$\frac{dx}{dt} = \cos\theta\cos\phi\frac{dr}{dt} - r\sin\theta\cos\phi\frac{d\theta}{dt} - r\cos\theta\sin\phi\frac{d\phi}{dt} \tag{7}$$

$$\frac{dy}{dt} = \sin\theta\cos\theta\frac{dr}{dt} + r\cos\theta\cos\phi\frac{d\theta}{dt} - r\sin\theta\sin\phi\frac{d\phi}{dt} \tag{8}$$

$$\frac{dz}{dt} = \sin\phi\frac{dr}{dt} + r\cos\phi\frac{d\phi}{dt} \tag{9}$$

Equation (7)-(9) is a dynamical system that will be estimated by using extended Kalman filter and its modification, with derive equations and made a stochastic discrete model the final results are as follow:

$$x_{k+1} = \frac{x_k}{\sqrt{x_k^2 + y_k^2 + z_k^2}} - y_k - \frac{x_k z_k}{\sqrt{x_k^2 + y_k^2}} + w_{1k} \tag{10}$$

$$y_{k+1} = \frac{y_k}{\sqrt{x_k^2 + y_k^2 + z_k^2}} + x_k - \frac{y_k z_k}{\sqrt{x_k^2 + y_k^2}} + w_{2k} \tag{11}$$

$$z_{k+1} = \frac{z_k}{\sqrt{x_k^2 + y_k^2 + z_k^2}} + \sqrt{x_k^2 + y_k^2} + w_{3k} \tag{12}$$

$$\begin{aligned} V_{k+1}^x = & \frac{-2y_k}{\sqrt{x_k^2+y_k^2+z_k^2}} - \frac{2x_k z_k}{\sqrt{x_k^2+y_k^2+z_k^2}\sqrt{x_k^2+y_k^2}} \\ & -2x_k + \frac{2y_k z_k}{\sqrt{x_k^2 + y_k^2}} + w_{4k} \end{aligned} \tag{13}$$

$$\begin{aligned} V_{k+1}^y = & \frac{2x_k}{\sqrt{x_k^2+y_k^2+z_k^2}} - \frac{2y_k z_k}{\sqrt{x_k^2+y_k^2+z_k^2}\sqrt{x_k^2+y_k^2}} \\ & -2y_k + \frac{2x_k z_k}{\sqrt{x_k^2 + y_k^2}} + w_{5k} \end{aligned} \tag{14}$$

$$V_{k+1}^z = \frac{2\sqrt{x_k^2 + y_k^2}}{\sqrt{x_k^2 + y_k^2 + z_k^2}} - z_k + w_{6k} \tag{15}$$

We define the following measurement model:

$$Z_k = \begin{pmatrix} \sqrt{x_k^2 + y_k^2 + z_k^2} + v_{1k} \\ tan^{-1}\frac{y_k}{x_k} + v_{2k} \\ tan^{-1}\frac{z_k}{\sqrt{x_k^2+y_k^2}} + v_{3k} \end{pmatrix}$$

where:

- x_k : target position at x-axes
- y_k : target position at y-axes
- z_k : target position at z-axes
- V_k^x : target velocity at x-axes
- V_k^y : target velocity at y-axes
- V_k^z : target velocity at z-axes

Equations (10)-(15) will be used to generate the real system, whereas the model used for estimation is the linearized version. The linearized model produces a system matrix **A**, that will be used in modified extended Kalman filter algorithm (see the next section).

IOP Conf. Series: Journal of Physics: Conf. Series **974** (2018) 012071 doi:10.1088/1742-6596/974/1/012071

3. Modified Extended Kalman Filter

Extended Kalman filter method is the extension of the classical Kalman Filter [12] for nonlinear systems. In reality, many models are nonlinear. Since the model is nonlinear, we use extended Kalman filter. Given a nonlinear stochastic model:

$$X_{k+1} = f(X_k, u_k) + w_k$$
$$Z_k = h_k(X_k) + v_k$$

where the nonlinear measurement model , $x_0 \sim N(\overline{X}_0, P_{x0})$, $v_k \sim N(0, R_k)$ has a normal distribution and assumed to be white that means there is no correlation between each other and also with initial condition X_0 [13].

Before the estimation process, linearization process is carried out since the system is not linear. Linearization process is done by defining it as follows:

$$X^*_{k+1} = f(\widehat{X}_k, u_k)$$
$$Z^*_{k+1} = h(X^*_{k+1})$$
$$\mathbf{A} = [A_{i,j}] = [\frac{\partial f_i}{\partial X_j}(\widehat{X}_k, u_k)]$$
$$\mathbf{H} = [H_{i,j}] = [\frac{\partial h_i}{\partial X_j}(X^*_{k+1})]$$

where $\mathbf{A}$ and $\mathbf{H}$ are the Jacobi matrix derived from a differential in f and h to the direction of X. A modification of Kalman filter algorithm is called extended Kalman filter algorithm [14].

From research conducted by Song Taek Park [1] modification extended Kalman filter is insert error covariance and mean-made into the algorithm as follows:

$$R^p_k \approx \begin{pmatrix} \sigma_r^2 + \frac{r_k^2(\sigma_\theta^4+\sigma_\phi^4)}{2} & 0 & 0 \\ 0 & \sigma_\theta^2 & 0 \\ 0 & 0 & \sigma_\phi^2 \end{pmatrix} \tag{16}$$

$$\mu^p_k \approx \begin{pmatrix} r_k[(\theta^m_k)^2 + (\phi^m_k)^2 - \sigma_\theta^2 - \sigma_\phi^2] \\ 0 \\ 0 \end{pmatrix} \tag{17}$$

where:
σ_θ: parameters of angle noise θ
σ_ϕ : parameters of angle noise ϕ
$\theta^m_k : \theta_k + v^\theta_k$
$\phi^m_k : \phi_k + v^\phi_k$
Error covariance R^p_k in Equation (16) and mean-made μ^p_k in Equation (17) which insert in modified extended Kalman filter algorithm, modified extended Kalman filter algorithm given by Table 1.

IOP Conf. Series: Journal of Physics: Conf. Series **974** (2018) 012071 doi:10.1088/1742-6596/974/1/012071

Table 1. Modified extended Kalman filter

	Description
Model system	$x_{k+1} = f(x_k, u_k) + w_k$
Measurement model	$Z_{k+1} = h(x_{k+1}) + v_k$
Assumption	$x_0\ N(\overline{x}_0, P_{x_0}), w_k\ N(0, Q_k)$, $v_k\ N(0, R_k)$
Initialization	$\widehat{x}_0 = \overline{x}_0, P_0 = P_{x_0}$
Predict	$\mathbf{A} = [\frac{\partial f_i}{\partial x_j}]$ Estimation: $\widehat{x}_{k_{\overline{k}}} = f(\widehat{x}_k, u_k)$ Error Covariance: $P_{k+1} = \mathbf{A}P_k + P\mathbf{A}^T + G_k Q_k G_k^T$
Update & Gain Kalman:	$K_{k+1} = P_{k+1}\mathbf{H}^T[\mathbf{H}_k P_{k+1}\mathbf{H}^T + \overline{R}_{k+1}^p]^{-1}$ Estimation: $\widehat{x}_{k+1} = \widehat{x}_{k+1} + K_{k+1}(z_{k+1} - h(\widehat{x}_{k+1}) - \overline{\mu}_{k+1}^p)$ Error Covariance: $P_{k+1} = [\mathbf{I} - K_k\mathbf{H}]P_{k+1}$

4. Results

Simulation is given initial conditions and parameters (see Table 2) with measurements model r, θ,and ϕ.

Table 2. Initialization

No	Variables	Values
1	x	15 kilometers
2	y	12 kilometers
3	z	5 kilometers
4	V^x	-60 meters/sec
5	V^y	-70 meters/sec
6	V^z	-40 meters/sec
	Parameters	
7	σ_r	0.003 kilometers
8	σ_θ	0.0261799 rad/sec
9	σ_ϕ	0.0261799 rad/sec
10	P_0	0.05
11	Q_k	0.0001
12	R_k	0.00002
13	dt	0.0001

In Figure 2-7 the green color of the graph shows the real value, the estimated value of eKf is blue and red color is value estimation results meKf. In Figure 2, the results showed the difference

IOP Conf. Series: Journal of Physics: Conf. Series **974** (2018) 012071 doi:10.1088/1742-6596/974/1/012071

in value between the real value and eKf estimation is 0.0094249, and the difference in value between the real value and meKf estimation is 0.0088085. For Figure 3, the results show the difference in value between the real value and eKf estimation is 0.0150247, and the difference in value between the real value and meKf estimation is 0.0073667. In Figure 4, the results show the difference in value between the real value and eKf estimation is 0.0111491, and the difference in value between the real value and meKf estimation is 0.0093434. In Figure 5, the results show the difference in value between the real value and ekf estimation is 0.01824580, and the difference in value between the real value and meKf estimation is 0.0114886. In Figure 6, the results obtained show the difference in value between the real value and eKf estimation is 0.0107461, and the difference in value between the real value and meKf estimation is 0.01008190. In Figure 7, the results show the difference in value between the real value and eKf estimation is 0.0115598, and the difference in value between the real value and meKf estimation is 0.0105467.

Figure 8 shows a graph of the error between the real and estimated values of all variables. The smallest error in the estimated value of the modification extended Kalman filter (red line) in each variable with the indicated value of each variable RMSE.

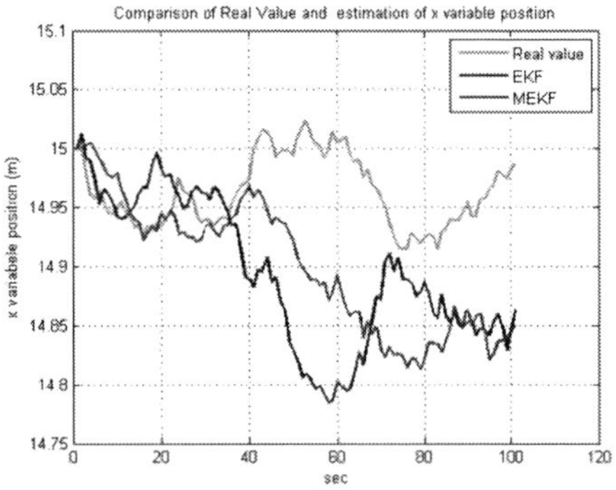

Figure 2. Comparison graphs real value and estimated x variable position

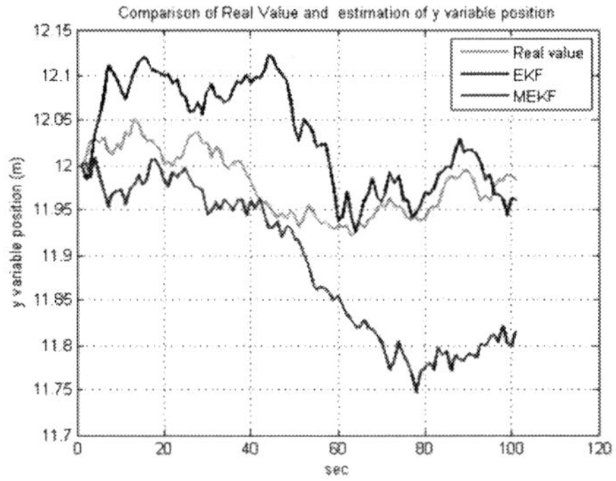

Figure 3. Comparison graphs real value and estimated y variable position

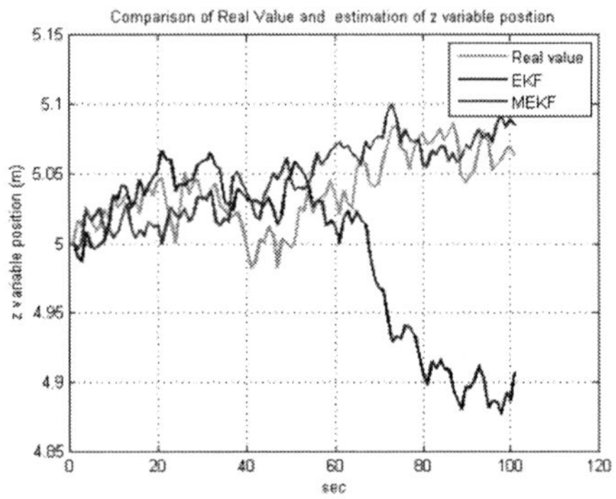

Figure 4. Comparison graphs real value and estimated z variable position

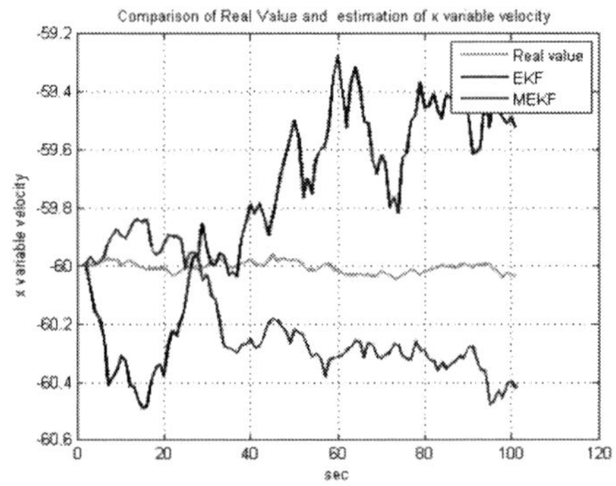

Figure 5. Comparison graphs real value and estimated x variable velocity

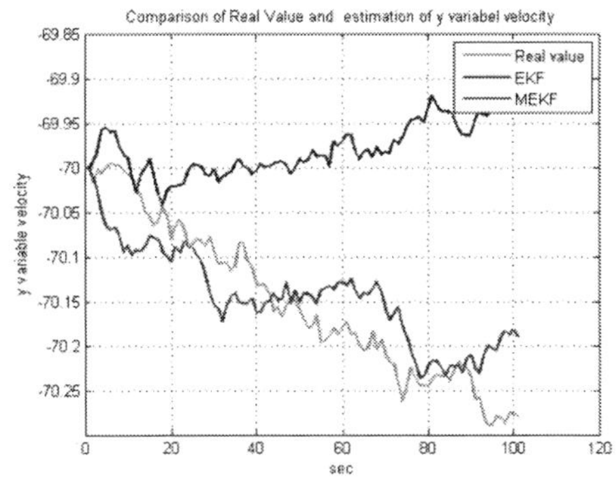

Figure 6. Comparison graphs real value and estimated y variable velocity

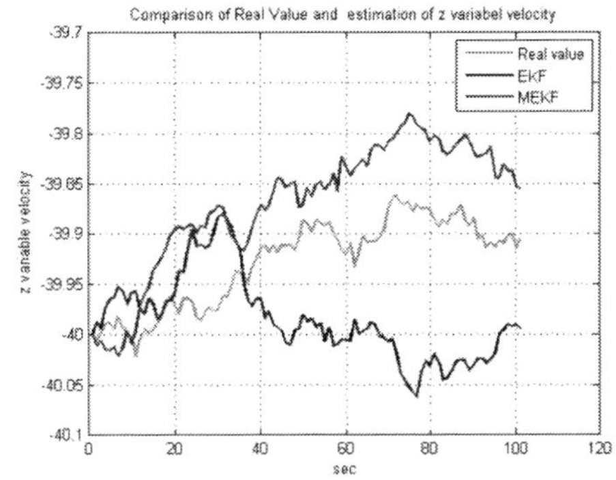

Figure 7. Comparison graphs real value and estimated z variable velocity

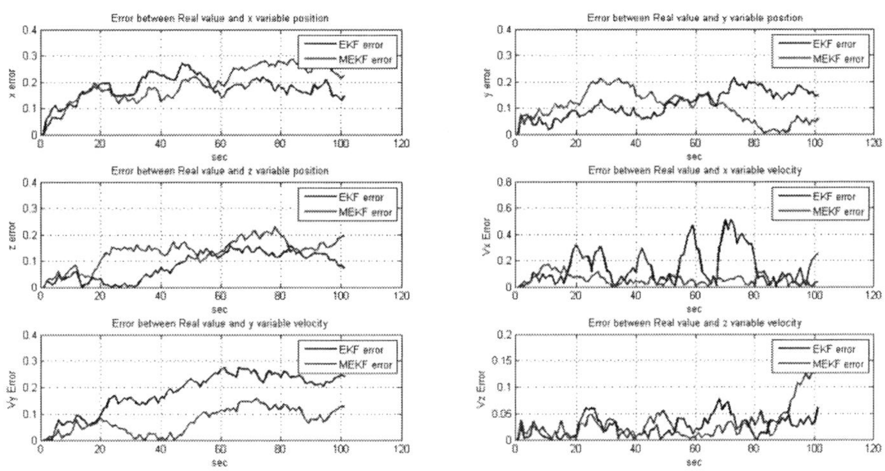

Figure 8. Comparison graphs error each variable

Table 3. Average value of rmse each variable

	RMSE of EKF	RMSE of MEKF
x position	0.00942490	0.00880850
y position	0.01502470	0.00736670
z position	0.01114910	0.00934340
V^x velocity	0.0182458	0.0114886
V^y velocity	0.0107461	0.0100819
V^z velocity	0.0115598	0.0105467

In Table 3 shows that the RMSE value of each variable is relatively small in the error value at intervals of $0.00736670 < RMSE < 0.0115598$, or it can be said mistakes by 0.77% up to 1.15% to modification of Extended Kalman Filter and $0.00942490 < RMSE < 0.01824580$, or it can be said mistakes of 0.94% to 1.82% for the extended Kalman filter. So overall it can be said that the modified extended Kalman filter method is suitable for estimating the three-dimensional radar tracking system in this case specifically for the measurement is not linear.

5. CONCLUSIONS

Based on the analysis and discussion presented in the previous chapter, we can conclude:

1. The dynamic model of three-dimensional radar tracking is formed as showed in chapter 2.
2. The estimation results indicate that each variables (position and velocity x, y, and z) of the tracking radar three dimensions by modification of extended Kalman filter better than methods extended Kalman filter with indicated by the error rate of only 0.77% up to 1.15% for modification of extended Kalman filter, while the error rate of extended Kalman filter is of 0.94% to 1.82%.
3. Based on the computation time show that the combined method of eKf and its modification takes 4.5770319 seconds.

References

[1] Einicke, G.A., "Smoothing, Filtering and Prediction: Estimating the Past, Present and Future" Rijeka, Croatia: Intech, 2012.

[2] R. Sani, Estimasi variabel keadaan gerak longitudinal pesawat terbang menggunakan metode fuzzy Kalman filter, Jurnal Sains dan Seni ITS, vol. 5, no. 2, 2016.

[3] S.T. Park and J. Lee, Improved Kalman filter design for threedimensional radar tracking, IEEE Transactions on Aerospace and Electronic Systems, vol. 37, no. 2, pp. 727739, 2001.

[4] N. Fahmedha, P. Prakash, A. Pooja, and R. Rachana, Estimation of system parameters using Kalman filter and extended Kalman filter, International Journal of Advanced Technology and Engineering Exploration, vol. 2, no. 6, p. 84, 2015.

[5] Z. Ermayanti, E. Apriliani, H. Nurhadi, and T. Herlambang, Estimate and control position autonomous underwater vehicle based on determined trajectory using fuzzy Kalman filter method, in International Conference on Advanced Mechatronics, Intelligent Manufacture, and Industrial Automation (ICAMIMIA), pp. 156161, 2015.

[6] E. Apriliani, F. Yunaini, and S. Hartini, Estimation and control design of mobile robot position, Far East Journal of Mathematical Sciences, vol. 77, no. 1, p. 115, 2013.

[7] D. Arif, M.Widodo, E. Apriliani, and Salmah, Distribution estimation of heat conduction using Kalman filtering which implemented on reduction model, in 3rd International Conferences and Workshops on Basic and Applied Sciences, p. M007, 2011.

[8] E. Apriliani, D. Arif, and B. Sanjoyo, The square root ensemble Kalman filter to estimate the concentration of air pollution, in International Conference on Mathematical Applications in Engineering (ICMAE10), 2010.

[9] R. Fitria and D. Arif, State variable estimation of nonisothermal continuous stirred tank reactor using fuzzy Kalman filter, in International Journal of Computing Science and Applied Mathematics, vol. 3, no. 1, pp. 1620, 2017.

[10] D. Leskiw, The Extended Preferred Ordering Theorem for Radar Tracking Using the Extended Kalman Filter. PhD thesis, Syracuse University, 2011.

[11] D. Arif, Widodo, Salmah, and E. Apriliani, Construction of the Kalman filter algorithm on the model reduction, International Journal of Control and Automation, vol. 7, no. 9, pp. 257270, 2014.

[12] F. Lewis, Optimal estimation: with an introduction to stochastic control theory. Wiley New York, 1986.

[13] L. Kleeman, Understanding and applying Kalman filtering, in Proceedings of the Second Workshop on Perceptive Systems, Curtin University of Technology, Perth Western Australia (25-26 January 1996), 1996.

[14] G. Bishop and G. Welch, An introduction to the Kalman filter, Proc. of SIGGRAPH, Course, vol. 8, no. 27599-23175, p. 41, 2001.

IOP Conf. Series: Journal of Physics: Conf. Series **974** (2018) 012072 doi:10.1088/1742-6596/974/1/012072

THE CONVERGENCE PROBLEMS OF EIGENFUNCTION EXPANSIONS OF ELLIPTIC DIFFERENTIAL OPERATORS

Anvarjon Ahmedov
Faculty of Industrial Sciences & Technology, Universiti Malaysia Pahang
E-mail: anvarjon@ump.edu.my

Abstract. In the present research we investigate the problems concerning the almost everywhere convergence of multiple Fourier series summed over the elliptic levels in the classes of Liouville. The sufficient conditions for the almost everywhere convergence problems, which are most difficult problems in Harmonic analysis, are obtained. The methods of approximation by multiple Fourier series summed over elliptic curves are applied to obtain suitable estimations for the maximal operator of the spectral decompositions. Obtaining of such estimations involves very complicated calculations which depends on the functional structure of the classes of functions. The main idea on the proving the almost everywhere convergence of the eigenfunction expansions in the interpolation spaces is estimation of the maximal operator of the partial sums in the boundary classes and application of the interpolation Theorem of the family of linear operators. In the present work the maximal operator of the elliptic partial sums are estimated in the interpolation classes of Liouville and the almost everywhere convergence of the multiple Fourier series by elliptic summation methods are established. The considering multiple Fourier series as an eigenfunction expansions of the differential operators helps to translate the functional properties (for example smoothness) of the Liouville classes into Fourier coefficients of the functions which being expanded into such expansions. The sufficient conditions for convergence of the multiple Fourier series of functions from Liouville classes are obtained in terms of the smoothness and dimensions. Such results are highly effective in solving the boundary problems with periodic boundary conditions occurring in the spectral theory of differential operators. The investigations of multiple Fourier series in modern methods of harmonic analysis incorporates the wide use of methods from functional analysis, mathematical physics, modern operator theory and spectral decomposition. New method for the best approximation of the square-integrable function by multiple Fourier series summed over the elliptic levels are established. Using the best approximation, the Lebesgue constant corresponding to the elliptic partial sums is estimated. The latter is applied to obtain an estimation for the maximal operator in the classes of Liouville.

1. Introduction

The mathematical models of the various vibrating systems are partial differential equations elliptic type and such equations, in the case of self adjoint elliptic operators, can be represented as a limit of eigenfunction expansions of the elliptic operators. The problem of funding the solution of partial differential equations will be reduced to the determining the conditions under which eigenfunction expansions of the functions from certain classes, which are used as a boundary values for the solution of the model equations, are convergent. Most complete solution of the expansion problems can be obtained by proper development of the spectral theory of elliptic differential operators. It is well known that the spectral theory of differential operators also used to describe slow flows of viscous incompressible fluids. Many physical process taking place in real space can be described using the spectral theory of elliptic differential operators. Many of the equations of physical sciences and engineering involve operators of elliptic type. Most important among these is non-relativistic quantum theory, which is based upon the spectral analysis of second order elliptic differential operators. Applications of second order elliptic operators to geometry and stochastic analysis are also now of great importance. Spectral theory of the elliptic differential operators is an extremely rich field which has been studied by many qualitative and quantitative techniques like Sturm-Liouville theory, separation of variables, Fourier and Laplace transforms, perturbation theory, eigenfunction expansions, variational methods, microlocal analysis, stochastic analysis and numerical methods including finite elements.

2. Eigenfunction expansions of the elliptic operators

We define Torus as a cube $[-\pi, \pi]^N$:

$$T^N = \{x = (x_1, x_2, \cdots, x_N) \in R^N : -\pi < x_i \leq \pi, i = 1, \cdots, N\},$$

naturally isomorphic to R^N/Z^N. Precisely, for $x, y \in R^N$ we say that

$$x \equiv y,$$

if $x - y \in 2\pi Z^N$. Here $\equiv$ is an equivalence relation that partitions R^N into equivalence classes, where $2\pi Z^N$ is the additive subgroup of R^N and Z^N is integer coordinates.

Let consider an arbitrary differential operator with constant coefficients:

$$A(D) = \sum_{|\alpha| \leq m} a_\alpha D^\alpha,$$

where $D^\alpha = D_1^{\alpha_1} \dots D_N^{\alpha_N}$, $D_j = \frac{1}{i}\frac{\partial}{\partial x_j}$, α is a multi-index.

The polynomial $A(\xi)$ is associated with differential operator $A(D)$ by replace D with $\xi \in R^N$ and it is called a symbol of operator $A(D)$, and the homogeneous polynomial

$$A_h(\xi) = \sum_{|\alpha| = m} a_\alpha \xi^\alpha$$

is called its principle symbol. The operator $A(D)$ is said to be elliptic of order m if its principle symbol is positive, i. e. $A_h(\xi) > 0$ for all $\xi \in R^N$, $\xi \neq 0$.

The operator $A(D)$ is considered in the Hilbert space $L_2(T^N)$ as an unbounded operator (we refer for the details [2]) with domain of definition $C^\infty(T^N)$ the class of infinitely differentiable functions on T^N, this class is dense in $L_p(T^N)$.

In case the coefficients are real, the $A(D)$ will satisfy the symmetric condition:

$$(Au, v) = (u, Av), \ \ \forall u, v \in C^\infty(T^N).$$

In addition, since the operator $A(D)$ is elliptic, then by Gårding's inequality the operator $A(D)$ is semi-bounded

$$(Au, u) \geq \lambda_A(u, u), \quad \forall u \in C^\infty(T^N),$$

where nonnegative constant λ_A is called lower bound of A. Hence, thanks for Friedrichs's theorem, asserts that for every symmetric semi-bounded operator there are at least one self-adjoint extension with the same lower bound, then there is a self-adjoint extension $\bar{A}$ in $L_2(T^N)$

of operator $A(D)$ which, indeed, its closure, and they are coincided on the domain of definition i. e. $\bar{A}u = Au,\ u \in C^\infty(T^N)$.

By von Neumann's spectral theorem, the operator $\bar{A}$ has a spectral decomposition of unity $\{E_\lambda\}$, and then it can be represented in the following form

$$\bar{A} = \int_{\lambda_A}^{\infty} \lambda\, dE_\lambda,$$

the projections E_λ increase monotonically, and continuous on the left, moreover

$$\lim_{\lambda\to\infty} \|E_\lambda u - u\|_{L_2(T^N)} = 0, \quad u \in L_2(T^N).$$

The operator $\bar{A}$ has a complete orthonormal system of eigenfunctions $\{(2\pi)^{-N/2}e^{inx}\}$ in $L_2(T^N)$ corresponding to the eigenvalues $A(n),\ n \in Z^N$. Thus, the spectral decomposition of $f \in L_2(T^N)$ coincides with partial sums of the multiple Fourier series of function f related to $A(n)$. A interesting fact that the lower order coefficients $a_\alpha, |\alpha| < m$ of $A(D)$ do not influence the convergence of the spectral decomposition $E_\lambda f$, provided the function f is sufficiently smooth. Then one can reduce the study of convergence for partial sum of eigenfunction expansions to the study of simpler case, that is, applying the summation over expanding its principle symbol $A_h(n)$.

The spectral decomposition E_λ can be written as an integral operator:

$$E_\lambda f(x) = \int_{T^N} \Theta_\lambda(x,y) f(y)\, dy,$$

where the kernel

$$\Theta_\lambda(x,y) = (2\pi)^{-N} \sum_{A(n)<\lambda} e^{i(n,x-y)}$$

is called the spectral function of operator $\bar{A}$. Indeed, the operator E_λ is a convolution operation $\Theta_\lambda * f$, such that

$$E_\lambda f(x) = \int_{T^N} \Theta_\lambda(x-y) f(y)\, dy.$$

The following is proved in the paper [12]:

Theorem 1.1 Let $\alpha > \frac{N-1}{2}$. Then the maximal operator E_* is bounded from $L_1^\alpha(T^N)$ to $L_1(T^N)$:

$$\|E_* f(x)\|_{L_1(T^N)} \le C\, \|f\|_{L_1^\alpha(T^N)}, \quad \forall f \in L_1^\alpha(T^N), \quad \alpha > \frac{N-1}{2},$$

where the constant C depends only on N and α. Furthermore

$$\lim_{\lambda\to\infty} E_\lambda f(x) = \lim_{\lambda\to\infty} \sum_{A(n)<\lambda} \hat{f}(n)\, e^{i(n,x)} = f(x)$$

almost everywhere in T^N.

In this paper we extend this result for the case of general liuoville classes $L_p^\alpha(T^N)$.

Theorem 1.2 Let numbers s and p be related by the condition

$$s > (N-1)(\frac{1}{p} - \frac{1}{2}), \quad 1 \le p \le 2.$$

Then the maximal operator E_* is bounded from $L_p^s(T^N)$ to $L_p(T^N)$:

$$\|E_* f\|_{L_p(T^N)} \leq C_p(N,s)\ \|f\|_{L_p^s(T^N)},\ \ \forall f \in L_p^s(T^N),$$

where the constant $C_p(N,s)$ depends on the degree p of integrability of f, dimension N and smoothness s.

The necessary and sufficient conditions for the almost everywhere convergence of the multiple series and Fourier integrals by spherical Riesz means in the classes $L_p(T^N)$ are obtained in [17],[18]. The papers [5], [6] are devoted to the problems on the almost everywhere convergence of multiple Fourier series and integrals by Riesz method corresponding to the elliptic differential operators. The methods in these papers based on the application of the Poisson formula to estimate the maximal operator for the Fourier series, where the conditions for the order of the Riesz means s: $s > (N-1)/2$ is required. The almost everywhere convergence of the partial sums of the Fourier integrals in the classes of Liouville $L_p^\alpha(R^N)$ are investigated in the paper [8], where the condition $\alpha > (N-1)/2$ guarantees the almost everywhere convergence of the spectral expansions. So far no results on the almost everywhere convergence of multiple Fourier series in the classes of $L_p^\alpha(T^N)$. In such case the difficulty arises because of not applicability of Poisson formula. In the current work the almost everywhere convergence problems of multiple Fourier series in the Liouville classes are investigated by using the estimation of Lebesgue constant and applying Jackson theorems. The almost everywhere convergence of the Fourier-Laplace series on the unit sphere is obtained in [4]. The almost everywhere convergence of spectral expansions of the general elliptic differential operators in the Liouville classes are investigated in [10], and the summability is established for the expansions of the function from L_p^α under the conditions $\alpha > N(\frac{1}{p} - \frac{1}{2}), 1 \leq p \leq 2$. The extension of the latter result to the case of pseudo-differential operators is carried out in [7].

3. Liouville Space $L_p^s(T^N)$

We deal with the functions from a class which are termed as a Liouville class L_p^s where $s > 0$ (coincide for s integers with Sobolev class). We study the differentiability and smoothness of functions . There are several ways to interpret smoothness and numerous ways to describe it and quantify it. A fundamental fact is that smoothness can be measured and fine-tuned the Fourier transform. Indeed, the investigation of the subject is based on this point. Certain spaces of functions are introduced to serve the purpose of measuring smoothness, for instance, Lipschitz, Sobolev and Hardy spaces. However, the main function spaces we introduce here are Sobolev space and its generalization Liouville space.

The Liouville space is considered as generalization of Sobolev Spaces, where the classes $W_p^k(T^N)$ can also be described in terms of Fourier transform. By that one can extend the definition of Sobolev space to the case in which the index k is real. We explain the idea by putting $p = 2$. By using the Plancherels Theorem we obtain that the norm for $f \in W_2^k$ can be introduced in the form

$$\|f\|^2_{W_2^k} = \sum_{n \in Z^N} \sum_{|\alpha| \leq k} |n^\alpha|^2 |\hat{f}(n)|^2.$$

The following inequality

$$c_k(1+|\xi|^2)^{\frac{k}{2}} \leq \sum_{|\alpha|\leq k} |\xi^\alpha|^2 \leq C_k(1+|\xi|^2)^{\frac{k}{2}},$$

which holds for all $\xi \in R^N$ with some positive constants c_k, C_k, gives a motivation for the following definition:

Let s be a real number and let $1 < p < \infty$. The Liouville space $L_p^s(T^N)$ is defined as the space of all functions u in $L_p(T^N)$, such that

$$\sum_{n \in Z^N} (1+|n|^2)^{\frac{s}{2}} \hat{f}(n) e^{inx} \in L_p(T^N). \tag{1}$$

Thus the norm of f in $L_p^s(T^N)$ has a form

$$\|f\|_{L_p^s(T^N)} = \Big\| \sum_{n \in Z^N} (1+|n|^2)^{\frac{s}{2}} \hat{f}(n) e^{inx} \Big\|_{L_p(T^N)}.$$

We observe that when $s = 0$ then $L_p^s = L_p$. The space L_p^s coincides with the space W_p^k when $s = k$ a positive integer number. We also note that when $s \geq 0$ the elements of L_p^s are always L_p, i.e. $L_p^s \subseteq L_p$.

If $0 < s_1 < s$ then $L_p^s \subset L_p^{s_1}$. Moreover we have the embedding

$$L_p^s \to L_q^{s_1}, \quad L_p^s \subset L_q^{s_1}$$

provided

$$s - \frac{N}{p} = s_1 - \frac{N}{q}.$$

4. Fractional Powers of Elliptic Operators

Let A be an elliptic differential operator of order m on the torus T^N. Let Λ be a subset of the complex plane (in the applications this will, as a rule, be an angle with the vertex at the origin). In spectral theory it is useful to consider operators depending on a parameter $\lambda \in \Lambda$. An example of such an operator is the resolvent $(A - \lambda I)^{-1}$.

Let proceed with the definition of the elliptic differential operators. For this reason we consider a polynomial on $\xi \in R^N$ of even order $m = 2m'$ with coefficients from $C^\infty(T^N)$

$$A(x, \xi) = \sum_{|\alpha| \leq m} a_\alpha(x) \xi^\alpha.$$

The formal differential operator

$$A(x, D) = \sum_{|\alpha| \leq m} a_\alpha(x) D^\alpha$$

is said to be an elliptic of order m, if $\forall x \in T^N$ and $\xi \in R^N$ for $\xi \neq 0$ the principal symbol $A(x, \xi)$ is positive:

$$A(x, \xi) = \sum_{|\alpha| = m} a_\alpha(x) \xi^\alpha > 0.$$

The polynomial $A(x, \xi)$ is called a symbol of the operator $A(x, D)$. We denote by A an operator acting in the Hilbert space $L_2(T^N)$ with the domain of definition $D(A) = C^\infty(T^N)$:

$$Au = A(x, D)u(x), \quad u \in C^\infty(T^N)$$

.

We assume that the operator A is symmetric

$$(Au, v) = (u, Av), \quad \forall u, v \in C^\infty(T^N).$$

Furthermore the operator A is called semi-bounded if there exists positive number κ such that the following inequality holds

$$(Au, u) \geq \kappa(u, u), \quad \forall u \in C^\infty(T^N).$$

It is proven that there exists $\lambda_0 > 0$ such that the resolvent $R_\lambda = (A - \lambda I)^{-1}$ is defined for $|\lambda| \geq \lambda_0$. The spectrum $P(A)$ of A is a discrete subset of the complex plane. We choose a number ρ satisfying the condition that the disk $|\lambda| < 2\rho$ does not intersect with the spectrum $P(A)$. Now we select a contour of the form $\Gamma = \Gamma_1 + \Gamma_2 + \Gamma_3$:

$$\Gamma_1 = \{\lambda \in C : \lambda = re^{i\pi}, +\infty > r > \rho\}$$

$$\Gamma_2 = \{\lambda \in C : \lambda = \rho e^{i\phi}, \pi > \phi > -\pi\}$$

$$\Gamma_1 = \{\lambda \in C : \lambda = re^{i\pi}, \rho < r < +\infty\}.$$

Consider the integral

$$A_z = \frac{i}{2\pi} \int_\Gamma \lambda^z (A - \lambda I)^{-1} d\lambda,$$

where λ^z is defined as a holomorphic function of $\lambda \in C\backslash(-\infty, 0]$, equal to $e^{z \ln \lambda}$ for $\lambda > 0$.

Note that the integral on the $\int_\Gamma \lambda^z (A - \lambda I)^{-1} d\lambda$ converges in the operator on $L_2(T^N)$ for $Re(z) < 0$ and also A_z is a bounded operator on $L_2(T^N)$.

With the help this operator we can define fractional power of the elliptic operator as follows: Let $z \in C$ and $k \in Z$ be such that $Re(z) < k$. Put, on $C^\infty(T^N)$

$$A^z = A^k A_{z-k}.$$

Such defined operator is independent of the choice of integer k and for arbitrary $k \in Z$ and $s \in R$, the function A^z is holomorphic operator function of z in the half-plane $Re(z) < k$ with values in the Banach space $L(H^s(T^N), H^{s-mk}(T^N))$ of bounded linear operators from $H^s(T^N)$ to $H^{s-mk}(T^N)$.

Let the distribution $f \in D'(T^N)$ and $f = \sum_{n \in Z^N} \hat{f}(n) e^{i(n,x)}$ be the Fourier series of f. Let denote eigenvalues of A by λ_n. Then

$$A^z f(x) = \sum_{n \in Z^N} \lambda_n^z \hat{f}(n) e^{i(n,x)},$$

which follows from the following identity

$$A^z \{e^{i(n,x)}\} = \frac{i}{2\pi} \int_\Gamma \lambda^z (A - \lambda I)^{-1} e^{i(n,x)} d\lambda =$$

$$\frac{ie^{i(n,x)}}{2\pi} \int_\Gamma \lambda^z (\lambda_n - \lambda)^{-1} d\lambda = \lambda_n^z e^{i(n,x)},$$

which is obtained by using the Cauchy formula.

5. The Lebesgue constant for the multiple Fourier series summed over elliptic levels

Let B denotes a closed, balanced set in R^N. We denote by B_ϵ the union of the balls of radius ϵ with centers in the boundary of B, and by $\mu(B_\epsilon)$ the measure of B_ϵ. We assume that for the set B the upper surface measure in the sense of Minkowski is bounded:

$$\lim_{\epsilon\to 0}\frac{\mu(B_\epsilon)}{2\epsilon}<\infty.$$

For convex, centrally symmetric bodies with bounded surface satisfy above condition, in the paper [20] the following estimation is obtained:

$$\int_{T^N}\Big|\sum_{n\in\lambda B\cap Z^N} e^{i(x,n)}\Big|\,dx = O(1)\lambda^{\frac{N-1}{2}},\quad \lambda\to\infty.$$

The partial sum $E_\lambda f(x)$ of the Fourier series of f summed over elliptic levels is the convolution $f*\Theta_\lambda$ with the corresponding Dirichlet kernel (spectral function)

$$\Theta_\lambda(x)=\sum_{A(n)\le\lambda} e^{i(x,n)}.$$

The Lebesgue constant corresponding to the self adjoint operator A can be defined as follows

$$L_\lambda=\int_{T^N}|\Theta_\lambda(x)|\,dx=\int_{T^N}\Big|\sum_{A(n)\le\lambda} e^{i(x,n)}\Big|\,dx.$$

Lemma. For the Lebesgue constant one has

$$L_\lambda=\int_{T^N}|\Theta_\lambda(x)|\,dx=O(1)\lambda^{\frac{N-1}{2m}},\quad \lambda\to\infty.$$

In the estimation in [19] as a set B we take $B=\{x\in R^N : A(x)\le 1\}$. The homothetic transform of the S is $\lambda S=\{x\in R^N : A(\frac{x}{\lambda})\le 1\}$, $\lambda>0$. Using the condition that the $A(x)$ is homogeneous operator of order m: $A(tx)=t^m\lambda^{-m}A(x)$, we obtain:

$$\int_{T^N}|\Theta_\lambda(x)|\,dx=\int_{T^N}\Big|\sum_{A(n)\le\lambda} e^{i(x,n)}\Big|\,dx=\int_{T^N}\Big|\sum_{A(n/\sqrt[m]{\lambda})\le 1} e^{i(x,n)}\Big|\,dx=\int_{T^N}\Big|\sum_{\sqrt[m]{\lambda}B\cap Z^N} e^{i(x,n)}\Big|\,dx.$$

by taking into account the definition B and From the estimation [] we derive that

$$\int_{T^N}\Big|\sum_{n\in\sqrt[m]{\lambda}B\cap Z^N} e^{i(x,n)}\Big|\,dx=O(1)\lambda^{\frac{N-1}{2m}}.$$

This estimation will be used to establish the almost everywhere convergence of the Fourier series in the classes of $L_p^s(T^N)$.

6. The almost everywhere convergence of the multiple Fourier series

In order to apply interpolation theorem of Stein we need to have appropriate estimation for the maximal operator in the classes of $L_1^s(T^N)$ and $L_2^s(T^N)$. We conclude that if $f\in L_2^s(T^N)$, $s>0$, then

$$\sum_{n\in Z^N}|\hat f(n)|^2\log^2(1+|n|)\le\|f\|^2_{L_2^s(T^N)},$$

and the elliptic partial sums of f

$$E_\lambda f(x) = \sum_{A(n)<\lambda} \hat{f}(n) e^{i(n,x)}$$

converges almost everywhere on the torus T^N. Moreover, the maximal operator $E^* f(x)$ is estimated as follows

$$\|E_* f\|_{L_2(T^N)} \leq C\Big(\sum_{n\in Z^N} |\hat{f}(n)|^2 \log^2(1+|n|)\Big)^{\frac{1}{2}} \leq C\|f\|^2_{L_2^s(T^N)}.$$

Now we are ready to apply the interpolation theorem of Stein. But the interpolation theorem is valid for linear operator. We need to linearize the operators T_λ^s. Let $\mu(x)$ be a measurable function on T^N that take at most finitely many distinct values and satisfy the condition $0 \leq \mu(x) \leq \mu_0 < \infty$ and $s(z) = \frac{N-1}{2} z + \delta$, $0 \leq \Re(z) \leq 1$, $\delta > 0$. We define an analytic family of linear operators:

$$T^{s(z)}_{\mu(x)} = \sum_{A(n)\leq \mu(x)} (1+|A(n)|^2)^{s(z)} \hat{g}(n) e^{inx}, \quad g \in C^\infty(T^N).$$

In the case $p=2$, which corresponds to the value of z when $\Re(z) = 0:$

$$\|T^{s(i\,\Im(z))}_{\mu(x)} g(x)\|_{L_2(T^N)} \leq C_1 e^{\pi|\Im(z)|/2} \|g\|_{L_2(T^N)}, \quad \Re(z)=0.$$

Secondly on the line $\Re(z)=1$, we have

$$\|T^{s(1+i\,\Im(z))}_{\mu(x)} g(x)\|_{L_1(T^N)} \leq C_2 e^{\pi|\Im(z)|/2} \|g\|_{L_1(T^N)}, \quad \Re(z)=1.$$

Thus, by the interpolation we have for the values of $t: 0<t<1:$

$$\|T^{s(t)}_{\mu(x)} g(x)\|_{L_p(T^N)} \leq K_t \|g\|_{L_p(T^N)},$$

where $s(t) = \frac{N-1}{2} t + \delta$, $\frac{1}{p} = \frac{1-t}{2} + \frac{t}{1}$. The definition of $T^{s(t)}_{\mu(x)}$ and $L_p^s(T^N)$ gives

$$\|E_*^{s(t)} f(x)\|_{L_p(T^N)} \leq K_t \|f\|_{L_p^{s(t)}(T^N)},$$

where K_t satisfies $\log K_t \leq C$. And if note that

$$\frac{1}{p} = \frac{1-t}{2} + \frac{t}{1}$$

then

$$s(t) = \frac{N-1}{2} t + \delta = \frac{N-1}{2}\left(\frac{2}{p} - 1\right) + \delta > (N-1)\left(\frac{1}{p} - \frac{1}{2}\right), \ \delta > 0.$$

This means for $s > (N-1)\left(\frac{1}{p} - \frac{1}{2}\right)$, $1<p<2$ we have

$$\|E_*^s f(x)\|_{L_p(T^N)} \leq C\|f\|_{L_p^s(T^N)}.$$

The statement of the Theorem 1.2 is established. As a consequence of the estimation for maximal operator we can prove the almost everywhere convergence of the multiple Fourier series summed over elliptic levels.

Acknowledgments
The current research is supported by Universiti Malaysia Pahang under Universiti Research Grant Scheme RDU170364.

References
[1] S. A. Alimov and V. A. Il'in, *Conditions for the convergence of spectral decompositions that correspond to self-adjoint extensions of elliptic operators.* I, II, Differential Equations 7 (1971), 516-543; 651 -667.
[2] Sh. A. Alimov, R. R., Ashurov, & A.K. Pulatov, *Multiple Fourier series and Fourier integrals. In Commutative Harmonic Analysis IV*, Springer Berlin Heidelberg, (1991), pp.1-95.
[3] Sh. A. Alimov, *Fractional powers of elliptic operators and isomorphism of classes of differentiable functions*, Differ. Uravn., 8:9 (1972), 1609-1626
[4] A.A. Akhmedov, *About the almost everywhere convergence of spectral expansions from* $L_1^\alpha(S^N)$, Uzbek Mathematical Journal, N 4 (1996), pp. 17-25. arXiv:0808.0352v1
[5] R. R. Ashurov, *Summability almost everywhere of Fourier series in* L_p *with respect to eigenfunctions*, Math. Notes, 34:6 (1983), pp. 913-916.
[6] R. R. Ashurov,*Summation of multiple trigonometric Fourier series*, Math. Notes, 49:6 (1991), pp.563-568
[7] R. R. Ashurov, *Conditions for the localization and summability of spectral resolutions of elliptic operators*, Doctoral dissertation, Moscow State University, Moscow, Russia (1992).
[8] R. R. Ashurov, K. T. Buvaev, *On the convergence almost everywhere of spectral expansions of functions in* $L_1^\alpha(R^N)$, (in Russian) Uzbek. Mat. Zh., 4, (1998), pp. 14-23.
[9] K. I. Babenko, *On the mean convergence of multiple Fourier series and the asymptotic behaviour of the Dirichlet kernel of the spherical means*, Preprint No. 52, Inst. Prikl. Mat. Akad. Nauk SSSR, Moscow, 1971.
[10] A. Y. Bastys P. *On some problems of convergence of spectral expansions of the elliptic differential operators*, PhD thesis, Moscow State University, Moscow, Russia (1983).
[11] L. Colzani and P. M. Soardi, L_p *norms of certain kernels of the N-dimensional torus*, Trans. Amer. Math. Soc. 266 (1981), pp. 617-627
[12] Anvarjon Ahmedov, Ehab Materneh, Hishamuddin Zainuddin, J. Phys.: Conf. Ser.(2017) 890 012112
[13] L. Gårding, *Eigenfunction expansions connected with elliptic differential operators.* C. r. ll congress Mah. Scandinaves, Lund (1953), pp. 44-55.
[14] H. Hadwiger, *Vorlesungen ber Inhalt, Oberfldche und Isoperimetrie*, Die Grundlehren der Mathematischen Wissenschaften in Einzeldarstellungen, vol. XCIII. Berlin-Gttingen-Heidelberg, Springer-Verlag, (1957), pp 325.
[15] S. M. Nikol'skii, *Approximation of Functions of Several Variables and Embedding Theorems*, Nauka, Moscow (1969).
[16] H. S. Shapiro, *Lebesgue Constants for Spherical Partial Sums*, Journal of Approximation Theory, 13, 40-44 (1975)
[17] E.M. Stein, *Localization and summability of multiple Fourier series*, Acta Math. 100:1-2, (1958), pp. 93-147.
[18] E.M. Stein, *On certain exponential sums arising in multiple Fourier series*, Ann. Math., 73, No. 2, (1961), pp.87-109.
[19] V. A. Yudin, *Behavior of Lebesgue constants*, Math. Notes, 17:3 (1975), pp. 233-235
[20] V. A. Yudin, *The multidimensional Jackson theorem*, Math. Notes, 20:3 (1976), 801-804

Institute of Physics
Dirac House, Temple Back
Bristol BS1 6BE UK

ISSN: 1742-6588
ISBN 978-1-5108-6042-1

CURRAN ASSOCIATES INC.
proceedings.com